Wolves

Wolves

Behavior, Ecology,
and Conservation

Edited by L. David Mech and Luigi Boitani

The University of Chicago Press   *Chicago and London*

The University of Chicago Press gratefully acknowledges the support of Valerie Gates, who kindly subsidized production of the color plates in this book.

The University of Chicago Press, Chicago 60637
The University of Chicago Press, Ltd., London
© 2003 by The University of Chicago
All rights reserved. Published 2003
Paperback edition 2006
Printed in the United States of America
12 11 10 09 08 07 06   3 4 5

ISBN: 0-226-51696-2 (cloth)
ISBN-13: 978-0-226-51697-4 (paper)
ISBN-10: 0-226-51697-0 (paper)

Library of Congress Cataloging-in-Publication Data

Wolves : behavior, ecology, and conservation / edited by L. David Mech and Luigi Boitani.
    p.   cm.
Includes bibliographical references and index.
ISBN 0-226-51696-2
I. Mech, L. David.   II. Boitani, Luigi.

QL737.C22 W6477 2003
599.773 — dc21

                    2003005231

WE DEDICATE THIS BOOK to Ulysses S. Seal, world-renowned scientist and author, and conservationist extraordinaire. Colleague to all of us, and close friend to many of us, Ulie personified a true genius, breadth of knowledge, enthusiasm, dedication, and accomplishment that inspired us all. While long active in pursuing his interest in the ecology, behavior, and conservation of wolves, he also applied his keen mind and abundant energies to numerous other species, fostering their chances of survival and thus bettering the world. We thank him for this and for the unique opportunity to have worked with him.

Contents

Foreword

George Rabb

FORWARD IS THE NATURE of this book. It brings forward an immense, scattered literature of research on wolf biology, ecology, and behavior, coupled with salient materials on the conservation issues that surround this charismatic creature. Evident from the outset of this comprehensive volume is concern that we assemble credible information. The intensity of such efforts over the past thirty years is reflected in the record of research by the chapter authors. I see this intensity and the sheer number of scientists and natural resource managers and agents cited throughout the text as evidence that many of those involved deeply care about the fabled subject of their studies. And the final chapters and conclusion make it very apparent that caring for the wolf's future requires social and political skills and sensitivity as well as scientific credibility.

Editor-authors L. David Mech and Luigi Boitani show very persuasively that crucial to conservation of this fascinating animal is the realization that the wolf is no longer an animal of the wilderness, symbolic as it may be of wilderness for many. They assert simply that people must therefore come to accept the necessity for management control of wolf populations if the animal's survival is to be assured. What a sobering conclusion this is for many people who extend their empathetic feelings for dogs to the ancestral wolf. Likely supporting bases for this biophilic empathy are the social nature of the species, its exceptional abilities to communicate, its caregiving behavior, and the transferability of its bonding capacity to people. The wolf is thus well constituted to command our attention to its survival. And its survival is important in some measure to the survival of biologi-

cal diversity in the environments it occupies. While no longer an icon for pristine wilderness, the wolf is a symbol for conscientious caring for the environment, for conservation that is enduring. And the admirable effort that the contributors to this volume have made to provide the information on which such a conservation commitment should be based is to be celebrated and emulated.

Why should we care so deeply about another species? The question comes to what values predominate in our concerns and prevail in our behavior, culturally diverse as we are. While there are ecological and economic aspects that we might consider, more significant are positions on ethical and biophilic values that we hold and manifest. Ethically, we owe other people and future human generations the opportunity to share environments with the wolf. That we also owe coexistence to the wolf and all other species is an extension of this human societal consideration, which is widely accepted today as a moral obligation. Biophilic value, or deep innate appreciation for another living entity, is little understood. However, qualities of plants and animals that are attractive to us appear to reinforce this affiliative instinct. Beauty, power, and mimetic behavior are examples, and the wolf has such qualities in full measure.

From my own studies long ago of wolf behavior, and my contacts then and since with several of Durward Allen's students, I have great appreciation for the determination needed to document fully the nature of the wolf and its complex of relationships with people, their domesticated animals, and the environment in general. Impressively emergent from the accounts in this volume is the extraordinary flexibility or adaptability of the wolf, rivaling our own plasticity in many ways. As people take away such impressions and extract detailed information from this great compendium, they too will surely come to have greater appreciation for the wolf and, I hope, greater commitment to its survival as a significant part of diverse environments throughout the northern hemisphere. These outcomes will well honor our friend Ulie

Seal, collaborator with Dave Mech and others on metabolic studies of wolves. Ulie, as chair of the Conservation Breeding Specialist Group of the Species Survival Commission, IUCN, was a man who dedicated much of his professional life to helping conserve biological diversity around the world by applying rigorous science along with his own very persuasive style of communication.

Preface

MORE BOOKS MAY have been written about wolves than about any other wildlife species. A quick survey of our bookshelves yields a count of forty-one. So why another one? Because none of the others presents the comprehensive, up-to-date, and documented treatment that we strove for in this one. None of them combines the 350 person-years of research experience with, and knowledge of, wolves that our authors bring to the project. We have attempted here to synthesize as much of the scientific and scholarly literature on the wolf as we could.

And the time for this project is right. Not only has the study of wolf biology blossomed during the last few decades, but wolf populations themselves have proliferated, expanding their ranges into areas that have not heard their howls for over half a century. Thus new generations of humans are viewing the wolf in new contexts, and with fresh perspectives.

Science, too, has now scrutinized the wolf in a revolutionary way. The new tools and techniques that became available just after the most recent book that synthesized wolf information was published (Mech 1970) yielded a treasure trove of information undreamed of at that time. Radio-tracking, first applied to wolves in the late 1960s (Kolenosky and Johnston 1967; Mech and Frenzel 1971a), has since unlocked multitudes of puzzles about the wolf's life. In Minnesota alone, about a thousand wolves have been radio-tracked (Mech 1973, 1979a, 1986, 2000c; L. D. Mech, unpublished data; Van Ballenberghe et al. 1975; Fritts and Mech 1981; Berg and Kuehn 1982; Fuller 1989b). In Alaska, over a hundred radio-tagged wolf packs had been studied by 1991 (Stephenson et al. 1995),

and such studies are ongoing. Nine monographs based on radio-tracking of wolves have been published (Mech and Frenzel 1971a; Van Ballenberghe et al. 1975; Fritts and Mech 1981; Peterson, Woolington, and Bailey 1984; Ballard et al. 1987, 1997; Gasaway et al. 1983; Mech et al. 1998; Theberge and Theberge 1999). Radio-tracking produced detailed information about wolf movements and spacing, but also fostered and aided in many other dimensions of wolf research.

Although most of the above work was done in North America, field research also began on wolves in other parts of the world at about the same time and is now flourishing. Pulliainen (1965) in Finland, Zimen and Boitani (1975) in Italy, and Bibikov (1975) in Russia paved the way, and the formation of the World Conservation Union's (IUCN) Wolf Specialist Group in 1975 (Pimlott 1975) fostered international exchange of information about wolves and wolf research techniques. At present, wolf studies have been conducted or are under way in Saudi Arabia, Israel, India, Mongolia, Romania, Croatia, and several other countries. These investigations have added considerably to our knowledge about the wolf.

So both wolf populations and knowledge about wolves have expanded so much in the last 30 years that a new synthesis of information was imperative. When we pondered this problem, we concluded that the only way such a project could get done was as a collaborative effort. As the amount of information about wolves expanded, certain scientists specialized in and became authorities on various aspects of wolf biology. We chose each of them to cover these specialties in separate chapters. A total of twenty-two authors have contributed to this book.

To launch the project, we held a two-day meeting of the chapter authors in Tuscany, Italy, in 1994. There each chapter's senior author presented his or her proposed chapter outline for all of us to critique. Areas of overlap were identified and resolved and missing material was added. Enthusiasm flourished, and we all dispersed and set about to tackle our assignments.

Then reality set in. Even when the total task was divvied up, it proved formidable. Each author had to superimpose his or her contributions on an already full schedule of teaching or research. Instead of taking two years to complete the project, it took seven. Meanwhile, more wolves were radio-tagged, more data came flooding in, and more papers were published.

Of course, there is no end to the data stream, so all we could do is work in the latest findings as they appeared for as long as the production process allowed. The ongoing studies of the biology of the reintroduced wolf population in Yellowstone National Park are particularly productive and will greatly promote our understanding of the wolf. We have tried to incorporate as many of the new findings from Yellowstone as we could, but certainly more will be forthcoming.

We hope this book will provide a lasting foundation on which to build new studies. We also hope it will promote a much better understanding of the wolf and foster ecologically sound wolf management. Together, research, public understanding, and proper management should help minimize the inevitable conflicts between wolves and humans and better the chances for wolf conservation worldwide. To this end, the authors and editors of this book are donating all of its royalties to the International Wolf Center.

Acknowledgments

MANY PEOPLE other than the authors contributed to this book directly and indirectly, and all deserve our hearty thanks.

The book was conceived as a way of thanking Valerie Gates for her strong interest in educating the public about wolves and her generous contributions to the International Wolf Center to foster that work; she inspired this project.

Also integral to the whole process were the outline and chapter reviewers. In keeping with science's peer review process, each chapter was critiqued by at least two colleagues. We promised to thank them publicly without indicating what chapter they reviewed, so that their critiques could remain anonymous. Several of the reviewers were themselves authors of other chapters. We very gratefully acknowledge the important contributions of the following reviewers in helping to improve this book: Layne Adams, Cheryl Asa, Warren Ballard, Marc Bekoff, Troy Best, Tim Caro, Bruce Dale, Stephen Forbes, Todd Fuller, Robert Haight, Fred Harrington, Phil Hedrick, David Hunter, Mark Johnson, Jane Packard, Rick Page, Michael Phillips, Douglas Smith, Robert Stephenson, and John Theberge.

Throughout the manuscript preparation process, our editor, Christie Henry, of the University of Chicago Press, was most patient and tolerant. As each new deadline for delivery of the manuscript passed unfulfilled, she understood and never tried to rush us. We suspect that her experience with other such projects served us well. We are most grateful to her.

Many other people contributed to this project. W. A. Fuller translated Russian manuscripts; U. Banasch, D. Dekker, W. A. Fuller, D. Boyd, and H. A. Whitelaw provided valuable editorial reviews of the chapter on wolf relations with non-prey; S. Fritts and M. Jimenez allowed access to their files concerning wolf-bear interactions; and numerous volunteers from the Yellowstone wolf project contributed many observations from the field, especially R. McIntyre and K. Murphy. Kinard Boone prepared the figures for chapter 9. Nick Federoff helped develop the account of the origin and hybridization of the domestic dog. Richard Tedford provided up-to-date information on the early evolution of canids, and David Waddington assisted with statistical analysis for chapter 9. Part of the work in that chapter was supported by U.S. Fish and Wildlife Service grant 1448-40181-98-6-073. Several biologists contributed information to a questionnaire used in chapter 4. Several individuals made outstanding contributions to the red wolf program and deserve specific recognition: Curtis Carley, Warren Parker, John Taylor, Chris Lucash, Michael Morse, Art Beyer, John Windley, and Jennifer Gilbreath.

Last but not least, we appreciate the support of the agencies and institutions that employ us and support and facilitate our work, including the Biological Resources Division of the U.S. Geological Survey, the North Central Research Station, the University of Minnesota, and the University of Rome.

We sincerely thank all of the above.

Introduction

L. David Mech and
Luigi Boitani

THE WOLF IS TRULY a special animal. As the most widely distributed of all land mammals, the wolf, formally the gray wolf *(Canis lupus),* is also one of the most adaptable. It inhabits all the vegetation types of the Northern Hemisphere and preys on all the large mammals living there. It also feeds on all the other animals in its environment, scavenges, and can even eat fruits and berries. Wolves frequent forests and prairies, tundra, barren ground, mountains, deserts, and swamps. Some wolves even visit large cities, and, of course, the wolf's domesticated version, the dog, thrives in urban environments.

Such a ubiquitous creature must, as a species, be able to tolerate a wide range of environmental conditions, such as temperatures from $-56°$ to $+50°$C $(-70°$ to $+120°$F). To capture its food in the variety of habitats, topographies, and climates it frequents, the wolf must be able to run, climb, lope, and swim, and it performs all these functions well. It can travel more than 72 km (43 mi)/day, run at 56–64 km (34–38 mi)/hr, and swim as far as 13 km (8 mi) (P. C. Paquet, personal communication), no doubt aided by the webs between its toes.

The wolf leads a feast-or-famine existence, gorging on as much as 10 kg (22 pounds) of food at a time, but able to fast for months if necessary. Nevertheless, if all goes well (which for most wolves it does not), wolves can live 13 years or more in the wild (Mech 1988b) and up to 17 years in captivity (E. Klinghammer and P. A. Goodman, personal communication).

As might be expected, a widely distributed animal like the wolf varies physically all around its circumpolar home. The desert-inhabiting variety of Israel can weigh as little as 13 kg (29 pounds), whereas its northern tundra cousin can reach over 78 kg (172 pounds). The wolf's color varies across the entire black-white spectrum, with most wolves tending to be a mottled gray (Gipson et al. 2002).

Wolves live in packs of up to forty-two, but can survive even as lone individuals. Although wolf packs are usually territorial, where necessary, they can migrate hundreds of kilometers between where they raise their pups and where they take those pups in winter to follow their prey. The great variation in the wolf's environment, and in the creature's behavior and ecology as it contends with that environment, makes generalizing difficult. This problem can lead to false generalizations and misunderstanding about the animal.

Although wolves have provoked human beings by sharing their domestic and wild prey without permission, these predators have withstood humankind's resulting assault on them everywhere except where the ultimate weapon, poison, has been used. Given half a chance, wolves have responded by repopulating suitable areas with remarkable success. In the process, wolves have gained the support of many humans and have become a conservation challenge—one that is rapidly being met. This animal that sits on its haunches at the top of the food chain has become a symbol of the wilderness, an icon to environmentalists, and a poster child for endangered species recovery efforts.

Because of the strong feelings that both wolf haters and wolf advocates hold, it has been hard to sell the truth about the wolf—folks of each viewpoint resist accepting information they believe supports views opposite their own. Yet we must at least present the information as best we can. In this respect, we have tried throughout this book to draw valid generalizations where possible while indicating where lack of information precludes them.

In the following pages, we and the chapter authors have assembled and synthesized the considerable amount of information available about this fascinating animal. This book, however, is not a collection of all works on the wolf, but rather a compendium of basic information. To have included all of the worthy published material about the wolf, or even to have referenced it, would have been too much for writers and readers alike.

We start by discussing in chapter 1 the wolf's basic social ecology; that is, its pack structure and spacing, natural history, and movements. Why do wolves live in packs? The answer is not as simple as once thought. What triggers dispersal? How do wolves defend their territories? These and many other questions are examined and discussed.

Chapter 2 focuses in more intimately on the wolf and dissects the details of individual behavior, courtship and reproduction, parental care, pack social dynamics, competition, aggression, rivalry, leadership, and rank order. How important and pervasive is the wolf dominance hierarchy for which wolves are so famous in the popular literature? This chapter explores this and many other questions in depth.

Focusing even more tightly, chapter 3 discusses wolf communication. Here wolf howling and other vocalizations, scent marking, body postures, and other significant signals are covered in detail. The composition of wolf urine, the influence of hormones on communication, the nature of wolf hearing, seeing, smelling, and tasting, and the role of wolf senses in the animal's dealings with its environment constitute much of the chapter.

Once these basics are dealt with, we turn to the behavior that has brought the wolf so much infamy—its food habits—in chapters 4 and 5. Chapter 4 approaches the subject with an overview of the wolf's predatory adaptations, its digestive system, general feeding habits, specific foods, and basic hunting behavior. The wolf is shown to be superbly adapted to its carnivorous lifestyle, from its forty-two teeth and massive jaw muscles to its large stomach capacity and thorough digestion.

Picking up where chapter 4 leaves off, chapter 5 examines the challenges wolves face in trying to procure their prey. With quarry ranging from the fleet and alert white-tailed deer to the massive moose, muskox, and bison, the wolf must find, catch, and kill regularly. How does it go about overcoming all the many antipredatory adaptations these animals have evolved? The many details of this vast subject constitute a large part of this chapter. Just as important, however, is the chapter's other main topic—assessing the influences of wolf predation on prey populations. Lack of agreement characterizes several aspects of this topic.

Less controversial, but still a dominant aspect of wolf behavior, ecology, and conservation, is the subject of chapter 6: wolf population dynamics. Wolf productivity, density, survival, mortality, population change, and population regulation are all discussed and analyzed. Relationships between food supply (prey biomass) and several demographic and ecological factors are explored, and the pervasive effect of food supply becomes apparent.

After all the above extensive and external topics are covered, chapter 7 turns to look inside the wolf. "The Internal Wolf: Physiology, Pathology, and Pharmacology" first examines hormonal aspects of wolf reproduction, then development, basal metabolic rate, and nutrition. Then it details internal threats to the wolf's health, including parasites and diseases of all sorts. Because wolf research and health management of captive wolves are becoming increasingly important, the chapter also includes a discussion of drugs used for anesthesia, capture, and treatment of wolves.

Also delving deeply into the wolf is chapter 8. Featuring the relatively new but highly enlightening techniques of molecular genetics, this chapter presents some novel perspectives on wolves. Through modern biochemical analyses, wolf DNA can be parsed, and various inferences can be drawn from the results. Some of these conclusions challenge existing ideas: was the dog domesticated from the wolf about 10,000–15,000 years ago, as has long been thought? Or much longer ago, as molecular analyses suggest?

While chapter 8 uses new technology to get at questions of wolf taxonomy and evolution, chapter 9 uses long tried and tested methodology to examine many of the same issues. Fossil evidence and skull measurements are used to trace the long path of evolution the wolf's ancestors followed, the relationships among wolves of different geographic regions, and the possible hybridization of closely related forms. Nicely complementing chapter 8, this chapter provides one more important perspective on these subjects. We hope that interested readers will examine chapters 8 and 9 closely and judge for themselves where the weight of evidence falls in any of the areas where their conclusions differ. The conclusions of both chapters depend greatly on inference, and in many cases the data base for the inferences is necessarily more meager than science would like.

Chapter 10 deals with the many interactions wolves have with prominent animals in their environment other than their regular prey. Species such as bears, tigers, and cougars not only compete with wolves for prey, but also sometimes kill wolves or are killed by them. Other smaller creatures, such as coyotes, foxes, ravens, and eagles, scavenge from wolf kills and can consume large amounts of the wolf's hard-earned bounty. Chapter 10 outlines these interactions that form such an important part of the wolf's life.

Chapter 11 differs from the others in that it does not cover a specific aspect of the wolf, but rather is devoted to describing the recovery of the red wolf (Canis rufus). We devote a whole chapter to this subject, while covering other wolf reintroductions as parts of other chapters, because the red wolf is a special case. The red wolf is the only long-recognized species of wolf other than the gray wolf, and scientists disagree on its taxonomic identity. Is the red wolf really a separate species, or might it be a subspecies of the gray wolf? Some scientists claim that the red wolf is a hybrid between the gray wolf and the coyote, while others dispute that conclusion. All workers agree, however, that the red wolf is endangered. Thus, after several years of raising remnant members of its

population in captivity, the United States Fish and Wildlife Service reintroduced the red wolf into part of its former range. Chapter 11 follows the progress of that historic endeavor and summarizes the lessons learned from it.

That topic leads nicely into the larger subject of chapter 12, the wolf's interactions with human beings. It is that very subject that makes this book possible and necessary. The chapter tracks human attitudes toward wolves through history and many various cultures. It documents the effects of wolves on human activities and vice versa, including wolf attacks on human beings and human extermination of entire wolf populations.

Concluding the book after the thought-provoking chapter 12 is our chapter on conservation of the wolf. Although human beings were responsible for the demise of the wolf throughout much of western Europe and most of the contiguous United States, the human species suddenly realized its mistake and began trying to conserve the wolf and restore it to parts of its former range. Chapter 13 details these efforts and projects an optimistic future for the wolf—a fitting conclusion to this book.

Down through the ages, the wolf has never had a neutral relationship with humanity. It has either been hated, despised, and persecuted or revered, respected, and protected. It has been, and continues to be, a subject of myth and legend, folklore and fairy tale. This book is meant to temper those misrepresentations by presenting a scientific view of the animal—one that we think is even more interesting, and certainly more accurate.

1

Wolf Social Ecology

L. David Mech and
Luigi Boitani

THE FIRST REAL BEGINNING to our understanding of wolf social ecology came from wolf 2204 on 23 May 1972. State depredation control trapper Lawrence Waino, of Duluth, Minnesota, had caught this female wolf 112 km (67 mi) south of where L. D. Mech had radio-collared her in the Superior National Forest 2 years earlier. A young lone wolf, nomadic over 100 km^2 (40 mi^2) during the 9 months Mech had been able to keep track of her, she had then disappeared until Waino caught her. From her nipples it was apparent that she had just been nursing pups.

"This was the puzzle piece I needed," stated Mech. "I had already radio-tracked lone wolves long distances, and I had observed pack members splitting off and dispersing. My hunch was that the next step was for loners to find a new area and a mate, settle down, produce pups, and start their own pack. Wolf 2204 had done just that."

During the decades since, we have seen this process many times, and it represents one of the primary ways in which wolves become breeders (Rothman and Mech 1979). However, there are several other ways, and it is only now, after 25 years of study and the wedding of wolf radio-tracking with biochemical analyses of wolf genetics (see Wayne and Vilà, chap. 8 in this volume), that we seem to have a reasonably complete picture of wolf social ecology (Meier et al. 1995; D. Smith et al. 1997; Mech et al. 1998).

Wolf Packs and Pairs: The Basic Social Units

The basic social unit of a wolf population is the mated pair. Known variations include a mature male and two

mature females; a mature male, his yearling son from a previous mating, and a new mate; and a mature female with a new mate and his younger brother (Mech and Nelson 1990b). There is no reason to believe that other similar combinations of a mated pair with various relatives of one or both members are not also possible.

There are two reports of packs of males, but these packs are not well documented or understood, and presumably are temporary until a mate is found. Ballard et al. (1987) reported without documentation that a pack of three males occupied a 3,077 km^2 (1,200 mi^2) area of Alaska for over a year. Two radio-collared males split off from a Montana pack and lived together from June to September before being joined by a third animal of unknown age and sex (Ream et al. 1991).

The most unusual type of pack ever recorded formed in Yellowstone National Park 7 years after wolf reintroduction (D. W. Smith, unpublished data). During winter 2001–2002, three packs were formed of various assortments of at least twelve dispersers from four packs. Each new pack included a Druid Peak pack female born in 1997. Individuals moved among these packs, sometimes daily. By late spring, one pack contained two males from the Chief Joseph pack and four Druid Peak females. These wolves produced two litters in separate dens, merged in midsummer into six adults and four pups, and remained such at least into winter. Less is known about the other two new packs.

Mech also once recorded an adult male, his yearling son, and his three pups remaining together for 10 weeks after his mate (wolf 5091) was killed by other wolves (Rothman and Mech 1979). This situation can be considered a temporary exception; a new mature female (5079) joined the pack after 10 weeks and remained with it, producing pups the next spring.

The natural extension of the mated wolf pair is the pair with its collection of offspring, or family, as earlier workers surmised (Olson 1938; Murie 1944; Young and Goldman 1944) and numerous radio-tracking studies have documented. In a thriving population, a wolf pair produces pups every year (Fritts and Mech 1981; Mech and Hertel 1983; Peterson, Woolington, and Bailey 1984; but cf. Mech 1995d). The offspring usually remain with their parents for 10–54 months, but except under special circumstances, all offspring disperse (Gese and Mech 1991; Mech et al. 1998). Packs therefore may include the offspring of as many as 4 years. A wolf pack, then, is some variation on a mated pair, and packs have contained as many as forty-two members, although most include far fewer (see table 1.1).

Adoptees

One poorly understood exception to the above basic rule is that strange wolves sometimes join packs already containing a breeding pair, at least temporarily (Fritts and Mech 1981; Peterson, Woolington, and Bailey 1984; Messier 1985b; Ballard et al. 1987; Mech 1991b; Boyd et al. 1995; Meier et al. 1995). We will refer to these animals as "adoptees" (Meier et al. 1995) to distinguish them from wolves that enter a pack to replace a lost breeder (see below). Most adoptees are males, and most adoptions take place from February through May (Messier 1985b; Meier et al. 1995).

One of the main mysteries of this behavior is why strange wolves are sometimes allowed to join packs, whereas in so many other cases they are chased, attacked, or killed (Mech 1993a, 1994a; Mech et al. 1998). A clue may be the fact that most adoptees are 1–3 years old (Messier 1985b; Meier et al. 1995), whereas a high percentage of wolves killed by other wolves are adults (Mech 1994a; Mech et al. 1998). Tests with captive wolves confirm that degree of aggressiveness depends on the rank, age, and residency status of the wolves involved (Fox et al. 1974).

The incidence of packs adopting strange wolves would be very difficult to measure without sampling each wolf in every pack of a population and resampling over time. Based on genetic determinations, nine of twenty-seven packs from three study areas included ap-

parent adoptees (Lehman et al. 1992). However, most members of most packs were not sampled, and the sampling was done over several years. In an Alaskan population subject to harvesting by humans, over 21% of the wolves that dispersed over a 7-year period were accepted into other packs (Ballard et al. 1987). These diverse sampling schemes, plus the fact that adoptees remain in packs for periods of only a few days to over a year, preclude an estimate of the proportion of adoptees at any given moment. A rough guess might be 10–20%, and this proportion could well vary by time and place. (Additional information about adoptees can be found in the discussion of multiple breeding below.)

Pair Formation

As in the case of wolf 2204, described above, one of the main methods of pair formation is for dispersing wolves of the opposite sex to find each other. However, there are several other methods ("strategies") of pair formation.

To understand the various breeding strategies wolves use, we must first make it clear that every wolf is a potential breeder, and as each begins to mature (see Kreeger, chap. 7 in this volume), its tendency will be to try to breed. This idea is contrary to earlier views that some wolves relinquish breeding "for the good of the species" (Rabb et al. 1967; Woolpy 1968; Mech 1970; Van Ballenberghe et al. 1975; Haber 1977).

Detailed studies of captive (Packard and Mech 1980; Packard et al. 1983, 1985) and wild wolves (Mech 1979a; Fritts and Mech 1981) show that many young wolves merely defer reproduction while still in their natal packs. In the basic social life of the wolf, this strategy can now be seen as merely a natural result of breeding competition, much like the failure to breed of many young male ungulates that lose in their competition with mature bulls.

The wolf population is comprised of tight, territorial social groups. To breed successfully, individual wolves must find a mate and a territory with sufficient food resources (Rothman and Mech 1979). In a saturated population, all territories are occupied, so the only local breeding possibilities will be to (1) wait until the established breeding position opens (A) in the natal pack or (B) in a neighboring pack, (2) become an extra breeder within the pack, (3) carve out a new territory from the established mosaic, or (4) usurp an active breeder.

Local Breeding Strategies

Wolves attempt all the above strategies and more. In Minnesota, a 2-year-old female bred with her stepfather after her mother was shot (Fritts and Mech 1981), illustrating strategy 1A above. The immigration of neighboring wolf 5079 into the pack described above after its breeding female (wolf 5091) was killed by other wolves illustrates strategy 1B; in this case, 5079 had produced pups in a neighboring pack the year before and apparently had lost them (L. D. Mech, unpublished data). Other cases of outside lone wolves joining existing packs to replace lost breeders have been documented by Fritts and Mech (1981), Mech and Hertel (1983), Peterson, Woolington, and Bailey (1984), and Stahler et al. (2002).

In some cases, wolves leave their pack but remain in the pack territory as "biders," presumably waiting for a chance to breed (Packard and Mech 1980). Such wolves have solved one of the two parts of their breeding problem, finding a territory with resources. However, they may have to wait for a parent to perish before they can breed. Lindstrom (1986) believed that in red foxes, biding might be the only type of breeding option for a weak individual.

Multiple Breeding

Rather than replacing a pack breeder, some maturing wolves breed in addition to the pack's established breeders while remaining in their natal pack. Such multiple breeding is favored by close genetic relatedness among the pack members (see below). Although some pertinent details about this behavior are still lacking, the behavior itself is well documented (Murie 1944; Rausch 1967; Clark 1971; Haber 1977; Harrington et al. 1982; Van Ballenberghe 1983b; Packard et al. 1983; Peterson, Woolington, and Bailey 1984; Ballard et al. 1987; Meier et al. 1995). In no case was the relationship between or among the multiple-breeding females known, but one suspects that the breeding females were mother and daughter, because the known structure of wolf packs (see above) suggests that strange females are adopted into packs only rarely (Meier et al. 1995) unless the breeding female is lost (see above).

The important unanswered question when more than one female in a pack breeds is, which male bred the extra (nondominant) female? The likely suspect would be the dominant male, even if the extra female were his daughter, since close inbreeding is well known in captive wolves (Medjo and Mech 1976; Packard et al. 1983; Laikre and Ryman 1991) and has long been considered common for wild wolves (Haber 1977; Woolpy and Eckstrand 1979; Theberge 1983; Shields 1983; Peterson, Woolington, and Bailey 1984). However, recent genetic studies of mated wolf pairs from the Superior National Forest (Minnesota) and Denali National Park (Alaska) populations indicated that inbred pairings were probably rare (D. Smith et al. 1997). This means chances are good that extra matings in a pack may be with immigrants from other packs, or even with outsiders through temporary liaisons. Or, if daughters of the dominant female breed, this could explain the role of adoptees (see above) and why most adoptees are males (Peterson, Woolington, and Bailey 1984; Messier 1985b; Meier et al. 1995).

Adoptee males may become interested in maturing females, which would explain their attraction to new packs. Sometimes such adoptees remain in their new pack from days (Peterson, Woolington, and Bailey 1984) to months (Fritts and Mech 1981) to over a year (Meier et al. 1995; M. E. McNay, personal communication). In Denali, an adoptee left his new pack after a year and was observed just outside the pack's territory with another wolf (a maturing female from the pack?); the adoptee and his mate produced pups in an adjacent territory the next year (Meier et al. 1995).

On the other hand, interest in a maturing female is not always an apparent motive for adoptees joining a pack, or for breeding pairs allowing them to do so. A male wolf radio-collared as a 10-month-old in Alaska during 1995 remained with his natal pack until June, then joined a breeding pair and their pups 58 km (36 mi) away in July, and remained with them at least through January and in their territory until the next July (M. E. McNay, personal communication). The pack had no maturing female when the adoptee joined, and when the pack's pups began maturing, the adoptee left.

Another situation in which multiple females in a pack could breed without inbreeding is when the father of maturing females is lost and replaced by a new male. This stepfather is then unrelated to any pack female and could breed any of them without inbreeding (Stahler et al. 2002).

For two reasons, it seems logical to suggest that multiple breeding is possible only when food supplies are flush (Mech et al. 1998), a hypothesis similar to the suggestion that multiple breeding is fostered by heavy exploitation (Ballard et al. 1987). First, ample food would be required for more than one female to gain sufficient nutrition to produce pups; young pack members receive less food when it is scarce (Mech 1988a; Mech et al. 1998). Second, as will be discussed, maturing members are more likely to remain with the pack when food is more plentiful, whereas aggression increases when food is scarce.

Previous workers have emphasized the importance of social and behavioral factors in prompting dispersal (Haber 1977; Harrington et al. 1982). While these factors may be involved, we believe that nutrition stress underlies them, as social competition is very much a function of food abundance (see below).

Regardless of the uncertainties about various aspects of extra litters per pack, multiple breeding represents a viable strategy by which some wolves succeed in the breeding arena.

Budding and Splitting

Another breeding strategy is for a dispersed wolf and its new mate to try to set up a territory along the edges of its natal pack territory; this approach can involve either a male or a female from the natal pack. The animal frequents one end of the territory, presumably pairs with a floater (see below) or a similar member from a neighboring pack, and forms a territory adjacent to, or overlapping with, its natal territory, a process known as "budding" (Fritts and Mech 1981; Fuller 1989b; Meier et al. 1995; L. Boitani, unpublished data). Budding conforms to the territory inheritance hypothesis, which attempts to explain why group living in carnivores is a stable strategy (Lindstrom 1986).

A variation on this strategy is pack splitting. Pack splitting differs from budding in that, rather than a single wolf budding off a pack with a mate, a group of wolves splits off and assumes a new territory. Pack splitting in this sense is not the same as the temporary splitting of large packs during winter (Mech 1966b, 1970; Haber 1977; Carbyn et al. 1993). Rather, pack splitting as a form of budding is a permanent phenomenon.

Several cases of permanent pack splitting have been reported, all involving larger-than-average packs during or around the breeding season (Mech 1986; Meier et al. 1995; Hayes et al. 2000). In Denali, a pack of twenty split into two packs of eleven and nine and split the territory; at least one of the new packs produced pups that year (Meier et al. 1995; Mech et al. 1998). In one recolonizing population, packs split when they averaged twelve ($\pm$ 1.5) wolves, and after 4 years of recolonization,

nine of twenty-eight (32%) packs were the products of pack splitting (Hayes and Harestad 2000a; Hayes et al. 2000).

It is probably when two related breeding pairs are present that packs split, perhaps after an immigrant male breeds a pack daughter. Presumably the additional members of the subunits are the previous offspring of each pair. Because breeders control the feeding of their offspring (see Packard, chap. 2 in this volume), they may compete too aggressively with other pack breeders as food needs peak in winter because of maximal pup weights. A solution that circumvents mortal competition among kin is to split the territory and resources (Mech 1970). This may be necessary only when food is scarce, thus explaining why large packs do not split every year.

Carving Out New Territories

Dispersers can also breed locally by carving new territories out of the existing pack territorial mosaic. Dispersers using such a strategy wander around the population ("floaters"), frequent areas along the interstices among territories (Mech and Frenzel 1971a; Rothman and Mech 1979; Fritts and Mech 1981; Meier et al. 1995), meet members of the opposite sex, mate, and attempt to set up a new territory (Rothman and Mech 1979). In some areas, however, such as parts of Quebec (Messier 1985b) and Denali (Mech et al. 1998), lone wolves do not seem to frequent pack territory edges and interstices. In any case, lone floaters may circulate over areas of 10,500 km^2 (4,100 mi^2) or more, many times the size of local pack territories (Mech and Frenzel 1971a; Fritts and Mech 1981; Berg and Kuehn 1982; Merrill and Mech 2000; Wabakken et al. 2001).

Often loners frequent two or three areas along various pack territory edges and float long distances among them until they meet a mate in one of them; then they settle (Mech and Frenzel 1971a; L. D. Mech, unpublished data). In a recolonizing population in northwestern Minnesota, three floaters that were monitored for more than 4 months all paired, and at least two of the pairs produced pups; in the same population, seven of eight dispersers paired, usually within 20 days of dispersal (Fritts and Mech 1981). Although most lone wolves float independently, three pairs in this recolonizing population formed and then floated together, exploring areas until they found one to settle in (Fritts and Mech 1981). The only other area where this strategy seems to have been reported was Scandinavia (Wabakken et al. 2001).

Whether any of the pairs that attempt to carve out territories in an established population succeed depends in part on food abundance in the population. In the re-colonizing population of northwestern Minnesota, these pairs tended to succeed (Fritts and Mech 1981), whereas 250 km eastward in the saturated, food-stressed Superior National Forest (SNF) population in the early 1970s (Mech 1977b), they tended to fail (L. D. Mech, unpublished data).

In the SNF during 1969–1989, a time that included periods both of food stress and of improved conditions, 65% of those wolves that dispersed as adults, 26% as yearlings, and 8% as pups succeeded in pairing and denning (Gese and Mech 1991). In an increasing wolf population in Denali National Park during 1986–1991, nine (56%) of sixteen new pairs succeeded in founding new packs that lasted a year or more (Meier et al. 1995).

Usurping a Breeder

A last way in which maturing wolves can breed in their own population is to usurp an established breeding position. An example of this approach was seen in the SNF, where a 3-year-old female bred with her stepfather a year after her mother bred with him and left (Mech and Hertel 1983); whether the mother was ousted or left voluntarily is unknown. On Ellesmere Island, a 3-year-old daughter took her mother's breeding role while the mother remained in the pack as a helper (Mech 1995d). In this case, the male had been the mother's mate for 2 years; he could have been the daughter's older sibling or an unrelated wolf, but probably was not the daughter's father.

No doubt the most dangerous strategy for gaining a breeding position would be to challenge an established breeder. Such challenges have been observed in captive situations, where yearling sons challenged their fathers and bred with their mothers (Zimen 1976; Packard et al. 1985). However, such fights that could become mortal in captivity might never take place in a wild pack, where a beaten contender can escape; furthermore, the best evidence so far is that close inbreeding does not occur where outbreeding is possible (D. Smith et al. 1997).

Nevertheless, wolves do often fight to the death in the wild (see below), and the losers are usually wolves encountered near a territory edge or inside a neighbor's territory (Mech 1994a; Mech et al. 1998). A disproportionate number of the dead wolves are adult breeders, but subordinate, maturing animals are also killed. There is a strong possibility that some of these fights result

from potential breeders challenging established breeders. The best such record was Messier's (1985b) observation in Quebec that a presumed breeding male was killed one March at the time his pack adopted a young immigrant male.

An incident that L. D. Mech (unpublished data) observed in the SNF during the breeding season (Mech and Knick 1978) also suggests such a challenge. The SNF Greenstone pack (four members) trespassed south of its southern neighbor, the Pagami Lake pack (five members), on 15 February 1972, then returned to its territory. The next day, Mech watched as the Greenstone pack entered the Pagami pack's territory from the south and attacked the sleeping five. At least one wolf from the Pagami pack was wounded, and the Greenstone pack returned to its territory. The only radio-collared Pagami wolf was alone the next eight times it was seen during the next month, and then dispersed. The one radio-collared Greenstone wolf was not seen with more than two others during the next twelve observations through 13 March; then her signal was lost. By fall, however, a newly radio-collared pup was part of a pack of six living in the former territories of both packs. Did the neighboring breeders form one pair after the fight, oust the others, and usurp both territories?

In Denali, the McKinley River pack (ten members) invaded the territory of the Bearpaw pack (also ten members) and, between January and March 1988, killed all three radio-collared members of the Bearpaw pack, wounded at least one other member, and may have killed two others (Meier et al. 1995). Two McKinley River wolves and two new wolves (former Bearpaw members?) then usurped the Bearpaw pack territory, even using the Bearpaw pack den.

Distant Dispersal

Besides the several strategies described above for obtaining a breeding position in the local population, wolves also use a strategy that takes them into a new population or to the very edge of the species' range. This strategy, called directional dispersal (Mech and Frenzel 1971a; Mech 1987a), is a tendency to move a long distance in more or less a single direction. Wolves of both sexes have dispersed to areas up to 886 km (531 mi) away (Fritts 1983; Ballard et al. 1987; Boyd et al. 1995), and some have crossed four-lane highways and open areas and circumvented large lakes and cities (Mech, Fritts, and Wagner 1995; Merrill and Mech 2000; Wabakken et al. 2001; L. Boitani, unpublished data). When long-distance dispersers settle, they may attempt to squeeze into the territorial mosaic of a distant population, join an existing pack, or pair with a member of the opposite sex in an area uninhabited by breeding wolves (Rothman and Mech 1979; Fritts and Mech 1981; Berg and Kuehn 1982; Peterson, Woolington, and Bailey 1984; Messier 1985b; Ballard et al. 1987; Fuller 1989b; Meier et al. 1995; L. D. Mech, unpublished data).

Frequency of Various Strategies

The relative proportions of potential breeders that use these various breeding strategies have not been measured (but see below). Those proportions must vary over space and time and depend a great deal on food supply and whether the population is increasing, decreasing, or stable (see Fuller et al., chap. 6 in this volume). However, a general idea of those proportions can be obtained from the proportions of wolves of various ages that disperse and the distances they move. Near-dispersers would include those wolves that attempt to breed with neighbors through biding, budding, or replacing established breeders. Distant-dispersers would be those that chance finding or founding new populations.

Some information on proportions of breeding strategies can be gleaned from both the SNF and Denali studies. In the SNF population, which between 1969 and 1989 declined, stabilized at a low level, and then increased, the pairing success of some seventy-five wolves that dispersed from their packs was examined (Gese and Mech 1991). A significantly greater proportion of maturing animals dispersed during the declining and increasing phases than during the stable phase, probably reflecting the least competition during the stable phase. Most of the wolves dispersing at less than 1 year of age traveled more than four territories away, whereas most yearlings and adults remained within a radius of three territories. (More details about dispersal are presented below.)

In the increasing Denali population, sixteen new pairs formed in 1986–1991 (Meier et al. 1995). Two of these pairs died out without producing pups; five produced pups, but failed to hold their territory beyond a year, in most cases because the adults were killed by other wolves; and nine produced pups and held territories for a year or more. Of the nine successful pairs, it is significant that at least seven succeeded through "budding," or carving out a territory partly inside or just adjacent to their natal territory.

The Breeding Flux

Competing with maturing wolves for new breeding positions are lone adults that have left or lost their mates or breeding positions. Individuals such as wolf 5079 in the SNF, mentioned above, as well as examples recorded by Fritts and Mech (1981), Peterson, Woolington, and Bailey (1984), Mech (1987a), L. D. Mech (unpublished data), Ream et al. (1991), and Meier et al. (1995), indicate that many adults join the floating members of the wolf population to compete with the maturing members. These adults tend to remain within 50 km of the area they leave, at least in Minnesota (Gese and Mech 1991).

Given all the above breeding strategies, a wolf population can be viewed as a highly dynamic system in which breeding pairs hold territories and pump out numerous offspring that travel about, criss-crossing the population and striving to gain their own breeding positions. In this flux, each pack tries to hold its position while competing with neighbors that try to expand their territories (see below) as well as with new breeding pairs, local lone wolves, and immigrants that are all trying to leverage themselves into the population structure.

The flexibility in the sizes of wolf packs and territories helps buffer the constant fluctuations in social and ecological factors that wolves face. Wolf populations are constantly churning, and a high proportion of their members are temporary. In the Denali National Park population, which is one of the least human-disturbed wolf populations anywhere, only 15% of wolves under 3 years of age remained in the population for more than 5 years (Mech et al. 1998). Thus at least some of the population's long-term breeders must be immigrants, another indication of the constant genetic mixing of the population.

Why Do Wolves Live in Packs?

The wolf and the wolf pack are as closely linked in the human mind as a child is linked to a family, and rightly so. The human family is a good analogy for the wolf pack. The basic pack consists of a breeding pair and its offspring, which function in a tight-knit unit year-round. As with humans, male wolves generally are larger than their mates, about 20% heavier in general (Mech 1970).

The offspring of the breeding pair often include members of more than one litter. Wolf pups reach adult size by winter, so the presence of pups then gives the pack the appearance of a group of adults. Because at least some young often remain with the pack for a year or more, when new pups are born, the social group constantly appears to contain more than a pair of adults.

Why do wolves remain with their parents for as much as 10–54 months while many other mammals leave sooner? At least some wolf pups can survive without their parents when as young as 4 months of age (Fritts et al. 1984, 1985). Their permanent canine teeth are in place by 7 months (Van Ballenberghe and Mech 1975), their long bones cease growth by 12 months (Rausch 1967), and at least some males and females are capable of breeding at 10 months (Medjo and Mech 1976).

The Pack as Nursery

One answer might be that there is great variation in wolf maturation. Some wolves are not reproductively capable even at 3 years of age (Mech and Seal 1987). Physiologically, wolves may not be completely "mature" until about 5 years of age. U. S. Seal et al. (unpublished data) found that wolf androgen and estrogen levels increased until this age. Thus the continued association of young wolves with their natal pack may simply be a way for the young to mature while still being subsidized by their parents. From the parents' standpoint, caring for young until they are mature may be the best way to ensure their original investment. In addition, long association with parents would increase the opportunity for offspring to learn the more subtle components of hunting and foraging behavior that are not innate (Leyhausen 1965, cited in Eaton 1970).

Pack Size and Prey Size

On the other hand, there is some evidence that wolf pack sizes may be influenced by other factors. There has been much theoretical discussion of carnivore group sizes (Murie 1944; Mech 1970; Kleiman and Eisenberg 1973; Zimen 1976; Bekoff and Wells 1980; Rodman 1981; Bowen 1981; Lamprecht 1981; Brown 1982; D. W. Macdonald 1983; Packer and Ruttan 1988; and others). Theory holds that pack size should vary with prey size up to some optimum number; this optimum should be that which allows predation with the least energy expenditure and the most energy return (D. W. Macdonald 1983).

Wolf pack sizes tend to be largest where wolves prey on the largest ungulates. Despite records of hundreds of wolf packs from many areas, however, the relationship of pack size to prey size is not definitive (see Fuller et al., chap. 6 in this volume). This is partly because of the extreme variation in pack size within each area and because in many of the areas studied the wolves were subject to harvesting or control.

Pack size data are available for relatively unexploited wolf populations in Minnesota, Denali National Park, Alaska, Wood Buffalo National Park, Alberta, and Yellowstone National Park. Other data from exploited populations tend to support these data, but are less definitive because of the possible effect of exploitation. The smallest packs tend to feed on garbage and small animals, and the largest on moose and bison (table 1.1).

However, this pattern is only a very general tendency (Mech 1970). For example, in 1971–1991, the mean pack size for Isle Royale, in Lake Superior, Michigan, where moose are the only ungulate prey, was 7.5, whereas for north-central Minnesota, where white-tailed deer were the exclusive prey, pack size averaged 7.3 (see table 1.1). Average pack sizes for wolves feeding on deer and moose are significantly smaller than for those feeding on elk and caribou (see Fuller et al., chap. 6 in this volume). Nevertheless, the largest packs where moose and bison were preyed on were twice as large as the largest packs from deer areas (see table 1.1).

Complicating Factors

As discussed above, it is reasonable to try linking group size to prey size. Some of the earliest wolf biologists assumed that wolf packs exist because they may promote greater hunting efficiency (Murie 1944), and this conclusion seems logical (Mech 1966b, 1970; Zimen 1976; Peterson 1977; Nudds 1978; Carbyn et al. 1993). Several important factors, however, complicate the picture.

If large numbers of wolves were necessary to prey on large ungulates, it would be difficult for lone wolves and pairs to survive and produce the offspring that enlarge the pack. In fact, large numbers of wolves are not necessary to kill large prey. Single wolves have been recorded to kill even the largest of the wolf's major prey species, including adult moose (Cowan 1947; A. Bjärvall and E. Isakson, personal communication; Thurber and Peterson 1993; Mech et al. 1998), muskox (Gray 1970), and bison (D. Dragon, cited in Carbyn et al. 1993).

TABLE 1.1. Distributions of wolf pack sizes primarily from unexploited populations using prey of different sizes

Main prey	N[a]	Pack size															$\bar{x}$[b]	Largest[c]	Source
		2	3	4	5	6	7	8	9	10	11	12	13	14	15	>15			
White-tailed deer	78	21	7	9	13	10	6	2	6	2	—	1	—	—	1	—	4.9	(17)	Mech 1986
White-tailed deer	35	3	1	2	2	6	6	2	6	2	2	2	1	—	—	—	7.3	(13)	Fuller 1989b
Moose	48	3	6	2	7	10	3	4	7	1	1	2	1	1	—	—	6.6	14	Mech 1986
Moose	50	7	4	4	3	3	6	5	1	5	4	1	2	2	—	3	7.5	18(22)	Thurber and Peterson 1993
Moose/caribou	106	8	6	4	9	7	13	13	7	8	4	8	2	1	4	12	9.1	29	Mech et al. 1998
Bison[d]	206																9.4	(42)[e]	Carbyn et al. 1993
Garbage, etc.[f]	24	12	2	2	3	2	3	—	—	—	—	—	—	—	—	—	3.6	(7)	Boitani and Zimen 1979
Garbage, small animals[f]	21	6	7	4	2	1	1	—	—	—	—	—	—	—	—	—	3.4	7	Mendelssohn 1982

[a]Pack-years, so many packs are represented during several years.
[b]Weighted.
[c]Numbers in parenthesis indicate pack sizes not reported as part of distribution of pack sizes but rather reported independently.
[d]Light to moderate exploitation.
[e]Fau and Tempany 1976; cited in Carbyn et al. 1993.
[f]Exploited population.

Even when a wolf pack attacks prey, not every pack member contributes significantly to the attack. Because of the general assertiveness and experience of the breeding pair, they tend to take the lead in chasing and attacking prey, and it is unclear how much the younger pack members contribute (see Mech and Peterson, chap. 5 in this volume). In a pack consisting of a breeding pair and their 7-month-old pups hunting for the first time, for example, it seems unlikely that the pups would assist very significantly in the kill.

Another factor that few workers have considered is that wolf pack size in the usual sense is not necessarily the same as hunting group size. Most pack size observations are made in winter, when the pack is nomadic. Thus the adults usually bring the whole family with them when they hunt. In summer, however, the den is the social center, and adults radiate out from it in foraging groups of various size (Murie 1944; Mech 1970, 1988a; Ballard, Ayres, Gardner, and Foster 1991). Even in winter, wolf packs do not always hunt at full size, especially when they are large. Most packs vary in the numbers traveling together throughout the winter (Stenlund 1955; Mech and Frenzel 1971a), as various members lag behind during travels, some visit old kills, or others disperse temporarily (see below).

In addition, packs sometimes split temporarily (but for days at a time) into smaller hunting groups, similar to the way African lion prides split (Packer et al. 1990). A pack of fifteen wolves on Isle Royale split into two groups about half the time during the 1961 winter study (Mech 1966b), and split again in 1963 and 1965 (Jordan et al. 1967). Similar pack splitting during winter has also been reported for Denali (Haber 1977), Italy (Boitani and Zimen 1979), and Wood Buffalo National Park (Carbyn et al. 1993). Therefore, published pack sizes, which are almost always stated as the maximum number of wolves observed over winter, are not necessarily hunting group sizes, thus complicating any analyses that do not consider this.

Pack Size and Hunting Efficiency

It certainly seems reasonable that, at least to some extent, hunting in groups would increase hunting efficiency even if no cooperative strategy were used. Multiple hunters, even if inept or inexperienced, would seem to yield greater sensing, chasing, restricting, attacking, and killing power than single hunters.

However, possibly offsetting this advantage is the fact that multiple hunters must also share the proceeds (Brown 1982). This and numerous other theoretical and empirical considerations have led some workers to the conclusion that "cooperative hunting is more often a consequence of gregariousness than its evolutionary cause" (Packer and Ruttan 1988,189).

A good test of the hypothesis that larger groups of wolves are more efficient at hunting or killing prey is to determine amount of food obtained per wolf for packs of various sizes. On Isle Royale during 1959–1961, the pack of about fifteen wolves mentioned above preyed on moose, but in 1961, when this pack split into two about half the time, the amount of food obtained was greater than during the previous 2 years, when the pack hunted as a unit (Mech 1966b). Similarly, lone wolves in Minnesota killed more prey per wolf than a pack of five (Mech and Frenzel 1971a), and pairs killed more prey per wolf than packs (Fritts and Mech 1981; Ballard et al. 1987, 1997; Thurber and Peterson 1993; Hayes et al. 2000).

When this hypothesis was tested more rigorously with wolves and moose on Isle Royale, the result was the same: the larger the pack, the less food obtained per wolf (fig. 1.1). Synthesizing data from many studies including most wolf prey gave the same result (Schmidt and Mech 1997).

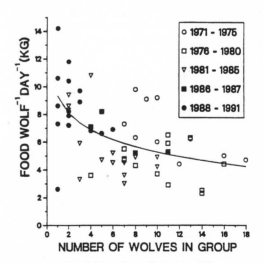

FIGURE 1.1. Food availability (kg/wolf/day) for different group sizes of gray wolves preying on moose in Isle Royale National Park during winter, 1971–1991 ($Y = 9.31 - 1.76 \log_{10}\bar{X}$). (From Thurber and Peterson 1993.)

Sharing the Surplus

What other factors might cause wolves (or other carnivores) to live in groups, then? And why do the largest packs seem to be those preying on the largest quarry? Put another way, why don't all young wolves disperse from their natal pack as soon as they are fully developed, at about 7–12 months?

It turns out that they do, at least in some areas, and these cases are instructive. During the early 1970s in Italy, when few ungulates were present, most packs consisted of little more than pairs in winter (see table 1.1). Similarly, packs are small in Israel, where wolves feed extensively on garbage and small animals (Mendelssohn 1982). That both the Italian and Israeli wolves were also subject to human exploitation confounds attempts to definitively relate small pack size to small scattered food sources, but the data are suggestive. With coyotes, pack size does relate to food source size (Bowen 1978).

If wolf pack size is related to food source size, but increased pack size does not necessarily yield greater hunting efficiency, then why live in packs? The answer seems to be that the evolution of grouping in wolves has facilitated subsidy of young wolves by their parents through the sharing of large prey (Mech 1970, 1991b; Schmidt and Mech 1997). Since adults prey on large animals, a surplus of food suddenly becomes available periodically. Making this surplus available to kin is the most efficient approach adult wolves can take, except for eating it and caching it. Without a sufficient number of feeders, this surplus can be lost to competitors, scavengers, insects, and bacteria. Ravens can remove up to 37 kg (17 pounds) of a carcass per day, and can usurp some 66% of a lone wolf's kill, compared with only 10% of the kills of a pack of ten (Promberger 1993; see also Stahler 2000).

The kin selection explanation of why wolves live in packs (Schmidt and Mech 1997) fits the resource dispersion hypothesis. This theory holds that food quantity and distribution is the primary cause and determinant of group size (D. W. Macdonald 1983; von Schantz 1984). The types of prey wolves rely on have unique characteristics of richness (a large amount of food per prey), renewal (slow turnover), and heterogeneity (highly patchy distribution and low density), which are the key conditions that the hypothesis predicts would foster group living (D. W. Macdonald 1983).

Wolf parents allow their young to remain with them so long as their food supply can support more individuals than themselves. From the offspring's standpoint, if the food supply is secure, it is advantageous for them to stay with their parents rather than trying to find resources on their own, at least until the urge to breed compels them to seek a mate outside the natal pack. Although there are no experimental results confirming this theory, the fact that pack size tends to correlate with food supply (Mech 1977a; Messier 1985a) lends support to the theory.

Clearly wolf packs that prey on smaller animals such as deer would have less surplus food available per kill than packs that prey on moose or bison. Packs preying on moose or bison could afford to include a larger number of offspring, thus improving the inclusive fitness of the family (Rodman 1981). An efficient pair of adult breeders in a moose area, then, could feed members of two or three of their last litters of offspring. This would enhance the survival of those offspring and increase the chances of the parents' own genes being disseminated. Inclusion of these maturing wolves on hunting forays would also give them practice and experience in hunting.

If maturing wolves accompany their parents in packs to gain easy forage, this may explain why large packs are not necessary to take large prey, yet the largest packs are usually found in areas with the largest prey. Simply put, large prey allow large packs, but do not require them.

When Mech (1966b) watched a pack of fifteen wolves lined up to feed around a moose carcass, he was impressed with the fact that not many more could have fit around it. Had there been any more wolves, some would have to have gone hungry. Long before, Adolph Murie (1944) had suggested that prey size might limit pack size in this way.

Such a relationship could also explain why large packs are occasionally found temporarily even among wolves hunting smaller prey. If enough smaller prey could be killed either concurrently or in close sequence, more individuals could accompany a wolf pack than otherwise. During 1990, when the East Fork pack in Denali numbered up to twenty-nine, they often killed more than one sheep or caribou at a time (Mech et al. 1998). This behavior conforms to the theory that when feeding constraints are relaxed, hunting group size should increase (Caraco and Wolf 1975).

Pack Size Regulation

Besides the general factors discussed above that affect pack size, other specific factors are also important. If the reason young wolves stay with their natal pack is to use their parents' provisioning skills to maximize their food intake during growth and maturation, this strategy would also explain certain aspects of dispersal. As indicated above, wolves mature at varying rates, probably because of varying nutrition. Thus it would be adaptive for them to do whatever possible to maximize their food intake. Because their parents have nurtured them throughout their lives, their tendency probably would be to remain with their parents until something forces them away.

However, because there are usually new offspring annually, with a greater need for parental nurturing, as the previous litters age, they must begin to compete with younger siblings for food. The parents' priority is to feed the youngest offspring; if there is enough to go around, then the older offspring are allowed to feed (L. D. Mech, unpublished data). (In the rare year when there are no pups, the adults continue to provision the yearlings, as would be expected [Mech 1995c,d].)

Some wolves disperse when as young as 5 months of age (Fuller 1989b), whereas others may remain with the pack for up to 3 years (Gese and Mech 1991), or occasionally longer (Ballard et al. 1997). As will be discussed below, intense food competition may be one of the main triggers for dispersal. If so, then perhaps when food is scarce, adults stop provisioning young as early as 5 months of age. By then, the young would be physically able to survive on their own (see above). This also would be a time when the adults would have to maximize their own intake to prepare for the next litter of pups.

It is probably only when food is sufficient that adults share it with their older offspring, and those offspring might then remain with the pack. Such offspring even provision the new litter of pups at the den, although they also sometimes usurp the pups' food as well (see Packard, chap. 2 in this volume).

Thus food competition could be the feedback mechanism that regulates pack size through dispersal. Prey size, and at times prey abundance, would set the upper limit to the number of individuals that could share without undue competition; any excess would disperse. If food were sparse, the young would disperse earlier; if abundant, they would remain longer, ideally until they were sexually mature. At that point, sexual competition and aggression might be the factor triggering dispersal.

A finer adjustment factor to this system could involve the pack dominance hierarchy (see Packard, chap. 2 in this volume). Presumably, as food competition increases, it is not only the lower-ranking classes of pack members (e.g., yearlings) (Messier 1985b) that must leave, but also the lower-ranking members within a class. Food competition has long been seen as a factor affecting the lowest-ranking pack members most adversely (Zimen 1976). In coyotes, it is also the most subordinate individuals that have least access to pack food resources and leave the pack soonest (Bekoff 1977b; Gese 1995).

This system of determining wolf pack size would explain why the age of offspring dispersal is so variable (Fritts and Mech 1981; Peterson, Woolington, and Bailey 1984; Ballard et al. 1987, 1997; Mech 1987a; Potvin 1988; Gese and Mech 1991) and why that age varies from year to year, with entire litters remaining with a pack in some years or dispersing in others (Mech 1995d).

The best evidence that food competition does affect dispersal comes from southwestern Quebec. There, yearling and "adult" wolves (which could have been as young as 2 years old) in an area of low moose density made significantly more excursions of 5 km (3 mi) or more from their territories than did yearlings and adults in a nearby area of high moose density; furthermore, more females than males made such excursions (Messier 1985b). These excursions lasted from a few days to a few months, averaged more than 22 km (13 mi) in straight-line distance, and eventually culminated in dispersal. On Isle Royale, more wolves also left packs during periods of lower food supply (Peterson and Page 1988).

Dispersal

As indicated above, most wolves disperse from their natal packs. Unless it assumes a breeding position within the pack, which is rare, any wolf born into a pack will leave it. In fact, each wolf pack can be viewed as a "dispersal pump" that converts prey into young wolves and spews them far and wide over the landscape. On the average, then, a thriving pack of three to nine members producing six pups each year (see Fuller et al., chap. 6 in this volume) thus "pumps out" about half its members annually.

In some circumstances, dispersal is more like a pulsating of members back and forth from the pack, for members may leave temporarily (see above) and return one to six times before finally dispersing (Fritts and

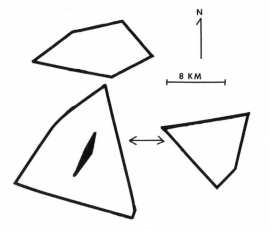

FIGURE 1.2. Territories of female wolf 6433 in the Superior National Forest of Minnesota. Upper polygon, 6433's natal territory; lower polygons, territories between which 6433 and her mate alternated from 11 January 1984 through 11 June 1984, after which they broke up and 6433 returned to her natal territory; blackened area, where the pair localized 5–29 April 1984.

Mech 1981; Van Ballenberghe 1983a; Peterson, Woolington, and Bailey 1984; Messier 1985b; Ballard et al. 1987; Mech 1987a; Potvin 1988; Fuller 1989b; Gese and Mech 1991). On the other hand, some wolves disperse without any known preliminary forays (Mech 1987a; Boyd et al. 1995; Mech et al. 1998; L. D. Mech, unpublished data).

Extraterritorial forays by wolves can even involve pairing, territorial establishment, and localizing during the denning season, followed by a return to the natal pack half a year later (Mech 1987a; Mech and Seal 1987) (fig. 1.2). In fact, wolf dispersal is probably most accurately viewed as a continuum, from single, short departures from the natal pack through intermittent and multiple extended forays to permanent, distant emigration. These movements appear to be motivated by attempts to maximize food input and opportunities to breed. However, they may also be underlain by a predisposition in some individuals to travel long distances, as we will see below.

Although predispersal forays might be viewed as "trial" or "exploratory" dispersals, they might also be merely movements that internal state plus food supply and social circumstances force on young wolves, and might not have any trial or exploratory function. Data from red foxes, however, tend to support the exploratory nature of some predispersal movements. Predispersing foxes moved much faster and spent little time resting or foraging during exploratory trips (Woollard and Harris 1990).

Sex and Age of Dispersers

Wolves of both sexes disperse, and there seem to be few consistent male-female differences in dispersal characteristics. In some regions or times, males apparently disperse farther or at a higher rate (Pulliainen 1965; Peterson, Woolington, and Bailey 1984; Wabakken et al. 2001). However, at other times or places females disperse farther on average, even though the longest-distance dispersers were males (Fritts 1983; Ballard et al. 1987). Nevertheless, the record dispersal lengths of males and females tend to be about the same (see below).

In south-central Alaska, males dispersed at a higher rate than females (Ballard et al. 1987). Perhaps such a difference has some ecological significance, because males showed the same propensity on Alaska's Kenai Peninsula during 1976–1980, whereas during 1980–1981, female dispersal tended to balance out the sex ratio of dispersers (Peterson, Woolington, and Bailey 1984). In southwestern Quebec, female pre-dispersers spent more time away from their packs than did males (Messier 1985b).

As already discussed, wolves disperse from their natal packs at a wide variety of ages, and this variation is probably related to food competition within individual packs. Wolves as young as 5 months and as old as 5 years have dispersed from natal packs, but the commonest age of dispersal in many areas is 11–24 months (Fritts and Mech 1981; Peterson, Woolington, and Bailey 1984; Messier 1985b; Ballard et al. 1987, 1997; Mech 1987a; Potvin 1988; Fuller 1989b; Gese and Mech 1991; Hayes and Harestad 2000a; Mech et al. 1998). Older adults that disperse from packs are often individuals that had immigrated into those packs (Meier et al. 1995; L. D. Mech, unpublished data). Fritts and Mech (1981) and Boyd et al. (1995) suggested that high rates of yearling dispersal were related to a high potential for colonization in the immediate area, but Gese and Mech (1991) found the same rate of yearling dispersal in a saturated population (63% of dispersers) as in the populations discussed by Boyd et al. 1995).

Notable exceptions to the usually high rate of yearling dispersal were reported under two conditions. In northwestern Alaska's exploited population, the average age of dispersal was about 3 years (Ballard et al. 1997). In the reintroduced Yellowstone population (see Boitani, chap. 13 in this volume), the mean age of the thirty wolves that had dispersed by October 2001 was 2 years and 1 month, with a range of 1 year and

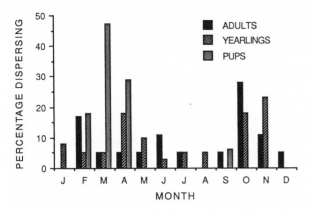

FIGURE 1.3. Month of dispersal for three age classes of dispersing wolves in the Superior National Forest, Minnesota, population, 1969–1989. (From Gese and Mech 1991.)

5 months to 3 years and 7 months (D. W. Smith, personal communication).

Dispersal Season and Triggering Mechanisms

The fact that wolves disperse primarily while beginning to mature sexually tends to implicate reproductive development (puberty) as a factor helping to trigger dispersal, or at least necessary for it. Such is the case with many other species (Howard 1960).

Although wolves have dispersed at every time of year, those in most areas leave during autumn and early winter or around the spring denning season (fig. 1.3). Most of the studies cited above agree with this conclusion, although not those in northwestern Alaska (Ballard et al. 1997). Pups that disperse in their first year usually leave from January to May (Fuller 1989b; Gese and Mech 1991).

The spring and fall peaking of dispersal in most areas suggests that one of the triggers for dispersal is social competition. During spring, aggression related to breeding is maximized (Rabb et al. 1967; Zimen 1976), and adults are presumably building reproductive fat stores. In fall, pups begin traveling with the adults and become nomadic with the pack. Their food needs peak at this time (Mech 1970), yet the pups are still dependent on the adults, so food competition also begins to peak.

Because food availability is variable throughout the year, dispersal could be expected during any season, but should peak when food and social competition peak. In Quebec, more yearling and adult wolves dispersed from packs living on a low prey base than from nearby packs living on a high prey base (Messier 1985b). When food stress increased on Isle Royale, so did dispersal rate, and dispersal rate was inversely related to pack size (Peterson and Page 1988). Furthermore, more dispersed lone wolves were present there during periods of low prey availability (Thurber and Peterson 1993).

In the Yukon, wolves repopulating an area from which wolves had been extirpated dispersed at increasing rates each year, and dispersal rates were positively correlated with pack size and negatively correlated with the ungulate biomass/wolf ratio (Hayes and Harestad 2000a). The larger the packs, the more competition and potential dispersers there are, and the greater the biomass of prey available, the less competition. The relatively high amount of food available to wolves in northwestern Alaska (Ballard et al. 1997), as well as to those in Yellowstone National Park (Mech et al. 2001), probably explains the high age of wolf dispersal in those areas (D. W. Smith, personal communication).

Thus competition and aggression, usually centered on food, can be considered a primary trigger for wolf dispersal. Postulating that aggression fostered by food and breeding competition helps trigger dispersal may seem to contrast with Bekoff's (1977b) view of the role of aggression in dispersal. However, Bekoff believed that aggression was not the "immediate cause" of dispersal, meaning that dispersers do not seem to be actively chased away. He did stress that avoidance of social interactions was characteristic of several species just prior to dispersal.

Bekoff's emphasis on social avoidance in dispersers accords with findings in both red foxes (Harris and White 1992) and coyotes (Gese 1995) that dispersers associate less with their social groups than do nondispersers. This lack of sociality, however, does not necessarily rule out aggression as a factor. In all the above situations, the aggression could merely be covert, or at least less perceptible, to a human observer. Schenkel (1947) emphasized that just the fixed stare of a dominant wolf wields great power over the behavior of subordinates.

In any case, wolves are often aggressive toward low-ranking wolves (Mech 1966b; Jordan et al. 1967), including their relatives, at least in captivity (Rabb et al. 1967; Packard et al. 1983). Thus chances are good that overt aggression, at least during some seasons, and possibly covert aggression at other times, is a strong factor in wolf dispersal (Zimen 1976).

TABLE 1.2. Frequency of wolves dispersing alone and with associates

Area	N	Alone	With 1	With 2	With > 2	Reference
Minnesota	9	9	—	—	—	Fritts and Mech 1981
Minnesota	28	23	1	1	—	Fuller 1989b
Minnesota	75	75	—	—	—	Gese and Mech 1991
Minnesota	7	7	—	—	—	Berg and Kuehn 1982
Alaska	21	15	3	—	—	Peterson, Woolington, and Bailey 1984
Quebec	11	7	2	—	—	Messier 1985b
Quebec	15	15[a]	—	—	—	Potvin 1988
Alaska	38	?[a]	1	?[a]	?[a]	Ballard et al. 1987
Wisconsin	16	16	—	—	—	Wydeven et al. 1995
Montana	13	11[a]	1[b]	—	—	Boyd et al. 1995

Note: The figures given here are probably minimum numbers because often it is not known whether a dispersing wolf has associates.

[a]Not explicitly stated.   [b]Mentioned by Ream et al. (1991) for same study as Boyd et al. (1995).

An intriguing piece of information that seems to contrast with the food competition theory of dispersal is the observation in northwestern Alaska that the highest annual dispersal rate was found following a rabies epizootic (Ballard et al. 1997). However, as the authors of the study indicated, this high dispersal rate may have been due to the resulting breakup of pack structure.

Individual versus Group Dispersal

Most wolves disperse alone, but there are notable exceptions, although little is known about such group dispersal (table 1.2). Among twenty-one wolves recorded dispersing from the Kenai Peninsula, in three cases, the animals dispersed as duos, but then split (Peterson, Woolington, and Bailey 1984). Seven of nine dispersers in Quebec dispersed alone (Messier 1985b), and in north-central Minnesota, one trio of wolves and one pair dispersed, whereas twenty-three other wolves dispersed alone (Fuller 1989b). Wolves also dispersed or made long forays in dyads in northern Montana (Boyd et al. 1995), and in Alaska, an adult female and five pups traveled some 72 km (43 mi) out of their territory in August (Ballard et al. 1997).

When groups of wolves permanently leave an area, their movements may not really be dispersal, which is defined as movement from a natal to a breeding site (Bekoff 1977b). Rather, the wolves may be emigrating. The most unusual group emigration reported was that of the Little Bear pack of eleven wolves, which moved 250 km (150 mi) from their territory in Denali National Park (Mech et al. 1998) (see below).

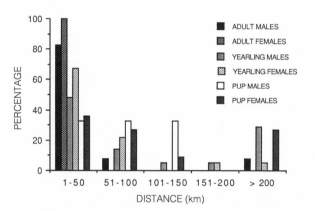

FIGURE 1.4. Distance traveled by three age classes of dispersing wolves in the Superior National Forest, Minnesota, population, 1969–1989. (From Gese and Mech 1991.)

Dispersal Distances

The distances wolves disperse reflect the great variation in types of dispersal, from merely moving to an adjacent territory through floating around the local population to dispersal up to 886 km (532 mi) distant (Mech and Frenzel 1971a; Van Camp and Gluckie 1979; Fritts and Mech 1981; Fritts 1983; Mech 1987a; Ballard et al. 1983; Messier 1985b; Gese and Mech 1991; Wabakken et al. 2001). The data suggest that the younger the disperser, the farther it disperses (fig. 1.4; see also Wydeven et al. 1995). This relationship might relate to the growing familiarity with the area or with the local population gained by a wolf as it remains with its natal pack. Perhaps older dispersers perceive more local opportunities, whereas younger dispersing animals feel less secure once they leave the fa-

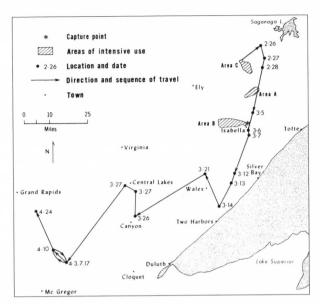

FIGURE 1.5. Dispersal of male wolf 1051 in the Superior National Forest of Minnesota. Lines merely indicate the sequence of locations. Only selected lakes are shown. (From Mech and Frenzel 1971.)

miliarity of their immediate social and physical surroundings. A 10-month-old wolf, for example, would only have had about 5 months of familiarity with its pack's territory and its immediate neighbors. Its naivete as it wanders through unfamiliar regions might drive it increasingly farther as it continues to seek security in wolf-free areas.

Wolves that disperse long distances appear to travel intently (Mech and Frenzel 1971a) in a manner that seems goal-directed (fig. 1.5). Whether the goal is to reach a particular kind of area, find a certain set of conditions, or travel a certain distance before settling is unknown. This type of travel is similar to that of translocated wolves when homing, a behavior that adults are good at if translocated less than about 130 km (80 mi) (Weise et al. 1979; Fritts et al. 1984; L. D. Mech, unpublished data). One possible deduction from this mode of travel is that wolves unable to find a breeding position locally are predisposed to proceed in a certain direction, possibly for a certain time or distance, before looking for a place to settle.

That the predisposition (Howard 1960) to distant dispersal might be genetic is an intriguing possibility. Gese and Mech (1991) found little evidence for genetic similarities in dispersal tendencies among the seventy-five dispersers they studied. However, consider two anec-

dotal cases. In Montana, two wolves 1 year apart in age dispersed 9 months apart and ended up 150 km (90 mi) away in the same pack 3 years later (Boyd et al. 1995). In Minnesota, two wolves caught 5 km (3 mi), but 12 years, apart were killed some 272 km (163 mi) away, 10 years, but only 11 km (7 mi), apart (Mech 1995e).

Some wolves disperse long distances as soon as they leave their pack, while others float around the natal population first (Mech and Frenzel 1971a; Fritts and Mech 1981; Peterson, Woolington, and Bailey 1984; Messier 1985b; Mech 1987a). Both floaters and distant dispersers can be found in the same litter (Mech 1987a), but why some wolves disperse long distances without first floating is unknown.

In addition, some wolves just float without ever dispersing long distances. A good example is male wolf 75 from the SNF study, who spent his life in an area of about 1,288 km² (L. D. Mech, unpublished data). Born in 1987, this wolf paired serially with females in two neighboring packs from 1989 through 1995. Even though each female survived, male 75 left each one, then serially paired again with two females that eventually were killed; wolf 75 was then killed himself when with his sixth female in late 1996 (all the deaths were human-caused). Other wolves in this area have floated around a region of 2,550 km² (996 mi²) (Mech and Frenzel 1971a).

Dispersal Direction

Care must be taken in analyzing wolf dispersal directions because dispersal data, especially final locations of dispersers, are often obtained from harvesting programs or other wolf studies. Such information is greatly biased toward areas where wolf retrieval is most likely. The dispersal directions reported for Denali wolves may suffer from such a bias (Mech et al. 1998).

In homogeneous habitat types, wolves would probably disperse equally in all directions. However, no habitat type is homogeneous, and topography, wolf density, and areas of human development no doubt play varying roles in steering dispersal direction. For example, most northwestern Montana dispersers settled north-northwestward in a narrow swath along the Rocky Mountain chain where there were other wolves (Boyd et al. 1995). It seems significant that few wolves inhabited the areas south, east, and west of this dispersal corridor.

We might conclude from the Montana data that wolf dispersal tends to be adapted toward maximizing the dispersers' chances of breeding rather than toward locating maximal resources. Thus, instead of dispersing toward areas of few wolves but much prey (e.g., south), the Montana wolves dispersed toward established wolf populations to the north (Boyd et al. 1995). Similar information from Wisconsin seems to show the same pattern (Wydeven et al. 1995). In other regions, however, both male and female wolves have dispersed long distances into wolf-free areas. For example, the Norway-Sweden wolf population apparently was begun by dispersers from a breeding population in Finland or Russia, more than 1,000 km (600 mi) away (Promberger, Dahlstrom et al. 1993, but cf. Sundqvist et al. 2001). In North and South Dakota, dispersed wolves have been found as far as 561 km (337 mi) from known breeding packs (Licht and Fritts 1994). A radio-collared male wolf from Michigan dispersed to Missouri, a 720 km (450 mi) move that took him hundreds of kilometers from any wolf population (J. Hammill, cited in Hutt 2002).

In the SNF, where the wolves studied were part of a larger surrounding population, yearling and adult dispersers initially headed in all directions. Pups and females settled significantly more to the southwest, and males to the north, for reasons unknown (Gese and Mech 1991). Lake Superior, lying some 35 km to the southeast, probably biased dispersal direction.

The wolves that recolonized France about 1994 no doubt dispersed from central Italy through the Apennine Mountains (Lequette et al. 1995). The relative isolation of the habitat type along the mountains, compared to the areas of high human disturbance surrounding it in the plains, probably helped funnel dispersing wolves toward France and sped up the recolonization.

Finding a Mate and Territory

As indicated above, dispersing wolves must ultimately find and acquire three things to succeed in life: a mate, food resources, and an exclusive area. A disperser can meet these needs by killing or usurping an established breeder, but it risks getting killed itself. It can also join a pack and lure out a mate, but it must then either disperse again to an unoccupied area or "carve out" a territory from the existing territorial mosaic, another risky strategy. Or it can disperse to the edge of the population range, locate a mate doing the same thing, set up a territory, and expand the species' range (Fritts and Mech 1981; Wabakken et al. 2001).

Where wolves are harvested, territories are left vacant and pack social structure is fragmented. This greatly enhances opportunities for budding wolves, floaters, and near and distant dispersers to succeed in meeting their life requirements. Thus it is common for harvested or controlled wolf populations to recover within a few years after harvesting stops (Ballard et al. 1987; Hayes and Harestad 2000a).

When most of the recent wolf studies using radio-tracking were being conducted, from the late 1960s through the mid-1990s, wolves in many areas were living in remnant populations, were legally protected, and were expanding their range. Thus considerable information is available about wolves dispersing to the edges of their range. However, under natural conditions, when wolves inhabited all of their range, such dispersal would have been rare. Floating, usurping, and other local breeding strategies would have predominated, as in many of the longer-established wolf populations in Alaska, Canada, and Minnesota. With such strategies, the actual detection of any potential mates, unused resources, and unoccupied territories would be prompt and direct.

Along the edges of the species' range, wolves might resort to more indirect means to evaluate their chances in an area. Through their daily hunting, they would learn whether the area provided enough catchable prey to support them and their offspring. In addition, they could determine through checking for scent marks and howling whether wolves occupied the area and whether there were potential mates there (see Harrington and Asa, chap. 3 in this volume).

Lone wolves in wolf-free areas tend to scent-mark and howl (Ream et al. 1985; R. P. Thiel, unpublished data), whereas those traversing wolf-inhabited areas tend not to (Rothman and Mech 1979; Harrington and Mech 1979). Presumably loners on the edges of wolf range need to advertise their presence, whereas those in wolf-inhabited areas must conceal their presence for fear of being harassed or killed.

Little is known about how wolves find mates, but the process probably takes only days once two predisposed individuals frequent the same region. The first wolf Mech ever radio-tracked, a dispersing male, in 1 week traveled more than 74 km (44 mi) and located a probable female along the edge of the species' range (Mech and Frenzel 1971a).

Information from northwestern Minnesota's expanding wolf population, where most wolves were settling locally, is instructive here (Fritts and Mech 1981). Six of seven dispersers paired within 8–30 days of leaving their packs, whereas three wolves radio-collared as loners that had already dispersed, possibly from outside the area, took longer (95–148 days). This difference could mean that pre-dispersers might be influenced to leave their pack by the presence of potential mates hanging around their area.

Pairing Success

Generally, dispersing wolves of both sexes have a high rate of success in settling and pairing in new areas (see above). Most studies indicating this involved populations that were expanding or were harvested (Fritts and Mech 1981; Peterson, Woolington, and Bailey 1984; Ballard et al. 1987). Nevertheless, the same was true even in the relatively protected central SNF population of Minnesota. Although dispersal rate there varied with phase of population trend, the pairing success of dispersers remained the same whether the population was expanding, declining, or stable (Gese and Mech 1991). In north-central Minnesota, where the population was subject to moderate human harvesting, all seventeen dispersers in one study settled within 267 days of leaving their pack and found potential mates (Fuller 1989b).

Dispersal beyond the Frontier

Some wolves become true pioneers by dispersing far beyond the frontier of their population. Examples include thirteen wolves killed in the northwestern United States from 1941 to 1978 (Nowak 1983), ten killed in North and South Dakota from 1981 to 1992 up to 561 km (337 mi) away from any breeding wolves (Licht and Fritts 1994), and a Minnesota disperser killed by a car in Wisconsin at least 80 km (48 mi), and probably 200 km (120 mi), from the nearest other wolves (Mech, Fritts, and Wagner 1995; see also Wabakken et al. 2001).

These pioneering wolves do not necessarily travel nomadically over a large area seeking mates. They may just disperse a long distance and then settle. In the Glacier National Park area of northwestern Montana, some 160–400 km (96–240 mi) from the nearest known wolf population, a lone female was radio-tracked over a region of about 1,100 km² (430 mi²), but concentrated her movements in two smaller areas, rather than moving nomadically over this extensive region (Ream et al. 1985).

Although the chances of distant dispersers finding a mate are small, if they do finally pair, they can begin a new population far from any source (Wabakken et al. 2001). About 20 months after the Montana female's radio collar expired, tracks of a pair were found in the area (Ream et al. 1985).

Multiple Dispersal

Although wolf dispersal typically occurs when animals are maturing, and involves a single move to a new area, some individuals may disperse and settle twice or more (Boyd et al. 1995). It is even conceivable that multiple dispersal is far more common than realized, but merely undetectable with radio collars that typically last no longer than 4 years.

In the SNF, for example, male wolf 75 dispersed from his natal pack as a yearling, lived in three other packs during the next 6 years, then successively paired with (traveled consistently with) two lone females in the next year, drifted as a loner again, then paired with another female before finally being killed by a vehicle at the age of 9.5 years (L. D. Mech, unpublished data). Two examples of male wolves dispersing and moving to successive packs were seen in Denali National Park (Mech et al. 1998).

Multiple Pack Affiliation

A few wolves associate with more than one pack more or less at the same time, but little is known about this behavior (Van Ballenberghe 1983a). It is best documented in a recolonizing population in northwestern Montana. There, "two individuals [both sexes] traveled freely between two packs and were observed caring for pups in two packs during one denning season" (Boyd et al. 1995, 139). As discussed below, such cases may involve related packs.

Colonization

The history of the northwestern Montana wolf population lends much insight into the colonization process. A few months after the tracks of a pair of wolves were found in the area, a litter of pups was born there in 1982 (Ream et al. 1991). Within the next few years, a total of about seven more founders arrived in the area, and the population had reached seventy wolves by 1996 (E. E.

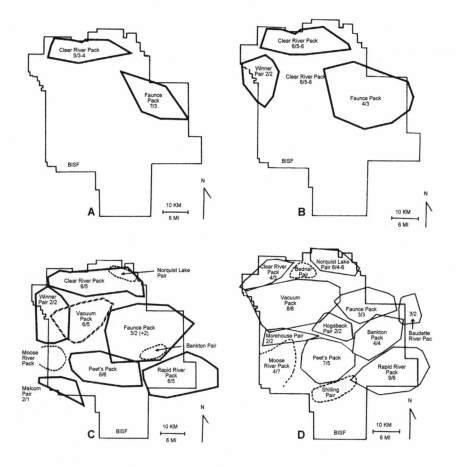

FIGURE 1.6. Proliferation of a colonizing wolf population in northwestern Minnesota. Numbers above the slashes indicate pack size in winter; numbers below the slashes indicate pack size in spring. Solid lines indicate the minimum area occupied by a pack; dashed lines indicate the approximate locations of non-radio-collared packs. (A) Size of and minimum area occupied by Clear River pack and Faunce pack in 1972–1973. (B) Size of and minimum area occupied by Clear River pack, Faunce pack, and Winner pair in 1973–1974. (C) Sizes of and areas occupied by ten social units of wolves in 1974–1975. (D) Sizes of and areas occupied by thirteen social units of wolves in 1975–1976. (From Fritts and Mech 1981.)

Bangs, personal communication). A similar process was seen in Scandinavia (Wabakken et al. 2001).

As wolves colonize or recolonize an area, the first pack soon begins to proliferate if conditions are favorable. The process of pack proliferation in Montana was similar to that in northwestern Minnesota, where, from 1972 through 1976, the number of packs increased by at least eight as a wolf population recolonized that area (fig. 1.6). Similar processes took place during population increases on Isle Royale (Peterson 1977; Peterson and Page 1988) and on Alaska's Kenai Peninsula (Peterson, Woolington, and Bailey 1984).

At least part of the pack proliferation process is fostered by dispersers from the original pack or packs (Fritts and Mech 1981). Conceivably, the first new breeders may be related animals, as they are on Isle Royale (Wayne et al. 1991). However, it is also possible, and we believe probable, that the mates of local dispersers from a colonizing pack are usually new immigrants (Forbes and Boyd 1996) that have gravitated to the territories of the original founders and helped them proliferate into

adjoining territories, as Hayes and Harestad (2000a) also suggested.

In Sweden (Bjärvall 1983; Wabakken et al. 2001) and in Wisconsin (Wydeven et al. 1995), the recent wolf recolonizations proceeded in several disjunct areas as well as proliferating from one core. This may have been because both these ranges contain "islands" of wilderness interspersed with agricultural and settled areas, unlike northwestern Minnesota and northwestern Montana. The recolonization of the French Alps, while still under way, shows a similar pattern (L. Boitani, unpublished data).

A year after a female wolf in Sweden reached a male that had been there for a few years, the two began producing pups in 1983 and disseminating long-distance dispersers, which then settled in new areas far from the core (Promberger, Dahlstrom et al. 1993). Despite considerable human-caused mortality, wolf numbers in Sweden and neighboring Norway had reached about thirty animals by 1994 (Wabakken et al. 2001).

In northwestern Wisconsin, which wolves began recolonizing in the mid-1970s (Thiel 1978; Mech and Nowak 1981), a cluster of four contiguous territories had been formed by 1979 (Wydeven et al. 1995). From 1979 to 1991, the wolves settled widely separated suitable areas (areas of low human accessibility), then gradually began filling in areas among them (Wydeven et al. 1995).

Because the wolves colonizing Sweden and Wisconsin had to have originated in wilderness, they probably selected disjunct wilderness areas in which to settle. Or, conceivably, any that tried to settle in unsuitable areas near existing packs were killed by humans. In both Sweden (Wabakken et al. 2001) and Wisconsin (Wydeven et al. 1995), humans caused much of the wolf mortality that occurred during colonization.

Of considerable interest is the observation that, during colonization, some wolves disperse out of the colonizing population foci and travel long distances through both wolf-free and wolf-inhabited regions (Fritts and Mech 1981; Wydeven et al. 1995; Boyd et al. 1995). It would seem more logical for them to settle locally, given the abundance of seemingly suitable areas available, where in fact other wolves do settle. Their failure to do so may evince a genetic predisposition for distant dispersal. The possibility that these wolves disperse far to seek unrelated mates tends to be negated by their passing through wolf-inhabited areas, where presumably unrelated candidates would be available. (See Mech 1987a for further discussion of distant dispersal.)

The Yellowstone National Park wolf reintroduction of 1995 provided interesting observations of wolf dispersal into an area with no breeding wolves. Male wolf R12 from the Soda Butte pack dispersed in January 1996, after the pack had been settled in a territory for 9 months. He then traveled in a semicircle that reached areas about 100 km (60 mi) southeast, 175 km (105 mi) south-southeast, and 240 km (144 mi) south of his pack territory before he was illegally killed (M. K. Phillips and D. W. Smith, personal communication). Conceivably, the animal was seeking out other wolves, and after not finding any after a certain distance, was continuing in a broad circle around his dispersal point.

Territoriality

Wolves generally are highly territorial (Mech 1973, 1994a; Van Ballenberghe et al. 1975; Fritts and Mech 1981; Jordan et al. 1967; Peterson 1977; Peterson, Woolington, and Bailey 1984; Messier 1985a; Ballard et al. 1987; Fuller 1989b; Ream et al. 1991; Meier et al. 1995; Mech et al.

1998). The development of territoriality is thought to depend on the influence of competition, the economic defensibility of resources, and the adaptive value of aggressiveness (see below) in relation to variation in these factors (Brown 1964).

The adaptiveness of territoriality has been explained as follows: "Territoriality is a very special form of contest competition, in which the animal need win only once or a relatively few times. Consequently, the resident expends far less energy than would be the case if it were forced into a confrontation each time it attempted to eat in the presence of a conspecific animal" (Wilson 1975, 268).

This explanation seems appropriate for wolves, for their territories encompass large areas replete with high numbers of prey. As wolves circulate about their territories seeking prey they can catch and kill, they rarely encounter neighbors, even along the edges of their territories. One of the main reasons they do not is that their territories are often so large (tens to thousands of square kilometers) that chance alone minimizes the possibility (see below).

A case has been made that wolf territorial patterns "arise naturally as steady state solutions" to the mediation of wolf movements and behavior by the presence or absence of foreign scent marks (Lewis and Murray 1993, 738). According to this theory, wolf movement is primarily dispersive, and marking frequency is low, in the absence of foreign marking, but with foreign marking, movement is toward an organizing center, and scent marking increases. The theory assumes that marks lose strength with age (Peters and Mech 1975b). A new model yields distinct home ranges through interaction between scent marking and movements in response to familiar marks (Briscoe et al. 2002).

Implied in the concept of territoriality is the need for defense, for a territory, by definition, is a defended area (Burt 1943). In theorizing about the implications of territoriality from an evolutionary perspective, various workers have uncovered several problems that a territorial species must solve. For example, defending a territory must be energetically efficient (Brown 1964), and defense must not take so much time or energy as to hamper courtship, copulation, and care of young (Wilson 1975).

The need to solve these problems is especially acute for a species with such large territories as the wolf's. However, the wolf has evolved very successful physical and behavioral solutions. The key to the wolf's solutions

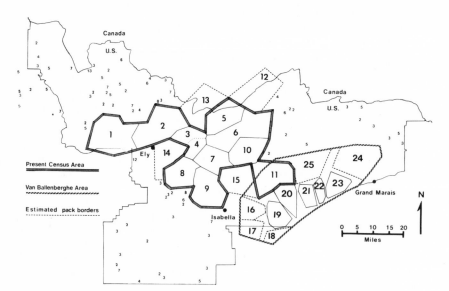

FIGURE 1.7. Wolf pack territories as first delineated in the Superior National Forest of Minnesota. Large numerals identify wolf packs, and lines around them indicate approximate pack territory borders. Small numerals represent the sizes of packs, or numbers of tracks, observed outside of the intensive study areas (lone wolves were not plotted). (From Mech 1973.)

is the ability and inclination to travel far and wide. The animal's physical stature, long legs, blocky feet, and powerful muscles allow it to travel tirelessly at about 8 km (5 mi)/hr for many kilometers per day in all sorts of climatic conditions (Mech 1966b, 1970, 1994a).

Widespread and regular travel functions both to help the wolf secure prey (see Mech and Peterson, chap. 5 in this volume) and to mark its territory (see below). Even a territory as large as 1,600 km² (625 mi²) (Mech 1988a; Meier et al. 1995) has a diameter of only about 40 km (24 mi). Wolves can cover this distance in less than a day (Mech 1966b, 1970). Since wolves both hunt and mark as they travel, and since marks are effective for long periods (Peters and Mech 1975b), this behavior allows efficient territorial defense. Howling at various locations along their routes, including homesites, complements this defense (see below).

In a well-established wolf population, a territorial mosaic develops (fig. 1.7). Each pack competes with neighbors for space and resources, and considerable territorial tension characterizes the population. The natural expansion tendency of each individual wolf has long been recognized (Schenkel 1947), and this trait expresses itself among packs as well. This expansion tendency allows the wolf population to adjust constantly, as a "flexible strategist" (von Schantz 1984), to variations in prey availability.

Each pack territory can in some respects be considered a mini-ecosystem, although not to the extreme degree Haber (1977) postulated. The size of a territory, the abundance of prey there, and the size of the pack occupying the territory are interdependent, consistent with the resource dispersion hypothesis (D. W. Macdonald 1983) discussed above, and these characteristics are unique for each territory.

Territory Size

A wolf pack's territory and home range are the same, since the territory is the defended home range. Home range size generally is related to an animal's size (McNab 1963), and body size explains about 75–90% of the variation in carnivore home range size (Harestad and Bunnell 1979). For wolves, however, this postulated relationship predicts a territory four times the size of the average home range used ($n = 30$) in helping derive the relationship. Thus the postulated relationship applies in only a very general way to wolves and provides little information about why wolves use the size territories they do.

Most mammals have little trouble finding and processing food, so it is not surprising that there is a general relationship between body size and home range size. In general, the larger the animal, the more food it requires, and the larger the area it needs to acquire that food. However, because wolves must hunt far and wide for prey they can catch and kill, it takes a disproportionately large area to support enough vulnerable prey (see Mech and Peterson, chap. 5 in this volume). Furthermore, both prey populations and prey vulnerability can fluctuate by orders of magnitude, and some prey are migratory or nomadic, greatly influencing the area a pack may need to cover.

Wolf pack territories vary in size by orders of magnitude. Estimated territory size depends considerably on the number of points used to define the territory, the period over which the points were derived (Fritts and Mech 1981; Scott and Shackleton 1982; Bekoff and Mech 1984; Mech et al. 1998), and the method used to analyze the points. Thus data from various studies are only roughly comparable. Nevertheless, the comparisons that can be made are instructive (Okarma et al. 1998).

The smallest well-documented territory reported seems to be that of the Farm Lake pack of six in northeastern Minnesota, which occupied an estimated 33 km^2 (13 mi^2) (88 locations) (L. D. Mech and S. Tracy, unpublished data). At the other extreme, Denali National Park's McKinley River pack of ten inhabited a 4,335 km^2 (1,693 mi^2) territory (51 locations year-round) in 1988 (Mech et al. 1998), and another Alaskan pack of ten covered some 6,272 km^2 (2,450 mi^2) in a 6-week period (calculated from Burkholder 1959).

Some wolves make grand excursions over areas of over 100,000 km^2 (38,000 mi^2) (P. C. Paquet, personal communication), but they are not considered here because the areas they cover do not seem to be defended. In addition, migratory wolves that follow migratory caribou herds (see below) occupy areas averaging 63,058 km^2 or 24,600 mi^2 annually (Walton et al. 2001).

Territory Size and Pack Size

In establishing a territory, a pair of wolves must select an area far larger than they themselves would need to gain a living (Peterson, Woolington, and Bailey 1984), because they can expect to produce an average of five or six pups per litter (Mech 1970), which they must feed. When pups are only 6 months old, they consume as much as adults (Mech 1970), which means that pack size and resource needs suddenly quadruple. Furthermore, some packs include not only a pair of parents and a litter of pups, but also offspring of earlier years, increasing pack size by a factor of up to fifteen. Thus a pair must either establish a territory up to fifteen times as large as they require to sustain themselves, or they must later expand their territory by this much.

Wolf pairs colonizing unoccupied habitat could resort to either approach, whereas those trying to carve out territories in an existing mosaic would have to start smaller and try to expand. In the many wolf populations that are exploited by humans, the exploitation sometimes leaves large gaps in the mosaic. There, pairs could assume full-sized territories whose occupants were recently destroyed.

Wolf pairs seem to resort to all the above approaches. On Alaska's Kenai Peninsula, where wolves are moderately exploited, two newly formed pairs established territories with areas per wolf three to four times larger than those of larger packs (Peterson, Woolington, and Bailey 1984), then maintained them as their packs grew. Hayes et al. (1991) found the same pattern in the Yukon. In Denali, the Headquarters pack maintained a territory of about 600 km^2 (234 mi^2) as a pair in 1987 and also when there were fourteen members in 1989 (Mech et al. 1998). When pack size declines, the remaining adult pair continues to maintain a large territory (Mech 1977a).

In a newly colonizing population in northwestern Minnesota, wolf pack territories began large and were compressed considerably as more and more packs formed and began filling the available space. Not only did individual pack territories shrink by 17–68%, but one territory that began as 555 km^2 (217 mi^2) eventually was occupied by four packs (Fritts and Mech 1981). In the Yukon, wolves recolonized an area in a similar pattern, with pack territories even overlapping after saturation (Hayes and Harestad 2000a).

It appears that the general competitiveness or aggressiveness of a pack increases with its size (Zimen 1976). By expanding a territory only slightly around the periphery, a pack could gain a considerable amount of space. For example, a territory of 250 km^2 (100 mi^2) would have a radius of 8.9 km (5.3 mi). The addition of only 1 km more to its radius would add 58 km^2, or 23% more space.

However, because pairs usually establish territories large enough for a full-sized pack from the beginning, the degree of expansion necessary is not great. Thus, in most relatively intact (saturated) wolf populations, there is only a minor relationship between pack size and territory size (Potvin 1988; Fuller 1989b; Mech et al. 1998). On the other hand, where human harvesting is high and the region is not saturated, pack size and territory size may be related (Peterson, Woolington, and Bailey 1984; Ballard et al. 1987, 1997).

Territory Size and Prey Biomass

One would expect that, on average, the greater the amount of prey (prey biomass) in an area, the smaller a

TABLE 1.3. Relationship between latitude and wolf pack territory size for wolves in North America

Latitude (°N)	Mean territory size (km²)	ml²	Source
46	137	54	Wydeven et al. 1995
46	199	78	Potvin 1988
47	625	244	Messier 1985[b]
48	198[a]	77	Mech 1973
48	285	111	Fuller 1989b
48	344	134	Fritts and Mech 1981
49	69	27	Scott and Shackleton 1982
51	260	102	Carbyn 1980
56	1,028	402	Fuller and Keith 1980a
60	638	249	Peterson, Woolington, and Bailey 1984
60	795	311	Carbyn et al. 1993
61	1,645	643	Ballard et al. 1987
62	1,478	577	Hayes 1995
64	1,330[a]	520	Mech et al. 1998
65	1,868	730	Ballard et al. 1997
70	1,225	479	Stephenson and James 1982
80	≥2,600	1,016	Mech 1988a

Note: See also Okarma et al. 1998.

[a]Calculated from data presented.

wolf pack territory would need to be. Many variables would affect this relationship, such as pack size, prey size and distribution, population lags, and differences in prey vulnerability. However, overall, such a relationship should exist (Walters et al. 1981), and for wolves feeding on widely varying densities of moose (Messier 1985a) and white-tailed deer (Fuller 1989b; Wydeven et al. 1995), there is good evidence that it does. In general, about 33% of the variation in wolf pack territory size is explainable by prey biomass (see Fuller et al., chap. 6 in this volume). Further confirmation comes from a strong relationship between latitude and territory size (see next section), since prey biomass density declines with latitude. Nevertheless, this relationship is complicated by patterns of variation in prey biomass over time (both within and between years) and space (clumped or dispersed), as well as by social and other ecological variables.

Territory Size and Latitude

Even though, due to methodological differences among studies, only gross relationships between territory size and latitude are detectable, one of the strongest links that does appear is that the higher the latitude, the larger the territory ($r^2 = .83$; $P < .00001$; table 1.3). This relationship probably results from the fact that productivity, and thus biomass density (standing crop), decreases with latitude (Rosenzweig 1968). In reality, then, this relationship is probably an extension of that between territory size and prey biomass (see above).

Territory Shape and Boundaries

Theoretically, if territory holders are competing maximally with neighbors, territorial mosaics should resemble the hexagonal cells of honeybee hives (Grant 1968; Wilson 1975). This spacing allows the maximum number of territories with the least space among them (Wilson 1975). The earliest published wolf territorial mosaic fits this model (see fig. 1.7), as do most later reports including sufficient numbers of territories (Fritts and Mech 1981; Messier 1985a; Ballard et al. 1987; Fuller 1989b; Mech, Meier, and Burch 1991; Mech et al. 1998; Hayes and Harestad 2000a).

Of course, landscape features also influence this basic mosaic structure. On Isle Royale, for example, which is 72 km (43 mi) long and 14 km (8 mi) wide, wolf pack territories tend to lie along the length of the island, but include its entire width (Peterson and Page 1988). This pattern may indicate that wolves grasp the idea of an easily defended boundary (the shoreline) and possibly some notion of the extensiveness of their territory. Otherwise, they might have divided the island up, for example, into parallel strips.

In mainland areas, topographic features such as long lakes also seem to be used as boundaries, as in the case of the Skilak Lake pack on the Kenai Peninsula (Peterson, Woolington, and Bailey 1984, 23). In the SNF Mech tracked a pack of seven to nine wolves for 2 km southwestward along the length of frozen Mahnomen Lake, which apparently formed their northwestern boundary. Thirteen times they had approached to within several meters of the opposite shore; each time, they had veered back toward the middle of the lake, presumably having detected their neighbor's' scent marks along the shore (Peters and Mech 1975b). In other areas, wolf pack territory boundaries adjoin extensive marshes and conifer swamps where ungulate prey are not present (Fritts and Mech 1981).

Territory Shifts

The degree to which territorial boundaries are stable is an intriguing question that is not easily answered (but cf. Haber 1977). One of the main problems involves methodology. Most estimates of wolf territory boundaries are based on sampling of wolf movements by radio-tracking. However, because this is usually done via aircraft and is expensive, only a tiny proportion of a wolf pack's movements are ever sampled.

Assuming that wolves travel 20 km (12 mi) per day, for example, and that a location data point represents their location to within 200 meters, then one location per day represents 1% of that day's locations. Most radio-tracking studies gather one or two locations per week. Thus the wolf pack territories described by biologists are only gross approximations of reality. Territory sizes can be reasonably estimated by determining when any additional points in a given sample contribute insignificantly to the calculated area (Fritts and Mech 1981; Scott and Shackleton 1982; Bekoff and Mech 1984).

However, determining precise territorial boundaries

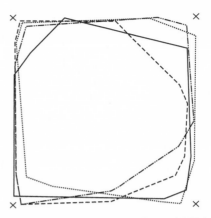

FIGURE 1.8. Varying minimum convex polygons (MCPs) resulting from four random selections of 50 points each from the same matrix of points. Note apparent differences in borders of MCPs due to sampling error. (From M. Bekoff and L. D. Mech, unpublished data.)

is impossible with standard radio-tracking. Given even a known or simulated territory, different sampling schemes produce grossly different boundaries (fig. 1.8). Thus researchers' perceptions of seasonal or yearly boundary shifts, for example, must be viewed very cautiously (Mech et al. 1998). Also, the use of old location data along with new may give an outdated or distorted picture of the current territory. (The use of Global Positioning System collars may help solve this problem: Merrill et al. 1998; Merrill 2002.) Even snow tracking cannot be relied on to yield a full understanding of a pack's territory, because rarely, if ever, does weather allow one to track wolves in the snow for months on end and thus learn the full extent of their territory.

Furthermore, there is no reason to believe that a pack territory is a constant, stable area. Rather, wolf packs in saturated populations are always competing with neighbors, defending their own areas and probably jockeying for advantages along whatever their current borders are. In exploited populations, such turbulence is no doubt accentuated.

Given all the above considerations, something can still be said about the spatial dynamics of wolf pack territories. As expected, the greatest shifts in wolf pack territories occur in colonizing or recolonizing populations. No doubt this is because of the lack of constraint by any neighboring packs. The Montana Magic pack shifted its territory 50 km (30 mi) south (Ream et al. 1991), and the Soda Butte wolves of the reintroduced Yellowstone population apparently shifted their territory back and forth

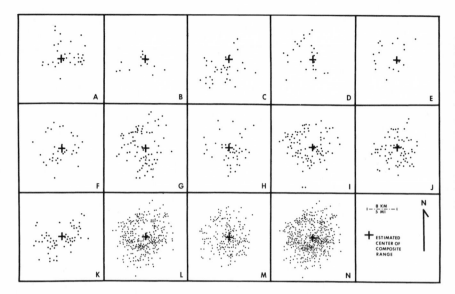

FIGURE 1.9. Distribution of locations of the Harris Lake pack in the Superior National Forest for various periods, as shown by radio-tracking: (A) winter 1968–69; (B) summer 1969; (C) winter 1970–71; (D) summer 1971; (E) winter 1971–72; (F) summer 1972; (G) winter 1972–73; (H) summer 1973; (I) winter 1973–74; (J) summer 1974; (K) winter 1974–75; (L) composite of all winters; (M) composite of all summers; (N) composite of all locations, summer and winter. (All trespasses are excluded.) (From Mech 1977a.)

over a large area (M. K. Phillips and D. W. Smith, personal communication). In both cases, these shifts could have been seasonal movements, since they occurred in autumn and could have been related to prey migrations. In the Yukon, such shifting by packs into the wolf removal area was more important in early years than later in the recolonization (Hayes and Harestad 2000a).

In the colonizing population of northwestern Minnesota, several packs shifted their territories grossly enough that the observed shifts were probably real (Fritts and Mech 1981). In some cases, the shifts were related to the formation of new packs that began filling up the available area. Hayes (1995) believed that his data showed that, during the first 2 years of his population's recovery in the Yukon, pack territories were exclusive, but that after enough territories developed, they began to overlap. Other gross shifts in wolf pack territories have been reported on the Kenai Peninsula (Peterson, Woolington, and Bailey 1984), in central Alaska (Ballard et al. 1987), in north-central Minnesota (Fuller 1989b), and in Wood Buffalo National Park (Carbyn et al. 1993).

In saturated wolf populations, of course, each pack territory is surrounded by others, so that any territorial shifts involve neighbors. Our impression is that the borders of most territories in the mosaic are constantly shifting, but that the center of each territory remains approximately the same over the years (fig. 1.9). Superimposed on this dynamism are the interactions among packs described earlier that snuff out some packs or create new territories as a wolf population fluctuates (Meier et al. 1995).

Seasonal Shifts in Territories

Because ungulates shift their movements seasonally to varying degrees, so too do the wolves that prey on them. Some territory shifts can be relatively minor, such as those within small territories where white-tailed deer are the primary prey (Van Ballenberghe et al. 1975; Mech 1977a). However, when SNF deer numbers were low during the 1970s and no deer remained in certain wolf territories during winter, those packs migrated some 50 km (30 mi) out of their usual territories. They then became nomadic around areas of up to 1,500 km^2 (585 mi^2) before returning to their territories in spring and denning there (L. D. Mech, unpublished data). Similarly, in central Italy, wolves made small-scale shifts during summer to the sheep-grazing areas (Boitani 1986).

Migration

In some regions, wolves migrate altitudinally as prey such as elk or moose that spend the summer in high areas migrate to valleys for the winter (Cowan 1947; Carbyn 1974; Ballard et al. 1987; Ream et al. 1991). In such cases, it appears that the packs remain territorial, but just shift their territories with the movements of the prey.

Where wolf prey are highly migratory, the wolves themselves must also migrate, unless alternative prey can tide them over until the migratory prey return. The longest wolf migrations are those in which wolves follow caribou herds, at times for up to 508 km (305 mi) (Kuyt 1972; Parker 1973; Miller and Broughton 1974; Miller

1975; Stephenson and James 1982; Ballard et al. 1997; Walton et al. 2001). In Kazakhstan, wolves follow the saiga antelope herds to their wintering grounds, and in central Canada, they track the long seasonal habitat type shifts of bison (Carbyn et al. 1993).

Little is known about the extent to which wolves are territorial during migrations, or when several packs are focused around prey concentrations during winter. Although pack ranges in some such areas do overlap, wolves still kill other wolves in such areas (Carbyn et al. 1993). In Yellowstone, the newly established packs maintain different territories in summer and winter, even around the high elk concentrations that inhabit the area (M. K. Phillips and D. W. Smith, unpublished data).

Territory Buffer Zones

Most studies of wolf pack territories indicate a certain amount of overlap among territories (see also Mech and Peterson, chap. 5 in this volume). The degree to which this overlap is spatial or spatio-temporal (Mech 1970; Ballard et al. 1987) has not been analyzed, and given the problems described above with radiotelemetry sampling, such an analysis may not be forthcoming any time soon. However, biologists have tracked wolf packs in snow enough to demonstrate that, at least along the immediate territory edge, movements of neighboring packs may overlap during short periods (Peters and Mech 1975b; Peterson and Page 1988).

Given this overlap, Mech (1977d) proposed that the overlap area is a kind of buffer zone between packs. Based on the fact that during a drastic deer decline, the wolves in the SNF eliminated deer first from the cores of their territories and only last from the edges, Mech (1977a,d) deduced the existence of the buffer zone, thought to be about 2 to possibly 6 km (1.2–3.6 mi) wide (Peters and Mech 1975b; Mech 1994a). He believed that the reason deer survived longer along the territory edges might be that neighboring packs felt more threatened there, so spent less time there, and thus deer bore less hunting pressure in these areas (Mech 1977a,c) (see Mech and Peterson, chap. 5 in this volume).

Evidence that buffer zones are areas of contest comes from an analysis of locations where wolves were killed by other wolves. In the SNF, in 1968–1992, 23% of twenty-two wolf-killed wolves perished along the estimated edge of their territory; 41% were killed along the edge or within 1 km of the edge, and 91% within 3.2 km of the edge (Mech 1994a). In the Yukon, 35% of wolf-killed wolves were killed within 2.5 km of their territory edge (Hayes 1995). In Denali, eight of twelve wolves killed by other wolves within their own territories (75%, a significant disproportion) died within 3.2 km of their estimated territory border, an area constituting only 29% of the average territory there (Mech et al. 1998). There is also theoretical evidence that buffer zones may be prey refuges (Lewis and Murray 1993) and that territorial stability in such zones would require inter-pack aggression by wolves (Taylor and Pekins 1991), which has been demonstrated (see above).

Territorial Defense

The concept of territoriality implies defense of an area (Burt 1943), and that defense theoretically should require less energy than that gained as a result of the defense (Brown 1964; Wilson 1975). The degree to which territorial defense is seasonal is unknown; however, territorial advertisement and defense tend to peak during the breeding season (Peters and Mech 1975b; Harrington and Mech 1979), as does aggression (Zimen 1976).

For a wide-ranging animal like the wolf, the problem of defending its entire home range is great. Wolves have solved this problem through a combination of at least three types of defensive behavior: scent marking, howling, and direct attacks. The first two behaviors are detailed by Harrington and Asa in chapter 3 in this volume, and will be discussed only briefly here. Both behaviors are indirect, and they complement each other in their application.

Scent Marking

Wolf scent-marking behavior used for territorial advertisement includes raised-leg urination (RLU) and perhaps standing urination (STU) by males, flexed-leg urination (FLU) and possibly squat urination (SQU) by females, and perhaps defecation (SCT) and ground scratching (SCR). Products of these behaviors are left, on average, every 240 meters throughout wolf territories, but especially along regular travelways and at junctions (Peters and Mech 1975b). Scats may carry anal gland secretions, and scratching may distribute secretions from interdigital glands (Peters and Mech 1975b; Asa et al. 1985).

Both wolf and coyote packs leave twice as many marks along the edges of their territories as in the core, resulting in an "olfactory bowl" (fig. 1.10, Peters and Mech 1975b; Bowen and Cowan 1980; Paquet 1991a; but

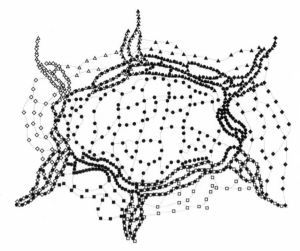

FIGURE 1.10. In this model of the distribution of raised-leg urination (RLU) scent marks, RLUs are indicated throughout the territory of one wolf pack (black dots) and for the areas where six neighboring packs border this territory. Each different symbol represents the marks of a different neighboring pack. Travel routes are simulated, but mean RLU density and territory size are to scale for a territory approximately 20 km (12 mi) wide. Note the bowl-shaped effect caused by heavier scent-marking by both the resident pack and its neighbors at the edges of the territory (Peters and Mech 1975).

cf. Barrette and Messier 1980). Scent marking appears to serve as a long-lasting (2–3 weeks) means of advertising a pack's presence.

Three aspects of scent marking imply that one of its most important functions is to deter neighbors from intruding (Peters and Mech 1975b): First, territorial packs mark, but nonterritorial animals do not. Second, packs that trespass into neighboring territories suspend marking until they return to their own territories, at least in the SNF (L. D. Mech, unpublished data). Finally, wolves are intimidated and deterred by neighbors' marks (Peters and Mech 1975b; Peterson 1977; Mech 1993a; L. D. Mech, unpublished data). There may be several other effects and functions of scent marking in relation to territorial defense, but these have yet to be explored.

Howling

The main disadvantage of scent marking as a means of advertising a territory is that it has little effect over long distances. Thus howling nicely complements scent marking. Although howling has several functions, at least one of them seems to be to inform neighboring packs that a territory is occupied (Joslin 1967; Harrington and Mech 1979).

Indications are that in forested areas, wolves can hear

howling at distances of up to 11 km (6.6 mi) (Harrington and Mech 1979) and on open tundra at up to 16 km (9.6 mi) (R. O. Stephenson, unpublished data, cited in Henshaw and Stephenson 1974; L. D. Mech, unpublished data). Observations of wolf packs howling to each other over expansive territories are necessarily rare, but territorial packs reply to human howling (Pimlott 1960; Joslin 1967; Harrington and Mech 1979).

Furthermore, wolves seem to be able to pinpoint the precise location of human howlers from distances of 2.7 km (1.6 mi), at least on tundra (L. D. Mech, unpublished data). Breeding animals tend to approach the howler (Joslin 1967; Peterson 1977; Harrington and Mech 1979), sometimes minutes after howling has stopped (L. D. Mech, unpublished data). Although no one has witnessed two packs interacting as a result of howling, the usual reaction of wolves encountering non-pack members is fighting and chasing (see below).

Direct Territorial Defense

Howling and scent marking must minimize the chances of neighboring wolves encountering one another. On the other hand, packs do sometimes meet up with each other, and these encounters often result in wolves being killed (Mech 1994a). Because the consequences of territorial encounters are so severe, and because systems are in place to avoid them, there is reason to believe that wolf territorial encounters are a result of either desperation (i.e., a wolf or wolves taking a chance for some kind of temporary gain), or deliberate aggressiveness (i.e., wolves seeking out others to kill or displace).

The main reason wolves might be desperate is probably hunger. Certainly in the SNF, the most trespassing by territorial packs deep into the territories of neighbors to kill prey was seen during a severe prey decline (Mech 1977a,b; L. D. Mech, unpublished data), and in Quebec, packs that made extraterritorial excursions were those in areas of low prey density (Messier 1985b). On Ellesmere Island, three of the five encounters that Mech and associates observed between a resident pack and outsiders involved food competition (fig. 1.11); one encounter ended in the death of the stranger (Mech 1993a; L. D. Mech, unpublished data). In Wood Buffalo National Park, a wolf pack that killed a bison in another pack's territory soon saw at least two of its members killed by the resident pack (Carbyn et al. 1993).

Deliberate attacks on neighbors represent more than territorial defense, but the tendency toward such behavior certainly must help enforce territoriality, as with

FIGURE 1.11. A breeding male wolf (right) attacks a strange wolf (left) near the edge of the breeder's territory. (From Mech 1993a.)

coyotes (Gese 2001). Forays into neighboring territories (fig. 1.12) and attacks on neighbors must be a result of a certain aggressiveness, to be discussed below. Many such deliberate attacks have been recorded (Haber 1977; Mech 1977a; L. D. Mech, unpublished data; Peterson 1977; Meier et al. 1995). In both northwestern Minnesota and Riding Mountain National Park (Manitoba), at times when there appeared to be abundant prey, wolf packs displaced neighboring packs (Fritts and Mech 1981; Carbyn 1981).

Whether encounters between packs are due to deliberate attacks or food competition, the end result is often death. As indicated above, most wolf deaths resulting from wolf attacks take place near territory boundaries or within buffer zones (see below), and killing by other wolves is one of the commonest causes of natural wolf mortality (Mech 1977b; Mech et al. 1998). Thus wolf pack territorial defense can be considered the most important feature of wolf spatial ecology.

Competition and Intraspecific Strife

"The predominant single factor tending to increase aggressiveness through natural selection should be competition" (Brown 1964, 161). Viewed on a population scale, wolf competitiveness is a pervasive phenomenon. Social competition in wolves is always intense. This contention may imply that the "ecological requisites" of wolves always exist at less than optimal levels (Brown 1964).

However, even where food is abundant and the wolf population is low—conditions that Brown (1964) claimed should minimize territorial defense—wolves are keenly competitive. Such conditions characterize colonizing populations, but even in those populations, fatal attacks are known to occur (Fritts and Mech 1981; Wydeven et al. 1995; R. R. Ream et al., personal communication). Furthermore, the three packs of reintroduced Yellowstone wolves released concurrently restricted their ranges within a few weeks and intermingled their areas

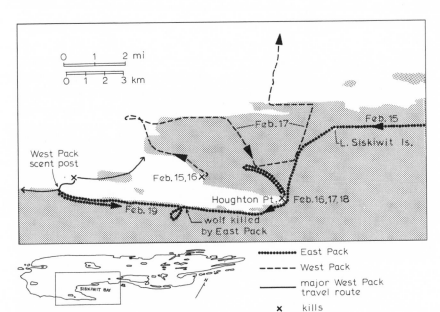

FIGURE 1.12. Movements of Isle Royale's East pack into traditional West pack territory in 1974 (Peterson 1977).

only minimally (M. K. Phillips and D. W. Smith, personal communication). Fatal aggression between members of two adjacent packs released in 1996 was observed within 6 months of release. The prey supply in the area numbered in the thousands.

It is easy to see why, in a declining wolf population with decreasing resources, the remaining members would be highly competitive. In an increasing population, too, competition should be keen because of the greater wolf numbers. The main difference between these two situations is that in an increasing population, the competitors are more likely to be kin than in a decreasing population (see above). Thus perhaps competition with close neighbors should be less intense in an increasing population.

The only observed relationship between the annual rate of wolves killed by wolves in the SNF from 1970 to 1989 (L. D. Mech, unpublished data) and population phase (Gese and Mech 1991) was that from 1978 through 1985, when the wolf population was lowest on average, intraspecific mortality was also lowest (L. D. Mech, unpublished data). This information, as well as observations of wolf packs tolerating each other when meeting (Pimlott et al. 1969), suggests that there may be periods when competition is relatively low. Most likely such times would be either when prey is surplus to the wolves' immediate needs or when relatedness among adjacent wolf packs is high. On the other hand, even when food appeared to be in surplus on Isle Royale, West pack II sought out and attacked other wolves (Peterson and Page 1988).

There is no absolute limit on competition among wolf kin, however. This is apparent within a pack when food is short; adults compete fiercely with yearlings, yearlings with pups, and pups with one another (see Packard, chap. 2 in this volume). Furthermore, on Isle Royale, where all wolves are related as closely as siblings (Wayne et al. 1991), wolves space themselves territorially and behave toward each other like any outbred population, and still kill each other (Peterson 1977; Peterson and Page 1988). This does not necessarily mean that wolves would kill close kin under other circumstances, for when all competitors are closely related, kin competition would be maximized. The only record we know of wild wolves killing close kin other than on Isle Royale involves apparent sisters in Yellowstone National Park (McIntyre and Smith 2000), although admittedly gathering such data is difficult.

Wolf aggressiveness might stem from food competi-

tion, breeding competition, or both. Nothing is known about the possibility of packs competing for breeding opportunities; however, the incidents described above in the section titled "Usurping a breeder" could be examples of such competition. In any case, analyses of deaths of wolves due to wolf attacks over a 22-year period in the SNF (Mech 1994a) and a 9-year period in Denali National Park (Mech et al. 1998) suggest that such intraspecific strife primarily represents territorial competition that reduces competing breeders and increases opportunities for packs to expand their territories, while indirectly tending to hold each pack in its territory.

This conclusion is supported by several lines of evidence. First, it is primarily maturing or mature wolves, which are the territory holders, that are killed by other wolves (L. D. Mech, unpublished data). Second, killings are concentrated in the few months before and after the breeding season (L. D. Mech, unpublished data), when chances are greatest of interfering with a neighboring pack's annual reproductive increase. Successful interference would reduce the pressure the neighboring pack would place on a pack's food supply and thus on its spatial needs. Third, some deaths involve individuals killed in their established territories by invaders. This last evidence tends to rule out strict territorial defense as the sole motive for intraspecific strife, although the territorial competition explanation obviously encompasses territorial defense.

Fates of Fractured Packs

Fractured packs are packs whose key members are lost to various sources of mortality. The fates of these packs depend a great deal on just which members are lost. Since the core of the pack is the breeding pair, the loss of any or all of their offspring means merely that the pair continues to hold the territory. Even with the loss of one member of the pair, the other member may hold the territory until a new mate arrives (see above). In Wisconsin's colonizing population, some single adult wolves that had lost a mate remained in their territories for years before finding another mate (R. N. Schultz and P. C. Wilson, unpublished data).

When the breeding pair is lost, the remaining members of the pack may disperse and join the floaters in the population (Meier et al. 1995; L. D. Mech, unpublished data), just as young members that are removed from packs and translocated behave like dispersers (Fritts et al. 1984, 1985). In one case, however, pups in their first winter whose parents were killed by other wolves eventually starved (Meier et al. 1995). In another case, both members of the breeding pair left or were lost, and the single remaining daughter paired with a dispersed neighbor (Mech 1987a).

An instructive case involves Montana's Ninemile pack. After producing a litter of pups, both adults were killed by midsummer. Government workers artificially fed the pups. The litter remained together all summer and autumn and eventually dispersed as yearlings (Jimenez 1992).

Spatial Structure and Population Change

Because wolf population size depends so much on the amount of vulnerable prey biomass available (Packard and Mech 1980), and because that figure varies widely each year, wolf populations can also fluctuate greatly (see Fuller et al., chap. 6 in this volume). Moderate fluctuations can change pack sizes without changing the structure of the territorial mosaic (Fuller 1989b). However, large population increases are accompanied by attempts at new pack formation, as described above. Much of the dynamism in the territorial mosaic described above results from population fluctuations.

Spatial Changes during Population Increases

When a wolf population is increasing, it produces large numbers of pups (see Fuller et al., chap. 6 in this volume), which increases food competition. As those pups begin to mature, they begin competing for breeding space. This competition increases the potential for biding, budding, splitting, dispersing, challenging, floating, and carving out new territories. Because these local territory formation strategies usually succeed during increases in wolf populations, one can conclude that established packs with adequate food can afford their territory size being reduced by competitors, so are less competitive with new packs than are food-stressed packs. This idea conforms to the elastic-disc view of territoriality (Huxley 1934) and makes sense from an evolutionary perspective.

Packs with sufficient resources should allow room for offspring so long as their own survival is not jeopardized. In fact, the above reproductive and territory proliferation strategies were observed primarily in increasing populations (Fritts and Mech 1981; Peterson and Page 1988; Meier et al. 1995), except for pack splitting, which

took place at the bottom of a population decline (Mech 1986) and at the top of an increase (Meier et al. 1995).

Based on two studies (Fritts and Mech 1981; Hayes and Harestad 2000a), the spatial changes that take place as wolf populations colonize or recolonize an area are as follows: First, pairs form large territories and breed; then pack size increases, territory size declines, and the number of territories increases. This progression of spatial and group size changes is similar to that of a fox population facing increasing food, which Lindstrom (1986) postulated was the basis for promoting group inheritance of territories. There appears to be some merit in this theory as applied to wolves in that pack budding, splitting, and multiple breeding (see above) would promote offspring remaining at home longer (philopatry). Philopatry, in turn, would further the inheritance of local resources by offspring of the original breeders.

Spatial Changes during Population Decreases

When wolf populations decrease, one might expect pack territory numbers to decrease and individual pack territories to enlarge. Although too few studies have been done on this subject to provide many details, the biology of the wolf tends instead to promote the quick proliferation of territories, but retard their decrease.

Here is why. Wolves can produce offspring at 2 years of age, so under favorable conditions, many potential breeders quickly become available seeking breeding positions. However, because breeders are productive for 8–10 years (see Fuller et al., chap. 6 in this volume), pairs established in territories must try to hold their territories for long periods, even during resource declines. Resource availability is greatly dependent on weather (see Mech and Peterson, chap. 5 in this volume), and weather can vary annually, so the chances of a pair experiencing a resource decline during their lifetime are good.

However, a wolf population can easily adjust to huge decreases in resource availability through increased dispersal of young and a reduction in productivity. For example, a pack of sixteen (two breeders, two 3-year-olds, six yearlings, and six pups) could drop almost 90% within a few months to just the two breeders merely through dispersal, and the pair could still hold its territory (see below). If necessary, the pair could then refrain from producing young until resource availability improved. Although we know of no such drastic decline, Mech (1986) has seen the Ensign Lake pack in the SNF drop from ten to two in 2½ years and Denali's East Fork pack drop from twenty-nine to eighteen in a few months (Mech et al. 1998).

Thus, once a wolf pair establishes a territory, it strongly resists losing that area. Some individual wolves and their offspring have held their territories for 8–12 years or more in the SNF (Mech and Hertel 1983; L. D. Mech, unpublished data) and on Ellesmere Island (Mech 1995d). (This does not necessarily mean that territory boundaries remain the same for the entire period, but the territory does cover the same general area.) On the other hand, there is some evidence that only a small percentage of packs hold their territories any longer than this (see below).

The history of the Isle Royale wolf territorial mosaic perhaps illustrates more dramatically than any other case the tendency of established breeders to continue holding territories. Wolves colonized the Lake Superior island in 1949. Little information about pack territories was recorded before 1959, but in that year one large breeding pack of fifteen dominated the island (Mech 1966b). Not until 1965 did a second breeding pack bud from the first (Jordan et al. 1967). However, ever since then, there have been two to four pack territories on the island, even after the population fell from fifty wolves to twelve (Wolfe and Allen 1973; Peterson 1977; R. O. Peterson, unpublished data; Peterson and Page 1988).

The only other population under primarily natural regulation that has been studied throughout a drastic decline in wolf numbers is the central SNF study population. It fell from eighty-seven wolves in 1969–70 to forty-four in 1974–75, and during this period the number of packs dwindled only from thirteen to eleven (Mech 1986). One of the lost packs resulted from the fight between and apparent merging of the Pagami and Greenstone packs, described above.

Relatedness among Packs

As should be apparent from the preceding sections, neighboring wolf packs tend to be genetically related. The closer one pack lives to another, the greater its chance of being related to the other. This tendency results from the budding and splitting processes constantly under way in a vigorous population, as well as from attempts by dispersed offspring to fill in interstices among pack territories. Molecular genetic data (Lehman et al. 1992) confirm the field data based on known wolf demographics (Fritts and Mech 1981; Peterson, Woolington, and Bailey 1984; Ballard et al. 1987; Meier et al. 1995).

Conceivably, the phenomenon of wolves living with more than one pack (Peterson, Woolington, and Bailey 1984; Mech 1987a; Ream et al. 1991; Boyd et al. 1995; see above) is explained by the close genetic relationships between certain packs in a population. On the other hand, it may be the cause of that close relationship. One male wolf that was known to live intermittently in his natal pack and in a neighboring pack over a 20-month period eventually moved into the neighboring pack and paired with a female there (Mech 1987a).

Nevertheless, the constant churning of the population resulting from strong competition and intraspecific strife (see above), as well as from the immigration of dispersers from distant populations, continues to ensure a certain level of unrelatedness.

One could theorize that expanding populations tend to be more closely related, on average, than contracting ones, for the reasons discussed above. Certainly the wolves in the best position to increase their genetic contribution to the population would be those occupying territories when prey availability increases. Because of the wolf's tendency to outbreed (D. Smith et al. 1997), however, a countertendency would develop, guaranteeing a constant influx of new genes and diluting the relatedness of the population (Kennedy et al. 1991; Lehman et al. 1992). The direct killing of breeders by neighbors (see above) tends not only to increase genetic heterogeneity, but to accelerate its increase (Mech 1977b).

Movements within Territories

Wolf travel within territories serves two main functions: foraging (hunting, scavenging, and food delivery) and territory maintenance. Travel would be most efficient if it were used for both functions, and there is every indication that this is the case. Whenever we have watched wolves traveling, they were both foraging and marking.

Wolves generally follow trails, shores, gravel bars, frozen waterways, ridges, roads, and other types of terrain that are easy to traverse. Even in closely cropped pastures and on frozen lakes, any part of which might seem to be an easy route, wolves still tend to follow trails or tracks of other animals. We have often thought this tendency may facilitate travel by allowing the wolves to concentrate on their surroundings rather than constantly having to focus on where to place their feet. Wolves generally travel single file, and in deep snow, this pattern allows more efficient travel by the younger individuals, which usually follow their parents.

Wolf movements within a territory differ between the pup-rearing season, spring to early fall (see Packard, chap. 2 in this volume), and the rest of the year (Mech 1970; Mech et al. 1998; Jedrzejewski et al. 2001). When rearing pups, pack members radiate out from the den or rendezvous site where the pups are to other areas of the territory, returning periodically to feed and care for the pups (Murie 1944; Chapman 1977; Haber 1977; Mech 1988a). Once the pups are developed well enough to join the adults on their hunts, the pack moves as a unit and becomes nomadic around the territory (Burkholder 1959; Mech 1966b; Peterson 1977; Musiani et al. 1998; Jedrzejewski et al. 2001). However, wolves occasionally use rendezvous sites even when offspring are as old as 13 months (Mech 1995c), and in central Italy, where wolves coexist with extensive human activities, they tend to maintain rendezvous sites year-round, radiating out from them at night (Boitani 1986).

Locations of Homesites in Territories

One might expect that wolf dens and rendezvous sites (see Packard, chap. 2 in this volume) would be located toward the center of the pack territory (Banfield 1954; Ballard and Dau 1983). Such a location would maximize the ability of adults to forage efficiently in all directions and minimize exposure to neighboring packs.

However, this hypothesis was not fully supported in the SNF (Ciucci and Mech 1992). Instead, wolves denned more or less randomly throughout their territories except for the outer 1 km, which seems to have been avoided. For wolves denning in the central 60% of their territories, however, there was evidence that the larger the territory, the closer to the center the wolves denned. Similar analyses are needed for other areas.

Where humans persecute wolves, the animals tend to locate their dens far from human disturbance. However, where wolves have not been persecuted for many years, they may den close to areas of high disturbance (Thiel et al. 1998). Rendezvous sites are usually located in the general denning region, so den location is the strongest determinant of their location in a territory.

Movements during the Pup-Rearing Period

Once wolves have denned, the social center of the pack is usually the pups (Murie 1944; Mech 1970, 1988a; Clark 1971; Haber 1977; Jedrzejewski et al. 2001). The reason for this is simple: The breeding pair's entire annual repro-

ductive investment is in the pups, which require regular care and feeding. Thus the parents must return to the pups as frequently as possible after foraging. Other pack members are also tied to the den area not only because they contribute to the care and feeding of the pups (see Packard, chap. 2 in this volume), but also presumably to maintain their bonds with the breeding pair and one another.

There are two exceptions to this generalization. First, maturing pack members sometimes leave the pack for varying periods and return much later (see above). Second, occasionally a large contingent of the pack, sometimes including the breeding male, lives nomadically while the breeding female, or the breeding female and a maturing offspring, feed and care for the pups (Mech et al. 1998). (Females have been known to raise pups alone, although no doubt they would not always be as successful as two parents; see Packard, chap. 2 in this volume.)

Wolves may travel as far as 48 km (29 mi) from the den or pups to obtain food (Mech 1988a). When Mech (1988a, 1995c,d) accompanied a habituated pack of wolves on Ellesmere Island during their hunts away from the den, he noticed no difference in their travel rates or patterns from those of nomadic packs in winter (Mech 1966b). The wolves basically traveled from prey concentration to prey concentration until they killed something. The difference, then, was that wolves with pups returned to the den soon after gorging.

Little is known about the patterns of post-denning territory use. However, there is little reason to think that wolves would use their territory any differently once they have left the den, although there is evidence of seasonal changes in territory use (Jedrzejewski et al. 2001; Merrill 2002). The relative extensiveness of territory use during the pup-rearing season versus the rest of the year seems to vary by study, and one must be cautious in interpreting these data because of the sampling problem discussed earlier (see "Seasonal Shifts in Territories").

Homesite Shifts

As the pups grow and develop, the adults may move them from one den or rearing site to another over the summer (see Packard, chap. 2 in this volume). When the pups are young, these moves may be as short as 0.25 km, whereas as they get older, such moves may be as far as 8 km (5 mi) (Joslin 1966; L. D. Mech, unpublished data). In one case, which may be an exception because it involved a pack with only a single 8-week-old pup, the Ellesmere Island pack shifted rearing sites some 32 km (19 mi) (L. D. Mech and L. Boitani, unpublished data).

Movements during the Nomadic Phase

Generally, wolves are nomadic during about half the year, after their pups have grown and developed enough to move with them. At least one study has suggested that the first movements of the nomadic phase are perhaps the most extensive of all (Fritts and Mech 1981). The most instructive information about movements during the nomadic phase derives from aerial snow tracking of wolves (Burkholder 1959; Mech 1966b; Jordan et al. 1967; Haber 1977; Peterson 1977). Much of this work was done on Isle Royale in Lake Superior, which has a long and narrow configuration. Although this fact should have little effect on such parameters as rate and distance of travel, it would affect the rate of doubling back, for example, and possibly other parameters.

Speed

Wolves usually travel at a lope. Since they are narrow-chested, and since their elbows are turned inward and their feet outward (Iljin 1941; Young and Goldman 1944), they put their feet one almost directly in front of the other as they walk. They can maintain this tireless gait for hours at a rate of about 8–9 km/hr (Burkholder 1959; Mech 1966b, 1994b; Shelton 1966). At times they break into an exuberant run at perhaps twice this speed, presumably in anticipation of something ahead. When returning to the den, their average speed increases to 10 km/hr (Mech 1994b). The wolf's elongated muzzle and the shape of the inner nose ensure optimal oxygenation and an efficient cooling system even in hot climates.

Distance

As already indicated, wolves are capable and inveterate travelers. Some of the claims for their travel distances may be exaggerated, such as J. Magga's that, when hunted, wolves can travel 200 km (120 mi) in a day (Pulliainen 1965); on the other hand, at only their usual rate of 8 km/hr, they could do so. What we do know is that in winter, packs can travel up to 56 km (35 mi) overnight (Stenlund 1955) and up to 72 km (45 mi) in 24 hours (Burkholder 1959; Pulliainen 1965; Mech 1966b; Pimlott et al. 1969). On average, wolves on Isle Royale traveled

14.4 km (9 mi) per day in winter (Mech 1966b). Even in territories of 172–294 km² (67–115 mi²) in Poland (Okarma et al. 1998), wolves traveled a mean of 22.8 km (13.7 mi) per day (Jedrzejewski et al. 2001). In Italy, wolves averaged 27.4 km (16.4 mi) per day (Ciucci et al. 1997).

The usual pattern of winter wolf movement includes travel for a long distance while hunting, making a kill, feeding, resting, and local movement near the kill, abandoning the kill, and repeating the cycle. When all elements of this pattern are considered, wolves cover their ranges at an average of about 2.4 km/hr (Mech 1970). This rate includes an actual travel rate while hunting of about 50 km (30 mi)/day, and an average of about 30% of the wolves' time is spent hunting (Mech 1970).

If wolves fail to produce pups, or lose them, the adults remain nomadic during summer (Mech 1995c), and their rate and distance of travel is similar to that during the rest of the year (Mech 1988a, 1994b, 1995c). One would expect that during summer wolves would need to rest less, since they would not be wading through snow. If so, they probably can cover more distance, although the only information available on this subject is from the SNF, where wolves averaged 19 km (11 mi) per day, with a range of 7–46 km (4–28 mi) per day (D. J. Groebner and L. D. Mech, unpublished data).

Differential Use of Habitat Types

Wolves gravitate to areas within their territories where prey live. Each prey species uses habitat types differently. White-tailed deer, for example, space themselves widely over a variety of habitat types in summer, but yard up in winter in protected lowlands or on south-facing slopes. Dall sheep tend to frequent the steepest mountain terrain and venture into the lowlands only to get to other mountains. Thus, during their routine hunting trips, wolves tend to travel wherever the prey reside in their territory and to avoid prey-free areas, such as extensive conifer swamps (Fritts and Mech 1981) and mountains in winter when prey inhabit valleys (Ream et al. 1991). In the Caucasus Mountains of Europe, an observer following wolf routes counted five to fifteen deer and several wild boar per hour, whereas a random route showed no more than five or six animals per hour (Kudaktin 1979, cited in Bibikov et al. 1983).

On the other hand, wolves will take advantage of easy travel routes, such as frozen lakes and shorelines, through prey-free areas (Stenlund 1955; Mech 1966b; Jordan et al.

1967; Peterson 1977) to get to where the prey is. In certain areas of Europe where there are no large prey, wolves hide in sterile, isolated habitat types such as mountainsides by day, then venture around villages and garbage dumps at night (Zimen and Boitani 1979).

Spatial Characteristics of Travel

The winter travels of wolves whose routes have been mapped (Burkholder 1959; Mech 1966b; Jordan et al. 1967; Pimlott et al. 1969; Haber 1977; Peterson 1977; Jedrzejewski et al. 2001) show three characteristics: long, linear routes, rather than, for example, zigzagging; repeated use of some routes; and a tendency to cover their territory extensively in short periods. That is, instead of searching one end of their territory thoroughly before moving on, the wolves tended to travel linearly from one end of their territory to the other. This was also true of the summer travels of wolves on the arctic tundra (Mech 1988a 1995c).

The extensive, rather than intensive, nature of wolf travel can be seen most dramatically on Isle Royale, where wolves follow the shoreline, trails, ridges, and strings of lakes and bays along the narrow lay of the island (Mech 1966b; Jordan et al. 1967; Peterson 1977). They could have cut across the island instead, but this would have made travel more difficult. Cutting across the island might have exposed more prey to them, however, for they would have had access to prey on both sides of their travel routes, rather than just one.

Even in the SNF on the mainland, wolves travel long strings of frozen lakes and rivers rather than cutting overland (Stenlund 1955; Mech and Frenzel 1971a). They might choose these routes because of the relative ease of travel, but if wolves travel primarily to find prey, and it is faster to find prey by traveling overland, then there may be some other reason why wolves travel linearly. Furthermore, on the arctic tundra during the snow-free summer, wolves also traveled linearly (Mech 1988a, 1995c). Traveling extensively instead of zigzagging over a smaller area would further territorial maintenance, and would add surprise as an advantage in hunting (Mech et al. 1998).

Rotational Use of Territory

Evidence is emerging that wolves may not revisit specific herds of prey for several days after a previous visit. This behavior might serve to reduce the prey's vigilance and perhaps increase the wolves' chance of catching them off guard (Mech et al. 1998, 105).

Indications of such movement patterns were apparent in travel data presented by Burkholder (1959) and Mech (1966b, 55–57), and possibly Haber (1977). Weaver (1994), however, quantified this behavior by measuring times between visits to individual prey herds. His study pack did not revisit a given elk herd until 12–16 days later, and did not revisit the same bighorn sheep until 36–37 days later. Wolf packs in Poland tended to visit any given area about every 6 days on average (Jedrzejewski et al. 2001).

Trappers' lore and common sense suggest that wolves (and other species) tend to use the same routes repeatedly throughout their lives, and that even after an absence of wolves, new wolves should use the same landscape features. Neither lifelong use nor interrupted sequential use has been documented, however.

Extraordinary Pack Movements

There are a couple of cases of pack movements that are rare exceptions to the patterns described above. Little is known about these movements except their rarity. The first such case was documented in interior Alaska and involved a pack of ten animals that was aerially tracked in the snow for 45 days. This pack covered an area of approximately 128 km (77 mi) by about 72 km (43 mi) (Burkholder 1959). Because these wolves were not radio-collared, nothing is known of where they lived before or after they were tracked. Thus it is not clear whether this was a pack that possessed an extra-large territory, or whether they were more or less nomadic during the study and settled down only if pups were born.

When this information was published, it was not known just how unusual such wide-ranging travels were, for no wolves had yet been radio-tracked. Since then, some two hundred packs have been radio-tracked in Alaska and the adjacent Yukon alone (Stephenson et al. 1995), and many others have been tracked in Canada and elsewhere. Only one other case of such widespread pack movements has been recorded.

This case involved the Little Bear pack, which was radio-tracked from 1988 to 1992 in Denali National Park (Mech et al. 1998). In fall 1991, this pack numbered twenty-three (including at least eleven pups), but it then split during the winter of 1991–1992. In May 1992, a group of eleven wolves, including three radio-collared adults, left and moved some 250 km (150 mi) to the southwest, where they eventually settled (Mech et al. 1998). They had not produced pups before they left, and it is unclear whether they produced any after they settled. This appeared to be a mass emigration. A shorter emigration was seen in Minnesota (Fuller 1989b). Nothing more is known about wolf emigration, but presumably it takes place in response to a food shortage.

———

It should be clear from the above discussion that the wolf has a highly adaptable social ecology that is flexible enough to contend with a wide variety of living conditions. Where food is small and scattered, offspring dis-

perse early, and packs are small. Where prey is much larger than the wolf itself, breeding pairs bring young from two or three litters with them as they travel and hunt, and packs are much larger. Changes in prey availability are met with changes in dispersal rates. Similar dynamics affect a population's social structure, with longer-term prey fluctuations translated into adjustments in the territorial mosaic. Such social fluctuations help wolves contend with the ever-changing nature of their dynamic economy—a live prey base that is itself subject to the vagaries of more basic environmental perturbations.

2

Wolf Behavior: Reproductive, Social, and Intelligent

Jane M. Packard

A LARGE WHITE WOLF appeared on the bank of a drainage flanking a meadow on Ellesmere Island in Canada's High Arctic. Through binoculars, I identified him as the breeding male "Left Shoulder" (Mech 1988a; Mech 1995d). Several hours earlier, he had left four pups playing in this meadow of gently rolling heather and willow tundra. Later, however, they had followed their mother out of view and toward their den. What would Left Shoulder do when he did not find the pups there?

Raising his chin, he released a long, low howl. We both waited. I was expecting to hear a chorus of high-pitched puppy howls in response, but heard none. Apparently the wolf didn't either. Looking toward where he had last seen the pups, Left Shoulder made his way down the slope, crossing their trail. Would he pick up the scent that was 15 minutes stale? Nose to the ground, he ambled up the drainage. However, the pups had gone the other way. Left Shoulder back-tracked the pups for about 3 meters, paused, reversed direction, and headed after the pups! I wondered if he would track the pups and his mate directly to the den.

Trotting briskly, Left Shoulder kept his nose to the ground for about 30 meters. Then, he raised his head, veered away from the scent, and headed directly toward the den, where he often met his mate and two yearlings, who also tended the pups. He seemed to have expected the behavior of the pups to be predictable.

Intelligence is demonstrated when a canid anticipates the behavior of a social companion (e.g., the pups) or solves a novel problem in obtaining food. Left Shoulder used several senses in solving the problem of how to find the pups: he scanned the landscape visually, he listened

for replies to his howl, and he sniffed the ground. He appeared to combine this sensory information with expectations based on what he had learned about the behavior of family members. Such integration of information will be emphasized by Harrington and Asa with regard to wolf communication in chapter 3 in this volume.

This chapter focuses on interactions between nature (genotype) and nurture (phenotype). Each wolf inherits genetic propensities (instinct; also called "neuroendocrine programming") that lead it into situations in which it learns from its environment (Fentress 1983). Its genotype enables it to solve ancient problems predictably encountered in the history of the species. Communication during social interactions helps fine-tune these responses, enabling individuals to solve new problems and respond to the unpredictable nature of their environment. (For a different perspective, see Mech and Boitani, chap. 1 in this volume.)

Wolves live in diverse and changing environments (see Boitani, chap. 13 in this volume). Variation in the environment is probably one of the keys to understanding why several species have evolved flexible problem-solving behavior (Byrne 1995). Not only does the wolf's physical environment pose challenges, but its social environment also provides both a challenge and a support to individuals learning to maneuver within the dynamic complexity of their physical environment.

This chapter will examine the three-way interaction among wolves, their social environment, and their physical environment (fig. 2.1). First, I will look at problem solving in courtship and reproduction: how are pups produced and cared for until they can start their own

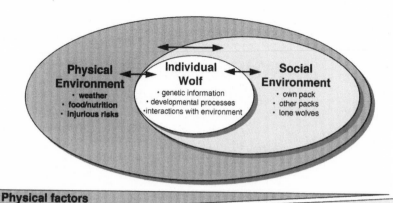

Physical Environment
• weather
• food/nutrition
• injurious risks

Individual Wolf
• genetic information
• developmental processes
• interactions with environment

Social Environment
• own pack
• other packs
• lone wolves

Physical factors

Social factors

Continuum of Environmental Effect

Fluctuating Predictability

Adult & Juvenile Traits
• learning & intelligence
• activity patterns
• leadership & travelling

Courtship & Reproduction

Adult Traits
• seasonal courtship
• denning
• pup care

Juvenile Traits
• dependent phase
• transition phase
• prolonged maturation

Social Cohesion & Conflict

Adult & Juvenile Traits
• temperaments & relationships
• access to food & mates
• interactions between groups

CLUSTERS OF BEHAVIORAL ISSUES

FIGURE 2.1. A conceptual map of this chapter, illustrating how cross-cutting themes (interactions among individual behavioral traits, social environment, and physical environment) overlap in the three sections addressing the categories of reproductive problem solving, physical problem solving, and social problem solving.

FIGURE 2.2. Behavior in a family group of wolves includes both (A) peaceful coexistence and (B) conflict over access to resources, such as food.

families? Second, I will explore the additional social problems of wolves living in families and extended families, sometimes sharing resources peacefully (fig. 2.2A) and sometimes competing for them (fig. 2.2B). Third, I will focus on how wolves survive in the midst of fluctuating predictability (e.g., obtaining and conserving energy while avoiding risks of death).

However, before addressing reproductive, social, and physical problem solving by wolves, I must (1) review the basics of what is known about wolf sociality, (2) examine what is meant by social and physical environments as they affect wolf sociality, and (3) explain the ethological approach I will use.

Wolf Sociality

Left Shoulder's pack can help us understand several key aspects of wolf sociality (fig. 2.3). Wolf packs are usually family groups that move within exclusive home ranges and are hostile to strangers from neighboring packs, although there are exceptions to this generalization due to the dynamics of social and physical environments (see Mech and Boitani, chap. 1 in this volume). Left Shoulder's pack, known as the Ellesmere pack, was a family whose breeding members remained together for over 10 years (Mech 1995d). As a group and alone, the wolves defended food resources from other wolves (Mech 1993a). In some years, the breeding female produced

FIGURE 2.3. Wolf packs are usually family groups that move within exclusive home ranges and are hostile to strangers from neighboring packs.

FIGURE 2.4. A litter of four pups born in a cave den on Ellesmere Island, which had probably been used by a series of family groups for centuries. (From Mech 1991b.)

pups in a cave den that had probably been used intermittently for centuries (Mech and Packard 1990) (fig. 2.4). In other years, pups were born in a pit scraped into the ground (Mech 1993b). When no pups were produced, the adults fed offspring from the previous year (Mech 1995e). Family membership and reproductive status changed over the years due to deaths, dispersal, and births (Mech 1995d). For example, when Left Shoulder's first mate, Mom, stopped producing pups, she switched to caring for pups produced by Left Shoulder and her daughter, Whitey (Mech 1995d).

How did Left Shoulder's pack compare with wolf packs in other environments? The following description of the structure of wolf populations provides a general overview. This basic framework will aid in understanding variation in the spacing, group size, and movements of wolves in different populations, topics detailed by Mech and Boitani in chap. 1 in this volume.

In low-density populations such as that of Ellesmere Island, wolves are generally monogamous, although there are exceptions. Wolf packs usually are founded by an unrelated male and female (D. Smith et al. 1997). After dispersing from the families where each was born, the members of a new pair travel together in an area not defended by other hostile packs (Rothman and Mech 1979; Fritts and Mech 1981). The chance of a new pair starting a family is relatively low in high-density populations where all suitable areas are already defended (i.e., "saturated populations") (Packard and Mech 1980). Occasionally, new packs consist of other combinations of members (Mech and Nelson 1990b), but groups without a breeding pair rarely persist for more than a few months (see Mech and Boitani, chap. 1 in this volume).

Each pack expands as it produces litters, averaging 5–6 pups, and shrinks as offspring disperse at 10–36 months of age (see Mech and Boitani, chap. 1 in this volume). For example, the Ellesmere pack's four 1988 pups were attended by a yearling female, "Whitey," and male, "Gray Back," in addition to Mom and Left Shoulder (Packard et al. 1992; Mech 1995d). Although Mom remained with the family when Whitey replaced her in 1990 as the breeding female, no other offspring remained with the pack for more than three winters (Mech 1995d).

As in most other packs, Left Shoulder, his mate, and the two yearlings delivered food to the pups, interacted congenially with them, rested near them, chased intruders, and were highly attentive to the pups as the family moved together during the summer (Packard et al. 1992; Mech et al. 1999). In a variety of ways, the social environment of the family buffered the pups from threats in the physical surroundings. However, there are variations on this theme; for example, even lone females have successfully raised pups (Boyd and Jimenez 1994).

Variation in Pack Structure

The tenure of breeders in wolf packs varies from 1 to 8 years, typically lasting only 3 to 4 years (Mech et al. 1998). For example, Left Shoulder was not the breeding male in the Ellesmere pack when it was first studied (Mech 1995d). We do not know whether Left Shoulder was born into the family or whether he joined it, as has been observed elsewhere (Van Ballenberghe 1983a; Mech et al. 1998). Outsiders are most likely to be accepted into a family by a widowed breeder seeking a new mate (Rothman and Mech 1979; Fritts and Mech 1981; Mech and Hertel 1983), although that is not always a requirement (Meier et al. 1995).

In a small percentage of wolf packs, more than one female may reproduce in a given year (see Mech and Boitani, chap. 1 in this volume). Breeding by two or more closely related females in the same pack has been noted after loss of one or both members of the original breeding pair (Packard et al. 1983; D. W. Smith, unpublished data), although whether such a loss is necessary for multiple breeding in a pack is unknown. In unsaturated populations, it is unlikely that two or more sisters will remain reproductively active in the same pack for more than 1–2 years (Packard 1980; D. W. Smith, unpublished data). In contrast to sisters, the relationships between aging mothers (over 7 years old) and their daughters appear more tolerant (Packard 1980; Mech 1999).

Although the Ellesmere pack has averaged four to five members over 10 years, the individual members have changed over the years (Mech 1995d; Mech 1999). At 3 years of age, Whitey replaced her mother as breeding female. Why didn't Whitey produce her own pups as a yearling? Wild female wolves usually don't ovulate until their second, third, or fourth winters (see Mech and Boitani, chap. 1 in this volume), and first deliver pups at 2–5 years of age (Rausch 1967; Mech et al. 1998). Physiological maturation may be delayed a year or two by nutritional or other stress (Packard et al. 1985; Mech and Seal 1987; Packard 1989). Under extremely good nutritional conditions, physiological maturation may also be accelerated a year, both in captivity (Zimen 1976; Medjo and Mech 1976; Packard et al. 1983) and in the wild (D. W. Smith, unpublished data).

Thus, reproductive characteristics vary within and between wolf populations, as well as within the lifetimes of individuals. For example, Left Shoulder and Whitey produced pups in only 5 of the 7 years when they were the breeders in the Ellesmere pack (Mech 1995d; L. D. Mech, unpublished data). In the Denali wolf population, packs failed to produce pups in 15% of 91 pack-years (Mech et al. 1998). Successful reproduction was lower in younger wolves (and in those breeding for the first time), as has also been reported for captive wolves (Packard et al. 1983).

Although no analyses of the variation in reproductive success among wolves in various packs and populations have been done, the Ellesmere pack helps illustrate general patterns. Left Shoulder produced seventeen pups over 9 years, and Whitey produced nine pups in 7 years.

Of these pups, seven (41%) of Left Shoulder's and three (33%) of Whitey's survived for at least 14 months (possibly longer); at least one survived for at least 8 years (Mech 1995d; L. D. Mech, unpublished data). Such information is difficult to obtain for free-ranging carnivores (Mech 1987a; Packer et al. 1988), which places a major constraint on the testing of hypotheses about the ecological and evolutionary functions of the variation observed in wolf reproduction. Therefore, this chapter focuses primarily on reviewing the literature on proximate mechanisms of wolf behavior, with only brief reference to hypotheses about its ultimate function and evolution.

Factors Affecting Wolf Sociality

Wolves are affected by both social and physical factors in their environment (see fig. 2.1). "Social environment" refers to the conspecifics with which a wolf interacts, including the members of its own pack, other territorial packs, and lone wolves not associated with a pack. "Physical environment" refers to abiotic factors (e.g., weather and landscape) as well as biotic factors other than wolves in the wolf's environment (e.g., prey species and animals that threaten wolves; see Mech and Peterson, chap. 5, and Ballard et al., chap. 10 in this volume).

Why distinguish between the effects of physical and social environmental variation on the behavior of wolves? To understand variation among wolves, we must understand the variation in both aspects of their environment (Packard and Mech 1980). Most packs within a local population usually experience similar environmental variation due to ecological cycles involving climate and the dynamics of predator and prey populations (see Mech and Peterson, chap. 5, and Fuller et al., chap. 6 in this volume). However, the wolves in any one pack share social experiences that differ from those of neighboring packs. For example, social relationships in packs may vary as diagrammed below in figure 2.7.

From the northern tundra to the southern desert mountains, the physical environment of wolf populations varies greatly (see Fritts et al., chap. 12 in this volume). Furthermore, changes in prey abundance affect each wolf population over the lifetime of family groups (Mech 1970; Packard and Mech 1983; Peterson, Woolington, and Bailey 1984). Thus, social and physical factors are both distinct and overlapping influences on individuals and on their reproductive fitness relative to the pop-

ulations in which they live and reproduce (see fig. 2.1). The additive effects of social and physical factors contribute to the variation observed between populations and across years within each population.

To understand wolf sociality in general, we must understand how it varies within and between populations in different physical environments. In part, the variation *within* each population results from the different histories of the packs that form the reproductive units in that population (i.e., social environment). Wolf families appear and disappear for reasons not directly linked to ecological cycles (Mech et al. 1998), although there are, of course, indirect linkages because nutritional condition can affect social interactions within and between packs (Packard and Mech 1983).

To the extent that wolves share genetic propensities for certain kinds of social behavior (Fentress 1983, 1992), the basic social reasons for pack dynamics are likely to be similar across wolf populations. To understand this neuroendocrine programming, we look for similarities in behavior across wolf populations that differ in latitude, prey species, and the phase synchrony of wolf-prey population dynamics (i.e., their physical environment). From an ethological perspective (Bekoff 1981), we seek to understand how much of that similarity results from shared information in the genome and how much from similar experiences that all wolves encounter growing up within families (i.e., their social environment).

Researchers are just beginning to develop and test hypotheses about the similarities and differences among wolf populations. One of the major problems has been the degree to which behavior observed in captive populations might differ from that in non-captive populations; for practical reasons, many more studies of wolf social behavior have been done in captivity than in the field (table 2.1).

Some field researchers discount certain aspects of captivity-based studies as being analogous to studying human behavior in refugee camps (Mech 1999). Other researchers point out that specific factors can be held relatively constant in captivity to tease out the independent effects of social, physical, and genotypic variation. For example, Bekoff (1972) compared play among wolves, dogs, and coyotes while holding the social grouping and physical conditions constant. In free-ranging populations, both the social and physical environments fluctuate simultaneously in an interactive and uncontrolled manner (Packard and Mech 1983). Thus, our under-

TABLE 2.1. Variation in structure of captive wolf families whose behavior has been studied

Structure	Description	Number of packs (pack-years) [a]
Nuclear family	Parents and their offspring, as in a newly formed pack in which unfamiliar breeders pair and produce a litter. The offspring may be of several ages.	6 (23) [b]
Extended family	Parents plus one or more of their siblings, and their direct offspring.	3 (6) [c]
Disrupted family	A family in which one or both of the original parents is missing. (Variations include disrupted nuclear family, disrupted extended family, disrupted foster family.)	11 (29) [d]
Step-family	A disrupted family that has accepted immigration of an outside breeder.	1 (6) [e]
Foster family	A family that has accepted immigration of nonbreeding individual(s) not born into the pack. (Foster pups are more likely in captivity than in the wild.)	5 (10) [f]
Complex family	A group of wolves with a history that does not fit into the categories defined above; e.g., a group of hand-raised siblings to which additional hand-raised pups have been fostered.	1 (5) [g]

[a] Abbreviations in notes: BeP = Berlin pack (Altmann 1987); SoP = South pack and NP = North pack (Packard 1980); CP = Connecticut pack (Jenks and Ginsburg 1987; Schotté and Ginsburg 1987); BuP = Burgers Zoo pack (Derix 1994; van Hooff and Wensing 1987); BrP = Brookfield Zoo pack (Rabb et al. 1967), WP = Washington Zoo pack (Paquet et al. 1982); OP = Oregon pack and ShP = Shubenacadie pack (Fentress and Ryon 1982; Fentress et al. 1987); RP = Rickling pack, KP = Kiel pack, and BaP = Bavarian pack (Zimen 1981); PI = pack I, PII = pack II, and PIII = pack III (Fox 1973).

[b] BeP (1959–66, 1974–78, 1978–81), NP (1977), CP (1975–80), BuP (1969–73).

[c] BrP (1960–63), WP (1968–70, 1973–74), BuP (1978–79).

[d] BeP (1959–66), BrP (1959, 1964–66), NP (1978), SoP (1977), CP (1981–82), ShP (1975–79, 1980–81), KP (1970), WP (1967, 1971–72), PI, PII, PIII.

[e] BuP (1981–85).

[f] SoP (1978), WP (1975–76), OP (1973–74), RP (1968–69), BaP (1970–72).

[g] BaP (1973–77).

standing of wolf social behavior will advance more rapidly when we integrate the information from both captive and field populations, so long as we take great care to recognize the limitations of each perspective and to understand the whole as the sum of the parts.

An Ethological Perspective

An ethogram is a catalogue of behaviors that functions like a dictionary of the meanings of all the actions of a particular species. However, rarely is it possible to document all possible behavioral acts, so such catalogues prepared for the purpose of specific studies are necessarily incomplete (Bekoff 1979a). A catalogue of the basic be-

havioral traits of wolves, as defined by diverse researchers, is compiled in table 2.2.

Many more hypotheses have been developed to explain wolf behavior than have been systematically tested. Students of wolf behavior have come from a wide range of disciplinary backgrounds (e.g., physiology, psychology, behavioral genetics, behavioral ecology, ethology, sociobiology, wildlife management). They have defined terms and questions in ways that are not always coherent, yet these workers provide a rich diversity of perspectives. Haber (1996), for example, described wolves as "eusocial," a term reserved by evolutionary biologists for species that live in colonies (e.g., social insects) in which some phenotypes cannot reproduce due to the

TABLE 2.2. Abbreviated ethogram of wolf behavior; conceptual model of internal states associated with behavioral assays (see also table 3.9)

Category	Subcategory code	Typical action patterns (behavioral assay of state)
Activity[a]	A1	Inactivity: lying sitting, or standing; without changes in angle of neck
	A2	Moderate activity: walking, interacting, exploring, feeding, grooming; neck angle changes
	A3	Strenuous activity: directional rapid movement (e.g., trotting, galloping, running)
Care[b]	C1	Care-solicit: suckle, whimper, lick-up, paw, roll-on-back, solo-howl
	C2	Care-ritual: grovel, over-the-muzzle-bite, roll (passive submission), curl (active submission)
	C3	Care-give: nurse, carry-pup, carry-food, regurgitate, lick-other, watch, follow, lead
Flight / Fight[c]	F1	Flight: avert-gaze, avoid, crawl, head-down, low-posture-retreat, ignore, leave, run, slink, refuge
	F2	Defensive (mixed fight/flight): bark, crouch, gape, growl, hackles, snap, snarl, whirl
	F3	Fight: chase, face-off, holding-bite, lunge, jaw-spar, nip, pin, sidle, stand-high, wrestle-fight
Humbleness[d]	H1	Humble-low: pricked-ears, high-posture, high-tail, flexed- and raised-leg-urination, scrape-back
	H2	Humble-neutral: ears-side, hanging-tail, squat- and stand-urinate, wait
	H3	Humble-high: brows-together, ears-back, low-posture, tuck-tail, long-mouth-line, hunchback
Ingest[e]	I1	Hunt: directional trot, chase, zigzag, sniff-ground, give-eye, dig
	I2	Handle: sprint, grab, hold, lunge, knock-down, pounce, neck-shake, nose-stab, cache, pluck
	I3	Consume: chew, swallow, rip, drink, lick
Maintenance[f]	M1	Maintenance-low: apathetic, foot-drag, disgust-mouth, head-hang, glazed-eyes, disheveled
	M2	Maintenance-normal: lick-self, scratch-self, head-shake, rub-body, urinate, defecate
	M3	Maintenance-high: repetitively-lick-injury, vomit, scoot-rear, limp, diarrhea
Proximity[g]	P1	Proximity-brief: approach, touch-nose, lick, lie-near, pass
	P2	Proximity-moderate: chorus-howl, carry-object, rally, stand-near, look-over-shoulder, wag-tail
	P3	Proximity-prolonged: bow, bounce-follow, gallop, play-wrestle, circle-wag, play-jaw-spar, roll
Sexual[h]	S1	Bonding: follow, mark-over (double-mark), nuzzle, parallel-walk
	S2	Courtship: chin-rest, prance, dart, ears-together, head-flick, hug, sniff-rear, T-formation, wrestle
	S3	Copulatory: escort, tail-avert, mount, thrust, ejaculatory-contraction, dismount, tie, tooth-clack

Note: This behavioral catalogue was compiled for observer reliability training at the Houston Zoo. A more complete ethogram compiled from several sources (Bekoff 1972; 1979a; Derix 1994; Fox 1971e; Goodmann and Klinghammer 1990; Zimen 1971, 1982) is available from the author. Categories are not mutually exclusive; subcategories are mutually exclusive within categories. See also Harrington and Asa, chap. 3, table 9 in this volume.

[a]Model A: Indicator of low (A1) to high (A3) cardiovascular activity; measured in field by variation in radiotelemetry pulses.

[b]Model C: C1 indicative of generalized state of need in juveniles (hunger, cold, full bladder); C2 indicative of low serotonin and moderate adrenal activity in juveniles, subadults, and adults; C3 indicative of high prolactin.

[c]Model F: indicative of adrenal activation above the individual's set-point range: F1 more effect of adrenaline than noradrenaline (Sapolsky 2002; Watkins 1997, 11); F3 more effect of noradrenaline than adrenaline (probably associated with high androgens); F2 indicative of both noradrenaline and adrenaline activity.

[d]Model H: continuum of serotonin above the group norm (H1) to serotonin below the group norm (H3).

[e]Model I: I1 indicative of low blood glucose, low fatty acids, empty gut in context of no food; I2 indicative of burst of epinephrine and endorphin activity; I3 indicative of low blood glucose, low fatty acids in context of available food, caching when stomach is full.

[f]Model M: M1 indicative of one or more diagnostic blood parameters outside the normal range; M2 indicative of diagnostic blood parameters within normal range; M3 may be indicative of active healing processes (e.g., cortisol, histamines).

[g]Model P: indicative of serotonin activity at or above the individual's set-point range and a continuum of endorphin activity below (P1) to above (P3) the individual's set-point range.

[h]Model S: S1 indicative of low steroid hormones (Seal et al. 1987); S2 indicative of estrogen above 2.5 pg/ml in females, baseline testosterone above 240 mg/dl in males; S3 indicative of declining estrogen, rising progesterone above 10 ng/ml and LH above 3 mg/ml in females, peak response to LRH above 600 mg/dl in males.

type of nutrition received at specific developmental stages (Lacey and Sherman 1997). In contrast, all wolves retain the readiness to breed when the social environment permits, and most disperse by 3 years of age (see Mech and Boitani, chap. 1 in this volume).

In part, the diversity of perspectives voiced by wolf biologists is related to the schools of thought encountered in their academic backgrounds. Botkin (1990) traced the roots of two schools of thought, the deterministic and the stochastic perspectives, back to the

Greek philosophers. According to the deterministic perspective, events in the natural world unfold according to certain predictable or predetermined rules and return to steady states. In contrast, from the stochastic perspective, order is apparent only at a specific time as a reflection of chance events. The dynamic changes between relatively steady states are characterized in terms of probabilities.

Although they are contrasting viewpoints, the deterministic and stochastic perspectives are not mutually exclusive. The basic research questions may be rephrased in terms of "those aspects that are relatively unchanging" (deterministic) and "those aspects that do change" (stochastic). Both perspectives should be integrated in predicting the responses of wolves to their ever-changing environment.

These contrasting viewpoints (stochastic and deterministic) are both reflected in the ethological approach of this chapter. Such a balanced perspective on canid behavior is reflected in the ethological writings of Fentress (1982, 1983, 1992) and Bekoff (1981, 1989). Other wolf researchers have tended to emphasize one perspective over the other, as may be apparent in other chapters.

According to the deterministic perspective, sociality evolved in wolves as an adaptation to their habit of hunting and feeding on large prey (see Peterson and Ciucci, chap. 4, and Mech and Peterson, chap. 5 in this volume). In contrast, the stochastic model examines the sources of variation within and between individuals, social groups, and populations (Botkin 1990; Creel and Waser 1997; Lucas et al. 1997); it allows for the hypothesis that behavioral traits of social mammals may not be currently adaptive (Solomon and French 1997), or may be neutral by-products of other traits that were adaptive in the species' history (Pusey and Packer 1994).

In this chapter, my approach to examining wolf sociality will be to (1) identify behavioral mechanisms from captive and field studies, (2) place these mechanisms in the context of limited information about the variation observed in field studies, and (3) clarify hypotheses that might be tested in the future. In short, I intend to explain the biological basis of wolf social behavior by relying on the four basic ethological concepts, two at the proximate level of individuals (causation, ontogeny) and two at the ultimate level of populations (function, phylogeny) (Solomon and French 1997), with an emphasis on the former. Causation addresses the relatively static aspects of behavioral traits, whereas ontogeny addresses the dynamics of behavioral variation with age.

Courtship and Reproduction

The behavior of adult wolves and juvenile wolves may be viewed as having coevolved over millions of years. For example, one can view care-giving behavior (in adults) as having coevolved with care-soliciting traits (in juveniles, subadults, and some nonreproductive adults). A deterministic perspective would posit that the benefits of group foraging shaped both juvenile and adult wolf traits in an optimally beneficial manner. According to a stochastic perspective, however, the factors affecting production of offspring (i.e., mating and care-giving traits) may be distinct from the factors affecting the survival of juveniles to reproductive age (i.e., survival and care-soliciting traits) (Caro 1994; Packard et al. 1992; Pusey and Packer 1994).

A coherent model of wolf reproductive behavior must explain the variation in both adult and juvenile traits. I will attempt this by addressing the following subjects: (1) seasonal courtship, (2) phases of the reproductive cycle, (3) behavior at dens and homesites (indirect care), (4) pup birth and stages of development, (5) direct biparental care of pups, and (6) the familial hunting school.

Seasonal Courtship

Domestic dogs (*Canis lupus familiaris*) can breed year-round (Haase 2000); why not wolves, from which dogs were derived (see Wayne and Vilà, chap. 8, and Nowak, chap. 9 in this volume)? In most of the Northern Hemisphere, wolf pups are born early enough in spring (Mech 1970; Fuller 1989a; Servín-Martínez 1997) that their nutritional needs coincide with a birth pulse of herbivores (May and June), providing relatively easy prey for wolf parents to catch. By autumn, the pups are large enough to follow adults on hunts for larger prey, which are more difficult to kill. Pups are born after the worst of the winter weather and grow to almost adult size before it returns (Van Ballenberghe and Mech 1975). Thus, wolves copulate in winter (Mech 1970), gestation being 61–64 days. However, new pairs form at all times of the year (L. D. Mech, unpublished data), and existing pairs remain together year-round, their tandem urination conveying essentially the same information a wedding ring does (Rothman and Mech 1979; Mertl-Millhollen et al. 1986).

Most of our knowledge of canid courtship behavior (see table 2.2) has been provided by captive studies (Schenkel 1947; Rabb et al. 1967; Packard 1980; Zimen 1981, 1982; Packard et al. 1983; Jenks and Ginsburg 1987; Schotté 1988; Derix et al. 1993; Asa 1995; Servín-Martínez 1991, 1997). Wolf researchers use these detailed studies to interpret the glimpses of reproductive behavior reported from the wild (Mech 1966b; Haber 1968, 1977; Peterson 1977; Mech and Knick 1978).

Phases of the Reproductive Cycle

Seasonal peaks in wolf reproductive behavior are correlated with seasonal changes in reproductive hormones (Seal et al. 1979; Zimen 1982; Asa and Valdespino 1998). In autumn, testosterone in males and estrogen in females begin to rise, priming the reproductive organs for a predictable sequence of behavioral and physiological phases: proestrus, estrus ("heat"), metestrus (pregnancy or pseudopregnancy), pup care, and anestrus (Packard et al. 1983). However, the duration of each phase and the magnitude of hormonal changes within each phase vary among individuals, possibly depending on interactions among such factors as genotype, age, experience, body condition, latitude, and the social environment (Packard et al. 1983, 1985).

The phases of the canid reproductive cycle can be recognized in the field. For example, during the 2 months prior to estrus, paired wolves sleep within 1 meter of each other, significantly closer than after mating (Mech and Knick 1978; Knick and Mech 1980). Usually, the breeding female in each pack is followed more closely by her mate than by other pack members (Mech 1966b; Peterson 1977). Each courting pair engages in reciprocal nuzzling, prancing, genital investigation, and scent marking (Schenkel 1947; Rabb et al. 1967). By back-tracking known radio-collared wolves during winter, Rothman and Mech (1979) learned that established pairs scent-mark more frequently than lone wolves and that newly formed pairs scent-mark more frequently than established pairs.

Observations of captive wolves help detail how the courtship behavior of a pair becomes synchronized when male and female are sexually naive. Wolves hand-raised by Erik Zimen (1981, 133) bred successfully although they had no opportunity to learn from experienced breeders. For example, one sexually naive male, Näschen, "took a great interest in the places in the snow where Finsterau urinated. . . . Näschen kept almost perpetual skin contact with Finsterau. At the end of February, Finsterau for the first time stopped, though only briefly, when Näschen tried to mount her. A few days later things progressed to a point at which she presented herself by approaching Näschen sideways and then in front of him, with her tail turned to one side. Näschen mounted her and, after a number of vigorous thrusts of the pelvis the two 'locked' in the fashion typical of canids."

Each mated pair of wolves progresses through the same fixed sequence of behavioral phases. The duration of each phase is determined by the rate at which ovarian follicles develop and mature within each female each season (Seal et al. 1979, 1987; Packard et al. 1983, 1985). The basic endocrinological patterns of wolves are described in more detail by Kreeger in chapter 7 in this volume; the physiological correlates of wolf reproductive behavior are outlined below.

Pre-proestrus

Pre-proestrus occurs in late autumn or early winter, before adult females show a bloody vaginal discharge (Seal et al. 1979). During pre-proestrus, it is not unusual for either a male or a female wolf to express unreciprocated interest in a potential mate. For example, in Zimen's (1981, 132) captive Bavarian pack, "Finsterau kept pressing against Wölfchen and whimpering, rolling on her back in front of him, and pulling his coat . . . but the more importunate she became the more he withdrew."

One theory is that flirtatious female behavior during pre-proestrus is affected by the hormonal changes associated with rising gonadotropin levels and waves of incomplete follicular development (Packard 1989). At this time, plasma estradiol rises above the 10 pg/ml typical of anestrus (Packard 1980, 75). More frequent scent-marking by males during this period may be correlated with elevated testosterone levels prior to proestrus in females (Hart and Haugen 1971; Packard et al. 1985; Packard 1989). In captive wolves, fights between males appear more likely in autumn and early winter prior to proestrus in females (Zimen 1975; Packard 1989). For adult male wolves, average baseline values for plasma testosterone ranged from 106 to 408 ng/ml, and average testosterone response to injection of LHRH ranged from

384 to 716 ng/ml, during this period prior to February 1 (Packard 1980, 91).

Proestrus

The proestrous phase begins when a bloody vaginal discharge appears, associated with squamous cells shed during rapid growth of the uterine lining. Plasma estrogen rises during proestrus, and adult males usually become very attentive to odors in the urine and vulva of their mates (see Kreeger, chap. 7 in this volume). Most likely, this olfactory communication functions primarily in behavioral synchronization of sexually naive, newly formed pairs. Experienced males copulate even if they cannot smell their mates (Hart and Haugen 1971; Asa et al. 1990; Asa 1995).

Typically, a proestrous female will prance, body-rub, paw, nuzzle, place her chin on her mate's back, or present her rear near his nose (Schenkel 1947). Such courtship behaviors (see table 2.2) are referred to as "active solicitation" or "proceptivity" (Beach 1976). However, the frequency of active solicitation varies greatly among individuals (Bernal and Packard 1997), as does female attractiveness to males (Packard et al. 1985). The most proceptive (solicitous) females are not necessarily the most attractive (Zimen 1981; Packard et al. 1985).

Estrus

In *Canis,* behavioral estrus is the phase in which a female is receptive to copulating (Concannon et al. 1975, 1977). Two diagnostic behavioral changes occur in a receptive estrous female: (1) she averts her tail to the side of her vulva (flagging), and (2) she stands still when a male mounts her. If a male is inattentive, an estrous female may paw at him, rub against him, straddle him, or even mount him (J. M. Packard, unpublished data). Cornified cells in vaginal smears and a soft or swollen vulva also define this phase (Concannon et al. 1975; Packard 1980, 83).

Due to their phylogenetic similarities, neurophysiological studies of estrous behavior in dogs provide an appropriate model for wolves (Seal et al. 1979, 1987; Asa and Valdespino 1998). The hormonal determinants of estrous behavior are complex (Hart 1970; Concannon et al. 1975; Thun et al. 1977). Receptivity appears to be correlated with rising plasma progesterone after priming by estrogen during proestrus.

The courtship behaviors of wolves described above are similar in form to those of dogs. Actions associated with play and conflict may also occur during courtship interactions. For example, a courting pair of wolves I watched in the Houston Zoo rose up on their hind legs with forelegs entwined in brief wrestling matches.

The male may respond to the female's visual and olfactory stimuli by licking her genitals, then mounting her. Inexperienced males may direct mounting behavior to the head or side of the female before learning to mount at the rear. An unreceptive female may snap, growl, pull away, lie down, roll over, or shove the male away. When receptive, females avert the base of the tail to one side, exposing the swollen vulva. Experienced females may spread the rear legs slightly, enhancing their stability as the male mounts and the penis is inserted into the vulva. In a successful copulatory sequence, rapid pelvic thrusts follow, while the male's forelegs clasp the female behind the ribcage.

When a wolf ejaculates, his final thrust is prolonged a bit and his chin and/or rear legs may be raised slightly. During pelvic thrusting, the bulbous gland at the base of his penis engorges with blood and locks the pair in a copulatory tie (Fuller and DuBuis 1962; Rabb 1968, cited in Mech 1970). Usually, the male dismounts, and the two stand or lie rear-to-rear until the swelling declines in 5–36 minutes (Mech 1970; J. M. Packard, unpublished data). The tie is shorter if the female struggles and tries to pull away, or if other wolves interact with the tied pair.

Ejaculation followed by expansion of the penile bulb to form the copulatory tie in *Canis* are spinally mediated reflexes facilitated by androgens (Hart 1968, 1974b). Female canids respond to stimulation by the penile bulb with rhythmic contraction of the smooth muscle of the uterus, such that sperm are squeezed toward the ovaries (Evans 1933), presumably in response to a short-term pulse of oxytocin. The copulatory tie may function in postcopulatory sperm competition (Dewsbury 1972) or in reinforcement of the pair bond (Mech 1970).

The total number of copulations per estrus varies among individuals, ranging from one to eleven and averaging six for five captive females observed continuously during estrus (Packard 1980). Although estrus usually lasts less than a week in experienced captive pairs (J. M. Packard, unpublished data), estrous periods of up to 15 days (Zimen 1976), and even multiple peaks of estrous activity, have been reported in captive wolves (Bernal and Packard 1997; Zimen 1981).

Metestrus

During metestrus in *Canis,* high levels of progesterone (10 to 19 ng/ml: Packard 1980, 75) are maintained whether pregnancy occurs or not (Asa and Valdespino 1998). Because these species are spontaneous ovulators, polyestrus does not occur in *Canis* (Asa 1995, 1997). Metestrous females that are not pregnant are said to be "pseudopregnant" (Johnston 1986), since some individuals show physical (slight growth of mammary tissue, loss of belly hair) and behavioral changes (den construction, pup care) usually associated with pregnancy (Packard et al. 1983; Mech, Phillips et al. 1996).

Despite much speculation (Moehlman 1986, 1987; McIntyre 1995; Lewis and Pusey 1997; Moehlman and Hofer 1997), there is no published evidence that pseudopregnant female wolves have nursed pups. Although milk can be expressed from the nipples of some pseudopregnant females during metestrus (Mech and Seal 1987), the secretion is probably nonfunctional. Reported cases of cooperative nursing in wolves (Packard 1980; Paquet et al. 1982; Fentress and Ryon 1982), have all involved females that were both pregnant.

Because estrus in a wolf population probably lasts only about a month (see Kreeger, chap. 7 in this volume), male wolves have less incentive than male dogs do to abandon their pregnant mates in search of other estrous females. Without much opportunity to inseminate other females, theoretically there is little ultimate cost to males that stay with their mates and help care for pups. Male care of infants is associated with obligate monogamy in other mammals in which males have few opportunities to inseminate other females and females are unlikely to succeed in raising young alone (Kleiman 1977).

Male care of pups may be indirect (den preparation, pup defense, food delivery to the breeding female) or direct (food delivery to the pups). This distinction is important, because all pups in a litter may share equally in the benefits of indirect care, while direct care may be unequally distributed (Malcolm and Marten 1982).

Dens and Homesites: Indirect Care

Preparations for pup care by family members may start before wolf pups are born. Dens may even be dug in autumn (Thiel et al. 1997). Adults and yearlings of both sexes participate in den digging (Ryon 1977; Mech, Phillips et al. 1996) and provisioning the pregnant female (Fentress and Ryon 1982).

Experience may influence choice of a den site (Fuller 1989a); however, experience is not a prerequisite for successful denning in wolves. For example, during her first pregnancy, a hand-raised female in Zimen's (1981, 134) captive Bavarian pack "dug small holes at several places in the enclosure, preferably in the sandy soil immediately under a tree stump. . . . The sand was easy to dig in, and the roof of the den was protected by the widespreading roots of an old pine tree which assured it against collapse."

Pregnant females may "localize" near a den for up to a month before parturition (Harrington and Mech 1982c; Fuller 1989a; Boyd, Ream et al. 1993), although they do not always do so (L. D. Mech, unpublished data). Usually, other pack members accompany the pregnant female near the den, but family members vary highly in their degree of association with the breeding female (Murie 1944; Clark 1971; Haber 1977; Harrington and Mech 1982c; Mech 1988a; Ballard, Ayres, Gardner, and Foster 1991). Reported dates of den localization correspond to the range of parturition dates recorded in captivity (Servín-Martínez 1997). As expected, the same latitudinal difference is seen in denning dates as is apparent in breeding dates (Mech 1970, 2002).

Wolf dens usually are located away from peripheral zones of the territory, where hostile encounters with neighboring packs are most likely (Ballard and Dau 1983; Fuller 1989a; Ciucci and Mech 1992). In the Superior National Forest of Minnesota, only 11% of twenty-nine dens were located within 1 km (0.6 mi) of territory boundaries (Ciucci and Mech 1992), a zone where 56% of the wolves killed by other wolves died (Mech 1994a). Dens in larger homogeneous territories tended to be more central (Ciucci and Mech 1992) unless there were attractive geographic features in the territory, such as a river or road (Mech et al. 1998, 104).

Distances between the active dens of neighboring packs vary with territory size; for example, in a south-central Alaska population, inter-den distances averaged 45 km (Ballard and Dau 1983). Several dens within each home range may be used (Joslin 1967; Chapman 1977; Mech et al. 1998), and females vary in the probability that they will reuse a previous den (Ballard and Dau 1983; Fuller 1989a; Ciucci and Mech 1992; Mech 1995d).

The characteristics of dens vary across diverse locations (Mech 1970; Mech et al. 1998), depending on what is available to the wolves, although most natal dens are located near water (Joslin 1967; Mech 1970). In the frozen tundra above the Arctic Circle, one pack used crevices on a rocky ridge, a shallow scrape, and a rock cave (Mech 1988a, 1993b). North of the tree line in the Alaskan Arctic, many dens are located on sandy bluffs (Stephenson

1974; Ballard and Dau 1983; Lawhead 1983; Williams 1990; Heard and Williams 1992). Where the thick root mat of vegetation on these sandy bluffs is intact, the roof of the burrow may be held fast; however, where the vegetation is sparse, the roof may cave in, resulting in multiple entrances to the burrow (Clark 1971; Murie 1944; Peace River Films 1975). In forested areas, dens may be dug under the roots of trees (Criddle 1947; Scott and Shackleton 1982). Mech et al. (1998) summarized the various kinds of dens that wolves use.

Hypotheses regarding the proximate mechanisms associated with pre-denning behavior include the following: (1) that rising prolactin is associated with den digging in males as well as both pregnant and nonpregnant females (Mech, Phillips et al. 1996; Asa and Valdespino 1998); (2) that the breeding female is often the focus of attention for her mate and offspring (Zimen 1976, 1981, 1982), who may be stimulated to dig where she digs; and (3) that den digging and homesite attendance by yearlings and subadults may be a by-product of a slow maturation rate and prolactin cycles in prereproductive wolves of both sexes. Insufficient data are available to test these hypotheses, but reports of den digging in autumn tend to refute the prolactin hypotheses (Thiel et al. 1997).

From a deterministic perspective, hypotheses about the ultimate reasons why nonreproductive pack members provide indirect pup care include the following: (1) to pursue the best option for promoting the genes they share with their kin while they are nonreproductive (kin selection), (2) to enhance pack social bonds, and (3) to gain experience for when they become breeders themselves (Macdonald and Moehlman 1983; Moehlman and Hofer 1997; Asa and Valdespino 1998). This "alloparental" care by nonbreeders could allow breeding females to remain with the pups continually when prey are not readily available near the den (Harrington et al. 1983). These hypotheses are also untested, and the issues associated with such babysitting behaviors are complex (Clutton-Brock et al. 2001).

Wolf pups live in and around a den during their first 8 weeks (Mech 1970), but their mother might move them from one den to another during this period (Mech et al. 1998). From about 8 to 20 weeks of age, pups inhabit an area above ground that includes a "nest" or nests where they huddle together, a network of trails, and various play areas. Known as loafing sites (Young and Goldman 1944) or rendezvous sites (Murie 1944; Joslin 1967; Theberge and Pimlott 1969), these areas, along with dens, are considered "homesites" (Harrington and Mech 1978a).

Wolf packs vary widely in the amount of time that pups remain at the natal den, in the number of activity centers within a homesite, and in the number of homesites used during the summer season of pup care. Variation in a pack's homesite use may be related to annual variation in the movements of prey between winter and summer feeding grounds (Scott and Shackleton 1982).

At homesites, pack members provide indirect care to pups in at least two ways: through general defense (Mech 2000a), and by provisioning lactating females (Mech et al. 1999). Aggressiveness toward intruders increases in both reproductive and nonreproductive males during denning (Mech 1970; Zimen 1981).

Prolonged bark-howling is a defensive response to unexpected movements of an intruder near a homesite. One day when I changed my 1988 observation site of the Ellesmere Island pack, Gray Back (a yearling male) bark-howled eighty-five times as the breeding female and her four pups calmly returned to the cave den from an open meadow. Sometimes the movement of pups to a new homesite is not associated with bark-howling, and the circumstances of disturbance are more ambiguous (Chapman 1977; L. M. Thurston, J. M. Packard, and D. W. Smith, unpublished data).

Indirect care is provided by all non-pup pack members, although not all participate equally. Harrington et al. (1983) suggested that larger packs may be more likely to successfully defend homesites from predators. In the reintroduced population of Yellowstone wolves (Phillips and Smith 1996), Thurston (2002) confirmed that gender, age, and pack affiliation influenced trends in homesite attendance. Furthermore, patterns of attendance changed over the pup-rearing season.

Pup Birth and Stages of Development

Birth and development in wolves are similar to those processes in dogs (Scott 1967), so it is useful to adopt Scott's and Fuller's (1965) classifications of pup developmental periods for wolves (Mech 1970). These authors recognized four developmental periods: (1) the neonatal period, from birth to the age of eye opening (12–14 days); (2) the transition period, from the age of eye opening to 20 days; (3) the period of socialization, from 20 to about 77 days; and (4) the juvenile period, from 12 weeks to maturity.

Within the socialization period, Packard et al. (1992) referred to phases of (a) milk dependency, (b) transition to solid food, and (c) independence from milk. They hypothesized that the duration of these phases might be affected by food availability, the nursing female's condition, and pup demand (litter size).

I once peeked in on the birthing behavior of a hand-raised wolf named Sitka. As I cautiously lifted the lid of her den box, I saw a pup emerging. Curled around the other pups nestled against her belly, Sitka licked the new arrival and nibbled gently at the membrane surrounding it. When the membrane slipped off, she swallowed the tissue, breaking the placenta in the process. The pup sprawled on its back, and Sitka nudged it gently until it was near the other pups at her belly. When Sitka looked at me and growled, I was afraid she might redirect her maternal defensiveness and bite the pups, so I withdrew.

Compared with smaller canids, wolves bear relatively large pups in relatively small litters (Moehlman 1986). There has been much speculation about this correlation (Bekoff 1989; Moehlman 1989; Geffen et al. 1996; Moehlman and Hofer 1997). Larger pups may be more resistant to wet, cold weather (Mech 1993b). However, larger pups also need more nutrition from their mother.

Neonatal Period

With eyes closed, newborn wolf pups look much like German shepherd pups. They are dark, and some even have a white star on the chest. Their tiny drooping ears and pug faces add to the round appearance of their heads. Holding a tiny pup in my hands, I used to marvel that such a little creature would someday look like its mother. The pups' small, uncoordinated legs are good for little more than crawling and kneading their mother's belly while nursing. When cold, they crawl toward warm objects. It is amusing to watch wolf pups as some squirm and crawl on top of one another, while others fall and tumble off the pile. When sleeping in a pile, sometimes a leg will twitch as if the pup is dreaming.

Within hours of birth, canid pups suckle in a reflexive response to a nipple-like object touching their lips (Fox 1971b). When their mother crawls into the den and lies with them, young wolf pups struggle to reach her belly, and move their heads from side to side until they encounter a nipple and start suckling. Some pups may doze off, tipping the head, then wake up with a start and continue suckling. Their mother may lick at the rear or inguinal area of the pups, stimulating them to urinate and defecate. She consumes eliminations, keeping the den clean, until the pups are large enough to walk to the entrance and eliminate outside.

At birth, wolf pup behavior is little more than a simple set of reflexes (e.g., heat seeking, nuzzling, suckling, elimination in response to maternal licking, crying when hurt, whimpering when cold, hungry, or isolated) (Scott and Fuller 1965; Fox 1971b). However, as the pups' senses and coordination develop, simple reflexes expand into interactive routines (Fentress 1983; Bekoff 1989; McLeod and Fentress 1997). For example, reflexive urination in response to maternal licking develops into the "belly-up" response, often referred to as "passive submission" in yearlings and adults (Schenkel 1967; Fox 1971e, 1972a).

Transition Period

Wolf pups' eyes open at about 12–14 days (Mech 1970), when the pups also become coordinated enough to stand and walk. At first the pups explore the natal chamber in the den, gradually moving farther each time before collapsing again in a puppy pile. Eventually they stumble out to the den entrance and stand looking at the outside world. It is not unusual for them to startle and duck back into the den after their first peek. Gradually, the growing pups explore farther, and soon they are playing, lying, nursing, and eliminating around the mouth of the den.

As indicated above, newborn wolf pups are relatively helpless, and their visual and auditory senses are poorly developed compared with their olfactory and tactile senses (see Harrington and Asa, chap. 3 in this volume). Their sensory systems, size, and muscular coordination develop rapidly during the transition period (Fox 1971b; Bakarich 1979; McLeod 1987, 1996; McLeod and Fentress 1997). The manner in which neuronal connections develop in the brain during the first few weeks of life may be determined by the pups' experience with their mother and siblings in the den (Klinghammer and Goodmann 1987). During this early window of development, pups learn to recognize familiar individuals, usually family members (Bekoff 1989). Rapid learning during the transition and socialization periods (Scott and Fuller 1965; Scott 1967; Fox 1971b) has important implications for the social context of learning later in life; for example, vocalizations become differentiated and associated with specific contexts.

Socialization Period

About 20–24 days after birth, the pups become mobile enough to explore as far as the mouth of the den (Mech 1970; Ryon 1977; Ballard et al. 1987; Fuller 1989a). They begin to elicit care from other pack members (Murie 1944; Ryon 1977; Fentress and Ryon 1982), and they start ingesting solid food (Mech 1970). In another 2 weeks, they are spending a lot of time outside the den and interacting with the adults.

I recall a typical scene at this stage from around the Ellesmere pack's den in 1988. Left Shoulder was lying like a sentinel near the den entrance when we first approached. As the 5-week-old pups milled around the head of their reclining mother, she arose and escorted them out of sight into the cave den, followed by the yearling female. Later, when the pups saw us, they disappeared into the den on their own, a highly effective response to intruders. When the yearling male, Gray Back, entered the den briefly, a pup followed him out, then waddled back in to rejoin its siblings. The pups appear rather indiscriminate of which family members they approach during this sensitive period of socialization.

While the pups are still small and immobile, they suckle from a lying position within the den (Coscia et al. 1990; Lawrence 1990). At about 3–5 weeks, they are large enough to reach their mother's nipples in a standing position, and will approach her outside the den, often in response to soft squeaks (Crisler 1958; Fentress 1967; Coscia 1989; Goldman 1993). At this stage, suckling bouts average 3 minutes in duration and occur at an average of 5-hour intervals (Packard et al. 1992).

Up to 5 weeks of age, wolf pups are relatively uncoordinated, and their range of movement is usually less than 0.5 km (0.3 mi). When a small pup is away from the safety of a den, its mother may pick it up gently in her mouth and move it to a safer site (see table 2.2). Mother wolves also carry pups when they lag behind while following adults between sites, or when the adults move the entire litter from one den to another. Pup carrying has been reported for a non-nursing female (D. W. Smith, personal communication), but not for male wolves (Mech 2000a).

The pups are very likely to follow their mother if she interrupts a nursing bout and trots away in response to a disturbance (Packard et al. 1992). Their behavior is quickly shaped to follow a departing adult moving in an intent, directional manner. This following response is effective in moving 5-week-old pups between homesites (Mech 1988a, 44). It could be a precursor to heeling, as in dog training.

At about 5 weeks of age, the pups are sufficiently coordinated to seek shelter from inclement weather and potential predators. They are still small enough to be carried by adult females and large enough to follow adults for short distances. Their sensory systems are fully developed, and their gastric system has developed to a stage at which solid food can be digested. Although their teeth have erupted, they do not have sufficient bite strength to chew large pieces of meat. The full capability of shifting to a meat diet develops later.

Between 5 and 10 weeks of age, wolf pups advance from dependent toddlers to active individuals engaged in learning from their physical and social environments (Mech 1970, 1988a; Havkin 1977; Havkin and Fentress 1985; Packard et al. 1992; Jensen 1993; McLeod 1996; McLeod and Fentress 1997). If there is a shortage of food during this period, the pups still have access to milk as a nutritional alternative, although their growth may be slowed. Under poor food conditions, individual variation in body size begins in the transition period and is likely to be accented by the end of summer (Van Ballenberghe and Mech 1975).

In the Ellesmere pack (in 1988), suckling bout duration declined to 1 minute, on average, at about week 9 (Packard et al. 1992). The intervals between bouts increased to an average of 10 hours, until the pups no longer solicited nursing during week 10. Weaning in the Ellesmere pack was not associated with agonistic interruptions (Packard et al. 1992), as reported for dogs (Scott and Fuller 1965). Theoretically, the degree of weaning conflict would be negatively correlated with food supply and positively correlated with litter size (Packard et al. 1992; Malm and Jensen 1996). We believe that food was abundant during our observations because pups received a daily average of one arctic hare and two regurgitations during their tenth week, when they stopped soliciting milk (Packard et al. 1992).

With increasing independence from milk, the following response of pups generalizes to whichever family member feeds them. When hungry pups spot an approaching pack member, they rush over and poke their muzzles around the adult's mouth, an action called "lick-up" (fig. 2.5). If the care-giver has a full stomach, this stimulus seems irresistible: the adult regurgitates food to the pups. In 76% of 115 regurgitation bouts we observed on Ellesmere, regurgitators delivered food

FIGURE 2.5. Pups solicit regurgitation by a behavior called licking-up, in which they poke their muzzles around the mouth and chin of an adult. This behavior persists into adulthood, when it serves an additional function of appeasement of conflict.

where they were met; in 11% of the bouts, pups followed the regurgitator 10–50 meters before it regurgitated; and in the rest, they followed up to 800 meters (Mech et al. 1999). Usually the pups were so excited and active that it was hard to see what food reached the ground in the midst of the confusion. By the age of weaning, pups are sufficiently mobile and have enough endurance to follow adults to carcasses (Gray 1993; L. D. Mech, unpublished data).

On days of food surplus, the pups cached food around the homesite, essentially filling "the pantry" with snacks that they later sought out or encountered haphazardly when the adults delivered little food and the pups were active. Caching routines initially are highly stereotyped in pups (Phillips et al. 1990), implying neuroendocrine programming. By measuring three cached regurgitations, Mech et al. (1999) estimated that an average of 1.25 kg (2.75 pounds) of meat was stored per cache.

The duration of activity bouts increases during the socialization period (Packard et al. 1992). Pups learn the contingencies of their interactions with other pups, and their combative routines increase in complexity during "play-fighting" (Moran 1978; Moran et al. 1981; Havkin and Fentress 1985; McLeod 1996; McLeod and Fentress 1997). Theoretically, cardiovascular conditioning increases with activity at this stage (Bekoff 1972, 1974b), and individual differences may become more pronounced (Folk et al. 1970; Fox and Andrews 1973; K. B. MacDonald 1983, 1987; Zimen 1978). During this stage, pups also become familiar with the identity of

family members, acquiring information that will influence how their social behavior is directed later in life (Bekoff 1981, 1989), particularly when they encounter hostile neighbors.

Play

Based on his observations of the Ellesmere pack during the pup-rearing season, Mech (1988a, 61) emphasized the importance of play in the socialization of wild wolves. He reported very few dominance interactions outside the context of food contests. He saw less conflictive than cohesive behavior, with much of the latter occurring in the form of social play.

Not all researchers agree on definitions of play. Typical characteristics of play include (1) actions also observed in other contexts (e.g., stalking, pouncing, chasing, face pawing), (2) metacommunication signals (e.g., bowing, tail wagging, grinning, head tossing), (3) repeated and exaggerated movement indicating a pleasurable quality (e.g., approach/withdrawal, leaps, bouncy galloping) and (4) exchange of roles (e.g., the "chaser" becomes the "chasee"), maintaining mutual participation (Bekoff 1974a, 1984). Object play also may occur when pups encounter novel stimuli, such as water (Coscia 1993).

While watching wolf pups play in the Ellesmere and Yellowstone packs, I was eager to see whether certain pups learned to avoid others during play. We were entertained by hours of play, including interactions analogous to keep-away, tag, wrestling, and king-of-the-mountain. Repetitive routines were interspersed with novel events resulting in new combinations of actions. I saw only two episodes in which elements of conflictive behavior occurred during play. However, the same frequent "play-wrestling" interactions that I saw were interpreted by another observer as "fighting." Clearly, interpretation depends on each observer's experience and mind-set.

According to a deterministic model of wolf behavior, there must be some sort of "social glue" that allows wolves to cooperate in caring for the young (Fox 1980). Fox (1975) compared the play behavior of pups in asocial, semi-social, and social canids (e.g., foxes, coyotes, and wolves). He hypothesized that wolves would be the most playful and the least aggressive (Fox 1969, 1970). Bekoff (1974a,b) obtained evidence from single litters of each of these species that supported Fox's hypothesis. He observed a stereotyped play bow in all species, presumably a trait inherited from a shared ancestor (Bekoff 1977c).

Detailed studies of play routines in wolf pups (Havkin 1977) have opened researchers' minds to the importance of examining relative probabilities as well as absolute rules (Fentress et al. 1987). For example, in 3-week-old pups, a side approach is more likely to knock another pup down, due to simple mechanical advantage, in contrast to a frontal approach (Havkin and Fentress 1985). Via trial and error, pups learn the consequences of their actions. With experience, older pups develop effective counter-tactics, including somersaulting.

The ability to learn counter-tactics during play may provide a basis for the learning of complex social relations later in life (Bekoff 1984). Other hypotheses about the function of play in canids include (1) physical exercise related to aerobic conditioning, (2) development of muscular routines, (3) practice of instincts useful for hunting later in life, and (4) "animals that play together, tend to stay together" (Bekoff 1974a, 338).

Direct Biparental Care

During the first month after birth, mothers generally contribute directly to pup care in the form of milk and body warmth, as well as choosing and maintaining a dry, clean environment. Fathers contribute indirectly in the form of defense of homesites, hunting, and provisioning the lactating female.

For newborns, pup care by a lactating female is quite different from care by her mate, who only occasionally enters the natal den (Fentress and Ryon 1982; L. D. Mech, unpublished data). Father wolves contribute to the feeding of their newborn pups only indirectly by feeding the nursing mother. According to the "division of labor" model (Mech 1999), the male spends most of his active time hunting while the female attends the pups during their first 3–4 weeks (Harrington and Mech 1982c; Ballard, Ayres, Gardner, and Foster 1991). Unless a focal-follow technique is used, however, it is hard to tell whether a male is hunting or resting when away from the homesite (Thurston 2002).

In Yellowstone packs, the relative probability of homesite attendance differed more between the breeding male and female during the first month than during the second month of pup care (Thurston 2002). In 44% of nine pack-years, the difference in den attendance between mother and father declined to zero by 5 weeks of

pup age, on average. In a third of the cases, no difference could be detected between mother and father throughout the monitoring period 1–12 weeks after birth. In two unusual cases involving communal denning by two sisters, the breeding male initially spent more time at the den than the mothers, although this difference was negligible by the second month of pup care. In nine of twelve cases, alloparental care was less than or equal to parental care when cases were matched for sex, pack, and year.

Regurgitative Provisioning

When the father wolf obtains food, he returns to the den and presents food to his mate, either by carrying it in his mouth or by regurgitating it to her from his stomach (summarized by Mech et al. 1999 and Mech 1999, 2000a). When the pups are out of the den, the breeding male and any other adults regurgitate food to the pups. The mother wolf may try to usurp whatever portion of this food she can get, later delivering some of it to the pups. When the mother joins the family on hunts, she similarly brings food back to her pups.

In the Ellesmere pack, the ratio of regurgitations by the breeding male compared with the breeding female varied by year (Mech et al. 1999). Relative female effort was not related to pack size, although it was positively correlated with litter size (Mech et al. 1999).

How much does the food provided by male wolves help offset the lactational drain on a female? No quantitative information on this subject is available. Mech (1999) proposed that the mother wolf probably can usually maintain her nutritional condition throughout the summer because prey is usually abundant then; furthermore, she travels little during the month or so after parturition and is fed mostly by the male and other pack members.

For wolves, monogamy does not appear to be obligate in the sense defined by Kleiman (1977), meaning that care by the father (fig. 2.6) is not essential under all conditions. At least one female wolf raised pups, apparently from birth, without help from other pack members (Boyd and Jimenez 1994). Other examples include situations in which mothers, and in one case a father, raised pups after losing a mate (Boyd and Jimenez 1994; D. W. Smith, unpublished data).

In their second month, when the pups can ingest solid food, biparental care may be more symmetric; that is, the difference between the sexes in homesite atten-

FIGURE 2.6. Adult and subadult wolves of both sexes care for and show tolerance toward pups within the family, although they may attack pups from another pack. Here, a father babysits pups.

dance and provisioning is less, but still apparent (Ballard et al. 1981; Harrington and Mech 1982c; Mech et al. 1999; Mech 2000a). Both male and female parents hunt and bring food to the pups. The female, however, continues to nurse them and still spends much of her time near the homesite. In the Ellesmere pack, the breeding male, Left Shoulder, was occasionally absent for more than a day at a time. Homesite attendance may be an indirect measure of parental care, correlated with direct care (Thurston 2002).

When parents return to the homesite, they regurgitate to the pups one or more times. On Ellesmere, adult wolves, including parents and helpers, regurgitated only once during each of 61% of the food deliveries, although they occasionally regurgitated up to five times during a single delivery, within 5–35 minutes after arrival (Mech et al. 1999). The food delivered per regurgitation bout was tentatively estimated to range from 1.10 to 7.25 kg (2.4–16 pounds). Since the average number of regurgitations per bout was 1.5, the average amount of food delivered per bout would have been 1.9 kg (4.2 pounds). This estimate assumes, however, that the amount of meat regurgitated to pups was the same as that regurgitated into caches (since that was the only time it could be measured)—a tenuous assumption. Theoretically, the amount actually received by each pup would vary with litter size, size of meat chunks, and individual skill at competing for meat.

Pups successfully competed with their mother and older siblings for regurgitated meat, receiving 81% of 171 regurgitations recorded over 6 years in the Ellesmere pack (Mech et al. 1999). The nursing female received

14% of the regurgitations, mostly from her mate, and primarily during the neonatal and transition periods, before the pups could reach the den entrance or digest solid food. Family members other than the parents were half as likely as parents to regurgitate to the pups, and they themselves received only 6% of the regurgitations during this study. The overall pattern was similar in a captive pack (Fentress and Ryon 1982).

Use of the Homesite

The mother wolf remains with the pups for most of the time during their first 3–4 weeks of life (Ballard, Ayres, Gardner, and Foster 1991). After that, the amount of time pups are left alone varies (Chapman 1977; Harrington and Mech 1982c; Ballard, Ayres, Gardner, and Foster 1991). Chapman (1977) estimated that pups were unattended 40–73% of the time that he monitored three packs in Denali National Park. Under conditions of abundant food, pups were left alone 5–15% of the time that two packs were monitored in south-central Alaska (Ballard, Ayres, Gardner, and Foster 1991). It is difficult to interpret this variation among populations, since the presence of a carcass near the homesite influences den attendance (Harrington and Mech 1982c; Ballard, Ayres, Gardner, and Foster 1991; Jedrzejewski et al. 2001).

On the few occasions when Murie (1944, 29) watched a lactating female leave the den with pack members, she "ran as if she were in high spirits, seeming happy to be off on an expedition with the others." However, she returned earlier than the other family members. Other researchers have reported a similar pattern (Harrington and Mech 1982c; Ballard, Ayres, Gardner, and Foster 1991; Vilà et al. 1995; Mech 1999, 2000a).

The Familial "Hunting School"

The association of pups with pack members between weaning and dispersal from the natal group is of particular importance to the pups' opportunities to learn hunting techniques. We watched one of the first times that 3-month-old pups followed adults away from the natal homesite at Ellesmere in 1988. Left Shoulder disappeared behind the crest of a hill. Shortly thereafter, the screams of a dying arctic hare echoed across the barrens. I naively expected the pups to run over and receive one of their first lessons in killing prey. Instead, they turned tail and ran for the shelter of the closest rock pile! However, when Left Shoulder brought the hare carcass to the pups, they readily converged around him and consumed the meal. Thus, the pups could have learned the association between food and the screams of the hare.

The propensity to chase and capture small moving animals appears to be genetically programmed in wolves, since it occurs without practice in hand-reared individuals (Sullivan 1979; Zimen 1981). I have watched 3-month-old wolf pups in Yellowstone repeatedly "mouse-pounce" in the stereotyped canid pattern (Fox 1971a, 1975). Such innate behavior gives young wolves a head start in practicing the skills that take learning, such as where to find prey, how to kill it, and how to avoid risks (Mech 1988a, 1991b; Fentress 1992).

At 3 months of age, wolf pups are full of energy, more likely to follow departing adults or to explore on their own, and less likely to remain at a homesite. Typically, they may move among sites where pack members are likely to return individually and in groups. For example, some tundra packs return to the site of a muskox kill, even after the carcass no longer provides food (Gray 1993). In Minnesota, where prey are not migratory, homesites may continue to function as activity centers in autumn and early winter (Harrington and Mech 1982b).

Between 4 and 10 months of age, juvenile wolves are sufficiently mobile to join adults on hunts, even though they have not attained full body size. Mech (1991b) has described the function of family groups at this stage as a "finishing school" for juvenile wolves, implying that they have opportunities to hone their hunting skills while traveling with the family. I suggest that the name "hunting school" would be better than "finishing school," since juveniles have already learned the "manners" of social interaction.

Most wolves disperse from their natal pack between the ages of 9 and 36 months (see Mech and Boitani, chap. 1 in this volume). In the Denali population, only 8% of dispersers were older than 3 years (Mech et al. 1998), although in northwestern Alaska, the average age of dispersal was about 3 years (Ballard et al. 1997). The complex and often subtle interactions within each family influence when offspring disperse. Conflictive behavior is tempered by gentle cohesive interactions in both captive and free-ranging families (see references in table 2.1).

Cohesion and Conflict

In the popular literature (e.g., Fox 1980; Savage 1988; McIntyre 1993), wolf packs are often portrayed as a dom-

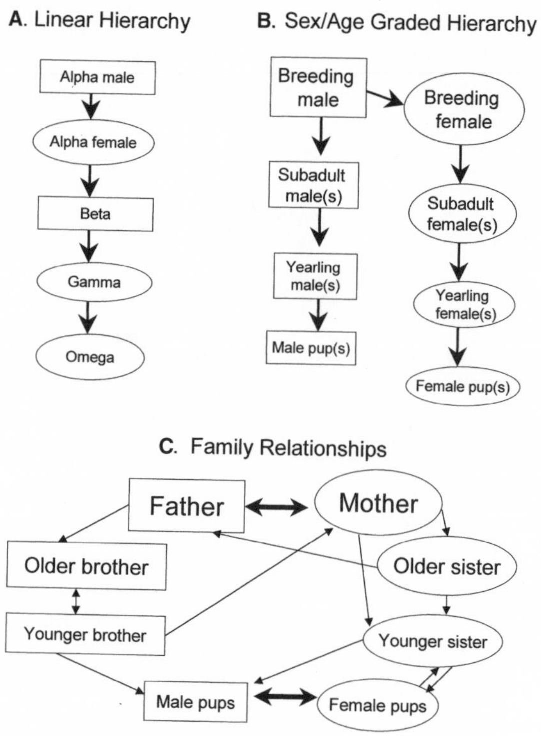

FIGURE 2.7. Three models that have been proposed for understanding wolf social structure.

inance hierarchy (fig. 2.7A), in which underlings are kept in line by dominant individuals. At first I was confused about how to describe social relations in the captive wolf packs I watched, because I did not see the amount of aggression I expected. Pups in the captive South pack spent endless hours in chase games, often joined by the yearling male and less often by their father. Adults in the North pack bedded near one another, tenderly licked one another's wounds, rubbed shoulders in exuberant rallies after naps, and trotted off together to explore the woods.

After reading an intriguing discussion by Lockwood (1976, 1979), I reread the observations by Murie (1944) and Schenkel (1947). All of these authors perceived life in a wolf pack as a balance between cohesive and conflictive behaviors. They described "submissive" behavior as the persistence of care-soliciting by offspring who remained in the family as adults. Perhaps the relative importance of dominance varies with pack composition, food availability (and thus competition), and even the eyes of the observer.

What is meant by cohesive and conflictive behavior?

Cohesive behavior is whatever brings wolves closer together, and conflictive behavior is whatever drives them apart from sharing within a family (Fentress et al. 1987). In table 2.2, actions in the categories "Care," "Proximity," and "Sexual" often serve a cohesive function; actions in "Flight/Fight" and "Humbleness" categories are often associated with conflict. However, any one action may occur in several contexts. For example, a "stare" might function in both cohesion and conflict, depending on its context.

We have seen that since wolf packs are families, basic family ties promote cohesion. However, from a deterministic perspective, each individual must also ensure its own survival first, or it cannot assist relatives. Thus, during food scarcity, competition and conflict may be viewed as a way of ensuring family survival over the long term. In this respect, the following aspects of social cohesion and conflict in wolf populations bear discussion: (1) the dominance hierarchy concept, (2) variation in individual temperaments and relationships, (3) access to food and mates within packs, (4) leadership, and (5) interactions among packs.

Age-Graded Dominance Hierarchy

A linear dominance hierarchy is the simplest way of describing conflict behavior in a wolf pack (see fig. 2.7A). According to that concept, the most dominant wolf is the one that wins fights over all others, and is called the "alpha." The "beta" loses fights with the alpha, yet wins over all others, and so on down the line. The wolf least likely to win any fights is called the "omega."

These terms (alpha, beta, omega) have been used in describing interactions among orphaned siblings in captive groups (Rabb et al. 1967; Folk et al. 1970; Zimen 1975, 1981; Fox 1980) as well as free-ranging wolves (Murie 1944; Haber 1968, 1977; Mech 1970, 1977c, 1993a; Peterson 1977). The terms may be appropriate in ambiguous situations in which the relatedness among pack members is unknown or complex (e.g., more than a single pair of breeders). However, I agree with Mech (1999) that the terms are inappropriate for typical packs consisting of parents and offspring (Packard 1980). Simon Gadbois (2002) has reviewed the complex, often ambiguous terms used in the literature on dominance.

The linear dominance hierarchy concept has been adopted and perpetuated by popular educational materials about wolves (Savage 1988; Lawrence 1993; McIntyre

1993). Wolf enthusiasts can usually distinguish easily between an alpha and an omega wolf by noting which carries the tail higher, as described in by Harrington and Asa in chapter 3 in this volume. However, in most wolf packs, family dynamics are more complex. Thus it is important to review the development of scholarly thought about the dominance hierarchy in wolf packs.

The classic work on this subject was conducted by Rudolf Schenkel (1947) on a pack of up to ten wolves kept in a 200 m^2 pen. His article was published in German, but an English translation by Agnes Klasson has long circulated among wolf biologists in North America. In that work, Schenkel wrote of "two sex orders of precedence," a "status order," a "rank order," a "dominance status order," and a "clear-cut hierarchy."

In the first book summarizing wolf natural history published since Schenkel's article, R. J. Rutter and D. H. Pimlott (1968, 43) wrote the following:

> Normally, a wolf population is divided into *packs,* and a pack is an organization within which every wolf knows its social standing with every other wolf. Each pack has its own territory and operates as a unit in its relations with neighboring packs.
>
> Dr. Niko Tinbergen, an internationally known authority on animal behavior, described a similar social order among Eskimo dogs in Greenland: "Within each pack the individual dog lived in a kind of armed peace. This was the result of a very strict 'pack order': one dog was dominant and could intimidate every other dog with a mere look; the next one avoided this tyrant but lorded it over all the others; and so on down to the miserable 'under dog.'"
>
> We believe this describes the fundamental organization of a wolf pack. . . .

Mech (1970, 69), citing Schenkel (1947), also implied that wolves showed a linear dominance order within each sex (fig. 2.7B). This conceptual model, largely structured according to age, was reinforced by Zimen's (1976, fig. 5) study of a captive wolf pack that he formed by assembling unrelated wolves from three sources. (It is important to note that this was not a natural wolf pack consisting of a pair of parents and their offspring [Mech 1999].) Zimen (1981, 128) later acknowledged that any model of rank order is an oversimplification.

Several researchers who have observed larger wolf packs over several years in a wider range of contexts (e.g., competition over food and mates) have rejected the hypothesis that all wolf packs fit the model of a linear dominance hierarchy (Lockwood 1976, 1979; Packard 1980; Zimen 1981; Mech 1999). Furthermore, K. B. MacDonald (1983, 1987) found that consistent temperament differences measured in captive wolf pups were not correlated with the probability of breeding in a nuclear family (Jenks and Ginsburg 1987; Schotté and Ginsburg 1987; Jenks 1988; Schotté 1988).

Based on studies of a larger sample of packs over multiple years (see table 2.1), I tend to think of the alpha behavioral profile as an internal state (or mood). Moods are subject to change as the health and environment of an individual change. According to a model of human behavior, moods are influenced by both temperament (heritable propensities) and character (learned styles of coping) (Cloninger 1986). This hypothesis is consistent with the stochastic perspective developed by behaviorists to understand conflictive behavior in other species (Appleby 1993; Barrette 1993; Moore 1993). Zimen (1982) expressed a similar notion in terms of the "age-graded dominance hierarchy" model of conflict within wolf packs.

What is this age-graded model? In interactions with adults, juveniles typically are more humble; thus older wolves effectively intimidate younger wolves. Littermates may squabble over food or during rough play, and pups are disciplined by older family members. As juveniles mature, conflict is more likely between members of the same sex; that is, females fight females and males fight males. This model has been presented in two ways: first, simply as separate linear hierarchies within each sex, influenced but not absolutely determined by age (Schenkel 1947; Zimen 1982), and second, as male dominance over females within each age class (Rabb et al. 1967; Fox 1971a, 1980; Lockwood 1979; Zimen 1982; van Hooff and Wensing 1987; Savage 1988). I hypothesize that the former is more likely in young nuclear families and the latter in disrupted or complex families (see table 2.1).

Theoretically, a predictable linear hierarchy is more probable in a social group when (1) individuals are added to the group one by one, (2) additions occur after each dominant-subordinate relationship has stabilized, and (3) there is a clear difference in fighting ability between the two individuals in each relationship (Chase 1974). For example, in the captive North pack (Packard 1980), there was a clear linear hierarchy among the mother and her two daughters of different ages: the mother interrupted squabbles between her daughters, and the older sister was more likely to chase her younger

sister than vice versa. However, the relationships among the females in the North pack were not reinforced by constant fighting, fear, or control by the breeding female. She was the mother, and her daughters had deferred to her ever since she started muzzle-biting them when they licked-up to her chin too insistently. Logically, the conditions favoring stable relationships are more likely to be met in small, young nuclear families like the North pack (fig. 2.7C) than in larger, older disrupted families (Packard 1989).

Although the typical wolf pack is a nuclear or extended family (Murie 1944; Mech 1970), field studies point to much turnover in packs and populations. These studies show that wolf populations consist of dynamic packs that are continually forming and dissolving, with a high annual turnover of offspring (Mech 1977b, 1987a, 1995d; Fritts and Mech 1981; Peterson, Woolington, and Bailey 1984; Messier 1985b; Fuller 1989b; Mech et al. 1998). By 3 years of age, most wolves have dispersed from their natal packs (see Mech and Boitani, chap. 1 in this volume), and deaths from disease (Ballard et al. 1997), fights with neighboring packs (Mech 1977b, 1994a; Mech et al. 1998), and hunting by humans (see Fritts et al., chap. 12 in this volume) further disrupt the stability of wolf families.

To understand the dynamics of social relationships within natural wolf packs, it is useful to distinguish among families with different histories (see table 2.1). The social relationships of captive wolves have been studied in detail in groups at five research centers and five zoos, representing a variety of different types of social groupings. However, of thirty wolf groups studied (92 group-years), only 34% of the groups were nuclear or step-families, and 50% were extended or disrupted families. L. D. Mech (personal communication) considers packs in the categories of "nuclear family" and "stepfamily" to be closest to those most frequently observed in stable wolf populations. Under the dynamic conditions of a colonizing or declining wolf population, the incidence of other types of families (e.g., "extended family" and "disrupted family" categories) may increase.

Often, the results of initial studies are perpetuated in popular literature, creating impressions that are difficult to dislodge. For example, observations of two dominant males that apparently allowed beta males to breed in captive packs have been repeated in several popular books (Lawrence 1993; McIntyre 1993; Savage 1988). Breeding by a beta male occurred in two of the first studies of captive wolves. Both groups were in the category of "complex family"—that is, siblings without parents (Rabb et al. 1967; Zimen 1975)—a situation rarely, if ever, found in wild populations (Mech 1970, 1999; Mech et al. 1998). Without the stabilizing influence of parents, the siblings fought, and their behavior was described in terms of competition for the social roles of "alpha," and "beta." However, there is no objective way of measuring a "social role" independently of the behavioral profile of its occupant, so this explanation is both circular and anthropomorphic.

Variation in Individual Temperaments

Individual behavioral profiles are a way of describing variation in temperament, analogous to "personality" in humans. One of the most challenging aspects of studying wolf social behavior has been understanding this variation among individuals. For example, Zimen (1975) observed distinct changes in the personalities of two hand-raised wolves as the result of a fight. Before the fight, Wölfchen was confident and assertive, carrying his tail high, challenging dogs outside the fence, and escalating conflict when challenged by his brother, Näschen. After being injured during the fight, Wölfchen's tail hung low, he no longer challenged outsiders, and he avoided conflict with Näschen. Wölfchen assumed an intimidated profile (high humbleness in table 2.2). In contrast, his brother Näschen assumed the confident profile previously typical of Wölfchen (low humbleness). A similar switch in behavioral profiles in the Ellesmere pack preceded daughter Whitey's assumption of the pup-producing role from her mother (Mech 1999).

In a debate about the heritability of dominance in social species (Appleby 1993; Barrette 1993; Moore 1993), three basic issues were identified: (1) which animal is assertive (individual temperament), (2) which is assertive with which (relationships), and (3) in what context conflict occurs. From a stochastic perspective, the claim is that it is not dominance, but rather the predisposition of each individual to escalate or reduce conflict in specific social contexts, that is heritable (Appleby 1993; Barrette 1993).

From this perspective, what we call a humble mood is associated with a low probability of conflict escalation in a specific interaction, in contrast to a confident mood (low humbleness), which is associated with a high probability of escalation. In wolves, it has been proposed that each individual has the potential for self-assertion, and

that this potential is realized when a wolf becomes a breeder (Schenkel 1947; Zimen 1981; Mech 1970, 1999).

Prior to Zimen's (1975) work, researchers had hypothesized that there were "born alphas" in each litter of wolf pups; in other words, measurable physiological variation within litters of pups was thought to predict later variation in their individual temperaments as adults (Folk et al. 1970; Fox and Andrews 1973). According to this deterministic model, it was logical that there would be polymorphism in temperaments within each litter so that some individuals would become breeders (alpha temperament) and some would cooperate in the care of the breeders' offspring (Rabb et al. 1967; Woolpy 1968; Woolpy and Eckstrand 1979). However, the hypothesis of "born alphas" has been tested and rejected for both captive (Lockwood 1976, 1979; K. B. MacDonald 1983; Ginsburg 1987; Schotté and Ginsburg 1987; Jenks and Ginsburg 1987; Ginsburg and Hiestand 1992) and wild wolves (Mech 1999).

Why isn't a dominant individual destined to always act dominant? According to a stochastic model, individual temperaments may change due to each individual's social experience (e.g., the predictability of interactions within a family) and mood state (e.g., the activity of neuroendocrine systems) (Packard 1980; Fentress 1982; Packard and Mech 1983; Packard et al. 1983; Fentress et al. 1987; McLeod et al. 1991, 1996).

Although our understanding of the neuroendocrine basis of mood state is still incomplete (Cloninger 1986; Overall 2001), sufficient information is available to formulate theories about which behaviors are indicators of neuroendocrine activity in canids. According to the models outlined in table 2.2, I would hypothesize that high serotonin is associated with the action patterns interpreted as a confident profile (e.g., high posture, tail high, ears forward) and low serotonin is associated with an intimidated profile (e.g., low posture, tail low, ears back). The communication function of these postures will be described by Harrington and Asa in chapter 3 in this volume. However, their association with serotonin-related neural networks needs to be tested.

Neuroendocrine systems, which affect moods and motivation, interact in complex ways (Cloninger 1986). According to one idea, when serotonin is low, the adrenal system is more likely to be activated. Fight behaviors are more likely to appear when noradrenaline-related neural circuits are activated, and avoidance behaviors (e.g., flight) are more likely when adrenaline-related circuits are activated (Sapolsky 2002; Watkins 1997, 11). In a finding consistent with this model, aggressiveness in dogs was reduced by treatment with fluoxetine, which inhibits the breakdown of serotonin (Dodman et al. 1996).

In captive wolves, assertiveness changes with age, reproductive state, nutritional condition, traumatic experience (fights), and resource context (pups, mates, or food) (Zimen 1975; Fentress 1982; Fentress et al. 1987). During the few field studies in which wolves have been observable interacting for any length of time, few fights have been seen among pack members (Murie 1944; Clark 1971; Haber 1977; Mech 1988a). Mech (1999) has been impressed with the peacefulness of interactions among pack members, at least in summer. Most serious aggression in wolves is directed at non-pack members (see Mech and Boitani, chap. 1 in this volume).

The issue of "which wolf fights" within a family is better phrased as "which wolf is more likely to be in an assertive versus avoidance state more often." The personality changes that occur with age and reproductive experience (Mech 1970; Packard et al. 1983) suggest that the patterns described as "temperament" are a function of the shifting internal state of an individual. Behavioral profiles may be predictable as long as internal states are relatively stable; however, fluctuations occur when external conditions or internal states change.

In his review of the literature on dominance and stress, Gadbois (2002) clarified that researchers in the field of human personality distinguish between temperament and character. "Temperament" refers to the aspects of personality that change little over a lifetime and many be highly heritable. "Character" refers to the aspects of personality that change as individuals mature and learn styles of coping with stressful situations. I agree that this is an important distinction that should be considered in future studies of the individual variation in wolves.

Several researchers have attempted to use multivariate statistical techniques to determine the basic dimensions of variation in personality among captive wolves (Bekoff et al. 1975; Colmenares Gil 1979; Lockwood 1979; Packard 1980; van Hooff and Wensing 1987; Derix 1994). Although the variables and results differed among these studies, all the researchers agreed on one conclusion: both cohesive and conflictive behaviors need to be considered in describing the variation in the personalities of individual wolves. This conclusion is consistent with the observations of wild wolves mentioned above, which

suggested that aggressive interactions are infrequent except during competition over food and mates.

To date, each research group studying wolves has used different terms to define and categorize wolf action patterns (Zimen 1971, 1982; Fox 1971b; Bekoff 1972, 1979a; Goodmann and Klinghammer 1990; Derix 1994). This diversity is appropriate, given the different questions studied (Bekoff 1979a). However, this variation also limits our ability to compare wolf temperament and its heritability across captive packs.

Behavioral indicators of social stress are the subject of one of the unsettled debates about conflictive and cohesive behaviors in wolves (Schenkel 1967; Packard 1980; Zimen 1982; Fentress et al. 1987; McLeod et al. 1996; Haber 1996; Moger et al. 1998; Gadbois 2002). Schenkel (1967) pointed out that inexperienced observers tend to interpret an "over-the-muzzle-bite" as adult aggression, stressful for the pup that receives it. However, he interpreted this behavior differently, because pups do not avoid an adult after muzzle-biting. They come back for more, indicating a cohesive function for the behavior. Indeed, the pups' behavior seems to solicit more interaction. Schenkel (1947, 1967) noted that such juvenile behavior persists into adulthood, when grown offspring continue solicitous behaviors that elicit muzzle-biting or pinning in such a manner that they acquire a new function of appeasement (i.e., conflict reduction). Based on his field observations, Mech (1999) agreed with Schenkel's interpretation of these behaviors as nonstressful conflict within the context of cohesive relationships within a social group. However, determination of stress without an independent physiological measure such as urinary metabolites is problematic (Creel 2001; Creel et al. 1997, 2002).

Thus, the social environment within a wolf pack may at first persist as a stable, predictable set of relationships. However, two main dynamic factors can eventually affect the relationships among relatives: competition for food and sexual tensions resulting from maturation of offspring. To better understand cohesion and conflict within wolf packs, it helps to look separately at each of these contexts in which conflict occurs.

Conflict within Packs

Are subordinate wolves stressed? Although extreme cases of individuals with elevated aggressiveness and stress hormones (urinary cortisol) have been reported for captive packs (McLeod et al. 1991; Moger et al. 1998; Gadbois 2002), researchers are only beginning to explore this question in wild populations (Creel 2001; Creel et al. 2002). Clearly not all wolves in each pack have equal access to food and mates. Studies of conflict over food will be reviewed below, followed by a discussion of conflict over mates and the apparent "incest taboo."

Access to Food within Packs

As Mech (1988a) watched the Ellesmere pack feed from a newly killed muskox carcass, the yearlings approached their father with highly exaggerated movements, curling their bodies dramatically, lowering their ears, and pawing widely into the air in a "groveling" manner (National Geographic Explorer 1988). Although the carcass potentially provided enough food for all, the breeding pair intimidated their offspring. They limited access to the meat until they had gorged enough to feed their pups and had torn off enough chunks to cache and eat at a distance in relative peace.

Without the exaggerated groveling, I have seen similar behavior in captive wolves feeding on a deer carcass (see fig. 2.2B). In the North pack, two adult daughters approached their father in a low crouch, and he pinned each to the ground with an inhibited muzzle-bite typical of a care-giver interrupting begging pups. The grown daughters stayed away from the carcass until their father was finished. In this nuclear family, order of feeding at the carcass was correlated with appeasement, not conflict, behaviors (Packard 1980, 163). Appeasement gestures reduce the probability of escalation of conflict after it has started.

In the captive South pack (a disrupted family), I watched two 6-month-old juveniles approach their foster father while he was feeding from a deer carcass. When he snarled at them, they lay down and looked the other way until he returned to chewing on the carcass (fig. 2.8). Not completely intimidated, the juveniles crept closer while the male was feeding and eventually fed beside him without conflict. In this family, in contrast to the North pack, the order of feeding at the carcass was correlated with conflict interactions, not with appeasement (Packard 1980, 163). Apparently the juveniles had not yet learned to use ritualized appeasing gestures, as their small size conveyed little threat to their foster father. The most easily intimidated were two older sisters, who also had the lowest blood indicators of nutritional condition (blood urea nitrogen and hemoglobin).

FIGURE 2.8. Juveniles that behave solicitously, like pups, are more likely to gain access to food defended by adults, which may affect their nutritional condition and dispersal during times of food scarcity.

Thus, elements of both cohesion and conflict can be associated with interactions around a carcass large enough to be shared by family members. The individuals least likely to be intimidated (i.e., most likely to act like pups) are the ones most likely to gain access to the carcass. The "appeasement" gestures derived from juvenile care-soliciting behavior provide clear signals that a younger individual is unlikely to escalate conflict, and elicit care-giving tolerance from other family members. Those individuals most likely to be intimidated in the context of food are the older siblings in poorer nutritional condition, which show fewer cohesive interactions with parents and more conflict with siblings.

Appeasement gestures reduce conflict only within relationships in which interactions have been patterned by repeated food solicitation and delivery (Schenkel 1967). Individuals in such cohesive relationships have been able to learn the contingencies of one another's actions and to modify their own behavior accordingly. In contrast, fighting between a confident wolf and an intimidated individual (e.g., siblings or strangers) involves jockeying for position to bite without being bitten (Schenkel 1967; Golani and Moran 1983; Mech 1993a).

By 2 months of age, pups in the Ellesmere pack had learned complex tactics for defending and acquiring pieces of arctic hare carcasses (Packard et al. 1992). The hare carcasses were small enough (about 1–3 kg, or 2.2–6.6 pounds) that one pup could initially monopolize the meat, while siblings waited and watched for an opportunity to grab a piece beyond lunging distance of the owner. Several researchers have noted that the wolf in possession of a food object is more likely to escalate conflict than an onlooker, even if the onlooker is bigger, older, or more assertive in other contexts (Mech 1970, 1999; Lockwood 1979; Zimen 1981; Townsend 1996). Likewise, it is very difficult to teach a hand-reared wolf to give up a bone or refrain from stealing steaks off the table!

According to a deterministic perspective, parents defend their right to monopolize food in order to continue producing pups, for which their immature older offspring will help care (Harrington and Mech 1982c, 1983). However, it is unclear whether alloparenting is a help or a hindrance to breeding wolves (Harrington and Mech 1982c; Harrington et al. 1983; Ballard, Ayres, Gardner, and Foster 1991). Indeed, its proximate costs and benefits are likely to vary with the wolf-prey dynamics of each population (Harrington et al. 1983; Mech et al. 1998).

A stochastic perspective would posit that if the physical environment is harsh, then strategies involving food sharing among relatives, high variation in rates of maturation, and a low probability of dispersal are likely to become fixed in mammalian populations (Honeycutt 1992; Lacey and Sherman 1997). Wolves certainly show a high degree of variation in rates of sexual maturation (see Kreeger, chap. 7 in this volume). Furthermore, food sharing in wolves appears to be conditional, varying with complex interactions among prey availability, maturation rates, and social conditions (Packard and Mech 1983; Ballard et al. 1997; Mech et al. 1998). Because of varying food conditions, the proportion of offspring dispersing also varies (see Mech and Boitani, chap. 1 in this volume), reflecting the dynamic competitive milieu.

When food is scarce, breeders may selfishly maintain their own nutritional condition at the expense of other family members, especially non-pups. Competition among siblings escalates in the context of food, until those that are more intimidated follow at a distance, dropping farther behind as they become more malnourished. Eventually they disperse (Mech et al. 1998).

Access to Mates
What about conflict in the context of mating? While offspring remain with their parents, there is no issue of sexual rivalry as long as the offspring are reproductively immature or defer to their parents (Packard et al. 1983, 1985). However, sexual rivalry can occur when offspring remain in the pack past the age of reproductive maturity

(Derix et al. 1993). In most packs, dispersal relieves this competition (see Mech and Boitani, chap. 1 in this volume). However, in larger packs enjoying a surfeit of food, as in the reintroduced Yellowstone population, sexual competition may occur among reproductively mature siblings (D. W. Smith, unpublished data).

Exclusive breeding by the dominant pair in a large nuclear family (e.g., having more than one mature member of each sex) apparently results from a delicate balance of asymmetric mate choice and same-sex rivalry (Packard et al. 1983, 1985; Schotté and Ginsburg 1987; Packard 1989). Monogamous relationships are likely in nuclear families only as long as (1) offspring are not reproductively mature, (2) the breeders are more attracted to each other than to their offspring, and (3) courtship between sibs is interrupted.

If these conditions are not met, plural breeding may occur (e.g., more than one female may reproduce within a pack in a given season). The term "plural breeding" is used by researchers studying cooperative breeding in other mammals (Solomon and French 1997) and is synonymous with the term "multiple breeding" used by Mech and Boitani in chapter 1 and elsewhere in this volume. A possible example of the apparent release of plural breeding was observed in the West pack on Isle Royale. Only one pair in the pack was reproductive until the breeding male disappeared prior to the breeding season in 1971 (Peterson 1977). When the breeding female accepted a mate from within the family, a second pair also successfully copulated (Peterson 1977, 80–84). In 1972, a subordinate female copulated with the West pack dominant male. Even though the subordinate was twice chased off the island by the dominant female and her companions, she returned each time (Peterson 1979). These observations illustrate the interaction of cohesion and conflict in determining the proportion of monogamous relationships in wolf populations.

Polygamous relationships have been reported in a few wolf step-families in which one of the breeding pair has died and a new breeder has immigrated into the pack (Haber 1977; Smith 1997, 2000), although polygamy was not reported in several other step-families (Rothman and Mech 1979; Fritts and Mech 1981; Mech and Hertel 1983). When a widowed breeder accepts an unrelated mate, the new step-parent may be less likely to interrupt rivals and more likely to be attracted to younger pack members in addition to the widowed breeder.

Occasionally, a cohesive relationship may persist between two adult males even though the less assertive individual appears to be more reproductively active (Murie 1944; Rabb et al. 1967; Haber 1977). In such cases, the assumption has been made that the more reproductively active wolf was the breeder, although the actual paternity of offspring is difficult to assess without genetic testing.

As indicated above, the proportion of monogamous packs (i.e., only one pair breeding per season) in the wild is high (Harrington et al. 1982). In captive packs in which more than one mature female was present and no dispersal was possible, plural breeding occurred in 30% of fifteen pack-years (Packard 1980). However, multiple litters were successfully raised in only 13% of these fifteen cases (Packard 1980); the losses were due to infanticide by females (Altmann 1974, 1987). Infanticide was also reported by McLeod (1990).

In captive wolves, mate choice may be influenced by sexual rivalry (Derix et al. 1993). For example, I was puzzled by the unsuccessful attempts of the North pack female, Chenequa (F60), to solicit sexual interest from an older brother, Negaunee (M64), after their father died. She stood near her brother, pranced coyly, rested her chin on his back, flagged her tail near him, and marked near his urine without eliciting a response. Why was Negaunee unresponsive to such a solicitous female? Often his older brother was looking over his shoulder (fig. 2.9)! Subsequently, Negaunee and Chenequa copulated during the night, when their older brother was sleeping. When the brother awoke and found his siblings in a copulatory tie, he rushed over and lunged at them. Big brother was too late, however. One tie was enough, and 9 weeks later, Chenequa produced pups in the same den as her mother.

Incest Avoidance

Is incest common in wolf packs? Incest is not common in wild wolves when they can choose mates other than close relatives (D. Smith et al. 1997). Packard (1980) hypothesized that close relatives in natural populations would be unlikely to form a pair bond that endured longer than a season or two. Genetic studies tend to confirm this hypothesis; in Denali National Park and the Superior National Forest of Minnesota, breeders are more distantly related to each other than neighboring packs are to one another (D. Smith et al. 1997). However, on Isle Royale, where there is no choice, wolves do breed with their close relatives (Wayne et al. 1991). During re-

FIGURE 2.9. Although a young male is being courted by his sister (on the left), he is unlikely to reciprocate as long as his older brother (on the right) is looking over his shoulder.

colonization of the Rocky Mountains, extensive dispersal should tend to reduce the probability of inbreeding (Boyd and Pletscher 1999; Forbes and Boyd 1996).

Does conflict in the context of courtship in wolves (Derix et al. 1993) serve a function analogous to that of parental manipulation in eusocial species (Lacey and Sherman 1997)? Currently, there is no evidence that nonbreeding adult wolves are physiologically suppressed, even under extreme conditions in which there is no option for dispersal (e.g., in captive packs) (Packard et al. 1985). In the captive North and South packs I studied, the nonreproductive adults all showed hormonal cycles typical of individuals that had reproduced. Thus, they retained the physiological readiness to breed (Packard et al. 1983, 1985), unlike African wild dogs (Creel et al. 1997). Although urinary indicators of stress may vary with aggressiveness (Gadbois 2002; McLeod et al. 1996), they may not be correlated with reproductive status.

Most nonbreeding adults within wild packs are merely in a transitory nonreproductive state (Mech et al. 1998). They may move into a breeding position within the natal pack if an opposite-sex parent dies and is replaced by a step-parent, they may disperse and start a new family, or they may enter a disrupted family as a step-parent (Rothman and Mech 1979; Fritts and Mech 1981; Mech and Hertel 1983; Packard and Mech 1983; Ballard et al. 1987; Mech et al. 1998). Similarly, senescent breeders that have lost a mate may continue to associate with the pack in a nonreproductive role (Mech 1995d; Thurston 2002). The probability of transition between these social roles will depend not only on the developmental history of each individual (possibly slowed by poor nutrition), but also on the dynamics of prey availability (Mech et al. 1998).

In summary, parents are more likely to escalate, and offspring to reduce, conflict in the usual nuclear family pack. In disrupted families, step-families, and complex packs with multiple breeders, the outcome of conflict for food and mates is more difficult to predict, as it depends on a complicated interaction of intrinsic (nutritional condition, reproductive state, temperament, relationship, relative attractiveness of potential mates) and extrinsic factors (prey and mate availability).

Leadership

Mech (1970, 73) defined leadership as "the behavior of one wolf that obviously controls, governs, or directs the behavior of several others." The first pack that he watched on Isle Royale contained a dynamic male who was consistently out in front of his family (Mech 1966b). Mech (1977c, 530) concluded that "the male leader guides the activities of the pack and initiates attacks against trespassers. Pack underlings can, however, sometimes effectively protest their leader's actions." Zimen (1981, 173) agreed that leadership in wolf packs is a "qualified democracy." This conclusion accords with a deterministic perspective.

The degree of authority the dominant male has over the pack has been disputed, however. Based on observations of a captive pack, Fox (1980, 128) asserted that there is "an alpha male who not only rules over the males but is the leader of the pack. He is the decision-maker. Other wolves, even older ones, respond to him submissively and affectionately as would cubs to their parents. Allegiance to the leader helps keep the pack together." Nevertheless, as Zimen (1981, 173) wrote, "no member decides alone when an activity is to begin or end, or which way or at what speed the pack is to move, or exercises sole power of command in any other activities that are vital to the cohesion of the pack. The autocratic leading wolf does not exist."

So, is the breeding male the leader of a wolf pack or not? Mech (2000a) has emphasized the concept of one-way, autocratic control by both parents in each family, consistent with the deterministic perspective. Taking a stochastic perspective, I like to emphasize the probability that parents influence offspring, but offspring also influence their parents. Wolf families can be so diverse

(see Mech and Boitani, chap. 1 in this volume) that both models probably have merit, depending on how the term "control" is defined and on the history of relationships within the pack.

Similar questions about leadership have been addressed by students of primates, and Rowell (1974) has suggested that researchers test the following hypotheses: (1) that one individual receives attention at each activity change, (2) that one individual is consistently in front of a line during directional movement, (3) that one individual defends the group, and (4) that the actions of others are policed by one individual.

Let us first examine the hypotheses that breeding male wolves are more likely to be the "attention center" and "at the front of the line." The notion that the breeding male is the center of the pack's attention may be traced back to Schenkel's (1947) description of behavioral expression in the two wolf packs he watched at the Zurich Zoo. He described the breeding pair as the center of attention for family members and the breeding male as the lead wolf. However, he was not able to document leadership during directional movement as it is observed in the field when wolves travel single file.

On a similar theme, Murie (1944, 28) enlivened his account of the East Fork pack in Denali National Park by describing a certain male as "lord and master of the group although he was not mated to any of the females." What he meant by "lord and master" is not clear, and he could not have known whose mate the male was. However, in his description of the interactions within the family, Murie gave the impression that the center of attention shifted among several individuals as they departed from and arrived at the den.

The "attention center" hypothesis has not been tested systematically for wolves in captive or field studies. Lockwood (1979) proposed that it might be a better model than a "dominance hierarchy" to describe the social interactions among wolves. However, the two measures of attention that he used (social proximity and looking across a barrier) did not account for the variation in social structure that he measured in seven captive packs over 3 years. Lockwood recommended that relations among wolves be analyzed in terms of dyads (e.g., which wolf looks at which more often?). In the few places where one can watch wolves in the field, it would be feasible to note the direction in which pack members look during a change in activity state (e.g., from resting to traveling). Although I know of no field studies in which such data have been collected, future research should focus on testing the attention center hypothesis regarding leadership in wolves.

The object of an individual's attention is likely to change with its mood (i.e., its internal state). When an individual's prolactin level is high, for example, pups should be the center of its attention. When a female is in estrus, she is likely to receive the attention of males. Thus, without knowledge of internal state and external stimuli, researchers are unlikely to be able to test how the center of attention shifts in a wolf pack.

Offspring are clearly more likely to follow their parents than vice versa (Mech 2000a). This pattern is consistent with the following response that develops in 2-month-old pups (Packard et al. 1992) and is reinforced in juveniles that gain access to food by following adults during their first year (Mech 1966b; Haber 1977; Peterson 1977; Mech et al. 1998).

Are breeding males more likely to be at the head of the line during travel? Generally, yes (Mech 1966b; Mech 2000a), except during the courtship season, when the breeding female is usually followed by her mate (Mech 1966b; Peterson 1977). Even when younger pack members forge ahead of a breeder (Murie 1944; Haber 1977; Mech 2000a), they appear uncertain about direction and take their cues from the breeders (Mech 1966b; Peterson 1977).

The Ellesmere breeding female followed Left Shoulder on 76% of twenty-nine occasions when departing from the den and on 67% of seventy occasions when the wolves were already traveling (Mech 2000a). In 1993, when no pups were present, and three yearlings begged more often from Left Shoulder than from the breeding female, the female followed Left Shoulder significantly more often than vice versa. Apparently Left Shoulder was a good source of information about the location of food. In 1996, when no yearlings were present, there was no difference between Left Shoulder and his mate in their order of travel (Mech 2000a).

Leadership during travel was studied in three Yellowstone packs during early and late winter sampling periods from 1997 to 1999 (Peterson et al. 2002). The factors influencing leadership included the pack, the season, and the type of measure used (frequency or duration). In March, breeding males were significantly more likely than their pregnant mates to be in the lead in the Rose Creek and Druid packs, but not in the Leopold pack. In November/December, there was no significant difference between the breeding male and female in their frequency of being in front. The Rose Creek pack, however, came very close to showing a significant difference ($P = .056$), with the male leading. Compared with nonbreeders, the breeders were more likely to be in front in the two smaller packs (Leopold and Druid), but not in the larger Rose Creek pack. These data were collected during a period of social transition for breeding females in the Druid and Rose Creek packs.

In nuclear families, the breeders are the most likely to trot directly and confidently toward a goal, to be solicited by hungry juveniles, and to be followed by family members. However, variations on this theme would be predicted in step-families or extended families. Consider a pack that accepts an immigrant step-father who is less familiar than the rest of the pack with the activity and movements of prey within the home range. The stochastic model predicts that pack members would be likely to continue to follow the experienced breeding female more often than the inexperienced step-father (at least until he became a predictable source of food). This pattern was documented in the Druid pack when they accepted an immigrant male from the neighboring Rose Creek pack (Peterson et al. 2002). Likewise, if the movements of a brother became a better predictor of food than the movements of an aging father, then one would predict that the younger male would attract more followers than the older male.

Conflict between Packs

As Rolf Peterson (1977) watched from an airplane over Isle Royale, the East pack encountered a scent post. First the breeders investigated the urine odor, then the younger pack members checked it. The breeding female reversed her direction and led the pack out of the edge of its territory and back toward the center. The scent was from the pack's neighbors, the West pack.

The West pack's range had covered the entire island until two changes occurred: (1) the pack's original breeding male died, and (2) their major prey (moose) became localized to the western end of the island (Wolfe and Allen 1973; Peterson 1977). A pair had then split from the West pack, found an eastern area where they could subsist on beavers during the summer, and produced pups of their own, becoming the East pack.

The reaction of the East pack breeders to the West pack's scent was predictable, given that direct encounters during territorial trespasses escalate rapidly, resulting in chases, injury, and death (Murie 1944; Mech 1966b, 1993a, 1994a; Marhenke 1971). Fights probably escalate most rapidly between breeders of the same sex. Are male wolves more likely than females to lead the pack against hostile neighbors? Unfortunately, few territorial encounters have actually been witnessed (Mech 1993a), but breeders and immatures of both sexes have been killed during territorial trespasses (Mech et al. 1998).

The conflict between packs is probably related to competition for food and mates. In contrast to conflicts within packs, which are inhibited by cohesive family relationships, fights between packs are likely to be injurious. Mortality due to conflict with hostile neighbors varies among wolf populations (see Mech and Boitani, chap. 1 in this volume).

Flexibility of Behavior: Physical Factors

Wolves can adjust to a wide range of physical factors in addition to the variation in social factors described in the previous section. This flexibility of behavior is illustrated in studies of activity patterns and intelligent problem solving.

Activity Patterns

How does one's dog know it is about 3:30 P.M. each day, such that he barks at his food bowl if he is not fed as usual by 4:00? The ability of wolves to learn predictable correlations among prey activity, temperature, and light intensity is not surprising to anyone who has watched a captive canid develop expectations of feeding time. When predictive cues fluctuate with season and weather, canids are flexible enough to learn new patterns and to adjust their activity accordingly. It takes my dog about a week to adjust to going off daylight savings time. The activity patterns of wolves can shift to adjust to changes in predictable patterns of temperature (Harrington and Mech 1982c; Fancy and Ballard 1995), prey activity (Oosenbrug and Carbyn 1982; Scott and Shackleton 1982; Fuller 1991; Theuerkauf et al. 2003), probability of encountering humans (Boitani 1986; Vilà et al. 1995), presence of a carcass (Harrington and Mech 1982c; Ballard, Ayres, Gardner, and Foster 1991; Mech and Merrill 1998), and reproductive season (Harrington and Mech 1982b,c; Ballard, Ayres, Gardner, and Foster 1991; Vilà et al. 1995).

Do wolves have an innate circadian rhythm? Mech and Merrill (1998) found that summer departures from the Ellesmere den occurred at predictable times (between 2200 and 0400). They proposed that this departure pattern was due to an internal rhythm because external triggers such as daybreak and nightfall were absent in summer at 80° N latitude. They cited supporting evidence from four denning studies in Alaska and Canada at latitudes ranging from 46° to 64° N (Kolenosky and Johnston 1967; Haber 1977; Scott and Shackleton 1982; Ballard, Ayres, Gardner, and Foster 1991). However, Mech and Merrill were not able to test alternative hypotheses related to several covariates (e.g., light intensity, temperature, food availability, and human activity). Variation in activity peaks has been reported from four other studies at latitudes also ranging from 42° to 64° N (Murie 1944; Harrington and Mech 1982c; Vilà et al. 1995; Theuerkauf et al. 2003).

In northwestern Alaska, the activity of twenty-three wolves (monitored continuously by satellite telemetry) was correlated with temperature in all seasons (Fancy and Ballard 1995). During summer, activity peaks occurred at 0600 and 2200. Activity was more likely in the morning than in the afternoon during summer, but occurred at a lower rate throughout the day in winter than in summer. On winter days in the Superior National Forest, wolves were more likely to be inactive (65% of the time) than traveling (28%) or engaged in activities such as feeding or socializing (Mech 1992). In summer, lower activity levels during midday may have been due to heat intolerance (Harrington and Mech 1982c). Whether such seasonal shifts in activity patterns are due to the wolves' own intolerance of heat or the activity patterns of their prey is hard to determine (Oosenbrug and Carbyn 1982; Scott and Shackleton 1982; Fuller 1991; Jedrzejewski et al. 2001).

In Spain, the time of day when two females were most active changed with the age of their pups (Vilà et al. 1995). During lactation, these females were most active during daylight, although their total activity was low (15%). When their pups were 2–5 months old, the females were more active at night, like their mates. During winter, the females' activity increased to 24% of the time and was relatively more evenly distributed across daylight (12%), dusk (18%), night (35%), and dawn (32%) (see also Theuerkauf et al. 2003).

Activity patterns also vary in captive packs. In one pack, daylight activity peaked in the morning and evening (MacDonald 1980). Morning activity associated with a cleaning routine was greater in a zoo enclosure than in a secluded naturalistic enclosure in Mexico (Bernal and Packard 1997). The overall amount of activity did not differ between wolves housed in outdoor chain-link kennels and in large (0.3–0.5 hectare) enclosures with natural vegetation in Minnesota (Kreeger, Pereira et al. 1996).

The seasonality of courtship and reproduction may also influence wolf activity patterns. The frequency of overall interactions and of aggression peaked during winter in several captive packs (Zimen 1981; Fentress et al. 1987; Servín-Martínez 1991). The occurrence of such peaks during autumn in another captive pack may have been specific to the instability of the relationship between an aging father and a maturing son (Packard 1989).

Thus, wolves develop predictable activity patterns, which may vary across seasons, locations, and individuals. Insufficient evidence is available to tease apart the effects of correlated external variables such as light intensity, temperature, and prey activity. If wolves are like other carnivores, we would expect the variation in their activity patterns to be a complex function of (1) daily internal cycles, (2) seasonal shifts in external stimuli from the physical environment (e.g., temperature, light, prey activity, human activity), (3) variation in the internal states of individuals (e.g., nutritional condition, reproductive phase), and (4) expectations of the actions of companions.

Learning and Intelligence

Why are wolves—or their domesticated descendants, dogs—so intelligent? Or are they? How do they learn to respond appropriately to changes in predictable patterns within their physical environment? To what extent does the social environment of the pack assist young wolves in learning what to expect from their physical environment?

Intelligence is the ability to apply knowledge gained from experience to novel problems (Byrne 1995). For example, a captive wolf learned to open his cage door by pulling a pulley rope that his caretakers had used predictably (Fox 1971d). From a human perspective, the "novel" problem for the wolf was how to open the cage door, obviously a challenge not encountered in the history of the species. By watching his caretakers, the wolf could have learned to associate the rope with the open-

ing of the cage door. Grabbing and pulling on the rope could be considered intelligent behavior because it opened the door. However, wolves also grab and pull on the noses, ears, and tails of exhausted prey (see Mech and Peterson, chap. 5 in this volume). To what extent was insight involved in this seemingly intelligent action, and to what extent was it motivated by reflex-like actions typical of all members of the species? Is learning any less intelligent if it is guided by programmed stimulus-response linkages shaped by problems solved in the history of generations of ancestors?

Questions related to wolf intelligence have been addressed in studies of (1) standardized problem-solving tasks in captivity (Cheney 1982; Frank and Frank 1984, 1985; Frank et al. 1989; Hiestand 1989; Hare et al. 2002), (2) the development of flexible hunting skills (Sullivan 1979; Crisler 1958; Mech 1988a, 1991b), (3) the modification of behavior contingent on the actions of social companions (Schenkel 1947; Moran and Fentress 1979; Moran et al. 1981; Lyons et al. 1982; Golani and Moran 1983; Havkin and Fentress 1985; Townsend 1996; McLeod and Fentress 1997), and (4) anecdotes in which wolves have solved unusual problems (Fox 1971d; Slade 1983; Fentress 1992; Ginsburg and Hiestand 1992; Mech 1999).

A standard approach to assessing wolf intelligence is to compare it with dog intelligence. Frank and Frank (1987) found that four juvenile wolves were better at standardized problem-solving tasks than were juvenile dogs raised in the same manner (cross-fostering to a wolf mother). As they had predicted based on models of canid domestication (Scott and Fuller 1965; Scott 1967; Frank and Frank 1987), the dogs were better at tasks that involved training by a human (Frank and Frank 1982; Frank et al. 1986, 1989). However, the wolves responded to the restraint of a leash or a box in an emotional way that seemed to interfere with their performance of the standardized tasks (Frank and Frank 1983). Was this really a fair test? Or did the physical restraint trigger an emotional response that has become neurally programmed in wolves? Does this neural programming exist because wolf ancestors that struggled against restraint (when grabbed by a bear or other predator, for example) were more likely to have survived than those that did not?

Addressing such questions about the interaction of emotional and learning systems in canid evolution, Hiestand (1989) tested spatial orientation in nine hand-reared wolves (Frank et al. 1986) and forty adult German shepherds using ropes hanging from a ceiling. She did not use restraint; rather, she called out "good" when the individual pulled the correct rope. She opened a gate and/or petted the individual lavishly when it learned to pull two or three ropes in a pre-established order. She was gruff with or ignored individuals that did not behave correctly. Apparently, social interaction with a human caretaker was as effective a reward in training wolves as it is for dogs (Frank and Frank 1988; Hiestand 1989). The two adult wolves had little difficulty learning the three-rope task. The seven juveniles learned the two-rope task, but had more difficulty with the three-rope task. The adult dogs showed high individual variation: 12% learned the three-rope task, 35% learned the two-rope task, and 13% failed to learn to pull even one rope.

The wolves acted differently from the dogs when initially placed in the testing arena: they spontaneously oriented to physical objects in their environment and learned the one-rope task in significantly fewer trials than the dogs (Hiestand 1989). The wolves also irreparably shredded more ropes than the dogs! Hiestand hypothesized that wolves are more attentive to physical objects in their environment than are dogs, as noted anecdotally by others (Fentress 1967, 1992; MacDonald 1980; Zimen 1981).

The wolves behaved as if they had both the neuroendocrine programming that attracted them to situations in which they were likely to learn and the attentiveness to social companions that shaped their behavior in ways that were effective in attaining their immediate goals. However, wolves also show high individual variation in learning ability in captivity (Sullivan 1979; Cheney 1982; Lyons et al. 1982; Slade 1983; Hiestand 1989). Hypotheses about heritable differences in emotional and learning systems still need to be tested with sample sizes adequate to account for individual variation in different rearing environments.

Claims that dogs are more skillful than wolves at reading human gestures must be evaluated critically. Hare et al. (2002) documented differences between wolves and dogs in a series of object-choice tests. The seven adult wolves less often used human gestures (gazing, pointing, and tapping) correctly to choose between two overturned food bowls, one baited out of view of the subjects. The subjects were tested in locations familiar to each, a small pen for the wolves and rooms for the dogs.

The degree to which such findings can be generalized is questionable for several reasons. First, the dogs were raised and housed continuously with humans in several households, whereas the wolves were raised with humans for 1.5 to 5 weeks, then housed outside with a wolf pack that had daily short-term contact with the experimenter (see supporting online material linked to Hare et al. 2002). No details were given about whether wolves and dogs had equal experience with humans delivering food in bowls. Second, the mean age of the wolves (6 years) was greater than that of the dogs (3.5 years), and age is important in the development of social cognition, at least in primates (Tomasello et al. 2001). Third, with such a small sample, a set of clever dogs or dull wolves may unintentionally have been used. Individual variation in how dogs and wolves respond to standard cognitive tests is high (Frank and Frank 1988; Hiestand 1989). Fourth, wolves respond to mechanical devices differently from dogs (Zimen 1981; Cheney 1982), as noted above. This fact accords with the observation (Hare et al. 2002) that wolf performance did not differ from that of dogs in a test involving food in a film canister in the experimenter's hands. Last, the use of standard cognitive tests across species can be problematic (Shumaker and Swartz 2002).

What are the indicators of learning and intelligence in wild wolves? Peters (1978, 1979) proposed that wolves learn to navigate in familiar landscapes by forming learned associations analogous to a cognitive map. However, because wolves are in no way unique in this ability (Tolman 1955; Hauser 2000; Santin et al. 2000), it is questionable whether the use of cognitive maps distinguishes wolf intelligence from that of any other mammal.

The wolf's diversity in hunting skills across a wide range of ecosystems (see Mech and Peterson, chap. 5 in this volume) could also be considered an indicator of flexible problem solving. Both spatial orientation and flexible hunting skills develop within the social environment of the pack, potentially providing a window of opportunity for individuals to learn about the specific physical environment in which they were born.

The social context of the pack has the potential to affect learning by pups in several ways. For example, in an episode of chasing play, I watched one pup "cut the corner" and grab a sibling that had evaded his chaser by making a sharp right turn. Later, such maneuvers were repeated, reminding me of a filmed hunting episode in which one wolf cut the corner and was able to grab a caribou that had evaded another wolf (Peace River Films 1975). The precursors to such combinations of actions in the caribou hunt could have been learned during the play routines of pups.

Another question that often arises in studies of intelligent behavior is the degree to which individuals modify their own actions contingent on the actions of others (McLeod 1987, 1996; McLeod and Fentress 1997). For example, adults in one captive pack were less likely to cache food when another wolf was present, consistent with predictions based on a model of deceptive communication (i.e., individuals withheld information about a food source when the goal was to keep the food for themselves) (Townsend 1996). However, 71% of the caches made in the same study were recovered by wolves other than those that made them, and 43% of these recoveries occurred in the presence of the cacher without retaliation. Much more study needs to be done on this subject.

The ability of wolves to solve novel problems often appears to draw on both instinct and experience. Left Shoulder was presented with a novel problem one day in 1988 when a muskox wandered near his pack's den, which was at the base of a cliff. The nursing female initially rushed over to the intruder, then gradually lost interest and lay down out of sight. Alarmed by the wolves, the muskox backed up to the cliff in a typical defensive maneuver also observed by Mech (2000a). The muskox stood directly in front of the den entrance, while the pups were a hundred meters away from the security of the den!

First, Left Shoulder picked up an abandoned arctic hare carcass and delivered it to the pups, capturing their attention for a short period. However, the satiated pups rapidly lost interest and started wandering. Left Shoulder trotted directly across a stream to a sandy bluff, dug

out a cache (at least 8 days old), and delivered it to the pups. The cache held the interest of the pups, and they stayed away from the den. With no wolves nearby, the muskox relaxed and wandered away from the den entrance. Left Shoulder's "problem" was solved. To what extent was his action insightful?

In a similar episode in which another muskox assumed a defensive position in front of the Ellesmere den, Mech (2000a) believed that Whitey behaved in an insightful manner by bark-howling and directing yearlings to a position off to the side so that the muskox could escape. From a deterministic perspective, wolves need to be insightful in order to cooperate in capturing prey and in defending the pack against enemies (Fox 1975, 1980; Haber 1996). However, the extent to which groups of wolves actually do cooperate in hunting has been questioned (Schmidt and Mech 1997; see Mech and Boitani, chap. 1 in this volume).

From a stochastic perspective, the ability of wolves to solve problems under a wide set of circumstances may be explained by basic learning processes shared by all mammals (Frank et al. 1989; Fentress 1992): (1) innate, simple decision rules that govern general learning situations encountered in the evolutionary history of the species, (2) the ability to learn specific predictable consequences of their actions (operant conditioning), (3) the ability to associate diverse sets of cues (e.g., smell, sound, and taste of a prey item) with specific predictable situations (associative learning), and (4) emotional thresholds associated with the social context of learning and response to novelty (affective state). At issue is the degree to which intelligent behavior in wolves reflects ancestral knowledge stored in the genome and the degree to which individual experience results in novel insight.

3

Wolf Communication

Fred H. Harrington and
Cheryl S. Asa

Overlooking Rookie Lake, Superior National Forest, Minnesota, U.S.A. 21 February 1974. Weather—cold, clear, and calm. I howled several times from a cliff overlooking the lake; a single wolf replied once from the northeast. Seconds later, a more distant pack howled from the west. When I howled again 20 minutes later, only the pack answered. Was the single wolf with the pack, or just keeping quiet?

Ten minutes later I got my answer when the single replied again from much closer. Ten minutes later it again replied. The pack remained silent. The single wolf's howls were beautifully modulated, long and haunting sounds. An approaching car forced me to end the session, but the next day the driver told me of a "large" adult wolf trotting down the highway just before the Rookie Lake lookout. Had I stayed, I likely would have had company!

I later followed every wolf track I could find nearby. The single wolf had nothing to do with the pack. At first, it had been about 600 m from the pack. Then it climbed a ridge toward me, stopping just over the crest where it left two beds of packed snow with scats [feces] in both. Not far away, it had urinated twice. After howling from one or both beds, it had moved toward the highway, where the driver had seen it.

This evidence, plus my knowledge of the local packs, led me to believe the single wolf was a lone adult female. The scat in the beds suggested she was fearful, excited, or anxious, when she howled. Her eventual approach indicated she may have been seeking contact. Was I being checked out, even courted? Why had the pack fallen silent after its first two replies? Did the animals consider the continuing serenade between two single animals of no importance to them?

The above notes of Fred Harrington portray what might be a typical night for a lone wolf during the breeding season. They also demonstrate the difficulties inherent in the study of communication. The howling interactions easily fit our expectations for communication (i.e., Smith 1977). One individual (the sender) produces a specialized signal (howl) in a specific context (breeding season), which has immediate (howl reply) and delayed (approach) effects on others (receivers) that perceive the signal. It is obvious that communication occurred, but what were the actual messages transmitted by the howling interaction?

And what about the olfactory signals left by the lone female? Were her defecations and urinations merely the involuntary result of fear or excitement? Or were they left as "calling cards" to announce her presence, sex, and reproductive state? What about the odor "rafts" she shed by the thousands and left wherever she traveled, or the smells left behind by her feet in the snow? All these olfactory signs can be perceived and acted on by other wolves, but should they be considered communication if they are entirely involuntary? At what point do these behaviors or physiological processes become communication?

Philips and Austad (1990, 258) propose that communication occurs when "an animal transfers information to an appropriate audience by use of signals." Signals are the behaviors and features of animals that have evolved to encode the information being conveyed. Information, itself, is the "property of entities and events that renders them, within limits, predictable" (Smith 1990, 235). Growling (the signal), for example, conveys aggressiveness (the information), and makes a wolf's subsequent behavior (likelihood of attacking) more predictable. Signals and information thus form the basis of communication: its media and its messages. A knowledge of signals and the information they convey is thus central to a fuller understanding of animal communication (Smith 1990).

In sharing information, a signaling animal influences the recipient's expectations about future events (Smith 1990). A growling wolf, for example, may cause one packmate to expect to be bitten painfully, should it pursue a certain course. However, another packmate might have a very different expectation of the likely outcome. These differing interpretations stem from the fact that three different sources of information are available to a recipient (Philips and Austad 1990). *The message* is the information present in the signal as a direct consequence of natural selection: growls evolved to convey aggressive intentions. *Metainformation* is additional information that is present in the signal not as a consequence of natural selection, but as a side effect of unrelated processes: the high pitch of the growls of an individual wolf might identify it as a pup. Finally, the context in which the signal occurs provides the last bit of information required as the recipient considers its options: "Although I am larger than the pup that is growling, its parent is lying nearby and will probably attack me if I try to steal food the pup is guarding." Thus, for wolves, communication involves the processing of a variety of information from numerous sources as individuals form and re-form their expectations about the future actions of packmates and other wolves.

The signals used in communication share a number of characteristics (Smith 1990). Within a species' repertoire, each signal is generally recognizable as a "discrete" entity. That is, despite variation in a signal, it is possible to draw bounds separating one signal from another. Thus individual signals vary within and among individuals, but within limits. As a consequence, each different signal may convey a qualitatively different kind of information. Within a signal, "rules" for the modification of form allow the message to change in specific ways, enabling an individual, for example, to convey not only "what" but also "how much." Two or more signals can be combined or sequenced to permit further modification. Finally, the signals of two (or more) individuals can be interwoven into a formalized interaction in which the behavior of one has meaning only in relation to the behavior of the other. All of these characteristics serve to make wolf communication both subtle and varied in its expression.

Communication in the Context of Wolf Sociality

In this chapter, we will consider both the signals used by wolves and the information those signals convey. In doing so, we must keep in mind that wolves communicate in a variety of contexts that range from competitive to cooperative. Our definition of communication requires that the sender, at least, have benefited from the evolution of the signal. Pack wolves generally share a common genealogy (see Mech and Boitani, chap. 1 in this volume) and, to a large degree, a common self-interest. Thus we expect that in activities of general benefit to pack members, such as hunting and pup rearing, signals will enhance the fitness of both sender and receiver. In these contexts, we expect to find honest and unambiguous signals. However, packmates also compete for food and for social status within the pack. In these competitive contexts, we expect signal evolution to have favored the sender rather than the recipient. When in the sender's best interests, deceitful or ambiguous signals may evolve in these contexts (Dawkins and Krebs 1978).

Interactions between packs ought to further this split in self-interest between sender and recipient. Most neighboring packs are only distantly related and are usually hostile to one another (see Mech and Boitani, chap. 1, and Wayne and Vilà, chap. 8 in this volume). Interactions involving lone wolves may range from competitive to cooperative, depending on the context and the identity of the interactants (i.e., former packmates, potential mate).

Communication as Integration

Most communication involves the transmission and receipt of simultaneous signals over multiple sensory systems. Indeed, most signals are composed of elements from two or more modalities; it is usually just simpler for us to separate them arbitrarily by sensory system. This multimodal aspect of communication provides signals with both emphasis and redundancy. Redundancy is of particular importance, as the relative salience of any one modality can vary tremendously as environmental conditions change. During the day, for example, wolves may attend more closely to visual signals, as slight variations in body posture may be sensitive indicators of a packmate's mood noticeable even at a distance. At night, however, attention may shift to vocal signals.

The integration of signaling systems has not been studied in wolves. In this review of communication, we will continue the usual practice of assessing each sensory system in isolation, but wherever possible, we will hint at possible interactions. Consider the case of a lone wolf locating and smelling a urine mark left by another wolf. After a brief sniff, the wolf turns and flees from the area. Observing from an aircraft, we would probably conclude that some message contained in the urine of the mark and decoded by the olfactory system stimulated the wolf's prompt exit. Yet perhaps it was a visual cue, such as the height of the mark above the ground, that so unsettled the loner.

Auditory Communication

The Importance of Audition in Wolf Communication

Wolves howl. Thus, of all the communication channels available to wolves, the auditory channel is the one best known to both the lay public and science. Much of wolf social behavior, from friendly greetings to vicious attacks, is accompanied by vocalization, thus providing, at the least, redundancy with other communication channels the wolves may also be using.

Auditory signals have certain advantages over other modalities, including the visual. They can instantly reach an audience from near to far at any time of day or night,

and they do not require a receiver's initial attention to be effective; in fact, they are excellent attention getters. However, they are highly dependent on suitable atmospheric and environmental conditions as distance between sender and receivers increases. Yet for all our preoccupation with howling, auditory communication in wolves most often involves relatively quiet, close-range signals that first emerge from pups in the darkness of the den.

Development of Vocal Communication

From birth on, wolves are very vocal. Newborn pups moan, squeal, and scream as they wiggle around and compete for their mother's nipples. Early studies of vocal development were largely anecdotal (summarized by Harrington and Mech 1978b). Coscia (1989, 1995; Coscia et al. 1990, 1991), however, described vocal development from birth to 6 weeks of age. By systematically sampling 100 hours of in-den videotapes, she described both the structure and the context of eleven discrete vocal signals (table 3.1). By 3 weeks of age, pups made all the adult sounds recognized by Schassburger (1993; see below), although some were rare.

Early Neonate Vocal Signals (Squeal / Scream / Yelp / Yawn)

Pups squeal and scream commonly during their first few weeks, but rarely by 5–6 weeks of age (Coscia 1995). Both of these sounds are short (< 0.5 s), harmonic, and relatively high-pitched, but screams are generally

TABLE 3.1. Acoustic properties of neonatal wolf pup vocalizations

Vocalization class	Gross spectral form	Duration (s)	Frequency modulation	Frequency range (Hz)	Peak age of occurrence
Moan	Richly harmonic	0.1–3.0	None to periodic	100–1,500	1–3 weeks
Whine	Richly harmonic	0.3–3.0	Aperiodic	200–2,500	—
Squeak	Harmonic	0.1–1.2, in series	Slow, aperiodic	5,400–11,000	4 weeks +
Yelp	Harmonic or noisy	0.1–0.3	—	500–2,600	Rare
Squeal	Harmonic or pure tonal	0.1–1.5	Aperiodic	200–2,400	1–3 weeks
Scream	Harmonic	0.1–1.0	Slow, aperiodic	700–2,900	1–2 weeks
Bark	Noisy and harmonic	0.1–0.3	—	200–800	Rare
Growl	Noisy	0.2–1.2	—	50–4,000	4 weeks +
Woof	Noisy	0.1–0.2	—	50–11,000	4 weeks +
Yawn	Complex	0.8–1.5	Slow, aperiodic	300–5,000	Rare
Howl	Harmonic	0.3–3.2	Slow, periodic	200–1,300	Rare

Source: After Coscia et al. 1991; Coscia 1995.

higher-pitched, much louder, and less frequent. Squeals are not associated with any particular pup or adult behaviors, and adults appear to ignore them. When pups scream, on the other hand, their mother grooms or repositions them, often eliciting more screams. Screams probably serve as mild distress calls.

Pups yelp when stepped on or being carried, and their mother responds promptly by attending to them (Coscia 1995). Yelps are rare, brief, loud sounds with a rapidly modulated fundamental frequency (Coscia 1995). Their chevron-shaped structure is similar to that of the bark, but is more rapidly modulated and much higher-pitched (Coscia 1995). Cohen and Fox (1976) believed that the yelp was a developmental precursor to the bark, but Coscia et al. (1991) recorded barks as early as day 1.

Yawns are sounds produced during yawning and have no apparent social role (Coscia et al. 1991). However, their presence as a possible auditory and visual signal in other species (Baenniger 1987; Provine 1989; Deputte 1994) suggests that they may also have signal value for wolves.

Early Adultlike Vocal Signals
(Moan/Whine/Growl/Bark)

Moans are the most common pup sound; they decline as pups grow older (Coscia et al. 1991; Coscia 1995). Both Frommolt et al. (1988) and Coscia et al. (1991) reported that moans were unique to pups, but Schassburger (1978, 1987, 1993) described moans in adults. Moans, like squeals, screams, and whines, are most common during the first 2 weeks of life (Coscia et al. 1991; Coscia 1995), when pups are deaf, suggesting that their target is the mother. Pups moan while nursing, asleep, or resting; their moans do not elicit any overt response from the mother (Coscia et al. 1991; Coscia 1995). Coscia felt that moans may be a passive consequence of respiration, but do provide a continuous (tonic) signal of the pup's location (and relative contentment?) in the dark den. Thus absence of moans may be more indicative of a pup's need for attention than their presence. Pups also moan when huddled together, providing a possible tactile (vibratory) signal to littermates (Coscia 1995).

Pups also whine from birth on. Whines are higher-pitched than moans and are modulated more slowly and variably (Coscia et al. 1991). Pups whine and moan in the same contexts, so the two vocalizations may carry similar information (Coscia 1995).

Pups growl as early as day 1, but not frequently until their fourth week (Coscia et al. 1991; Coscia 1995). During the first 3 weeks, pups growl mostly when huddled. Older pups growl during active interactions with littermates and occasionally when being moved by the mother.

Pups bark infrequently from day 1 (Coscia et al. 1991; Coscia 1995). Before 4 weeks of age, pups bark primarily when huddled together. Later, they bark most frequently during interactions with littermates. Pups also bark in response to sounds outside the den, and frequently as they leave the den.

Late-Appearing Adultlike Vocal Signals
(Woof/Squeak/Howl)

Pups woof infrequently from their third week on, most often during the mother's absence, or in response to adult movement or squeaking about the den entrance (Coscia et al. 1991; Coscia 1995). Woofs are associated with both hesitancy and approach and probably reflect uncertainty.

Pups do not squeak until day 15, but by week 4, squeaks become the most common sound in the pups' repertoire (Coscia et al. 1991; Coscia 1995). The mean *minimum* fundamental frequency of squeaks is higher than the mean *maximum* frequency of all other pup sounds. The average pitch of squeaks decreases between weeks 3 and 5 at a rate of 10% per week; during the same period, the mean fundamental frequency of squeals decreases at a similar rate (8% per week).

In weeks 3 and 4, pups squeak primarily while in contact with one another but not interacting (Coscia 1995). During the next 2 weeks, they squeak most commonly during interactions with littermates or with their mother. Pups also squeak in response to adult squeaks and howls from outside the den.

Pups howl only rarely and sporadically during their first few weeks (Coscia 1995), but soon after emerging from the den at about 3 weeks, they join in most of the daily howling of the adults (Mech 1988a; L. D. Mech, personal communication). Compared with other harmonic pup sounds, howls are much lower-pitched. Mech (1970), Schassburger (1978, 1987, 1993), and Frommolt et al. (1988) thought the howl may develop from a neonatal sound variously called the whine or cry, largely because this sound resembles the howl and because pup howls often begin with whines that lead into howls. Coscia (1995; Coscia et al. 1991) was skeptical about the evidence for this conclusion.

Mixed Series and Sounds

Of the wolf pup sounds recorded by Coscia (1995), 15% occurred with other sound types, as also reported by others (Cohen and Fox 1976; Schassburger 1987; Coscia et al. 1991). This was especially true of squeals, screams, and squeaks. Within these series, sound types were distributed nonrandomly, suggesting that squeaks, squeals, and screams represent increasing levels of arousal or distress in the pup. In addition, Schassburger (1987, 1993) and Coscia (1995) recorded a variety of mixed sounds (combinative or derived sounds) that may reflect complexity, uncertainty, or changes of motivational state (Schassburger 1978).

The Neonatal Vocal Repertoire

The vocal repertoire of newborn pups begins to give way to an adult-structured repertoire shortly after the pups can see, hear, and move about. The four most common sounds of newborns (moans, whines, squeals, screams) are made repeatedly during the first 2 weeks of life, then decline rapidly to either disappear entirely (squeals, screams) or continue in limited roles. The peak occurrence of these sounds before the pups can hear implies that the sounds are directed toward the mother. Together with the yelp, these four sounds provide a stepped series that permits the mother to continually monitor the status of her litter. The decline of these early sounds may indicate that pups rapidly outgrow the need for a multi-stepped distress signal system. By 3–4 weeks of age, pups can avoid or endure many situations that may have been life-threatening earlier. The loss of squeals and screams still leaves moans, whines, and yelps to indicate low, moderate, and high levels of distress, respectively.

Coscia's (1995) research suggests that each neonatal and adult vocalization is developmentally independent. Each is distinctive from its first appearance. That some adult sounds (growls, barks, squeaks) are made in a common context early in life suggests that the motor programming for the sounds is present before they are needed, as are human speech sounds in the babbling of infants.

The pup sounds recognized by Coscia provide the foundation from which the adult vocal repertoire emerges relatively early in life. At 1–3 weeks, before the development of directed social behavior, pup signals primarily serve to indicate distress. Other sounds probably exist in neural form early on, but are not expressed or, if expressed, occur out of context. After 3 weeks of age, however, these vocal signals quickly assume their adultlike roles. The switch between the neonatal scream and the adultlike howl is a case in point. Screams are effective within the den, where distress is typically caused by trampling by the mother, who is also the intended receiver. But several weeks later, when pups become increasingly mobile around the den, a scream, with its shorter range, is not as effective as a howl for a pup whose distress is caused not by trampling, but by separation.

Thus by the time it emerges from the den at about a month of age, a pup can produce the full range of vocal signals used by adults. However, it will take another 6 months or more before the pup "grows" into its voice and fully sounds like an adult. The fundamental frequency of howls, for example, drops from an average of about 1,100 Hz at 2 weeks (Coscia 1995) to about 350 Hz by 6–7 months of age (Harrington and Mech 1978b). Until then, the higher pitch (and shorter duration) of the pup's howls, and indeed all of its vocalizations, provide metainformation, revealing its age to its audience.

The Adult Vocal Repertoire

Schassburger (1978, 1987, 1993) has provided the most detailed description of the adult wolf vocal repertoire (table 3.2). He studied captive wolves in three enclosures, classifying sounds by ear. From this aural grouping he selected "representative" series, and then analyzed their context, in order to surmise the meanings and functions of each sound.

Schassburger divided the wolf's vocal repertoire into harmonic sounds and noisy sounds, with the whine and the growl being the focal sounds for the respective groups. These groups separate sounds used in friendly and submissive contexts (harmonic sounds) from those used in aggressive and dominance contexts (noisy sounds), paralleling Cohen's and Fox's (1976) division between attracting and repelling calls, and consistent with Morton's (1977) motivation-structural (M-S) rules.

Morton's M-S rules propose that the motivation behind a vocalization influences its physical structure. Sounds conveying nonaggressive motivation (affiliation, submission) should be higher-pitched and purer-toned than sounds motivated by aggression. These differences reflect the relationship between body size and both competitive ability and vocal capacity. Small individuals (i.e., young) are typically nonthreatening and possess small vocal cords; large individuals may be threatening and possess larger vocal cords. Small vocal cords under

TABLE 3.2. Descriptions of adult wolf vocal repertoires

| Sound type[a] | Description[b] | | | | | | | Acoustic properties of sounds (Schassburger 1993) | | |
	Young 1944	Joslin 1967	Theberge and Falls 1967	Mech 1970	Cohen and Fox 1976	Tembrock 1976a,b	Harrington and Mech 1978b	Nikol'skii and Frommolt 1989	Type	Duration (s)	Frequency range (Hz)
Whine	X	s	X	s	X	X	s	–	Harmonic	1–2+	440–7,500+
Whimper[c]	–	X	X	X[c]	X	X	X[c]	X	Harmonic	0.1–0.2	575–7,300+
Yelp	–	–	X	–	X	X	–	X	Harmonic	0.3–0.7	460–9,900+
Growl	–	X	X	X	X	X	X	X	Noisy	1–several	70–2,175
Snarl	X	–	–	–	–	s	s	–	Noisy	1–several	145–8,000+
Woof	–	–	–	–	X	s	–	X	Noisy	0.1–0.15	90–1,080+
Bark	X	X	X	X	X	X	X	X	Noisy	0.25–0.4	145–2,720+
Moan	–	–	–	–	X	–	–	–	Variable	tenths–5	75–2,795
Howl	X	X	X	X	X	X	X	X	Harmonic	up to 14	300–1,800

[a] Adapted from Schassburger 1993, with our modifications (see text).

[b] X = explicitly recognized as major category; s = explicitly recognized as subclass of major category; – = neither explicitly recognized nor implicitly described.

[c] Includes the "squeak," which Schassburger (1993) calls the "whistle whimper."

tension produce higher, purer sounds; large, relaxed cords produce lower, coarser sounds. Therefore, to convey friendliness, one should sound small; to convey aggressiveness, one should sound large. This pattern recurs throughout the wolf vocal repertoire, as indeed it does in most mammals and birds.

Schassburger's system is also graded rather than discrete, with subtle differences in message and arousal conveyed by variation among and within sound types. Mammals, in general, possess graded vocal systems, in which specific types of calls vary extensively (Marler 1965). This variation may result from individual differences (i.e., age, sex, mood, social status) as well as from context, and given sufficient study, may become predictable. But such variation may make it difficult to clearly categorize sounds, leading to the splitter-lumper dilemma. This has been true in studies of wolf vocal communication, with early workers (one of us included) taking the more conservative lumper's approach and recent researchers such as Schassburger making finer distinctions. This debate will ultimately be resolved when we learn where the animals themselves draw the line between signals. Until such time, the sound categorizations discussed below should be taken as hypotheses.

Short-Range Vocal Signals

Despite the preoccupation of humans with the howl, most wolf vocalizations are used at close range among packmates, where visual, tactile, and olfactory information is also available. One of the benefits of using auditory signals is the great degree of variation possible within each sound type, permitting wolves to convey subtle differences in mood and meaning. This is particularly so at close range because the effects of attenuation and distortion are nil; subtle differences can be conveyed clearly and unequivocally. A wolf can use the full range of its vocal cords, with each signal using a discrete portion of the possible vocal output.

Harmonic Sounds (Whimper/Whine/Yelp)

Wolves whimper, whine, and yelp in friendly or submissive contexts (table 3.3). There are two variants of each of these sounds: full forms with low tones, and whistle forms lacking low frequencies (Schassburger 1993). The whistle form of the whimper is often called the squeak (Crisler 1958; Fentress et al. 1978; Field 1978, 1979; Gold-

man 1993; Goldman et al. 1995). We do not know, however, whether these variants are discrete signals, or how they are produced.

Schassburger (1978, 1987, 1993) suggested that the high pitch of these sounds results from emphasis on the fifth or sixth harmonic (3,000–3,800 Hz) of a low-frequency (500–700 Hz) call. However, sonograms (Harrington and Mech 1978b, 111; Schassburger 1993, 16; Coscia 1995, fig. 9) indicate that the low and high components are not harmonically related, so they must be formed by independent mechanisms, perhaps vocally (low-frequency) and nasally (high-frequency) (Nikol'skii and Frommolt 1989) or by a laryngeal whistle mechanism similar to that used by rodents to produce ultrasound (i.e., Roberts 1975a,b). The low component often appears at the end of the sound or series of sounds rather than at the beginning (sonograms in Harrington and Mech 1978b; Schassburger 1987, 1993), suggesting that changes in vocal cord tension affect the relative content of the two components.

Schassburger (1978, 1987, 1993) proposed that the full forms of these sounds evolved to facilitate localization. Alternatively, whistle forms may be less threatening (Morton 1977), and may be used when maximizing attractiveness, as when an adult approaches a pup or a socially dominant animal approaches a subordinate.

Wolves often whimper and whine together in series and with overlap in context (see table 3.3). However, they whine more often during submission/appeasement and less often in greeting. They yelp largely in submissive situations involving physical contact, and probably in an attempt to terminate the interaction (see table 3.3). Wolves show long-term individual differences, as well as sex and age differences, in squeaks (Field 1978, 1979; Goldman 1993; Goldman et al. 1995; T. D. Holt and F. H. Harrington, unpublished data), and there is circumstantial evidence that young pups can distinguish the squeaks of their mother (Goldman 1993; Goldman et al. 1995).

Noisy Sounds (Growls/Snarls/Woofs/Barks)

Wolves growl and snarl in a wide variety of aggressive contexts, including assertion of dominance, during threat or attack, or as a warning or defense (Schenkel 1947; Joslin 1966; Fentress 1967; Fox 1971a, 1978; Schassburger 1978, 1987, 1993). Growls are the typical vocalization used during dominance interactions. Snarls, along with growls, occur more often in the latter two contexts

TABLE 3.3. Relative occurrences of short- and medium-range wolf vocalizations

Context	Yelp	Whimper	Whine	Whine-moan[b]	Moan	Growl-moan[b]	Growl	Snarl	Woof	Bark
					Occurrences[a]					
Pain, fear	484									
Greeting		355	7							
Frustration, anxiousness	3 ←	254 ←	143 ←	20 ←	14 ←	5				
				↕	↕	↕	↕			
Submission, appeasement		8	144 ←	11 ←	14 ←	7				
Dominance				1 →	11 →	25 →	443 →	1		
				↕	↕	↕	↕			
Threat, attack				3 →	17 →	55 →	432 →	80	5 →	6
				↕	↕	↕	↕	↕		
Warning, defense				25 →	38 →	54 →	180 →	14	70	18
Protest							1	52 →		51
Play		1	8 ←	11 ←	24 →	23 →	11			
Sexual arousal		71	34							

Source: After Schassburger 1993.

[a]Numbers represent occurrences in three packs observed for differing periods with varying degrees of sampling effort. Therefore, numbers should be used as rough guides to relative occurrence. Horizontal arrows represent changing "self-assertiveness," increasing to right, decreasing to left. Vertical arrows represent frequent overlap in context.

[b]These vocalizations were considered major classes by Schassburger (1993), and will be so distinguished here.

(see table 3.3). Growls constituted one-third of all the vocalizations Schassburger recorded in his captive studies.

Woofs are soft, low sounds that may be "nonvoiced" (i.e., not involving the vocal cords), providing a localizable short-range signal similar to a whisper. Wolves woof primarily in warning or defense (see table 3.3), as when humans approach summer homesites with pups (Nikol'skii et al. 1986). Pups respond to adult woofs by lying quietly in brush or returning to the den, whereas nearby adults may be alerted and recruited to defend against the threat (Schassburger 1978).

Barks are short, low-pitched sounds that clearly involve the vocal cords. Though described as harsh or noisy, barks recorded at close range and analyzed at high resolution have a rapidly modulated, chevron shape (Frommolt et al. 1988; Coscia et al. 1991; Coscia 1995). Barks probably vary in their relative harmonic/noisy quality with motivation, following Morton's (1977) motivation-structural rules: as an animal's aggressiveness increases, its barks should decrease in pitch and become more noisy.

Barks and woofs overlap greatly in context (see table 3.3). Wolves woof or bark when humans or other large animals approach dens or rendezvous sites closely (summarized by Harrington and Mech 1978b). In these situations, the wolves maintain distance, but do not flee, suggesting a motivational conflict between "fight" and "flight." The chevron form of the bark coincides with the sound structure expected by Morton (1977) in such a situation. Because it should also be relatively easy to localize (Marler 1955; Scott 1961; Harrington and Mech 1978b; Lehner 1978a,b), the bark probably evolved to draw visual attention to the vocalizer.

Intermediate Sounds (Moans)

Moans constitute a variable class of harmonic to noisy sounds that resemble a "low, prolonged sound of grief or pain" (Schassburger 1993, 19). They have not been reported from wolves in the wild. Schassburger distinguished three subclasses of the moan (harmonic, noisy harmonic, and intermediate) as well as two other related sounds (whine-moan and growl-moan). As no statistical evidence was presented to support these subdivisions as discrete signals, we will consider them all here either as graded variants of the moan or as mixed sounds (see below).

Wolves moan in many contexts (see table 3.3) that in Schassburger's (1978) view reflected ambivalence between self-assertion (aggression) and submission. He also thought the moan was both the developmental precursor of the wolf's vocal repertoire, as it dominates the newborn pup's repertoire, and the link between aggressive/noisy sounds and friendly/harmonic sounds. However, wolf pups make many sounds on their first day of life, and other adult wolf sounds show no developmental trends from simpler precursors (Coscia 1995).

Mixed Sounds

Schassburger (1978, 1987, 1993) also identified mixed sounds, continuous sounds with successive characteristics of two or three sound types (i.e., yelp-whine, whimper-yelp, bark-growl, woof-bark, bark-growl-snarl, etc.). Most combinations occurred among sounds within either the harmonic or the noisy groups, excepting the whine-bark. Schassburger believed that mixed sounds represented "functional summation," the expression of different but compatible messages, or served to grade "message intensity" (also proposed by Cohen and Fox 1976).

Long-Distance Vocal Signals

Long-distance vocal signals evolved under important constraints imposed by the environment through which the sound travels (i.e., Morton 1975; Marten and Marler 1977; Marten et al. 1977). Characteristics of the carrier medium (air), such as turbulence, temperature, humidity, precipitation, and noise, affect the fidelity and range of the vocal signal. Intervening structures, from landscape features (hills, valleys) to vegetation, further scatter and attenuate sounds. Although some of these effects may be independent of sound frequency, some of the most important causes of signal loss and distortion hit higher-pitched sounds harder than lower ones. Consequently, one general rule for maximizing signal distance while minimizing signal distortion and loss is to use lower and harmonically purer sounds.

The wolf's principal long-distance vocalization is the howl, which fits this rule as a lower-pitched, harmonically simple sound. The maximum range of detection by another wolf may be 10 km (6 mi) in forested habitat

(Harrington and Mech 1978b) and 16 km (10 mi) on tundra (Henshaw and Stephenson 1974), although the usual range is probably much less (Joslin 1967). But the gain in extended range comes at a cost. As we saw above, the wolf is able to exploit the full range of its vocal cords for short-range sounds, so that a rich variety of information can be clearly transmitted. Howls, on the other hand, are constrained in their structure and therefore are more limited in the variety or subtlety of information they can transmit. Variation in the message, following Morton's M-S rules, is still possible with howls; it is only that the range of variation is more limited.

Howls are harmonic sounds with a fundamental frequency of 150 Hz to more than 1,000 Hz for adults (Theberge and Falls 1967; Harrington and Mech 1978b; Harrington 1989; Tooze et al. 1990). The lowest frequencies usually occur briefly at the beginning or end of a howl, and the highest frequencies typically occur during the first half of the howl (Theberge and Falls 1967; Tooze et al. 1990). The mean pitch of adult howls varies between 300 and 670 Hz (Harrington and Mech 1978b; Tooze et al. 1990). Both pups and adults respond differentially to playback of adult and pup howls, replying to and approaching adult howls more readily (Harrington 1986).

The fundamental frequency and the first one or two harmonics of a howl are usually the strongest (Theberge and Falls 1967; Harrington and Mech 1978b; Tooze 1987; Tooze et al. 1990). The relative strengths within these bands may shift as pitch changes (Theberge and Falls 1967; Schassburger 1978), probably due to resonant features of the individual's vocal tract. Although wolf howls show up to a dozen harmonics (Theberge and Falls 1967), there is relatively little energy in the upper harmonics, so they attenuate rapidly with distance (Morton 1975; Marten and Marler 1977; Marten et al. 1977). "Spontaneous" solo howls are softer than elicited howls or howls given during choruses (Klinghammer and Laidlaw 1979).

Solo Howling

Single adult wolf howls vary from less than a second to 14 seconds in length (Theberge and Falls 1967; Harrington and Mech 1978b; Schassburger 1993), but mean durations range between 3 and 7 seconds (Theberge and Falls 1967; Harrington and Mech 1978b; Tooze 1987; Tooze et al. 1990). Single animals may howl in bouts up to 9 minutes long (Joslin 1966; Theberge and Falls 1967; Harrington and Mech 1978b; Schassburger 1978;

Klinghammer and Laidlaw 1979; Tooze 1987). Solo howling by captive wolves peaks during the winter breeding period and is most common during morning and evening (Schassburger 1978; Klinghammer and Laidlaw 1979).

The structure of a howl varies greatly within and among both individuals and social contexts (Theberge and Falls 1967; Harrington and Mech 1978b; Schassburger 1978, 1987, 1993; Tooze 1987; Harrington 1989; Tooze et al. 1990). Howl form (i.e., change in frequency modulation during the howl) and frequency characteristics (i.e., fundamental frequency, range of modulation, and harmonic content) differ consistently among individuals (Theberge and Fall 1967; Harrington and Mech 1978b; Tooze et al. 1990). Thus the information is available for wolves to identify one another, although we still need to determine how they do so. As the frequency composition of the howl changes with distance (Morton 1975), there may be a maximum distance over which individual recognition is possible.

Solo howls vary from scarcely modulated ("flat" howls) to highly modulated, sometimes with discontinuities in pitch ("breaking" howls) (Theberge and Falls 1967; Harrington and Mech 1978b; Tooze 1987). This variation may be related to general arousal, to whether the howl occurs early or late in a bout, or to whether the howling was elicited or spontaneous. Other examples of apparently systematic variation in wolf howls have been noted (Harrington and Mech 1978b; Harrington 1987; F. H. Harrington, unpublished data; Tooze 1987). For example, elicited howls from adult wolves that approached an apparent intruder were significantly lower and coarser than elicited howls from the same wolves when they did not approach (Harrington 1987). This finding suggests that changing levels of aggressiveness influence the structure of howls in the same way they do the structure of short-range vocalizations (Morton 1977).

Chorus Howling

Howls in choruses typically are even more variable than those from single wolves (McCarley 1978; Harrington 1989; Holt 1998). Wolves generally begin choruses with a relatively unmodulated howl, and other wolves join in at an accelerating rate. The fundamental frequencies of subsequent howls usually differ by at least 15 Hz from one another (Filibeck et al. 1982). Howls become shorter, higher, and more frequency-modulated as the chorus continues (Joslin 1966; Tooze 1987; Harrington 1989).

Joslin (1966) and McCarley (1978) described choruses ending with a bark, although highly modulated howls (i.e., yips and yaps) may sound like barking from a distance (F. H. Harrington, unpublished data). Holt (1998) found that in addition to howls (56% of all vocalizations), squeaks (36%), barks (7%), and growls (1%) were common vocalizations during choruses.

Wolf choruses last 30–120 seconds (Joslin 1966; McCarley 1978; Klinghammer and Laidlaw 1979; Harrington 1989), and possibly longer during the breeding season (Klinghammer and Laidlaw 1979; Servín[-Martínez] 2000). Their lengths may (Klinghammer and Laidlaw 1979) or may not (Harrington 1989) be related to pack size.

Schassburger (1978, 1987, 1993) distinguished two forms of chorus howling: harmonious, characterized by relatively little modulation of individual howls, and discordant, with a "more random, non-overlapping distribution of harmonics, as well as apparently random distribution of the sound energy" (1978, 95). Pup choruses, and choruses in association with pre- or post-chorus "rallies" (see Packard, chap. 2 in this volume) were frequently discordant, whereas "spontaneous" choruses without accompanying group activity were more harmonious. The above descriptions match McCarley's (1978) descriptions of spontaneous and elicited choruses in red wolves, as well as similar descriptions of chorus howling in coyotes (McCarley 1975; Ryden 1975; Lehner 1978a,b).

In our experience (Harrington and Mech 1978b; Harrington 1989), proximity to packmates seems to influence the structure of a chorus: when wolves were close together, their howling was discordant; when wolves were separated (by tens of meters or more), their howling was harmonious. This difference suggests that chorus structure may reflect the degree of physiological or behavioral arousal experienced by animals during the chorus, contrary to other hypotheses (Schassburger 1978, 1987, 1993; Lehner 1978a,b, 1982).

Functions of Howling

A variety of functions have been proposed for wolf howling: (1) reunion, (2) social bonding, (3) spacing, and (4) mating. Evidence for these functions ranges from speculative through anecdotal to quasi-experimental. It is obvious, however, that howling has functions both within and among packs.

Many researchers have noticed that wolves, when separated from their packmates, howl readily (Murie 1944; Young 1944; Fentress 1967; Theberge and Falls 1967;

Zimen 1981; Tooze 1987; Coscia 1995). Anecdotal accounts of separated wild wolves meeting after howling, or howling as they returned to summer homesites (Murie 1944; Mech 1966b; Rutter and Pimlott 1968; Peterson 1974, 1977; Dekker 1985), leave little doubt that howling helps coordinate movements among separated packmates. The likelihood of individual recognition (Theberge and Falls 1967; Tooze et al. 1990) would allow wolves to distinguish between packmates and strangers.

Evidence for a second commonly proposed function, the strengthening of social bonds (see Packard, chap. 2 in this volume), is more equivocal. This function has probably been proposed for intuitive reasons: chorus howls are highly contagious events within the pack (Crisler 1958; Mech 1970; Zimen 1981); wolves that have been expelled from the pack or are being "suppressed" by others usually do not join the chorus (Zimen 1981); packmates typically engage in a variety of often vigorous social activities before, during, or after a chorus howl (Murie 1944; Crisler 1958; Zimen 1981); and, since wolves are highly social, with strong ties to packmates, many people think that such a spirited activity must strengthen bonds among individuals.

Unfortunately, no objective evidence exists with which to test this assertion. Furthermore, individual wolf packs vary widely in their howling rates (Voigt 1973; Harrington and Mech 1978a; Joslin 1982). Does this variation result in differences in social cohesion within packs? Until evidence of a relationship between chorus howling and some objective measure of social bonding is produced, we must consider this role a hypothesis to be tested. However, whatever form of arousal wolves experience during chorus howling may serve to ensure that wolves howl when appropriate to fulfill other important functions, such as reunion or spacing.

The role of howling in spacing packs (see Mech and Boitani, chap. 1 in this volume) has been explored in some detail (Joslin 1967; Theberge and Falls 1967; Harrington 1975, 1987, 1989; Harrington and Mech 1979, 1983). In general, packs' responses to imitation wolf howls or playbacks of strangers' howls are consistent with a spacing function: packs that reply generally remain where they are; packs that do not reply often retreat (Harrington and Mech 1979). Packs are more likely to reply and stand their ground if they are at a relatively fresh kill or if they are accompanied by relatively small pups, both important resources critical to reproductive success. Furthermore, factors that stack the odds in favor of a pack in a confrontation, such as pack size, and factors correlated with increased aggressiveness, such as hormone levels during the breeding season, increase the probability of a reply (Harrington and Mech 1979). Observations of packs avoiding one another following vocal interactions (Rutter and Pimlott 1968; Harrington 1975) further support a role in spacing.

The precise role that howling plays in spacing, however, may be distorted by human preconceptions of the concept of territory. A pack's response to strangers' howling shows no relationship to territory boundaries (Harrington and Mech 1983). In addition, some tundra packs are nomadic much of the year (Stephenson and James 1982). Thus, it may be more appropriate to consider inter-pack howling an avoidance mechanism than a "territorial" mechanism (see also Jaeger et al. 1996).

Mech (1970) proposed that packs could assess one another's size through their chorus howls, but Harrington (1989) considered this unlikely on both theoretical and empirical grounds. As packs are generally competitors and vary in size relative to their neighbors from year to year and even day to day, and as they may have no prior knowledge about their competitors' pack size, we would expect them to benefit more often than not from not providing clear information on their true size.

An analysis of chorus howls found no reliable information on pack size (Harrington 1989). Furthermore, humans often overestimate the size of packs howling *at a distance* (Eckels 1937), probably due to the differential echoes of highly modulated howls in a chorus (Harrington 1989). Filibeck et al. (1982) reported a method to determine pack size from harmonious choruses recorded at close range in captivity. The technique failed for modulated choruses, however, and has not been tested in the wild. If a similar misperception occurs in the canid auditory system, then chorus howls may often provide information that leads to an overestimate of the pack's actual size, similar to the Beau Geste hypothesis (Krebs 1977).

Captive wolves not integrated into packs often howl "spontaneously" during the breeding season (Klinghammer and Laidlaw 1979; Zimen 1981). Harrington and Mech (1979), however, found that wild lone wolves never replied to simulated solo howling, although they sometimes did approach very closely. These findings suggest that lone wolves may use howling to locate potential mates, with the relative danger of vocalizing determining whether they reply or approach silently.

Variation in Vocal Signals

The way in which wolves express their changing state of arousal may vary among sound types, but in general it is expressed by changes in pitch, loudness, duration, or rate of frequency or amplitude modulation. For example, within growls, a decrease in pitch or an increase in duration or loudness may represent an increase in aggressiveness. Discontinuities between sound types underlying similar motivational states (i.e., growls-snarls, whimpers-whines-yelps, woofs-barks) may arise at specific thresholds in the same manner that specific gaits (i.e., walking, trotting, galloping) replace one another over a limited range of speeds (Stewart and Golubitsky 1992). Growls and squeaks, for example, represent opposite motivations, and their structures are antithetical. Moans, whines, and yelps represent steps along a pain/fear (distress) axis, becoming higher, shorter, and more modulated as distress increases. Growls and snarls represent increasing aggressiveness as conflicts escalate from threat to action. Woofs and barks indicate increasing aggressiveness or assertiveness as the wolf switches from a muted warning call to a louder confrontational sound.

For the wolf, the bark is an intermediate-range vocalization. The concentration of energy in harmonic bands should extend its range beyond that of a broadband noise (i.e., growl), while the extensive modulation makes the sound easier to localize (Marler 1955; Konishi 1973). With short-range vocalizations, attenuation and noise are unimportant, so the entire range of frequencies and durations permitted by the wolf's vocal tract can be exploited. Specific signals occupy discrete portions of this range with minimal ambiguity and maximal dynamic range. With increasing transmission distance, spreading and interference markedly attenuate sounds, especially higher or noisier ones, relegating long-distance calls to lower, purer-toned, harmonic sounds (Morton 1975; Marten and Marler 1977; Marten et al. 1977). The howl thus acts as the long-distance medium upon which varying messages may be carried, albeit with greater ambiguity and less subtlety, representing a trade-off between transmission range and information content.

The long, pure, low harmonic structure of howling is well suited for long-distance communication. For dogs (Baru 1971) and most mammals (Fay 1992), the ability to detect a pure tone increases as the duration of the sound increases ("temporal summation") up to an optimal du-

ration dependent on pitch (Green 1973). At 1,000 Hz, this optimum duration is about 100–200 ms in dogs, but at 250 Hz, near the fundamental frequency of adult wolf howls, it is at least 1,000 ms (Baru 1971). Thus howling wolves should achieve the greatest effective range by sustaining a frequency for at least a second before changing pitch, which many howls do (Harrington 1975; F. H. Harrington, unpublished data).

The frequency range of howling (150–780 Hz: Theberge and Falls 1967) neither coincides with the "best frequency" of the canid auditory system nor fits the frequency "window" for optimal range (Morton 1975). However, this frequency window is caused largely by a "ground effect" on sounds broadcast and heard less than a meter above the ground (Marten and Marler 1977; Marten et al. 1977). For calls broadcast more than a meter above the ground, this "window" disappears, and the lower the call, the longer the range (Marten and Marler 1977; Marten et al. 1977; Waser and Waser 1977). Wolves howl at this height threshold, so any means to avoid the ground effect (e.g., standing or sitting while howling rather than lying, pointing muzzles above the horizon, using higher elevations) should help maximize range. One unquantified aspect of howling involves the relations among motivation, posture, and immediate habitat and the structure, amplitude, and pitch of howls, all of which should be important influences on range.

When a wolf utters a growl or whimper or bark, there is usually little uncertainty about its intended target. But when a wolf howls, it usually does not know what wolves might be listening. This uncertainty has probably influenced the form of the howl in a variety of ways. For example, given the hostility that generally occurs between wolves of different packs, wolves might limit the amount of information made available by a howl. When packs see each other, a confrontation often follows rapidly, with one pack pursuing the other (see Mech and Boitani, chap. 1 in this volume). These confrontations may happen so quickly because visual contact removes most of the uncertainty about the other pack (i.e., its size, number of adults, confidence, etc.). If a chorus readily provided all this information, packs ought to be just as likely to seek out and attack one another following a howling interaction. The relative rarity of confrontations following howling interactions (Harrington and Mech 1979) suggests that uncertainty about the other group makes wolves more cautious.

Similarly, a wolf separated from its packmates might

reach an unintended, hostile audience by howling. The observation that solo howls are often relatively quiet (Klinghammer and Laidlaw 1979) suggests that a wolf may limit the range of its howls to avoid alerting strangers. Furthermore, several sets of evidence suggest that wolves separated from their pack, or wolves interacting with strangers, might modify their howls in advantageous ways. First, pack wolves that had been interacting vocally with an apparent stranger at a distance lowered the pitch of their howls when later approaching it (Harrington 1987). These wolves used a higher howl during the initial interaction, and resorted to a lower (and presumably more threatening) howl only when the stranger continued to howl. Second, on other occasions, the howls initially elicited from separated pack members were little modulated; these howls were sometimes extremely difficult to localize (F. H. Harrington, unpublished data). Finally, on one occasion, two pups howled for at least an hour from an abandoned rendezvous site. When a packmate eventually replied, their howls changed in form and seemed to become much louder than before (F. H. Harrington, unpublished data). These observations suggest that in times of uncertainty, wolves may use a "poker" howl that limits the amount of information that might be made available to unintended ears.

The Auditory System

Effective communication with sound requires an auditory system that can both receive and decode sound signals. Unfortunately, little is known about the wolf's auditory abilities. Dogs, however, appear to have relatively ordinary hearing abilities compared with other mammals (table 3.4) (Masterton et al. 1969). Their sensitivity at low frequencies (< 1 kHz) is generally similar to that of cats and humans (Dworkin et al. 1940; Lipman and Grassi 1942; Neff and Hind 1955; Masterton et al. 1969; Baru 1971; Heffner 1983). At higher pitches, dogs become increasingly more sensitive than humans, and cats become even more sensitive than dogs (table 3.4).

The upper frequency limit of hearing in four dogs did not vary with their size (4.3–45.5 kg or 9.5–100.1 pounds), indicating that high-frequency sensitivity may be a species-typical characteristic (Heffner 1983) (see table 3.4). Peterson et al. (1969) found upper frequency limits of 60 kHz for dogs, 80 kHz for coyotes, 65 kHz for red foxes, and 100 kHz for domestic cats, suggesting systematic differences among species even as closely related and similar in size as dogs and coyotes.

The role of the pinna (external ear) in the dog or wolf is not yet clear. Two studies found no systematic effects of pinna type (upright, partly upright, flopped) on sensitivity (Peterson et al. 1966; Heffner 1983). Its role in sound localization is also unknown, although Heffner and Heffner (1992c) have suggested that movable pinnae could serve as directional noise filters, allowing an animal to amplify a weak signal without moving its head. Wolves point their ears when alert and facing some point of interest. The spatial acuity (ability to localize sound) of dogs is similar to that of cats, but nearly an order of magnitude poorer than that of humans (Heffner and Heffner 1992a,b) (see table 3.4). Goats, horses, and cattle have very poor ability to localize sound (18°, 25°, and 30°, respectively), yet all have very mobile pinnae (Heffner and Heffner 1992a,b). Heffner and Heffner (1992a,c) suggested that the primary function of sound localization is to direct the field of best vision toward the sound source. Thus species with small areas of best vision (i.e., humans with foveae) require high spatial acuity, whereas species with broad bands of best vision (i.e., ungulates with pronounced visual streaks) need only mediocre spatial acuity. The modest abilities of the canid at localizing sounds may be sufficient to direct its best vision (area centralis, see below) toward the sound source.

A dog's ability to detect a change in pitch is better for immediate than for delayed comparisons, and worsens for sounds shorter than 100–200 ms (Baru 1971) (see table 3.4). At 1 kHz, dogs are capable of discriminating a change in frequency of 8–10 Hz (1%), as compared with the human threshold of 3 Hz (0.3%) (Sinnott and Brown 1993).

At birth, dog and wolf pups are deaf, evidently because the ear canals are closed (Strain et al. 1991). Once they open at about 12–14 days of age (Mech 1970; Foss and Flottorp 1974; Breazile 1978; Strain et al. 1991), hearing matures rapidly and reaches adult sensitivity (35 dB) by day 20 (Strain et al. 1991). The auditory cortex attains adult complexity at about 4–5 weeks (Breazile 1978).

Dog pups respond earliest (days 13–16) to frequencies between 250 Hz and 750 Hz, slightly later (days 13–23) to frequencies between 1,000 Hz and 4,000 Hz, and latest (days 20–22) to 8,000 Hz, the highest frequency tested (Foss and Flottorp 1974). Four-week-old beagle pups are sensitive to tones up to 50 kHz, whereas 4-month-old beagles are sensitive only up to 25 kHz (Peterson et al. 1966). Dog pups can localize pup whines and adult female barks as early as 13 days of age (Ashmead et al. 1986).

TABLE 3.4. Characteristics of the auditory systems of various mammals

Species	Auditory range (at 60 dB)	Range of best sensitivity	Best frequency	Lowest threshold	Low-frequency sensitivity (at 1 kHz)	Area of audible field	Spatial acuity	Pitch discrimination (at 1 kHz): Immediate	Delayed
Dog[a]	67 Hz–45 kHz (4)	500 Hz–32 kHz (4) 200 Hz–15 kHz (11)	2 kHz (9), 4 kHz (8) 8 kHz (4)	−7 dB (8)	8 dB (8)	430 dB by octave (8)	4–8° (3,6)	1% (1)	2–3% (1)
Cat	48 Hz–85 kHz (5,10)	250 Hz–45 kHz (10)	8 kHz (9)	−17 dB (8)	−6 dB (8)	700 dB by octave (8)	4–12° (3)	1% (2)	—
Human	20 Hz–18 kHz (9)	400 Hz–12 kHz (10)	1 kHz (10)	−7 dB (8)	6 dB (8)	480 dB by octave (8)	<1° (3,6,12)	0.3% (12)[b]	—
Ungulates[c]	Up to 42 kHz (9)	—	10 kHz (9)	−11 dB (8)	14 dB (8)	460 dB by octave (8)	5–30° (6)	—	—

Sources (in parentheses): 1, Baru 1971; 2, Elliott et al. 1960; 3, Fay 1988; 4, Heffner 1983; 5, Heffner and Heffner 1985; 6, Heffner and Heffner 1992c; 7, Heffner and Heffner 1992b; 8, Lipman and Grassi 1942; 9, Masterton et al. 1969; 10, Neff and Hind 1955; 11, Peterson et al. 1969; 12, Sinnott and Brown 1993.

[a] Presumably the wolf's system is similar to the dog's.

[b] at 500 Hz and 4 kHz.

[c] Includes pig, sheep, cattle, and horse.

Olfactory Communication

The Importance of Olfaction in Communication

Olfaction, the sense of smell, is probably the most acute of the wolf's senses. Wolves, unlike humans, are strongly reliant on odors to acquire information about the outside world—for example, about food or danger—and to communicate with other wolves. The strength of the wolf's drive to investigate the environment with its nose was demonstrated in a study of adult wolves made anosmic (unable to smell) (Asa et al. 1986). Both males and females continued for years to sniff objects as if seeking olfactory information, despite their failure to find it.

In many respects, acoustic and olfactory signals play complementary roles in wolf communication. At close range, and particularly at night, these two kinds of signals provide wolves with different "pictures" of one another: vocal signals best indicate the immediate state of an individual's changing mood and probable future actions over the next few seconds or minutes, whereas odors provide a deeper and broader window into the animal's history, present status, and future prospects over a much longer time frame. Thus, while a female's urine may inform a male that she is physiologically ready to mate and conceive, her voice may inform him that now is not the time. Over distance, the time differential between acoustic and olfactory signals is magnified. Howling permits wolves to know one another's locations, identities, and moods in real time, whereas scent marking permits the same information, and more, to be "archived" for periods of days or even weeks.

Odors may contain information on species or individual identity, gender, breeding condition, social status, emotional state, age, condition, and even diet. Olfactory signals are produced by the wolf's entire body, both inside and out. Although much has been learned about scent marking by wolves, there is still much we do not know. However, a vast literature on domestic dogs provides additional details and insight into the olfactory workings of these closely related animals.

Signal Sources

Olfactory signals are much harder for humans to discern than are sounds or visual displays, primarily because the human olfactory sense is relatively poorly developed. Odors are not easily captured and often leave little evidence of their presence that we can detect directly. Although some forms of olfactory communication are accompanied by conspicuous behaviors, such as a raised leg during scent marking, many are known only from the interest that wolves direct toward odor sources with their noses. In the following sections, we will survey the various odor sources most likely to play a role in wolf communication.

Skin Glands

The secretions of the wolf's skin glands not only keep the skin supple and hydrated, but also function in chemical communication. There are three basic types of secretory skin glands: sebaceous, apocrine, and eccrine. Sebaceous glands are typically found in hair follicles, except those in the hairless preputium (Sokolov 1982). They produce an oily, waxy substance called sebum that, when acted on by bacteria, emits distinct odors. Increased gland size and sebum production are stimulated by androgens, whereas estrogens retard production (Albone 1984). In dogs, large sebaceous glands associated with hair follicles are found along the dorsal part of the neck, back, and tail, especially in the supracaudal tail gland area. However, those present at the junction of the skin and mucous membranes of the lips, vulva, and eyelids are even larger (Al-Bagdadi and Lovell 1979).

Apocrine sweat glands have been detected microscopically in the skin of dogs (Aoki and Wada 1951), so wolves also are likely to possess them. They are most numerous on the face, lips, and back and between the toes. Dog apocrine glands are typically associated with hair follicles and, although they first become active at puberty, appear not to be under the control of sex steroid hormones after that time. Their watery secretion is not used for cooling, but is under autonomic control (Aoki and Wada 1951).

Eccrine glands, the true sweat glands that function primarily for cooling, are not associated with hair follicles, but secrete salty fluid directly onto the skin. In dogs they are located only on the footpads (Nielsen 1953). Their secretion is influenced predominantly by exercise or heat stress, but can be stimulated by the involuntary nervous system (Albone 1984).

Although fresh secretions are odorless, microflora in the ducts or on the skin act on the products of sebaceous and apocrine glands to produce odorous compounds (Nicolaides 1974). Individual differences in microflora and in diet result in the functional equivalent of chemical "fingerprints" (Halpin 1980). Thus, it should be possible for wolves to recognize one another by distinctive "odor fingerprints" (Brown and Johnston 1983). In fact, domestic dogs not only can identify individual humans by odor, but can even distinguish between identical twins that are eating different diets (Kalmus 1955). In addition, skin gland activity can vary seasonally, as changing hormone levels result in changes in gland size and in the composition of secretions (Adams 1980).

Feet

Apocrine sweat glands are present in the "webs" of the paws, in the form of small glandular pockets near the bases of the toes of dogs (Speed 1941; Nielsen 1953) and probably on those of other canids as well. Eccrine sweat glands, in contrast, are numerous in the footpads (Nielsen 1953; Lovell and Getty 1957). The stiff-legged scratching that sometimes follows urine marking and defecation by breeding wolves may lay down an additional scent mark from the paws (Peters and Mech 1975b; Paquet 1991a). A scent-marking function for such scratching has also been suggested for coyotes (Young and Jackson 1951; Barrette and Messier 1980; Paquet 1991a) and dingoes (Corbett 1995). In addition to scent, scratching leaves obvious visual marks that may draw attention to the associated urine or fecal deposit.

Back and Tail

Little attention has been paid to the skin on the backs of canids in terms of signal value. Wolves, for example, have contrasting color markings across the shoulders, and the relatively longer guard hairs covering that area are raised during periods of arousal. Also called "hackles," these raised hairs convey an obvious visual signal that can be recognized even by humans. There are clusters of sebaceous and apocrine glands at the bases of the follicles of these hairs (Al-Bagdadi and Lovell 1979). The skin tends to lie in folds when the hairs are not erect, providing a microclimate for bacterial action on the glandular secretions. Bacteria can break down larger molecules into smaller, more volatile forms (Brouwer and Nijkamp 1953). During piloerection, the skin folds are spread, forcing the gland contents up the ducts and allowing dispersion of accumulated volatiles (Al-Bagdadi and Lovell 1979; Albone 1984). This passive release of scent has been little studied and may indeed play a more important role in communication than previously suspected (see Eisenberg and Kleiman 1972). In the case of canid piloerection, the wolf's emotional state, especially alarm or aggression, may be signaled, perhaps as in the fright-induced release of the black-tailed deer's metatarsal scent (Müller-Schwarze 1971).

Clusters of sebaceous glands, together with apocrine glands, make up the supracaudal or dorsal tail gland found on the top surface of the tail about a third of the way from its base (Hildebrand 1952). Previously called the violet gland because of its scent in the fox, in which it was first described, this glandular area is distinguished by black-tipped guard hairs and the absence of underfur. Although wolves do not typically investigate this area on one another, Ewer (1968) contended that its secretion is rubbed onto the roof of den entrances. However, Hildebrand (1952) proposed that the secretion may contain information about individual identity.

Ears

The olfactory significance of sebaceous secretions from the ears is not clear. In a study of domestic dogs (Dunbar 1977), males showed little interest in ear wax samples, but investigated the ears of females twenty times as frequently as those of males. These results suggest that secretions other than, or in addition to, ear wax are involved, and that the information they contain is related to gender.

Anal Sacs

The paired anal sacs just inside the wolf's anus are invaginations of the skin that contain both apocrine and sebaceous glands (Montagna and Parks 1948). They are surrounded by a muscle layer that is under voluntary control. The components of anal sac secretions vary by season and between males and females whether they are castrated or ovariectomized, suggesting that this mode of communication can supply information about gender and reproductive state (Raymer et al. 1985). Microbial action influences the relative composition of volatile anal sac secretions (Raymer et al. 1985), as has been demonstrated for some other canid species (Albone and Perry 1976; Preti et al. 1976).

The roles of anal sac secretions in wolf communication may be varied. The common greeting position, in which two individuals stand head to tail, suggests an interest in anal sac odors. The more dominant animal holds its tail away from the body, and the more subordinate holds its tail close, covering the anal area (Mech 1970). However, it also is possible that the odorant involved in this behavior is generated by the circumanal glands, sebaceous glands associated with hair follicles surrounding the anus (Parks 1950). The possible involvement of circumanal gland or anal sac secretions in reproductive communication is suggested by the observation that female wolves rarely investigate another wolf's anal area except during the breeding season (C. S. Asa, unpublished data).

It was long thought that anal sac secretions were passively deposited on feces during defecation. However, in a study of wolves whose anal sac contents were labeled with an inert dye, anal sac secretions were found on a surprisingly small percentage of feces (Asa, Peterson et al. 1985; Asa, Mech, and Seal 1985). In addition, the presence of secretions was not related to the firmness of the stool. However, adult males, especially the breeding male, deposited anal sac secretions more frequently when defecating than did females or juveniles, indicating a scent-marking function.

Another function of anal sac secretions has been described for the domestic dog, in which the sac contents are forcibly expelled during acute stress (Donovan 1969). A cornered male captive wolf also sprayed anal sac contents, and the presence of anal sac secretions on the floor of the enclosure was confirmed (Asa, Peterson et al. 1985).

The odor of sprayed anal sac secretions seems much more intense than that of secretions on feces (C. S. Asa, unpublished data). It may simply be that more secretion is sprayed than is applied to feces, or the most pungent elements may be the most volatile and disappear rapidly. Indeed, the odor does dissipate rapidly (in about 15 minutes), as judged by the human nose, from spots sprayed by domestic dogs (C. S. Asa, unpublished data). Thus secretions on feces investigated after that amount of time may be less pungent. It is also possible that different components are contained in the spray expelled during acute stress. Thus the message in anal sac spray (alarm) may be different from that on feces.

Conceivably, the same type of tissue that comprises the anal glands may make up the mid-dorsal line on wolves, for a tumor in that area on a captive wolf was histologically similar to anal gland tissue (L. D. Mech, personal communication).

Preputial Glands

The preputial glands that line the opening of the penile sheath and the vulval folds are a likely source of sexual odors in wolves, as has been described for boars (Dutt et al. 1959). The production of preputial gland secretions in male dogs is stimulated by androgens and inhibited by estrogens (Sansone-Bassano and Reisner 1974; Van Heerden 1981), a finding that implicates these secre-

tions in reproductive communication. Their composition may be modified by bacterial activity as they adhere to the hairs at the tip of the penile sheath (Ling and Ruby 1978).

Females sometimes sniff and lick the preputial area during breeding activity, but more commonly, male wolves have been seen standing over subordinate males, apparently presenting this area for investigation (Mech 2001b; C. S. Asa, unpublished data). This behavior suggests that the more dominant male may be advertising his hormonal condition or reinforcing his social position relative to the subordinate male. Nevertheless, similar "standing over" has been observed between wolves of all combinations of gender and age (Mech 2001b).

Preputial secretions of male mice and pigs are attractive to females of those species (Signoret 1970; Bronson and Caroom 1971) and elicit aggression from other males in mice (Jones and Nowell 1973). There are no data regarding differential reactions by gender in wolves. Preputial secretions may also play a role in scent marking, because they are washed into the urine and so incorporated into the urine marks of dogs (Huggins 1946).

No studies of wolves or other canids have considered the possible significance of preputial gland secretions in females. In female rodents, however, estrogen stimulates preputial gland secretions that are attractive to males (Pietras 1981). The secretions produced by these glands (associated with the clitoris) in estrous female wolves may contribute to the attractiveness of the perineal area to males. Treatment of female wolves with testosterone resulted in clitoral enlargement and production of a copious exudate resembling male preputial secretions (C. S. Asa, unpublished data).

Although preputial gland secretions are most apparent in adult male wolves, these glands appear to be active in young pups as well. The preputial area is noticeably swollen in both male and female pups and secretes a creamy fluid held on long, wispy hairs. These secretions are attractive to female wolves, who lick the area and stimulate urination (C. S. Asa, unpublished data). The swelling gradually recedes during the first month or two of life, its size perhaps being inversely related to the frequency of spontaneous urination.

Vagina

The wolf vagina and uterus also secrete substances that play a role in reproductive communication. Under estrogen stimulation during proestrus and estrus (see

Packard, chap. 2, and Kreeger, chap. 7 in this volume), blood flows from the uterus through the vagina, where it incorporates vaginal secretions. This proestrous sanguinous discharge lasts an average of 6 weeks, as assessed by vaginal swabbing (Asa et al. 1986). This discharge communicates the reproductive state of the female wolf and is obviously attractive to males even from considerable distances. Males eagerly sniff and lick the vulval area of estrous females (see Packard, chap. 2 in this volume).

Male interest in urine may also be influenced by vaginal secretions. Because of the position of the urethra within the vaginal canal, vaginal secretions are undoubtedly incorporated into urine. Male dogs show an unequivocal preference for vaginal secretions of estrous versus nonestrous females (Beach and Merari 1970; Doty and Dunbar 1974). A compound isolated from the vaginal secretions of estrous female dogs, when applied to anestrous or spayed females, elicited sexual arousal and mounting attempts by males (Goodwin et al. 1979). However, even though the same compound was identified in the urine of a female wolf during the breeding season, the compound did not elicit a response when presented to a male wolf (E. Klinghammer, personal communication). This finding suggests that other components of the vaginal secretions or the context—that is, other features of an estrous female—are important sources of information in wolves.

Saliva

Saliva may also contain information on gender or reproductive state (e.g., Booth 1972; Booth et al. 1973; Doty et al. 1982). Male dogs more frequently sniff and lick the muzzles of females than of males (Dunbar 1977). Saliva contains high concentrations of hormones (progesterone: Luisi et al. 1981; testosterone: Baxendale et al. 1982; cortisol: Walker et al. 1978). The ritualized muzzle-licking performed by pack members (see Packard, chap. 2 in this volume) could not only convey olfactory information, but might also transfer small amounts of hormone. Although this behavior has not reported for wolves, red foxes draw branches through their mouths, presumably to anoint them with saliva (Macdonald 1979). Alternatively, muzzle-licking may relate to sebaceous secretions on the lips.

Social grooming also transfers saliva, possibly accomplishing olfactory marking of the partner. Transfer of saliva is perhaps most common between mother and young. The mother regularly licks her pups, both to clean them and to stimulate urination and defecation, leaving traces of her saliva behind. The pups also leave saliva on the mother's nipples, perhaps marking a nipple with an odor signature and providing the basis for "nipple fidelity." Further evidence for the role of saliva during the development of social behavior comes from studies of the Mongolian gerbil. Block et al. (1981) found that chemical messages contained in saliva facilitate the formation of social bonds between mother and young, among siblings and peers, and between sexual partners. During greetings that involve head contact, volatile substances excreted through the lungs may also mediate olfactory communication (Whitten and Champlin 1973).

Feces

Wolves may use feces, with or without streaks of anal sac secretions, in territorial marking (Peters and Mech 1975b; Vilà et al. 1994). Captive wolves left most feces near the gate where keepers entered their enclosure, which was possibly perceived as a point of intrusion (Asa, Peterson et al. 1985). In the wild, feces are more common along trails and roads, particularly at junctions (Peters and Mech 1975b; Vilà et al. 1994). This evidence, together with the observation that feces often are placed on conspicuous objects, implies that they are used in scent marking.

Similar conclusions have been reached for other canids (e.g., dingoes; Corbett 1995). A more dramatic example of intentional placement is shown by some male dogs (and fewer females) that use a type of "handstand" to elevate the hindquarters during defecation, ostensibly to deposit feces on a vertical object such as a tree or fence (Sprague and Anisko 1973).

Urine

Although glandular secretions, feces, and saliva may provide important modalities for information exchange, the best-studied means of olfactory communication in canids is urine marking. In fact, canid scent marking is often considered synonymous with urine marking. And though numerous reviews of urine marking have debated its function, most agree that spacing is its primary function for most species (for reviews see Ralls 1971; Eisenberg and Kleiman 1972; Johnson 1973; Gosling 1982; Doty 1986).

Kleiman (1966) postulated that scent marking may be a response to unfamiliar or frightening surroundings. According to this theory, the animal creates familiarity and reassures itself by permeating the area with its own scent. Such fear-induced urination may have provided the initial conditions for the evolution of a urine-based spacing system, and may still be an important proximate factor stimulating the production of scent marks.

Dogs (Hart 1974a; Bekoff 1979b), coyotes (Bowen and Cowan 1980; Wells and Bekoff 1981), and wolves (Peters and Mech 1975b) mark more frequently or vigorously in areas of intrusion where the scent of conspecifics, or even of other canid species (Paquet 1991a), is present. Male dogs also mark more frequently in unfamiliar than familiar areas (Bekoff 1979b). Wolves may also avoid areas with alien scent (Peters and Mech 1975b; Peterson 1977). Whether driven by uneasiness, fear, or aggression, scent marking can reduce the risk of injury and reduce the energetic cost of aggressive encounters by allowing individuals to assess one another or the situation and perhaps avoid combat (Gosling 1982). Wolves mark more frequently along the boundaries of their territories than in the interior, creating an "olfactory bowl" (see Mech and Boitani, chap. 1, fig. 10 in this volume; Peters and Mech 1975b).

The posture that canids use for marking with urine also conveys a visual message, making it easy to recognize, even for humans. Everyone has seen a male dog raise its leg to urinate, usually on a vertical object. For male dogs, just the sight of another male in the raised-leg urination (RLU) posture is sufficient to elicit urination (Bekoff 1979b). The function of the RLU in marking is probably to place the scent so as to maximize the chance of its detection by conspecifics (i.e., at nose level). For dogs, the height of the scent mark may reflect the stature of the marker (Lorenz 1954), which may explain why some dogs sometimes overreach themselves, rotating their bodies to raise their legs so high they almost topple over.

Well studied in the dog, the development of this raised-leg posture in the male is under testosterone control, first appearing during puberty (Beach 1974; Ranson and Beach 1985). Urine marking with a raised leg is less common among female than among male dogs (Sprague and Anisko 1973), although breeding female wolves regularly mark year-round with flexed-leg urinations (FLUs) (Peters and Mech 1975b; Asa 35 al. 1985, 1990). The typical raised-leg urination posture of the female, the FLU, is somewhat different from the RLU of the male. One leg is lifted slightly forward, or flexed, while squatting, a position that restricts placement of the scent

mark. In studying females, it may be more relevant to determine whether or not the urination is "oriented" than to rely on detecting the sometimes subtle FLU.

In a detailed study of the ontogeny of urinary patterns, domestic dog puppies required maternal stimulation to urinate during the first few weeks of life (Ranson 1981), corresponding to the wolf denning period. Dog pups began spontaneous urination at about 3 weeks of age, the age when wolf pups begin emerging from the den (Mech 1970). This pattern suggests that urinary development in the dog has been conserved and reflects the ancestral function of maintaining a clean den. The juvenile urination postures are similar in dogs and wolves: males stand, slightly stretched, and females squat.

The adult pattern, however, is much more complex in wolves. The onset of urine marking at puberty is similar to that of the dog, but is expressed only by dominant males. Subordinate male wolves continue to use the juvenile standing posture throughout adulthood (Asa et al. 1990), but exceptions may occur if they challenge the dominant male. This observation suggests that virtually all male dogs consider themselves dominant.

Testosterone is necessary, but not sufficient, to permit raised-leg urination in male wolves. That is, the interaction of social status and the male hormone is required to activate the behavior. Interestingly, though, the exposure to testosterone is required only during fetal development; even if castrated as pups, dogs (Beach 1974; Ranson 1981) and wolves (L. D. Mech, unpublished data) still show the raised-leg posture at the time of puberty. Testosterone is thus necessary for the organization, but not the activation, of the behavior.

The first expression of urine-marking behavior by male dogs, whether castrated or not, is preceded by an increased interest in sniffing vertical objects, suggesting that the pup's perceptions of his sensory world change as he reaches puberty, and that the RLU is a response to those perceptions (Ranson and Beach 1985).

Female wolves are much more likely to urine-mark (with FLUs) than are female dogs, but again, social status is pivotal. As with male wolves, only the dominant female urine-marks. Female wolves mark less regularly than males (Peters and Mech 1975b; Asa, Mech, and Seal 1985, Asa et al. 1990), but both sexes use their marking posture year-round (Mech 1991b, 1995d). Although female dogs are not as likely as female wolves to urine-mark, the frequency of dog squatting urination does increase dramatically during estrus (Beach 1974; Anisko 1976).

In both wolves and dogs, males urine-mark significantly more often than females. The strength of this drive in males can be deduced from several lines of evidence. For males, urine marking may not even be associated with the passage of urine. They will continue to raise a leg, as if to mark, even when the bladder is empty ("pseudo-urination": see Bekoff 1979b; Harrington 1982a). Furthermore, anosmic male wolves continued to urine-mark at typical levels in spite of their inability to smell (Asa et al. 1986).

Wolves of both sexes urine-mark more often during the breeding season. However, captive females tend to mark less during the summer, even though females in the wild continue FLUs year-round (Mech 1991b, 1995d). RLU rates in males are correlated with seasonal changes in serum testosterone. Interestingly, the urine-marking behavior of female wolves also appears to be influenced by testosterone, not estrogen (Asa et al. 1990). The wolf's seasonal pattern of urine marking contrasts with that of dogs, which show no seasonal trend in marking or in circulating testosterone. This aseasonality no doubt arises because female dogs can enter estrus at any time of the year (Christie and Bell 1971b).

The chemical composition of wolf urine differs between males and females and between breeding and nonbreeding seasons (Raymer et al. 1984, 1986). The administration of gonadal hormones to both male and female wolves demonstrated their important role in these differences (Raymer et al. 1986). Treating castrated male wolves with testosterone resulted in increases in many urinary components that are typically high during the breeding season (table 3.5). However, treating spayed female wolves with estradiol and/or progesterone resulted in fewer changes in urine composition. These studies suggest that wolf urine carries a message about gender and reproductive condition. Dogs and wolves can also discriminate between the urine of individual conspecifics (Brown and Johnston 1983).

Not only do the urine-marking rates of dominant male and female wolves increase during the breeding season, but their marks often overlay each other (Peters and Mech 1975b). First one wolf marks, then the other sniffs the mark and marks close to the other's mark, and sometimes each marks two or three times (L. D. Mech, unpublished data). Double or tandem marking appears to be related in some way to formation and maintenance of the pair bond. The frequency of double marking is highest in newly formed pairs (Rothman and Mech 1979). Furthermore, a captive, anosmic female did succeed in becoming dominant, but urine-marked with the dominant male at a significantly lower rate than in other pairs. She did not appear to bond with the dominant male, since she showed few of the proximity-related behaviors characteristic of other pairs (Asa et al. 1986). In another captive wolf colony, reduced rates of double marking were associated with failure to bond and reproduce (Mertl-Millhollen et al. 1986).

In dogs as well, investigation of the opposite sex or its urine stimulates further urinations (Dunbar 1977). Furthermore, the stimulus value of estrous female urine is progressively diminished by increasing quantities of male urine (Dunbar and Buehler 1980). Males show less interest in the urine of an estrous female if it also contains the urine of another male. This effect cannot be attributed to dilution, because addition of the same quantity of water rather than male urine does not have the same effect. This finding suggests that overmarking by a male reduces the chances of another male perceiving that an estrous female is in the vicinity, or announces that she is already paired. In wild wolves, the mere proximity of both adult male and adult female urine probably has the same effect, for usually male and female urine fall many centimeters apart (L. D. Mech, unpublished data).

Thus, it appears that urine marking in both males and females is primarily related to reproduction, advertising proestrus and estrus and establishing a pair bond in both sexes. A pair's double marks may serve a triple purpose: to a partner they convey a courting message, to single wolves they indicate a mated pair, and to other pairs they warn against territorial intrusion (Peters and Mech 1975b; Rothman and Mech 1979).

TABLE 3.5. Some chemical constituents elevated in the urine of intact male wolves during the breeding season and in castrated male wolves treated with testosterone

Ketones	Sulfur compounds
4-Methyl-3-heptanone	Methyl propyl sulfide
3,5-Dimethyl-2-heptanone	Methyl isopentyl sulfide
3,5,7-Trimethyl-2-nonanone	
3,5-Dimethyl-2-decanone	
2-Heptanone	

Source: Raymer et al. 1986.

A particular feature of female wolf reproductive physiology associated with urine marking that may facilitate pair bonding is the relatively long proestrous period (average 6 weeks: Asa et al. 1986) compared with that of the dog (1 week: Christie and Bell 1971a). Proestrus is preceded by a transient increase in testosterone and accompanied by elevated estrogen. During proestrus, vaginal secretions washed into the urine, coupled with the increased frequency of urination, render the female attractive and appear to stimulate double marking. This is probably an important time for the female in evaluating the suitability of the male as a mate and in cementing their commitment. Yet, although proestrus and estrus may be focal times for exchange of olfactory information between the male and female, wolves in the wild can pair-bond at any time of year and double mark year-round) (L. D. Mech, unpublished data).

The importance of odor in male mating behavior has been shown by experiments that eliminated the sense of smell in selected wolves (Asa et al. 1986). During three breeding seasons, neither of two sexually inexperienced, anosmic males sniffed or licked the urine or vaginal secretions of their female partners. Neither did they respond to the females' solicitations to mate. A sexually experienced male behaved similarly during his first year of anosmia, but in subsequent years he responded to the sexual invitations of his partner, copulated, and sired pups. He did not show interest in his mate's urine, but he did lick her genital area when she presented and deflected her tail. Thus, males require olfactory information from proestrous or estrous females to stimulate the full complement of sexual behavior. However, after sexual experience, the importance of olfaction is reduced and visual or social cues may suffice, demonstrating the role of learning.

The effect of anosmia on a female wolf was different. The female copulated normally despite her prior lack of sexual experience. Her maternal behavior appeared unaffected, although, because she was captive, we could not confirm that for all maternal behaviors (Asa et al. 1986).

Wolves also use urine marking as part of a "food inventory" system. As has been demonstrated for red foxes (Henry 1977) and coyotes (Harrington 1982b), wolves may keep track of the status of cached food by urine-marking caches that have been emptied, thus making more efficient use of their foraging time (Harrington 1981). It is notable, however, that wolves do not mark all of their caches after emptying them (L. D. Mech, unpublished data).

Scent Rolling

The tendency of wolves to roll on pungent substances is difficult to explain (reviewed by Reiger 1979). This ritualized behavior involves lowering the head and shoulders onto the substrate, followed by rubbing the chin, cheeks, neck, shoulders, and back on the odorous substance. Natural stimuli for scent rolling, such as rotten carcasses, typically have very strong odors that humans find offensive.

Tests with captive wolves have resulted in extensive lists of potential scent-rolling stimuli, both natural and unnatural (Goodman 1978; Ryon et al. 1986), but they have not conclusively demonstrated the function of this behavior. Possibilities include (1) familiarization with novel odors or changes in odors (Fox 1971a; Ryon et al. 1986), (2) strong attraction or aversion to particular odors (Ryon et al. 1986), (3) concealing one's own scent with something more pungent (Zimen 1981), and (4) making oneself more attractive by applying a novel odor (Fox 1971a). Female African wild dogs roll in the urine of males whose pack they are attempting to join, perhaps to coat themselves with an odor that would be familiar to the pack and increase the chances of acceptance (Frame et al. 1979).

The Message

As we have seen, glandular secretions and excretory products can carry a wide range of information. Our own olfactory limitations undoubtedly prevent us from fully appreciating the olfactory realm of canids. To have a wolf's nose for only a day would surely reveal a whole new world. But even from our limited perspective, studies have demonstrated that wolves' olfactory messages can contain information on species, individual identity, age (i.e., infant, pre- vs. post-puberty), gender, social rank, and sexual status. The odors involved in these messages are called either "releasing" or "signaling" pheromones (Wilson and Bossert 1963; Bronson 1968).

One characteristic of these olfactory messages is that they are honest signals. Most are the result of physiological and bacteriological processes that the wolf has little, if any, control over. A male under social stress will excrete elevated levels of corticosteroids in his urine for any other wolf to "read" (McLeod et al. 1996). A female in search of a mate may consequently find him unattractive, based on her investigation of his urine. Scent rolling may be an attempt to change an animal's olfactory "image" (see above), but if so, we have no data to indicate that it succeeds. However, if wolves cannot modify their scent signals, they can at least limit their distribution. Observations from both the wild (Peters and Mech 1975b) and captivity (F. H. Harrington, unpublished data) indicate that subordinate wolves urinate infrequently and in areas less likely to be visited by other wolves. Dominant wolves, on the other hand, tend to urine-mark frequently and widely.

Another aspect of olfaction that has been best studied in rodents involves the action of odors on the physiology of the recipient, especially on reproductive processes. Such odors, called primer pheromones, can accelerate or delay puberty, induce, synchronize, or suppress estrus, and interrupt pregnancy (reviewed by Vandenbergh 1988). Hradecky (1985) has suggested that these phenomena may also be at work in wolves.

However, the evidence from wolves does not support this hypothesis. The failure of reproduction in subordinate wolves can be explained by social suppression (Packard et al. 1983, 1985) and incest avoidance (D. Smith et al. 1997). Furthermore, reproductive parameters, including the gonadal hormone profiles of both males and females, testicular development, sperm production, and ovulation, are not affected by social status or by anosmia (Seal et al. 1979; Asa et al. 1986). Anosmic female wolves can even conceive, give birth to, and care for young (Asa et al. 1986). This independence of reproductive physiology from the direct influence of social status may give female wolves the ability to respond quickly to changes in reproductive opportunity that take place during the relatively short breeding season.

Similarly, there is no evidence to relate the physiological changes in reproductive systems that occur seasonally to pheromonal communication or coordination. That these changes occur so predictably at the same time each year, in widely spaced females, argues for an environmental cue (i.e., day length). However, anecdotal observations from dog colonies suggest an exception. Although dogs do not have a defined breeding season, colony females often have synchronous estrous periods (P. W. Concannon, personal communication). A new estrous female introduced into a colony can even trigger estrus in the others. Domestic dogs belonging to people working at our wolf colony began to cycle in synchrony with the wolves (C. S. Asa, unpublished data). These observations suggest that group-living female canids might have relatively synchronous estrous periods driven by pheromonal communication as well as environmental cues.

Signal Transmission

An advantage of olfaction over other modes of communication is that messages can be detected long after they are left, much like a bulletin-board posting. Information about the passage of time since deposition can also be incorporated, as odor intensity wanes and components evaporate at different rates. Urine, feces, and footpad odors can all be left for later detection. In contrast, transmission of information through saliva or skin glands is typically direct, with the recipient sniffing the source, not a scent mark. Preputial, vaginal, and anal sac secretions apparently can be used in either context; that is, left or sniffed directly.

Rates of evaporation of volatile molecules such as those found in wolf odors are affected by many factors, including the compound's molecular weight (MW). Because compounds with lower MW are more volatile, they are good candidates for olfactory signals ("pheromones"). However, larger, heavier molecules stimulate the olfactory system more efficiently (Dethier 1954). Thus, a balance must be struck: the signaling molecule must be small (i.e., volatile) enough to vaporize under local conditions, but large enough to cause an olfactory response. Wilson and Bossert (1963) have hypothesized that the MW range of pheromonal compounds is 80–300.

Interestingly, the average MW of gonadal steroid molecules, such as estradiol, progesterone, and testosterone, is about 250, suggesting that hormone molecules in glandular secretions might serve as pheromones. Even though hormones are excreted in urine as much larger complexes, enzymes in urine can return the molecules to their original form (Whitten and Champlin 1973). In fact, receptors for estradiol have been found in the olfactory epithelium of the rat (Vannelli and Balboni 1982). Furthermore, the rich capillary bed of the nasal epithelium can absorb hormones directly into the bloodstream (Chien 1985).

Environmental factors and odor substrates also play critical roles in the rate of transfer of olfactory messages. Increases in temperature, air currents, or relative humidity result in increased evaporation. However, surfaces that are porous or that have a polar charge can significantly slow vaporization, as can the presence of lipids or sebum from sebaceous glands (Regnier and Goodwin 1977). The deposition of urine on elevated vertical structures may both enhance its detection, by increasing the evaporative surface at nose height, and prolong its potency, by protecting it from dilution by rain and dew.

Signal Reception

Clearly, wolves have a wide variety of sources of olfactory information at their disposal, and these sources may provide them not only with current information about their packmates and competitors, but also with a historical record of the movements of both packmates and strangers during the recent past. It is also obvious that wolves spend much of their time investigating these odors, and they appear to use the information they contain as they decide on future behavior (Peters and Mech 1975b; Peters 1978). The ability of wolves to utilize odors with such apparent sophistication requires both a sensitive receptor system and a complex neural center.

The Olfactory System

Inside the nose of the dog, and presumably the wolf, the surface containing the olfactory receptors is much enlarged by extensive folding supported by a thin, bony structure (Hare 1975). This feature accommodates an estimated 280 million olfactory receptors, more than the number of visual receptors in the retina (Wieland 1938, cited in Moulton 1967). Nerve fibers from the olfactory receptors project to the brain's olfactory bulb.

The relative importance of olfaction to the dog as compared with various other species can be indirectly surmised from the percentage of nasal epithelium devoted to olfaction as opposed to respiration (table 3.6). Among several common species evaluated, only the cat has proportionately more nasal epithelium devoted to olfaction than the dog.

The main olfactory system just described is one of three olfactory systems possessed by the wolf. The others, the accessory and trigeminal systems, are less understood. The relative roles of the main and accessory systems in wolf communication have not been clearly distinguished. The accessory system, which receives signals from the Jacobson's or vomeronasal organ (VNO), is thought to mediate sexual response in some species (Estes 1972; Scalia and Winans 1975; Wysocki 1979).

Although the VNO of the dog is well developed (Adams and Wiekamp 1984), a definite function for this organ is not known. Many mammals perform a behavior known as "flehmen," or lip curl, which is thought to help transfer molecules from urine or glandular secretions into the VNO. Although wolves and other canids do not shown flehmen after contacting urine or vaginal secretions, they do rapidly and repeatedly press the tongue against the roof of the mouth just behind the front teeth (Asa et al. 1986), the site of the nasopalatine ducts that open into the nasal cavity near the VNO.

Pathways from the VNO via the accessory olfactory bulb reach areas of the brain associated with sexual behavior and stimulation of gonadotropic hormones. In sexually naive male wolves, surgical transection that blocks both the main and accessory olfactory systems interferes with the ability to court and mate (Asa et al. 1986). However, removal of only the vomeronasal organs does not affect sexual behavior (C. J. Wysocki et al., unpublished data). These results suggest that the main, but not the accessory, olfactory system is important in the sexual response of male wolves, especially those that have not learned to use other cues.

Scent rolling also appears to involve the main rather than the accessory olfactory system. Following VNO removal, wolves continued to respond to a fish odor by typical rolling, whereas wolves whose main olfactory systems had been blocked did not (C. S. Asa, unpublished data).

The trigeminal system may also be involved in scent rolling. With receptors in both the olfactory and respiratory epithelium, this system responds primarily to noxious chemicals and probably prevents the organism from inhaling something potentially dangerous by invoking protective reflexes (Allen 1937). This combination of triggering by pungent odors and a reflex reaction is consistent with scent rolling. However, trigeminal systems typically evoke evasive responses.

Olfactory Sensitivity

To gain insight into the sensitivity of a wolf's nose, it is useful to turn again to information about the dog and to

TABLE 3.6. Comparison of the absolute surface areas of the respiratory and olfactory epithelia and the ratio of respiratory to olfactory surface areas for various mammalian species

Species	Respiratory surface (cm^2)	Olfactory surface (cm^2)	Ratio[a]
Guinea pig	2.08	1.12	1.85
Rat	1.60	1.12	1.33
Cat	6.08	5.76	1.06
Dog	12.08	9.76	1.23
Monkey	9.76	4.16	2.34
Human	10.40	3.08	3.37
Sheep	33.00	8.84	3.73

Source: Adapted from Dieulafé 1906.

[a]The lower the ratio, the greater the importance of olfaction to the species.

assume that wolves are at least as sensitive as dogs. Dogs are a hundred to millions of times more sensitive than humans in perceiving odors (Neuhaus 1953; Moulton and Marshall 1976; Marshall and Moulton 1981). This superior olfactory ability may be explained by noting that, although the same proportion (1–2%) of the molecules that enter the nose reach receptors in each species, (1) dogs have a much greater olfactory receptor area; (2) dogs probably have proportionately more active receptors, so that an odor molecule is much more apt to reach a binding site; (3) the dog's olfactory bulbs are considerably larger; (4) dogs can better discriminate among different odors and resolve complex mixtures into components; and (5) the resulting increased reliance by dogs on olfaction probably results in more efficient sniffing and attention to odors (Marshall and Moulton 1981).

The Role of Olfaction in Hunting and Tracking

The sensitivity and discriminatory powers of the canid olfactory system are demonstrated not only by its important role in social communication, but also by its use in hunting (see Peterson and Ciucci, chap. 4, and Mech and Peterson, chap. 5 in this volume). Again, much of our information relating to the use of olfaction in hunting comes from the dog, particularly tracking dogs. Although little of this information is based on controlled scientific studies (e.g., Brisbin and Austad 1991, 1993), much trackers' lore (e.g., Budgett 1933; Syrotuck 1972; Pearsall and Verbruggen 1982) can provide insight into the mechanisms involved. Hypotheses to explain the cues used by tracking dogs include scent on the substrate or in the air, the odor of crushed vegetation, and odor "rafts" falling from the animal being pursued. These "rafts" consist of sloughed skin cells, sometimes with odor enhanced by bacteria, that continually fall from living animals (Marples 1969); they presumably can be detected by canids (Syrotuck 1972).

Whatever the odor source, dogs and wolves probably use similar strategies when tracking prey. For example, they may catch an airborne scent or a trail on the ground and follow it directly, but more commonly have to cast about to locate and stay on the scent. By criss-crossing scent trails, tracking dogs might avoid adapting to an odor, which could result in their losing it (Wilson and Bossert 1963). Casting might also allow them to keep assessing the odor gradient, with the concentration being strongest in the center of the trail.

Mech (1966b) reported that 43 of 51 hunts by a wolf pack on Isle Royale involved detecting moose by direct scenting. Lone wolves also seemed to locate prey by odor. Commonly, the wolves detected moose within 300 yards if the wolves were downwind. However, they once scented a cow with twin calves from about 2.4 km (1.5 mi) away. Wolves in Minnesota also seem to locate deer by odor (Mech 1970).

Indeed, the popular consensus regards olfaction to be of paramount importance for hunting (i.e., Asa and Mech 1995; Mech 1995e). However, it should be remarked that no experimental research has compared the senses used in hunting by wolves. Experimental research with captive red foxes (Osterholm 1964) and coyotes (Wells 1978; Wells and Lehner 1978) found vision to be most important during hunting, followed by audition or olfaction, depending on conditions. If olfaction is truly predominant for wolves, its importance probably reflects the low density and wide dispersal of their large ungulate prey in forested habitats. For wolves, visual information may predominate in open plains and tundra habitats, and at close quarters.

Visual Communication

The Importance of Vision in Communication

Visual communication is probably every bit as important to wolves as acoustic and olfactory communication. In fact, our knowledge of the latter two systems has usually been acquired by careful observation of the visual signals that accompany sounds and smells. Yet in some respects, visual communication has been the more difficult to describe objectively. Vocal signals are discrete, occur in temporal sequence, and can be physically "captured" and measured with relative ease. Olfactory signals, as we have mostly studied them to date, have been reduced to the level of their visual appearance: scent marks and scent-marking behavior. We know little about the chemical composition of these signals. Visual signals, however, come and go in less obvious fashion and might consist of the simultaneous movement of most body parts from nose to tail. Consequently, we have usually relied on our ability to detect patterns in these otherwise complex stimuli, which gives us a good "feel" for the basic patterns, but less than an intimate knowledge of the details—although that is changing.

Analyses of wolf visual communication have followed the lead of Darwin (1872), who felt that aggression and submission were the two central motivational states in animals. His principle of antithesis holds that when animals express opposing emotional states, the signal elements used to express those states are also opposites. As Morton (1977) did in his motivation-structural rules for vocalizations, Darwin organized expressive behavior along axes of fear and aggression, with each axis indicating the level of emotional arousal.

The "Elements" of Visual Communication and the Principle of Antithesis

Schenkel (1947) first described visual communication in the wolf. He noted that features, or "elements," of the face (ears, eyes, lips, teeth, nose, and forehead), the body (posture, hair), and the tail are important components of visual signals, while facial and body coloration often enhance a signal's value (Schenkel 1947; Fox 1969; Fox and Cohen 1977). Variation in each element was presumed to express variation in underlying motivation along a continuum from aggressive/confident to submissive/anxious (table 3.7).

Aggressive or self-assertive individuals are characterized by a high body posture, enhanced by raised hackles and general piloerection along the back and the tail (Schenkel 1947). The legs are held stiffly, and movements

are slow and deliberate. These dominance signals reflect the wolf's readiness to attack, with other visual elements enhancing the its size and fully displaying its teeth. Postures indicating submission or fear represent preparations for defense or flight, with visual features reducing the animal's apparent size and hiding its teeth. Submissive animals carry the body low, sleek the fur, and lower the ears and tail.

Schenkel (1947, 98) called the tail the "most dynamic" of the visual elements (fig. 3.1). It may increase an individual's apparent size, and may also give it a mechanical advantage in the event of a fight (Schenkel 1947). In addition, elevating the tail exposes the anal region. Wagging of the tail conveys friendliness, and submissive wolves may enhance this signal by moving the hindquarters as well; slow, stiff movements of the tail or tip indicate an aroused state that might lead to attack (Schenkel 1947; Zimen 1981).

Recently, van Hooff and Wensing (1987) distinguished three distinct postural attitudes used by wolves during social interactions: "high," "neutral," and "low." The high and low postures were based on opposite expressions of the head, ears, tail, and legs, following the above descriptions. Of their twenty-one behavioral measures, high and low postures were among the few that

TABLE 3.7. Expressive characteristics of visual features used during social interactions in wolves

Feature	Aggressive ⟵	⟶ Fearful
		Expression
Eyes	Direct stare	Looking away
	Open wide	Closed to slits
Ears	Erect and forward	Flattened and turned down to side
Lips	Horizontal contraction ("agonistic pucker")	Horizontal retraction ("submissive grin")
Mouth	Opened	Closed
Teeth	Canines bared	Canines covered
Tongue	Retracted	Extended ("lick intention")
Nose	Shortened (skin folded)	Lengthened (skin smoothed)
Forehead	Contracted (bulging over eyes)	Stretched (smoothed)
Head	Held high	Lowered
Neck	Arched	Extended
Hair	Erect (bristled)	Sleeked
Body	Erect/tall	Crouched/low
Tail	Held high	Tucked under body
	Quivering	Wagging

Sources: After Schenkel 1947; Fox 1970.

FIGURE 3.1. Schenkel's model of the motivational bases of tail position in wolves. (a) Self-assertion during social interactions; (b) assertive threat; (c) (with lateral wagging) intimidation; (d) normal position during conditions without social tension; (e) not-quite-certain threat; (f) normal position (similar to d), especially while eating or watching others; (g) "depressed" mood; (h) between threat and defense; (i) (with lateral wagging) active submission; (j) and (k) strong inhibition. (From Schenkel 1947.)

clearly revealed the rank relationships among packmates. This consistency between body attitude and social relationships indicates that signals of dominance and signals of submission honestly convey information about individual wolves at any given instant. Within a pack, being able to "read" another packmate's signals unambiguously benefits both individuals, confirming a

previously established relationship without the need for recurring conflict.

Following earlier work by Lorenz (1966), Zimen (1981) developed a visual model to indicate how the various elements of the face might change as the underlying motivation of a wolf changes along fear and aggression axes (fig. 3.2). Although recent work on motivation indicates that wolves express more than just aggression and fear in their signals (table 3.8; see Mech and Boitani, chap. 1 in this volume), Zimen's model does illustrate a portion of the range of wolf facial expression. Figure 3.3 depicts the greater range of expressions possible.

Recent study of the development of visual signals further underscores the complexity of this mode of communication. Fox (1969, 1970, 1971a,c,e, 1976) originally described the general development of visual signals in canids. McLeod (1987, 1996) and McLeod and Fentress (1997) studied the way these signals were expressed in wolf pups between 2 and 12 weeks old. Clustering analyses placed the twelve postural elements studied along two orthogonal axes, which were labeled dominant-submissive and playful-serious. In addition, four elements did not cluster with any others, suggesting that they may each express a different message. Tail wagging by older pups, for example, may serve as a metacommunication signal, indicating playful intentions. Thus an aggressive message indicated by the eyes, ears, and voice may be "voided" by a wagging tail. Bekoff (1995) has suggested the same role for the "play bow" in canids, including wolves. Thus McLeod's work indicates that even

TABLE 3.8. Results of behavioral studies that identified presumed motivational bases underlying wolf social behavior

Motivational clusters identified	Number of packs	Number of variables	Statistical method[a]	Reference
Attention/aggression; submission; play; sexual/friendly; defense/appeasement	1	29	pca	Colmenares Gil 1983
Aggression/dominance; defense; sexual; inspection; affiliation; play; submission	2	60	ca, pca	Derix et al. 1994
Dominance; affiliation; agonism; activity; deference	7	27	fa	Lockwood 1979
Dominant/submissive; playful/serious	1	12	ca, mds	McLeod 1996
Affiliative; dominant; defensive; sexual	2	13	pca	Packard 1980
Dominance/subordinance; sexuo-affiliative	1	21	ca	Van Hooff and Wensing 1987

Note: For each study, a number of behavioral variables (ranging from 12–60) were quantified during behavioral observations typically carried out over one to several years. The behavioral variables were then correlated using a number of statistical methods to identify groups or clusters of behaviors that are assumed to share similar underlying motivations. These groups are named to reflect the motivation expressed in the original behaviors.

[a]ca = cluster analysis; fa = factor analysis; mds = multidimensional scaling; pca = principal components analysis.

FIGURE 3.2. Zimen's model of the motivational bases of facial expression in wolves. Aggressiveness (the tendency to attack) increases from left to right, and fear increases from bottom to top. The facial expressions shown represent only a sampling of the many variations possible. Zimen believed that high levels of fear inhibit the tendency to attack, so the upper right portion of the model is blank. (From Zimen 1981.)

FIGURE 3.3. An aggressive interaction between two males competing for access to the breeding female. The wrinkled nose, bared teeth, and direct stare indicate that the male on the left is aroused aggressively, although the tightly clamped teeth suggest that the animal is not likely to bite (Zimen 1981). The ears indicate the low to moderate level of fear that is expected of adults when interacting aggressively with animals of similar size and fighting ability. The tongue, however, suggests a strong element of friendliness behind the animal's expression, in the form of a "lick intention" common to active submission. The male on the right responds to this complex expression by turning slightly aside, averting his gaze, and pulling back his ears, signs of fear or submission. (Photo by K. Hollett.)

in young wolf pups, more than aggression and fear lie behind visual signals.

Visual Signals

Consistent combinations of visual elements recur in patterns that are recognizable through time and across individuals. Schenkel (1947, 1967) described a number of these elements and the contexts in which they occur (table 3.9). In some cases, these patterns in adults may resemble behaviors shown by pups, but often in different contexts. These repeatable patterns comprise visual displays or signals that carry specific messages and meanings for wolves. In this section, we will indicate how recent studies have furthered some of Schenkel's pioneering work.

Submission

Schenkel (1947, 1967) described two primary visual signals that occurred in submissive contexts: active submission and passive submission (see table 3.9). In active submission, the submissive animal approaches another wolf in a low posture, slightly crouched, ears back and close to its head, and tail held low. The submissive wolf wags its tail or hindquarters and attempts to lick or mouth the other wolf's muzzle. Active submission occurs often in "greeting," during "group ceremonies" in which dominant wolves become the focus of nuzzling, licking, and mouthing about the face by other members of the pack, and as a "nose-push" given by submissive animals toward dominant ones when they are still a few meters away.

Passive submission is often a reaction to approach and investigation by a dominant animal (Schenkel 1947, 1967; Fox 1971e). The submissive animal lies partly on its side and partly on its back, with its tail curved between its legs and its ears flat and directed backward. If the approaching wolf sniffs the animal's anogenital area, the submissive animal may raise its upper leg and expose more of the belly area. Although he described and named these two quite different behaviors, Schenkel (1947, 1967, 319) emphasized that there were many variations of submissive behavior, representing subtle expressions of mood, between the "extremes" of passive and active submission.

Van Hooff and Wensing (1987) recently studied dominance and submission in a captive wolf pack, using cluster analysis to organize twenty-one behavioral variables.

Among them were Schenkel's categories of active and passive submission. Active submission was very common, constituting 3–10% of the 80,000 behavioral acts recorded; passive submission was less frequent (1–2%). Along with high and low body postures and retreat, however, active and passive submission were the best and most consistent indicators of dominant or submissive status, particularly for indicating the direction of the relationship between two wolves. A related but even more detailed study of two captive packs gave similar results (Derix et al. 1994), confirming Schenkel's original descriptions and justifying the labels he gave to them.

Active and passive submission in adults resemble two very common pup behaviors: the food-begging that elicits regurgitation and the response of pups to inguinal licking by the mother that elicits urination and defecation (Schenkel 1967) (see Packard, chap. 2 in this volume). This relationship in form illustrates how motor patterns used in one context early in life may acquire new meaning and function at a later stage. Food-begging, however, does recur in adulthood as well, particularly among mothers and yearlings during pup rearing (Fentress and Ryon 1982; Mech 1995c,d).

Active and passive submission also illustrate well the multimodal form of most wolf signals. The visual form that humans readily recognize, and that wolves perceive from a distance, is accompanied by vocalizations, touch, tastes, and odors. It is likely that the relative importance of these modalities changes developmentally (i.e., taste [food] and touch [licking] may be central to pups; visual and vocal stimuli more crucial for adults) and with context.

Conflict

Strife is a fact of life for wolves. In some wild populations, inter-pack hostility accounts for the majority of adult deaths by natural (non-human-related) causes (see Mech and Boitani, chap. 1, and Fuller et al., chap. 6 in this volume). Between packs, submissive displays do not "cut off" attacks; whether a confrontation leads to an attack with injury or death depends most on how the combatants are matched. In one case, a lone wolf behaved submissively after being caught by a pack, but was nevertheless attacked and killed (Marhenke et al. 1971). In another case, a lone wolf was able to escape his single aggressor before other members of the pack arrived (Mech 1993a). Zimen (1981) indicated that a captive wolf that was not inhibited by fear often attacked with little

TABLE 3.9. Behavior patterns used by wolves during social interactions

Behavior pattern[a]	Social context
During less intense dominance/submission interactions	
Active submission	Subordinate wolf approaches a more dominant individual with lowered (or crouched) body posture, often directing licking or "licking intention" to mouth of dominant. From a distance this pattern may be reduced to a "nose-push."
Anal presentation	Raising of tail by dominant wolf to expose anal region upon approach by other individuals of similar or lower rank.
Anal withdrawal	Submissive wolf tucks tail, lowers and moves away hindquarters in response to approach to hindquarters by more dominant individual.
Fixed stare	A dominant wolf directs its gaze toward a rival.
Passive submission	Submissive wolf rolls onto side or back in response to approach by dominant individual; rear leg may be raised and urine may be expressed upon closer inspection by dominant individual.
Riding up	Among adult males, a dominant wolf will mount the back of a rival, directing "threat bites" toward the other's neck. Both may growl. A third individual may also be "ridden on" by one male as he directs a threat across its back to another male on the other side.
During more intense dominance/submissive interactions ("ritualized fighting")	
Ambush threat	A dominant wolf assumes a low, stalking posture, preparatory to a pounce and sometimes from cover, oriented toward its subordinate rival.
Bite threat	Dominant wolf stares at rival, teeth bared, forehead and nose wrinkled, ears erect, body tense and hair raised.
Defensive snapping	An empty, snapping movement, often accompanied by barking, by a wolf under threat of attack. The relative orientation of these biting motions to the rival may vary.
During friendly social interactions	
Group ceremony (rally)	A group activity characterized by localized but active movement and by mutual muzzle nuzzling, body rubbing, and whimpering/whining vocalizations. Often precedes other group pack activities, such as chorus howling or movement away from the current site (i.e., hunting).
Play invitation	One wolf approaches another with forequarters lowered ("play bow") and hindquarters and tail raised and often wagging.
Standing over	One wolf stands over the forequarters of one that is lying down. The lying individual may lick the genital area of the other.
During courtship	
Anal presentation	Raising of tail to expose anal region by sexually receptive female upon approach (and inspection) by courting male(s).
Muzzle nuzzling	Head to head contact between a courting pair involving sniffing, rubbing, pushing, and seizing the other's muzzle.
Riding up	The female places her forelegs on the back of a male. Males other than the highest-ranking male may reject this behavior while growling.

Source: From Schenkel 1947, 1967.

Note: See also Packard, chap. 2, table 2 in this volume.

[a]Variation of body elements within each of these named behaviors is often extensive.

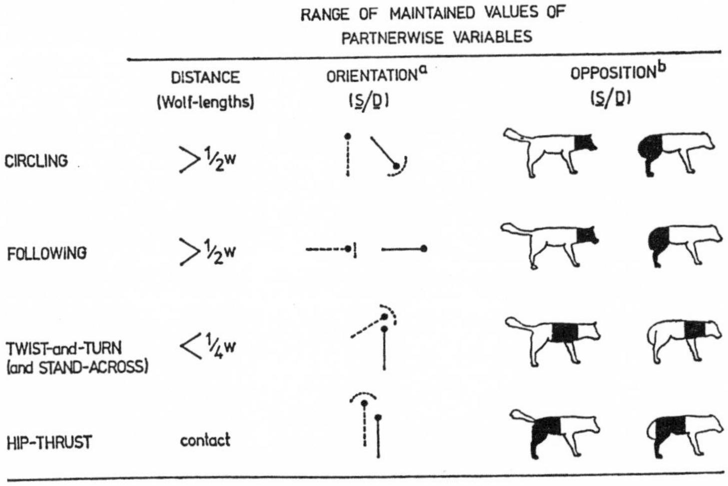

RANGE OF MAINTAINED VALUES OF
PARTNERWISE VARIABLES

	DISTANCE (Wolf-lengths)	ORIENTATION[a] (S/D)	OPPOSITION[b] (S/D)
CIRCLING	$> \frac{1}{2}$w		
FOLLOWING	$> \frac{1}{2}$w		
TWIST-and-TURN (and STAND-ACROSS)	$< \frac{1}{4}$w		
HIP-THRUST	contact		

a. Dotted lines represent range of stable orientations.
b. Shaded areas indicate range of stable points of opposition on each interactant.

FIGURE 3.4. Moran and colleagues' summary of the stable configurations (left column) that occur during conflict between two wolves. The three relationships recorded between the two wolves are distance apart (measured in wolf body lengths), relative orientation to each other, and nearest "point of opposition" (closest body area to opponent). D = displacing wolf (also termed "superior" or "dominant" wolf by others) on left; S = supplanted wolf (called "inferior" or "subordinate" wolf by others) on right. Dotted lines about heads of wolves (when present) represent the range of movement typical of these stable configurations. (From Moran et al. 1981.)

outward sign of aggressiveness. These observations suggest that the richness of information available in visual signals may not be as fully expressed or attended to during clashes between strange wolves.

Within a pack, however, conflict is restrained; although deaths and serious injuries do occur, they are rare considering the amount of time the animals spend together (Mech 1999). Most conflicts are settled rather quickly by avoidance, mediated by visual and vocal signals of dominance and submission (Schenkel 1947, 1967). For example, if one wolf encroaches too closely on another's "personal space," a growl through bared teeth may be all that occurs. In those circumstances in which conflicts escalate or are prolonged, as might occur between rivals during the mating season (see fig. 3.3), signals from a number of modalities may combine into what is akin to a dance between the opponents.

Schenkel (1947, 1967), Lorenz (1966), and Fox (1969) have described isolated aspects of these aggressive interactions (e.g., table 3.9), but Moran and colleagues (1978, 1987; Moran and Fentress 1979; Moran et al. 1981; Golani and Moran 1983) have described the interactions themselves. Moran took thirty interactions that fit Schenkel's (1947) descriptions of "ritualized fighting" between packmates and measured three variables: relative distance between individuals, relative orientation of the pair, and point of nearest opposition, much as one would do if describing the choreography of a dance. Within an interaction, the two wolves played different roles; one (the displacer) continually harassed the other (the supplanted

animal), which tried to escape while protecting its vulnerable flanks and hindquarters.

Moran found four relatively stable configurations between the two combatants and five transitions that rather abruptly led them from one configuration to another (fig. 3.4; table 3.10) (Moran et al. 1981; Moran 1987). When the two wolves were more than half a body length apart, the supplanted wolf usually moved forward with the displacing wolf in pursuit (*Following*). *Circling* involved the same body distances and oppositions, but now the animals seemed to pivot about a point, with the supplanted wolf's body curved to point its head toward its opponent's. The displacer, meanwhile, continued to approach its rival's hindquarters. At closer distances (< 1/4 body length), circling was replaced by *Twist-and-Turn* as the supplanted animal suddenly shifted to bring its forequarters to oppose those of its rival. Finally, *Hip-Thrust* occurred when the animals came into physical contact. The supplanted wolf kept its jaws opposite the shoulder and neck of the displacer, while the latter maintained contact with its rival's rear flank, where it could deliver hip-thrusts to unbalance its opponent. Throughout all these configurations, the displacer directed its attention toward its opponent's hindquarters, where most wounds are commonly found, while the supplanted wolf used its jaws to protect its flank. Although many interactions involved quite extensive movements, the physical relationships between the two animals changed little; the animals performed a "dance," with each constrained by a set of rules of movement (Moran et al. 1981).

TABLE 3.10. Stable behavioral configurations and the transitions that occur among them during ritualized fighting between two adult wolves

Previous stable configuration(s)	Transition involved	Subsequent stable configuration(s)	Change in relative orientation	Individual contributions to the transition[a]
Circling	Swivel/Stand-Across	Twist-and-Turn Hip-Thrust Stand-Across	From anti-parallel to parallel orientation, with a decrease in distance to near zero	Active movement by S to pull hindquarters away from D
Following	Lunge/Swivel	Twist-and-Turn Hip-Thrust Stand-Across	Point of opposition rapidly shifts from rear to head	Initiated by S with a sudden 360° turn, pulling rear away from D
Following	Walk-Up/Stop	Twist-and-Turn	Point of opposition slowly shifts from rear to head	S slows down and then stops forward movement
Twist-and-Turn Stand-Across	Turn-to-Rear/Turn	Follow Circle	Increase in relative distance and shift of opposition from front to rear	Variable—often involves S breaking away from D
Twist-and-Turn Stand-Across Hip-Thrust	Walk-Away/Walk	Follow (or another transition)	Increase in relative distance and shift of opposition from front to rear	S walks sideways away from D, then turns once distance is increased

Source: Moran et al. 1981.

[a]D = displacing wolf; S = supplanted wolf. Moran et al. referred to their wolves using the neutral terms "Supplanting" and "Displaced" to avoid using the terms "Dominant" and "Submissive," which may have wider connotations than they intended. Unfortunately "D" and "S" are used commonly to indicate dominant and subordinate wolves, respectively. To avoid confusion but to retain the more neutral tone of Moran et al., we have substituted the terms "Displacing" and "Supplanted" above and in the text.

Although we have chosen to discuss Moran's work in this section on visual signals, the two combatants are signaling and assessing each other using other communication channels as well. Growls and whines change to snarls and yelps as the intensity of the conflict escalates. Tactile information may be especially important during the hip-thrust, and olfactory signals may also give each animal further information about its opponent as the conflict evolves. Some of these interactions may end after the displacer succeeds in getting one or more bites into the other's hindquarters; on occasion the displacer may even be able to overcome the other's defenses and launch a mortal attack. Other interactions end after the supplanted wolf manages a good defensive parry to its tormentor's head, perhaps tearing an ear. Most, however, simply end without injury to either wolf.

The Visual System

To use the often subtle information made available by visual signals, as well as to permit quick and effective reactions to the movements of fleeing prey, wolves require a visual system that combines reasonable acuity with the ability to operate well under the low light available when wolves are often active. Consequently, the wolf's visual system differs markedly from that of humans, and possibly even from that of the domestic dog, on which our limited knowledge of canid vision is based (see Miller and Murphy 1995). Differences between wolves and humans probably reflect adaptations to different lifestyles or environments, whereas differences between dogs and wolves may be due to domestication. We will compare the visual system of wolves (or dogs) with those of humans and cats, which have provided most of our insights into the vision of terrestrial carnivores.

General Characteristics

Wolves (and dogs) possess "24-hour" (duplex) eyes, adapted to function both day and night (Walls 1942; Andreyev 1985). However, their eyes have modest capabilities for both collecting and excluding light (table 3.11). Cats can use a greater range of illuminations effectively than dogs or humans (Walls 1942).

The "eyeshine" of a dog's eye is caused by the tapetum lucidum (see table 3.11), a cellular layer that reflects light back through the layer of rods and cones (Walls 1942;

TABLE 3.11. Characteristics of the eyes of various mammals

Species	Eye diameter	Pupil diameter	Variation in pupil area	Tapetum lucidum Type	Tapetum lucidum No. layers
Dogs[a]	25 mm (1)	3–4 mm (1)	—	Cellulosum (4)	9–11 (5)
Cats	—	—	135-fold (3)	Cellulosum (4)	15–20 (5)
Humans	24 mm (2)	1.3–9 mm (2)	10-fold (3)	None	

Sources (in parentheses): 1, Andreyev 1985; 2, Duke-Elder and Wybar 1961; 3, Hughes 1977; 4, Walls 1942; 5, Wen et al. 1985.

[a] Presumably wolf eyes are similar to those of dogs.

Duke-Elder 1958; Andreyev 1985; Wen et al. 1985). The resulting increase in low-light sensitivity, however, comes with some loss in visual acuity due to scattering (Walls 1942). The dog's tapetum is roughly triangular, with the base of the triangle situated along the equator of the retina (Walls 1942; Parry 1953; Andreyev 1985), reflecting light gathered from at and below the visual horizon.

The wolf's visual field is directed somewhat laterally, giving it a relatively wide view of its surroundings but reducing the size of its binocular field (table 3.12). Hughes (1977) hypothesized that the binocular field aids "praxic" behavior: visually controlled activities involving the manipulation of the environment (e.g., use of paws in prey capture). The relatively small binocular field of canids may reflect their limited use of their forelimbs.

The ability to change focus (accommodation) may also be related to praxic skill (Hughes 1977). In dogs, accommodation has been described as "feeble" (Hughes 1977, 680); they may be able to focus on objects no nearer than 30 cm (Schmidt and Coulter 1981). Wolves may do slightly better (Duke-Elder 1958) (see table 3.12).

The Retina

Both canids and humans possess rods and cones, but differences in the distribution and density of these photoreceptors across the retina give dogs and wolves a fundamentally different vision from that of humans. Rods are highly sensitive to light, but do not permit good spatial resolution or color discrimination. Cones need relatively strong light to function, but permit sharper acuity and some degree of color vision.

Human vision is dominated by the fovea, a depression at the focal point on the retina that is densely packed with cone cells but devoid of rods (Hughes 1977) (table 3.13). Rods are found in greatest density about 20° from the center of the retina, and the density of both rods and cones drops off rapidly away from the center (Hughes 1977; Wassle and Boycott 1991). Thus human vision is dominated by the center of the visual field, which is adapted for daylight use, color discrimination, and fine visual acuity.

Dog (and wolf) vision has not been as extensively studied as that of humans or cats, but dogs and cats are similar in many respects (Peichl 1989; Loop and Martin 1991) (see table 3.13). Like cats, dogs and wolves lack a defined fovea and instead possess an area centralis, a broad, central region of higher rod and cone density (Peichl 1989). The density of cones falls gradually across the retina (Koch and Rubin 1972), whereas rod density changes little (Peichl 1989).

Dogs and cats have similar rod and cone densities and distributions, which parallel those found in humans (Steinberg et al. 1973). Cats and dogs differ from humans in the three- to fourfold higher densities of rods throughout the carnivore retinas, the complete lack of rods in the human fovea, and the five- to tenfold higher density of cones in and around the human fovea. Interestingly, the density of cones beyond about 12° of the center is nearly identical in all three species (Steinberg et al. 1973; Peichl 1989).

Peichl (1989, 1992a,b) found pronounced "visual streaks," bands of ganglion cells running along the horizontal (equatorial) axis of the retina, in all of eight wolves, but only eighteen of forty dogs, that he examined. Densities of ganglion cells in the area centralis of the visual streak were highest for wolves and quite variable in dogs (see table 3.13). Over the entire retina, two wolves studied had nearly twice as many ganglionic cells as did two dogs (see table 3.13), suggesting that wolves, on average, have sharper vision than dogs. Pronounced visual streaks would give the wolf relatively sharp vision along much of the visual horizon without having to shift its gaze.

TABLE 3.12. Spatial and temporal characteristics of the visual systems of various mammals

| Species | Size of visual field | | | | Accommodation | Visual acuity | Flicker fusion frequency | |
	Divergence of eye axis	Total	Binocular	Best vision			Rods	Cones
Dogs	15–25° (6, 16)	240–250° (9, 13, 16)	60–80° (5, 6, 16)	5–6° (4)	1 diopter (3, 6)	3–12 c/deg (8, 11)	25–40 Hz (2, 12)	75–90 Hz (2, 12)
Wolves	25° (6)	260° (1)	70° (1)	—	2.75 diopters (3)	15 c/deg (1)	—	—
Cats	4–9° (6, 16)	180–190° (6, 16)	90–105° (5, 6, 16)	5–6° (4)	2–6 diopters (6, 16)	5–18 c/deg (5, 6, 7, 15)	—	—
Humans	0° (6, 16)	180° (6, 16)	105–130° (6, 16)	1° (4, 5)	10–20 diopters (3, 6, 16)	56–67 c/deg (5, 6)	18 Hz (2)	40–60 Hz (10, 14)
Ungulates[a]	35–65° (6, 16)	300–350° (1, 16)	20–55° (5, 6, 16)	12–120° (4)	None (3)	8–12 c/deg (5)		

Sources (in parentheses): 1, Andreyev 1985; 2, Coile et al. 1989; 3, Duke-Elder 1958; 4, Heffner and Heffner 1992b; 5, Heffner and Heffner 1992c; 6, Hughes 1977; 7, Jacobson et al. 1976; 8, Loop and Martin 1991; 9, Magrane 1977; 10, Marler and Hamilton 1966; 11, Neuhaus and Regenfuss 1963; 12, Schaeppi and Liverani 1979; 13, Sherman and Wilson 1975; 14, Simonson and Brozek 1952; 15, Steinberg et al. 1973; 16, Walls 1942.
[a]Ungulates include pig, horse, cattle, elephant, and deer.

TABLE 3.13. Characteristics of the retinae of various mammals

| Species | Total rods | Rod densities (cells/mm²) | | | Total cones | Cone densities (cells/mm²) | | | Total no. ganglion cells |
		Maximum	In central area	In periphery		Maximum	In central area	In periphery	
Dogs	—	450,000 (5)	250,000 (5)	300–400,000 (5)	—	40,000 (5)	20–30,000 (5)	4–7,000 (3)	115,000 (6)
Wolves	—	—	250,000 (1)	185,000 (1)	—	45,000 (1)	—	3–5,000 (1)	200,000 (6)
Cats	204 million (2)	460,000 (7)	275,000 (7)	250–400,000 (7)	3 million (2)	27,000 (7)	15–30,000 (7, 8)	3–5,000 (7)	200,000 (2)
Humans	120 million (2)	160,000 (7)	0 (7)	80–120,000 (7)	6.4 million (4)	150,000 (7)	150,000 (2, 7)	4–5,000 (7)	1.2 million (4)

Sources (in parentheses): 1, Andreyev 1985; 2, Hughes 1977; 3, Koch and Rubin 1972; 4, Østerberg 1935; 5, Peichl 1991; 6, Peichl 1992b; 7, Steinberg et al. 1973; 8, Wassle and Boycott 1991.

Spatial and Temporal Acuity

The visual (spatial) acuity of canids has been little studied (Andreyev 1985). The fact that wolves (Peichl 1992b) and cats (Hughes 1977) possess similar numbers of ganglion cells suggests that wolf visual acuity may be like that of cats (see Loop and Martin 1991). Neither carnivore can match the extremely sharp foveal vision of humans (see table 3.12). Dogs, however, appear to be quite sensitive to motion (Duke-Elder 1958) and much better than humans at distinguishing among shades of gray (Pavlov 1927).

Flicker fusion rate, a measure of the eye's temporal resolution, improves as illumination increases (Marler and Hamilton 1966) and as cones become active (the rod/cone break) (Schaeppi and Liverani 1979). For both rod and cone systems, dogs have better temporal resolution than humans (see table 3.12), although the rod/cone break occurs at much brighter illumination in dogs (Coile et al. 1989). These results indicate that dogs, and probably wolves, see the world "faster" than humans and are thus able to make finer temporal use of visual information.

Color Vision

All canids studied to date have two cone systems, one sensitive in the blue range (429 nm) and the other in the green (555 nm) (Neitz et al. 1989; Jacobs 1993; Jacobs et al. 1993). Dogs can discriminate color best on either side of 480 nm (blue-green), the wavelength to which both cone types are equally sensitive. Color discrimination then deteriorates rapidly toward both blue and red. A large drop in color acuity above 520 nm indicates little ability to discriminate colors in the green to red range (Neitz et al. 1989). Reports that dogs (Rosengren 1969) and wolves (Asa and Mech 1995) can discriminate red and yellow as colors need to be evaluated with proper controls for brightness; in theory, red should appear as a fainter shade of green to a canid possessing only blue and green cones.

Although the rod systems of dogs and humans apparently have similar sensitivities to light intensity, the dog cone system is only 1/30 as sensitive as that of humans (Schaeppi and Liverani 1979). Thus the dog's color vision provides a coarser picture with fewer hues, and is limited to brighter illuminations, than human color vision.

Vision and Communication

The visual system of wolves is more than adequate to make use of the variety and nuances of their visual signals. Though wolves lack the high definition and color resolution of the human eye, they have the ability to use their visual system well into darkness. Although they lose color vision and suffer further deterioration in acuity at night, they can still discern the features of nearby packmates. In particular, the white facial patterns of many wolves may facilitate visual communication at night, and could also play an important role in coordinating group attacks on prey, enabling wolves to keep track of packmates without shifting their gaze from the prey. Thus vision probably remains an important communication medium throughout most of the night.

Tactile Communication

The Importance of Touch in Communication

Tactile communication in the wolf has been little studied (Fox and Cohen 1977; Zimen 1981), yet touch undoubtedly carries important information. Newborn pups are deaf and blind, yet are able to nurse and huddle (Fox 1971e). Although olfaction may cue these activities, the "rooting" reflex of young pups is also important. The huddling of pups around rocks and other debris in the den (Coscia 1995) is further evidence of their tactile orientation.

After a pup's eyes and ears open, tactile activities remain common. Pups continue to huddle, especially when cold, even when several months old, although this activity wanes over time (Coscia 1995). Socially directed actions, such as play, involve frequent body contact. Food-begging (see Packard, chap. 2 in this volume) involves licking and other direct contact, and active submission continues this snout-to-snout contact throughout life (Schenkel 1967; Fox 1971e).

For adults, body contact occurs often in "friendly" contexts, particularly during group ceremonies (Fox and Cohen 1977; Zimen 1981). While walking his captive wolves, Zimen (1981) found that they made brief muzzle-to-muzzle or muzzle-to-fur contact an average of six times per hour per wolf. Licking of fur occurs in a variety of situations; males, for example, lick the female's genital area during courtship (Fox 1972a; Zimen 1981) and also lick one another's wounds (Zimen 1981). Besides the olfactory or hygienic value of these activities, they may provide a tactile message as well.

Agonistic activity involves much body contact, from pushing against the flank of a rival to pinning another's muzzle to the ground (Schenkel 1947, 1967; Moran et al. 1981; Mech 1993a). Havkin (1977, 1981; Havkin and

Fentress 1985) described the development of "combative social behavior" ("play fighting") in wolf pups. He focused on the analysis of falls because agonistic interactions among adults often involve attempts by one wolf to throw its opponent and efforts by the other to avoid being thrown (Schenkel 1967; Moran et al. 1981).

Pups use three basic types of falls during play fighting (Havkin and Fentress 1985). *Side falls* occur earliest; one pup uses a bite attack coupled with a foreleg sweep to knock the other pup over. As pups grow older, they learn to avoid the foreleg sweep. At 40 days, the *rear fall* first occurs. The *front fall* emerges last; it involves the head landing first, followed by the rest of the body. This latter fall appears to be the most effective from a defensive standpoint, as the fall usually breaks the opponent's grip and carries the fallen pup's hindquarters out of the other pup's reach.

The best offensive position during play fighting is the stem of the so-called T configuration. In that position, the pup can bite its opponent virtually anywhere, whereas the latter must twist and turn to bite, making it likely to fall. During adult fighting, however, the animal on the offense stands at the top of the T (Schenkel 1967; Moran et al. 1981). This inherently unstable position has been likened to a "challenge" (Schenkel 1967). However, it may be that one of the most important lessons a pup learns from play fighting is how to protect itself effectively; thus a defensive wolf may quickly assume the more stable stem position when threatened at close quarters, denying that advantage to its opponent.

Communicative Roles of Touch

Two possible roles of tactile communication can be advanced. First, it may strengthen social bonds through reduction of stress. Studies of humans and their pet dogs have shown that tactile contact reduces heart rate and blood pressure in both humans and dogs (Lynch 1974; Baun et al. 1983; Vormbrock and Grossberg 1988). Many tactile contexts for pups lead to reduction of stress (i.e., huddling, nursing, defecating or urinating, and eating). Thus adult tactile behaviors may "tap into" earlier behavioral-sensory networks; when wolves receive a particular type of stimulation, in a specific context, they feel reassured.

Second, contact during aggressive behavior probably plays a role in assessing a rival. Information gained through physical contact during play and ritualized

fighting may indicate the strength or skill of an opponent. In fact, of all the ways of receiving information about a rival (i.e., visual—body size; vocal—pitch of voice; olfactory—hormonal profile), tactile information may provide the most reliable means of assessing an individual's current status. Strength and skill in combat change over time; thus assessment should be ongoing, particularly among young animals competing for social status. Once achieved, a strategy for maintaining status might be to avoid assessment, as age ultimately saps strength and dulls skills. Thus the involvement of individuals in behaviors that assess strength, skill, or stamina, including "play," may show predictable variation with age, rank, and health.

Development of Tactile Responses

Fox (1971e, 191) called a pup's head a "thermotactile sensory probe"; it moves side to side in a semicircular manner until it contacts an object. Stimulation of the face with a soft or warm object causes the pup to move toward the stimulus; this response strengthens during the first 2 weeks. When cold or hungry, pups may circle one another, maintaining contact all the while.

The initial movements of a newborn pup are limited to "swimming" motions until 2 or 3 weeks of age, when the pup can begin to support its weight (Breazile 1978). By 9–10 days of age, the pup can support itself with its forelegs, and its rear legs provide some support by 14–16 days.

Newborn pups defecate when stimulated in the anal area and urinate in response to stimulation of the vulva or preputial area (Fox 1964, 1971e; Breazile 1978). Both responses are typically elicited by the mother's licking of the pups during grooming. Earlier, we indicated that the preputial glands of newborn pups are swollen and exude a creamy fluid that the mother finds attractive. Thus an olfactory/gustatory signal provided by the pups elicits a tactile stimulus from the mother that stimulates urination or defecation in the pups. This important network operates during the same time frame as the pups' neonatal repertoire of squeals and screams, suggesting that there may also be a vocal component to the network (do pups vocalize more as the bladder fills?). If so, are the screams/squeals and preputial fluids part of the same developmental system, or are they independent?

Gustatory Communication

The role of taste in wolf communication has not been explored. Because investigations of taste can be confounded by the influence of smell, it can be difficult to properly evaluate the significance of taste as a sensory modality. Dogs, and thus possibly wolves, possess receptors for all four major categories of taste: salt, bitter, acid, and sweet (Appelberg 1958). Because wolves may consume fruits such as berries (see Peterson and Ciucci, chap. 4 in this volume), sweet taste receptors would be adaptive.

Taste may be involved in the transmission of pheromonal information contained in urine and various glandular secretions. Male wolves eagerly lick urine and secretions on the vulva of proestrous and estrous females as well as preputial secretions of other males. Females may lick preputial secretions of males as well as those of pups. Adults and pups lick the muzzles of other wolves. Although the primary pheromonal function of licking could be to facilitate the transfer of low-molecular-weight compounds into the nasal cavity or VNO, it may be important that these substances also taste good to perpetuate the behavior.

Grooming stimulated by blood on the muzzle or head of a packmate may be reinforced by taste. The methodical grooming of pups by their mothers suggests that their fur may contain a pleasant-tasting substance not present on older animals. Likewise, tastes associated with pups are apparently very appealing; the amniotic fluid, for example, stimulates the cleaning of newborns and consumption of the placenta (Dunbar et al. 1981). Similarly, the urine and feces of pups are readily consumed by their mothers during the first few weeks of life, a practice that gradually declines thereafter (Ranson 1981), reflecting the change in their contents from milk to meat waste products and thus possible changes in their stimulus value. However, wolves do commonly practice coprophagy, at least in captivity (L. D. Mech, personal communication).

Unfinished Business

Since Rudolph Schenkel (1947) launched the study of wolf communication a half century ago, we have made some progress in understanding how and what wolves signal to one another. Yet we still have much to learn. As should be clear from the last two sections, we know little about the signals and information wolves convey using touch and taste. This deficiency is regrettable. For an animal as large and powerful as a wolf, touch is probably a very important channel for both sending and receiving information about other wolves. The pressure applied during a hip-thrust, for example, could send an important message to the wolf on the receiving end, information that might even prompt an animal to disperse from its pack. Furthermore, how important is the physical contact that occurs during a social rally or while a pair of wolves is traveling?

And what about touch delivered and sensed via the jaws? "Mouthing" is a behavior that involves seizing the muzzle or head of another wolf, often in relatively relaxed contexts. How important is the amount of pressure, if any, that is applied during the mouthing? Captive wolves will mouth their human companions or acquaintances. In one small pack we studied in Minnesota, a subordinate male licked our hands when we placed them through the fencing. His dominant brother, on the other hand, would grasp a hand briefly in his mouth and then immediately release it; each time he relaxed his grip just as we began to feel pain. Did he simply not know his own strength, or was there a message in the pressure of his jaws, like that of a firm handshake? It is obvious that we have much more to learn about touch and taste.

Even the well-studied senses pose numerous questions for future work. Take the short-range vocalizations. Given Schassburger's (1993) and Coscia's (1995) research, it seems safe to say that wolf vocal communication is based on the dynamic and graded use of less than a dozen sound types or vocal signals. These vocal signals divide up the acoustic spectrum in a predictable manner that is consistent with Morton's (1977) motivation-structural rules and is therefore intuitively understandable to humans. Variation within the acoustic spectrum is continuous, yet our research suggests that each vocal signal occupies its own discrete portion of this spectrum. Are wolf vocal signals indeed categorical, and if so, are the boundaries between sound types a function of production or perception? There do seem to be no intermediates between sound types in pup vocalizations; is the same true in adults? If so, what rules of production (i.e., neural, mechanical) keep these sounds discrete?

Theoretically, short-range vocalizations appear to use the full dynamic range of the vocal cords. Indeed, it is possible that the squeak and other whistle forms greatly exceed the upper limit of the vocal cords by utilizing a whistle mechanism, thus giving larger animals the ability to sound even less intimidating or more submissive than otherwise. This possibility suggests that canids produce a sonic parallel to the ultrasonics of rodents, and that the only difference between wolf squeaks and rodent squeaks lies in the ears of humans. After all, most rodent squeaks are not ultrasonic to a wolf. In fact, the high upper frequency limit of wolves' hearing suggests a further question: What use do wolves make of this upper range, if any? Lehner (1978a) suggested that coyotes may use their upper limits for either intraspecific communication or the detection of rodents.

The longer-range vocal signals (barks and especially howls) appear to combine the motivation-structural rules with the basic structural features required for long-distance propagation. What sort of trade-offs are wolves making in this combination? What price in information content is paid to gain extended range? And from a social point of view, how much does the inability to vocalize without being overheard by an unintended audience influence the manner in which wolves howl?

With olfactory communication, we may never be able to fully understand how the world of scent appears to wolves, just as they will never share our experience of a spectacular sunset. We have made initial progress in describing and understanding the use of urine in marking from a functional perspective. Unfortunately, the signal itself, that cocktail of molecules that stimulates the olfactory bulb and association areas of the cortex, has yet to be fully described (cf. Raymer et al. 1984, 1986). How does this mixture correlate with sex, age, social status, reproductive condition, and health, how does it age over time since deposition, and how do wolves use it to discriminate among individuals and to find their way?

Finally, the visual channel offers challenges much like those presented by the olfactory channel, but is much more tractable to human study; we can see at least as well as wolves, and perhaps even better. Like olfactory signals, visual signals offer a "Gestalt" involving a collection of many simultaneous elements. How well do all these elements actually fit together as a package? Which elements covary and which are independent, and what are the rules for their combination? Schenkel (1947, 1967) hinted at the complexity of expression that is made available through visual signals. McLeod (1996) has shown how we might begin to understand this signaling system, and Moran (1987) has shown how it pays to look at the system from the point of view of relationships

rather than separate individuals. In the future, an increasing knowledge of the cognitive structures and rules used in visual perception will give us better insight into the importance of the various elements used in visual signaling.

Communication and Behavior

We have tried to present a thorough picture of the signals and information involved in wolf communication without delving too deeply in the details of the social fabric within which these signals are used, the topic of chapter 2 in this volume. Before we finish, however, we would like to return to an issue mentioned at the outset that we have touched on along the way: wolf sociality. Wolves are highly social animals. From its birth through perhaps its first or second year, a wolf is usually in the company of packmates. There may be some periods when it is alone, but those periods typically are short. For this reason, we should expect wolf communication to have several important features.

First, communication signals should have a predictable developmental history. Wolf pups, like human infants, ought to possess an initial signal repertoire that guarantees that their critical needs will be met. Shortly afterward, however, that repertoire should reveal important developmental plasticity as pups react and adjust to their littermates, mother, and other packmates. Thus the simple messages available in the initial signals begin to acquire further information (metainformation) as pups learn their packmates' identities and personalities. Finally, contextual information is factored in as pups become more experienced with settings and situations. Thus we should expect to see increasing sophistication in the use of signals, as well as a moderate degree of individual variation due to different developmental histories.

Second, given the amount of time wolves spend together, we should expect them to develop a great deal of sensitivity to signals as they learn to better predict the ensuing actions of packmates. Thus the signals required to coordinate effectively between two long-standing packmates ought to be relatively subtle, compared with those between animals with less mutual experience. However, most pups will emigrate before they realize their full potential for such coordination. During the ensuing period of wandering, the lone wolf may spend most of its time monitoring the howling or scent marking of pack wolves, while at the same time attempting to minimize any inadvertent sign to these packs. But while it is avoiding packs, it is also monitoring signals and other signs for a breeding territory and a potential mate. If it succeeds in finding both, it probably faces a period of uncertainty as it gets to know, understand, and predict the actions of its mate. Given time, it may reestablish the same sensitive understanding it acquired in its natal pack as a pup.

4

The Wolf as a Carnivore

*Rolf O. Peterson and
Paolo Ciucci*

FROM THE BEST-SELLING BOOK and popular movie, *Never Cry Wolf* (Mowat 1963), millions of people gained the impression that wolves eat mice, rarely caribou. Author Farley Mowat later admitted fabricating much of the story (Goddard 1996), originally billed as true, to gain public sympathy for the wolf. Mowat succeeded enormously, and decades later the misconception remains.

Wolves are flexible and opportunistic predators, but they usually rely on large ungulates for food. They do not always have to kill their prey. R. O. Peterson has seen wolf scats full of maggots after wolves consumed rotting carcasses of moose, and he has watched wolves in winter dig out year-old moose hides dried by the summer sun months earlier. Perhaps the most accurate statement that can be made about the diet of the wolf is that it is usually hard won and highly variable.

While the wolf is popularly viewed as a consummate carnivore, it belongs to a family of carnivores that is adapted to feeding on a diverse array of foods. Wolves and other canids obtain most of their food from prey, but they are not exclusive meat eaters, or hypercarnivores, like the many species of cats. Throughout their evolutionary history, wolves have been shaped as cursorial predators of large herbivores, their characteristic niche. Yet, as generalist carnivores, wolves can effectively hunt prey that range in size over three orders of magnitude, from 1 kg (2.2-pound) snowshoe hares to 1,000 kg (2,200-pound) bison, and they can even subsist on garbage.

In this chapter we focus on the living economy of the wolf, from the wolf's standpoint. Where does the wolf find food, and how does the animal make use of it? We have learned the characteristics of the wolf diet both directly, by aerial or ground tracking, and indirectly, especially through analysis of prey remains in wolf feces (i.e., scats). After reviewing the techniques used to discover patterns of prey use, we describe the digestive anatomy and physiology of the wolf, summarize geographic differences in wolf diet between Eurasia and North America, and discuss the ecology of feeding. The next chapter will explore the ecological implications of wolf predation, and chapter 7 will deal with nutrition itself.

The Food of Wolves

There has long been strong scientific interest in knowing what wolves eat, as the wolf's diet forms the core of human conflict with this large carnivore. Data on the food habits of wolves have accumulated since Murie's pioneering study (1944) in Mt. McKinley (now Denali) National Park. Prior to the aerial tracking of wolves initiated by Burkholder (1959), the only technique used to determine wolf diet was analysis of the contents of either stomachs or scats. More recently, with fewer wolf control programs in North America, the contents of wolves' stomachs have rarely been available for study, but more field studies have provided direct analyses of wolf-killed prey as well as more extensive reporting of scat contents.

While examining wolf scats and stomach contents provided basic data about the wolf's diet, these approaches afforded no information about wolf hunting behavior. In summer, and in southern ranges without seasonal snow, heavy reliance is still placed on scat studies to understand the food base of wolves.

Direct Recording of Kills

The travels and kills of wolves may be studied in winter by following tracks, either from aircraft or on the ground. Bjärvall and Isakson (1982) tracked wolves on skis in Sweden when wolves were rare and could be distinguished as identifiable individuals or packs. While wolves can be visually tracked from the air through snow when it is not windblown (Burkholder 1959; Mech 1966a,b; Pimlott et al. 1969; Peterson 1977), radiotelemetry has greatly furthered this approach (Mech and Frenzel 1971a; Kolenosky 1972; Fuller and Keith 1980a; Ballard et al. 1987, 1997; Fuller 1989a,b; Kunkel et al. 1999). Boyd (1994) and Weaver (1994) relied primarily on snowshoes to ground-track radio-collared wolves to find kills and determine their kill rates. The primary problem with any tracking technique is maintaining constant contact with the wolves' path so that no kills are missed.

In theory, radio-collared wolves could be located and observed from aircraft frequently enough so that all kills would be recorded (e.g., two kills found on 14 days suggests a kill rate of one per 7 days). However, even if packs are located daily, kills are often missed when the prey is deer-sized. In a 5-year study of wolf predation on deer in Minnesota, Fuller (1989b) found wolves on kills 96 times out of 840 locations of packs, but never on the same kill during two consecutive flights (even with two flights per day). No packs, regardless of size, stayed at a deer kill for an average of more than 12 hours, so with one flight per day, the proportion of flights when wolves were found on kills underestimated the kill rate by at least 50%. Fuller corrected the data by multiplying the daily kill rate by 2 and considered this a minimum estimate.

With large prey such as moose, one can find virtually all kills if packs are located one or two times daily (Fuller and Keith 1980a), particularly right after sunrise, when wolves are most apt to be sleeping near kills (R. O. Peterson, unpublished data). By locating radio-collared wolves at least daily and supplementing with two locations on about half the days, Fuller and Keith (1980a) believed they found all moose kills. However, if prey are smaller and abundant, locating wolves even twice daily may not be sufficient to record all prey killed. In northwestern Alaska, a pack of eight wolves killed and ate a yearling caribou and left the site, all within a 3-hour period (Ballard et al. 1997, 12). Only by following snow trails can complete records of kills be made, and then only for prey large enough that the wolves leave remnants of it behind.

Indirect Studies of Wolf Diet

Following the lead of Murie (1944), scat analysis (Korschgen 1980; Putman 1984) has been used worldwide to characterize the wolf diet. Large samples are possible at all seasons of the year, and this method leaves the wolves undisturbed. However, wolf scats may be indistinguishable from dog scats (Ciucci et al. 1996) or coyote scats (Weaver and Fritts 1979), and the contents may vary with the digestibility, size, or frequency of meals (Meriwether and Johnson 1980; Kelly 1991).

Problems also arise in identifying the scat contents themselves (Frenzel 1974; Weaver and Hoffmann 1979; Reynolds and Aebischer 1991). Macroscopic food items in scats (i.e., hair, bones, feathers, invertebrates, seeds, and vegetation) are usually identified by comparing them with a reference collection or manual (Adorjan and Kolenosky 1969; Debrot et al. 1982; Teerink 1991). Observer reliability should be evaluated either by reanalyzing a random subsample (e.g., Fritts and Mech 1981) or by testing observers with a "blind" sample (Carbyn and Kingsley 1979; Ciucci et al. 1996).

Prey importance in the diet can be ranked either by direct measurement (i.e., by frequency, dry weight, volume) of the undigested remains in scats or by converting prey frequency data to estimates of biomass ingested based on experimental data (Floyd et al. 1978; Weaver 1993). For large prey, however, such data are sparse. Additional fine-tuning of the regression equations used to convert frequency data to prey biomass (Floyd et al. 1978; Weaver 1993) could incorporate provisions for variable carcass utilization and wolf activity levels; more replicates of various prey types would also be useful. In addition, if the energy equivalents of various prey species and their components were better known, hypotheses on the feeding strategy of wolves could be tested (e.g., Crawley and Krebs 1992), which has rarely been done for carnivores as a whole (Haufler and Servello 1994). To determine kill rates for wolves throughout the year, Jedrzejewski et al. (2002) supplemented ground searches for radio-collared wolves and prey carcasses with detailed analyses of scats, including the proportion of amorphic mass from meat and soft tissues. More amorphic material corresponded with a shorter interval since a fresh kill. In this study, up to 41% of wolf kills were detected only from scats, representing mostly small-bodied prey.

Some aspects of wolf diet may be determined through analyses of various isotopes. Prey that consume lichens and mosses, such as caribou and black-tailed deer, may ingest high levels of radiocesium, ^{137}Cs, from nuclear bomb testing (in the 1950s and 1960s) or nuclear accidents (e.g., Chernobyl), which can be detected in muscle tissue from wolves (Holleman and Stephenson 1981). Naturally occurring heavy isotopes of carbon (C) and nitrogen (N) might also be employed to advantage. The relative abundance of ^{13}C and ^{12}C, expressed as a modified ratio termed δ^{13}C, is particularly useful in distinguishing marine and terrestrial food sources and establishing the relative importance of C_3 and C_4 photosynthetic products in food chains (Tieszen and Boutton 1988). Applied to wolves, the isotope technique could be used to establish whether wolves in coastal areas are supported by marine life, such as spawning salmon (Szepanski et al. 1999; Darimont and Reimchen 2002) or seal carrion (Gilmour et al. 1995), or to investigate whether individual wolves have been eating domestic animals that have corn in their diet. Corn, as a rare C_4 plant in temperate regions, leaves a distinctive isotopic signature (Schoeninger et al. 1983).

Wolf Feeding Habits

The adaptability of the gray wolf is best exemplified by its highly variable diet throughout the world. The economy of a wolf in Canada's Yukon is likely to be based on moose, while a wolf living in certain Mediterranean regions may subsist largely on garbage and domestic animals. Wolves also inhabit regions with large seasonal fluctuations in the environment, with consequences for prey availability. They may frequently kill juvenile Arctic hares in the brief High Arctic summer on Ellesmere Island (Mech 1988a, 1995c) or beavers along the well-watered valleys of Isle Royale National Park in Michigan when those waters are unfrozen (Mech 1966b, Peterson and Page 1988). In Alaska, wolves eat salmon (Mech et al. 1998).

The wolf is an opportunist with an amazing ability to locate food. A midwinter thaw might find a wolf abandoning its cursorial habit to curl up next to a hole in the ice where beaver have suddenly renewed their foraging (Thurber and Peterson 1993). Wolves might spend hours scavenging seal carcasses washed up on a beach after a storm (Klein 1995), hunting unfortunate prey displaced by a wildfire (R. O. Peterson, unpublished data), or carefully inspecting mounds of garbage for anything that might be nourishing (L. D. Mech, unpublished data). Wolves stay alive through attention to such details, yet

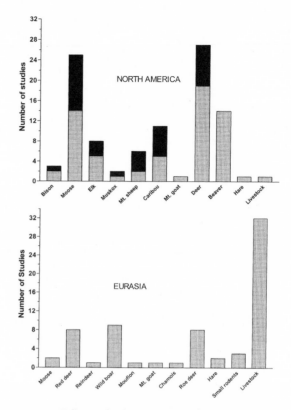

FIGURE 4.1. Reliance of wolves on various prey species in North America and Eurasia, as indicated by > 20% occurrence in wolf scats (stippled columns) or in > 10% of wolf-killed prey (solid columns). All studies reviewed are listed in appendix 4.1. Because this metadata set was not collected systematically, occurrences of individual prey species may be biased, depending on how much or how little wolf food habits were studied in various areas.

these usually disappear in the screening of information that accompanies scientific reporting of wolf diet. Nevertheless, it is clear that the diet of the wolf is as broad as its geographic range.

Studies of the wolf's diet are perhaps more common than any other kind of wolf research. By simply enumerating these studies (see appendix 4.1), one can gain a preliminary impression of variation in the food economy of wolves around the Northern Hemisphere, as well as broad-scale differences among continents (fig. 4.1). Of course, the geographic locations of such studies were not systematically established, so prey that predominate in areas of frequent study are overrepresented, and vice versa. White-tailed deer, for example, predominate in the midwestern United States, where wolf research has been conducted over many decades, so deer are probably overrepresented among North American studies.

The significance of moose in the diet of North American wolves, on the other hand, seems accurately reflected by the tally of studies. Beaver also emerges as a significant summer prey in North America. In more human-dominated Eurasian ecosystems, livestock appear to be the most common prey, with red deer, boar, and roe deer also important. Most wolves in Eurasia live in Siberia, where moose are the most common prey (Bibikov 1985), but these areas are underrepresented in the tally of studies for Eurasia.

Many aspects of wolf prey selection patterns might be understood in terms of prey size; other things being equal, one might think that wolves would prefer small ungulate prey that escape by running to those that use direct defensive behavior, such as large-bodied moose and bison. However, as indicated by Mech and Peterson in chapter 5 in this volume, a rigorous attempt to determine wolf preference for any kind of prey is fraught with methodological problems. Flexibility and opportunism dominate wolf foraging behavior, as wolves can thrive on any prey that is large enough, abundant enough, and catchable enough. In addition to abundance and size, relative species abundance, physical vulnerability, defensive behavior, and environmental conditions (snow parameters) all influence which prey species are most important in a local area for a given period (Mech et al. 1998). Where several prey species are available to wolves, fluctuating in density and vulnerability across time and space, the resulting pattern of prey selection may be complex, highly variable, and subject to change.

Eurasia

Much of the natural habitat of the wolf and its prey throughout Eurasia has been fragmented, altered, and destroyed by human activities, and many native prey species have been extirpated. Historically, native prey diversity declined from five or six species to just two or three (Okarma 1995). In areas most affected by humans, wolves have been forced to subsist on domestic animals and garbage, although native prey are important in the most remote portions of Eurasia (e.g., Russia and mountainous regions of eastern Europe), and where they have recovered. Variation in the relative abundance, vulnerability, and accessibility of wild and domestic prey produces a complex food economy for Eurasian wolves, a testimony to the animal's highly flexible and opportunistic feeding behavior.

Native wild ungulates important for wolves in Eurasia are moose, red deer, roe deer, and wild boar (Bibikov 1982, 1985; Filonov 1980; Okarma 1995). Other species that may be locally abundant and thus important to wolves include wild reindeer, mouflon, European bison, saiga, ibex, chamois, mountain goats, fallow deer, and musk deer. In the boreal zone of Scandinavia, moose are the primary prey in forested areas, but roe deer are important in agricultural areas (Bjärvall and Isakson 1981; Wabakken et al. 1983; Olsson et al. 1997). Wild reindeer are the primary prey in tundra regions of Siberia (Makridin et al. 1985; Kolpashchikov 1995), giving way to moose in the vast taiga zone (Labutin 1972; Bibikov 1982; Filonov 1980).

In temperate forests throughout Eurasia, wild boar may be abundant, providing a valuable prey for wolves. Near the Caspian Sea (Kyzyl-Agach Reserve), wild boar provide about two-thirds of the wolf diet (Litvinov 1981). In the Apennine range of Italy, the recent recovery of wolves corresponded to a large expansion of the wild boar population, which constitutes 12–52% of the wolf diet (Ciucci and Boitani 1998a).

The red deer is a commonly killed species in mixed and deciduous temperate forest regions, both in lowlands and in mountains (Kudaktin 1978; Okarma et al. 1995); it is supplemented by roe deer, sika deer, and wild boar. In Poland's Białowieza National Park, where Europe's richest community of wild ungulates inhabits pristine lowland forests, wolves' primary prey is red deer, followed by roe deer and wild boar (Reig and Jedrzejewski 1988; Jedrzejewski et al. 2000). Wolf predation accounted for nearly 75% of red deer natural mortality, but wolves virtually ignored European bison (Okarma et al. 1995). In France's Mercantour National Park, colonized by wolves from northern Italy in the mid-1990s, the primary prey, at least initially, were mouflon and chamois (Poulle et al. 1997).

Where wild ungulates have been reintroduced or historically conserved, these herbivores support wolf populations, as in portions of Italy and Spain. Wild boar and red deer compose a large portion of the wolf diet in the southern mountains of Spain (Estremadura and Sierra Morena). Especially in Sierra Morena, where big-game hunting reserves allow high ungulate densities, wolves rely almost exclusively on red deer and wild boar (Castroviejo et al. 1975, Cuesta et al. 1991). In Spain's Cantabrian Mountains, where otherwise domestic prey prevail in the wolf diet, scats collected in Sierra de Invernadeiro most frequently contained roe deer remains (Guitian et al. 1979). Along the northern Apennines in Italy, wild boar and roe deer generally appear as the most frequent prey, followed by red deer and locally available species such as fallow deer and mouflon (Mattioli et al. 1995; Meriggi et al. 1996; C. Matteucci, personal communication).

In India and China, where wolves frequently live on foods of human origin, they thrive almost exclusively on wild ungulates in some restricted localities (Gao 1990, Jhala 1993). Both wolves and their prey are endangered in Velavadar National Park in India, where blackbuck constitute 88% of the biomass consumed by wolves and wolf predation is the primary cause of blackbuck mortality. There is essentially no wild prey available in India outside relatively small and scattered nature reserves (Jhala and Giles 1991).

As in North America, Eurasian wolves appear to kill prey based on size, abundance, and vulnerability. Wolves rely on moose only where smaller prey are uncommon (Bibikov 1982). In Bialowieza Primeval Forest, wolves hunted in packs averaging 4.4 members and often killed wild boar piglets, young deer, and beavers, but generally, larger packs more often killed red deer (Jedrzejewski et al. 2002). Okarma (1995), after reviewing food habit studies of the wolf in Europe (including twenty-seven studies from the former Soviet Union), concluded that red deer were most often taken in most areas. Wolf tendency to kill red deer in ungulate communities of lowland mixed deciduous forests has also been documented by Filonov (1980) for the European part of the former Soviet Union. The significance of moose and red deer in the diet of wolves finds certain parallels in North America.

Prior to the restoration of wild prey in many areas of Eurasia in the 1980s and 1990s, wolves preyed on livestock. They continue to do so today where wild ungulates are absent or extremely rare. This is especially true where preventive measures are ineffective and domestic stock are easily accessible to wolves. In northern Finland, for example, herds of semi-domestic reindeer are important to wolves (Pulliainen 1993). In the former Soviet Union, domestic prey are important especially at lower latitudes and in less forested areas, where the grazing period is longer and wild ungulates are scarce (Bibikov 1982, 1985). Bibikov et al. (1985) found pigs, cattle, and sheep as the most frequent items in both scats and stomach contents of wolves from the eastern Caucasus and Voronez area.

In densely populated northwestern Spain (Galicia and Asturia), the wolf diet contains no wild prey, being composed of goats, sheep, and dogs (Castroviejo et al. 1975; Guitian et al. 1979; Cuesta et al. 1991; Reig et al. 1985; Llaneza et al. 1996). Similarly, in the Douro Meseta area of Spain, domestic ungulates, along with garbage and offal from domestic animals, represent the bulk of the wolf diet (Reig et al. 1985; Cuesta et al. 1991). Wolves from this region also rely on carrion of horses and cattle (Castroviejo et al. 1975; Barrientos 1993). A preponderance of livestock in the wolf diet was also reported for Portugal (Álvares 1995) and some localities of Italy during the 1970s (Macdonald et al. 1980; Boitani 1982; Ragni et al. 1985).

Where both wild and domestic ungulates are found, the feeding ecology of wolves appears to be driven by the relative availability of wild prey, rather than by human-related food. For example, in southeastern Poland (Bieszczady Mountains), wolves kill few domestic ungulates during the grazing season (Smietana and Klimek 1993). However, during fall and winter, they readily feed on livestock carcasses used to bait them in to be shot from towers (Lesniewicz and Perzanowski 1989; Smietana and Klimek 1993). In Greece, wolves subsist mostly on free-ranging livestock (Adamakopoulos and Adamakopoulos 1993) because wild prey are rare (Papageorgiou et al. 1994).

There may be seasonal differences in wolf reliance on wild versus domestic prey, depending on the influence of climatic (e.g., winter severity, snow depth) and demographic (i.e., production of newborns) factors on wild ungulate vulnerability as well as on livestock accessibility as determined by husbandry practices and length of the grazing season. In northern Sweden and in Siberia (Chukotka), wolves rely on moose and wild reindeer in winter, but turn to semi-domestic reindeer that are herded within range in summer and fall (Bjärvall and Isakson 1982; Zheleznov 1992). In Bulgaria, domestic prey predominate in the wolf diet in summer, but during winter wild ungulates are most important (Ivanov 1988).

Similar seasonal patterns in wolf predation have been reported in several other areas of Europe, where livestock depredation generally increases during the grazing period and predation is redirected to wild ungulates during the rest of the year. Where wild prey have been restored, wolf predation on domestic animals has sometimes been reduced. In Poland, for example, livestock losses to wolves were lower in the 1990s than in the 1950s, after a four- to fivefold increase in red deer (Okarma 1993; cf. Perzanowski 1993). Similarly, in Romania, the proportion of wolf stomachs containing domestic prey declined from 76% in 1954–1967 to 22% in 1991, while the proportion containing wild prey increased from 25% to 78%; an increase in wild prey may explain the change, but during the same period livestock grazing became prohibited in many forested areas where wolves live (Ionescu 1993).

As in North America, wolves may rely on medium-sized prey such as hares and beavers in some locales. However, the European beaver was reported in only trace amounts in the wolf diet in Sweden (Olsson et al. 1997) and as a secondary prey after red deer in the Ukraine (Tkachenko 1995); thus it seems much less important than the beaver in North America. Hares, however, were included in the wolf diet in twenty-two of thirty-one studies in Eurasia. In the northern taiga and tundra regions of Eurasia, where ungulate availability may be quite low, the blue hare may locally be one of the main prey (Okarma 1995). Hares are more important in the wolf diet in summer, when they are more available and ungulates may be less vulnerable (Bibikov 1982; Salvador and Abad 1987; Jedrzejewski et al. 1992; Smietana and Klimek 1993; Ciucci 1994).

In parts of Eurasia, wolves live in areas with relatively little wild prey, but subsist nevertheless on a wide variety of foods provided indirectly by humans. Foraging in garbage dumps, wolves eat meat scraps and various fruits, as well as inadvertently consuming nonfood debris. In Israel, the following items were found in wolf scats: human hair, plastic, tinfoil, cigarettes, matches, and eggshells (B. Shalom, personal communication). In Minnesota, long, sharp shards of glass were found in scats of garbage-dump-feeding wolves (L. D. Mech and H. H. Hertel, unpublished data).

Perhaps because of the greater availability of fruit, wolves in southern portions of Eurasia may feed on plant material more extensively than those in North America. Radio-collared wolves in the lowlands of central Italy have been monitored as they moved through mature vineyards (P. Ciucci, unpublished data). Fruit may provide vitamins for wolves in summer, as even in North America it is not uncommon to find seeds from raspberries and blueberries in wolf scats (Van Ballenberghe et al. 1975; Peterson 1977; Fuller 1989b). Cherries, berries (Hell 1993; Ciucci 1994), apples, pears, figs, plums,

grapes (Castroviejo et al. 1975; Guitian et al. 1979; Bibikov et al. 1985; Gao 1990; Cuesta et al. 1991; Papageorgiou et al. 1994), melon, and watermelon (Gao 1990) have been reported in wolf scats.

Grass (Graminae) merits brief mention, as it appears in wolf scats in North America as well as in Eurasia with 14–43% frequency (Ragni et al. 1985; Salvador and Abad 1987; Patalano and Lovari 1993; Ciucci 1994; Papageorgiou et al. 1994). Possibly grass acts as a scour or inducement to vomit, ridding the intestine of parasites or the stomach of long guard hairs that delay passage of food through the gut, or as a source of vitamins (Mech 1970; Kelly 1991).

North America

In North America, the wild prey of the wolf have largely continued to occupy suitable habitats with relatively low human density. An exception is the American bison, which was virtually extirpated in the United States, but survived as wolf prey in Canada's Northwest Territories (Carbyn et al. 1993), and has recovered in Yellowstone National Park (Smith et al. 2000). Wolves have generally subsisted on wild prey throughout their current North American range, and they currently feed on garbage and domestic stock only in isolated circumstances.

Wolves in North America experience strong seasonal shifts in environmental conditions, including a lengthy winter season in which ungulates are most vulnerable to wolf predation because of snow accumulation and elimination of water as a refuge (see Mech and Peterson, chap. 5 in this volume). In winter, the diet of North American wolves is dominated by ungulates. Juvenile ungulates are often the most common age class killed (see Mech and Peterson, chap. 5 in this volume), but because of their smaller biomass, they contribute proportionately less to the wolf diet in winter than adults do.

Wolf predation patterns were documented in at least eighteen studies within Canada during the 1980s (Hayes and Gunson 1995). Wolves in Canada's eastern region are supported by moose and white-tailed deer, with deer the commonest prey, while wolves in the western provinces exist in multi-prey systems with up to four or five ungulate species (elk, moose, mule deer or black-tailed deer, mountain sheep, and caribou). In these mixed-prey complexes, elk usually predominate in the winter diet of wolves, but the carnivores may also rely on a secondary prey species (Weaver 1994; Bergerud and Elliott 1998; Kunkel et al. 1999). North of continuous forest in mainland Canada, wolves rely on migratory caribou, while on the Arctic islands, muskoxen are the primary ungulate prey (Miller 1995).

Six telemetry-based studies in Alaska revealed that moose usually predominate in the diet of wolves in multi-prey systems lacking elk or deer (Stephenson et al. 1995). In south-central Alaska, Ballard et al. (1987) found that 70% of wolf kills were moose, with caribou a secondary source of prey. In northwestern Alaska, where migratory caribou were the most abundant prey, moose still composed 42% of wolf kills (Ballard et al. 1997). Caribou are the most common prey where they are abundant (Dale et al. 1994; Mech et al. 1998), and Dall sheep (Murie 1944; Mech, Meier et al. 1995) and black-tailed deer (Klein 1995) can be locally important prey within their Alaskan ranges.

The near elimination of bison on the North American continent late in the nineteenth century removed the most abundant ungulate prey of wolves in North America, both in numbers and biomass. Up to 60 million bison may have existed on the grasslands and plains of North America (Roe 1951), although Native American hunting may have kept bison numbers much lower than the carrying capacity of their range, perhaps at only 10–15 million animals (Shaw 1995) or even less (Kay 1998). Hampton (1997) opined that 200,000 wolves might have been supported by bison prior to European contact. Nowhere else was there such a concentration of prey and wolves (Van Ballenberghe et al. 1975).

Today we can only glimpse something similar when wolves accompany migratory caribou numbering in the tens of thousands on Canada's Barrenlands. In the 1970s, a single observer counted over fifty wolves in one day following a caribou herd across a frozen lake during the spring migration (D. Thomas, personal communication). Similarly, over thirty wolves have been observed at one time on a frozen lake in Wood Buffalo National Park (L. Carbyn, personal communication), one of the few areas in the western provinces of Canada where wolves today exist on a diet of bison. Bison were largely replaced by domestic herbivores, and throughout the original range of bison, wolves were eliminated by private and government control efforts (Young and Goldman 1944).

There have been ten studies of wolf predation in mixed-prey complexes including elk, providing data on wolf predation in winter from 868 kills (Weaver 1994). Wolf predation on deer equaled that on elk, as each prey species constituted about 42% of the diet. Considering food biomass available to wolves, elk emerged as the most important prey (56% of total biomass), while deer and moose each contributed about 20%.

Based on his study in Jasper National Park, Alberta, Weaver (1994) believed that deer and elk were about equal in terms of profitability (energy gain divided by handling time; see Mech and Peterson, chap. 5 in this volume), followed by moose and then bighorn sheep. While moose provide a large biomass per kill, they require longer chases than elk (mean chase distance 883 vs. 100–200 m, respectively: Weaver 1994), and they pose a greater danger during encounters because they often stand and pugnaciously defend themselves (Mech 1966b). Elk are still relatively large prey, but they may be less dangerous than moose and more hampered by deep snow because of their shorter legs. Living primarily in groups, elk probably have a relative advantage over more solitary prey by having better senses and a lower individual probability of capture. Deer are even less able to cope with deep snow and pose less threat to attacking wolves, but Weaver (1994) found that the average chase distance for deer exceeded that of elk. Bighorn sheep were taken least often, probably because of the difficult escape terrain they inhabit.

Studies of wolves in other multi-prey systems have revealed a multitude of factors that may influence types of prey killed (see Mech and Peterson, chap. 5 in this volume). Mech et al. (1998) examined 526 wolf-killed ungulates in Denali National Park. They found that wolves killed approximately equal numbers of moose and caribou (47% and 42%, respectively) and a much lower number of sheep (11%), even though the park contained approximately twice as many caribou (3,000–4,000) as moose and sheep (2,000 each). Wolves appeared to concentrate on the prey species in poorest condition, so patterns of prey selection varied among years and among seasons.

Dale et al. (1995) studied wolf predation patterns on the same three prey species in Gates of the Arctic National Park, Alaska. There, Dall sheep and moose were locally abundant, at 5.0/10 km^2 (3.9 mi^2) and 1.2/10 km^2 in suitable range, respectively, while caribou density varied fortyfold among four wolf pack territories (0.6–23.4/10 km^2). Wolves killed a disproportionate number of caribou, which accounted for 93% of 177 wolf kills. When caribou migrated away and their density dropped to less than 2/10 km^2, resident wolves usually did not follow the caribou, but rather preyed on the sparse moose population (Ballard et al. 1997).

In summer, wolves in North America have a more

Wolves in Yellowstone National Park's restored population often must compete with grizzly bears at the carcasses of their kills. Especially after their first feeding at a kill, wolves tend to relinquish their kills to bears. Photograph by Doug Smith.

The red wolf *(Canis rufus)* tends to possess larger ears, a narrower nose, smaller feet, and a more reddish coat than the gray wolf. The appearance of the animal is intermediate between that of a coyote and a gray wolf. Photograph by Barron Crawford.

Elk are one of the most common prey of wolves in the northern Rocky Mountains of North America, primarily because in most areas they tend to outnumber other prey. This large bull standing his ground escaped this pack. Photograph by Doug Smith.

Wolf pups nurse for up to 9 weeks in several bouts of 2–5 minutes each per day (Packard et al. 1992). However, if the nursing female is killed, pups as young as 5 weeks can survive without further nursing (D. W. Smith, personal communication). Photograph by L. David Mech.

Bottom left: Wolves attack every species of mammal in their range, including large carnivores such as cougars and bears. Like other animals pursued by wolves, bears sometimes take to water to escape. Photograph by Doug Smith.

Top and bottom right: "The wolf is kept fed by his feet," according to an old Russian proverb. To obtain sufficient vulnerable prey, wolves generally must travel 25–50 km (15–30 miles) per day. Their usual travel formation is single file. *Top:* Photograph by Isaac Babcock. *Bottom right:* Photograph by Doug Smith.

Often a wolf pack will attack a strange wolf and kill it. This behavior usually occurs when the strange wolf enters a pack's territory. However, sometimes an especially aggressive pack will trespass and attack residents of the territory they invade (D. W. Smith, personal communication). Photograph by Doug Smith.

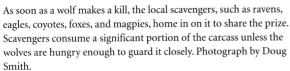

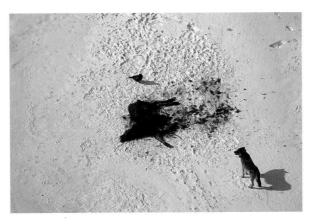

As soon as a wolf makes a kill, the local scavengers, such as ravens, eagles, coyotes, foxes, and magpies, home in on it to share the prize. Scavengers consume a significant portion of the carcass unless the wolves are hungry enough to guard it closely. Photograph by Doug Smith.

Middle and bottom: A moose provides several meals for the wolf pack that kills it. On Isle Royale in Lake Superior, lying between Canada and the United States, moose is the only large prey available. Photographs by Rolf Peterson.

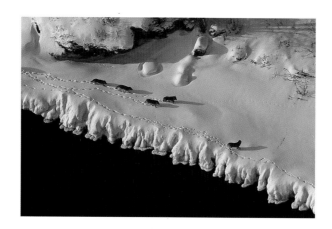

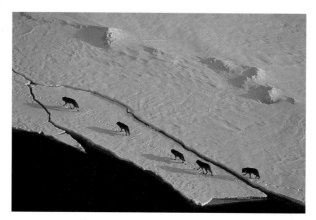

Wolves prefer to travel on hard-packed surfaces such as ice, shorelines, frozen rivers, rocky ridges, roads, and trails. Their usual rate of
travel is about 8 km (5 miles) per hour (Mech 1970, 1994b).
Photographs by Rolf Peterson.

Moose are one of the wolf's most formidable prey. If a moose flees from wolves, the wolves follow, sometimes for a kilometer or more. Photograph by Rolf Peterson.

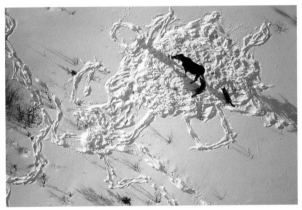

Moose often stand their ground and fight off attacking wolves. Rarely do wolves capture such a moose unless they can make it run. What makes some moose stand their ground and others flee wolves is unknown, but conceivably individuals in the best condition have the confidence to stand and defy wolves. Photograph by Rolf Peterson.

Wolves usually try to bite the rump of a moose. After these attacks slow the animal down, often one wolf will grab it by the nose while the others continue to bite at its rump. This process frequently takes wolves hours and sometimes days. Other times, wolves wound a moose, leave it, and return days later to kill it. Photograph by Rolf Peterson.

"Standing over" is a common behavior in which a wolf pack member stands over another that is lying down. The standing wolf positions its groin over the recumbent wolf's head. Often the recumbent wolf ignores the other, but sometimes sniffs its genitals (Mech 2001b). Photograph by Rolf Peterson.

Although prey often can escape wolves by heading for water, this ploy does not always work, as this old, arthritic Dall ram learned. Wild sheep and goats find their best refuge from wolves in steep mountain terrain. Photograph by L. David Mech.

Like dogs, mature, dominant wolves raise their legs when they urinate. Males perform a complete raised-leg urination (RLU), whereas females merely cock the leg similarly in a flexed-leg urination (FLU). Photograph by L. David Mech.

When wolves secure an abundance of prey, they bury some of it for future use. Sometimes large pieces of prey, such as half a newborn caribou calf, are so cached. Other times, wolves cache food that they regurgitate. Photograph by L. David Mech.

Howling seems to serve at least three functions: (1) letting packmates know where each one is, thus facilitating assembly; (2) advertising the pack territory and warning against intrusion; and (3) motivating packmates to join in on a hunt. Photographs by L. David Mech.

Top: Each wolf pack usually produces a litter of pups each year in a den, which may be a rock cave, a hole in the ground, an abandoned beaver lodge, a hollow log, or even a shallow pit. Photograph by L. David Mech.

Bottom left and right: Distinctively white, the arctic wolf *(Canis lupus arctos)* is one of five subspecies or geographic races of the gray wolf recognized in North America. Most arctic wolves live north of 70° N latitude, but some venture much farther south. Photographs by L. David Mech.

diverse diet than in winter, with significant contributions from beavers, snowshoe hares, and juvenile ungulates. However, ungulates still usually compose more than 75% of the biomass consumed by wolves in summer (Ballard et al. 1987; Fuller 1989b). Predation patterns of wolves in summer have been estimated primarily by scat analysis. Only Ballard et al. (1987) have reported summer predation patterns determined from direct observations; these workers studied radio-collared wolves in south-central Alaska. They also analyzed over 5,000 wolf scats at dens and rendezvous areas. Prey types were ranked similarly by direct observation and by scat incidence converted to prey biomass: adult moose > calf moose > adult caribou (cf. Ballard et al. 1997 and Spaulding et al. 1998). While Ballard et al. measured moose occurrence in the diet by direct observation of kills, Peterson, Woolington, and Bailey (1984) believed that most summer occurrence of adult moose in wolf scats on Alaska's Kenai Peninsula resulted from scavenging on moose killed in winter, as these authors observed wolves scavenging on moose kills from the previous winter, but rarely saw them feed on freshly killed adult moose in summer.

In Minnesota, reliance on deer dropped from 90% (frequency of occurrence in scats) in winter to 68% in summer (Fuller 1989b) as beavers and snowshoe hares became important prey. Potvin et al. (1988) found that deer occurrence dropped from 81% in winter to 14% in summer in Quebec, with moose increasing proportionately. Fritts and Mech (1981) described a similar pattern in northwestern Minnesota.

Calves and fawns are particularly important as wolf prey in summer in many areas (Mech 1966b, 1988a; Peterson 1977; Mech et al. 1998). Peterson (1977) found moose calf remains in wolf scats six times more frequently than remains of adult moose on Isle Royale. In northeastern Minnesota, studies of radio-collared deer indicated that wolves there rarely killed adult deer in summer (Nelson and Mech 1986b), while they did take fawns (Kunkel and Mech 1994).

Beavers are eaten by wolves in summer wherever the two species coexist (Mech 1970; Frenzel 1974; Van Ballenberghe et al. 1975; Fritts and Mech 1981), occasionally composing the most frequent food item in scats (Voigt et al. 1976). Fuller (1989b) found that in April and May beavers provided 16% of wolf food biomass in north-central Minnesota, similar to the 11% at Isle Royale in summer (Thurber and Peterson 1993). No one has ever described wolves preying on beavers, but at Isle Royale

wolves seem to concentrate their travels along beaver drainages, often resting for long periods near beaver dams where these aquatic animals briefly, but predictably, become vulnerable as they cross their dams.

Small prey are frequently eaten by wolves in summer, but relatively little of the wolf's total food comes from small prey. At Isle Royale, adult moose (especially yearling animals, R. O. Peterson, unpublished data) compose 72% of ingested prey biomass in summer, but based on calculated biomass and relative number of prey (Thurber and Peterson 1993), other individual prey animals are consumed more often (calf moose 1.7 times more often, with beaver and snowshoe hare 3.4 and 2.9 times more often, respectively). At Ellesmere Island, juvenile Arctic hares are the most frequent and reliable wolf prey in summer (Mech 1988a, 1995c).

Wolves in North America typically do not subsist on domestic animals and garbage except in small local areas, but these may become secondary food sources in summer. Where wolves had ready access to cattle in Alberta (pastures adjacent to continuous forest), Bjorge and Gunson (1983) found remains of cattle in 20% of wolf scats in summer. A Minnesota wolf pack that had a municipal dump in its territory made nightly foraging trips in summer for garbage (Hertel 1984).

The Wolf's Predatory Adaptations

The prey defenses described by Mech and Peterson in chapter 5 in this volume, and no doubt many others still unknown, form one part of a constant tension between wolves and their prey. The other part of that tension resides in the total of the wolf's abilities to locate, subdue, and kill its prey. Over the eons, the coevolution of wolf and prey (Bakker 1983) has resulted in an evolutionary "arms race" (Dawkins and Krebs 1979; but cf. Abrams 2000). Darwin (1896, 110–111) put it this way:

> Let us take the case of the wolf, which preys on various animals, securing some by craft, some by strength, and some by fleetness; and let us suppose that the fleetest prey, a deer for instance, had . . . increased in numbers, or that other prey had decreased in numbers, during that season of the year when the wolf was hardest pressed for food. Under such circumstances the swiftest and slimmest wolves would have the best chance of surviving and so be preserved or selected.

The tension or basic equality between wolf and prey explains the continued existence of prey in the face of wolf populations that survive by killing them. Under usual food and weather conditions, the abilities of a prey population to survive about equal the abilities of the wolf population to make a living from the prey. During unusual conditions favorable to prey, their numbers increase despite wolves; during unfavorable conditions, wolves foster their decrease.

Before the interactions of wolves and their prey are discussed in detail (see Mech and Peterson, chap. 5 in this volume), we will examine the wolf's abilities to overcome enough prey to survive and reproduce. The wolf is well adapted to predatory life physically, mentally, and behaviorally. Its digestive system, travel and sensory abilities, aggressiveness, speed, endurance, and intelligence are all advantageous to its predatory existence.

Digestive System

While it is said that the wolf is kept fed by his feet (Mech 1970), wolves possess other structures more directly involved in capturing prey animals and consuming them as quickly as possible. The wolf's adaptation to a variety of prey types and environmental conditions around the world is indicated by the animal's unspecialized carnivore dentition and digestive process.

Teeth and Skull

The hunting techniques of wolves are reflected in the evolutionary shaping of the skull, especially the teeth. Whereas felids often kill prey with a single penetrating bite to the head or neck, canids usually dispatch prey with numerous, shallower bites, delivered more opportunistically (Biknevicius and Van Valkenburgh 1996). Solitary felids, assisted by sharp retractile claws, must be able to kill even large prey with a single crushing bite, often to the head or neck region. Canids, with claws dulled by travel, deliver bites with less precision, relying instead on multiple slashes of teeth alone.

The variable tooth requirements for killing and consuming prey are evident in the wolf's heterodont type of dentition, comprising many different tooth types. Muscle and skin from prey can be effectively reduced by the shearing and compression actions of sharp, bladelike teeth, but bone-cracking is best done by stout, cone-shaped teeth (Lucas and Luke 1984; cited in Biknevicius and Van Valkenburgh 1996).

A wolf's mouth reflects a generalized carnivore pattern, without the extreme specialization seen in the more robust mandible, jaw-closing muscles, and premolars of

hyenas, adapted for bone-crushing, or the loss of bone-crushing teeth in felids, adapted for an all-meat diet (Biknevicius and Van Valkenburgh 1996). The one movable joint in the wolf's skull, the temporomandibular joint (TMJ), where the mandible (lower jaw) connects with the temporal bone of the cranium, is well grounded at the back by a postglenoid process that helps prevent dislocation while the mandible is severely stressed during prey capture and consumption. Although the wolf TMJ has better-defined structural boundaries than the flat and open joint of herbivores, it lacks specializations such as the extended preglenoid and postglenoid processes of the mustelids, which allow the jaw to be locked or heavily stabilized during closure around large and active prey (fig. 4.2).

The wolf's jaw is closed by several massive muscles that act sequentially, each inserted for best mechanical advantage as the mouth closes on its prey (fig. 4.3). The primary closure muscle, the temporalis, is larger in wolves and other carnivores than in herbivores, but it has relatively less mass and mechanical advantage in wolves than in felids, hyaenids, and mustelids (Radinsky 1981; Van Valkenburgh and Ruff 1987, cited in Biknevicius and Van Valkenburgh 1996).

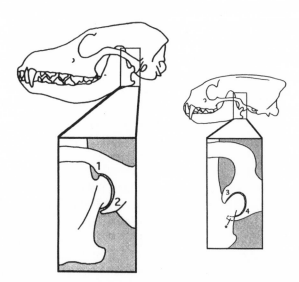

FIGURE 4.2. Lateral views of the skulls and details of the temporomandibular joints of a mustelid (wolverine) (right) and a canid (left) representing the wolf skull plan. The gray wolf has a small preglenoid process (1) and a modest postglenoid process (2) compared with the condition found in most predaceous mustelids, in which the mandibular condyle is surrounded by enlarged preglenoid and postglenoid processes (3 and 4). (From Biknevicius and Van Valkenburgh 1996.)

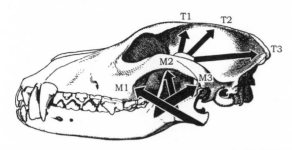

FIGURE 4.3. Orientation of the fibers of the temporalis (T1–T3) and masseter (M1–M3) muscles of the wolf, demonstrating the manner of attachment that conveys powerful mechanical advantage to the jaws.

Asymmetrical mechanical loading within the wolf's skull may be severe, as when bone is suddenly encountered by just one canine in a locking bite. Such torsional strain may result in spiral fractures, but it is resisted by buttressing of bone along diagonal lines that cross the skull, and also by the limited length of the skull. Covey and Greaves (1994) analyzed torsional strain in carnivore skulls and concluded that the skulls of wolves and other canids were as long as possible, given their width, allowing maximal gape or jaw-opening dimensions. Felids, ursids, and mustelids, on the other hand, have relatively shorter, more robust skulls.

Canines and Incisors. The anteriormost teeth, the incisors and canines, are the primary tools that wolves use to subdue their prey. The canines are designed to stab and hold, assisted in the latter task by the incisors. In the ensuing struggle, these teeth are undoubtedly subjected to enormous stress. The entire weight of the wolf, plus the force generated by movement of both predator and prey, must be borne by these anterior teeth and their bony anchorage in the mandible and skull.

Some appreciation of the magnitude of these forces can be gained from watching wolves, teeth locked onto a moose's nose, being tossed from side to side and lifted off the ground by their prey (Mech 1966b). Peterson has observed wolves with a lock on a rear leg of a running moose being dragged along for dozens of meters. In thick timber, a moose can swing a clinging wolf with full force against nearby trees, occasionally dislodging a wolf with a poor grip. (This may explain the frequent broken ribs seen in old wolves.)

Wolves use their canines to slash at the hide and muscle of prey, producing lacerations and extensive bleeding. This use contrasts with that of felids, which use their canines to stab and hold struggling prey. These different uses of the canines are reflected in their configu-

ration in the two types of carnivores. The circular cross section of felid teeth, for example, reflects the unpredictable direction of stresses on teeth deeply imbedded in the flesh of prey. In contrast, the elliptical cross section of wolf canines suggests adaptation to loads primarily applied along the long axis of the ellipse, front to rear, as when a wolf is pulled along by running prey (fig. 4.4). Nevertheless, 50% of all observed broken teeth in nine species of extant large carnivores were canines (Van Valkenburgh 1988). While Vilà, Urios, and Castroviejo (1993) found no broken canine teeth in 500 wolf skulls from the former Soviet Union, Phillips (1984) and Haugen (1987) found that broken canines were common in Alaskan wolves that fed on moose, but less common where smaller prey predominated.

The wolf's incisors play an important role in grasping and holding prey. The parabolic arrangement of the incisors, in front of the canine teeth, appears to be the primitive carnivore plan, but it may also reflect a need for teeth that can be selective during omnivory by canids. Among ungulate herbivores, a parabolic arrangement of the incisors is characteristic of selective browsers, while less selective grazers tend to have linearly arranged incisors (Biknevicius and Van Valkenburgh 1996). In any case, the arrangement of wolf incisors allows them to be used independently from the canines (Mech 1970) to nip or pull at live prey, to remove tissues from dead prey, or to ingest small, nonstruggling food items such as berries or small mammals (Biknevicius and Van Valkenburgh 1996).

Carnassials. While incisors and canines make up the prey-killing apparatus of wolves, the postcanine

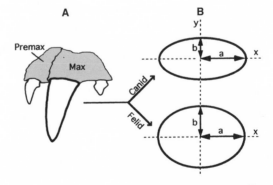

FIGURE 4.4. Cross sections of canine teeth reveal design differences between canids and felids. Canids have an elliptical cross section (large a/b ratio) that helps the tooth resist the strong front-to-back pull exerted when a wolf is clinging to running prey. Felids, in contrast, have a more circular cross section (a/b ratio closer to unity). (From Biknevicius and Van Valkenburgh 1996.)

teeth are used primarily during the consumption of carcasses. A diagnostic carnivore tooth adaptation is the carnassial pair, the upper fourth premolar and the lower first molar, used to slice through hide and meat. Among canids, these teeth are relatively enlarged in species such as the wolf that feed consistently on large prey (Van Valkenburgh and Koepfli 1993). Each tooth from this pair has two shearing edges separated by a V-shaped notch on each edge of the cutting blade (fig. 4.5). As the jaw closes, the upper and lower blades shear past each other, trapping and cutting food between the converging notches (Greaves 1974; Lucas and Luke 1984). The cutting edge is self-sharpening, maintained by mutual abrasion as the shearing edges pass (Mellett 1981).

The wolf carnassial pair is multipurpose, as the rear portion of the lower first molar (termed a "talonid basin") is adapted as a crushing or grinding surface, paired with a smaller upper first molar (see fig. 4.5). The comparable tooth in mustelids is further specialized, with crushing function enhanced by the presence of an enlarged inner lobe. Hyaenids, on the other hand, have lost all postcarnassial molars, and bone-crushing is accomplished by enlarged cone-shaped premolars anterior to the carnassials (Biknevicius and Van Valkenburgh 1996).

Nutritional Needs and Body Composition

Prey flesh and wolf flesh are similar in terms of elemental constituents, so they are nutritionally exchangeable —wolves are not likely to suffer specific nutrient deficiencies if they consume entire carcasses of prey. More than muscle tissue is required, however, and wolves obtain essential elements of their diet from various body organs and bones of their prey.

The body composition of a wolf gives a gross indication of the type of diet needed for physical maintenance. A whole-body analysis of wolves has not, to our knowledge, been performed, but dogs can provide a reasonable approximation. Average body composition (plus observed range) for adult dogs was as follows (Meyer and Stadtfeld 1980):

> Water, 56% (42–67%)
> Protein, 16% (11–20%)
> Fat, 23% (10–41%)
> Minerals, 3.5%
> Carbohydrates, 1.7%

The skeleton is distributed among these compartments, primarily as protein, fat (in bone marrow), and minerals. The skeleton of dogs constitutes 5–20% of body weight, averaging 12%; from allometric considerations, we would expect wolves to be at the high end of this range.

Water Requirements

In midwinter at northern latitudes, wolves drink little or no water, yet the fat-free wet mass of most mature wolves is about 73% water (Kreeger et al. 1997). Water is the universal solvent for a diverse array of chemical reactions, is the principal blood constituent for transport and the medium of the immune system, is required for temperature regulation and digestion, and helps eliminate toxins from the liver (Blaza 1982). Wolves regularly lose

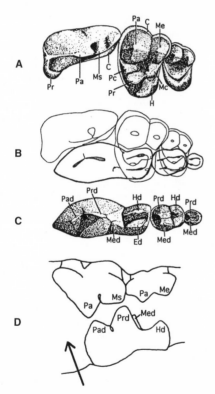

FIGURE 4.5. Functional morphology of the wolf carnassials and adjacent molars. (A) Occlusal view of upper left fourth premolar and first and second molars. (B) Occlusal relations of upper and lower teeth. (C) Occlusal view of lower right first through third molars. (D) Lateral outside view of the left carnassial teeth (fourth premolar and first molar) approaching centric occlusion. The labels refer to functional surfaces of the teeth. (From Biknevicius and Van Valkenburgh 1996.)

water via urine, expired air, feces, and lactation, although their kidneys are able to concentrate urine to a high degree (Afik and Pinshow 1993).

Wolves in warm climates usually require free water for thermoregulation. Desert wolves are not highly adapted to conserve water, as are many other desert mammals; rather, they evaporate considerable water in order to dissipate excess heat (Afik and Pinshow 1993). The mobility of wolves allows them to search out distant sources of water that are inaccessible to many desert mammals. In captive desert wolves, daily water turnover in summer was 1.7 liters, compared with 0.5 liters in winter (Afik and Pinshow 1993).

Except for thermoregulation, wolves can obtain their maintenance water requirements from their prey, both from water in prey tissues and from water produced by chemical oxidation of food. The primary ingredient of tissues eaten by wolves is water, which constitutes 55–75% of fresh meat (Blaza 1982). The proportion of water in muskoxen ranged from 51% to 75%, with adult animals generally at the low end of the range (Adamczewski et al. 1995). By oven-drying several tissues from a freshly killed moose, R. O. Peterson (unpublished data) found that the water content of various organs was as follows: liver 72%, heart 80%, kidney 81%, lung 77%, muscle 78%.

The water produced by digestion—so-called metabolic water—is also substantial. (Digestion is really a process of hydrolysis, or "splitting with water.") One hundred grams of fat actually yields 107 g water (with extra mass coming from inspired oxygen), and complete oxidation of 100 g protein yields 40 g water (Randall et al. 1997). The feces of wolves are relatively dry, reflecting the extraction of water in the intestine. The average water content of fresh wolf scats collected in winter was 28.9% (SE = 0.5) (Fox 2001), less than half the water content of food.

Free water intake is helpful, but not required, to process food with a low water content (bones and hide) or when a wolf is heat-stressed, as after a vigorous chase. Liquid water is not commonly available to wolves in winter in boreal or arctic regions. Even when water is obtainable, it is infrequently drunk in winter. In a 50-day field study on Isle Royale in January and February, in which wolves or wolf packs were observed 160 times, wolves were seen drinking water only twice, both times after the heavy exertion of a moose chase (R. O. Peterson, unpublished data). It is not uncommon to see wolves grab a mouthful of snow after an unsuccessful chase.

Lactation needs probably require nursing females to drink water frequently; this water requirement may limit den sites to locations near water (Mech 1970). Water intake in lactating females increases approximately in proportion to their food intake, which may rise two- or threefold over that of non-lactating individuals (Oftedal and Gittleman 1989), but much of their water need will be met by the food itself. We estimate, based on data provided by Oftedal and Gittleman (1989), that for a litter of five or six pups, a nursing female needs to produce about 1.2 liters (1.5 quarts) of milk each day (assuming an energy density for milk of 1.4 kcal/g, an energy need of 882 kcal/day, and a milk density of 1.0 g/ml).

R. O. Peterson and L. D. Mech (unpublished data) observed a lactating female with two pups at a den at Ellesmere Island in late June 1996; during this time, the pups were usually in the den. The water source for the female was outside the den, and she drank water for 413 seconds during 77 hours of observation. R. O. Peterson (unpublished data) found that two large dogs drank water at an average rate of 9.5 and 10.7 ml/sec, and the maximum he recorded was 15.3 ml/sec. If the female wolf drank water at this maximum rate, her intake would have been 2.0 l/day.

Ingestion and Digestion

Like other carnivores, wolves possess a relatively short gut because their diet of animal parts is highly digestible. The digestive tract is a differentiated tube with valves controlling the flow and movement of food through the gut, assisted by wavelike muscular contractions, or peristalsis. Our understanding of digestion in wolves is largely inferred from studies of the domestic dog, and the general plan of canid digestion presented here follows Blaza (1982). Because the diet of dogs is also a domestic product, we should draw parallels carefully between the wolf and its domestic derivative. Nevertheless, the basic anatomy and physiology of digestion is probably conservative—we might expect that the dog lacks some enzyme systems or physiological capacity that wolves retain, but it is unlikely that dogs would exhibit physiological innovations or capacities not shared with wolves (Meyer and Stadtfeld 1980).

Feeding responses in wolves begin even before food is ingested, as salivation in the mouth is prompted by the sight and smell of food (the gustatory response studied early in the twentieth century by Pavlov, the famous Russian behaviorist). Salivation is heightened by the presence of food in the mouth as taste is added to other sensations. The most potent compounds for stimulating the taste buds in dogs, and presumably wolves, are amino acids that taste sweet to humans, especially L-cysteine, L-proline, L-lysine, and L-leucine (Bradshaw 1992). Taste buds respond to nucleotides, and their constituent amino acids probably help distinguish among meats of various nutritional qualities.

Saliva, produced by the tongue, the orbital and parotid glands (dorsal), and the mandibular and sublingual glands (ventral) (fig. 4.6), is a slightly acidic secretion containing lubricating mucus that facilitates swallowing. This reduces the time it takes for wolves to fill the stomach, allowing maximum intake and minimum loss to scavengers and fellow pack members. There is little mechanical breakdown of soft food in the mouth before swallowing, but hide and bone may require considerable chewing before they can be swallowed.

The route to the stomach is through the short esophagus, where more mucus (but no enzymes) is added. The stomach serves as a site of digestion and as a reservoir that regulates flow to the small intestine (fig. 4.7). The stomach mucosa (epithelium and underlying tissue) secretes more mucus (which protects the stomach wall from digestive enzymes), together with hydrochloric acid and various proteases (protein-digesting enzymes), especially pepsin. These stomach secretions are released under nervous and hormonal control as a result of the gustatory response and the presence of food in the stomach. The hormone gastrin, for example, stimulates the anterior portion of the stomach to produce acid and digestive enzymes, and it promotes motility of the stomach. Gastrin is released by the stomach mucosa as stretch receptors detect distension of the stomach wall.

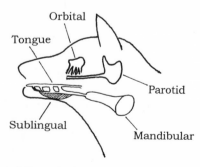

FIGURE 4.6. Salivary glands of the wolf.

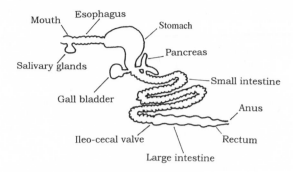

FIGURE 4.7. Digestive anatomy of the wolf.

In its distal portion, the stomach lining produces alkaline fluids low in enzymes. These fluids are thoroughly mixed with the stomach contents, partly neutralizing stomach acids and producing a thick, milky liquid called chyme. A pyloric sphincter muscle controls the flow of stomach contents into the small intestine. Many factors influence this flow, but the rate of passage is generally reduced by the presence of acids, irritants, fats, or chyme in the anterior portion of the small intestine.

The small intestine is the "workhorse" of the digestive tract, with an internal surface consisting of myriad fingerlike villae that give the tract an area the size of a small room. Here digestion is completed, aided by more secretions from the gall bladder and pancreas. The pancreas, under hormonal control, secretes more digestive enzymes, including proteases (for proteins), lipases (for fats), and amylase (for carbohydrates), plus a large volume of bicarbonate salts, which neutralize the acidic chyme. The gall bladder releases bile, a substance continuously produced by the liver, into the anterior small intestine. Bile salts emulsify particles of fat and also activate some lipases.

While some absorption of nutrients occurs from the stomach and large intestine, the small intestine is the site of most nutrient uptake. Protein products (and any carbohydrates) are absorbed by the bloodstream, while most fat derivatives enter the lymphatic system. Although newborn animals can absorb some intact proteins, such as maternal antibodies in colostrum, most protein uptake occurs via the derivative amino acids. Water is absorbed passively via osmosis, mostly in the small intestine, as are vitamins, both water-soluble (B group, C) and fat-soluble (A, D, E, and K).

The final portion of the wolf gut, the large intestine, conducts most of the minor water and electrolyte absorption and hosts the bacterial breakdown of protein and fiber residue (which give feces their characteristic smell and color). Undigested residues, together with water, minerals, and dead bacteria, are stored in the rectum for evacuation as feces. As explained below, the time required for these processes depends on the type of meal eaten.

Caching

When a wolf is sated, it begins caching food (Kreeger et al. 1997). Food caches might include anything from an intact caribou calf weighing some 6–8 kg covered with snow (Adams, Dale, and Mech 1995) to several food chunks regurgitated into a hole dug in the ground and then covered by movements of the nose (Mech 1970; Mech and Adams 1999). Mech (1970) and Peterson (1977) did not observe wolves caching food in winter on Isle Royale, although their observations were from aircraft. Mech (1970) suspected that wolves sometimes cached food in winter, yet there appear to be no documented cases of wintertime caching by wolves. Mech (1970) pointed out that in winter wolves would have little reason to cache, as a pack could easily clean up a kill in its entirety.

In summer, when wolves hunt singly or in small groups, caching appears to be an important behavior (Murie 1944; Cowan 1947) that helps secure excess food left from large prey or reduce loss to scavengers and maggots (Mech 1970). Caching helps buffer the irregularities of a predatory economy, for if prey are difficult to find or catch, a wolf can dig up caches and bide its time until it can capture prey again. On Ellesmere Island, where the ground is frozen for 9–10 months of the year, L. D. Mech (unpublished data) saw a wolf eating a hare that apparently had been cached for about a year.

After an initial feeding, and before resting, a female wolf on Ellesmere Island was observed to immediately commence caching trips, probably up to 5 km (3 mi) from the kill (Mech and Adams 1999). Two caches that Mech dug up weighed 0.65 and 0.66 kg, so up to sixteen caches could have been made from a full stomach weighing 10 kg (22 pounds). Wolves appear to have a high degree of control over the amount of food that may be regurgitated, either for packmates or into caches, as they often regurgitate multiple times after a single filling of the stomach.

Magoun (1976) observed six caches made by a wolf on a single trip after 20 minutes of feeding on a fresh carcass. In 1996 Mech and Peterson saw a breeding male on Ellesmere Island regurgitate five times in less than 2 hours, provisioning pups and his mate as well as making two caches in one hole that was 30 cm (11 inches) deep (Mech et al. 1999). The hole contained 2.5 kg (5.5 pounds) of muskox parts and a small bit of Arctic hare.

Magoun (1976) found that only one of twenty-eight food caches made by wolves in summer in the Arctic National Wildlife Refuge were in the vicinity of the kill (< 46 m). Wolves appeared to purposely distance themselves from the kill before caching, probably to reduce theft by other scavengers (Murie 1944; Magoun 1976). Murie (1944) related an instance in which a wolf appeared to purposely leave a confusing trail when carrying food away for caching, especially by intermittently walking in water. Nevertheless, food caches of wolves are raided by bears, ravens, and foxes, the same suite of scavengers that follow wolves at prey carcasses (Murie 1944; Mech 1970; Magoun 1976). Only five of fourteen simulated food caches were left undisturbed after 2 days (Magoun 1976).

Fat Storage and Fasting

The bodies of wolves may store surplus food energy as fat, which contains, gram for gram, over twice as much caloric energy as proteins or carbohydrates (Randall et al. 1997). Potentially, at least 15% of a wolf's weight can be composed of fat (Kreeger et al. 1997), and this is probably a conservative estimate.

On average, wolves are probably much leaner than most dogs. As total body components, fat and water are inversely proportional to each other, and together they make up 80% of the canid body: fat levels of 40% are associated with water levels of 40%; levels of 10% fat, with 70% water (Meyer and Stadtfeld 1980). Like many carnivores, wild wolves are often food-limited and typically exist near the low end of their fat content range. This can be seen in the depletion of their bone marrow fat, which is the last fat reservoir in the body, the final stored source of energy before muscle catabolism is required to sustain life (Mech and DelGiudice 1985). The bone marrow fat of free-ranging wolves is often partly depleted. Of forty-two wild wolves taken by trappers and hunters during winter in Ontario, 17% had used most of their marrow fat, and another 12% had begun marrow fat depletion (fat content < 70%) (LaJeunesse and Peterson 1993). Total body water for three desert-dwelling wolves living in captivity in Israel averaged 68% of body mass,

TABLE 4.1. Extent of total weight loss for Isle Royale wolves that died of starvation

Wolf	Age (yrs)[a]	Live weight and date of capture	Weight and time of death	Estimated weight loss
Male 2224	11	36 kg, 1 May 1988	29 kg, Jan 1991	24%
Male 2518	9	31 kg, 21 Aug 1990	25 kg, Jan 1994	19%
Female 2542	8	—	22 kg, Feb 1994	29%[b]
Male 1565	10	—	26 kg, Jun 1980	33%[b]

[a]Waite 1994.

[b]Assuming average female weight of 31 kg ($N = 7$) and male weight of 39 kg ($N = 7$), based on Isle Royale wolves weighed live during 1988–1994 (R.O. Peterson, unpublished data).

corresponding to fat levels of about 12% (Afik and Pinshow 1993).

Wolves possess a remarkable ability to survive long periods of low food intake. The actual period of fasting for wild wolves is unknown because monitoring of individuals is inevitably fragmentary. Peterson commonly observes single wolves on Isle Royale existing for several weeks in winter without killing a moose. They frequently scavenge old prey carcasses and may, for example, spend days at a time in winter at a single location while hunting snowshoe hares or beavers (Thurber and Peterson 1993).

In the absence of food, wolves and other animals lose weight daily. Captive wolves that were fasted for 13 days lost an average of 262 g (9 oz)/day (DelGiudice et al. 1987), while others fasted 10 days lost 400 g (14 oz)/day (Kreeger et al. 1997). In the latter case, 54% of the weight that males lost consisted of water, 28% of fat, and 18% of protein/ash; for females, the comparable figures were 58%, 20%, and 22% (Kreeger et al. 1997). Based on these figures, a 50 kg (110-pound) wolf composed of 15% fat (7.5 kg) could fast for about 67 days before losing all its fat. Of course, wild wolves without food would be spending much time traveling and hunting, and would use more energy than captives. The wolves that were fasted 10 days were able to regain all of their weight in 2 days of ad libitum feeding (Kreeger et al. 1997).

After exhaustion of fat reserves, muscle protein can be catabolized as a last resort to sustain life. If wolves typically have fat levels near the minimum for dogs (10%) and the same range of protein levels (11–20%), then as much as 21–30% of their weight might be catabolized before all internal energy reservoirs are exhausted. Captive wolves ranged from 9.0% to 13.1% fat (Kreeger et al. 1997), and when fasted for 10 days, lost mass in water, fat, and protein/ash in a manner similar to seasonal changes

in body mass observed in coyotes (Poulle et al. 1995). Four Isle Royale wolves that died of starvation had lost 19–33% of their initial live weight (table 4.1). Peterson boiled the carcass of one of these wolves to recover the skeleton, and found no fat whatsoever.

As long as water is available, fasting is not life-threatening for most animals until weight loss approximates about 50% of the normal body mass (Kleiber 1961). However, survival time for starving animals is probably highly variable, and several records exist of wolves, which normally have low fat reserves, succumbing to apparent starvation when about one-third of body mass was lost (see table 4.1). Mech et al. (1984) reported a wild female wolf weighing 18.6 kg (40.9 pounds), about 27% below the average weight for well-fed captive female wolves in Minnesota, that appeared near death from starvation. After 9 days of recuperation and ad libitum feeding, the wolf was released in the wild and survived.

Dogs, and presumably wolves, possess the ability to alter their enzyme systems to adapt to a diet low in protein and its constituent amino acids (Meyer and Stadtfeld 1980). With a low-protein diet, catabolic enzymes for diet amino acids decline, primarily in the liver, so that a lower proportion of the diet amino acids are catabolized. These enzyme adaptations occur on a scale of hours to days. Thus the animal conserves nitrogen from protein and survives in spite of a low protein intake. Of course, if a wolf has inadequate protein, it is probably starving, while a domestic dog might simply be responding to low protein in a diet that is energy-rich from carbohydrate additives. Nevertheless, dogs and wolves can adjust their enzyme systems in ways that cats cannot, as cat liver enzymes are permanently set for a medium- to high-protein diet.

The wolf, adapted to a feast-or-famine diet, can quickly

recover weight lost during fasting (Kreeger et al. 1997). The amazing record of a Scotch collie dog that survived two experimental fasts early in the twentieth century demonstrates this capability. After 45 days of fasting, the dog was near death, but recovered completely upon refeeding and went on to survive a second fast of 117 days, in which 63% of its original weight of 26.3 kg (57.9 pounds) was lost (Howe et al. 1912; cited in Kleiber 1961). Many physiological aspects of fasting in wolves are poorly understood. For example, we know nothing about the possibility that organs may diminish in fasting wolves, thus reducing metabolic demands (Hammond and Diamond 1997).

Hunting Behavior

While they are able to deal with food scarcity, wolves are also well equipped for minimizing it. As capable travelers, wolves can locate prey over large areas, even when it is scarce or less vulnerable (Mech et al. 1998). In winter on Isle Royale, packs of wolves traveled average distances between kills of 42.4 km (26.5 mi) (Mech 1966b) and 49.4 km (29.6 mi) (R. O. Peterson, unpublished data). Much of this travel involves searching for prey, and may occur in a single night of hunting. The wolf's keen hearing, acute sense of smell, and excellent eyesight greatly aid in these searches (see Harrington and Asa, chap. 3 in this volume).

When the wolf does find prey, it uses several other traits that serve it well, especially boldness, speed, and endurance. As already indicated, most prey species are potentially dangerous to wolves. Thus, in approaching them, a wolf must be careful enough to avoid harm, yet bold enough to attack. No doubt experience and learning come into play in helping wolves fine-tune their judgment at this crucial stage. This probably explains why wolves give up many hunts readily (Mech 1970).

On the other hand, wolves may also press the attack vigorously. They can run at 56–64 km or 34–38 miles/ hour and continue running for 20 minutes or longer, although not necessarily at this speed (Mech 1970). An 11– 13-year-old male chased a young arctic hare for 7 minutes, finally catching it when the hare slowed down (Mech 1997). While wolves most often give up chases within 1–2 km (Mech 1970), one wolf chased, trailed, and followed a deer for 21 km (13 mi) (Mech and Korb 1978). Wolves' bold persistence is most apparent when they attack their largest prey, moose (Mech 1966b; Peter-

son 1977; Hayes 1993; Nelson and Mech 1993), bison (Carbyn et al. 1993), or muskoxen (Mech 1987b, 1988a). Television documentaries such as National Geographic's *White Wolf* (1988) and the BBC's *Wolves and Buffalo: The Last Frontier* (1996) show how bold and persistent wolves can be. These videos, made with the assistance of biologists, accurately show wolves dashing into running herds of muskoxen and bison, respectively, and attacking despite the thrashing hooves and horns all around them.

These observations of wolves abandoning and pressing attacks suggest that wolves can reasonably estimate the vulnerability of individual prey in various situations. As they hunt day after day, gaining experience, their basic ability to learn quickly and perceive complexities (see Packard, chap. 2 in this volume) no doubt continually helps hone their hunting and killing prowess.

Wolves' almost universal concentration on young-of-the-year soon after they are born (Murie 1944; Mech 1966b; Pimlott 1967; Kunkel and Mech 1994) is another reflection of their learning ability. L. N. Carbyn (personal communication) even observed an instance in which wolves passed up an obviously debilitated adult bison to focus on killing calves. No doubt wolves also learn when to concentrate on other categories of vulnerable prey that may not be so obvious to humans (see below).

In addition to learning when certain prey classes are vulnerable, wolves probably also learn where to find these prey. That wolves can remember locations of vulnerable animals was demonstrated on Isle Royale, where in 1984 Peterson watched a pack head 20 km (12 mi) diagonally across the island straight to a moose wounded 6 weeks before.

There are several stages to the wolf's hunting behavior: (1) locating prey, (2) the stalk, (3) the encounter, (4) the rush, and (5) the chase (Mech 1970, but cf. MacNulty 2002). Depending on the species of prey and the individual situation, these stages may overlap or be telescoped to varying degrees.

Locating Prey

To locate prey, the wolf depends on its keen senses and its ability to travel long distances. Not only must the wolf find prey, but it must find prey it can catch and kill. Doing so usually requires scanning large numbers of prey. Thus wolves spend 28–50% of their time traveling, at least during winter (Kelsall 1957; Mech 1966b,

1977a, 1992; Peterson 1977; Peterson, Woolington, and Bailey 1984).

On Isle Royale, where the entire route of a pack of fifteen wolves was mapped from 4 February through 7 March, 1960, the animals covered 443 km (266 mi) and killed eleven moose (Mech 1966b). Assuming their average hunting success rate of 7.8% (Mech 1966b), this implies that the wolves actually tested an average of about one moose per 3 km (1.8 mi). On the other hand, where prey is much less dense, as on Ellesmere Island, wolves must travel much farther on average to find large prey (Mech 1988a).

Wolves use all their senses as well as luck to find prey, relying primarily on chance encounters, tracking (Peterson 1977), and scenting (Mech 1966b; Peterson 1977) in heavily wooded areas (Mech 1970) and on sight (Haber 1977; Peterson 1977) and scent in open areas. With young arctic hares that crouch in furrows on Canada's barren grounds, wolves sometimes spot one from several meters away, walk to it, and merely snatch it up (L. D. Mech, unpublished data).

When traveling, wolves generally file along one behind the other. However, sometimes when climbing a hill or heading through areas of thick cover where prey such as deer or smaller species may reside, they fan out, presumably to better scare up prey and position themselves to capture it (Stenlund 1955; Kunkel and Pletscher 2001).

Stalking Prey

Once wolves locate prey, they attempt to get as close to it as possible, unless their initial encounter results in the prey detecting them and fleeing (Kunkel and Pletscher 2001). "As they close the gap between themselves and their prey, the wolves become excited but remain restrained. They quicken their paces, wag their tails, and peer ahead intently. Although they seem anxious to leap forward at full speed, they continue to hold themselves in check" (Mech 1970, 199). An excellent example of wolves stalking muskoxen can be seen in the television documentary *White Wolf* (National Geographic 1988). On the barren grounds, wolves try to use gullies and other uneven terrain to make their stalk (Mech 1988a).

The Encounter

Next comes the encounter, when wolves and prey first see and confront each other, often at a distance. Three things can happen: The prey can remain in place, ap- proach the wolves, or flee. It is larger prey that most often stand their ground. However, even arctic hares sometimes hold still when they spot wolves and continue to hide until grabbed, or until they realize they have been seen (L. D. Mech, unpublished data). With larger prey, wolves generally wait until the prey flees before they respond. Our experience is similar to that of Crisler (1956, 340), who observed that "a wolf prefers not to be eyed when approaching prey."

When hunting Dall sheep, according to Murie (1944, 110), "the method is to get above a sheep and force it to run down, for a sheep running upward can quickly outdistance the wolves and escape. Sheep on somewhat isolated bluffs where space for maneuvering is limited are in danger of having their upward retreat cut off and of being forced to run down to the bottom." In a case in which young wolves had cornered a lamb on a precipice below which they could not reach, their mother helped them out by shinnying down a rock toward the lamb. However, she ended up plummeting 5 meters over a cliff, bouncing once, and landing in a boulder field. Meanwhile, to avoid the wolf, the lamb had climbed up and into the jaws of the waiting pack. The mother wolf apparently survived (J. W. Burch, cited in Mech et al. 1998).

Large prey that approach wolves or stand their ground upon seeing wolves present a dilemma. Even deer occasionally stand their ground or defy wolves (Mech 1984; Nelson and Mech 1993). With any such animals, wolves approach, usually cautiously, and try to threaten the prey face-to-face. Whether the threat includes any of the signals with which wolves threaten one another (i.e., bared teeth, growling, barking,) is unknown; such threats have not been reported and are not evident in photos (Mech 1987b, 1988a) and drawings (Gray 1987) of wolves approaching prey.

If the prey stands its ground long enough, the wolves eventually leave, although they may remain with the prey for 4 hours or longer, periodically trying to get them to run (Gray 1987). With bison, wolves may remain near herds for 6 days (Carbyn et al. 1993). When prey do flee, whether directly upon encounter or after harassment, the wolves immediately pursue. This is true even if they are feeding on a kill and nearby standing prey suddenly try to flee (L. D. Mech, unpublished data). Such behavior triggers the rush phase of the hunt (Mech 1970).

The Rush

The exact nature of the rush depends on the prey species and on the individual circumstances. With small single prey such as hares, deer, or caribou calves, which are easily killed or gravely wounded with a few bites, wolves attempt to catch up to them as soon as possible. With herds of prey such as caribou and elk, the wolves may just run along with the herds, apparently searching for a vulnerable individual (Landis 1998).

With moose (Mech 1966b), muskoxen (Gray 1987; Mech 1988a), bison (Carbyn et al. 1993), and elk (Mac-Nulty 2002), wolves attempt to get their quarry running. In one notable example of this strategy, a single wolf approached from behind twenty muskoxen that were lying down facing into the wind; it then poked its nose into a bull's rump. "The startled bull jumped up and whirled around to face the wolf. He gland-rubbed as the wolf circled, then charged the wolf. The wolf ran briefly, then walked away, yawning" (Gray 1987, 131).

The Chase

The chase is the continuation of the rush. Thus, if wolves do get a moose or a herd of muskoxen or bison running, they chase them and probably watch for weakness. Murie (1944, 109) wrote that the prey were "given a trial," and Mech (1966b, 124) believed the wolves were "testing" the prey. As discussed below, the victims of wolves are almost always vulnerable in some way. With individual moose, sometimes the wolves chase without attacking and finally give up; on other occasions the wolves attack and kill the moose (Mech 1966b, 1970; Peterson 1977; Haber 1977).

When muskoxen and bison attempt to keep their calves toward the center of the herd, the wolves try to rush in and grab them away. Sometimes the wolves try grabbing an adult out of the herd, presumably after having detected one with a weakness.

Generally chases are not long. With the exception described above of a wolf chasing, tracking, and following a deer for 21 km, the longest chase recorded, which involved caribou, was about 8 km (4.8 mi) (Crisler 1956). Most often the distance is much less (Mech 1970).

Solo, Pair, and Pack Hunting

Although it seems logical that one of the wolf's primary adaptations for hunting is its social nature, the evidence seems to contradict this notion (see Mech and Boitani,

chap. 1 in this volume). Indeed, single wolves are known to be able to kill even the largest prey, such as moose (Young and Goldman 1944; Thurber and Peterson 1993), bison (D. Dragon, cited in Carbyn et al. 1993), and muskoxen (Gray 1970).

It would seem that a pair of adult wolves would be more efficient at hunting and killing most prey than would single wolves. Pairs could maneuver better than a single wolf, and one wolf could distract the prey while the other attacked. With even more wolves in the group, the job should be even easier. However, packs larger than an adult pair include primarily young, less experienced wolves, which often just follow along behind their parents and contribute little to the attack (Mech 1966b; Haber 1977). On the other hand, emerging evidence from Yellowstone National Park indicates that when pack breeders are older, the maturing pack members may lead attacks on prey (D. R. MacNulty, unpublished data). In any case, packs kill less food per wolf than do pairs, and of all pack sizes, pairs are the most efficient per wolf (see Mech and Boitani, chap. 1, and Mech and Peterson, chap. 5 in this volume). Whether pairs are any more efficient per wolf than single wolves is unknown, although it seem likely that they are.

Strategic Cooperation

Although it is difficult to assess the intelligence of animals, wolves probably rank among the more intelligent species, as anyone who has raised dogs can attest, and they appear to learn quickly and show insightful behavior (see Packard, chap. 2 in this volume). Thus we would expect that wolves might develop cooperative hunting strategies. By "cooperation" we do not mean merely group chasing in which the pack is all strung out in line behind the prey (see fig. 5.2). Rather, we mean conducting the hunt or chase in such a way as to capture the prey more effectively than by merely running after it as a group. Examples of such strategies would be chasing into ambush, heading off fleeing prey, or taking turns chasing ("relay running").

Uses of teamwork, relay running, ambushing, or decoying by wolves have all been claimed (Young and Goldman 1944; Kelsall 1968; Rutter and Pimlott 1968); Haber (1977) considered the use of cooperative strategies "frequent." On the other hand, several workers who observed countless instances of wolves hunting Dall sheep and caribou (Murie 1944), moose (Mech 1966b; Peterson

TABLE 4.2. Wolf and ungulate biologists' opinions and observations about the possibility of wolves using cooperative hunting strategies

| | Reply | | | |
Question	Yes	No	Maybe	Species
1. Do you believe that wolves consistently or deliberately ever employ a strategy in which one or more wolves chase prey to one or more wolves waiting in ambush?	8	10	1	
2. If you answered "yes" to the above, do you think you have observed one or more such hunts, and if so, with what prey species?	7	4	1	Caribou (4), horse, white-tailed deer, moose, arctic hare
3. Do you believe that wolves consistently or deliberately use a strategy of running prey in relays; that is, first one wolf chasing an individual and then, while that wolf rests, a second wolf taking up the chase, etc.?	4	14	0	
4. If you answered "yes" to the above, do you think you have observed one or more such relay runs, and if so, with what prey species?	3	1	0	Caribou (3), moose
5. Do you believe wolves use any other form of cooperative hunting strategy?	15	0	2	

Note: Respondents were W. Ballard, D. Boyd, J. Burch, L. Carbyn, B. Dale, T. Fuller, B. Hayes, D. Heard, D. MacNulty, M. McNay, D. Mech, T. Meier, F. Messier, F. Miller, M. Nelson, R. Peterson, D. Smith, B. Stephenson, and V. Van Ballenberghe.

1977), deer (Mech 1966a; Mech and Frenzel 1971a), bison (Carbyn and Trottier 1987, 1988), muskoxen (Gray 1970, 1987; Mech 1988a), and elk (MacNulty 2002) made little mention of the wolves' use of cooperative maneuvers. One exception was an observation of parent wolves positioning themselves along the ends of rocky ridges and ambushing arctic hares as the pack yearlings chased them by (Mech 1995c).

Because hunts usually include several wolves and prey, all moving over a large area, and because most observations of hunts have been made from circling aircraft, it is possible to make many interpretations for each hunt. Apparent cooperation or high-order strategies might be only the incidental result of haphazardness in movements or of interpretation by the observer.

African lionesses do cooperate strategically in hunting (Stander 1992), so it is not far-fetched to think that wolves might do so. On the other hand, lionesses remain together for years, which would give them time to learn to cooperate. Most wolf pack members disperse before 2 years of age (Gese and Mech 1991), and even the tenure of most breeding adults is short (Mech et al. 1998). The fact that food acquisition per wolf decreases with pack size (Schmidt and Mech 1997), contrary to the findings for lions (Stander and Albon 1993), also implies that wolves may not often cooperate strategically in hunting, except possibly for mated pairs.

A survey of nineteen biologists who have observed many wolf-prey encounters reveals no unanimity in their beliefs about the wolf's possible use of ambushing or relay running (table 4.2). Most wolf biologists believed that wolves do sometimes use some forms of cooperative strategy, but the number of descriptions including convincing examples of it is low. Most described chases were simple and straightforward.

The Ecology of Feeding

There is surprising order to the natural frenzy of wolves killing and consuming prey. When wolves kill a large animal, there is usually more food available than the pack can consume during its initial feeding. Some parts of prey are readily eaten by wolves or by any of the important competing scavengers, so one would expect easily stolen components to be eaten first. Portions of prey that are high in essential nutrients should also be eaten as a first priority. Thus the order in which various organs and body parts are consumed tells us something about the ecological and nutritional constraints that wolves face.

Feeding Patterns

Small prey, up to the size of a 20 kg (44-pound) beaver, a deer fawn, or even a calf moose, can be eaten entirely within a few hours, especially if several wolves are present. Eight wolves in Alaska killed and consumed a yearling caribou and left, all within 3 hours (Ballard et al. 1997). Researchers may find nothing at all remaining from juvenile prey, or perhaps just a mat of hair at the kill site (Pimlott et al. 1969), even from a 150 kg (330-pound) moose calf in winter (R. O. Peterson, personal observation). Large prey, such as moose, reveal more about the systematic manner of wolves' feeding. The following are some general patterns, based on published studies (Mech 1966b, 1988a) and R.O. Peterson's field inspection of hundreds of moose killed by wolves.

Eating prey carcasses as quickly as possible is an obvious priority for wolves, but there are physical limits to intake even for these voracious animals. Scavengers descend on fresh wolf kills with vigor, removing substantial amounts of food that would otherwise be available to wolves. Clearly, the greatest gain (for both wolves and competing scavengers) is associated with the first feeding on a fresh carcass.

In a study of scavenging on potential wolf prey, C. Promberger (unpublished data) assumed that average per capita consumption of a carcass by wolves would be 10 kg (22 pounds) on the first day, then 6 kg (13 pounds) on successive days. In addition to removal by wolves, there is progressive loss of food to scavengers. Furthermore, in winter, even hide-covered portions of carcasses will freeze within 2 days (if the internal organs are removed) in very cold conditions (approximately −20°C) (C. Promberger, unpublished data), so the rate of intake from a single prey carcass will steadily decline.

Freezing of carcasses provides a food bank that reduces loss to other scavengers and preserves food for wolves. And carcasses of large prey, such as moose, are protected from small scavengers while still covered by thick hide. Most scavengers active in winter have smaller mouths than wolves and are less able to reduce a frozen carcass effectively. Wolverines may be powerful enough, and another exception may be the wild boar, a major scavenger of wolf kills in many European countries (P. Ciucci, unpublished data). Peterson once watched five frenzied red foxes trying to steal parts from an intact and unfrozen cow moose; in almost 3 hours they had succeeded only in removing the tongue, eyes (together with ravens), and outer portions of the anus.

Even wolves may be challenged by the frozen carcasses of prey. In January and February 1994, daily minimum temperatures on Isle Royale were rarely above −18°C (0°F) for 5 weeks. During this period, an elderly breeding pair in one pack, both 8−9 years old and with well-worn teeth, died of starvation, even though their pack was able to kill moose. Another pack, after a mid-February thaw that lasted several days, traveled directly across its territory and consumed a moose that had died of malnutrition more than a month before and had lain intact, but frozen solid.

It is the prerogative of the dominant male and female wolves to feed first, and, based on observations on Isle Royale in winter, these are usually the wolves to first open a fresh carcass. There is, of course, great variation in feeding patterns, depending on individual wolves and situations, and systematic data on the feeding order of wolves has not been gathered except in captive packs (Zimen 1976). Parent wolves are usually (although not always; see above) the hardest workers during the actual killing of prey (Mech 1988a); sometimes they lie down to rest once a kill is made, and other wolves begin to open the carcass.

The feeding order lies at the core of important behavioral dynamics within a wolf pack, probably influencing dispersal of individual wolves and ultimately regulating pack size in relation to prey size (Mech 1970). It also reflects the competition for food that may exist within a pack. Many wolves feeding centrally at a kill usually results in squabbles, and wolves with little social standing must wait their turn, especially if food is short or prey is small. High-ranking wolves may complete their initial feeding on a fresh kill in an hour or less (Murie 1944; Mech 1970). Magoun (1976) found that mean time for continuous feeding in summer was 14 minutes, although Mech and Adams (1999) observed a female wolf on Ellesmere Island feeding for 63 minutes on a freshly killed muskox. Some 81% of wolf feeding bouts in summer lasted less than 30 minutes (Magoun 1976).

Wolves usually tear into the body cavity of large prey and pull out and consume the larger internal organs, such as lungs, heart, and liver. The large rumen (weighing about 60 kg or 132 pounds for a moose) is usually punctured during removal and its contents spilled. The vegetation in the intestinal tract is of no interest to wolves, but the stomach lining and intestinal wall are consumed, and their contents further strewn about the kill site. Smaller internal organs, such as kidneys and spleen, are then exposed and eaten immediately. Sometimes the rumen contents, with or without the surrounding rumen, freeze and become one of the few signs left of a kill in winter.

Magoun (1976) provided details on the feeding behavior of two wolves that fed on a caribou carcass she provided in summer. In the first 15 minutes, the backstrap, some skin, haunch muscle, some ribs, a kidney, and a piece of small intestine were consumed. A caching trip of 20 minutes was then undertaken, and in the ensuing 2 hours, meat from the backbone, rib cage, haunch, and legs was eaten, and the lungs were removed. In the next 2 hours, meat from the ribs and a leg was eaten; then the carcass was left for 4 hours, presumably while the wolves slept. They returned to feed over a 3-hour period on muscle from the ribs and hindquarters plus mesenteries. After 10 hours, part of the heart was eaten and the remainder cached, the lungs were eaten, and another caching trip was undertaken.

After choice organs are consumed, the large muscle masses associated with each leg of a prey animal provide the bulk of the food consumed by wolves. At a fresh kill, wolves typically eat until their sides are distended and their stomachs are packed. Young and Goldman (1944) recorded a wolf stomach containing 9 kg (19 pounds) of food. The stomach of an Alaskan wolf weighed by Peterson contained 7 kg (15 pounds), and a Minnesota wolf had a stomach weighing 10 kg (22 pounds) (L. D. Mech, personal communication). These stomach weights, probably typical for wolves after first feeding, represent up to 25% of body weight.

In winter, wolves with full stomachs usually collapse and sleep for at least 5 hours, usually at midday, sometimes traveling some distance to find a suitable location. Such extensive rest after gorging probably aids digestion (Mech 1970). In the Alaskan Arctic, most wolf feeding was concentrated between the hours of 0600 and 1200, while resting predominated between 1200 and 1800 (Magoun 1976). Wolves with distended stomachs usually lie on their sides rather than curling up. If they have not already been satiated, subordinate wolves have ample time while the dominant wolves are resting to tear off parts of a prey carcass, and they may search out a nearby ridgetop or tree canopy to eat in relative seclusion. Wolves usually rest less than 100 m from their kills, where they can still scare away a gathering mob of scavengers, but parts of a kill may be distributed at den sites tens of kilometers distant or buried in caches away from the kill.

For a pack of fifteen or sixteen wolves, a second feeding on a moose carcass typically occurred about 6 hours after the first feeding (Mech 1970). After a second feeding, only bones would typically remain, but these would keep the wolves in the vicinity for a second day. In winter, smaller packs may remain around a kill for several days.

Wolves are not fed by meat alone; in fact, they require the less palatable or less accessible portions of their prey in order to maintain a balanced intake of nutrients. To grow and maintain their own bodies, wolves need to ingest all the major parts of their herbivorous prey, except the plants in the digestive system.

The liver of prey may be the most important organ for wolves to eat, based on the variety of vitamins and minerals it provides (table 4.3). Wolves can synthesize some vitamins, such as vitamin C (which humans cannot), if their diet is inadequate. Like the nearby liver, the lungs and heart are quickly consumed by wolves. The high palatability of liver and lungs from ungulate prey is exploited by several wolf-prey parasites, notably *Echinococcus* and *Taenia* tapeworms, which encyst in prey lungs and liver, respectively, preparatory to being

TABLE 4.3. Important nutritional derivatives from constituents of wolf-killed prey

Prey item	Nutritional benefits
Liver	B vitamins
Liver	Vitamin A, for integrity of epithelial and mucosal surfaces
Liver	Source of copper, zinc, manganese, required for many enzyme systems
Brain, liver, kidney, heart	Essential fatty acids, for skin and membrane integrity, nervous tissues, retinal pigments, prostaglandins
Muscle	Protein, important for body structures (nerves, muscle, blood) and as a source of energy; also a minor source of fat
Fat	Source of energy and essential fatty acids, and essential for absorption of fat-soluble vitamins (A, D, E, and K)
Bone	Calcium and phosphorus, the main components of the skeleton
Hair	Speeds passage through intestinal tract, possibly a source of minerals such as zinc and manganese (if digestible)
Blood	Source of protein and water
Digestive tract	Stomach and intestinal wall are sources of fatty acids; microflora may be source of biotin (B complex), required for maintenance of skin and hair

Sources: Meyer and Stadtfeld 1980; Rivers and Frankel 1980; Bradshaw and Thorne 1992.

consumed by wolves and hatching into adults (see Kreeger, chap. 7 in this volume).

Bones from prey are required by wolves as the major source of calcium and phosphorus for the maintenance of their own skeletons. Bones, in fact, are a surprisingly well-balanced food for canids. In the nineteenth century, dogs experimentally fed only tendons died, but dogs fed only fresh bones thrived for long periods (McCay 1949). Single wolves, which often scavenge to stay alive, may not see a fresh kill for weeks, yet they maintain themselves on a diet of bones from old kills. Fresh bone, if replete with fatty marrow, may contain 15–20% protein and an equivalent amount of fat (McCay 1949).

No soft portion of a prey carcass is ignored by hungry wolves, and seemingly obscure parts may be important sources of required nutrients or aid in processing other portions. Brain tissue contains the highest amount of polyunsaturated fatty acids (PUFAs), some of which are required for body maintenance. Thus brains of prey are inevitably eaten, so long as wolves are able to break open the skull. They invariably do so with calf moose, and in a case observed by L. D. Mech (personal communication) in Alaska, they even managed to open an adult cow skull.

If the stomach of a large prey animal freezes before wolves find the carcass (e.g., if it died from starvation, accident, or earlier wounding by wolves), the wolves commonly consume the entire stomach wall by laborious nibbling with their incisors. Mucous membranes, an important component of ungulate stomachs, intestines, and snouts, are a vital source of essential fatty acids. Muscle tissue contains only about 1% of its maximum metabolizable energy as isomeric D-EFA's (essential fatty acids), while the heart contains 3–4%, the liver 4–7%, and the brain over 11% (Rivers and Frankel 1980).

Food Passage Rate and Digestibility

If wolves were to be judged as domestic dogs are, they would probably rank as dysfunctional feeders. The irregular, very large meals ingested by wolves commonly result in very liquid, diarrhea-like feces. The rapid passage of chyme through the digestive system is associated with osmotic imbalance, stimulation of secretion and gut motility, and inhibition of nitrogen and water absorption, all of which lead to increased water content in feces (Blaza 1982).

The first scats after a kill are usually black and runny, reflecting the initial meal of only body organs and muscle (Mech 1970; Jhala 1993). Judging by the popularity of wolf scat consumption among attentive ravens near wolf kills, the dark, liquid scats reflect poor digestive efficiency in wolves after large meals. This inefficiency is thought to result from ingestion of large quantities of protein, the decomposition of which produces organic acids and ammonia, which may restrict sodium-dependent enzymes (ATPases) in the intestinal wall and lead to sodium upset and therefore reduction in water absorption (Meyer et al. 1980).

Easily digestible items pass through the digestive tract more slowly than do foods with indigestible components. When the proportion of raw fiber in the diet of dogs was experimentally increased from near zero to 8%, passage time was reduced from 60–70 hours to 35 hours, primarily because of more rapid passage through the large intestine (Meyer et al. 1980). Wolves may attempt to mix their food with indigestible components. Peterson has seen them shear off mouthfuls of hair from moose just after making a kill but before tearing into soft tissues. A rapid passage rate would allow wolves to empty the gut and feed again, minimizing losses to scavengers (see also Ballard et al., chap. 10 in this volume)

Gut clearance time for wild wolves has not been adequately documented. On Isle Royale, most kills are made during the night, and in the morning the wolves are found sleeping after their first feeding. They usually arise and feed again before the day is over, so passage time may often be as short as 12 hours (R. O. Peterson, personal observation). In captive wolves, 8 hours was the minimum time required for production of feces after feeding (Floyd et al. 1978), and all scats from a given feeding cleared the system within 48 hours (Kreeger et al. 1997).

Hide and bones are the final portions of a prey carcass to be consumed. The extent to which these portions are eaten is a rough indicator of prey vulnerability (Pimlott et al. 1969; Mech and Frenzel 1971a; Mech et al. 2001). If an adult moose carcass is left with the skeleton disarticulated, the leg bones heavily chewed on the ends, and all hide consumed from the distal portions of the legs down to the hooves, then moose vulnerability is relatively low (though normal). At the other extreme, if the skeleton is still articulated and hide remains on the skull, the lower half of the legs, and the underside of the carcass, then wolves are finding vulnerable prey without a great deal of difficulty. Of course, pack size will influence the degree of prey utilization. Large packs with more than ten wolves feeding on moose rarely leave much hide un-eaten, except on the lower portion of the legs.

The wolf's diet, except for hair and bone, is highly di-gestible—generally in excess of 90%, based on experi-ments with dogs (Meyer et al. 1980). Fat is the most thor-oughly digested component of the diet (97%), followed by protein (93%). As food intake increases, digestibility drops by a few percentage points. When dogs were fed daily on a balanced diet amounting to 2.6% of their weight, digestibility was 93%, but when their intake in-creased to 5.2% of their weight, digestibility dropped to 90%. The effect of doubling the intake of food, then, was a near-tripling of feces production.

Food Requirements

Three important calculations for converting docu-mented kill rates of prey into food consumption rates for wolves are the amount of food provided by prey car-casses of various types, the amount of food lost to scav-engers, and the amount of food cached to be eaten later. None of these subjects has received much attention. Pe-terson (1977) weighed the rumen contents from a moose carcass—its largest inedible component—and esti-mated the weight of the hide and skeleton that are typi-cally not consumed. He suggested that about 75% of an adult moose could be eaten by wolves (or scavenged by other species). C. Promberger (unpublished data) weighed the remains of seven wolf-killed moose the day after wolves had left them. Based on whole weights of Alaskan moose (400 kg or 880 pounds for adult males, 350 kg or 770 pounds for adult females, 250 kg or 550 pounds for yearlings), about 35% of the carcasses re-mained. Jedrzejewski et al. (2002) estimated that wolves consumed 65% of total prey mass for large prey with a body mass exceeding 65 kg (143 pounds). Additional measurements would be useful, particularly for other large prey species for which consumption is incomplete.

C. Promberger (unpublished data) pointed out that evaporation and loss of blood would further reduce the edible mass of prey. Assuming similarity to domestic cows, the blood volume of ungulate prey would make up 6–10% of whole weight (McCay 1949). For one Isle Royale moose weighing 360 kg (792 pounds), R. O. Pe-terson (unpublished data) estimated total blood and fluid loss during dissection (a process not unlike the feeding of wolves) to be 27 kg or 59 pounds (or almost 8% of whole weight). On the other hand, rarely is that

amount of blood found around a moose kill (L. D. Mech, personal observation). The largest variation should come from the degree to which hide and bone are consumed. The skin and skeleton together constitute about 25% of the whole weight of all mammals, regard-less of size (Calder 1984).

The extent to which caching could modify estimates of food consumption is unknown, but it could be quite important. As indicated above, wolves sometimes cache large quantities of food from fresh kills. In such cases, estimates of food consumption based on amounts re-moved from a carcass could be grossly inflated (Mech and Adams 1999).

Basal Metabolic Rate

We can estimate wolf food requirements from equations relating energy requirements to body mass for mam-mals. The daily basal metabolic rate (BMR) in kcal/day, measured for a fasting and resting animal within its ther-moneutral zone, is related to body mass (W) in kg by the following equation (Kleiber 1961):

$$BMR = 70W^{0.75}$$

For a 35 kg (77-pound) wolf, BMR is thus estimated to be 1,007 kcal/day, or 175.4 kJ/hr (1 kcal = 4.18 kJ). Experi-mental estimates for adult wolves (Okarma and Koteja 1987) range from 154 to 173 kJ/hr (see Kreeger, chap. 7 in this volume). Assuming an average energy content of 7.7 kJ/g for prey carcasses (Glowacinski and Profus 1997), a BMR requirement of 1,007 kcal/day corresponds to 0.55 kg (1.2 pounds) prey/day. However, even a resting animal will not be maintained by a diet equaling BMR, because some chemical energy is lost during ingestion, digestion, and assimilation; some is excreted in urine, expelled as methane, or lost undigested in feces; and some is used to produce additional heat in a fed animal (as compared with a fasting animal) (Kleiber 1961). As-suming digestibility of 93% and 10% loss of energy in urine (Gorman et al. 1998), then 0.65 kg (1.4 pounds) prey/day is required for wolf basal metabolism.

Food intake providing 2–4 times the BMR is re-quired for active wild mammals (Weiner 1989), but this figure may be low for a cursorial predator such as the wolf. Kreeger (see chap. 7 in this volume) estimated that wolves burn calories at a rate of 4.6 times BMR. The painted hunting dog of Africa, with a similar lifestyle, expends about 5.2 × BMR during its normal activi-ties (Gorman et al. 1998). Considering these findings, we estimate the minimum daily energetic requirement for a wild wolf to be about 5 × BMR, approximately 3.25 kg (7.2 pounds)/wolf/day (for a 35 kg animal), or 0.09 kg per kilogram of wolf per day. Wolves might survive on less, but withdrawal of body reserves would be necessary.

Our estimate amounts to 25,025 kJ/wolf/day, compared with 47,100 kJ/day expended by Alaskan sled dogs during a 3-day, 490 km race. Painted hunting dogs burned energy at a rate of 25 × BMR while actively chasing prey (Gorman et al. 1998). Additional indirect measurements of food consumption or energy requirements provide some comparisons with these estimates. Glowacinski and Profus (1997) summarized estimates of field metabolic rate (FMR) for carnivores of different sizes. For a wolf weighing 35 kg, an energy requirement of 21,300 kJ/day was estimated, corresponding to 2.8 kg (6.2 pounds)/day/wolf (assuming 7.7 kJ/g for prey).

Comparisons of food requirements among species are possible if energetic demand is expressed in terms of "metabolic body mass," or mass raised to the 3/4 power. As body size increases, energetic demand per unit mass declines. Our estimated requirement of 25,025 kJ/wolf/day then becomes 1,750 kJ/kg$^{0.75}$/day for a 35 kg wolf. In comparison, human bicyclists in the grueling 22-day, 3,826 km (6,122-mile) Tour de France race expended about 1,200 kJ/kg$^{0.75}$/day, and Robert Scott and his ill-fated colleagues burned about 860 kJ/kg$^{0.75}$/day while hauling sledges toward the South Pole (Hammond and Diamond 1997). Yet wolves seem almost lethargic compared with hummingbirds (1,900 kJ/kg$^{0.75}$/day) or laboratory mice with large litters of young (3,900 kJ/kg$^{0.75}$/day) (Hinchcliff et al. 1997).

Our minimum daily energy requirement estimate of 3.25 kg/wolf/day (5 × BMR) is higher than most early estimates of wolf food requirements based on estimated food consumption in captivity and in the wild. Mech (1970) reported that wolves in captivity could be maintained on 1.1 kg (2.4 pounds)/wolf/day, while Kuyt (1972; cited in Mech 1977a) found that 1.7 kg (3.7 pounds)/day was sufficient for maintenance and reproduction in captivity. In contrast, early estimates of food consumption for wild wolves ranged from 2.5 to 6.3 kg (5.5–13.9 pounds)/wolf/day (Mech 1966b; Mech and Frenzel 1971a). Our estimate of 3.25 kg/wolf/day compares favorably with the finding that when food availability for a Minnesota wolf pack fell below an estimated 3.2 kg (7.0 pounds)/wolf/day, productivity or survival declined (Mech 1977b). An important research need to allow bet-

ter expression of wolf food requirements on an energetic basis is to determine the caloric content of different prey, as well as of different body components of large prey.

Food Consumption Rates

With the help of radiotelemetry, food consumption has now been estimated many times for wolves in the wild, based on efforts to record all kills made by packs of known size (table 4.4). For eighteen studies in North America, average estimated daily food consumption (or, more correctly, availability) was 5.4 kg (11.9 pounds)/wolf/day, or 0.14 kg/kg wolf/day (SE = 0.01 kg or 0.02 pounds), similar to an estimate of food consumption in eastern Poland of 5.6 kg (12.3 pounds)/wolf/day (Jedrzejewski et al. 2002). The North American estimates correspond to a low of 0.06 kg/kg (0.06 pounds/pound)/wolf/day (Fuller 1989b) and a maximum of 0.29 kg/kg (0.29 pounds/pound)/wolf/day (Hayes 1995).

Another estimate of food consumption used "fallout cesium," ^{137}Cs, from atmospheric testing of nuclear weapons in the 1950s and 1960s. Holleman and Stephenson (1981) estimated a mean daily intake of only 0.62 kg (1.36 pounds) per wolf, although individual intake rates were as high as 3.0 kg (6.6 pounds) per wolf. In this case, the wolves probably had a more diverse diet than assumed, particularly in summer, thus diluting the assumed ^{137}Cs levels; also, non-muscle tissues have much lower ^{137}Cs levels than muscle tissues.

The proportion of a carcass lost to scavengers, or cached for later use, must be subtracted from the above consumption estimates. The loss to scavengers is primarily dependent on pack size—large packs lose little, while small packs may lose half or more of their kill (C. Promberger, unpublished data). This variation can be illustrated by examining the range of estimates of food availability for wolves on Isle Royale (fig. 4.8): the smaller the pack, the larger the estimate of food available (although not necessarily consumed) per wolf. This finding may mean that smaller packs lose much more of their food because they take longer to handle carcasses, so that scavengers have more time and opportunity to remove parts of the prey (C. Promberger, unpublished data).

Estimates of food availability for 158 wolf packs (see fig. 4.8), measured almost entirely in winter, confirm that food availability per wolf declines as pack size increases (Thurber and Peterson 1993; Schmidt and Mech 1997; Hayes and Harestad 2000b; but see Jedrzejewski

TABLE 4.4. Estimated winter food availability from wolf-killed ungulates in North America

Major prey species	Location	Mean availability		Reference
		kg/wolf/day	kg/kg of wolf/day	
Deer	Ontario	2.9	0.10	Kolenosky 1972
Deer	Minnesota	2.9	0.10	Mech 1977a
Deer	Minnesota	2.9	0.09	Fritts and Mech 1981
Deer	Minnesota	2.0	0.06	Fuller 1989b
Dall sheep	Yukon	3.0	0.08	Sumanik 1987
Caribou	Alaska	6.9	0.16	Dale et al. 1995
Elk	Manitoba	6.8	0.21	Carbyn 1983b
Elk	Yellowstone	6.1–17.1	—	Mech et al. 2001
Moose	Isle Royale	4.4–6.3	0.13–0.19	Mech 1966b
Moose	Alaska	4.5	0.11	Haber 1977
Moose	Isle Royale	7.2	0.22	Peterson 1977
Moose	Alberta	5.5	0.14	Fuller and Keith 1980a
Moose	Alaska	4.8	0.12	Peterson, Woolington, and Bailey 1984
Moose	Quebec	2.2	0.07	Messier and Crête 1985
Moose	Alaska	7.1	0.18	Ballard et al. 1987
Moose	Isle Royale	4.9	0.15	Peterson and Page 1988
Moose	Yukon	11.4	0.29	Hayes 1995
Bison	Alberta	5.3	0.14	Oosenbrug and Carbyn 1982

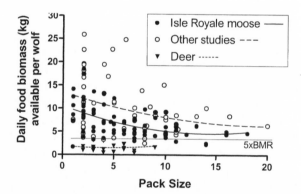

FIGURE 4.8. Food availability in relation to pack size for wolves preying on moose on Isle Royale ($N = 82$), wolves preying on white-tailed deer ($N = 27$), and all other packs studied ($N = 58$), with exponential decay equations fitted to each data set ($r^2 = .26, .10,$ and .57, respectively). Also shown is the estimated food requirement ($5 \times$ BMR, or basal metabolic rate; see text). Note that the deer predation rate may have been underestimated. (Data from Schmidt and Mech 1997.)

et al. 2002). It has been hypothesized that the ultimate explanation for group hunting in wolves is that breeding pairs can efficiently direct the short-term food surplus from their kills toward their offspring—food that would otherwise be lost to scavengers (Schmidt and Mech 1997; see also Mech and Boitani, chap. 1 in this volume).

Surplus Killing of Prey

When wolves kill prey frequently and eat little or none of the carcasses, they are said to be engaging in "surplus killing" (Kruuk 1972). Surplus killing occurs when prey are both abundant and highly vulnerable, a rare combination that, for wolves and wild prey, has been recorded only for neonates and prey that are relatively immobilized by deep snow. This subject is covered in more detail by Mech and Peterson in chapter 5 of this volume.

Loss to Scavengers and Other Predators

Wolf-killed prey are a food bonanza for many species besides wolves. Wolf-scavenger relationships are old and well established on evolutionary time scales, and have mutual influences on behavior and ecology. Stahler et al. (2002) confirmed an intimate relationship between wolves and ravens in Yellowstone that probably reflects a shared evolutionary history. Ravens preferred to associate with wolves, rather than being found with other species or at random points in the landscape, and they invariably found wolf-killed prey within a few minutes after the kill was made.

C. Promberger (unpublished data) documented the potential extent of scavenging on wolf kills and measured carcass removals by scavengers in the Yukon and

Northwest Territories of Canada. Of sixteen species of vertebrate scavengers, twelve were present in winter, typically showing up a day or more after the kill. During summer, both black bears and grizzly or brown bears are important scavengers. In winter, the raven emerged as the dominant scavenger in northwestern Canada (C. Promberger, unpublished data). Ravens and gray jays were consistently among the first scavengers to appear at test carcasses, and the importance of ravens grew as their numbers swelled to dozens of birds per carcass. While a territorial pair of ravens does not advertise the presence of a carcass, the carcass is soon discovered by other ravens on patrol, which attract more ravens with their calls and mob the local pair (Heinrich 1989). Up to 135 scavenging ravens were recorded on a wolf-killed elk in Yellowstone (Stahler et al. 2002), and they removed up to 37 kg (81 pounds) per day from fresh moose carcasses in the Yukon (fig. 4.9). As many as ten red foxes were observed simultaneously at a wolf-killed moose on Isle Royale (Peterson 1977), and six coyotes on wolf-killed elk in Yellowstone (L. D. Mech, personal communication). The scavenging significance of wolf-killed prey is not limited to vertebrate macrofauna; over 400 species of beetles were associated with elk carcasses in Yellowstone (Sikes et al. 1995).

C. Promberger (unpublished data) estimated the proportion of a moose carcass removed by scavengers, assuming that wolves each consumed 10 kg (22 pounds) on the first day and 6 kg (13 pounds) on subsequent days.

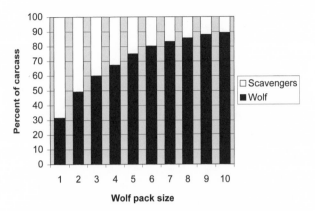

FIGURE 4.10. Proportion of prey carcasses lost to scavengers in relation to wolf pack size. (From C. Promberger, unpublished data.)

The larger the wolf pack, the less would be lost to scavengers. By this analysis, a lone wolf might lose two-thirds of a moose kill to scavengers, and a pair of wolves, half of their kill, while a pack of ten wolves would lose only about 10% (fig. 4.10).

Research Needs

While we have a good understanding of wolf food consumption (or at least availability) in winter, little is known about food intake in summer. Do the reduced weights of wolves in summer measured by Peterson, Woolington, and Bailey (1984) reflect reduced food availability, precisely at a time when the demands of growing pups reach a peak? Or are they merely a result of an internal seasonal weight cycle (Seal and Mech 1983)? Jedrzejewski et al. (2002) recently applied novel field techniques to estimate year-round food consumption of wolves.

The physiology of fasting in general is poorly understood in the wolf, a species well adapted to a feast-or-famine existence. The energy content of various prey and parts of prey should be better documented in an effort to put estimates of wolf energetics on more solid footing. Loss of food to scavengers needs to be documented across the range of prey types and wolf distribution. Likewise, biomass models for conversion of occurrence data in scats should be extended to include more prey types.

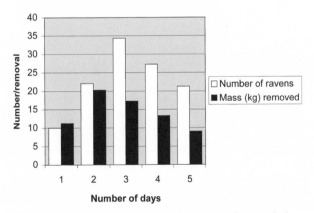

FIGURE 4.9. Numbers of ravens and their estimated removal of biomass from moose carcasses over time. (From C. Promberger, unpublished data.)

APPENDIX 4.1 Studies of wolf food habits in Eurasia and North America

Eurasia

Álvares 1995

Brangi et al. 1992

Ciucci 1994

Cuesta et al. 1991

Fernandez et al. 1990

Fonseca 1990

Guitian et al. 1979

Hell 1993

Ionescu 1993

Jedrzejewska et al. 1994

Jedrzejewski et al. 1992

Kochetkov 1988

Kudaktin 1978

Labutin 1972

Lesniewicz and Perzanowski 1989

Llaneza et al. 1996

Macdonald et al. 1980

Magalhaes and Fonseca 1982

C. Matteucci, unpublished data

Mattioli et al. 1995

Meriggi et al. 1991

Meriggi et al. 1996

Okarma et al. 1995

Olsson et al. 1997

Patalano and Lovari 1993

Poulle et al. 1997

Pullianen 1965

Ragni et al. 1985

Ragni et al. 1996

Reig et al. 1985

Salvador and Abad 1987

Smietana and Klimek 1993

Vilà et al. 1990

North America

Ballard et al. 1987

Bjorge and Gunson 1983

Boyd et al. 1994

Carbyn 1974

Carbyn 1980

Carbyn and Kingsley 1979

Clark 1971

Frenzel 1974

Fritts and Mech 1981

Fuller 1962, cited in Pimlott 1967

Fuller 1989b

Fuller and Keith 1980a

Gasaway et al. 1983

Hertel 1984

Huggard 1993a

Kelsall 1957, cited in Pimlott 1967

Marquard-Petersen 1998

Mech 1966b

Messier and Crête 1985

Murie 1944

Oosenbrug et al. 1980

Peterson 1977

Peterson, Woolington, and Bailey 1984

Pimlott 1967

Pimlott et al. 1969

Potvin et al. 1988

Schmidt and Gunson 1985

Scott and Shackleton 1980

Shelton 1966

Smith et al. 1987

Stephenson and James 1982

Theberge and Cottrell 1977

Thompson 1952

Thurber and Peterson 1993

Van Ballenberghe et al. 1975

Voigt et al. 1976

Weaver 1994

5

Wolf-Prey Relations

*L. David Mech and
Rolf O. Peterson*

AS I (L. D. MECH) watched from a small ski plane while fifteen wolves surrounded a moose on snowy Isle Royale, I had no idea this encounter would typify observations I would make during 40 more years of studying wolf-prey interactions.

My usual routine while observing wolves hunting was to have my pilot keep circling broadly over the scene so I could watch the wolves' attacks without disturbing any of the animals. Only this time there was no attack. The moose held the wolves at bay for about 5 minutes (fig. 5.1), and then the pack left.

From this observation and many others of wolves hunting moose, deer, caribou, muskoxen, bison, elk, and even arctic hares, we have come to view the wolf as a highly discerning hunter, a predator that can quickly judge the cost/benefit ratio of attacking its prey. A successful attack, and the wolf can feed for days. One miscalculation, however, and the animal could be badly injured or killed. Thus wolves generally kill prey that, while not always on their last legs, tend to be less fit than their conspecifics and thus closer to death. The moose that the fifteen wolves surrounded had not been in this category, so when the wolves realized it, they gave up. That is most often the case when wolves hunt.

Throughout the wolf's range (most of the Northern Hemisphere; see Boitani, chap. 13 in this volume), ungulates are the animal's main prey (see Peterson and Ciucci, chap. 4 in this volume). Ordinarily, ungulate populations include both a secure segment of healthy prime animals and a variety of more vulnerable or less fit individuals: old animals; newborn, weak, diseased, injured, or debilitated animals; and juveniles lacking the strength,

experience, and vigor of adults. Prey populations sustain themselves by the reproduction and survival of their vigorous members. Wolves coexist with their prey by exploiting the less fit individuals. This means that most hunts by wolves are unsuccessful, that wolves must travel widely to scan the herds for vulnerable individuals, and that these carnivores must tolerate a feast-or-famine existence (see Peterson and Ciucci, chap. 4, and Kreeger, chap. 7 in this volume).

When environmental conditions change, the relationship between wolves and prey shifts: conditions favorable to prey hamper wolf welfare; conditions unfavorable to prey foster it. With their high reproductive and dispersal potential (see Mech and Boitani, chap. 1, and Fuller et al., chap. 6 in this volume), wolves can readily adjust to changes in proportions of vulnerable prey. The result is that, under average prey conditions, wolf populations generally survive at moderate, lingering levels. All the while, they remain poised to exploit vulnerable prey surpluses, expand, and disseminate dispersers far and wide to colonize new areas (Mech et al. 1998).

Prey and Their Defenses

The dependence of wolves on ungulates implies that the entire original range of the wolf around the world must have been occupied ungulates, and that is indeed the case. Although ungulates vary considerably and may occupy highly specialized habitats, some representative of this large group of hoofed mammals lives almost everywhere throughout wolf range, from pronghorns on the

FIGURE 5.1. Healthy prime-aged moose can withstand wolves. These wolves left after 5 minutes.

prairie to mountain goats on the craggiest cliffs. And the primary predator on all of them is the wolf.

Each ungulate species is superbly and uniquely adapted to survive wolf predation. Most possess several defensive traits, while some depend on one or a few (table 5.1). In no case can a wolf merely walk up and kill a healthy ungulate that is more than a few days old.

All but a few ungulate species are highly alert and responsive to sight, smell, and sound (see table 5.1). The degree of such vigilance is affected by several factors (table 5.2). This fine-tuning of vigilance serves ungulates well in allowing them to feed relentlessly while still being able to suddenly choose their course of escape or defense should wolves threaten. With deer (Mech 1984), sheep (Murie 1944), goats, pronghorn, and even hares (L. D. Mech, unpublished data), all of whose other main defense is flight, either away from the predators or to safer terrain, alertness could make the difference. Furthermore, after wolves have hunted an area, local prey increase their vigilance (Huggard 1993b; Laundré et al. 2001; K. E. Kunkel et al., unpublished data). Deer, beavers, and probably most other prey species can even distinguish the odor of predator urine or feces (Muller-Schwarze 1972; Steinberg 1977; Ozoga and Verme 1986; Swihart et al. 1991; Smith et al. 1994), and probably use this ability to avoid their enemies (Adams, Dale, and Mech 1995).

Clearly speed combined with vigilance is an important defensive factor for smaller prey. R. O. Peterson (unpublished data) measured the speed of an arctic hare at 60 km (36 mi)/hr. White-tailed deer can run at 56 km (34 mi)/hr or more (Newsom 1926, 174, cited in Taylor 1956) and can leap hurdles as high as 2.4 m (Sauer 1984); these abilities facilitate their flight through the thick forested areas they often frequent and through deep snow (Mech and Frenzel 1971a). Although most chases of deer by wolves appear to be relatively short (Mech 1984), deer do possess the endurance to flee for 20 km (12 mi) or more (Mech and Korb 1978). Other relatively small ungulates such as sheep and goats combine alertness and speed with ability to outmaneuver wolves around steep, dangerous terrain, and thereby manage to evade wolves. When on level ground, these animals are almost defenseless (Murie 1944).

In addition to these obvious types of defense, prey animals use a variety of more subtle defensive and risk-reducing behaviors (Lima and Dill 1990); the precise manner in which many of these behaviors work is still unknown (see table 5.1). White-tailed deer, for example, flag their tails in response to disturbance. The most recent explanation for this behavior is that it signals the predator that its presence is known and that pursuit is therefore useless (Caro et al. 1995).

At the other end of the size spectrum, prey such as

TABLE 5.1. Antipredator characteristics and behavior of wolf prey species

Trait/behavior	Species	Reference
Physical traits		
Size	Moose	Mech 1966b
	Bison	Carbyn et al. 1993
	Muskoxen	Gray 1987
Weapons		
Antlers/horns	Male ungulates	Nelson and Mech 1981
	Some females	See text
Hooves	All ungulates	See text
Cryptic coloration	Most ungulate young	Lent 1974
Speed/agility	Pronghorn	Kitchen 1974
	Hares	Mech, unpublished data
	Blackbuck	Jhala 1993
Lack of scent	Deer neonates	Severinghaus and Cheatum 1956
Behavior		
Birth synchrony	Most ungulates	Estes 1966; Rutberg 1987; Ims 1990; Adams and Dale 1998b
Hiding	Deer neonates	Walther 1961; Lent 1974; Carl and Robbins 1988
	Pronghorn neonates	Lent 1974; Carl and Robbins 1988
Following	Caribou neonates	Walther 1961; Lent 1974
	Goat neonates	Lent 1974
	Sheep neonates	Lent 1974
	Moose neonates	Lent 1974
Aggressiveness	All ungulates	See text
Grouping	Caribou	Bergerud et al. 1984
	Elk	Darling 1937; Hebblewhite and Pletscher 2002
	Muskoxen	Gray 1987; Heard 1992
	Bison	Carbyn et al. 1993
	Deer (winter)	Nelson and Mech 1981
	Pronghorn	Kitchen 1974; Berger 1978
	Sheep	Berger 1978
	Goats	Holroyd 1967
	Hares	Mech, unpublished data
Vigilance	All species	Dehn 1990; Laundré et al. 2001
	Deer	Mech 1966a
Vocalizations	Deer	Schaller 1967; Hirth and McCullough 1977; LaGory 1987
	Sheep	Berger 1978
Visual signals	Deer	Smythe 1970, 1977; Bildstein 1983; LaGory 1986; Caro et al. 1995
	Elk	Guthrie 1971
	Sheep	Berger 1978
	Muskoxen	Gray 1987
	Arctic hares	Mech, unpublished data

(continued)

TABLE 5.1 *(continued)*

Trait/behavior	Species	Reference
Landscape use		
Migration	Caribou	Banfield 1954
	Deer	Nelson and Mech 1981
	Elk	Schaefer 2000
	General	Fryxell et al. 1988
Nomadism	Caribou	Bergerud et al. 1984
	Muskoxen	Gray 1987
	Bison	Roe 1951
	Saiga	Bannikov et al. 1967
Spacing		
Away	Caribou	Bergerud et al. 1984; Ferguson et al. 1988; Adams, Dale, and Mech 1995
	Deer	Hoskinson and Mech 1976; Mech 1977a,d; Nelson and Mech 1981
	Moose	Edwards 1983; Stephens and Peterson 1984
Out	Deer	Nelson and Mech 1981
	Moose	Mech et al. 1998
	Caribou	Bergerud et al. 1984
Escape features		
Water	Deer	Nelson and Mech 1981
	Moose	Peterson 1955; Mech 1966b
	Caribou	Crisler 1956
	Elk	Cowan 1947; Carbyn 1974
	Beavers	Mech 1970
Steepness	Sheep	Murie 1944; Sumanik 1987
	Goats	Rideout 1978; Fox and Streveler 1986
Shorelines	Caribou	Bergerud 1985; Stephens and Peterson 1984
Burrows	Wild boar	Grundlach 1968

TABLE 5.2. Factors affecting vigilance in wolf prey

Factor	Reference
Body size	Berger and Cunningham 1988
Herd size	Berger 1978; Lipetz and Bekoff 1982
	LaGory 1987; Berger and Cunningham 1988; Dehn 1990
Position in herd	Lipetz and Bekoff 1982
	Berger and Cunningham 1988
Maternal status	Lipetz and Bekoff 1982; Boving and Post 1997; Berger, Swenson, and Persson 2001
Cover	LaGory 1986, 1987
Degree of predator risk	Boving and Post 1997; Berger, Swenson, and Persson 2001

moose (Mech 1966b; Peterson 1977), bison (Carbyn et al. 1993), horses, muskoxen (Gray 1987; Mech 1988a), elk (Landis 1998), wild boar (Reig 1993), and even domestic cattle depend on their sheer size and aggressiveness for much of their defense. Although individuals of any of these species will flee if they detect wolves from far enough away, they will stand their ground and fight when confronted. They lash out with heavy hooves, and those with horns or antlers wield them well. Even deer hooves and antlers can be deadly weapons, and some deer will stand and fight off wolves (Mech 1984; Nelson and Mech 1994). Wolves have been killed by moose (MacFarlane 1905; Stanwell-Fletcher and Stanwell-Fletcher 1942; Mech and Nelson 1990a; Weaver et al. 1992), muskoxen (Pasitchniak-Arts et al. 1988), and deer (Frijlink 1977; Nelson and Mech 1985; Mech and Nelson 1990a).

The large ungulates are especially aggressive when defending their young. Cow moose are dangerous even to humans when their calves are newborn, and they will battle wolves fiercely to protect their young calves (Mech 1966b; Peterson 1977; Stephenson and Van Ballenberghe 1995). When the calves are several months old, a cow running from wolves remains close to her calves' rear ends (their most vulnerable area) and tries to trample any wolf coming close (Mech 1966b). In one case, a cow moose fended off wolves from her two dead 10-month-old calves for 8 days (Mech et al. 1998).

Muskoxen form a defensive line or ring to protect calves (Hone 1934; Tener 1954; Gray 1987; Mech 1988a). All the oxen press their rumps together in front of their young, and the calves press in close to the rumps of their mothers. Bison react similarly, with calves running to the herd and seeking protection from adults (Carbyn and Trottier 1987, 1988). Both muskoxen and bison, especially calves, are most vulnerable to wolves when running (see Gray 1983, 1987; Mech 1987b, 1988a for muskoxen; Carbyn and Trottier 1988 for bison).

Water as a Defense

One of the defensive techniques that most wolf prey resort to when possible is to run into water (Mech 1970). This tactic may provide the prey with several advantages, and it seems to hinder the wolves. Larger prey can stand in deeper water than a wolf can, so the wolf would have less leverage there. The prey can also stand still in the water, while the wolf and its companions must maneuver around through the water. Long-legged species such as moose probably could wallop a wolf with a hoof while the wolf is forced to swim around it. On the other hand, a swimming wolf has been known to kill a swimming deer (Nelson and Mech 1984).

Another common wolf prey species uses water in a different way to protect itself. By building dams, the beaver surrounds itself and its lodge with water deep enough to provide security from wolves most of the time (Mech 1970). It is vulnerable to wolves primarily when it ventures ashore or on top of the ice to cut food, or when its pond freezes to the bottom and the wolves dig the beavers out of the lodge (Mech 1966b; Peterson 1977). The propensity of wolves to travel on beaver dams, where crossing places used by beavers are quite obvious, suggests that waiting at such points at night when beavers are active would be a successful hunting strategy for wolves.

Safety in Numbers

Another defensive trait of many wolf prey species, small and large alike, is herding (Nelson and Mech 1981; Messier et al. 1988). Prey as diverse as wild boar, elk, muskoxen, saiga antelope, domestic animals, and arctic hares, as well as many others, live in herds, at least during certain seasons. The antipredator benefits of herding are well known (Williams 1966; Hamilton 1971): (1) increased sensory potential (Galton 1871; Dimond and Lazarus 1974), (2) dilution of risk (Nelson and Mech 1981), (3) greater physical defense, (4) increased predator confusion (McCullough 1969), (5) a reduced predator/prey ratio (Brock and Riffenburgh 1960), and (6) an increased foraging/vigilance ratio (Hoogland 1979).

Herding is so beneficial that some species go to great lengths to group together during their most vulnerable season, winter. White-tailed deer, for example, which live solitarily during summer, may migrate 40 km (24 mi) or more to herd, or "yard," on winter range (Nelson and Mech 1981). Elk sometimes join herds of 15,000 or more (Boyd 1978), although sometimes living in small herds reduces their rate of encounter with wolves (Hebblewhite and Pletscher 2002). Muskox herd size increases by 70% in winter, and the higher the wolf density, the higher the herd size (Heard 1992). Moose tend to aggregate in larger groups the farther they are from cover (Molvar and Bowyer 1994), probably because moose use woody vegetation as a tactical defense when attacked by wolves (Geist 1998).

Movements

Migration itself, aside from herding, also tends to reduce predation. Migration (seasonal movement between different ranges) can carry ungulates to more favorable areas away from wolves (Seip 1991) and increase wolf search time. Modeling of African ungulates suggests that migration confers such a strong antipredator benefit that migrants should always outcompete residents (Fryxell et al. 1988). By itself, migration may greatly increase an ungulate's short-term risk (Nelson and Mech 1991), but this fact only further supports the long-term benefit of migration. That migration is a general adaptation to enhance survival is shown by the tendency for cow elk that have calves to migrate farthest to escape deep snow in Yellowstone's Northern Range, both before and after the introduction of wolves in 1995 (Schaefer 2000). In some areas, elk migrate 64 km (38 mi) or more (Boyd 1978).

An increase in search time is also an advantage of the nomadism (constant movement over a large area) that several ungulates practice (see table 5.1). Mech was continually impressed with the difficulty of finding nomadic caribou every time he searched by helicopter for the Denali herd in Alaska. Despite the advantages of speed, broad visibility, and a general knowledge of past areas the caribou had frequented, it often took him hours to find them. A related type of wolf avoidance was documented for a bison herd of about ninety that fled 81.5 km (50 mi) during the 24 hours after wolves killed a calf in the herd (Carbyn 1997). L. D. Mech (unpublished data) has noticed that muskox herds also tend to disappear from a region after wolves have killed one.

Spacing

Caribou and other ungulates (Kunkel and Pletscher 2000) also space themselves in other ways that tend to thwart wolves. "Spacing out" (Ivlev 1961) is the tendency of prey to disperse themselves widely within their populations, which helps maximize wolf search time (e.g., deer in the Superior National Forest: Nelson and Mech 1981). A similar advantage is gained by the "spacing away" of caribou cows, the tendency to calve on steep mountain ridges, in extensive spruce swamps, or in other areas far from wolf travel routes such as rivers and from other potential wolf food sources ("apparent competition": Holt 1977) such as moose, which concentrate in lower areas with better nutrition (Bergerud et al. 1984, 1990; Bergerud 1985; Edmonds 1988; Bergerud and Page 1987).

Similarly, the spacing of calving caribou herds away from wolf denning areas or year-round wolf territories also increases wolf search and travel times, thus reducing predation risk (Bergerud and Page 1987; but cf. Nelson and Mech 2000). The Denali herd used this tactic to avoid any increase in wolf predation risk even when the wolf population doubled (Adams, Singer, and Dale 1995). A more dramatic example is the extensive spring migration of barren-ground caribou, which travel hundreds of kilometers from their winter range to calving grounds where wolf numbers are minimal (Bergerud and Page 1987 and references therein). By frequenting islands, peninsulas, shorelines, and other areas where exposure to approaching predators is minimized, prey reduce their chances of encounters with wolves (Edwards 1983; Stephens and Peterson 1984; Bergerud 1985; Ferguson et al. 1988).

These areas, along with mountaintops and extensive habitats such as spruce swamps that few prey, and thus few predators, regularly frequent, are especially important as birthing areas (Skoog 1968; Bergerud et al. 1984, 1990; Bergerud 1985; Bergerud and Page 1987; Adams, Dale, and Mech 1995). If using such areas improves the chances of a newborn's survival for just its first few days when it is most vulnerable, that might make the difference between whether the animal lives out a full life or not.

Wolf Territory Buffer Zones

A specialized type of spacing away involves wolf pack territory buffer zones, or overlap zones along the edges of territories (see Mech and Boitani, chap. 1 in this volume). During a drastic deer decline, wolves in the Superior National Forest eliminated deer first from the cores of their territories and only last from the edges. Based on this observation, Mech (1977a,d) proposed the existence of a buffer zone, or a "no-man's land," thought to be from 2 (Peters and Mech 1975b) to possibly 6 km wide (Mech 1994a). He felt that the reason deer survived longer along these territory edges might be that neighboring packs felt insecure in the buffer zone so spent less time there, minimizing hunting pressure on the deer there. In both summer and winter, deer were more abundant in buffer zones than in territory cores (Hoskinson and Mech 1976; Mech 1977a,d; Rogers et al. 1980; Nelson and Mech 1981). Similar wolf-deer relationships were observed in northwestern Minnesota (Fritts and Mech 1981) and on Vancouver Island (Hebert et al. 1982; Hatter 1984). Furthermore, theoreticians have found mathematical support for the buffer zone as a prey refuge (Lewis and Murray 1993), and others have described similar prey-rich zones between warring Indian tribes (Hickerson 1965, 1970; Martin and Szuter 1999). Carbyn (1983b) did not find disproportionate use by elk of pack boundary edges in Riding Mountain National Park.

"Swamping"

Another antipredator strategy pervasive among wolf prey species that helps promote survival of their young is the tendency toward synchronous births (Estes 1966; Wilson 1975). This phenomenon tends to "swamp" wolves with a short burst of vulnerable individuals of a given species. While wolves are occupied preying on some individuals, the others grow quickly and become less vulnerable by the day. For example, about 85% of caribou calves in Denali National Park are born within a 2-week period (Adams and Dale 1998b). During years of favorable weather, almost all the wolf predation on calves takes place during the calves' first 2 weeks of life (Adams, Singer, and Dale 1995). Similarly, white-tailed deer and arctic hares are born over a short period and tend to be vulnerable to wolves primarily during their first few weeks of life (Kunkel and Mech 1994 for deer; L. D. Mech, unpublished data, for hares). Because neonates of most ungulates are so vulnerable, but develop so quickly, it seems reasonable that swamping in some form helps minimize wolf predation on them as well.

Hunting Success

The many effective antipredator traits and strategies of most prey ensure that most hunts by wolves are unsuccessful (Mech 1970). Moreover, the actual hunting success of any predator varies considerably and depends greatly on many circumstances, such as season, time of day, weather, and terrain; predator experience; prey species, numbers, age, sex, associates, and vulnerability; past and immediate prey history; and no doubt many other factors. Furthermore, subtle factors, such as prey odor, prey behavior, and recent exposure of prey to attacks, may play important roles in the outcome of wolf-prey encounters (Haber 1977; Carbyn et al. 1993).

Measurements of wolf hunting success have been made primarily in winter, when hunting success for most large prey species is probably maximal because their vulnerability is greatest then (see below). In addition, the fact that many of the wolf's prey species live in herds complicates determinations of success. If wolves kill one elk in a single attack on a herd, but try to catch three of them, is their success rate 100% or 33%? Thus we have only a glimpse of the total picture. This glimpse shows both the relatively low success rate and its variation (10–49% based on number of hunts and 1–56% based on number of prey attacked) (table 5.3).

TABLE 5.3. Wolf hunting success rates based on number of hunts (encounters involving groups of prey) and on number of individual prey animals

Prey	Location	Number			% success based on		Reference
		Hunts	Individuals	Kills	Hunts	Individuals	
Winter[a]							
Moose	Isle Royale, MI	—	77	6–7[b]	—	8–9[b]	Mech 1966b
Moose	Isle Royale, MI	—	38	1	—	3	Peterson 1977
Moose	Kenai, AK	—	38	2	—	5	Peterson, Woolington, and Bailey 1984
Moose	Denali, AK	—	389	23	—	6[c]	Haber 1977
Moose	Denali, AK	37	53	7–14[b]	19–38	13–26[b]	Mech et al. 1998
Deer	Ontario	—	35	16	—	46[e]	Kolenosky 1972
Deer	Minnesota	—	60	12	—	20	Nelson and Mech 1993
Caribou	Denali, AK	—	16	9	—	56[c]	Haber 1977
Caribou	Denali, AK	26	303	4[d]	15	1	Mech et al. 1998
Dall sheep	Denali, AK	—	100	24	—	24[c]	Haber 1977
Dall sheep	Denali, AK	18	186	6	33	3	Mech et al. 1998
Bison	Alberta	31	—	3	10	—	Carbyn et al. 1993, table 46
Elk	Yellowstone, WY	102	1,532	21	21	1	Mech et al. 2001
Summer							
Bison	Alberta	86	—	28	33	—	Carbyn et al. 1993, table 48
Caribou	Denali, AK	110	1,934	54	49	3[c]	Haber 1977
Dall sheep	Denali, AK	14	108	4	29	4[c]	Haber 1977

[a]Results from Mech et al. 1998 include a few instances from spring, summer, and fall.

[b]Larger figures include wounded animals that may have died later.

[c]Results from Haber 1977 should be considered minimum estimates because he included prey that he believed the wolves "tested" from distances of "several hundred feet or more" (Haber 1977, 381).

[d]Includes two newborn calves in May.

[e]Probably biased upward because it was based on ground tracking where likelihood of interpreting kills is much greater than for failures (Kolenosky 1972).

One factor that might influence wolf hunting success rate is motivation based on time since last kill. However, wolves sometimes show interest in attacking prey within minutes of leaving a kill (Mech 1966b), or stop feeding on fresh kills to take advantage of new opportunities to catch prey (L. D. Mech, unpublished data). Thus it is not surprising that wolves seem to show no more intensity in attacking prey several days after feeding than just a day after.

Effects of Snow and Other Weather

Because wolves tend to kill prey that are vulnerable, and because prey vulnerability is greatly affected by weather conditions, weather is important to wolf-prey relations. The most significant weather factor is snow conditions, including snow depth, density, duration, and hardness.

Snow affects prey animals primarily by hindering their movements, including foraging and escape from wolves. The effect of snow on prey escape is mechanical: the deeper and denser the snow, the harder it is for prey to run through it. Most prey probably have a heavier foot loading than do wolves, so they would sink deeper and be hindered more than wolves. Estimates for foot loading in deer, for example, range from 211 g/cm^2 (Mech et al. 1971) to 431–1,124 g/cm^2 (Kelsall 1969), whereas for wolves, the estimate is about 103 g/cm^2 (Foromozov 1946). Ungulates are usually much heavier than wolves and possess hard hooves that puncture snow much more easily than the spreading, webbed toes of a wolf foot. This difference can tilt the balance toward wolves during predation attempts on animals from the size of deer (Mech et al. 1971) to bison (D. R. MacNulty, personal communication).

The condition of snow changes daily, even hourly, and wolves and their prey are very sensitive to subtle changes that might work to their advantage or disadvantage. R. Peterson (personal observation) has seen packs of wolves sleep through late afternoon and early evening during midwinter thaws, apparently waiting for the crusted snow that will follow when the temperature drops at night. During daily tracking of a pack of five wolves in upper Michigan during a 3-month period, B. Huntzinger (personal communication) documented three cases of the pack killing five to ten deer overnight; during two of these instances the kills were made during heavy blizzards, and in the third case wolves took advantage of a strong snow crust that supported them, but not the deer.

In addition to the acute effect of hindering prey escape, deep snow has a longer, more pervasive effect on prey nutrition. Snow resistance reduces foraging profitability for ungulates and causes them to lose weight over the winter, the amount depending on snow depth and density and duration of cover. During severe winters, prey often starve. The combination of reduced nutrition and poor escape conditions for prey can result in a bonanza for wolves (Pimlott et al. 1969; Mech et al. 1971, 1998, 2001; Peterson and Allen 1974; Mech and Karns 1977; Peterson 1977; Nelson and Mech 1986c).

However, severe snow conditions can also have indirect effects on prey animals that predispose them to wolf predation. These take the form of intergenerational effects and cumulative effects. Intergenerational effects result from the fact that ungulates are gravid over winter. Thus undernutrition or malnutrition caused by deep snow can affect fetal development and viability (Verme 1962, 1963), resulting in offspring with increased vulnerability to wolf predation (Peterson and Allen 1974; Mech and Karns 1977; Peterson 1977; Mech, Nelson, and McRoberts 1991; Mech et al. 1998). This intergenerational effect can even persist for a second generation. That is, animals with poorly nourished grandmothers can be more vulnerable to wolf predation even if their mothers were well nourished (Mech, Nelson, and McRoberts 1991; also see below).

The cumulative effects of snow conditions on prey vulnerability operate across winters. Ungulates must replenish their nutritional condition during the snow-free period each year. Thus if the replenishment period is too short, or if an animal reaches that period in too poor a condition, that creature may be vulnerable the next winter (Mech 2000d). If it survives, its condition may worsen, especially if the following winter is also severe or prolonged. In this manner, a series of severe winters can cumulatively reduce an animal's condition and increase its vulnerability (Mech, McRoberts et al. 1987; McRoberts et al. 1995; cf. Messier 1991, 1995a).

These same principles operate in the opposite direction if winters are mild and snow depth low or snow cover duration short. The result is prey in better condition and with lower vulnerability.

Although the effects of snow conditions on wolf-prey relations are the best-studied weather effects, drought and probably several other extreme conditions that affect prey nutrition no doubt similarly influence wolf-prey relations. For example, warm and dry weather during spring leads to heavy infestations of winter ticks

(*Dermacentor albipictus*) the following winter in North American moose, which cause direct mortality from starvation and probably make the moose more vulnerable to hunting wolves (DelGiudice, Peterson, and Seal 1991; DelGiudice et al. 1997).

The effects of weather, especially snow, so pervade wolf-prey relations that some workers believe that they actually drive wolf-prey systems (Mech and Karns 1977; Mech 1990a; Mech et al. 1998; Post et al. 1999). When snow conditions are severe over a period of years, they reduce prey survival and productivity, and wolves increase for a few years, whereas during periods of mild winters, the opposite happens. This bottom-up interpretation of driving factors may seem to conflict with a top-down interpretation (McLaren and Peterson 1994). However, ecosystems are complex and dynamic, with multiple food chains, so they can include both bottom-up and top-down influences (see Sidebar).

The Role of Tradition

Captive-raised wolves with no experience can hunt and kill wild prey and survive for years when released into the wild (Klein 1995) just as dogs, cats, and other species can hunt and kill instinctively. Captive-reared Mexican wolves (*Canis lupus baileyi*) reintroduced into Arizona in the spring of 1998 began killing elk within about 3 weeks of release (D. R. Parsons et al., personal communication). The wolves translocated from Canada to Yellowstone began killing elk within days after their release, despite no tradition of hunting in the area.

Nevertheless, it seems reasonable to suggest that naturally raised wolves gain a keen knowledge of the prey in their territory and that they develop habits, traditions, and search patterns that increase their hunting efficiency. Under good conditions (for example, in the case of the wolves reintroduced in Yellowstone) such an advantage may not be crucial, but perhaps with fewer or less vulnerable prey, it might make some difference.

This supposition has been extended to great lengths with the contention that tradition is critical to wolves and that packs are inbred groups that maintain long traditions of hunting routes and habits (Haber 1996). However, as indicated by Wayne and Vilà in chapter 8 in this volume, wolves generally outbreed (D. Smith et al. 1997), and the turnover of individuals in packs is high (Mech et al. 1998), so this extreme degree of reliance on tradition seems highly unlikely. Furthermore, the facts that dispersing wolves readily colonize new areas and prosper

(Rothman and Mech 1979; Fritts and Mech 1981; Ream et al. 1991; Wydeven et al. 1995; Wabakken et al. 2001) and that populations quickly recover following wolf control (Ballard et al. 1987; Potvin et al. 1992; Hayes and Harestad 2000a) demonstrate that hunting traditions are far from critical to wolf functioning under most conditions. The constant variation of wolf prey vulnerability (Mech, Meier et al. 1995; Mech et al. 1998) may force wolves to be flexible enough to deal with the conditions of the moment rather than relying heavily on traditions.

––––––––

The above overview of wolf-prey relations does not necessarily apply to wolf interactions with domestic prey. Domestication has left some prey, such as sheep, defenseless, and the ways in which humans restrain domestic animals—for instance, in wide-open, fenced fields—often makes them more vulnerable to predation. Thus wolf predation on domestic animals does not necessarily fit generalizations based on wild prey.

Characteristics of Wolf Predation

Prey Species Preferences

Do wolves prefer certain prey species? This is an interesting question and one not easily answered. Generally wolves eat whatever meat is available, including carrion and garbage (see Peterson and Ciucci, chap. 4 in this volume). There is probably not one potential prey species in wolf range that wolves have not killed. Furthermore, wolves in the ranges of several prey species kill them all. Single packs in Denali National Park, for example, kill moose, caribou, and Dall sheep, as well as many smaller species (Murie 1944; Haber 1977; Mech et al. 1998). The question can be broken down into two parts: First, do individual wolves or packs prefer to prey on certain species if given choices? Second, how readily do wolves that are accustomed to preying on certain species learn to prey on others, and under what circumstances will they switch prey?

Several observations spawn these questions. Cowan (1947) concluded that in the Canadian Rockies, wolves tended to forsake mountain sheep and goats for elk, deer, and moose. Carbyn (1974, 173) stated that, in the same area Cowan studied, "elk calves and mule deer are preferred prey, followed by adult elk, moose, sheep, small mammals, caribou and goat." In Riding Mountain National Park, Carbyn (1983c) found that wolves killed elk disproportionately to moose. Fritts and Mech (1981) noted that several wolf packs living among farms continued preying on wild prey and did not kill domestic animals. Potvin et al. (1988) learned that even when deer were scarce, wolves concentrated on them during winter despite the presence of moose. On the other hand, in Minnesota, wolf packs preying primarily on deer sometimes killed moose (Mech 1977a; L. D. Mech, unpublished data). Dale et al. (1995) recorded wolves preying primarily on caribou even though moose were more abundant. Kunkel et al. (in press) found that although wolves tended to hunt during winter in deer concentration areas, they killed disproportionately more elk and moose.

Speculating about this subject, Mech (1970, 205) wrote the following: "No doubt wolves in each local area become very skilled at hunting prey on which they specialize. But it is also possible that the same animals might be inept at hunting species they have never seen. It would be extremely interesting to take a pack that is accustomed to killing deer, for instance, and move it to an area where caribou and moose are the only prey available. Possibly such a pack would be at so great a disadvantage that it would fail to survive."

That experiment still has not been done; however, similar tests have been performed. Captive-reared wolves that had never killed any prey were released on Coronation Island, Alaska, and just about exterminated the deer there (Klein 1995). Similarly, captive-raised red wolves learned to kill deer and smaller prey upon release (see Phillips et al., chap. 11 in this volume). Captive-reared Mexican wolves began killing elk within 3 weeks of release, as we saw above. Bison-naive wolves reintroduced into Yellowstone National Park learned to kill bison within 21 days to 25 months after release (Smith et al. 2000). This evidence weakens the notion that individual wolves cannot learn to hunt and kill prey with which they have never had experience.

But this evidence still does not show that wolves highly experienced with one kind of prey can readily hunt and kill others. Although we strongly suspect they can, it is worth considering how to explain the observations of apparent prey preferences mentioned above. Those observations do not constitute definitive evidence for prey preferences because no study has compared the vulnerabilities of several prey species in a given area and thus ruled out the possibility that an apparent species preference was anything more than a temporary differential vulnerability.

Huggard (1993b) and Weaver (1994) illustrated some of the complexities involved in analyzing prey selection patterns by recording locations of deer and elk as well as the travel patterns and success rates of hunting wolves. If deer scattered across the landscape could not be located or killed, it became profitable to go to predictable elk locations, even when groups of elk were fewer.

While elements of learning, tradition, and actual preference may be involved in apparent prey species preferences, the most likely explanation for these patterns involves a combination of capture efficiency and profitability relative to risk, which boils down to prey vulnerability. In other words, we believe that as wolves circulate around their territories and encounter and test prey under constantly changing conditions, they gain information about the relative vulnerability of various types of prey to hunting (including finding, catching, and killing). Through trial and error they end up with whatever prey they can capture. Thus as conditions change, the wolves' prey changes in species, age, sex, and condition. This explains the seasonal and annual variation so apparent in any overview of wolf predation for any given area (Mech et al. 1998). It would also explain the finding that, in the Glacier National Park area, wolves killed disproportionately more elk and moose even though frequenting an area with more deer (Kunkel et al., in press).

Relying on whatever class of prey is currently vulnerable means that lags are inevitable because of the time it takes for wolves to gather the information about changing conditions. With the dramatic burst of vulnerable newborn caribou calves each spring, for example, it takes the wolves about a week to begin utilizing them (Adams, Singer, and Dale 1995).

Detailed analyses have been attempted to try to explain why wolves seem to specialize in killing more of some prey species when others are available (Huggard 1993b; Weaver 1994; Kunkel 1997). However, such studies must assume that equal proportions of each prey species are equally vulnerable at any given time—a critical condition that cannot be demonstrated and probably is rarely true. Therefore, we doubt that a more detailed explanation will be forthcoming than that wolves prey on whatever individuals of whatever species are vulnerable enough for them to kill with the least risk at any given time.

Vulnerability and Prey Selection

As we indicated earlier, wolves tend to kill the less fit prey. Evidence for this contention is considerable (summarized by Mech [1970], Mech et al. [1998], and in table 5.4); the main aspect of this issue that needs further study is the question of when or whether wolves *ever* take prey that are maximally fit. Given that it is almost impossible to gather enough evidence to prove that an animal is fit in every way (Mech 1970, 1996), this question may forever go unanswered. For example, even if a fresh, intact carcass of a wolf-killed animal could be examined, one could not determine enough about the animal's sensory abilities or keenness to draw conclusions about its fitness.

Our reasoning for claiming that wolves are heavily reliant on prey that are in some way defective is as follows (cf. Mech 1970). A complete examination of an animal for any traits that might predispose it to predation would require testing of live prey for various sensory, mental, behavioral, or physiological flaws as well as intact carcasses for detecting any anatomical or pathological conditions. Rarely are enough remains of wolf prey found to allow anything close to a complete carcass examination; most often the only remains are bones, and even then the complete skeleton is rarely available. However, based on even these partial remains of prey, a wide variety of predisposing conditions have been found (see table 5.4). Regardless of the approach used, including examination of prey before death (Seal et al. 1978; Kunkel and Mech 1994) and comparison of wolf-killed prey with the prey population at large (Pimlott et al. 1969; Mech and Frenzel 1971a), the results consistently indicate that wolves tend to kill less fit prey.

One possible exception to this tendency involves calves or fawns. Because remains of prey less than 6 months old are rarely found, it is usually impossible to determine the condition of such animals. Are they vulnerable just because they are young? Certainly some are debilitated, weaker than others, or otherwise inferior (Kunkel and Mech 1994), but are these the only individuals wolves kill? Or are all young-of-the-year more vulnerable?

The answer probably varies by species or even by year or population. Caribou calves in Denali National Park born after average or mild winters, for example, were

TABLE 5.4. Prey characteristics that may determine vulnerability to wolves

Characteristic	Remarks	Reference
Species	Some indication that in multi-prey systems, certain species may be "preferred" to others, but no definitive evidence (see text)	Cowan 1947; Mech 1966a; Carbyn 1974, 1983b; Potvin et al. 1988; Huggard 1993b; Weaver 1994; Kunkel et al. 1999
Sex	Males killed most often around the rut	Nelson and Mech 1986b; Mech, Meier et al. 1995
Age	Calves and fawns and old animals most often taken	Summarized by Mech (1970) and Mech, Meier et al. (1995)
Nutritional condition	Individuals in poor condition most often taken	Summarized by Mech (1970) and Mech et al. (1998); Seal et al. 1978; Kunkel and Mech 1994; Mech et al. 2001
Weight	Lighter individuals most often taken	Peterson 1977; Kunkel and Mech 1994; Adams, Dale, and Mech; 1995 [a]
Disease	Diseased animals most often taken	Summarized by Mech (1970) and Mech et al. (1998)
Parasites	Hydatid cysts and winter ticks may predispose prey	Summarized by Mech (1970) and Mech et al. (1998)
Injuries, abnormalities	Injured or abnormal individuals most often taken	Summarized by Mech (1970) and Mech et al. (1998); Mech and Frenzel 1971a; Landis 1998
Parental or grandparental condition	Offspring of malnourished mothers or grand-mothers most often taken	Peterson 1977; Mech and Karns 1977; Mech, Nelson, and McRoberts 1991
Defensiveness	Aggressive individuals taken less often	Mech 1966b, 1988a; Haber 1977; Peterson 1977; Nelson and Mech 1993; Mech et al. 1998
Parental age	Offspring of older parents taken less often	Mech and McRoberts 1990

[a] Adams, Dale, and Mech found a strong inverse relationship between caribou birth weight and wolf-caused mortality among, but not within, years.

rarely killed by wolves after they were about a month old (Adams, Singer, and Dale 1995b), so presumably they were not especially vulnerable as a class. On the other hand, deer and moose young are killed throughout their first year (Mech 1966b; Peterson 1977; Nelson and Mech 1986b), so possibly they are more vulnerable. We believe that probably wolves do kill some normal, healthy young prey that are vulnerable just because they are young, but the proportion of such animals in their total take of young probably varies considerably.

Other possible conditions that might make otherwise fit individuals vulnerable to wolves could include the sudden appearance of a strong crust over deep snow (Peterson and Allen 1974), as might follow a rainstorm in winter. Animals such as Dall sheep may suddenly be caught far away from cliffs (although Murie [1944] believed that this is most apt to happen to sheep in poor condition). Other chance circumstances involving environmental conditions might strongly disadvantage a prey animal.

Some of the conditions that predispose prey to wolf predation are dramatic, such as necrotic jawbones (Murie 1944), lungs filled with tapeworm cysts (Mech 1966b), arthritic joints (Peterson 1977), and depleted fat stores (Mech, Meier et al. 1995). However, others are more subtle, such as abnormal blood composition (Seal et al. 1978; Kunkel and Mech 1994) or even malnourished grandmothers (Mech, Nelson, and McRoberts 1991). While it may seem hard to explain how the nutrition of a deer's grandmother has anything to do with the deer's being predisposed to wolf predation, rats with poorly nourished grandmothers show learning deficits (Bresler et al. 1975), fewer brain cells (Zamenhof et al. 1971), and reduced antibodies (Chandra 1975). Any of these traits could predispose an animal to predation.

From a strictly logical standpoint, wolves could not kill every prey individual they wanted to, for given their high productivity and other characteristics, they would soon end up depleting their prey. The wide variety of antipredator traits that prey have evolved (see table 5.1) prevents this outcome. Thus generally wolves must strive hard in order to capture enough prey to survive.

Through constant striving, however, wolves are able to find and capitalize on the usually small proportion of their prey population that is vulnerable. Because of environmental changes and the natural history of prey, defective individuals are constantly being generated. Aging, accidents, progressing pathologies, birth, competi-

tion for food, and various other natural processes assure that. A high degree of buffering in the form of excellent mobility, fat storage, caching behavior, and variation in productivity, survival, and dispersal rates helps wolves survive most mismatches between their needs and the defensive capabilities of their prey (Mech et al. 1998).

Thus as wolves travel about among their prey, they try to catch whatever they can (fig. 5.2). Each attempt represents a test or trial of sorts (Murie 1944; Mech 1966b; Haber 1977). A parsimonious view of how these tests result in the wolves ending up with the inferior prey individuals is that the process happens mechanically. Prey that are not alert, fleet, strong, or aggressive enough simply end up being killed more often.

On the other hand, there may be more to it. A study using borzoi dogs as surrogates for wolves showed that the dogs actually detected inferior members of prey herds and targeted them (Sokolov et al. 1990). Film footage in real time of two wolves chasing a herd of elk clearly documented the wolves scanning the herd by coursing through it at restrained speeds, targeting an individual with an arthritic knee joint (fig. 5.3), and chasing it through the herd until they caught and killed it (Landis 1998). You could almost hear Charles Darwin cheering, "Yes, Yes!"

Kill Rate

The rate at which wolves kill prey has been measured many times and, as is to be expected, is highly variable. Because both prey size and pack size must influence kill rate, it is useful to express kill rate as biomass per wolf per day. The range runs from 0.5 to 24.8 kg/wolf/day (table 5.5). Given all the vagaries of a wolf's existence (countered by the various buffers discussed above), the only reasonably certain generalization that can be made is that wolves kill enough to sustain themselves.

How much does this amount to? Based on studies of dogs and of captive wolves, Mech (1970) concluded that the basic daily requirement for an active animal would be about 1.4 kg/wolf. Assuming about 7 kg of inedibles such as rumen contents and skull, this would amount to about one 45 kg deer per 27 days, or 13 such deer per year. This figure should be considered the minimal maintenance requirement because it is based on captive wolves that are much less active than wild ones. However, wild wolves will eat far more than this minimal requirement. Captive wolves will consume over 3 kg (7 pounds) of

FIGURE 5.2. Wolves usually try to attack any prey they can. When they are chasing prey, often young-of-the-year are strung out behind the adults.

FIGURE 5.3. Arthritic knee joint of an elk culled from a herd by wolves in Yellowstone. Observers filmed two wolves targeting the limping elk from among its herd and killing it (Landis 1998).

food per day (see Peterson and Ciucci, chap. 4 in this volume), and many of the reported kill rates reflect that (see table 5.5).

There is an interesting difference between kill rates for wolves preying on deer and those for wolves preying on larger species. Generally wolf kill rates for larger prey run about five times those for deer (Schmidt and Mech 1997). The highest kill rate reported for deer-killing wolves is 6.8 kg/wolf/day, whereas for wolves killing larger species, it is 24.8 kg/wolf/day (see table 5.5), While it is true that wolves preying on moose and caribou generally weigh about 40% more than those preying on deer, this difference could not account for the difference in kill rates.

So what does account for it? Conceivably, the kill rates for wolves killing deer are higher than have been measured, perhaps because a wolf pack can clean up a deer kill in a few hours and leave, so the kill goes undetected by researchers checking the wolves periodically by aircraft, the usual method (Fuller 1989b). However, even tracking wolves on the ground in the snow (Kolenosky 1972) yields much lower kill rates for deer than for moose or caribou. Possibly greater scavenging (Promberger 1992; Hayes et al. 2000) or caching (Mech and Adams 1999) around larger prey than was earlier realized explains the difference. However, the question remains unanswered.

Seasonal Variation in Kill Rate

The question of seasonal variation in wolf kill rates has been little studied, but, due to the extreme variation in

TABLE 5.5. Wolf kill rates during winter

Prey	Pack size	N	kg/wolf/day	Reference
White-tailed deer	3	1	4.5	Stenlund 1955
White-tailed deer	5	1	0.6	Mech and Frenzel 1971a
White-tailed deer	8	1	3.7	Kolenosky 1972
White-tailed deer	2–9	4	1.6–3.6	Mech 1977a
White-tailed deer	2–7	20	0.5–6.8	Fritts and Mech 1981
White-tailed deer	1–10	—[a]	2.0[b]	Fuller 1989b
Moose	15–16	36	4.4–6.0	Mech 1966b
Moose	4	1	1.8	Mech 1977a
Moose	6–11	6	4.1–12.1	Fuller and Keith 1980a
Moose	2–9	8	3.5–19.9	Ballard et al. 1987
Moose	2–17	5	5.5–14.6	Peterson, Woolington, and Bailey 1984
Moose	4–11	5	8.7–24.8	Dale et al. 1994
Moose	2–20	40	7.9[b]	Hayes et al. 2000
Caribou	2–20	20	2.5[b]	Hayes et al. 2000
Caribou	4–8	3	5.7–10.2	Ballard et al. 1987
Caribou	2–15	13	8.6–24.8	Dale et al. 1994
Dall sheep	6–13	3	8.7–17.9	Dale et al. 1994
Bison	7–13	8	3.5–7.4	Carbyn et al. 1993
Elk	2–14	106	2.3–22.0	Mech et al. 2001

Note: See also Mech 1970.

[a]Not given.   [b]Mean.

environmental conditions throughout the year, it is reasonable to expect much seasonal variation. Almost all kill rate studies have been conducted during winter, so sparse data are available for summer (see table 5.5, Peterson and Ciucci, chap. 4 in this volume). Furthermore, because all kill rate studies have been conducted in the northern part of the wolf's range, where daylight is short until late winter and spring, most such rates are for late winter and spring. That is also the period when ungulate nutritional condition is poorest and ungulates are most vulnerable. Thus published kill rates no doubt represent maxima for the year.

Only a few studies have sought to compare winter wolf kill rates by month. Although Ballard et al. (1987) did not make monthly comparisons, they did estimate that wolves killed about the same biomass of prey during summer as during winter. Three of the studies that did make monthly comparisons (Mech 1977a; Fritts and Mech 1981; D. W. Smith, unpublished data) showed that, as expected, kill rates peak in February and March. A fourth study (Dale et al. 1995) showed higher rates in March than in November, but indicated that the differences were not statistically significant. However, because the researchers' data consisted of all the kills their packs

made during their study, and were not samples, their packs actually did kill more in March than in November.

Surplus Killing

When prey are vulnerable and abundant, wolves, like other carnivores, kill often and may not completely consume the carcasses, a phenomenon known as "surplus killing" (Kruuk 1972) or "excessive killing" (Carbyn 1983b). The amount of each carcass wolves eat depends on how easy it is to kill prey at the time, but sometimes they leave entire carcasses (Pimlott et al. 1969; Mech and Frenzel 1971a; Peterson and Allen 1974; Bjärvall and Nilsson 1976; Carbyn 1983b; Miller et al. 1985; DelGiudice 1998). Surplus killing of domestic animals lacking normal defenses against wolf predation may not be unusual (Young and Goldman 1944; Bjärvall and Nilsson 1976; Fritts et al. 1992), but it is rare for wolves to kill wild prey in surplus. All cases of surplus killing of wild prey reported for wolves have occurred during a few weeks in late winter or spring when snow was unusually deep. In 30 years of wolf-deer study, Mech observed this phenomenon only twice (Mech and Frenzel 1971a; L. D. Mech, unpublished data), and in forty winters of wolf-moose studies, it was seen in only three winters (Peterson and Allen 1974; R. O. Peterson, unpublished data). DelGiudice (1998) recorded it during only a few weeks in one of six winters.

Presumably what happens when wolves kill more than they can immediately eat is that they respond normally to a situation that is drastically different than usual—prey are highly vulnerable, rather than being especially hard to catch. Programmed to kill whenever possible because it is rarely possible to kill, wolves automatically take advantage of an unusual opportunity.

This phenomenon has not been thoroughly studied. It has been dubbed "surplus killing" because individual carcasses are not eaten right away, contrary to the wolf's usual hungry habit. However, it stands to reason that, if scavengers did not consume these carcasses, eventually the wolves would return to them when prey was harder to kill, just as they do to caches (see above) or carrion (see Peterson and Ciucci, chap. 4 in this volume). In fact, a follow-up study supports that notion. In Denali National Park, six wolves killed at least seventeen caribou about 7 February 1991, and of course could not eat them all. By 12 February, however, 30–95% of each carcass had been eaten or cached (Mech et al. 1998); by 16 April, wolves had dug up several of the carcasses and fed on them again.

Number of Prey Killed

Actual numbers of individual prey killed per year cannot accurately be determined because of the lack of kill rate data from non-winter periods. Estimates could be made by projecting from late winter data, but besides almost certainly being overestimates, they would require using a sliding scale to account for the ever-growing fawns and calves that constitute much of the wolf's diet during summer. Supplementary prey such as beavers, hares, and other small animals taken in summer must also be considered (Jedrzejewski et al. 2002).

Nevertheless, attempts have been made to determine annual kill rates of individual prey, but they remain estimates. For deer, they ranged from 15 to 19 adult-sized deer (or their equivalents) per wolf per year, assuming that other prey constitute another 20% of the diet (Mech 1971; Kolenosky 1972; Fuller 1989b). For moose on Isle Royale, where the only other significant prey are beavers, taken mostly during warm periods, the annual estimate was 3.6 adult moose and 5.3 calves per wolf (Mech

1966b). In south-central Alaska, the year-round estimated kill rate, adjusted for prey type (adult and calf moose and caribou), averaged one kill per 8.3 days for a pack of six wolves (Ballard et al. 1987), or about 7.3 kills per wolf per year. For the Western Arctic caribou herd, where an estimated 55% of wolves' prey was caribou, some 1,740 wolves were estimated to be killing the equivalent of 28,000 adult cows annually, or 16 per wolf per year (Ballard et al. 1997).

Seasonal Vulnerability of Prey

Because of the extreme variation in size and natural history among ungulates, including differences between mature ungulates and their newborn offspring, the type of prey accessible to wolves varies throughout the year. This is especially true when one considers the need for wolf prey to be vulnerable in order to be accessible. For example, newborn ungulates are generally more vulnerable than adults, as we saw above.

An example of seasonal variation in the vulnerability of various age and sex classes, even of a single species, is the white-tailed deer in northeastern Minnesota (Nelson and Mech 1986b). Throughout the year, fawns are vulnerable as a class, although not every individual is (Kunkel and Mech 1994); during summer, adults are rarely taken, so fawns form most of the wolf's diet. In fall, adult bucks—occupied with fighting and the rut instead of eating—become vulnerable, and finally during late winter and spring, when pregnant does reach the nadir of their condition (DelGiudice, Mech, and Seal 1991), they become more vulnerable (Nelson and Mech 1986b).

This basic pattern varies among different ungulates and areas, and probably among years (Mech 1966b, 1970; Peterson 1977; Peterson, Woolington, and Bailey 1984; Nelson and Mech 1986b; Ballard et al. 1987, 1997; Carbyn et al. 1993; Mech et al. 1998; Kunkel and Pletscher 1999). However, several generalizations can be made. Young are most vulnerable in their first few weeks and remain relatively vulnerable throughout their first year, except for caribou calves (Adams, Dale, and Mech 1995; Adams, Singer, and Dale 1995). Adult males are most vulnerable immediately before, during, and after the rut, and adult females are most vulnerable in late winter. However, depending on the species, area, and year, some adults may be vulnerable year around. In the multi-prey systems of Denali (Mech et al. 1998) and Glacier (Kunkel and Pletscher 1999; Kunkel et al. 1999) National Parks, various ages and sexes of several ungulate species form different proportions of the wolf's diet during different seasons.

Influences of Wolves on Prey Numbers

Do wolves control the density of their prey, or does wolf predation merely substitute for other mortality? Probably no question has dogged wolf research more, or generated more disagreement among biologists. The influence of wolf predation on prey populations has been a subject of public controversy and scientific debate for decades. How is it possible that wolves introduced to Coronation Island, a small island in southeastern Alaska, almost wiped out the resident black-tailed deer (Klein 1995), yet on Isle Royale wolves coexist with the world's highest density of moose (Peterson et al. 1998)? Can both case studies be understood under a single scientific umbrella? Do they tell us anything useful about wolf predation in mainland systems? Since Mech's (1970) review, there has been a wealth of fieldwork on this subject, as well as much effort to place wolf predation in the context of general ecological theory.

As the complexity and unique features of real-world ecosystems have become more evident, it has also become clear that simple platitudes about whether or not wolves control prey populations are naive (Mech 1970). Under some circumstances, wolves can dramatically reduce, even locally extirpate, some prey species (Mech and Karns 1977). At other times, wolf predation may only compensate for other mortality that takes over in the absence of wolves (Ballard et al. 1987).

Important determinants of wolf-prey relationships include whether or not multiple prey species or other predators (especially humans and bears) are influential in a system, the relative densities of wolves and prey, the responses of wolf and prey populations to prey density, and the effects of environmental influences such as winter severity and diseases on both wolves and prey. All of these factors may affect the rate of increase for prey, the number of wolves present, and the kill rate of prey by wolves.

To discuss this subject, it is first necessary to distinguish among the many terms used to describe the effects of wolf predation. The alleged "control" of prey populations by predators, for example, might be interpreted in at least six ways, depending on the definition used (Tay-

lor 1984). Several recent reviews have used definitions by Sinclair (1989), who proposed that "limiting" factors include all mortality factors that operate in a prey population, and that "regulating" factors are those that act in concert with prey density (i.e., are density-dependent) to maintain prey populations at equilibrium, or within a usual range. Density-dependent mortality, for example, would be proportionally higher when a population is above an equilibrium than below it, while reproduction would follow an opposite trend. The result of such relationships would be a strong tendency for a prey population to stabilize.

While all populations are limited, not all are regulated. Similarly, all regulating factors are limiting factors, but not all limiting factors are regulating factors. Eberhardt (1997) applied yet another definition of "regulation" as a phenomenon involving two-way actions of the predator-prey system: prey density affects wolf numbers, and wolves affect prey populations. While we endorse the general truth of this concept, we will use terms as defined by Sinclair (1989). After reviewing theoretical concepts of predator-prey dynamics, we will try to apply them to the real world through comparison with field studies of wolves and prey.

Predator-Prey Theory

Two perspectives are necessary to understand wolf-prey interactions: the reproductive potential of the prey, or its annual increment, and the prey-killing potential of the wolf population. The latter is commonly understood as a set of two responses of wolves to their prey: the "numerical response," or change in wolf population size, and the "functional response," or change in individual wolf kill rate. Important features that make each wolf-prey system unique can be examined in the theoretical context of prey reproduction plus wolf numerical and functional responses (Seip 1995), assuming that wolf-caused mortality predominates.

Potential Prey Increment

The annual increment to a prey population is usually expressed in relation to prey density. This is best illustrated by a graphic in which the potential increment (vertical axis) to a population appears as a dome-shaped curve (fig. 5.4) that drops to zero, thus reaching the horizontal axis (corresponding to population density) at the population's carrying capacity (K). At this point, the prey

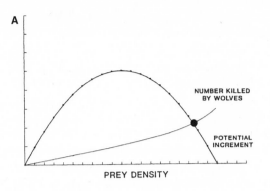

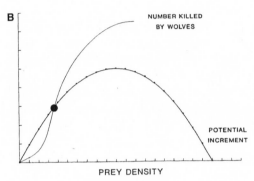

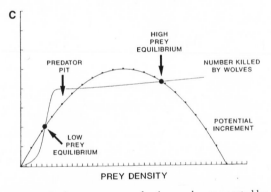

FIGURE 5.4. In theory, prey reproduction can be represented by a hump-shaped "recruitment curve," here labeled "Potential increment." A stable prey equilibrium is possible where this curve intersects the "total response" curve of wolves ("Number killed by wolves"). A variety of prey equilibria are possible, depending on the shape of the total response curve. If wolf predation is density-dependent at low prey densities, prey density may be regulated within a "predator pit." (From Seip 1995.)

population should remain stationary; as prey density approaches this level, the population growth rate is slowed by poor nutrition. (Such curves have been termed yield curves or stock-recruitment curves [Ricker 1954; Caughley 1977].) Potential annual increment is also low when prey density is low, because the population contains few

individuals. The highest annual increment is usually at some intermediate population density at which a herd has grown to substantial size, but not to a size at which nutrition begins to suffer.

At carrying capacity, prey population density is high, and the population is limited by resource scarcity. Evidence of nutritional limitation will be common. This is the state of an ungulate population absent natural predation or hunting mortality. If carrying capacity is overshot by the prey population, there will be no annual increment, and the population will fall back to carrying capacity. If a prey population at carrying capacity is harvested, whether by humans or wolves, prey numbers will decline and annual increment will be positive. If the additional production is not harvested or taken by other mortality factors, the population will increase back to carrying capacity.

Numerical Response of Wolves

The response of wolf populations to increased prey density will obviously influence their effect on prey. Keith (1983) and Fuller (1989b) found a linear correlation between wolf density and prey abundance; an increase in prey is associated with an increase in wolves (see Fuller et al., chap. 6 in this volume).

Messier's (1994) review of nineteen studies suggested that, where wolves preyed on moose, wolf density increased nonlinearly as moose density rose, and that wolf density plateaued at 58 ± 19 per 1,000 km^2. However, seven of his nine data points corresponding to high prey density were derived from Isle Royale, and included two periods when wolves were probably limited by disease and its aftermath (Peterson et al. 1998). An eighth point came from Kenai, Alaska, where wolf density was limited by harvest (Peterson, Woolington, and Bailey 1984). Messier did not propose any mechanism that might cause wolf density to stabilize at about 60 per 1,000 km^2. Isle Royale wolves actually reached a density of 92 per 1,000 km^2 in 1980 before the likely advent of canine parvovirus retarded wolf numbers; projections of vulnerable prey numbers suggested that wolves could have increased to about 110 per 1,000 km^2 (cf. Peterson et al. 1998). It has not been demonstrated that any social or territorial restrictions limit wolf density to a level lower than that allowed by food supply (Packard and Mech 1980).

Nevertheless, prey density is not necessarily synonymous with wolf food supply (Packard and Mech 1980).

Especially when prey density is high in a complex, multi-prey system, wolf numbers may not increase in proportion to total prey density. Wolves may rely primarily on one prey species (Dale et al. 1995), at least temporarily, and therefore may not benefit if other prey species increase. For example, wolves in Riding Mountain National Park rely on elk and deer (Carbyn 1983b); they might not respond numerically if moose increased.

On the other hand, moose are a common prey for wolves, providing most of the food for wolves in many areas. Bergerud and Elliott (1998) argued that, if moose (or elk) density were relatively high in such an area, then wolves would increase and sheep and caribou would decline until they equilibrated at fewer than 0.25/km², at which level they would be adequately spaced to avoid wolves. For example, Sumanik (1987) found a high-density Dall sheep (0.68/km²) system in the Yukon, where moose were so scarce (0.06/km²) that wolves supported by moose could not exert much predation pressure on sheep. As a result, sheep were limited by scarce forage and severe winters, not by predation (Hoefs and Cowan 1979; Hoefs and Bayer 1983). Bergerud and Elliott (1998) predicted that if moose were to increase in such a system, wolves would likewise increase, but then Dall sheep would be reduced by wolf predation.

Areas with high prey density often contain multi-prey systems with one or more highly social prey species such as elk or caribou. Wolf encounters with group-living prey are based on the frequency of groups, not of individuals (Huggard 1993b; Weaver 1994). Therefore, increased prey density in such areas would not lead to increased encounters with prey, so wolf response to increased prey density may be lessened for social prey. Bergerud and Elliott (1998) pointed out that the difference between observed wolf numbers and those predicted by prey biomass increased with prey species diversity. They interpreted this finding as evidence of "destabilization" of wolf numbers caused by high prey diversity. However, we believe that the difference more likely results from wolves concentrating their predation on only one or two of the available species.

Despite the rough, large-scale correlation between wolf density and prey abundance, there is much about wolf numerical response that remains unknown. Spatial refuges or migration may make increasing numbers of prey inaccessible to wolves (Krebs et al. 1999), and, depending on patterns of prey selection by wolves, the response of wolf populations to changes in a single prey

species in a multi-prey system may be complex (Dale et al. 1994). Even though most prey biomass for Denali wolves consisted of moose, increased caribou vulnerability arising from several winters with unusually deep snow allowed the wolf population to flourish briefly. The wolf population finally declined as caribou crashed, but the wolf decline was proportionately less because the wolves were supported by other prey (Mech et al. 1998).

The linear relationship between wolf density and prey density is simply a correlation, commonly interpreted as showing the response of wolf numbers to changes in prey numbers. But the general correlation between prey and wolf numbers does not necessarily tell us anything about how a wolf population responds to changes in prey density. This claim is documented by the tortuous pathway actually followed by the wolf and moose population relationship on Isle Royale (fig. 5.5) and the often inverse relationship between wolf and moose numbers there (fig. 5.6). Wolf population change may lag behind that of prey simply because of demographic inertia. At Isle Royale, wolf density closely tracked the abundance of moose at least 9 years old, rather than the total moose population (Peterson et al. 1998), so a decade

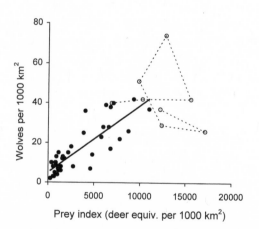

FIGURE 5.5. Linear relationship between wolf density (Y) and prey density (X), based on forty-one studies in North America, shown here as a straight line of the form $Y = 5.12 + 0.0033X$ ($P < .0001$, $r^2 = .71$). Data points (solid circles) were summarized by Fuller (1989b) and Messier (1994). Wolf and moose fluctuations in Isle Royale National Park, shown here as open circles corresponding to 5-year population averages from 1960 to 1999 (Peterson et al. 1998; R. O. Peterson, unpublished data), were excluded from the regression analysis, but are shown here to illustrate the actual path followed by wolf and prey in a single system. The linear regression is commonly used to represent the numerical response of wolves to changing prey density (see also fig. 6.2).

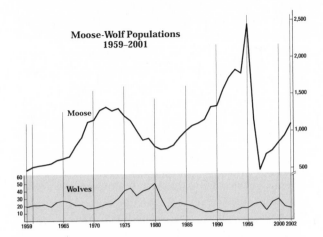

Moose-Wolf Populations 1959–2001

FIGURE 5.6. Fluctuations of wolf and moose populations in Isle Royale National Park from 1959 to 2002 illustrate the generally inverse trends in wolf and moose populations over time. (Data from Peterson et al. 1998; R. O. Peterson, unpublished data.)

may pass between successive changes in moose and wolf populations.

Of course, human persecution and disease may limit wolf numbers quite apart from any influence of prey populations. For example, canine parvovirus emerged in the 1980s as an often lethal disease for wild wolves, at times reducing wolf density in several areas of North America (Mech et al. 1986; Johnson et al. 1994; Wydeven et al. 1995; Mech and Goyal 1995; Peterson et al. 1998).

Wolf Functional Response

Ever since the pioneering work of Holling (1959) on the kill rate of invertebrate prey by deermice, change in the per capita kill rate of predators with change in prey density (functional response) has been a core feature of predator-prey theory. Holling described three basic types of predator functional responses to increasing prey density: a linear (type I), an asymptotic (type II), and a sigmoidal (type III) increase in the per capita kill rate (fig. 5.7C). While these different types of functional response have important implications for theories about predator-prey stability, the differences may not be of overriding importance in real-world wolf-prey systems (Dale et al. 1994; Van Ballenberghe and Ballard 1994). Conceivably, as prey populations increase and wolves remain constant, the number of prey killed per wolf might tend to increase. Under such circumstances, wolves might simply eat less of each prey animal (Mech et al. 2001), or peripheral members of a pack might be able to

increase their food intake. If prey density continued to increase, however, the individual kill rate would eventually begin to level off as each wolf became satiated.

There are more aspects of functional response than wolf satiation. Broken into its component parts, functional response depends primarily on the search time required to locate a vulnerable prey animal plus the handling time associated with eating it. The time required to actually kill a vulnerable prey animal is usually short. According to theory, as prey density increases, there is a

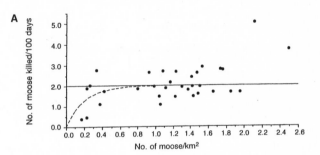

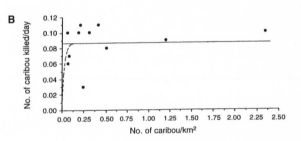

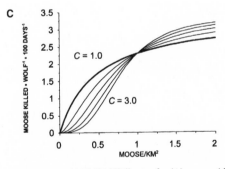

FIGURE 5.7. (A,B) Wolf kill rates for (A) moose (data from Messier 1994) and (B) caribou (data from Dale et al. 1994) were redrawn and interpreted by Eberhardt (1997) as unrelated to prey density except at very low prey levels. (C) Three types of predator functional response, represented by the equation $Y = 3.36X^C/(0.46 + X^C)$, where $C = 1.0$, 1.5, 2.0, 2.5, and 3.0. $C = 1.0$ is a type II functional response (thick line), and $C > 1.0$ is a type III functional response (thin lines). (From Marshall and Boutin 1999.)

progressive reduction in search time (except with prey that herd), allowing the kill rate to increase until handling time alone dictates the kill interval. Handling time comprises feeding time and rest to allow digestion. It, in turn, can be further compressed if prey carcass use is incomplete and feeding time is thus shortened. At the extreme, the kill rate is limited by the time required for an engorged wolf to digest its meal and sleep; thus at this point, wolf functional response must level off. (For actual kill rates, see above.)

Eberhardt (1997), Mech (cited in Ballard and Van Ballenberghe 1998), and Person et al. (cited in Ballard and Van Ballenberghe 1998) argued, and Ballard and Van Ballenberghe (1998) tended to agree, that the functional response concept was inappropriate for application to wolf-prey systems. Because of the inherent difficulties with the concept, there have been relatively few studies of wolf functional response. In a single study area where different wolf packs had access to various numbers of caribou, Dale et al. (1994) found that wolf kill rates were relatively constant across a wide range of caribou densities, although the kill rate tended to decline when caribou density was very low (fig. 5.7B). Eberhardt (1997) found the same with moose, contrary to Messier's (1994) analysis (fig. 5.7A). Although per capita kill rates on Isle Royale increased asymptotically with moose density, only 17% of the variation in kill rate is explained by moose density (Vucetich et al. 2002). In contrast, the ratio of moose to wolves explained 34% of the variation in kill rate. The prospect that moose/wolf ratios better predict kill rates than does prey density has important implications for understanding the strength of top-down influences of wolves on moose (Vucetich et al. 2002).

One of the problems in assessing wolf functional response as a per capita kill rate is that the killing unit for wolves is the pack, not the individual. High kill rates and high pack sizes usually coincide (Hayes et al. 2000). Added members of wolf packs eat portions of prey that would be lost to scavengers if pack size were small, so it seems reasonable to expect pack size to increase faster than kill rate. Finally, the hunting behavior of the pack is probably not dictated by per capita satiation as much as by the satiation of the breeding pair—usually the wolves that take precedence in feeding from kills. Subordinate members of a large pack, with the poorest opportunities to feed, usually remain in their natal pack only if supported by adequate food. Large packs, then, consistently kill more prey, while small packs kill at a disproportionately higher rate (Thurber and Peterson 1993; Schmidt

and Mech 1997). A pack of three wolves on Isle Royale killed, on average, nine moose in 100 days, while a pack 5 times larger killed only 2.8 times as many moose (twenty-five in 100 days) (Thurber and Peterson 1993).

While the functional response concept is a critical part of wolf-prey theoretical models, it has poor predictive power. This is because actual kill rate probably depends more on pack size and on prey vulnerability (which varies with snow depth, population age structure, and nutritional plane) than it does on prey density (see above). Thus we agree with the workers cited above that the concept of functional response, established in laboratory experiments with small mammals and invertebrate prey, is poorly suited to describing wolf predation.

Total Predation Rate

The total number of prey killed by wolves is the product of the number of wolves present and their per capita kill rate. At low prey densities, the total kill is usually small because wolves are scarce. Theoretically, as prey density increases, the number killed by wolves increases disproportionately faster because both wolf numbers and functional responses are increasing, with a multiplicative effect. As a result, the total loss to wolves should be density-dependent, increasing faster than prey density, and thus wolves might be able to regulate prey density (Messier 1994).

According to theory, if prey density continues to increase, wolf numbers or per capita kill rates usually plateau. In reality, we believe there is little reason to expect wolf numbers to plateau if prey density increases (see above). Nevertheless, if the kill rate of wolves does not keep pace with rising prey density, total predation losses may be inversely density-dependent, or "depensatory." The term "depensatory" implies that predation is not density-dependent or regulatory; predators either drive prey to extinction or prey erupt despite predation.

The actual outcome of this contest between predator and prey depends on the extent of predation losses compared with the annual increment of prey. Either one may exceed the other, so prey populations may increase, decline, or, if annual prey increment matches predation losses (plus other mortality), stabilize. Graphically, it is easy to see how prey might stabilize at high or low densities, depending on the height and shape of total predation (total response) and prey increment curves (see fig. 5.4).

Eberhardt (1997) felt it was difficult to assess the effect

of wolf predation on ungulates, owing to the limited quality of the data, the pervasive harvesting of wolves and their prey, and the fact that prevailing wolf-prey theory was based on studies of invertebrates, not wolf-prey systems. Ungulates are often difficult to census, and many assessments of wolf-prey dynamics are based on indices of abundance, or merely informed opinion of likely trends in populations. Furthermore, Eberhardt argued that the use of differential equations in wolf-prey theory is inappropriate, because neither wolves nor their prey reproduce instantaneously, as assumed by these equations. Thus in nonequilibrium systems there will be lags in wolf numerical responses to prey, and, as Holling (1959) pointed out, his "total response" model will be an oversimplification.

Eberhardt (1997) also felt that the equation usually used to describe the functional response of wolves commonly did not fit the actual data on wolf predation. He used a constant kill rate (functional response), a method that Messier and Joly (2000) criticized but Eberhardt (2000) defended (see fig. 5.7). Incorporating this approach into generalized difference equations of the Lotka (1925)-Volterra (1928) genre, he explored whether a wolf-prey model based on wolf-prey ratios instead of prey density might be more suitable.

Eberhardt and Peterson (1999) reexamined wolf abundance and rate of increase in relation to prey biomass and the conclusion of Eberhardt that wolf and prey numbers usually were proportional, with an average of over 200 deer-equivalents per wolf (Eberhardt 1998). Eberhardt and Peterson (1999) revised this figure to 122 deer-equivalents per wolf (equivalent to 40 elk or 20 moose per wolf, based on Keith's [1983] estimates of relative biomass). When this figure is combined with an average wolf kill rate (estimated at 6.9 moose/wolf/year) and a productivity of 7.8 moose/wolf/year (assuming an even sex ratio, 13% yearlings, and a 90% pregnancy rate and single calves for moose over 2 years old; see Peterson [1977] and Schwartz [1998]), it is apparent that wolves can harvest moose at a rate close to their maximum annual increment (Mech 1966b, 163). Thus wolves could potentially regulate prey abundance (Eberhardt 1997), and the combined effects of predation by both wolves and humans may lead to prey declines (Eberhardt et al. 2002).

Multiple Equilibria: A Theoretical Possibility

Much has been made of the concept of "equilibria" in wolf-prey systems, sometimes in the context of wolf control programs (Haber 1977; Walters et al. 1981; Messier and Crête 1985; Seip 1995). A graphic model illustrates that if the total predation curve intersects the annual increment curve in a certain way relative to prey density, then three potential equilibria between wolves and prey might result (see fig. 5.4). One of the equilibria would be unstable, and prey would not remain at this level, while the other two would be stable.

Theoretically, if a wolf-prey system existed at a low equilibrium (called a "natural enemy ravine" [Southwood and Comins 1976] or a "predator pit" [Walters 1986]) and if the total predation curve were lowered temporarily (as when wolf numbers are reduced by control programs), prey could escape predation and increase to a high equilibrium. Alternatively, prey could reach a high equilibrium if prey productivity improved dramatically. Having achieved a high equilibrium, prey would, in theory, remain there even if wolves were allowed to recover. This conclusion would be attractive to wildlife managers, suggesting that a long-term increase in prey might result from short-term predator control. But does it really work this way?

A theoretical condition for the existence of multiple equilibria or "two-state systems" is that total losses to predation be density-dependent at low prey densities (Messier 1995b). In addition, if a wolf reduction allows prey to escape from a low to a high equilibrium, then prey should remain at the high-density equilibrium even after predator numbers are restored (Skogland 1991). Finally, dramatic prey population "outbreaks" should occur when there is an increase in herd productivity, as might occur following substantial habitat improvement (Van Ballenberghe and Ballard 1994; Kunkel and Pletscher 1999). During the 1980s and 1990s there were extensive efforts, primarily through wolf control in Alaska and the Yukon, to induce prey populations to increase to a high stable equilibrium, but none was successful, except possibly that in interior Alaska between 1976 and 1982 (Boertje et al. 1996) (see below).

The degree to which wolf predation is density-dependent at low prey densities should affect the persistence and stability of prey populations. A type II functional response, with increasing slope as prey density decreases, would allow less prey persistence and stability than a sigmoid type III curve because predation would be more apt to drive prey to extinction. Efforts to distinguish between type II and III curves underlie recent efforts to assess wolf predation effects (Messier 1994; Dale et al. 1994).

Marshal and Boutin (1999), however, point out that the statistical power to distinguish among these curve types is very low because of the low sample sizes and high variability typical of field studies. They suggest bypassing this analysis and directly measuring mortality rates for moose at low and intermediate densities. They proposed two possible ways to tell whether wolf predation is density-dependent, and thus regulatory, at low prey densities (Walsh and Boutin 1999): first, if moose density can be induced to increase by removing wolves, then the proportion of moose killed by wolves should increase from before wolf removal to after wolf recovery; second, if moose density is reduced, the proportion of moose killed by wolves should decrease. Regardless of whether wolf predation regulates prey, wolf predation can still be considered a limiting influence on prey density.

Most studies of wolves and prey have involved relatively simple systems with one to two prey species. As the prospect and reality of wolves in the northern Rocky Mountains emerged in the 1980s, interest grew in the nature of wolf-prey dynamics in systems with as many as five prey species. Building on the pioneering work of Cowan (1947) and Carbyn (1974), recent studies by Huggard (1993a,b) and Weaver (1994) in Jasper and Banff National Parks have sought to understand the apparent preference of wolves for certain prey species.

Huggard (1993b) assessed prey abundance and predation patterns for two packs in the Bow River Valley of Banff National Park. Occupying lowland habitats were elk, mule deer, and moose, while bighorn sheep and mountain goats inhabited primarily steep slopes and higher elevations. The sheep and goats overlapped little with the wolves and were infrequently killed. Elk biomass exceeded deer biomass by an order of magnitude, and moose were uncommon. As in the earlier studies, elk predominated among wolf kills, and based on the number of encounters with prey, there was no apparent preference for any prey species.

Nevertheless, Huggard (1993b) revealed greater complexities in this system. Wolves encountered many elk in groups in predictable locations, while they encountered deer more randomly. For elk, the herd was the basis for wolf encounters; with many animals in a herd, the chance of a successful kill was higher than for an individual prey encounter. Hunting wolves appeared to key in on predictable elk herds, with a high probability of making a kill, and they also killed elk and deer during random encounters while traveling between predictable

elk herds. Huggard (1993b) argued that prey encounter rate was the most important determinant of wolf diet, because as prey density changes, the unique grouping tendencies and habitat selection patterns of each prey species result in different responses by wolves. Based on a simple model of functional response, Huggard showed that the changing pattern of encounter rates, by itself, would generate different patterns of selectivity by hunting wolves as prey density changed. For example, with constant deer density but declining elk density, selection for elk would increase as wolves concentrated on predictable elk herds.

In the real world, additional factors contribute to more complexity: wolves may have inherent preferences for certain prey based on experience (see above), and capture success may vary as prey density and vulnerability change. Snow conditions affect each prey species in a unique manner, and even the carcass use patterns of wolves (which vary with pack size and the presence of scavengers) affect their response to changing prey density (Mech et al. 2001).

From the foregoing discussion, it is apparent that the numerical and functional responses of wolves are grossly oversimplified when modeled simply against prey density, and our limited understanding usually prevents realistic elaboration of the existing models. Consequently, trying to predict wolf responses in a multi-prey system is quite a primitive business.

Wolf Predation in the Real World

It has been repeatedly stressed that critical features of wolf-prey dynamics will differ between wolf populations that are naturally regulated and those that are harvested by humans; additionally, simple systems with a single predator and prey will be fundamentally different from those with either alternative prey or additional predators (such as bears) (Filonov 1980; Gasaway et al. 1992; Van Ballenberghe and Ballard 1994). For any given wolf-prey system, there will always be unique characteristics that must be understood before the effect of wolf predation can be predicted. For example, factors that limit wolf populations, including control programs, may greatly influence wolf-prey dynamics (Seip 1995).

Wolf control by wildlife managers has always been controversial. In 1994, Governor Tony Knowles of Alaska suspended that state's wolf control program and asked the National Academy of Sciences to undertake a scientific review and economic analysis of wolf and bear management in Alaska. The resulting committee report (National Research Council 1997) reviewed eleven case histories of wolf control in Alaska, the Yukon, and British Columbia. The committee concluded that wolves and bears in combination could limit prey at low numbers for many years, and that predator reduction might hasten the recovery of prey. An increase in prey density was demonstrated in only three of the eleven cases, but increased juvenile survival after predator reduction was a common finding.

The three cases of increased prey density (National Research Council 1997) involved Game Management Unit (GMU) 20A in east-central Alaska; Finlayson, Yukon Territory; and northern British Columbia. They illustrated the committee's conclusion that wolf control is unlikely to result in increased prey populations unless a very high proportion of resident wolves are killed annually over a large area for at least 4 years. Such a high level of wolf control is necessary to prevent local reproduction and rapid recolonization from surrounding areas from bringing the wolf population rapidly back to its previous levels (Boertje et al. 1996; Bergerud and Elliott 1998; Hayes and Harestad 2000a).

In GMU 20A, wolves were killed from aircraft for 7 years, after a combination of overharvest and severe winters had reduced moose to a low level (0.2/km²). Over 17,000 km² (6,640 mi²), 337 wolves were killed during 1976–1982, reducing wolf density to about 44% of its pre-control level for 6 years (fig. 5.8; Boertje et al. 1996). During the 7 years after official control ceased, another 190 wolves were killed by private hunters, but wolf density nevertheless increased to 80% of the pre-control level. During the 7 years of wolf control, moose density increased from 0.2 to 0.5 moose/km², and in the next 15 years to 1.3 moose/km². This increase was high enough to cause concern that a severe winter might cause a moose die-off (Boertje et al. 1996).

In fact, beginning in 1989–1990, 4 years in 5 brought deep snow (90 cm or 36 in), but the moose population continued to increase. Caribou, on the other hand, declined with the severe winters, after increasing from 0.2 to 0.9 caribou/km² in the 14 years during and after wolf control, when winter weather was favorable. While wolf control apparently led to impressive increases in caribou and moose herds, it must also be realized that hunting of both prey species was also greatly curtailed (National Research Council 1997), and it remains un-

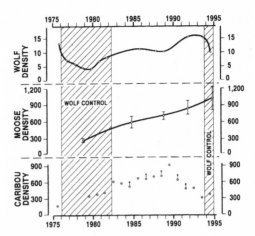

FIGURE 5.8. Population densities for wolves, moose, and caribou in Game Management Unit 20A in interior Alaska, 1975–1994. Moose estimates include 90% confidence intervals. Caribou and moose density increased during and after 7 years of wolf reduction when winter weather was benign. Caribou declined, but moose continued to increase during four consecutive severe winters from 1989–1990 to 1992–1993. (From Boertje et al. 1996.)

known what would have transpired if winter weather had not cooperated.

A 6-year wolf reduction experiment conducted in Finlayson, Yukon Territory, was also followed by increases in moose and caribou (National Research Council 1997). However, an upper prey equilibrium was not maintained; when wolf control ended, prey populations began to decline. Between 1983 and 1989, over 23,000 km² (9,000 mi²), 454 wolves were removed, mostly shot from helicopters, producing an 85% reduction in wolf density. As in central Alaska, harvest of prey by human hunters was also greatly restricted.

Caribou density rose from about 0.1/km² in 1983 to 0.3/km² in 1990 as the proportion of calves almost doubled (from 26 to 50 calves/100 cows). Moose density was not estimated before wolf control began, but in 1987, after 4 years of wolf control, there were 67 calves/100 cows, and, based on hunting statistics, moose density was increasing. In 1996, 6 years after wolf control ended, the proportion of moose calves had dropped to about 30/100 cows. Likewise, in the 4 years after wolf control ended, there were about 32 caribou calves/100 cows, and caribou density declined to 0.2/km². This study revealed wolf predation as a major limiting factor for these prey species, but it raised little hope for an upper equilibrium for prey in the absence of continued wolf control.

Similar results were reported from northern British

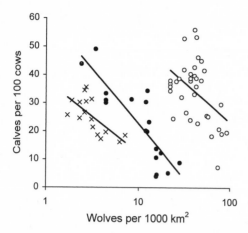

FIGURE 5.9. The relationship between moose calf abundance and wolf density is consistently negative, yet differs geographically (Xs, data for Alaska from Ballard et al. 1987; solid circles, data for British Columbia from Bergerud and Elliott 1998; open circles, data for Isle Royale from R. O. Peterson, unpublished data). Wolf control caused variation in wolf density in the Alaska and British Columbia study areas, but on Isle Royale wolf density was correlated with the number of old moose (Peterson et al. 1998). Wolf density varied between study areas probably because of differences in prey density. Note that moose populations usually increase when calf abundance exceeds 24–26 calves per 100 cows (Peterson 1977; Bergerud and Elliott 1998). Linear regression lines all had negative slopes ($P < .01$) and $r^2 = .47$, .73, and .20 for data shown for Alaska, British Columbia, and Isle Royale, respectively.

Columbia (Bergerud and Elliot 1998). In several study areas, wolves were reduced by 60–86% for 3–4 years. The proportion of juvenile prey (at least 5 months old) increased twofold to fivefold, and population densities increased for all four large ungulates in the area: moose, caribou, elk, and Stone's sheep. Interestingly, this study suggested that, for all four prey species, an average recruitment of 24 juveniles/100 females was sufficient to balance average mortality (fig. 5.9). Where wolves were not reduced, average recruitment for moose and sheep was 14–23 young/100 females, while in areas of wolf control there were 32–45 young/100 females. Projections suggested that, without wolves, average recruitment for all four prey would be 53–57 young/100 females. As in the Yukon experiment, the British Columbia data did not suggest that an upper equilibrium could be maintained without continued wolf reduction.

Seven of the eleven case studies (National Research Council 1997) involved reduction of only wolves, not bears, yet bears prey heavily on newborn ungulates. In the Nelchina Basin, a 61,600 km² (24,000 mi²) area in

south-central Alaska, 60 wolves were killed in an experimental area of 7,262 km² (2,837 mi²) during 1976–1978. Public wolf harvest outside the experimental area also increased, reducing wolf density throughout the Nelchina. Ballard et al. (1987) concluded that for this and other reasons it was not possible to fully evaluate the effect of wolf predation on the moose population. Nevertheless, autumn moose calf/cow ratios were negatively related to wolf density (see fig. 5.9).

A companion study of moose calf mortality conducted in 1977–1978 involved determining the cause of death for 120 moose radio-collared soon after birth (Ballard et al. 1979, 1981). In the first 6 weeks of life, 55% of the moose calves died. Predation accounted for 86% of natural deaths, and brown bears accounted for 91% of those deaths. Wolves, reduced to a low density (2.7/1,000 km²), were responsible for only 4–9% of the predation deaths, and estimated recruitment greatly exceeded the proportion removed by wolves. After brown bear density was reduced 60% by moving bears away, calf survival increased. Most of the bears returned, however, and calf survival returned to pre-bear-removal levels.

Van Ballenberghe and Ballard (1994) listed another four areas where predation was judged to be a major limiting factor during specific periods. However, because bears coexist with wolves throughout wolf range, the difficulty of evaluating the effects of wolf predation alone has bedeviled scientists and game managers alike.

The wolves secluded in Isle Royale National Park, probably the world's safest wolf sanctuary, provided one of the most impressive natural wolf control experiments by their population crash during 1980–1982, which was circumstantially linked to canine parvovirus (Peterson et al. 1998). In 1981, coincident with the wolf crash, the proportion of moose calves shot from an average of about 22/100 cows to, briefly, 60/100 cows (Peterson and Page 1988). Over the next 15 years, with wolves unexpectedly few, moose increased to over 4/km², about ten times higher than usual moose densities in mainland areas of North America (Messier 1994). Thus the limiting nature of wolf predation was revealed (Peterson 1999).

The high level of moose on Isle Royale led to reduced growth of balsam fir (McLaren and Peterson 1994), which moose eat in winter, demonstrating that the indirect effects of wolf predation in an ecosystem can be significant. This cascading relationship from wolf to moose to fir recalls Aldo Leopold's (1949) essay, "Thinking like a mountain," in which he proposed that the integrity of mountains themselves was influenced by wolves in this manner. Of course, at the top of this cascade was the lowly canine parvovirus.

Wolf predation appears to fit the generalization (Hairston and Hairston 1993, 379) that "predation is a major source of herbivore mortality in terrestrial communities, and grazing on plants is held at a lower level than would otherwise be the case." Similarly, Krebs et al. (1999, 447) concluded that "all vertebrate herbivores are limited primarily in abundance by predation unless they have evolved an escape mechanism in space or time."

Limiting Effect of Wolves on Prey

There has been much attention to theoretical models in attempting to explain the effects of wolves on prey populations (Mech and Karns 1977; Walters et al. 1981; Van Ballenberghe 1987; Crête 1987; Skogland 1991; Boutin 1992; Gasaway et al. 1992; Van Ballenberghe and Ballard 1994). The primary scientific debate centers on whether wolf predation regulates prey at low-density equilibria, with predation rate increasing faster than prey density, or whether it acts more simply as a limiting factor that, when combined with bear predation and other limiting factors, leads to prey densities far below the carrying capacity set by food supply. As in the above case studies, while prey have been induced to increase via predator control, they tend to decline again after predators recover (Gasaway et al. 1992). This outcome supports the notion that wolf predation limits, but does not regulate, prey populations.

The moose population on Isle Royale is highly dynamic, and wolves may well contribute to this instability (Peterson, Page, and Dodge 1984). Statistical analysis suggests that the observed moose dynamics arise from dynamics that alternate between periods of wolf increase and decrease (Post et al. 2002). Specifically, during years of wolf decline, moose exhibit strong direct density dependence, and during years of wolf increase, moose exhibit only weak direct density dependence and strong delayed density dependence. These patterns suggest that moose are strongly attracted to an equilibrium during wolf decreases and exhibit unstable dynamics, characteristic of a cyclic population, during wolf increases.

At low prey densities, the distinction between regulation and limitation hinges on whether wolf predation is density-dependent. Two studies have claimed that wolf predation is density-dependent at low prey densities. Messier and Crête (1985) estimated losses to wolf predation at three low moose densities in Quebec and argued that they had evidence of density dependence. However, others found the evidence equivocal (Van Ballenberghe 1987; Boutin 1992). Pooling data from several studies in North America, Boutin (1992) showed that wolf predation rates were density-independent and were remarkably constant over a wide range of moose densities.

Bergerud (1992) also argued that wolf predation is density-dependent, based on his analysis of correlations between calf survival, wolf density, and prey density for caribou and moose. Bergerud's hypothesis is that a major strategy to reduce predation is "spacing out" (see above), and he relies heavily on the logic that predation at low prey densities *must* be density-dependent; it seems reasonable that predators should be able to kill a higher proportion of young animals if they are clumped instead of spaced out (e.g., Miller's [1983] surplus killing of caribou calves by wolves). While Bergerud's analysis provides evidence that predation by both wolves and bears can be strongly limiting, his claim that wolf predation is generally regulatory is based more on reasoning than on actual evidence.

Most reviewers have stressed that the rather scarce empirical data used to evaluate alternative hypotheses have serious limitations. Regardless of whether wolf predation is density-dependent or not, bear predation seems to be additive and density-independent (Boertje et al. 1988; Van Ballenberghe and Ballard 1994), and wolves coexist with bears throughout their North American range, except on Isle Royale, and in many areas of the Old World (Filonov 1980). Even if wolf predation is density-dependent, it is usually overlain by bear predation, which apparently is not (Gasaway et al. 1992; Van Ballenberghe and Ballard 1994).

Limited by predation, prey populations will rise and fall at irregular intervals based on demographic and environmental factors that influence losses to predators (Van Ballenberghe 1987). These factors include relative numbers of predator and prey (Mech 1970, 277; Eberhardt 1997); snow depth, which influences wolf kill rate (Mech and Frenzel 1971a; Mech and Karns 1977; Peterson 1977; Nelson and Mech 1986c; Mech, McRoberts et al. 1987; Mech et al. 1998; DelGiudice 1998; Post et al. 1999; Jedrzejewski et al. 2002; Hebblewhite et al. 2002; Kunkel et al., in press); and fluctuations in other predator and prey species in the system (Kunkel et al. 1999; Kunkel and Pletscher 1999). Thus a severe winter, a string of mild winters, or habitat rejuvenation by fire may induce prey populations to fluctuate by altering prey reproductive output or losses of prey to predators.

While many factors may influence prey density, the basic conclusion is that wolves and bears always help limit prey numbers, as demonstrated by the study in GMU 20 (fig. 5.10). An extreme example is seen in the 3,000 km² (1,170 mi²) area of poor habitat in northeastern Minnesota where, during a series of severe winters, wolves decimated a white-tailed deer population (Mech and Karns 1977), and deer did not repopulate the area for at least the next 30 years (L. D. Mech and M. E. Nelson, unpublished data).

Gasaway et al. (1992) argued that, for moose, each additional predator species resulted in a stepwise reduction in density. If we consider wolves, brown bears, black bears, and humans as the potential predator guild for moose and caribou, it is clear that prey density depends on the number of predator species (fig. 5.11). Prey density can be quite high if the wolf is the only carnivore present, as in Isle Royale National Park. However, throughout their global range, wolves everywhere else coexist with human hunters or bears.

Although wolves do help limit or retard the growth of their prey populations, it is also clear that these predators do not necessarily hold prey numbers down. Mech (1970, 268) distinguished between systems where wolves controlled their prey (and where removing wolves would allow the prey population to increase) and where they did not, and concluded that they did not where prey/wolf ratios were greater than 25,000 pounds (11,364 kg) of prey/wolf. More recent cases in which prey populations increased despite the presence of wolves include the moose on Isle Royale (Peterson et al. 1998), caribou in Denali National Park (Adams and Dale 1998a; Mech et al. 1998), and deer in northeastern Minnesota (Mech 1986; Nelson and Mech 1986a, 2000; Mech, McRoberts et al. 1987). In the first two cases, the prey/wolf ratio exceeded the above level, but in the last case it did not. In all three cases, the prey population trends were related to snow depths, which affect prey nutrition and thus the degree to which prey are vulnerable to predation (see also Jedrzejewski et al. 2002).

Additive versus Compensatory Mortality

In trying to assess whether wolves are controlling a prey population in any given situation, it would be helpful to know the extent to which wolf predation is compensatory (Errington 1967) to other mortality factors and the extent to which it is additive (Mech 1970, 268). When wolf predation is compensatory, it is only substituting for other mortality factors. For example, if wolves took only deer that would have starved to death otherwise, then wolf predation would be compensatory.

Usually the situation is more complex, however, with wolves killing some of the prey individuals that would have died from other causes and some that would not have. As indicated above, bear predation usually seems to be additive to wolf predation, although when wolves are removed from a system, bear predation may compensate for wolf predation on calves (Ballard et al. 1987). In certain Russian nature preserves, prey mortality shifted among various predators in a compensatory way

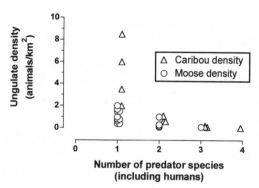

FIGURE 5.11. Ungulate density in relation to the number of predator species present, including black bears, brown bears, wolves, and humans (from Peterson 2001).

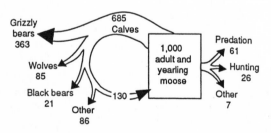

FIGURE 5.10. From a hypothetical pre-calving population of 1,000 moose in east-central Alaska (Game Management Unit 20E), an average of 685 calves are born, and about 19% of those calves survive to the age of 1 year. Most mortality is caused by predators, especially brown bears. For moose older than 1 year, average mortality was 9.4%, and predation by bears and wolves was the largest source of mortality. Mortality from hunting was less than 3% annually. (From Gasaway et al. 1992.)

as various carnivores were controlled by humans (Filonov 1980).

As discussed earlier, it is with young-of-the-year that the least is known about the degree to which wolf predation takes inferior animals and thus the extent to which it is compensatory. How many of the calves that wolves kill would have lived otherwise? The answer to this question would help us considerably in determining the effect of wolves on prey populations, so this is an area that needs considerably more research.

Disagreement about Wolf Effects on Prey Numbers

Why is there still no scientific consensus on the significance of wolf predation in prey population dynamics? One reason is that scientists have studied a wide range of wolf-prey systems, each with a combination of ecological factors that renders it unique (Mech 1970, 268). Factors of importance include different combinations of prey species (wolves supported by one prey species may have a disproportionate influence on alternative prey, as with caribou affected by wolves that subsist primarily on moose [Bergerud 1974; Seip 1995]); other predator species (mountain lion [Kunkel et al. 1999], grizzly bear, black bear); a wide range of human effects on both predators and prey (confounding any understanding of predator-prey interaction); differences in the inherent productivity of habitats and of prey populations (Seip 1995; National Research Council 1997); and regional differences in the importance of winter snow conditions (Coady 1974; Mech and Karns 1977; Mech et al. 2001; Hebblewhite et al. 2002). Any of the above factors may influence the degree to which prey are limited by wolf predation, both in different geographic regions and at different times in the same area (Mech, McRoberts et al. 1987; Mech et al. 1998; DelGiudice 1998).

A second reason that disagreement persists is that wolf-prey systems are inherently complex, with population dynamics affected by nonlinear predator-prey linkages, multi-trophic-level interactions (Bergerud 1992; McLaren and Peterson 1994; Hayes and Harestad 2000b; cf. Krebs et al. 1995 for another predator-prey system), and even predator and prey mental states (Brown et al.

1999). Finally, data on wolf and prey population densities often are inherently neither precise nor accurate, and measured predation rates by wolves also show great variation (Schmidt and Mech 1997; Marshal and Boutin 1999), leaving much room for differing interpretations of field data.

In summary, although considerable debate still rages over several theoretical issues related to wolf-prey interactions, we find general agreement on a few key points. First, wolf predation can be an influential limiting factor for prey populations, especially where wolves themselves are not limited by harvest. Second, when wolves coexist with grizzly bears, black bears, or both, the combined effects of these predators are usually sufficient to reduce primary prey populations to levels below that which could be supported by their forage base. (That is not to deny that food and other environmental factors may also influence prey dynamics.) Third, wolves have their greatest demographic effects on prey via predation on young-of-the-year (Pimlott 1967; Mech 1970).

———————

In this chapter, we have tried to discuss the very essence of the wolf: how the animal interacts with its prey in order to eat, survive, and reproduce. The coevolution of the wolf and its prey, an ongoing contest during which the prey must survive in the face of constant threat by the wolf, and the wolf must succeed in overcoming specialized prey defenses often enough to survive, is scientifically one of the most intriguing aspects of wolf biology. Likewise, it is most captivating—and disconcerting—to members of the lay public who are interested in the wolf, negatively or positively.

The innate need of the wolf to attack large prey is what most often brings the creature into conflict with humans (see Fritts et al., chap. 12, and Boitani, chap. 13 in this volume). In addition, the wolf's wide geographic distribution and diverse prey base result in great variation in interactions between the wolf and its prey. Thus, perhaps it is understandable that, even after much study, scientists still disagree on the precise nature of several aspects of this fascinating topic.

Ecosystem Effects of Wolves
L. David Mech and Luigi Boitani

Wolves form a major force in the ecosystems of which they are a part. On a circumpolar basis, these animals represent probably the single most important predator on large mammals (but cf. Murphy et al. 1999).

Like any other species, wolves inevitably influence other ecosystem components and processes, but they may do so in a more obvious way because their effects on other vertebrate species are more prominent and more easily observed. Several primary effects of wolves have long been recognized (Mech 1970): (1) the sanitation effect (culling of inferior prey individuals), (2) control or limitation of prey numbers, (3) stimulation of prey productivity, (4) an increase in food for scavengers, and (5) predation on non-prey species.

We have covered some of these primary effects in detail in various chapters of this volume. However, these primary effects cascade or "ripple" (N. A. Bishop, personal communication) through the ecosystem, causing other important changes. Unfortunately, the further down the ecological cascade from the wolf itself, the less we know about its effects, and the less certain we are about what we know. Thus, for now, all we can do is highlight a few of the secondary or tertiary effects of the wolf, which we will label "indirect" effects to distinguish them from its primary, or "direct" effects.

There is a tendency to view the ecological effects of wolves as "good" or positive, probably because one of the main postulates of conservation biology views ecological complexity as good and simplification of ecosystems by humans as bad. From this perspective, many of the primary effects mentioned above can be considered positive. However, in the interest of science and objectivity, we advise against this outlook, for two reasons. First, science does not really know enough about many of the cascading effects to judge the extent of their positive or negative effects on other elements of ecosystems—and many of them may well be both. Second, claims about positive or negative, good or bad, are human value judgments and differ depending on which humans make them. The fact that wolves tend to spread the hydatid tapeworm,

which occasionally is detrimental to humans (see Kreeger, chap. 7 in this volume) is an example.

The reintroduction of wolves into Yellowstone National Park is a useful case study to show the complexity of the role of wolves in ecosystems. Two main types of cascading effects have gained scientific attention: (1) those resulting from the wolf's interference competition with coyotes, and (2) those resulting from predation on ungulate herds. Although there has been little time yet for studies of these phenomena in Yellowstone, there has been considerable speculation about them.

Effects of Coyote Reduction
Wolves have reduced coyote numbers in part of Yellowstone (see Ballard et al., chap. 10 in this volume), and predictions are that this reduction will result in an increase in red foxes, which without wolves are more subject to interference competition from coyotes (Crabtree and Sheldon 1999a,b; Singer and Mack 1999). Reductions in coyotes could also lead to increases in their prey, which in turn could lead to increases in several other mesocarnivores (Buskirk 1999).

On the other hand, an increase in the year-round carrion supply in the form of numerous wolf-killed ungulates might support an increased mesocarnivore population quite aside from the direct killing of coyotes by wolves. That increase, in turn, might dampen small mammal numbers, which could have a differential effect on various of the mesocarnivores. Such effects would not necessarily be constant, but most probably would vary over time and space depending on other ecological conditions, including weather. Whether an increase in mesocarnivores is positive or negative depends on one's viewpoint. Mesocarnivores are considered "important ecologically" for various reasons (Buskirk 1999, 166), but may also contribute to the "loss of ground-nesting birds and probably other small vertebrates" (Terborgh et al. 1999).

Although the indirect effects of wolves on mesocarnivores and their prey are complex and dynamic, the influence of wolves on more diverse taxa, such as birds (Stahler 2000) and insects (Sikes 1994), is much more so. Most effects on these taxa are probably related to their interactions with carrion from wolf kills and with mesocarnivores and their prey. Certainly with wolves in the system, there will be changes in the numbers and distributions of these creatures and, in turn, of those they interact with. But accurately predicting the direction of those effects and any net benefit or detriment may forever be impossible.

Wolf Effects on Prey

Indirect effects of wolves on prey could be caused by structural changes in prey herds, such as changes in the species, age, sex, and condition of the standing crop; by changes in prey distributions, behavior, and movements; and by changes in prey numbers.

Because wolves tend to kill primarily the oldest, youngest, and most debilitated and undernourished members of prey herds, such herds under the influence of wolves tend to be made up of individuals of prime age, condition, and health, and therefore of highest productivity (Mech 1966b; Bubenik 1972; Schwartz et al. 1992). Over tens of thousands of years of such natural selection, the antipredator defenses and adaptations of prey should tend to become keener as wolf and prey continue their arms race. Antipredator behavior affects the movements and distributions of prey. Thus, for example, in Yellowstone National Park, some predict that elk may increase their use of forest cover, bison may seek tree cover during snow-free periods, and bighorn sheep may abandon the gentler slopes (Singer and Mack 1999).

Probably the most pervasive and yet diffuse indirect effects of wolf predation come from reductions in prey numbers. Whether or when wolves control, limit, or regulate their prey, and under what conditions, is discussed in this chapter. There certainly are documented cases in which wolves have led to fewer prey. Numbers of prey can influence vegetation and cause ripples further down and out from the trophic cascade; that much is obvious. The question is, just what are these effects, how do they work, and what are their consequences?

Controversy and confusion have arisen over whether such effects tend to flow from the bottom trophic level (vegetation) upward or from the top level (wolves) downward (Schmitz et al. 2000). Long-term studies of the wolves, moose, and vegetation on Isle Royale (Mech 1966b; Jordan et al. 1967; Peterson 1977) have served as the catalyst for debate on this issue. Two views of the workings of the Isle Royale ecosystem have emerged.

Isle Royale wolf numbers tend to follow the trend in moose numbers with a lag of 7–10 years (Peterson, Page, and Dodge 1984). Since herbivore numbers depend on vegetation quality and quantity, and the latter is modified by snow conditions (Mech, McRoberts et al. 1987; McRoberts et al. 1995; cf. Messier 1991), this finding suggests a bottom-up flow of effects. On the other hand, it also appears that high moose numbers tend to follow low wolf numbers, and that moose browsing on balsam fir, and thus fir growth, is related to moose numbers (McLaren and Peterson 1994). This finding suggests top-down flow (but cf. Schmitz et al. 2000).

In other wolf-prey systems, evidence has accumulated that snow depth, modulating vegetation availability, drives the systems, a bottom-up flow. Furthermore, there is no reason to believe that forces in all food chains act in the same direction. The Isle Royale top-down flow was postulated only for the wolf-moose-balsam fir chain, but fir is eaten during only part of the year, and even during that time, it constitutes only 59% of the moose diet (McLaren and Peterson 1994). Food chains involving other plants could evince a bottom-up influence; Boyce and Anderson (1999) postulated that such would be the case in Yellowstone. Discussion on the prevalent direction of flow of effects will never end and may well be a trivial issue, as it focuses on only a few interactions of selected trophic chains within an ecosystem. Recent findings of a long-term study of northern ecosystems in the Yukon show the largely unpredictable interaction between simple species relationships and complex stochastic events that affects ecological processes at a variety of spatial and temporal scales (Krebs et al. 2001).

Regardless of the direction of forces in wolf-prey ecosystems, wolves must exert some indirect influence on the vegetation through predation on the main herbivores in a system. Some historical evidence has been offered that successful aspen recruitment ceased after wolves were removed from the Yellowstone ecosystem in the 1920s (Ripple and Larsen 2000) and that aspen growth increased after wolf restoration (Ripple et al. 2001).

On a broader scale, Crête et al. (2001) have shown what they consider to be negative effects of herbivores on 197 plant taxa eaten by white-tailed deer, moose, and caribou/reindeer and positive effects on only 24. Presumably, then, wolf predation on these ungulates would bring reverse effects on the plants by reducing the ungulates. However, assigning positive and negative values to these effects is, as mentioned earlier, controversial. For example, claims have been made by some (Wagner 1994) that biodiversity—generally considered positive ecologically—is reduced by ungulate feeding, while others claim the opposite (Boyce 1998).

Wolves do affect ecosystems through multiple interacting ecological processes whose nonlinear effects confuse the superficial observer. It is possible that much of the discussion on the role of wolves within ecosystems is due to the mismatch of data collected under different sampling scales. We do not claim to know whether the wolf's effects are positive or negative, what its net effect is, or whether its effects are of any great consequence ecologically. We thus favor continued research into these issues to help solve the unknowns about this interesting and complex subject.

6

Wolf Population Dynamics

*Todd K. Fuller, L. David Mech, and
Jean Fitts Cochrane*

A LARGE, DARK WOLF poked his nose out of the pines in Yellowstone National Park as he thrust a broad foot deep into the snow and plowed ahead. Soon a second animal appeared, then another, and a fourth. A few minutes later, a pack of thirteen lanky wolves had filed out of the pines and onto the open hillside.

Wolf packs are the main social units of a wolf population. As numbers of wolves in packs change, so too, then, does the wolf population (Rausch 1967). Trying to understand the factors and mechanisms that affect these changes is what the field of wolf population dynamics is all about. In this chapter, we will explore this topic using two main approaches: (1) meta-analysis using data from studies from many areas and periods, and (2) case histories of key long-term studies. The combination presents a good picture—a picture, however, that is still incomplete. We also caution that the data sets summarized in the analyses represent snapshots of wolf population dynamics under widely varying conditions and population trends, and that the figures used are usually composites or averages. Nevertheless, they should allow generalizations that provide important insight into wolf population dynamics.

What Is a Wolf Population?

Trying to define a wolf population is problematic. As chapter reviewer Bruce Dale (personal communication) reminded us, two adjacent wolf packs may each depend on separate prey bases and thus respond independently to prey changes. In that respect, they could be regarded as separate populations. However, in regard to a disease outbreak that might affect many adjacent packs, the entire group affected could be considered a population. Or, genetically, all the wolves in the contiguous range from northern Alaska through Canada into Minnesota, Wisconsin, and Michigan could be thought of as one population.

There is no convention applicable here: a wolf population can be whatever interacting conglomeration of wolves one wants to consider for a particular reason. For example, in the Yellowstone Wolf Reintroduction Environmental Impact Statement (USFWS 1994b, 6–66), the following operational definition of a wolf population was adopted: "A wolf population is at least 2 breeding pairs of wild wolves successfully raising at least 2 young each (until December 31 of the year of their birth), for 2 consecutive years in an experimental area."

Studies of wolf population dynamics cover wolf density and distribution, population composition, the rates of births, deaths, and dispersal of wolves, and in particular, the means by which these parameters vary and change and the factors that affect them. Numerous scientific and popular articles and books deal with wolves, and most cover some aspects of wolf population dynamics. Several recent works have made important points concerning wolf conservation (e.g., Peek et al. 1991; Fritts et al. 1994; Fritts and Carbyn 1995; Mech 1995a), and central to all of them is information on wolf populations.

Since Mech's (1970) comprehensive summary of wolf biology, thousands of wild wolves have been radio-collared and monitored intensively (Mech 1995e), and many others have been studied in captivity (Frank 1987). These studies have allowed the collection of information critical to understanding wolf population dynamics. Radio-tracking has not only helped produce better data on wolf population size and trends, but has also yielded important new data about wolf mortality and survival, birth rates, and dispersal. In addition, radio-tracking studies have shed much light on wolf interactions with their prey, another key to understanding wolf population dynamics (Nelson and Mech 1981; Peterson, Woolington, and Bailey 1984; Ballard et al. 1987, 1997; Fuller 1990; Mech et al. 1998).

In 1983, Keith suggested that four factors dominate wolf population dynamics: wolf density, ungulate density, human exploitation, and ungulate vulnerability. Subsequent studies of wolf population dynamics (e.g., Fuller 1989b, 1995b) show that to understand wolf population ecology and conservation in a general way, we can reduce these factors to three key elements: food, people, and source populations. These are complex elements, to be sure, but they are clearly the most important to understand.

The abundance and availability of food (i.e., hoofed prey such as red deer or moose; see Peterson and Ciucci, chap. 4, and Mech and Peterson, chap. 5 in this volume) determine the potential for wolves to inhabit areas. Given higher ungulate populations, wolves should have more opportunities to catch prey, and food accessibility ultimately affects nutritional levels and thus wolf reproduction, survival, and behavior (Mech 1970; Van Ballenberghe et al. 1975; Zimen 1976; Packard and Mech 1980; Keith 1983; Mech et al. 1998). Prey accessibility is related not only to the abundance, but also to the vulnerability of prey (Mech 1970; Peterson and Page 1988; Mech et al. 1998; Peterson et al. 1998). Deep snow, age, or disease may make some prey more vulnerable, and thus more "accessible," than others (see Mech and Peterson, chap. 5 in this volume).

Second, human behaviors that result in the direct or indirect killing of wolves may influence where wolves live and in what numbers. In designated wilderness areas, national parks, and wildlife refuges, wolves are generally protected from human-related deaths. Wolf populations also seem little affected by snowmobiles, vehicles, logging, mining, and other human activities outside of these areas (Thiel et al. 1998; Merrill 2002), except as these factors facilitate accidental or intentional killing by humans or change prey density (e.g., logging). Even then, once a wolf population is large enough, such human take of wolves affects the population level little (see below) except along the frontier of the wolf's range.

In the past, of course, adverse human attitudes and traditions played a significant role in reducing wolf populations, especially in North America and western Europe when extensive poisoning and deliberate government persecution were applied (Young and Goldman 1944; Boitani 1995). In much of eastern Europe and Asia today, human attitudes toward wolves are still important determinants of wolf killing and hence population trends. Poland, for example, is experiencing its second wolf population resurgence in the last century (fig. 6.1) as a result of more tolerant public attitudes (Okarma 1992).

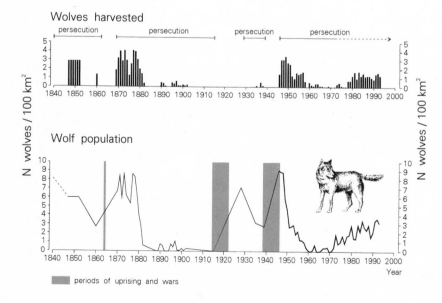

FIGURE 6.1. Long-term dynamics of wolf hunting harvest and wolf density in Bialowieza Primeval Forest (BPF), Poland, 1847–1993. Thin line, density reconstructed based on regression between numbers shot and population size between 1946 and 1971. Thick line, density determined by snow tracking surveys; numbers recorded in the exploited forests of the Polish part and in the Belarusian part of BPF were summed. Wolves recorded in Bialowieza National Park (BNP) were not added (with the exception of 1961, when wolves were recorded in BNP only) because they were most likely already counted in either of the two parts. (From Jedrzejewska et al. 1996.)

Finally, source populations of wolves are crucial to the establishment of new populations and to the maintenance of populations that are heavily controlled or harvested. For example, wolf populations in marginal habitats (i.e., where food resources are poor or human-caused mortality is high) are often successfully augmented by dispersal from adjacent source populations (Mech 1989; Lariviere et al. 2000). This is also true for small populations within larger regions of wolf abundance (Hayes and Harestad 2000a). Wolves are great dispersers and can move to new areas fairly easily (see Mech and Boitani, chap. 1 in this volume, and below). Thus, the distance of one wolf population from the next nearest one plays yet another primary role in wolf population ecology (Wydeven et al. 1995).

Below we try to summarize and synthesize what is known about wolf populations and the way they behave. We draw on data from perhaps the most comprehensive set of population literature there is for a large mammal to detail how various factors interact to affect wolf population dynamics. We will begin by looking at the bigger picture of how wolf populations are distributed geographically and by examining the role of packs in population change. Then we will continue through discussions of wolf density and how variables such as food affect it; the critical factors of reproduction, survival, mortality, and dispersal; rates of wolf population change; natural regulation of wolf populations; the role of cumulative effects on populations; and how well wolf populations persist. We conclude by assessing future needs for studying wolf population dynamics.

Wolf Distribution

Large-Scale Patterns

Historically, wolves occupied every habitat containing large ungulates in the Northern Hemisphere from about 20° N latitude (mid-Mexico, southern Saudi Arabia, and India) to the polar ice pack (Young and Goldman 1944). Vegetation type makes little difference to wolves as long as populations of hoofed prey are available. Wolves inhabit deserts, prairies, woodlands, swamps, tundra, and "barren lands" from sea level to mountaintops.

In general, wolves are very adaptable: they enter towns or villages at night (Zimen and Boitani 1979), cross four-lane highways and open landscapes (Merrill and Mech 2000), and den near logging sites, open-pit mines, garbage dumps, and military firing ranges (Thiel et al. 1998; Merrill 2002). They have few, if any, natural predators

(see Ballard et al., chap. 10 in this volume), but persecution of wolves by humans, primarily by poisoning, long ago eliminated wolves from many portions of their historical range (see Fritts et al., chap 12, and Boitani, chap. 13 in this volume).

Minimum Spatial Requirements

Wolf distribution on a small scale is limited mostly by the amount of land available containing enough prey with high enough productivity to support at least one pack. Even at the highest imaginable average prey densities (e.g., a biomass equal to 15 deer or 3 moose/km^2), it would seem that an individual pack of four wolves probably requires a territory of about 75 km^2 (30 mi^2) to meet its nutritional requirements (see fig. 6.2; see also Peterson and Ciucci, chap. 4, and Kreeger, chap. 7 in this volume). Few territories that small have been documented outside of small islands, although a pack of six wolves in 39 km^2 (15 mi^2) has been recorded in northeastern Minnesota (L. D. Mech and S, Tracy, unpublished data). Mean territory sizes of wolf packs on the mainland whose major prey occur at the highest measured densities (equivalent to 7–10 deer/km^2) actually average 100–200 km^2 (39–78 mi^2) (see below). In places where prey are at very low densities, average pack territories may measure more than 1,000 km^2 (390 mi^2) each (Mech 1988a; Mech et al. 1998).

An important consideration regarding the minimum area required by a wolf population is that a single, isolated pack should have a lower chance of persisting than a group of several adjacent packs. Theoretically, the chances of an isolated pack avoiding some catastrophe or difficulties from inbreeding vary inversely with its distance or degree of isolation (e.g., distance to a natural travel corridor) from the next nearest pack or packs (see the section on dispersal below). However, with an abundant food supply and no human-caused deaths, a population of 12–50 wolves on Isle Royale resulting from a single pair survived for 50 years (Peterson 2000), even after having lost an estimated 50% of its genetic variability (Wayne et al. 1991).

Nevertheless, if we were prescribing a formula for the smallest demographically viable wolf population, we might include two to three adjacent packs (cf. USFWS 1992, 18) of four wolves each, 40–60 km (24–36 mi) from other wolves. At average ungulate densities (e.g., 8 deer/km^2), pack territories might each cover 300 km^2 (117 mi^2). Such a population could persist anywhere ungulate prey occurred at the specified biomass density and at reasonable productivity, and where wolf mortality was less than net reproduction.

Studies of several small wolf populations add insight to this question. In the wildlife reserves of Quebec, human harvesting of wolves averaged 2–74% of the populations annually; populations persisted in reserves larger than 1,500 km² (585 mi²), but tended to be unstable in smaller reserves (Lariviere et al. 2000). This finding was similar to that in Poland's 1,538 km² (600 mi²) Bieszczady National Park, with a population of 26–33 wolves in five packs (Smietana and Wajda 1997).

Packs

Wolf populations are composed of packs and lone wolves, but as indicated earlier, packs form the basic units of a population. Most lone wolves are only temporarily alone as they disperse from packs and either start their own packs or join existing packs (see Mech and Boitani, chap. 1 in this volume, and below).

Origin of Packs

Packs originate when a male and female wolf meet, pair up, and produce pups (Rothman and Mech 1979). There are many variations on this method of pack formation and pack maintenance, but basically packs are composed of a mated pair of wolves and their offspring (see Mech and Boitani, chap. 1 in this volume).

Pack Size and Composition

Packs vary in size from two to forty-two wolves (table 6.1) (Rausch 1967; Fau and Tempany 1976, cited in Carbyn et al. 1993), and average pack sizes range from three to eleven. As indicated by Mech and Boitani in chapter 1 in this volume, the size of a given pack can vary by many multiples of the basic founding pair. Furthermore, when prey availability is reduced, large packs can be reduced in size through lower reproduction and/or survival or through dispersal. In addition, as packs enlarge, they sometimes split or proliferate. Therefore, we do not view pack size as a serious constraint on wolf population increases or decreases, but change in pack size is one of the primary mechanisms through which wolf population size changes (Rausch 1967).

Pack size does not necessarily differ among wolf pop-

ulations whose major prey are different. That is, average sizes of wolf packs feeding mainly on moose are not larger than those of packs feeding on deer, although mean pack sizes for those feeding on caribou and elk are larger than for those feeding on deer or moose (see table 6.1) (but also see Mech and Boitani, chapter 1 in this volume). Pack size also does not vary with relative prey biomass; packs are just as large at high prey densities as they are at low prey densities (tables 6.1 and 6.2).

Seasonal Changes in Pack Size

Packs are obviously largest just after pups are born; this is the major annual increment to wolf populations. As summer progresses, some pups and a few adults die, reducing overall pack size, and mortality of adults typically peaks during fall and winter (see below). Fall and winter are also major times of wolf dispersal, so pack sizes diminish further as members leave. However, a few wolves also join packs, single wolves pair with others, and young wolves often make predispersal trips away from packs for periods of days to months (see Mech and Boitani, chap. 1 in this volume). Thus pack sizes can fluctuate through the year.

"Observed" pack sizes may seem to follow a somewhat different pattern than outlined above, because during summer pack members more often travel alone. Pack members are more often together during winter, but even then packs may split apart for days to weeks before getting together again (see Mech and Boitani, chap. 1 in this volume). Thus most studies estimate pack sizes from the maximum number of wolves observed in a pack during winter, as recommended by Mech (1973, 1982b).

Pack Composition

In most wolf packs, pups, or young-of-the-year, form the single largest age class, followed by yearlings. Some packs may include one or more 2- or 3-year-olds. These wolves are usually all offspring of the breeding pair. Some packs also contain a postreproductive female or a wolf "adopted" from another pack (see Mech and Boitani, chap. 1 in this volume). Wolf packs in national parks such as Denali (Mech et al. 1998) and Yellowstone (Bangs et al. 1998), where human-related mortality is minimal, usually typify this type of age composition.

Where wolves are subject to human taking, estimates of pack composition in midwinter indicate that adults and yearlings usually constitute 54–76% of all pack

TABLE 6.1. Estimated early- to midwinter size and composition of wolf packs and proportion of nonresident wolves in various populations

Location	Main prey	Pack size[a]			% adults and yearlings[b]	Percent non-residents	Reference
		Mean	Maximum	N			
Northern Wisconsin	Deer	3.6	—	46	—	—	Wydeven et al. 1995 (1985–1991)
Northwestern Minnesota	Deer	4.3	9	24	—	14	Fritts and Mech 1981
Voyageurs Park, Minnesota	Deer	5.5	11	23	—	10	Gogan et al. 2000
Southern Quebec	Deer	5.6	10	19	—	—	Potvin 1988
East-central Ontario	Deer	5.9	9	54	69	20	Pimlott et al. 1969
Algonquin Park, Ontario	Deer	6.0	13	44	—	—	Forbes and Theberge 1995
North-central Minnesota	Deer	6.0	8	4	—	—	Berg and Kuehn 1982
North-central Minnesota	Deer	6.7	12	33	54(54)[c]	7	Fuller 1989b
Northeastern Minnesota	Deer	7.2	10	11	58(50)[c]	—	Van Ballenberghe et al. 1975
West-central Yukon	Sheep	4.6	8	5	—	—	Sumanik 1987
Northwestern Alaska	Caribou	8.6	24	34	—	—	Ballard et al. 1997
Northern Alaska	Caribou	9.5	15	12	—	—	Dale et al. 1995
Southwestern Manitoba	Elk	8.4	16	13	—	—	Carbyn 1980
Central Rocky Mts, MT, BC	Elk	10.7	19	38	—	—	Boyd and Pletscher 1999
Jasper Park, Alberta	Elk	11.5	14	4	—	—	Carbyn 1974
Southwestern Quebec	Moose	3.7	6	16	—	9	Messier 1985a,b (low prey area)
Pukaskwa Park, Ontario	Moose	3.8	7	39	—	—	Bergerud et al. 1983
Southwestern Quebec	Moose	5.7	8	11	—	9	Messier 1985a,b (high prey area)
Southern Yukon	Moose	5.8	25	103	66(55)[c]	10	Hayes et al. 1991
Northwestern Alberta	Moose	6.0	12	22	76	12	Bjorge and Gunson 1989
East-central Yukon	Moose	6.8	20	146	—	—	Hayes and Harestad 2000b
South-central Alaska	Moose	7.5	20	59	—	—	Ballard et al. 1987
Northeastern Alberta	Moose	7.6	7	7	60	13	Fuller and Keith 1980a (both study areas)
Denali Park, Alaska	Moose	8.9	29	91	57	—	Mech et al. 1998
Kenai Peninsula, Alaska	Moose	9.8	29	32	65	—	Peterson, Woolington, and Bailey 1984
Northern Alberta	Bison	8.4	14	20	—	—	Carbyn et al. 1993
Isle Royale, all years Isle Royale, Michigan	Moose	5.8	22	135	75[d]	—	Mech 1966b; Jordan et al. 1967; Peterson 1977; Peterson and Page 1988; Peterson et al. 1998; R. O. Peterson, personal communication (1959–1994)
Isle Royale, by wolf trend Isle Royale, Michigan	Moose	4.4	11	39	—	—	Peterson and Page 1988; Peterson et al. 1998; R. O. Peterson, personal communication (1983–1994)
Isle Royale, Michigan	Moose	7.8	18	37	—	—	Peterson 1977; Peterson and Page 1988 (1973–1980)

(continued)

TABLE 6.1 *(continued)*

Location	Main prey	Pack size[a]			% adults and yearlings[b]	Percent non-residents	Reference
		Mean	Maximum	N			
Isle Royale, Michigan	Moose	6.2	22	46	—	—	Mech 1966b; Jordan et al. 1967; Peterson 1977; R. O. Peterson, personal communication (1959–1972)
Isle Royale, Michigan	Moose	3.9	—	21	—	—	Peterson and Page 1988 (1980–1982)
Isle Royale by ~stable biomass index							
Isle Royale, Michigan	Moose	6.5	—	34	—	—	Peterson 1977; Peterson and Page 1988; R. O. Peterson, personal communication (1968–1976)
Isle Royale, Michigan	Moose	3.1	—	18	—	—	Peterson et al. 1998; R. O. Peterson, personal communication (1987–1991)
Isle Royale, Michigan	Moose	4.7	—	30	—	—	Peterson and Page 1988 (1980–1985)
Isle Royale, Michigan	Moose	7.4	—	24	—	—	Mech 1966b; Jordan et al. 1967; R. O. Peterson, personal communication (1959–1966)
Northeastern Minnesota, all years							
Northeastern Minnesota	Deer	5.8	15	198	—	16	Mech 1986 (1967–1985)
Northeastern Minnesota by wolf/deer trend							
Northeastern Minnesota	Deer	6.5	—	—	—	—	Mech 1986 (1967–1970)
Northeastern Minnesota	Deer	6.2	—	—	—	—	Mech 1986 (1971–1975)
Northeastern Minnesota	Deer	5.2	—	—	—	—	Mech 1986 (1976–1984)

Summary and statistical test results
Average pack size and principal prey species

Species	Deer	Moose	Elk	Caribou
No. studies	10	11	3	2
Mean pack size	5.66	6.49	10.2	9.05

Two-sample, two-tailed *t* tests assuming equal variance

Test	d.f.	P	t
Deer vs. moose	19	.24	−1.22
Deer vs. elk	11	< .001	−5.89
Deer vs. caribou	10	.002	−4.31
Moose vs. elk	12	.01	−3.08

Biomass index/wolf and % adults + yearlings in the fall population: $r^2 = .25$; d.f. = 8; $P = .17$.

[a]Including all groups ≥2 wolves.
[c]Percentage of females within age class.
[b]Percentage of wolves ≥1 year old in the population.
[d]Average percentage of wolves ≥1 year old in the fall when the population was stable between 1971 and 1995 (Peterson et al. 1998).

TABLE 6.2. Mean ungulate and wolf densities and ungulate biomass/wolf ratios during winter in North America

Location	Years	Prey species	Number/1,000 km²				Reference(s)
			Ungulates	Ungulate biomass index[a]	Wolves	Ungulate biomass index per wolf	
Northeastern Minnesota	1970–1971	Deer	5,100	9,900	42	236	Van Ballenberghe et al. 1975
		Moose	800				
Voyageurs Park, Minnesota	1987–1991	Deer	8,370	9,150	33	277	Gogan et al. 2000
		Moose	130				
Southwestern Manitoba	1975–1978	Elk	1,200	8,740	26	336	Carbyn 1980, 1983b
		Moose	800				
		Deer	340				
Northwestern Alberta	1975–1980	Moose	1,165	7,332	24	306	Bjorge and Gunson 1989
		Elk	114				
Northern Wisconsin	1986–1991	Deer	7,200	7,200	18	400	Wydeven et al. 1995
Northwestern Minnesota	1972–1977	Deer	5,000	6,800	17[b]	400	Fritts and Mech 1981
		Moose	300				
East-central Ontario	1958–1965	Deer	5,769	6,645	38	175	Pimlott et al. 1969
		Moose	146				
Southern Quebec	1980–1984	Deer	3,000	6,600	28	236	Potvin 1988
		Moose	600				
North-central Minnesota	1980–1986	Deer	6,160	6,280	39	161	Fuller 1989b
		Moose	20				
North-central Minnesota	1978–1979	Deer	6,170	6,170	10	617	Berg and Kuehn 1980
Northeastern Minnesota	1946–1953	Deer	3,475	5,791	23	252	Stenlund 1955
		Moose	386				
Kenai Peninsula, Alaska	1976–1981	Moose	800	4,826	14	345	Peterson, Woolington, and Bailey 1984
		Caribou	13				
South-central Alaska	1945–1982	Moose	665	4,612	7[c]	659[c]	Ballard et al. 1987; Davis 1978
		Caribou	311				
Algonquin Park, Ontario	1969	Deer	3,100	4,024	36	112	Kolenosky 1972; Pimlott et al. 1969
		Moose	154				
Jasper Park, Alberta	1969–1972	Elk	500	2,730	8	364	Carbyn 1974
		Sheep	470				
		Goat	120				
		Moose	80				
		Deer	80				
		Caribou	40				
Algonquin Park, Ontario	1988–1992	Deer	395	2,615	27	97	Forbes and Theberge 1995
		Moose	370				
East-central Yukon	1989–1994	Moose	353	2,609	6[c]	435[c]	Hayes and Harestad 2000a,b
		Caribou	238				
		Goat	11				
		Sheep	4				
Northern Alaska	1989–1990	Caribou	510	2,240	7	320	Adams and Stephenson 1986; Singer 1984; Dale et al. 1995
		Moose	120				
		Sheep	500				
Southwestern Quebec	1980–1984	Moose	370	2,200	14	159	Messier 1985a,b (high prey area)

(continued)

TABLE 6.2 *(continued)*

Location	Years	Prey species	Number/1,000 km²				Reference(s)
			Ungulates	Ungulate biomass index[a]	Wolves	Ungulate biomass index per wolf	
Denali Park, Alaska	1966–1974	Moose	164	2,002	6	334	Haber 1977
		Sheep	478				
		Caribou	270				
Pukaskwa Park, Ontario	1975–1979	Moose	296	1,789	12	149	Bergerud et al. 1983
		Caribou	13				
Interior Alaska	1975–1978	Moose	206	1,560	9	173	Gasaway et al. 1983
		Caribou	162				
Southern Yukon	1983–1988	Moose	207	1,556	8[c,d]	207[c,d]	Hayes et al. 1991
		Sheep	260				
		Caribou	27				
		Goats	29				
Denali Park, Alaska	1986–1992	Caribou	300	1,531	6	255	Meier et al. 1995
		Moose	133				
		Sheep	133				
Southwestern Quebec	1980–1984	Moose	230	1,380	8	173	Messier 1985a,b (low prey area)
Northwestern Alaska	1987–1991	Moose	166	1,324	5	267	Ballard et al. 1997
		Caribou	164				
Northern Alberta	1979	Bison	153	1,224	8	152	Oosenbrug and Carbyn 1982
West-central Yukon	1985–1986	Moose	62	1,143	7	153	Sumanik 1987
		Caribou	45				
		Sheep	681				
Northeastern Alberta	1975–1977	Moose	180	1,114	6	186	Fuller and Keith 1980a,b, Gunson 1995 (AOSERP area)
		Caribou	17				
Denali Park, Alaska	1984–1985	Moose	94	865	3	288	Singer and Dalle-Molle 1985
		Caribou	106				
		Sheep	89				
Isle Royale, all years							
Isle Royale, Michigan	1959–1994	Moose	2,096	12,576	44	286	Mech 1966b; Jordan et al. 1967; Peterson 1977; Peterson and Page 1988; Peterson et al. 1998; R. O. Peterson, personal communication
Isle Royale by wolf trend							
Isle Royale, Michigan	1983–1994	Moose	2,399	14,394	31	465	Peterson and Page 1988; Peterson et al. 1998; R. O. Peterson, personal communication
Isle Royale, Michigan	1973–1980	Moose	2,247	13,482	71	190	Peterson 1977; Peterson and Page 1988

TABLE 6.2 *(continued)*

Location	Years	Prey species	Number/1,000 km²		Wolves	Ungulate biomass index per wolf	Reference(s)
			Ungulates	Ungulate biomass index[a]			
Isle Royale, Michigan	1959–1972	Moose	1,844	11,064	41	270	Mech 1966b; Jordan et al. 1967; Peterson 1977; R. O. Peterson, personal communication
Isle Royale, Michigan	1980–1982	Moose	1,485	8,910	58	154	Peterson and Page 1988
Isle Royale, by ~stable biomass indicator							
Isle Royale, Michigan	1968–1976	Moose	2,678	16,068	49	328	Peterson 1977; Peterson and Page 1988; R. O. Peterson, personal communication
Isle Royale, Michigan	1987–1991	Moose	2,558	15,348	25	614	Peterson et al. 1998; R. O. Peterson, personal communication
Isle Royale, Michigan	1980–1985	Moose	1,490	8,940	51	175	Peterson and Page 1988; Mech 1966b; Jordan et al. 1967; R. O. Peterson, personal communication
Isle Royale, Michigan	1959–1966	Moose	1,321	7,926	50	158	
Northeastern Minnesota, all years							
Northeastern Minnesota	1967–1993	Moose[e] Deer[e]	560 1,212	4,572	28	163	Mech 1973, 1986; Mech and Nelson 2000; Peek et al. 1976; Fuller 1989b
Northeastern Minnesota by wolf/deer trend							
Northeastern Minnesota	1967–1970	Moose[e] Deer[e]	600 3,380	6,980	38	186	Mech 1973, 1986; Peek et al. 1976; Fuller 1989b
Northeastern Minnesota	1971–1975	Moose[e] Deer[e]	570 1,800	5,220	33	161	Mech 1973, 1986; Mech and Nelson 2000; Peek et al. 1976; Fuller 1989b
Northeastern Minnesota	1976–1984	Moose[e] Deer[e]	550 600	3,900	23	170	Mech and Nelson 2000; Peek et al. 1976; Fuller 1989b

Summary and statistical test results

Test	r^2	d.f.	P	Regression
BMI/wolf and mean pack size	.06	24	.23	
Total BMI and mean pack size	.004	24	.76	
Total BMI and mean density	.64	31	< .001	$y = 3.5 + 3.27x$

(continued)

TABLE 6.2 *(continued)*

	BMI/wolf summary statistics
Mean	271
SE	23
SD	131
Range	97–659
No. studies	32

Source: Adapted from Keith 1983; Fuller 1989b.

[a]Relative biomass values were assigned as follows (similar to Keith 1983); bison, 8; moose, 6; elk, 3; caribou, 2; bighorn sheep, 1; Dall sheep, 1; mountain goat, 1; mule deer, 1; white-tailed deer, 1.

[b]Wolf population newly protected and expanding.   [c]Wolf population heavily exploited.   [d]Wolf population recovering from heavy exploitation.

[e]Ungulate densities extrapolated between estimates for 1970 (Peek et al. 1976) and 1975 (Fuller 1989b), then assumed constant after 1975.

members (see table 6.1). The limited data do not indicate any particular bias in sex ratios of adults and yearlings; there is either an equal sex ratio or one slightly biased toward females (Mech 1970).

Populations with the highest proportion of pups in packs are usually those whose numbers have been reduced substantially through control efforts, thus leaving only small packs or pairs. When these groups produce an average litter of pups (4–6; see below), surviving pups can clearly make up a high proportion of the pack. Similarly, populations of wolves recolonizing areas have ample opportunity to form new packs made up of only a pair of wolves, so newborn pups form a large part of populations in such areas.

Lone Wolves

At any given time, some wolves that have dispersed from packs are traveling alone. These wolves may be either temporarily away from their pack or permanently dispersed and looking for mates. The proportion of these nonresident wolves in a population probably varies seasonally, as do dispersal rates and the rates at which individuals settle into territories (see Mech and Boitani, chap. 1 in this volume), but a variety of studies have documented or surmised that these wolves compose about 10–15% of a wolf population in winter on average (see table 6.1).

Density

Variation

Wolf densities naturally vary tremendously. It is common for studies in the far north to record healthy wolf populations with densities of less than 5/1,000 km^2

(391 mi^2) (see table 6.2), whereas on Isle Royale in Lake Superior (Canada-U.S. border) wolf density reached 92/1,000 km^2 in 1980 (Peterson and Page 1988). Furthermore, studies of wolf density have varied in the precise methods used to derive the area involved, so often their results are not strictly comparable (Burch 2001). In general, however, maximum midwinter wolf densities documented for mainland populations over a number of years have rarely measured more than 40/1,000 km^2 (see table 6.2).

Pimlott (1967) suggested 30 years ago that some intrinsic control on wolf numbers limited density to a maximum of about 40 wolves/1,000 km^2 in most areas. This conclusion was based on his own observations in Algonquin Provincial Park in Ontario, Canada, and on limited observations of others. Mech (1973) concurred with this assessment, but noted exceptions where prey densities were extremely high. In addition to the findings on Isle Royale noted above, Fuller (1989b) recorded maximum densities in north-central Minnesota during the 1980s of 69 wolves/1,000 km^2 in early winter and 50/1,000 km^2 in late winter. The work of Peterson and Page (1988) on Isle Royale, and the evidence presented by Keith (1983), convinced Peterson and Page (1988) that the ultimate limit on wolf density is that imposed by food, as many other workers had also concluded (Mech 1970; Van Ballenberghe et al. 1975; Packard and Mech 1980; Keith 1983).

Food

In fact, 64% of the variation in wolf density in all North American studies was directly accounted for simply by variation in prey biomass. This relationship (Keith 1983; Fuller 1989b) is now based on thirty-one intensive stud-

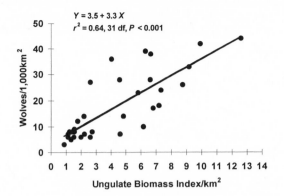

FIGURE 6.2. Relationship between ungulate biomass index and wolf density, plotted from data in table 6.2. (Adapted from Keith 1983 and Fuller 1989b.)

ies that measured total average ungulate biomass (often more than one prey species) and average wolf populations for a period of several years (see table 6.2). The relationship between prey abundance and wolf numbers may vary for areas with migratory versus nonmigratory prey, or where prey concentrate seasonally. However, there are no indications that, over time, wolf numbers are mainly limited by anything other than food (usually ungulate numbers and accessibility), given the above considerations. A plot of the relationship between food abundance (i.e., ungulate biomass index; see table 6.2) and wolf density (fig. 6.2) does not "level off," and thus suggests that even at prey densities higher than have been recorded thus far, this relationship should be valid.

Effect of Long-Term Mortality

The actual ungulate biomass index per wolf varies among studies (mean = 271; median = 254; range = 97–659; see table 6.2), as indicated by the deviation of data points from the regression line in figure 6.2. This ratio, however, is highest for heavily exploited (Peterson, Woolington, and Bailey 1984; Ballard et al. 1987; Hayes and Harestad 2000a,b) and newly protected wolf populations (e.g., Berg and Kuehn 1980; Fritts and Mech 1981; Wydeven et al. 1995), and lowest for unexploited wolf populations (Oosenbrug and Carbyn 1982; Bergerud et al. 1983) and those where ungulates are heavily harvested (Kolenosky 1972).

It seems clear that newly protected wolf populations would have the potential to grow until food was a limiting factor; thus the relative number of ungulates initially and for some time would be high. In addition, it

also makes some sense that perpetually harvested wolf populations, despite compensatory reproduction, might never "catch up" with prey densities and thus would fail to achieve some maximum density. Gasaway et al. (1992, 39) demonstrated for numerous regions in Alaska that wolf populations that they believed were limited by harvesting occurred at much lower densities in relation to prey availability than did populations that were lightly harvested.

Conversely, completely unexploited or completely protected wolf populations are probably making the most of their food supply and achieving the highest densities possible. This should be especially true where, in addition, ungulates are harvested by humans, thus holding their numbers low.

Over the long run, however, we would expect that the average ratio of wolves to ungulate biomass in a system unaffected by humans might reach some median value that reflects the bioenergetic balance of predator and prey. In fact, Isle Royale's unexploited population seems to have done just that; the mean ungulate biomass per wolf there over a 36-year period was 286, almost identical to the mean for all areas (see table 6.2).

The relationship between food or prey density and wolf density is sufficiently strong that, given specific conditions, one can make reasonable predictions concerning the average density of wolves. For example, a lightly to moderately harvested wolf population whose only prey is moose occurring at a density of $1/km^2$ (6 "deer-equivalents"$/km^2$) would probably have a density of 23 ($\pm$ 5 SE) wolves/1,000 km^2. As will be discussed below, however, other factors determine the specific wolf numbers and population trends in various areas.

Temporal Variation

Changes in wolf density due to varying prey density have been documented by long-term studies in northeastern Minnesota (Mech 1977b, 1986, 2000c) and on Isle Royale (Peterson et al. 1998), in areas of varying moose density in southwestern Quebec (Messier and Crête 1985), and in Denali National Park, Alaska (Mech et al. 1998). The numerical response of an individual wolf population to a change in food supply or prey biomass may be like that for other cyclic mammals (Peterson, Page, and Dodge 1984), and thus for any one year, the ratio of prey biomass to wolves may differ from other years and from other areas in the same year. When ungulate numbers fluctuate from year to year, changes in wolf density may lag for up to several years in a single-prey system (Mc-Laren and Peterson 1994); we will discuss the reason for this finding below. In multi-prey systems, wolf numbers may respond more quickly to changes in prey vulnerability (Mech et al. 1998; see below).

Wolf densities also vary where wolves are heavily harvested, have the opportunity to recover from overharvest, or are newly protected. In some areas wolves have been intentionally harvested more heavily in one or more years to reduce their effect on prey populations (e.g., Ballard et al. 1987; Bjorge and Gunson 1983; Gasaway et al. 1983; Hayes and Harestad 2000a), and their numbers have declined precipitously. Conversely, some of these same populations have then been allowed to recover, and their numbers have increased to a similar degree (e.g., Bjorge and Gunson 1989; Hayes and Harestad 2000a). In other cases, wolves have recolonized areas from which they were extirpated many years earlier, and these populations, too, have increased rapidly (e.g., Fritts and Mech 1981; Peterson, Woolington, and Bailey 1984; Wydeven et al. 1995; see below).

Territory Size

Wolves usually occupy exclusive, defended territories, although there are several exceptions to this generalization (see Mech and Boitani, chap. 1 in this volume). Territoriality is generally thought to help stabilize population dynamics by tightening the feedback loop to local resources. This theory has not been tested in wolf populations. About all we can add to this discussion is that, as indicated above and by Mech and Boitani (chap. 1 in this volume), wolf pack sizes adjust considerably to food supply or vulnerable prey biomass within territories, but factors affecting prey vulnerability, such as winter severity, usually are pervasive across many territories.

Wolf pack territory sizes vary, on average, fourteenfold among areas (table 6.3). Average territory size and, more particularly, the average area per wolf vary most directly with food resources or prey abundance, as well as with prey type and the mean annual rate of population change. On average, about 33% of the variation in mean territory size ($r^2 = .33$, $P < .001$, d.f. = 32) and 35% of that in mean territory area per wolf ($r^2 = .35$, $P < .001$, d.f. = 32) can be attributed to variation in prey biomass; in general, the higher the prey density, the smaller the territory (table 6.3). In Wisconsin, a similar relationship ($r^2 = .59$; $P < .01$) has been documented for individual wolf territories and their corresponding deer densities (Wydeven et al. 1995).

However, territory sizes still vary considerably, even among areas where total prey biomass is about the same.

TABLE 6.3. Ungulate biomass index, mean territory size, mean pack size in winter, and mean territory area per wolf for wolf populations utilizing different primary prey

Primary prey	Location	Ungulate biomass index[a]	Territory size (km^2) $\bar{x}$	N	Pack size	Territory area per wolf (km^2)	Finite rate of increase	Reference
Deer	Northeastern Minnesota	9,900	143	11	7.2	20	—	Van Ballenberghe et al. 1975
Deer	Voyageurs Park, Minnesota	9,150	152		5.5	28	—	Gogan et al. 2000
Deer	Northern Wisconsin	7,200	176	41	3.5	50	1.16	Wydeven et al. 1995 (1986–1991)
Deer	Northwestern Minnesota	6,800	344	8	4.6	80	1.13	Fritts and Mech 1981
Deer	Algonquin Park, Ontario	6,645	259	47	5.9	25	—	Pimlott et al. 1969
Deer	Southern Quebec	6,600	199	21	5.6	36	—	Potvin 1988
Deer	North-central Minnesota	6,280	116	33	5.7	20	1.02	Fuller 1989b
Deer	North-central Minnesota	6,170	230	4	6.0	46	—	Berg and Kuehn 1980, 1982
Deer	Algonquin Park, Ontario	4,024	224	1	8.0	28	—	Kolenosky 1972
Deer	Algonquin Park, Ontario	2,615	149	44	6.0	25	1.01	Forbes and Theberge 1995
Sheep	West-central Yukon	1,143	754	5	4.6	164	—	Sumanik 1987
Elk	Southwestern Manitoba	8,740	293	12	8.4	35	0.86	Carbyn 1980, 1983b
Moose	Northwestern Alberta	7,332	424	9	6.0	71	1.29	Bjorge and Gunson 1989

TABLE 6.3 *(continued)*

Primary prey	Location	Ungulate biomass index[a]	Territory size (km²)		Pack size	Territory area per wolf (km²)	Finite rate of increase	Reference
			$\bar{x}$	N				
Moose	Kenai Peninsula, Alaska	4,826	638	18	11.2	57	1.03	Peterson, Woolington, and Bailey 1984
Moose	South-central Alaska	4,612	1,645	—	7.5	219	0.88	Ballard et al. 1987
Moose	East-central Yukon	2,609	1,478	17	6.8	217	1.49	Hayes and Harestad 2000a,b
Moose	Southwestern Quebec	2,220	397	14	5.7	68	1.06	Messier 1985a,b (high prey area)
Moose	Interior Alaska	2,080	665	—	9.3	72	0.76	Gasaway et al. 1992 (1972–1975)
Moose	Pukaskwa Park, Ontario	1,789	250	—	2.8	89	0.84	Bergerud et al. 1983
Moose	Southern Yukon	1,556	1,192	—	5.5	193	0.97	Hayes et al. 1991
Moose	Denali Park, Alaska	1,531	1,330	15	8.9	133	1.20	Mech et al. 1998
Moose	Southwestern Quebec	1,380	255	16	3.7	69	1.11	Messier 1985a,b (low prey area)
Moose	Northwestern Alaska	1,324	1,372	14	9.0	152	0.88	Ballard et al. 1997
Moose	Northeastern Alberta	1,114	834	7	7.7	110	1.21	Fuller and Keith 1980a (AESORP area)
Bison	Northern Alberta	1,224	1,352	3	12.3	110	—	Carbyn et al. 1993; Oosenbrug and Carbyn 1982
Isle Royale, all years								
Moose	Isle Royale, Michigan	12,576	145	135	5.8	25	1.00	Mech 1966b; Jordan et al. 1967; Peterson 1977; Peterson and Page 1988; Peterson et al. 1998; R. O. Peterson, personal communication (1959–1994)
Isle Royale by wolf trend								
Moose	Isle Royale, Michigan	14,400	167	39	4.4	38	0.97	Peterson and Page 1988; Peterson et al. 1998; R. O. Peterson, personal communication (1983–1994)
Moose	Isle Royale, Michigan	13,480	118	37	7.8	15	1.11	Peterson 1977; Peterson and Page 1988 (1973–1980)
Moose	Isle Royale, Michigan	11,070	166	46	6.2	27	1.01	Mech 1966b; Jordan et al. 1967; Peterson 1977; R. O. Peterson, personal communication (1959–1972)
Moose	Isle Royale, Michigan	8,910	78	21	3.9	20	0.42	Peterson and Page 1988 (1980–1982)
Isle Royale by ~stable BMI periods								
Moose	Isle Royale, Michigan	16,070	144	34	6.5	22	1.09	Peterson 1977; Peterson and Page 1988; R. O. Peterson, personal communication (1968–1976)

(continued)

TABLE 6.3 *(continued)*

Primary prey	Location	Ungulate biomass index[a]	Territory size (km²)		Pack size	Territory area per wolf (km²)	Finite rate of increase	Reference
			$\bar{x}$	N				
Moose	Isle Royale, Michigan	15,350	151	18	3.1	49	0.93	Peterson et al. 1998; R. O. Peterson, personal communication (1987–1991)
Moose	Isle Royale, Michigan	8,940	109	30	4.7	23	0.85	Peterson and Page 1988 (1980–1985)
Moose	Isle Royale, Michigan	7,920	181	24	7.4	24	1.04	Mech 1966b; Jordan et al. 1967; R. O. Peterson, personal communication (1959–1966)
Northeastern Minnesota, all years								
Deer	Northeastern Minnesota	4,572	198	198	5.8	34	0.99	Mech 1973, 1986; Mech and Nelson 2000; Peek et al. 1976; Fuller 1989b (1967–1985)
Northeastern Minnesota by wolf/deer trends								
Deer	Northeastern Minnesota	6,980	172	48	6.5	26	1.00	Mech 1986 (1967–1970)
Deer	Northeastern Minnesota	5,220	184	56	6.2	30	0.87	Mech 1986 (1971–1975); Mech and Nelson 2000
Deer	Northeastern Minnesota	3,900	219	94	5.2	42	1.00	Mech 1986 (1976–1985); Mech and Nelson 2000

Summary and statistical test results

Test[b]	r^2	d.f.	P	Regression
Biomass Index (BMI) and territory size	.33	31	< .001	$y = 900 - 0.07x$
BMI and pack size	.04	31	.29	
BMI and wolf density	.35	31	< .001	$y = 124 - 0.01x$
BMI and rate of increase	.008	21	.7	
Rate of increase and wolf density	.33	21	.005	$y = -104 + 181x$
Rate of increase and mean territory size	.30	21	.008	$y = -891 + 1407x$

Mean territory size and wolf density

Prey species	Deer	Moose
No. studies	11	13
Mean territory size	199	817
Mean wolf density	36	113

2-sample, two-tailed t-tests assuming equal variance

Test	d.f.	P	t
Mean territory size (deer v moose)	22	< .001	3.87
Mean wolf density (deer v moose)	22	< .001	3.89

Source: Adapted from Fuller 1989b.

[a]Relative biomass values were assigned as follows (similar to Keith 1983): bison, 8; moose, 6; elk, 3; caribou, 2; bighorn sheep, 1; Dall sheep, 1; mountain goat, 1; mule deer, 1; white-tailed deer, 1.

[b]Isle Royale and northeastern Minnesota data entered by phase of population trend. Other variations yielded similar results.

This variation may be related to prey type. Irrespective of ungulate biomass, all but two of twenty-four average wolf pack territory sizes, and two values for territory area per wolf, are higher ($P = .001$, two-tailed t test, d.f. $= 22$ for both territory size and area/wolf) where wolves prey mainly on moose than where they prey primarily on deer (see table 6.3). In areas of similar prey biomass, this relationship probably reflects the amount of prey biomass "accessible" to wolves. If moose are, on average, less vulnerable to wolf predation (i.e., harder to catch) than are deer, then we would expect a wolf pack of a particular size living on moose to need relatively more living biomass, and thus a larger territory, in order to provide enough prey that it can catch and kill.

There still remains much unexplained variation in territory size. Even in areas with the same major prey species and a similar total prey biomass, wolf pack territory sizes can differ markedly. For example, in southwestern Quebec boreal forest, moose (230–370/1,000 km^2 or 590–950/1,000 mi^2) compose 100% of total ungulate prey biomass (see table 6.2), and wolf territories average 250–400 km^2 (98–156 mi^2). In the Yukon, moose (62–353/1,000 km^2 or 160–900/1,000 mi^2) compose 75% of total ungulate prey biomass, generally inhabiting forest patches and tundra, and wolf pack territories average 1,300 to 1,500 km^2 (508–586 mi^2). Perhaps moose in particular, and ungulate prey in general, are less "vulnerable" when co-occurring with several other species in open habitats.

Reproduction

Age

Although there are recorded instances of captive wolves breeding at age 9–10 months (Medjo and Mech 1976), the earliest that breeding in wild wolves has been documented is 2 years (Rausch 1967; Peterson, Woolington, and Bailey 1984; Fuller 1989b), except for some equivocal evidence of first-year breeding in the restored Yellowstone population (D. W. Smith, personal communication). In some areas, females do not usually breed until age 4 (Mech and Seal 1987; Mech et al. 1998). As with other species, age of first breeding in wolves probably depends on environmental conditions such as food supply. In addition, because wolves must find a vacant territory before rearing young, those in saturated populations may have to wait longer.

This considerable flexibility in age of first breeding could have important effects on population change.

Thus, when food is abundant, such as during severe winters that make prey more vulnerable to wolves or in low-density reintroduced or heavily controlled wolf populations, wolves could rear pups when younger, quickly making use of the newly available resources to increase their numbers.

Few wolves live longer than 4 or 5 years, but female wolves as old as 11 years have been known to produce pups in the wild (Mech 1988c). There is no evidence that females reach reproductive senescence before they die, as coyotes do (Crabtree 1988). However, old females may be replaced as breeders by their daughters (Mech and Hertel 1983) and, if they remain in the pack, become postreproductive (Mech 1995d) (see also Kreeger, chap. 7 in this volume).

Breeding Frequency

Female wolves are capable of producing pups every year, and in most areas except the High Arctic (Mech 1995d), packs usually produce pups each year. Most wolf packs produce only a single litter per year (Harrington et al. 1982; Packard et al. 1983), although two litters from two females per pack have been reported (Murie 1944; Clark 1971; Haber 1977; Harrington et al. 1982; Van Ballenberghe 1983a; Peterson, Woolington, and Bailey 1984; Ballard et al. 1987; Mech et al. 1998), and in Yellowstone National Park there were three litters in one reintroduced pack (D. W. Smith, personal communication). Except for these unusual packs, if there are more than two female wolves older than 2 years in a pack, usually some do not breed, or if they do breed, they may resorb their fetuses (Hillis and Mallory 1996a) or fail to rear the pups. Thus populations with larger packs contain a lower proportion of breeders (Peterson, Woolington, and Bailey 1984; Ballard et al. 1987). Increased human harvest of wolves may result in smaller packs and territories and in the establishment of new packs in vacated areas, so that breeders then compose a higher proportion of the population and the rate of pup production increases (Peterson, Woolington, and Bailey 1984).

There is not yet a good explanation as to why packs in some areas more frequently include two females that produce pups (e.g., the East Fork pack in Denali National Park, Alaska) (Murie 1944; Haber 1977; Mech et al. 1998). Two founding packs in Yellowstone National Park have produced multiple litters in several consecutive years (D. W. Smith, unpublished data). Because these packs have a maximal food supply, this observation suggests that a surfeit of food fosters multiple breeding in a pack. Surplus food would certainly minimize competition and thus delay dispersal (Mech et al. 1998), so perhaps the founding breeding female would become more tolerant of her daughters breeding (see Mech and Boitani, chap. 1 in this volume).

Litter Size

Wolf litter sizes tend to average about five or six (Mech 1970; table 6.4) except in the High Arctic, where fewer pups are produced (Marquard-Petersen 1995; Mech 1995d). Litter size was small for an unexploited population in Ontario ($\overline{x}$ = 4.9; Pimlott et al. 1969) but large for exploited populations in Alaska ($\overline{x}$ = 6.5; Rausch 1967) and northeastern Minnesota ($\overline{x}$ = 6.4; Stenlund 1955), leading Van Ballenberghe et al. (1975) and Keith (1983) to suggest that litter size may increase with ungulate biomass per wolf. More recent data strongly confirm this assertion (Boertje and Stephenson 1992), with litter sizes across studies increasing an average of 31% with a sixfold increase in ungulate biomass available per wolf (r^2 = .38, P = .01, d.f. = 16, table 6.4).

Survival

Age and Sex

Wolf pups in most areas survive well through summer (table 6.5), probably because of a temporary abundance of a greater variety of food (Mech et al. 1998). Where canine parvovirus is prevalent, however, summer pup survival can be quite low (Mech and Goyal 1995). Pup survival is directly related to prey biomass (table 6.5), for the greater the biomass, the greater the chance that more will be accessible. Summer pup survival was almost doubled (0.89 vs. 0.48) where per capita ungulate biomass was four times greater (table 6.5). In northeastern Minnesota, pup condition and survival decreased during a decline in the deer population (Van Ballenberghe and Mech 1975; Seal et al. 1975; Mech 1977b). The percentage of pups in the population or in packs (see table 6.4) was highest in newly protected (Fritts and Mech 1981) and heavily exploited populations (Ballard et al. 1987), and probably reflected both larger litters and higher pup survival where ungulates were abundant (Pimlott et al. 1969; Keith 1974, 1983; Harrington et al. 1983), as well as a higher percentage of the population being reproductive (see above).

Mean prey biomass/wolf ratios and mean percent-

ages of pups in fall and winter populations are not clearly correlated (see table 6.4), in contrast to the findings of Keith (1983) and Boertje and Stephenson (1992). Prey biomass/wolf ratios and percentages of pups in packs are somewhat correlated, however (see table 6.4); the percentage of pups in packs on the Kenai Peninsula increased from 26% to 46% when wolf harvest was high and available biomass per wolf increased (Peterson, Woolington, and Bailey 1984). Autumn can be a critical period as pup food requirements are maximized (Mech 1970), but prey supply and vulnerability diminishes. Thus, where food is insufficient, it is usually fall, rather than summer, when pups starve (Van Ballenberghe and Mech 1975).

During winter, pup survival may differ from that of yearlings and adults in the same area. Sometimes it is higher (Ballard et al. 1987; Potvin 1988; Gogan et al. 2000); at other times, it is lower (Mech 1977b; Peterson, Woolington, and Bailey 1984; Fuller 1989b; Hayes et al. 1991). Overall, documented yearling and adult wolf annual survival rates where humans have not purposely tried to eliminate a high proportion of wolves (e.g., Bjorge and Gunson 1983; Gasaway et al. 1983) vary from about 0.55 to 0.85 (table 6.6). There is no evidence that female wolf survival differs from that of males.

Residency Status

In some studies, dispersing wolves seem to have had lower survival than wolves of the same age that remained in packs (Peterson, Woolington, and Bailey 1984; Messier 1985b; Pletscher et al. 1997). Dispersing wolves travel through new areas, where they are not familiar with the distribution of prey, and must work harder to maintain their condition. They also are less familiar with the distribution of other wolves that may kill them, and they may be more likely to be struck by a vehicle or to meet humans that may kill them (see below). Elsewhere, mortality did not differ by residency status (Fuller 1989b; Ballard et al. 1997; Boyd and Pletscher 1999), and in a population disrupted by control mortality, dispersing wolves survived better than residents (Hayes et al. 1991).

Mortality

Natural Factors

Wolves die of a variety of natural causes, including starvation, accidents, disease, and intraspecific strife (table 6.7). On Isle Royale, where no human-caused deaths occur,

TABLE 6.4. Ungulate biomass/wolf ratio, litter size, and percentage of pups in wolf populations during late fall to early winter for several areas of North America

Location	Ungulate biomass per wolf[a]	Litter size[b] $\bar{x}$	Litter size[b] N litters	Percentage of pups in packs	Number of pups per pack $\bar{x}$	Number of pups per pack N packs	Reference
Central Alaska	101[c]	4.6	7	—	—	—	Boertje and Stephenson 1992 (low prey density)
North-central Minnesota	161	6.1	5	46	3.2	36	Fuller 1989b
Northeastern Minnesota	164	—	—	49	2.6	24	Harrington et al. 1983 (Superior National Forest)
Interior Alaska	173	4.4	12	29	—	—	Gasaway et al. 1983
Algonquin Park, Ontario	175	4.9	10	32	1.9	—	Pimlott et al. 1969
Northeastern Alberta	186	4.8[d]	5	40[d]	—	—	Fuller and Keith 1980a
Southern Yukon	207	4.4	18	34	2.1	—	Hayes et al. 1991
Southern Quebec	236	5.6	10	—	—	—	Potvin 1988
Northeastern Minnesota	236	—	—	43	3.4	5	Van Ballenberghe et al. 1975
Isle Royale, Michigan	243	—	—	45	2.4	9	Peterson and Page 1988 (1984–1986)
Northeastern Minnesota	252	6.4	8	—	—	—	Stenlund 1955
Denali Park, Alaska	255	4.2[d]	23	43	3.8	91	Meier et al. 1995; Mech et al. 1998
Northwestern Alaska	267	5.3	22	—	—	—	Ballard et al. 1997
Central Alaska	285[c]	5.7	12	—	—	—	Boertje and Stephenson 1992 (medium prey density)
Northwestern Alberta	306	6.2	5	29	—	—	Bjorge and Gunson 1989
Denali Park, Alaska	334	—	—	39	5.4	5	Haber 1977
Kenai Peninsula, Alaska	345	5.0	5	36	3.8	15	Peterson, Woolington, and Bailey 1984
Jasper Park, Alberta	364	—	—	45	5.2	5	Carbyn 1974
Northwestern Minnesota	400	5.6[e]	8	44	2.7	21	Fritts and Mech 1981; Harrington et al. 1983
East-central Yukon	435	5.7	19	—	4.3	—	Hayes and Harestad 2000a,b
North-central Minnesota	617	—	—	45	3.3	3	Berg and Kuehn 1980, 1982
South-central Alaska	659	6.1	16	67	5.4	28	Ballard et al. 1987
Central Alaska	675[e]	6.9	15	—	—	—	Boertje and Stephenson 1992 (high prey density)

Summary and statistical test results

Test	r^2	d.f.	P	Regression
BMI/wolf and litter size	.38	16	.008	$y = 4.5 + 0.003x$
BMI/wolf and % pups in packs in fall	.32	15	.02	$y = 31 + 0.03x$
BMI/wolf and no. pups in packs in fall	.32	13	.04	$y = 2.15 + 0.004x$

Fetal litter sizes

Mean	5.5
N	164
No. studies	14

Source: Adapted from Fuller 1989b.

[a]From table 6.2, unless noted otherwise.

[b]Litter sizes are based on fetal observations unless noted otherwise.

[c]Average ungulate biomass estimate from Boertje and Stephenson 1992.

[d]Based on May–June observations.

[e]Based on May and July observations.

TABLE 6.5. Summer wolf pup survival and ungulate biomass in various areas of North America

Location	Summer pup survival rate[a]	Ungulate biomass per wolf[b]	Annual finite rate of increase	Annual adult survival rate	Reference
Northern Wisconsin	0.39[c]	400[d]	1.16	0.82	Wydeven et al. 1995
North-central Minnesota	0.48	161	1.02	0.64	Fuller 1989b
Southern Yukon	0.48	207[e]	0.97	0.56[e]	Hayes et al. 1991
Northwestern Minnesota	0.57[f]	378[d]	1.13	0.72[d,g]	Fritts and Mech 1981
Northeastern Alberta	0.69[f]	231	1.21	0.86[g]	Fuller and Keith 1980a (AOSERP area)
East-central Yukon	0.75[c]	435[h]	1.49	0.84[g]	Hayes and Harestad 2000a,b
Kenai Peninsula, Alaska	0.76	345	1.03	0.67[g]	Peterson, Woolington, and Bailey 1984
South-central Alaska	0.89	659[e]	0.88	0.59[e,i]	Ballard et al. 1987
Denali Park, Alaska	0.91[j]	334	1.06/1.20[k]	0.73	Mech et al. 1998

Summary and statistical test results

Test	r^2	d.f.	P	Regression
BMI/wolf and rate of increase	.003	8	.9	
BMI/wolf and adult survival rate	<.001	8	.94	
BMI/wolf and summer pup survival	.26	8	.16	
BMI/wolf and summer pup survival[l]	.69	6	.02	$y = 0.40 + 0.0008x$

"Summer" pup survival summary statistics

Mean	0.66
SE	0.06
SD	0.19
Range	0.39–0.91
No. studies	9

[a]Calculated from average litter size (fetal unless noted otherwise) and average number of pups in fall, from table 6.4.

[b]From table 6.2.

[c]Survival to or through winter.

[d]Wolf population expanding.

[e]Wolf population heavily exploited.

[f]Based on summer, not fetal, litter size.

[g]Survival rate for all ages combined.

[h]Wolf population recovering from heavy exploitation.

[i]Excludes mortality due to control program.

[j]Pup survival from May observations (not fetal) to average number of pups in August.

[k]Rate of increase based on late winter and early winter population estimates respectively.

[l]Omits Mech et al. 1998 and Wydeven et al. 1995.

annual mortality due to starvation and intraspecific strife (mostly related to relatively low food availability) ranged from 0 to 57% and averaged 23.5% (± 3.3 SE) from 1971 to 1995 (Peterson et al. 1998). In the Superior National Forest from 1968 to 1976, annual wolf mortality rates ran from 7% to 65%, and 58% of that mortality was natural, primarily due to fall pup starvation and intraspecific strife (Mech 1977b). In Denali National Park, Alaska, annual mortality averaged 27% and varied from 13% to 41% from 1986 through 1994; most (81%) of the mortality was natural (Mech et al. 1998). Elsewhere, average annual natural mortality has varied from 0% to 24% (average 11% ± 2% SE) in populations also subject to 4–68% human-caused mortality (see table 6.8 and below).

Diseases such as rabies, canine distemper, and parvovirus and parasites such as heartworm and sarcoptic mange might be important causes of death for wolves, but documentation is somewhat lacking (see Kreeger, chap. 7 in this volume).

TABLE 6.6. Age-specific dispersal rates of wolves and annual survival rates of nonresident wolves

Location	Dispersal rate Adult	Yearling	Pup	$\bar{x}$	Pack size	Survival rate Resident	Non-resident	Ungulate biomass per wolf[a]	Finite rate of increase	References
Northeastern Minnesota	3	83	35	—	—			145	1.04	Gese and Mech 1991 (1985–1989); Mech and Nelson 2000
Southern Quebec	—	—	—	—	5.7	(0.65)[b]		159	1.06	Messier 1985a,b (high prey area)
North-central Minnesota	17	49	10	35	6.7	0.67	0.52	161	1.02	Fuller 1989b
Northeastern Minnesota	7	70	19	—	6.4	(0.58)[b,c]		168	0.91	Gese and Mech 1991; Mech 1977a, 1986; Mech and Nelson 2000 (1969–1975)
Northeastern Minnesota	5	47	4	—	5.2			171	1.02	Gese and Mech 1991; Mech 1986; Mech and Nelson 2000 (1975–1985)
Southern Quebec	9[d]	76[d]	13[d]	—	5.6	(0.64)[b]		236	—	Potvin 1988
Northwestern Alaska	17	15	—	18	8.6	(0.55)[b,e]		267	0.88	Ballard et al. 1997
Kenai Peninsula, Alaska	(19)[f]		—	22	11.2	<10	0.38	345	1.03	Peterson, Woolington, and Bailey 1984
Northern Wisconsin	9	23	13	—	3.5	(0.82)[b,g]		400	1.16	Wydeven et al. 1995 (1986–1991)
Non-age-specific dispersal:										
Northwestern Alaska				13	8.4	(0.60)[b]		236	1.22	Ballard et al. 1997 (1987–1989)
Voyageurs Park, Minnesota				37	5.5	(0.75)[b]		277	—	Gogan et al. 2000
Denali Park, Alaska				28	6.9	(0.73)[b]		320	1.06/1.20[b]	Mech et al. 1998
East-central Yukon				25	6.8	(0.84)[b]		426	1.49	Hayes and Harestad 2000a,b

Summary and statistical test results

Test	r^2	d.f.	P	Regression
BMI/wolf and adult dispersal	.05	6	.64	
BMI/wolf and yearling dispersal	.44	6	.10	$y = 94 - 0.19x$
BMI/wolf and pup dispersal	.06	5	.65	

[a] From table 6.2, unless noted otherwise.
[b] Combined survival rate for all wolves > 6 months old.
[c] Apparent survival rate from Mech (1977a).
[d] Calculated from number of age-specific dispersals per month monitored (Potvin 1988, fig. 4).
[e] Includes period with rabies epidemic.
[f] Combined yearling and adult dispersal rate.
[g] Wolf population expanding.

TABLE 6.7. Known causes of deaths of wolves

Cause	Reference[a]
Accident	
Avalanche	Mech 1991b; Boyd et al. 1992
Starvation	Mech 1977a
Cliff fall	Child et al. 1978
Human (accidental)	
Train	L. D. Mech, personal observation
Vehicles	de Vos 1949
Human (purposeful)	
Aerial hunting	Stenlund 1955
Corrals	Young and Goldman 1944
Deadfalls	Young and Goldman 1944
Den digging	Young and Goldman 1944
Dogs	Young and Goldman 1944
Eagles (falconry)	Kumar 1993
Edge traps	Young and Goldman 1944
Fish hooks	Young and Goldman 1944
Guns	Young and Goldman 1944
Ice box trap	Young and Goldman 1944
Lassoing and hamstringing	Young and Goldman 1944
Piercers	Young and Goldman 1944
Pitfalls	Young and Goldman 1944
Poison	Young and Goldman 1944
Ring hunts and drives	Young and Goldman 1944
Salmon poisoning	Young and Goldman 1944
Set guns	Young and Goldman 1944
Snares	Young and Goldman 1944
Spears	Young and Goldman 1944
Steel traps	Young and Goldman 1944
Wolf knife	Young and Goldman 1944
Wildlife	
Bear, black	Joslin 1966
Bear, brown	Ballard 1980; 1982
Deer	Frijlink 1977; Nelson and Mech 1985
Moose	MacFarlane 1905; Stanwell-Fletcher and Stanwell Fletcher 1942
Muskox	Pasitchniak-Arts et al. 1988
Wolves	Murie 1944; Mech 1994a
Disease	
Canine parvovirus	Mech et al. 1997
Distemper	Grinnell 1904
Encephalitis	Young and Goldman 1944
Mange	Young and Goldman 1944
Rabies	Young and Goldman 1944; Chapman 1978

[a]First (and other significant) reference(s) in the scientific literature.

Human-Related Factors

Over the years, humans have devised many ways to kill wolves (see table 6.7). With focused wolf reduction programs, populations have been reduced over 60% in some years (table 6.8). In a few cases, site-specific control programs have eliminated entire packs (Fritts et al. 1992; Hayes et al. 1991; T. K. Fuller, unpublished data). Since wolves were legally protected in Minnesota and Wisconsin in 1974, human-caused wolf deaths have taken 13–31% of the studied populations there annually (Mech 1977b; Fritts and Mech 1981; Berg and Kuehn 1982; Fuller 1989b; Gogan et al. 2000). In Wisconsin, human-caused mortality declined after 1986, from 28% to 4%/year on average (Wydeven et al. 1995).

Many of the human-caused deaths in protected wolf populations occur because of depredations on livestock (see Fritts et al., chap. 12 in this volume). The government control program in Minnesota, for example, accounted for the deaths of 161 wolves there in 1998 (Mech 1998b), or about 7% of the population. Private citizens also kill wolves illegally to protect livestock, pets, and even deer (Fritts and Mech 1981; Berg and Kuehn 1982; Fuller 1989b; Corsi et al. 1999), or for other reasons. Wolves also are killed accidentally when hit by cars or trains, and are captured in traps or snares set for other wildlife species. Some are mistakenly shot as coyotes, but historically this source of mortality has been lower than intentional killing (Berg and Kuehn 1982; Fuller 1989b).

In examining factors correlated with the historic demise of wolves in Wisconsin, Thiel (1985) found that, in the era when wolves were persecuted by people, wolf populations did not survive where road densities exceeded about 1 km/km², because the roads made these areas accessible to people who killed wolves illegally or accidentally. Other studies supported that conclusion (Jensen et al. 1986; Mech, Fritts, Radde, and Paul 1988; Fuller 1989b). However, after public attitudes toward wolves changed (Kellert 1991, 1999) and wolves greatly increased and expanded their range, wolf populations have been able to survive even where road densities are higher than 1 km/km² (Mech 1989; Fuller et al. 1992; Berg and Benson 1999). Wolves are successfully occupying areas where road and human densities were thought to have been too high 10 years ago (Berg and Benson 1999; Merrill 2000).

Dispersal

Dispersal is a major means by which maturing wolves of both sexes leave their natal packs, reproduce, and expand their population's geographic range. Dispersers also fill any gaps in a population's territorial mosaic left by packs that have died or been killed out (see Mech and Boitani, chap. 1 in this volume). They also serve as sources for "sink" populations that could not sustain themselves without immigration from elsewhere (Mech 1989; Lariviere et al. 2000). Most often, dispersing wolves establish territories or join packs located anywhere from near their natal pack to some 50–100 km (30–60 mi) away (Fritts and Mech 1981; Fuller 1989b; Gese and Mech 1991; Wydeven et al. 1995). However, they sometimes move much longer distances; one disperser traveled at least 886 km (532 mi) away from its home area (Fritts 1983).

Several factors affect the timing and age of dispersal (Mech et al. 1998). Whether wolves pair and settle in a vacant area (Rothman and Mech 1979; Fritts and Mech 1981; Ballard et al. 1987) or join already established packs (Fritts and Mech 1981; Van Ballenberghe 1983b; Peterson, Woolington, and Bailey 1984; Messier 1985a; Mech 1987a) probably depends on relative prey abundance, the availability of vacant territories, and survival rates of breeding pack wolves.

Across populations, annual dispersal rates range from 10% to 40%, with most variation due to the irregular dispersal of nonbreeding wolves older than 1 year (see table 6.6). When food is sufficient, few yearlings may be driven to disperse ($r^2 = .44$, $P = .10$, d.f. = 7), although in unsaturated populations nonbreeding wolves may leave at younger ages to take advantage of breeding opportunities (Fritts and Mech 1981). Thus dispersal age is what varies most. Most adult dispersal (see table 6.6) consists of nonbreeding wolves 2 years old or older; these animals disperse at rates similar to those of yearlings (once breeding wolves are removed from the analysis).

Rates of Population Change

Potential

L. D. Mech once saw a vacant wolf territory in the Superior National Forest colonized by a new pair of radio-collared wolves one summer, and a year later the pair had produced seven pups. Wolves in that territory thus increased from two to nine, or 450%, in one year. Small

TABLE 6.8. Mean rates of population increase and annual mortality rates of exploited wolf populations in North America

Location	Number of years	Population increases		Annual mortality rate		Reference
		Finite rate	Exponential rate	Total	Human-caused	
Northwestern Alberta	2	0.40	−0.92	0.68[a]	0.68	Bjorge and Gunson 1983
Interior Alaska	4	0.76	−0.27	0.58	0.50	Gasaway et al. 1983; Ballard et al. 1997
Southwestern Manitoba	4	0.86	−0.15	0.56	0.32	Carbyn 1980
South-central Alaska	8	0.88	−0.13	0.45	0.36	Ballard et al. 1987
Northwestern Alaska	5	0.88	−0.13	0.45	0.27	Ballard et al. 1997
Northeastern Minnesota	6	0.89	−0.12	0.42	0.18	Mech 1977a, 1986 (1970–1976)
North-central Minnesota	3	0.93	−0.08	0.31	0.31	Berg and Kuehn 1982
Southern Yukon	6	0.97	−0.03	0.60	0.40	Hayes et al. 1991
Isle Royale, Michigan	4	0.95	−0.05	0.34	0.00	Peterson and Page 1988 (1983–1986)
Isle Royale, Michigan	9	1.01	0.01	0.21	0.00	Peterson et al. 1998
Algonquin Park, Ontario	5	1.01	0.01	0.37	0.24	Forbes and Theberge 1995
North-central Minnesota	6	1.02	0.02	0.36	0.29	Fuller 1989b
Kenai Peninsula, Alaska	6	1.03	0.03	0.33	0.28	Peterson, Woolington, and Bailey 1984
Denali Park, Alaska	8/9[b]	1.06/1.20	0.06/0.18	0.27	0.05	Mech et al. 1998
Southwestern Quebec	4	1.06	0.06	0.35	0.30	Messier 1985a,b (high prey area)
Northwestern Minnesota	5	1.13	0.12	0.28	0.17	Fritts and Mech 1981
Northern Wisconsin	6	1.16	0.15	0.18	0.04	Wydeven et al. 1995 (1986–1992)
Northeastern Alberta	3	1.21	0.19	0.15	0.15	Fuller and Keith 1980a (AOSERP area)
East-central Yukon	6	1.49	0.40	0.16	0.02	Hayes and Harestad 2000a,b

Summary and statistical test results

Test	r^2	d.f.	P	Regression
Total mortality and rate of increase	.7	18	< .001	$y = 1.4 − 1.17x$
Human-caused mortality and rate of increase	.6	18	< .001	$y = 1.2 − 0.93x$
BMI/wolf[c] and rate increase	.004	18	.80	
BMI/wolf and total of mortality	.07	18	.29	
BMI/wolf and human-caused mortality	.03	18	.46	
Human-caused mortality and wolf density	.04	18	.43	
Total mortality and wolf density	.003	18	.83	
Human and total mortality	.72	18	< .001	$y = 0.2 + 0.73x$
Human and natural mortality (no Isle Royale)	.14	16	.15	

Mortality summary statistics	Total	Human	Natural
Mean	0.37	0.24	0.11
SE	0.04	0.04	0.02
SD	0.15	0.18	0.08
Range	0.15–0.68	0–0.68	0–0.24
No. studies	19	19	17

[a]Mortality rate of early winter population; assumes all mortality is human-caused and summer survival of adults = 1.00.

[b]For spring and fall estimates, respectively. [c]Biomass Index from table 6.2 and Peterson et al. 1998; Mech 1977a; 1986.

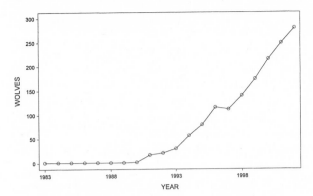

FIGURE 6.3. Trend of a colonizing wolf population in Michigan. Wolves spread from Minnesota into Wisconsin and by 1990 from Wisconsin into Michigan. A small proportion of Michigan wolves may also have immigrated from Ontario. This population trend represents an expanding population, not a density change. (From Michigan Department of Natural Resources 1997 and unpublished data.)

wolf populations have increased as much as 90% (from 30 to 57) from one year to the next (Michigan Department of Natural Resources 1997).

Populations that increase at such high rates are usually those that (1) have recently colonized or recolonized new areas (e.g., in Wisconsin, Michigan, and Yellowstone National Park), (2) have rebounded after deliberate removal of a subpopulation from within a much larger population (Ballard et al. 1987; Boertje et al. 1996; Hayes and Harestad 2000a), as Keith (1983) postulated, or (3) have been heavily harvested (see table 6.8) or devastated by disease (Ballard et al. 1997).

The population of wolves recolonizing the Upper Peninsula of Michigan increased by 90% in 1993 and at a mean rate of about 58%/year from 1993 through 1996 (fig. 6.3) (Michigan Department of Natural Resources 1997). In Bieszczady National Park, Poland, where wolves were heavily harvested, annual increase ranged from 15% to 53% (Smietana and Wajda 1997). The recolonizing Scandinavian wolf population increased an average of 29% from 1991 through 1998 (Wabakken et al. 2001). Given such high potential rates of increase and adequate food, wolf populations can more than double in 2 years.

Reproduction

The main component of dramatic increases in wolf numbers is reproduction, especially pup survival to fall. Because the single largest age class of wolves in a pack and in a population is the young-of-the-year, it is easy to see that annual change in pack or population size is most dependent on the fate of pups. In north-central Minnesota, annual wolf population change was highly correlated ($r^2 = .79$; $P < .02$) with the average number of pups per pack the previous fall (Fuller 1989b). Similarly, in Denali National Park, Alaska, from 1986 through 1993, 80% of the annual variation in spring-to-spring percent wolf population change was attributable to percent pup production and survival to the previous fall (Mech et al. 1998). In the Superior National Forest, percent change in the winter wolf population was correlated ($r^2 = .39$; $P = .05$) with an index of pup production in the previous summer (Mech and Goyal 1995).

It is interesting that in the unexploited wolf population on Isle Royale, where neither immigration nor emigration is a factor, the relationship between pup percentage (combined reproduction and pup survival to winter) and population change was only 35% (Peterson et al. 1998). Probably mortality influenced the dynamics of this isolated population more than did reproductive success because mortality rates varied more among years (Peterson et al. 1998).

Immigration

Depending on the reproductive status of wolf populations in surrounding areas, immigration could also provide a major component of population increase in areas where the potential for wolf density is relatively high. Especially in areas where intensive wolf control has been conducted, dispersal from adjacent populations can quickly resupply breeding pairs, which then produce large litters, recolonize the control zone, and within 2–4 years refill the area where wolves had been almost eliminated (Gasaway et al. 1983; Ballard et al. 1987; Potvin et al. 1992; Hayes and Harestad 2000a).

Mortality

For a wolf population, like any other wild population, mortality is a year-round process. Theoretically, as soon as wolf pups are born, mortality can begin, and no doubt this sometimes occurs. Because newborn pups remain in the den for their first 10–24 days (Young 1944; Clark 1971; Ryon 1977; Ballard et al. 1987), however, it is almost impossible to measure early pup mortality.

Most often the best that can be done, without disturbing the pups and the adults by invading the den—and thus possibly affecting the study results—is to count the pups when they first emerge from the den. By then, of course, some might already have died. Even regularly observing pups around a den is difficult or impossible in many areas. Thus data on wolf pup mortality often are based on a comparison of pup numbers around a den or rendezvous site in summer versus fall, when they can be seen and distinguished from the air (Fritts and Mech 1981; Fuller and Keith 1980a; Mech et al. 1998). Some pups can be identified from the air even in winter, but workers disagree on how consistently that can be done (cf. Van Ballenberghe and Mech 1975 and Peterson and Page 1988). An alternative approach is comparing fetal litter sizes (from carcasses) with average fall litter sizes in the same area (Peterson, Woolington, and Bailey 1984; Hayes et al. 1991).

Causes and rates of pup mortality were discussed above. However, we wish to emphasize here that most reported wolf mortality rates (see table 6.8) pertain to the population of wolves aged about 4–8 months and older. Mortality rates for younger pups usually remain unknown for two reasons: first, wolf pups are usually not large enough to be caught and radio-collared until they are at least 4–6 months old (Van Ballenberghe and Mech 1975), and second, many mortality studies depend on aerial observation of wolves, which is usually not feasible until winter. Reported annual mortality rates, then, are likely to be lower than if pups younger than 6 months old were included because pup mortality generally exceeds that of older wolves during late spring and summer.

Because of their high reproductive potential, wolf populations can withstand a high rate of mortality. On Isle Royale, where pups constitute a smaller percentage of the population than usual and wolves do not disperse to and from the island, annual natural mortality of adult-sized wolves averaged 15% when numbers were increasing or stable, 41% during population declines, and 24% when the population was stable (Peterson et al. 1998). Of course, in most populations, in which litters average five or six pups (Mech 1970), sustainable mortality can be even higher because this mortality keeps a higher percentage of the population breeding (Peterson, Woolington, and Bailey 1984) (see above).

If exploitation rates are too high to be fully compensated for by reproduction, however, the population should decline. Observed rates of increase are, as expected, negatively correlated with both total mortality ($r^2 = .70$; $P < .001$, d.f. = 19) and human-caused mor-

tality ($r^2 = .60$; $P = < .001$, d.f. = 19) (see table 6.8). These relationships suggest that, on average, wolf population size should stabilize ($r = .00$, $\lambda = 1.00$) with a mortality rate of 0.34 ± 0.06 SE, or a human-caused rate of 0.22 ± 0.08 SE, in late autumn populations of wolves (i.e., excluding mortality from birth to autumn). The slope of this relationship between intrinsic rate of increase and mortality, however, is fairly gentle. Thus even a considerable amount of additional mortality does not necessarily reduce the population so much that it cannot compensate or rebound through increased reproduction and/or immigration (Lariviere et al. 2000).

As recovering wolf populations continue to grow (see Boitani, chap. 13 in this volume), managers and the public will become increasingly interested in both sustainable levels of wolf harvest and the percentage take necessary to reduce a population or keep it stable (Mech 2001a). Because the above figures represent a general average, it is also useful to examine the results of specific studies dealing with the subject in order to better understand the high degree of variation that is possible.

Mortality Rates for Control and Sustainable Harvest

The maximum percentage of a wolf population that can be harvested annually on a sustainable basis is just short of the percentage that must be taken to control a wolf population. Thus we will discuss these two figures as one. By "control" we mean keeping a wolf population below the level to which it would rise without human-caused mortality.

Mech (1970, 63–64) suggested that over 50% of the wolves over 5–10 months old must be killed each year to control a wolf population, basing his estimate on Rausch's (1967) age structure data on over 4,000 harvested Alaskan wolves. Because these wolves were killed in fall and winter, the 50% kill figure would have been in addition to natural mortality from birth to 5–10 months of age. Keith (1983) reevaluated the proposed 50% kill figure by assembling data from several field studies. He concluded that the figure should be less than 30%, including a precautionary hedge. However, the data he used (Keith 1983, table 8) included populations that may have been stationary when 41% were taken, and declining populations with a 58–70% take. These data do not conflict with the 50% figure.

Other studies have directly measured the effects of various harvest rates. Gasaway et al. (1983) reported stable wolf populations after early winter harvests of 16–24%, but declines of 20–52% after harvests of 42–61%. On Alaska's Kenai Peninsula, wolf density dropped following two annual kills of over 40%, but increased 58% after a harvest of 32% (Peterson, Woolington, and Bailey 1984). Elsewhere in Alaska, Ballard et al. (1987) estimated that a 40% human take of the fall wolf population caused a decline. By reanalyzing their data, however, Fuller (1989b) concluded that the population would stabilize with a total overwinter mortality of 34%, including a fall harvest rate of 27%.

Fuller (1989b) also concluded that, in north-central Minnesota, a human-caused annual mortality rate of 29% resulted in a stable or slightly increasing wolf population. This finding is supported by similar work in Poland's Bieszczady National Park. There, annual mortality of 21–39% ($\bar{x}$ = 29%) of the 26–33 wolves in five packs, in a population with little or no immigration, resulted in a stable or slightly decreasing population (Smietana and Wajda 1997).

Additional evidence that human take of wolves can sometimes exceed 35% without permanently reducing a population comes from the annual rates of increase of the colonizing Michigan wolves discussed earlier. The figures imply that from 1993 to 1996, if humans had killed 58% of the wolves each year, the population would only have remained stable rather than continuing to increase.

These latter figures are much lower than one derived from a 5-year study in northwestern Alaska. There, wolf numbers remained stable at an annual winter mortality rate of 53%, including a minor amount of natural mortality (Ballard et al. 1997). The harvest in this study was biased toward nonreproductive animals, which may typify human-caused mortality; variation in this proportion probably helps explain the variation found among studies (Fuller 1989b).

The highest mean annual sustained human take of wolves was 74%, reported from the Portneuf Wildlife Reserve in Quebec, Canada, from 1990 to 1997 (Lariviere et al. 2000). The authors believed that the population there, and in nearby reserves, was being maintained by wolves immigrating from surrounding areas.

Causes of Variation in Sustainable Mortality Rate

Why all the variation in this important figure? Fuller (1989b, 25) noted that "these values may vary with the age and sex structure of the population. For example, a population with a high proportion of pups may be able to withstand somewhat higher overall mortality because pups (non-reproducers) may be more vulnerable to some harvest techniques and make up a disproportionate part of the harvest. Also, net immigration or emigration may mitigate effects of harvest." Ballard et al. (1997, 24) agreed, adding that "relatively small packs can sustain high mortality rates so long as reproductively active adults are not killed." These authors also stressed that multiple denning within individual packs (Harrington et al. 1982; Ballard et al. 1987; Mech et al. 1998) could significantly affect rates of increase and sustainable mortality rates.

Boiled down to its essence, the factor most critical to the annual percentage of a wolf population that can be killed by humans without reducing the population is the population's productivity. Clearly, if productivity is low, or immigration limited, then allowable harvest must be low as well, and field studies confirm that conclusion (Peterson, Woolington, and Bailey 1984; Fuller 1989b; Ballard et al. 1997). However, where productivity is average or high, a much higher take can be sustained, especially if the harvested or controlled population is surrounded by a population with a lower human take that can serve as a source population (Gasaway et al. 1983; Ballard et al. 1987; Hayes and Harestad 2000a,b; Lariviere et al. 2000).

Compensatory Mortality

As in other populations, the principle of compensation (Errington 1967) operates in wolf populations (Mech 2001a). This principle, simply stated, means that wolves killed by one factor cannot be killed by another. Thus, for example, if some wolves are killed by humans, there are fewer wolves that can starve or be killed by other wolves, the two main sources of natural wolf mortality (see above). Also, survival prospects may improve for the remaining wolves due to greater food availability or fewer conflicts, thus further reducing natural mortality. In addition, a population reduction can lead to increased reproduction through higher litter sizes and/or higher pup survival (see above). However, human-caused mortality can compensate for natural mortality even if it does not affect the rate of natural mortality (R. G. Haight, personal communication).

In Minnesota, where wolves were legally protected from human hunting by the federal Endangered Species Act and illegal human-caused mortality was 17–31% (Mech 1977b; Fritts and Mech 1981; Berg and Kuehn

1982; Fuller 1989b), and in Denali National Park, Alaska, where wolves in much of the area are protected by the National Park Service, some 10% of the population each year was killed by other wolves (Mech 1977b; Mech et al. 1998). However, in parts of Alaska where wolves are legally hunted and trapped by humans at a rate of 28–38%, very few wolves are killed by other wolves (Peterson, Woolington, and Bailey 1984; Ballard et al. 1987, 1997). Another indication of how natural and human-caused mortality compensate for each other can be found in the relationship between rates of total mortality and human-caused mortality, where human take replaces about 70% of mortality that would have occurred otherwise ($r^2 = .72$, $P < .001$, d.f. = 19; see table 6.8).

Because of the compensatory nature of various mortality factors, if humans wish to control a wolf population (keep it stable or reduce it), they must kill a higher percentage of wolves than would be expected to die of natural causes in a stable or increasing population. In addition, control measures must be carried out for several consecutive years, or the population bounces back.

A good example can be seen in the Tanana Flats area south of Fairbanks, Alaska. During a 7-year period, a population of 239 wolves was reduced to about 143 animals, but 337 wolves had to be killed to effect that reduction (Boertje et al. 1996). A take of 61% of the population in the first year and 42–43% of the remaining number in each of the next 2 years reduced the population, but a take of 38% in the fourth year then affected it little. A 19% kill in the fifth year was followed by a 51% population increase. In a review of wolf control in Alaska and elsewhere, a U.S. National Academy of Sciences committee concluded that wolf control is likely to be successful only if, among other things, "wolves are reduced to at least 55% of the pre-control numbers for at least 4 years" (National Research Council 1997, 184).

No doubt some of the resistance of the Tanana Flats wolf population to reduction came from dispersing wolves from the surrounding area (Hayes and Harestad 2000a). However, in addition, much of the high human kill, especially in the first 2 years, merely compensated for any natural mortality that might have taken place and fostered an increase in the percentage of breeders, as detailed earlier.

The relationship between dispersal and compensatory mortality involves two main aspects. First, an important factor in wolf dispersal is food competition. The greater the food competition, the more likely maturing wolves are to disperse (see above and Mech and Boitani,

chap. 1 in this volume). Human-caused mortality, especially when heavy, reduces food competition, which in turn reduces dispersal. Thus wolves that might have been lost from the population through dispersal remain, helping to compensate for the human-caused mortality. This mechanism operates, of course, only if the dispersal from the population would have exceeded the dispersal into it from the surrounding area.

The second aspect of the dispersal-compensation relationship involves the flux of lone, nonresident wolves circulating through the population. These animals are searching for opportunities to take up breeding positions by inserting themselves and a mate among the existing pack territorial mosaic, by joining an existing pack, or by colonizing areas at the edge of the population's range (see Mech and Boitani, chap. 1 in this volume). If a wolf population is subjected to human control, that creates vacancies both in packs and in territories that these floaters can fill. Thus the controlled population becomes a sink for wolves immigrating from as far as hundreds of kilometers away. These wolves then help compensate for the wolves being killed.

Wolf Population Models

Wolf populations have been the subject of several attempts to understand and predict their trends by mathematically modeling their dynamics. Efforts have ranged from the simple correlating of wolf density and prey biomass to highly complex computerized models that include consideration of age-specific mortality rates, varying reproductive rates, immigration, dispersal, spatial organization, and various life history relationships.

The first wolf population model was Keith's (1983) correlation of wolf density with prey biomass, which Fuller (1989b) extended and Dale et al. (1995) refined. While it is valuable for describing general relationships, this correlation's wide confidence intervals limit its value in predicting wolf density or population trends at specific locations and times (Mech et al. 1998). Population viability analysis (PVA) models based on computerized demographic simulations (Soulé 1980, 1987; Seal and Lacy 1989) have also been applied to wolves (USFWS 1989; Ciucci and Boitani 1991; D. R. Parsons, personal communication, cited in Fritts and Carbyn 1995). However, for several reasons, they have proved unsatisfactory or even misleading (Caswell 1989; Boyce 1992; Fritts and Carbyn 1995; White 2000). Similarly, a stochastic population model to predict wolf numbers in Yellowstone Na-

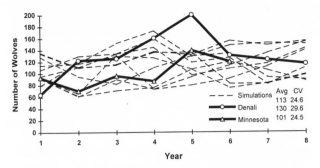

FIGURE 6.4. Annual variation in population sizes from ten random simulations of the Cochrane (2000) wolf population model compared with actual data from north-central Minnesota (Fuller 1989b) and Denali National Park, Alaska (Mech et al. 1998) for average fall / winter pack sizes extrapolated to equivalent fifteen-territory areas. The coefficient of variation for the model was based on thirty random simulations. (From Cochrane 2000.)

tional Park after reintroduction (Boyce 1990) proved problematic (see Boitani, chap. 13 in this volume), even though it incorporated prey dynamics.

More recent wolf population models have included consideration of wolf social structure, and one of them (Vucetich et al. 1997, 957) predicted that "demographic stochasticity may pose the greatest threat to small, isolated wolf populations," an interesting conclusion that has yet to be tested. Modern models include consideration not only of wolf social structure, but also of wolf population territorial structure (Haight and Mech 1997; Haight et al. 1998, 2002; Cochrane 2000). Obviously the greater the number of critical factors a model includes, the greater the chance that it will faithfully simulate reality. The Cochrane (2000) model, for example, tests well in generating wolf population trends similar to those actually described (fig. 6.4).

Natural Wolf Population Regulation

Although most wolf populations worldwide are strongly influenced by humans through control, harvesting, or illegal or incidental taking, valuable insight into wolf population dynamics can be gained by examining a few wolf populations under natural regulation. The key question to be asked about these populations is what is driving or regulating them.

Intrinsic Population Control

The idea that wolves might regulate their own numbers has been entertained by researchers as far back as

Adolph Murie (1944, 15) who wrote that "intraspecific intolerance may hold a population in check." Stenlund (1955), Mech (1966b, 1970), Pimlott (1967, 1970), Woolpy (1967), and Van Ballenberghe et al. (1975) have added to the speculation. As indicated by Mech and Boitani in chapter 1 in this volume, wolf populations are characterized by various mechanisms that might contribute to intrinsic regulation of their numbers: territoriality, intraspecific strife, high dispersal rates, and reproductive inhibition in subordinate pack members and lone wolves. Mech (1970) discussed how these various intrinsic mechanisms might work, and Pimlott (1970) concluded that such mechanisms operate to regulate wolf numbers at about 40/1,000 km^2 (102/1,000 mi^2).

However, as more and more data accumulated, it became increasingly clear that, while social factors might play some role, it was available food that ultimately limited wolf populations. Mech (1970, 317) mentioned this possibility—"Of course, if there were no other factors controlling a wolf population, ultimately it would be limited by a shortage of food"—and stressed that "food" meant "vulnerable prey." Van Ballenberghe et al. (1975, 36) stated similarly that "environments rich in food lower the threshold of such [intrinsic] mechanisms and are the ultimate factor accounting for the existence of dense wolf populations." Packard and Mech (1980, 1983) viewed the intrinsic limitation theory as "outdated" and reiterated the importance of vulnerable prey biomass. Keith's (1983) synthesis nailed the coffin of the intrinsic regulation theory shut with his findings of the importance of per capita prey biomass to wolf population dynamics.

Vulnerable Prey Biomass

Thus, although the intrinsic social characteristics of wolves modulate the way in which wolf populations react to their vulnerable prey biomass (Packard and Mech 1980), ultimately wolf numbers depend on the food supply, except when limited by disease. The combination of reproduction, mortality, immigration, and dispersal determines wolf population levels at any given time (see above). Changes in numbers from year to year depend on how these factors are affected by food, and that can vary over time or space.

Although the general relationship between food supply (prey biomass) and wolf numbers is strong (Keith 1983; Fuller 1989b; see table 6.2), it is also highly variable. Thus, for a given prey biomass, wolf numbers can vary as much as fourfold (Fuller 1989b). As indicated by Mech and Peterson in chapter 5 in this volume, not all prey animals are accessible to wolves. Rather, it is the older, weaker, younger, and otherwise vulnerable individuals in the prey population that wolves generally kill. Thus, although on average a large prey herd should contain more vulnerable members than a small one, it is possible for a large herd to include fewer vulnerable members than a small one, and vice versa. A large, increasing herd, for example, will be younger on average, and thus will include fewer vulnerable individuals, than a small, decreasing, and thus older, herd. Because prey condition is highly dependent on weather conditions (Mech and Peterson, chap. 5 in this volume), and weather is so variable, the annual percentage of a herd that is vulnerable is also highly variable.

Therefore, we agree that the proper unit of prey biomass to consider in analyzing wolf-prey interactions is vulnerable prey biomass. Although vulnerable prey biomass is an ever-changing proportion of a prey herd and is seldom measurable, the concept is critical to an understanding of wolf-prey relations and wolf population dynamics.

Fortunately, sometimes a single vulnerability factor is so overwhelmingly important that vulnerable prey biomass can be measured. For example, in one of the most elegant findings of any wolf-prey study done anywhere, the trend in numbers for the long-studied Isle Royale National Park wolf population (fig. 5.6) was found to depend on the number of moose (their sole year-round prey) 10 years old or older (Peterson et al. 1998). From 1959 to 1980 and from 1983 to 1994, the number of wolves was related to the number of old moose ($r^2 = .80$ and $.85$, respectively).

In Denali National Park, Alaska, where humans also have little effect on the wolf population, the trend in wolf numbers from 1986 through 1994 (fig. 6.5) was driven by snow depth, which influenced caribou vulnerability (Mech et al. 1998). Although Denali wolves fed primarily on moose, caribou, and Dall sheep, the vulnerability of caribou was the main determinant of wolf population change during the study. As snow depth and caribou vulnerability increased, adult female wolf weights also increased, followed by increased pup production and survival and decreased dispersal (Mech et al. 1998).

A more complicated situation existed in the east-central Superior National Forest of Minnesota (fig. 6.6). There wolves were protected by the Endangered Species

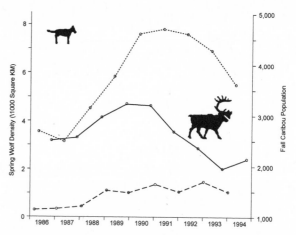

FIGURE 6.5. Wolf and caribou population trends in Denali National Park, Alaska, 1986–1994, in relation to snow-depth trend (bottom graph). Other important prey of these wolves are moose and Dall sheep, but wolf numbers changed in relation to caribou numbers. (From Mech et al. 1998.)

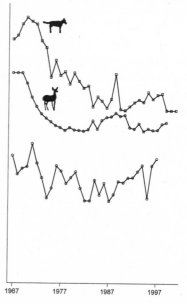

FIGURE 6.6. Wolf and white-tailed deer trends in the east-central Superior National Forest of Minnesota (Mech 1973, 1977b, 1986, 2000c) in relation to trend in cumulative 3-year snow depth (Mech et al. 1987 and unpublished). Deer population for 1967–1974 subjectively based; for 1975–1985 based on Nelson and Mech (1986a); and for 1986–1999 based on projections from correlation ($r = .31$; $P = .09$) between reported buck kill in Isabella area (M. Lennarz, personal communication) and winter deer counts in the same area (Nelson and Mech 1986a) for 1975–1976 through 1984–1985. Note that the wolf population trend followed the deer population trend through about 1984, when canine parvovirus affected the wolf population (Mech and Goyal 1995; L. D. Mech and S. M. Goyal, unpublished data).

Act of 1973 in August 1974. Although poaching by humans continued (Mech 1977b), it was not enough to reduce the population. From about 1966 to 1983, the wolf population trend (Mech 1973, 1977b, 1986) followed that of the white-tailed deer herd (cf. Mech 1986 and Nelson and Mech 1986a; Mech and Nelson 2000), which was related to winter snow depth (Mech and Frenzel 1971a; Mech and Karns 1977; Mech, McRoberts et al. 1987). Thus snow was seen as the driving force in the wolf-deer system (Mech 1990a). However, canine parvovirus (CPV), a new disease of domestic dogs that apparently began as a laboratory artifact, began to spread to the wolf population in the late 1970s, and by 1984 began influencing the wolf population (Mech and Goyal 1995), thereby at least partly unlinking wolf and deer numbers (Mech and Nelson 2000, fig. 2).

From the above three long-term investigations of wolf population trends, we can conclude that the factors that determine the annual changes in natural wolf populations are usually those affecting the availability of wolf prey. Prey availability is determined by prey density and vulnerability, so theoretically habitat quantity and quality, weather conditions, and competing predators (including humans) all can ultimately affect wolf numbers. The role of CPV in the Superior National Forest population can be considered an artifact. Wolves are well adapted to most diseases, and their populations are not usually affected by most of them (but cf. Carbyn 1982b), except perhaps by rabies (Chapman 1978) in the far north (see Kreeger, chap. 7 in this volume).

Cumulative Effects

Although there are no experimental studies per se on the accumulated effects of a variety of potentially negative factors on wolf populations (sensu Salwasser and Samson 1985; Weaver et al. 1987), none may be required. While it is difficult to test specific factors alone and in combination and then determine their joint effects on wolf demography (but see Cochrane 2000), the significant factors affecting wolf population trends are well studied (see above). The relative impact of concurrent effects can be deduced from the knowledge we currently have (e.g., relations between food abundance and productivity and survival, and human propensity to kill wolves). Simple demographic models (Keith 1983; Fuller 1989b) account for most observed differences in wolf population levels, and Cochrane's (2000) more compre-

hensive model allows more sophisticated exploration of the effects of multiple factors. Also, we know that wolves are very adaptable (i.e., can live under a great variety of circumstances) and can pass on adaptive behaviors to their offspring; few disturbances short of extensive killing affect wolf population demography.

Given the limits on our ability to assess the distribution, density, and mortality of wolves and their prey, our knowledge of the significant aspects of wolf biology is reasonably detailed and is unlikely to get much better. Even Geographic Information System-based landscape analyses (Mladenoff et al. 1995, 1999) of the Lake Superior Region, while confirming and refining earlier findings (Thiel 1985) about the relationship of wolf distribution to road densities and other landscape features, added little new information. Because of the overwhelming effort that has already gone into wolf studies, it seems unlikely that more complex landscape-explicit models (e.g., Weaver et al. 1987) will greatly improve our accuracy in predicting cumulative effects.

Persistence of Wolf Populations

Wolf populations possess a remarkable ability to persist so long as food supply is adequate, despite being subject to a number of possible mortality factors (see above). Even small populations of wolves have persisted and increased in several areas of the world during the last three decades. Because wolves were exterminated across almost all of the forty-eight contiguous United States, Mexico, and most of western Europe, many people think of the species as being fragile. However, it was primarily through poisoning that wolves were extirpated (Fritts et al., chap. 12, and Boitani, chap. 13 in this volume). Now that poison has been outlawed or greatly restricted in many areas, wolf populations are rebounding vigorously.

Examples of the wolf's ability to persist are many (table 6.9). Even the thoroughly inbred (Wayne et al. 1991) Isle Royale wolves, whose population once dropped to 12, have persisted for 50 years (see fig. 5.6). Italy's 100 wolves of the early 1970s have quadrupled and are recolonizing France (Poulle et al. 1999). Norway and Sweden's 1 or 2 wolves of the early 1970s numbered 90–100 in 2002 (Vilà, Sundqvist et al. 2002). Some wolf populations have been beset by canine parvovirus, depredation control, sarcoptic mange, lice, poaching, hunting, trapping, snaring, snowmobile pursuit, or aerial hunting.

TABLE 6.9. Persistence histories of small wolf populations

| Location | Lowest population | | Current numbers | Reference |
	Year	No.[a]		
Isle Royale, Michigan	1949		29[b]	R. O. Peterson, personal communication
Mainland Michigan	1991		216	J. Hammill, personal communication
Wisconsin	1975		266	A. Wydeven, personal communication
Minnesota	1953	450–700	2,450	Berg and Benson 1999
Montana	1985		80–100	USFWS 2000
Italy	1970	100	400–500	Chapter 13 in this volume
Norway/Sweden	1978		80–95	Chapter 13 in this volume
Riding Mountain National Park	1930		40–120	Fritts and Carbyn 1995
Kenai Peninsula, Alaska	1960		150–180	Fritts and Carbyn 1995

[a] A blank cell in this column indicates that the population began in the given year.
[b] In 2000.

Nevertheless, to our knowledge, humans have not caused any wolf population to permanently decline in the last 30 years.

Important Knowledge Gaps

Despite the thousands of scientific and popular articles that have been written about wolves (Fuller 1995c; Fuller and Kittridge 1996), and despite the fact that enough information is available to formulate general guidelines for their management, many aspects of wolf biology remain to be thoroughly described (Mech 1995e). However, given financial constraints and the nature of wolf conservation problems, we have identified a more limited set of research goals that, if carried out, would improve our understanding of wolf population change. These are vital areas of investigation precisely because they are difficult to study, but advances in technology and accumulation of anecdotal information leading to testable hypotheses will greatly assist research efforts.

Dispersal and Immigration

We do not have sufficient description and quantification of movements of dispersing wolves to predict when and where wolves will go (Merrill and Mech 2000). We need to know what constitutes barriers to dispersal, and whether for wolves there are such things as dispersal corridors.

Effects of Prey Types

Wolf density and territory size seem to be affected, in part, by prey type (see table 6.3). These effects probably result from differences in vulnerability due to prey behavior, but may also be related to the habitats in which certain prey reside.

Effects of Multiple Prey

Many wolves have been studied in essentially single-prey systems, and some information is available on functional responses of wolves to changes in relative prey densities (Dale et al. 1995). However, numerical responses to such changes in multi-prey systems have only begun to be studied (Mech et al. 1998).

Multiple Breeding Females

We do not fully understand why in some packs with more than two females of breeding age, two or more produce pups, while in others, only one does (Ballard et al. 1987; Mech et al. 1998), although food abundance probably plays a strong role. The wolf reintroduction to Yellowstone National Park (Bangs et al. 1998) affords an excellent opportunity for such studies.

Role of Disease

The effects of disease on the short-term and long-term status of wolves need to be investigated. Disease is a potentially great (e.g., Mech and Goyal 1995), but understudied, mortality factor affecting wolf populations (see Kreeger, chap. 7 in this volume). Additional collaborative work with veterinary scientists should prove invaluable in the future.

Wolf-Human Relationships

Continual assessment of human attitudes, beliefs, knowledge, and reactions to wolf recovery and control (Kellert 1985, 1999) are essential to successful wolf conservation programs because all wildlife management is, in essence, people management. In addition, better documentation of the lack of significant population effects on wolves caused by anthropogenic disturbances (e.g., snowmobile traffic [Creel et al. 2002], hiking near den or rendezvous sites, and other recreational activities) is needed. These disturbances often are proposed as being important, but probably influence populations only when they are very widespread and intensive, if at all (Thiel et al. 1998; Blanco et al. 1992; Merrill 2002).

Population Assessment

Standardized, accurate, and cost-effective methods of assessing wolf distribution and abundance need to be identified and implemented. Future planning for and monitoring of wolf recovery, harvest, and control depends critically on unassailable population assessment techniques.

Effects of Wolves on Low-Density Prey

In contrast to our knowledge of moose-wolf population dynamics (Gasaway et al. 1992), the precise role that wolves, and other predators such as bears or humans, play in limiting deer populations at relatively low densities (e.g., Mech and Karns 1977) is poorly known (see Mech and Peterson, chap. 5 in this volume). Experiments to assess this role are difficult, and long-term studies (e.g., Mech and Nelson 2000) in several study areas may be needed.

Pup Survival

Almost 30 years ago, Keith (1974) concluded that "the factors which produce [wolf pup] mortality during the first 5 months are almost wholly unknown. This is probably the single greatest enigma in wolf biology today." Though some strides have been made toward identifying these factors, this is still a much needed area of research.

7

The Internal Wolf: Physiology, Pathology, and Pharmacology

Terry J. Kreeger

IT WAS FEBRUARY IN MINNESOTA, the onset of wolf mating season. From a blind in an experimental enclosure, I observed a female curled up asleep on a snow-covered hillside. The breeding male of the pack appeared at the edge of a clearing and walked over to the female. The two sniffed noses. No sound was made, no tails were wagged, and no hair was ruffled, yet the male walked quickly away.

Any other observer would be left with a variety of hypotheses to explain this behavior, but I had an inside clue. Both of these wolves were equipped with internal radio transmitters that allowed me to monitor their heart rates. I saw that the female's heart rate was normal, whereas the male's was exceedingly high. I believe that the male was testing the sexual receptivity of the female and was anticipating success, whereas the female flatly wasn't interested. Thus, we can see that what goes on inside a wolf can tell us a great deal about what we observe on the outside.

Exploring inside the wolf is not a trivial task. It takes handling the wolf and either extracting a sample or attaching a device that provides data about the wolf's internal functioning. In the field, single samples can often be obtained when the wolf is captured for other purposes. The use of a remote-controlled recapture collar has allowed multiple recaptures of wild wolves (Mech and Gese 1992), improving our ability to study their physiology. In general, however, the need to handle wolves to obtain sufficient data almost mandates that captive wolves be used.

Wolves have been studied for decades, and much of what we know about their behavior and the biochemical basis for it comes from captives. Some argue that captivity somehow alters the wolf's physiology. However, years of studies comparing captive and wild wolves have failed to produce any evidence of a difference. Conversely, a strong argument can be made that the study of captive wolves is the best method to obtain sound physiological data. Science requires observation, developing a hypothesis, and testing that hypothesis. Hypothesis testing usually involves statistical analyses, which can be complicated by data that are highly variable. Captivity helps control such variability. In captivity, one can study animals based on age, sex, reproductive state, or time of day or season. All these factors can change physiological samples. One can also test the same animal repeatedly or many animals at the same time. Such control over sampling helps yield sound results.

In this chapter, I will examine the life cycle of the wolf and identify what we know about wolf physiology as it applies to the various stages of reproduction, birth, growth, and survival. Diseases and other pathological threats to the wolf's survival will also be discussed. Interwoven will be information about pregnancy diagnosis, nutrition, chemical restraint, and health evaluation. Behavioral aspects of wolf reproduction are covered in chapter 2, and communicative aspects in chapter 3 of this volume.

Reproductive Physiology

Female Reproduction

A great deal is known about the reproductive physiology of the domestic dog (Anderson 1970; Concannon et al.

1989), and much of this knowledge can be extrapolated to the dog's progenitor, the wolf. However, a few differences between the two canids exist. Whereas the female dog may come into estrus twice a year and at any time during the year, the female wolf is strictly monestrous and highly photoperiodic (Hayssen et al. 1993).

Some wolves have bred in captivity at 10 months of age (Medjo and Mech 1976; Zimen 1976; Seal et al. 1979), as do many dog breeds, but the only possible record of breeding so early in life in the wild is from the restored wolf population in Yellowstone National Park, where prey is unusually abundant (D. W. Smith, personal communication). Most wolves do not come into estrus or breed until 22 months of age or older (Murie 1944; Rabb et al. 1967; Rausch 1967; Lentfer and Sanders 1973; Hayssen et al. 1993), and one wolf had not ovulated by 34 months of age (Mech and Seal 1987).

The maximum breeding age for female wolves is not known. The oldest female recorded breeding in the wild was at least 10 years old (Mech 1988c). Females in captivity have whelped at 14 years of age (M. Callahan, personal communication), but litter sizes decline from age 9 on (C. S. Asa, personal communication). In the wild, older breeding females are sometimes replaced by their daughters. In one such case, the displaced mother left the pack (Mech and Hertel 1983); in another, she remained for two more breeding seasons and then disappeared (Mech 1995d).

The female reproductive cycle is characterized by several distinct stages: anestrus, proestrus, estrus, and metestrus (see Packard, chap. 2 in this volume). Each stage is characterized by cellular changes in the vagina (table 7.1). In North America, wolves generally come into estrus between late January and early April, with those far-

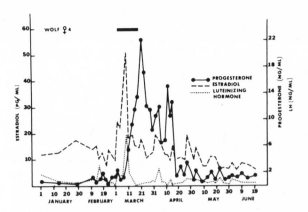

FIGURE 7.1. Reproductive endocrine profile of the wolf. (Adapted from Seal et al. 1979, 1987.)

ther north tending to cycle later (Mech 1970, 2002). In India, the wolf reproductive cycle occurs about 3 months earlier (Jhala and Giles 1991; Kumar and Rahmani 2001). When dogs are housed near wolves, they can become as photoperiodic as wolves, coming into estrus in January–February and again very predictably in August–September (Kreeger and Seal 1992). Estrus is synchronized in bitches housed together (Kreeger and Seal 1992), so the dogs in this study may have been receiving cues for the onset of estrus from the wolves.

Proestrus

Proestrus length in captive wolves averages 15.7 ± 1.6 days (mean ± SE), about double the average for dogs (Seal et al. 1979; Concannon et al. 1989). However, in wild wolves, this stage can last up to 45 days (Young and Goldman 1944), and based on cellular changes in the vagina, it can last 60 days (see Packard, chap. 2 in this volume). Behaviorally, proestrus is characterized by increased attractiveness to males and proceptive behavior, but refusal to allow mounting. This behavior is attributed to increased blood concentrations of estrogen. Vaginal bleeding is commonly seen, but close inspection or vaginal swabs may be required to confirm this in some animals.

The proestrous hormone profile of the wolf is similar to that of the dog. There is a rise in estradiol-17β varying between 10–20 pg/ml initially and peaking at 30–50 pg/ml late in proestrus (fig. 7.1). Progesterone remains low, generally below 1 ng/ml, but it occasionally increases to 3 ng/ml. Like dogs, wolves may demonstrate minor luteinizing hormone (LH) surges 9–24 days prior

TABLE 7.1. Vaginal cytology of the wolf

Stage	Description
Proestrus	Cornifying cells; increased erythrocytes; decreased leukocytes
Estrus	Uniform cornification of epithelial cells with pyknotic nuclei; disappearance of erythrocytes and leukocytes
Metestrus	Abundance of leukocytes; reappearance of round, noncornified epithelial cells and neutrophils
Anestrus	Noncornified epithelial cells having light blue cytoplasm with distinct, uniformly sized nuclei; leukocytes in relatively high numbers.

to the major preovulatory LH surge during estrus (Seal et al. 1979, 1987).

Estrus

In one study of captive wolves, estrus averaged 9.0 ± 1.2 (mean ± SE) days (Seal et al. 1979), but it lasted 15 days in another (Zimen 1976). Estrus in dogs averages 1 week (Concannon et al. 1989). Little is known about the length and variation of estrus seasons in wild wolves. However, it is known that the same female in different years can breed over a span of a month (Boyd, Ream et al. 1993). This finding suggests that estrus in a population, although not necessarily in any individual, might last at least a month, as also implied by Rausch (1967).

As in dogs, estrus in the wolf is characterized by positive sexual behavior toward the male, including standing firmly in place, presentation of the vulva in a lordosis-like manner, reflex deviation of the tail to one side, and permitting mounting and pelvic thrusting. This behavior is attributed to a decline in estrogen and a rise in progesterone (Concannon et al. 1977).

Preovulatory LH surges in wolves measure 5–15 ng/ml (Seal et al. 1979), somewhat lower than in dogs (Concannon 1986b), and can last 1–3 days in both canids. The LH surge enlarges and luteinizes mature ovarian follicles, resulting in ovulation. In the process, it transforms the estrogen-secreting follicles into progesterone-secreting corpora lutea (Concannon et al. 1977). In dogs and wolves, the preovulatory LH surge usually occurs within a day of the transition from behavioral proestrus to estrus. However, aggressive male dogs may mount females as early as 4–5 days before the LH peak, and some females may refuse to mate for 6–7 days after the peak (Concannon et al. 1989). After the LH peak, estradiol-17β concentrations fall precipitously to about 10–20 pg/ml, and progesterone begins rising rapidly above the baseline of 1–3 ng/ml (see fig. 7.1).

Metestrus

Metestrus (or diestrus) generally encompasses the luteal phase of the pregnant and nonpregnant wolf and lasts until parturition or the decline of progesterone to basal concentrations. Estradiol-17β fluctuates from 10 to 30 pg/ml (see fig. 7.1), but in contrast to dogs, no prepartum rise in estradiol-17β has been observed in wolves (Seal et al. 1979). Progesterone peaks 11–14 days after the LH peak at 22–40 ng/ml, which is lower than in dogs (Concannon 1986b). Elevated progesterone is main-

tained for 56–68 days, the length of gestation (Concannon et al. 1989). Prolactin (PRL) increases slowly throughout metestrus in both the pregnant and non-pregnant wolf.

Anestrus

Wolves are anestrous from June to December (Seal et al. 1987), except in India (see above), and this period is generally one of endocrine quiescence.

Male Reproduction

Like the female, the male wolf may sometimes be physiologically capable of breeding at 10 months of age, but rarely does so (Medjo and Mech 1976; Mitsuzuka 1987; Hayssen et al. 1993). The testicles are small at 10 months and may still appear undeveloped at 22 months (Mitsuzuka, 1987). Semen samples from 10-month-olds show a high percentage of immature spermatozoa (Mitsuzuka, 1987).

The age of reproductive senescence is not known for the male. A captive male sired pups at 15 years of age, the last time he had a female available (E. Klinghammer and P. A. Goodmann, personal communication), but one proven breeding male no longer was producing sperm at age 14 (C. S. Asa, personal communication). In the wild, a male known to be at least 11 years old sired two pups (Mech 1995d).

Whereas the male dog is considered to be reproductively viable throughout the year, spermatogenesis is seasonal in the male wolf. The male wolf demonstrates a photoperiodic reproductive cycle relative to LH and testosterone secretion and testicular morphology. In North America, testosterone fluctuates from 10 to 560 ng/dl during the year, with zeniths occurring from December to March and nadirs from June to September; LH stimulated by luteinizing hormone-releasing hormone (LHRH) shows a similar cycle (Asa et al. 1987; Seal et al. 1987).

This cyclic testosterone production explains why sperm production also is cyclic, reaching a maximum during the breeding season (Gensch 1968; Zimen 1971; Mitsuzuka 1987). During January through March, samples collected from one 5-year-old wolf showed volumes averaging 1.6 ± 0.8 ml (range = 0.2–4 ml), total sperm counts averaging 372.1 ± 222.0 × 10⁶/ml (range = 125.0–1005.4 × 10⁶/ml), and motility from 90% to 95% (Mitsuzuka 1987). Sperm abnormalities noted included

TABLE 7.2. Testicular measurements of a 5-year-old wolf before, during, and after the breeding season

	September	January	April
Length (cm)	4.0	3.3	5.2
Width (cm)	3.0	3.5	2.5
Volume (cm³)	19.55	35.29	10.90

Source: Mitsuzuka 1987.

bent neck and curled tail, protoplasmic droplets, and acrosomal abnormalities (Mitsuzuka 1987). (Studies that have examined sperm characteristics in red wolves [see Phillips et al., chap. 11 in this volume] include those of Koehler et al. 1998 and Goodrowe et al. 1998.) Measurements of testicular lengths for this animal during the breeding season were significantly different from those during summer (table 7.2, Mitsuzuka 1987). Testicular sizes in two other captive males also showed a strong circannual rhythm when measured over a 3-year period (Asa et al. 1987).

Fertilization, Gestation, and Parturition

Little is known about oocyte maturation and fertilization in the wolf, but what is known about the domestic dog probably applies to the wolf. Dog spermatozoa may readily penetrate immature oocytes and form male pronuclei, but actual fertilization (i.e., fusion of pronuclei) must await oocyte maturation. Oocytes do not mature until 2–3 days after ovulation, which occurs about 2 days after the LH surge (Concannon et al. 1989). The fertile life of mature oocytes probably lasts another 2–3 days, since matings in late estrus, 7–8 days after the LH peak, are often fertile.

Dog matings 9–10 days after the LH peak rarely result in pregnancy, and when they do, produce litters of only 1–2 pups after an apparent gestation of 55–57 days (Concannon 1986a). Matings more than 2 days before the LH peak are rarely fertile and result in apparent gestations of 68 days or longer. Dog spermatozoa can remain fertile in the female reproductive tract for 6–7 days (Concannon et al. 1983). Peak fertility for natural matings occurs 0–5 days after the LH peak (Holst and Phemister 1975).

Pregnancy lasts 60–65 days in wolves and 65 ± 1 days in dogs (Seal et al. 1979; Concannon et al. 1989; Hayssen et al. 1993). A day or two before parturition, progesterone falls below 3 ng/ml, and PRL increases (Seal et al. 1979; Kreeger, Seal et al. 1991). Prolactin concentrations are much higher in the postpartum lactating female (20–50 ng/ml) than in the postluteal nonpregnant female (10–20 ng/ml; Kreeger, Seal et al. 1991). In the dog, and presumably the wolf, PRL is a required luteotropin, and its suppression by bromocriptine will cause luteolysis and terminate pregnancy (Concannon et al. 1987). After weaning, PRL falls rapidly to basal concentrations of around 3 ng/ml.

Wolf litter sizes average 6 pups (range = 1–13) that weigh 300–500 g (11–18 oz) at birth (Gavrin and Donaurov 1954; Mech 1970; Hayssen et al. 1993). No comprehensive studies detail the sex ratio of wolf pups at birth, but there is no reason to suspect a skewed ratio (Crisler 1958; Rutter and Pimlott 1968; Mech 1970; but cf. Mech 1975). There does appear, however, to be a bias toward males in adults (Mech 1970).

Pregnancy Diagnosis

Pregnancy can be diagnosed in dogs by a variety of mechanical tests: uterine swellings are generally palpable by days 20–25 of gestation; ultrasound can detect amniotic cavities at days 19–22 and fetal heartbeats at days 22–25; and radiography can detect fetal skeletons at days 45–48 (Concannon et al. 1989). These methods and periods generally apply to wolves as well (T. J. Kreeger, unpublished data).

Because pregnant and nonpregnant wolves in the luteal phase have similar endocrine profiles, pregnancy diagnosis based on hormone concentrations, such as progesterone, is complicated. Pregnancy can be diagnosed in wolves within a week of breeding by stimulating PRL with the opioid antagonist naloxone. In pregnant wolves, PRL increases twice as much as in nonpregnant wolves in response to naloxone (Kreeger et al. 1993).

Pregnancy, but not viable fetuses, can also be inferred retrospectively by examining the uterine horn for placental attachment sites ("uterine scars"). Pre- and postimplantation losses resulting in resorption sites are common in dogs (Tsutsui 1975) and wolves (Rausch 1967; Hillis and Mallory 1996a).

Nipple (*papilla mammae*) measurements can also be used to differentiate between female wolves that have bred and those that have not. Inguinal nipple width plus height in wolves that have pups of the year, or have had pups previously, measures 1.6 ± 0.018 cm (range 1.34–2.12); in yearlings or adults that have never bred, this measurement is 0.6 ± 0.009 cm (range 0.44–0.72; Mech et al. 1993). Although not a fail-safe index, nipple measurements appear to be useful for assessing breeding status in the field.

Reproductive Success

Although a pack of wolves may contain several reproductively mature females (Rausch 1967), usually only one female reproduces (see Mech and Boitani, chap. 1, Packard, chap. 2, and Fuller et al., chap. 6 in this volume). Studies of captive wolves to determine the causes of such reproductive "failure" implicated social behavior in the inhibition of breeding by subordinate animals (Packard et al. 1985). Physiological suppression of subordinate wolves was ruled out because five adult females that did not reproduce had ovulated, and when social circumstances were altered, as by the death of a parent, these subordinate offspring copulated with parents or siblings.

However, physiological suppression of reproduction apparently can occur under high stress. Wild, mature female wolves brought into captivity have remained anestrous for over 2 years (Seal et al. 1979; T. J. Kreeger, unpublished data). Such stress suppression of reproduction has been observed in a variety of species, including humans (Whitacre and Barrera 1944; Christian 1971), and hormones elaborated by the hypothalamic-pituitary-adrenocortical axis have long been implicated as mediators of this suppression. Adrenocorticotropin hormone (ACTH) and glucocorticoids such as cortisol can block LH release in a variety of species, although not in wolves (Kreeger et al. 1992). However, LH in wolves is responsive to the opioid antagonist naloxone. Thus it was hypothesized that stress suppression of wolf reproduction could be explained by endogenous opioids suppressing LHRH in the hypothalamus, resulting in depressed LH secretion and, thus, failure to ovulate (Kreeger et al. 1992). This mechanism may have implications for the captive breeding of wolves.

Stress can also cause progesterone secretion, presumably from the adrenal gland, in the wolf and other species (Kreeger et al. 1992). The function of this progesterone is unknown, but it could serve to maintain pregnancy under adverse conditions. Cosecretion of adrenal progesterone with glucocorticoids could neutralize the pregnancy-terminating effect of the latter (Plotka et al. 1983).

Nutrition also plays a role in the suppression of reproduction in subordinate wolves. Furthermore, social and nutritional factors may interact considerably. In some populations, several pack members may become pregnant (Rausch 1967), but apparently through fetal resorption due to nutritional constraints, the subordinate females fail to produce pups (Hillis and Mallory 1996a). During years of better nutrition, subordinates succeed in carrying their litters to term, and multiple litters per pack are born (see Mech and Boitani, chap. 1, Packard, chap. 2, and Fuller et al., chap. 6 in this volume).

Sterilization

Recently sterilization has been suggested as a possible complementary means of wolf control (Mech 1995e; Mech, Fritts, and Nelson 1996; Haight and Mech 1997; Haight et al. 2002), and some experimentation with surgical sterilization has been conducted (Spence et al. 1999). Vasectomy is fairly simple to conduct in the field, but depending on the goal of the control program, there may be reasons to attempt female sterilization as well. Any of the sterilization methods used with dogs can be applied to wolves, although ovariohysterectomies may be more difficult with immature wolves because of the difficulty of locating the ovaries.

The Physiology of Growth and Development

Energy is required to survive, grow, and reproduce. Some of the energy consumed as food is used merely to maintain the basic processes of life, such as respiration, circulation, and temperature regulation. Surplus energy can be used for growth and eventually for reproduction. In this section, we will explore what is known about factors that affect nutritional intake, energy expenditure, and metabolism in wolves throughout their lives, from the nutritional requirements and care of pups through the biochemical bases of the feeding strategies of adult wolves. Wolf digestion and nutrition are also discussed by Peterson and Ciucci in chapter 4 in this volume.

Lactation and Weaning

Wolf pups are weaned from their mother's milk at 5–9 weeks of age (Young 1944; Mech 1970; Hayssen et al. 1993; Packard et al. 1992), although they will continue to solicit

TABLE 7.3. Milk composition of the wolf (n = 4)

Total solids	23.50%
Water (by difference)	76.60%
Fat	6.60%
Solids-not-fat (by difference)	16.90%
Ash	1.35%
Lactose	2.95%
Protein	12.40%
Specific gravity, 22°C	1.012
pH	6.23
Calories	144 kcal/100 g

Source: Lauer et al. 1969.

nursing from their mother for several weeks thereafter (Packard et al. 1992; see Packard, chap. 2 in this volume).

In general, wolf milk is similar to that of the domestic dog, but it is higher in protein and lower in fat than in some dog breeds, such as the husky or beagle (table 7.3) (Lauer et al. 1969; Anderson 1970). Since wolf milk contains more arginine than some commercial puppy milk replacers, wolf pups raised by humans and fed a milk formula may develop cataracts if the formula is not supplemented with arginine (1 g/100 g formula). A lactose supplement (15 g/100 g formula) can be substituted for the arginine because lactose apparently assists in the absorption of arginine (Vainisi et al. 1981). Gelatin is a common arginine source that can be mixed with puppy milk replacer.

Role of Prolactin

Prolactin (PRL) from the pituitary gland is required not only for duct and lobule growth in the mammary gland, but also for maintenance of lactation (Anderson 1970). Prolactin, however, may have other roles in the development of wolf pups. In wolves and other wild animals, as well as dogs, PRL has a definite circannual rhythm in both sexes, rising in spring with the onset of long days and decreasing during the short days of winter (Bubenik et al. 1983; Kreeger, Seal et al. 1991; Kreeger and Seal 1992).

Prolactin can induce or maintain parental behavior in mammals (Dixson and George 1982; Bridges et al. 1985). All wolves in a pack, including males, demonstrate parental behavior in the form of pup feeding (see Packard, chap. 2 in this volume). Expectant mothers even elicit feeding behavior from their mates prior to parturition (Fentress and Ryon 1982). The circannual PRL rhythm

in wolves may not only serve as a cue for breeding activity (Mirachi et al. 1978), but may also induce parental behavior in all pack members, which enhances pup survival (Kreeger, Seal et al. 1991).

The photoperiodic rhythm of PRL in wolves may also be associated with coat shedding (Martinet et al. 1984). Wolves shed their thick, insulating undercoat in spring as day length and PRL concentrations increase. Conversely, in fall, when day length and PRL decrease, the summer coat is replaced by the thick winter pelage. Any reproductive or pelage effects that PRL may have, however, are most likely a function of a complex relationship with several other hormones, such as melatonin. Since melatonin is also highly photoperiodic, it could be acting as a primary messenger mediating the effect of photoperiodicity on seasonal events, in which PRL acts as a modulating, secondary messenger (Bubenik et al. 1986).

Pup Growth and Development

Most of the following information about wolf pup growth and development is based on material presented by Mech (1970:123–125). Information about the social development of wolf pups is presented by Packard in chapter 2 in this volume.

Newborn pups weigh 300–500 g (see above), are blind and deaf, and are darkly furred. Their ears are small, and their noses are blunt or pugged. Their eyes open at about 10–14 days of age, and often are bluish, gradually turning to brownish or yellow, but sometimes remaining blue (L. D. Mech, personal communication).

Based on measurements of well-fed captive pups (Pulliainen 1965), three growth periods are recognized: (1) 0–14 weeks, maximum growth rate (1.2 kg or 2.6 pounds/week for females and 1.5 kg or 3.3 pounds/week for males); (2) 15–27 weeks, rapid growth (0.6 kg or 1.3 pounds/week for both males and females); and (3) 28–51 weeks, slow growth (0.03 kg or 0.07 pound/week for females and 0.20 kg or 0.40 pound/week for males). Growth ends at about 12–14 months of age, when the growing points (epiphyses and diaphyses) of the radius and ulna fuse (Rausch 1967). Of course, wolves can gain weight by adding fat and muscle throughout their lives (Kreeger et al. 1997).

The rates of growth and development of wolf pups in the wild depend on how well the pups are fed. In northeastern Minnesota during a period of prey decline, for example, growth rates of 0.35–1.6l kg (0.77–3.54 pounds)/week were observed in pups aged 8–28 weeks

(Van Ballenberghe and Mech 1975). These pups weighed from 13.2 kg (29.0 pounds) to 24.5 kg (54 pounds) in early November, although one weighed 32.7 kg (72.0 pounds) in October. In the reintroduced Yellowstone National Park population, for which food is abundant (see Fritts et al., chap. 12 in this volume), a pup born in 1997 weighed 52.3 kg (115 pounds) on 12 March 1998 (D. W. Smith, personal communication).

The rates of physical developmental processes, such as tooth replacement, generally parallel those of weight gain (Van Ballenberghe and Mech 1975). In the northeastern Minnesota population, most pups caught before mid-September retained their deciduous upper canines, but eruption of permanent canines was seen as early as 18 August. Upper canines were fully in place by November. Wolves with sharp upper canines less than 20.5 mm (0.8 in) long can be considered pups (Van Ballenberghe and Mech 1975).

Once adult dentition is in place at about 6–7 months of age, a wolf's teeth begin to wear. However, the amount and type of wear is highly variable, depending on what the wolf eats (bone, stomach contents, etc.). Neverthe-

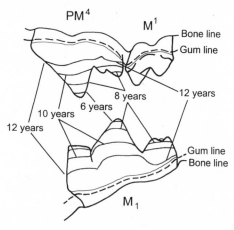

FIGURE 7.3. Progressive wear on wolf carnassials (upper premolar 4 and lower molar 1) in 2-year increments from less than 6 to more than 12 years of age. Wear is visible on tips of major prominences at 5 years of age, and profiles flatten slightly by 6 years. Deep wear on the posterior cusp of the lower carnassial after 10 years of age results from occlusion with the first upper molar, not the upper carnassial. (From Gipson et al. 2000.)

less, as a wolf ages, the amount of wear generally follows a pattern that allows a wolf's age to be estimated (fig. 7.2 and 7.3) (Gipson et al. 2000).

Basal Metabolic Rate

How fast a wolf grows and how large it eventually becomes is a function of many factors, such as genetics (mature body size and growth rate have a high degree of heritability), nutrition (how much and what quality of food is available), and disease (parasites compete with the wolf for nutrients). Underlying all these factors is how well a wolf can transform food into energy (see Peterson and Ciucci, chap. 4 in this volume). This energy is necessary to sustain all the processes of life; when the energy of absorbed nutrients is insufficient, body tissues are used for this purpose, and the animal literally consumes itself to make energy available for vital life processes.

Metabolism is the sum of all the chemical changes occurring in tissue, consisting of anabolism and catabolism. The energy of carbohydrates, fats, and proteins is liberated by oxidation. The "joule" (J) is the standard unit of energy used in metabolic studies. The average values for oxidation of various classes of foods by mammals are 0.98 J/g for carbohydrates, 2.22 J/g for fats, and 1.03 J/g for proteins.

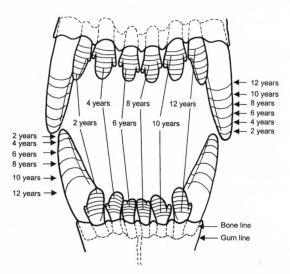

FIGURE 7.2. Progressive wear on wolf incisors and canines in 2-year increments from less than 1 to more than 12 years of age. Wear on incisors typically progresses beyond the lobes on the first two upper and lower incisors at 8 years of age, leaving approximately 5 mm of enamel. At 10 years of age, 2–4 mm of enamel remain on the first and second incisors. Length of canines is reduced 30–50%, with 10–16 mm of enamel remaining. Beyond 12 years of age, incisors may be worn to the roots, with a few peglike stumps projecting above the gum line, or the gums may cover the roots. Length of canines is reduced 50% or more, with less than 10 mm of enamel remaining. (From Gipson et al. 2000.)

An animal's basal metabolic rate (BMR) gives insight into its energy requirements. Knowledge of a species' BMR allows us to analyze predator-prey interactions quantitatively as well as understand how an animal responds to its environment. The BMR essentially represents the metabolism of an animal at rest in a thermally neutral environment. Under these circumstances, BMR, as a function of metabolic weight ($kg^{0.75}$, see below), is reasonably uniform.

Wolf BMR averages 158.8 ± 17.9 kJ/hr (converted from liters of oxygen per hour as measured by Okarma and Koteja [1987]). Adult males (172.5 ± 9.4 kJ/hr) have a higher BMR than adult females (153.8 ± 11.8 kJ/hr) or pups (142.3 ± 21.2 kJ/hr). These BMR values are about 27% higher than predicted by theory for placental mammals, but 15% lower than predicted for carnivores as a group (Hayssen and Lacy 1985). The BMR of desert wolves (C. l. pallipes) is also lower than predicted (Afik and Pinshow 1993).

Factors Affecting BMR

Body size can affect metabolic rate. Although a mouse produces less total heat than an elephant, the mouse produces significantly more heat on a per weight basis because the mouse has a much higher BMR. The metabolic level of adult warm-blooded animals, from mice to elephants, averages 293 $kJ/kg^{0.75}$/day. Since animals expend energy in proportion to their weight raised to the 3/4 power, they require dietary energy on the same basis; that is, small animals need more calories per kilogram of weight than do large animals. Thus, if wolves had the metabolic rate and energy requirement of shrews, they could not possibly catch enough prey to survive.

Heart rates and body temperatures reflect BMR. Small animals have higher heart rates and body temperatures than do large animals. Wolf heart rates (45–115 beats/min; Folk and Hedge 1964) are similar to those of large dogs (Kirk and Bistner 1981). To obtain values unbiased by physical restraint or drugs, Kreeger, Kuechle et al. (1990) used implanted radio transmitters to measure both heart rates and body temperatures in wolves (table 7.4). Other studies have examined the electrocardiographic consequences of the wolf's peripatetic lifestyle (Constable et al. 1998a).

Merely feeding on protein increases BMR and heat production (Kleiber 1961) and elevates body temperature (see table 7.4). In wolves, BMR increases after feeding to an average of 233.8 kJ/hr (converted from figures by Okarma and Koteja [1987]) and remains at that level for approximately 15 hours after feeding (Okarma and Koteja 1987).

Muscle tone contributes about 20% of total heat production. During strenuous exercise, such as running, oxygen consumption may increase twentyfold, while energy expenditure may increase a hundredfold. The difference between these two measures represents an "oxygen debt," which is "repaid" by a continued elevated rate of oxygen consumption after the exercise ends. When BMR is combined with maximum oxygen consumption, a metabolic index (VO_2max/BMR) can be calculated. The energy consumption of running wolves is 5,070 kJ/hr (Weibel et al. 1983), which, when combined with Okarma's and Koteja's (1987) data, results in a metabolic index of about 32 (5,070/158); this is about three times higher than the average for mammals as a whole (Lechner 1978). This high metabolic index in running wolves may reflect the wolf's adaptation for high running speeds and endurance while chasing prey and for saving energy when no prospective prey is available (Okarma and Koteja 1987).

Temperature extremes also affect BMR. Wolves inhabit ecosystems that vary from the frigid Arctic to hot, arid deserts. The average body temperature of a wolf, which it attempts to maintain through a combination of physical and chemical processes, is 39.6°C (103°F) (Kreeger, Kuechle et al. 1990). Dogs are unable to regulate their temperature at birth, but are able to do so by 4 weeks of age (Anderson 1970); wolves may be able to thermoregulate even sooner (Mech 1993b).

TABLE 7.4. Heart rates and body temperatures of four wolves (2 male, 2 female) measured remotely by radiotelemetry

Activity	Heart rate (bpm)	Temperature (°C)
Sleeping	56 ± 3	39.61 ± 0.11
Resting	84 ± 5	39.72 ± 0.08
Active	123 ± 7	39.69 ± 0.13
Feeding	138 ± 5	40.00 ± 0.09
Interacting[a]	159 ± 11	39.60 ± 0.13
Chased	253 ± 10	40.17 ± 0.14

Source: After Kreeger, Kuechle et al. 1990.

Note: Heart rates for each category were significantly different ($P = .001$) from each other; there were no differences among body temperatures. Means are reported with standard errors.

[a] This category represents data collected when two or more wolves were engaged in behaviors directed at each other.

As ambient temperatures fall, wolves maintain their body temperature by curling up to reduce the surface exposed to the cold or by increasing the insulating effectiveness of their fur through piloerection. Additionally, wolves can constrict peripheral blood vessels, which reduces the thermal gradient between the skin and the environment to reduce heat loss (Henshaw et al. 1972). However, since the limbs of animals are farthest from the central body core of heat generation and have a large surface relative to their mass, limbs still tend to lose significant heat to the environment. To reduce heat loss from limbs, a process known as countercurrent exchange has evolved. Within the limbs, deep arteries and veins run close together. Cooled blood returning from the limb surface picks up heat from the warm arterial blood. The arterial blood thus conserves heat by rewarming the venous blood before it returns to the body core. The arterial blood meanwhile loses less heat when it perfuses the skin.

While countercurrent exchange may be adaptive in temperate climates, it could be maladaptive for Arctic animals exposed to extreme cold. When an animal is exposed to tissue-freezing temperatures, a countercurrent system might exclude so much heat from the appendages that they would freeze. Wolves have evolved protective mechanisms to prevent this problem (Henshaw et al. 1972). They have unbranched arteries that carry blood directly through the footpad to a cutaneous plexus in the pad surface. This plexus keeps foot temperature just above the tissue freezing point (about −1°C). Maximum energetic efficiency is achieved because the unit of heat exchange is located in the pad surface that contacts the cold substrate, rather than throughout the pad, where tissue damage could occur if the cooled blood fell below freezing (Henshaw et al. 1972).

The external temperature at which these heat-retaining mechanisms are insufficient to maintain a constant body temperature and at which heat production must be increased is known as the lower critical temperature. Below this temperature, wolves can generate internal heat through exercise and shivering. During prolonged exposure to cold, however, the endocrine system increases metabolic heat production. Although this effect is primarily mediated by the thyroid hormones, others, such as growth hormone, insulin, and adrenal hormones, exert a regulatory effect.

When ambient temperatures rise, wolves may seek cooler environments, such as shade or a den, in which to cool down. If this strategy is unavailable or insufficient, cutaneous vasodilation increases and raises the skin temperature to foster convective heat loss to the environment. Heat is also dissipated by sweating, which dogs do via glands in the footpads. Wolves, however, have 80% fewer sweat glands in their pads than do dogs (Sands et al. 1977), which may serve to minimize lameness from frozen sweat. Panting, which is usually accompanied by increased salivation, also expels body heat through evaporative cooling.

Wolves in deserts have a higher BMR in summer than in winter (Afik and Pinshow 1993). To dissipate excess metabolic heat and maintain a constant body temperature, these wolves must evaporate a substantial quantity of water. Fortunately, wolves are highly mobile, which allows them to find water. Nonetheless, wolves are better adapted to cold or temperate climates than to warm ones, and historically did not live south of about 12° N latitude (Young and Goldman 1944; Kumar and Rahmani 2001).

Field Metabolic Rate

Whereas BMR reflects energy expenditure in a uniform state (e.g., a sleeping wolf metabolizing only fat), field metabolic rate (FMR) is an estimate of total energy expenditure, including basal metabolism, maintenance, thermoregulation, and activity. After these essential energy requirements are met, extra energy may then be allocated to growth, fat deposition, or reproduction. Thus the daily food requirements (see Peterson and Ciucci, chap. 4 in this volume) of a mammal such as the wolf are determined by its FMR (Nagy 1994).

The FMR of a 37 kg wolf is 17,700 kJ/day (Swain, Costa, and Mech, cited in Nagy 1994). The average daily BMR, based on the estimate of Okarma and Koteja (1987), is 3,811 kJ/day (158.8 kJ/hr × 24 hr). The ratio of FMR to BMR in wolves, then, is 4.6 (17,700 kJ/3,811 kJ). This FMR/BMR ratio is higher than for mammals as a whole (approximately 3.0; K. A. Nagy, personal communication).

Another way of looking at relative energy expenditures is to compare the FMR of wolves to that of other eutherians (placental mammals). The FMR for a typical eutherian having the same body mass as a wolf (37.3 kg, or 82 pounds) would be 14,100 kJ/day ($37.3^{0.772} \times 4.63$) (Nagy et al. 1999). Thus, the wolf's FMR is about 26% (17,700/14,100 = 1.26) higher than that of a typical eutherian of similar body mass. All of these data suggest that wolves work harder than the average mammal to earn their daily bread.

Nutrition

Chapter 4 dealt with the wolf's digestive system. Here I will discuss the physiology of wolf nutrition (see also table 4.3). Wolves kill prey sporadically, often going days between feedings (Mech 1970). During this period, wolves must rely on catabolism, which results in loss of weight. It may be several days before the wolves eat again, and often they must travel far to find food (Mech 1970). When food is found, wolves gorge, sometimes eating 15–19% of their weight in a single feeding (Mech 1970; Kreeger et al. 1997). In fact, L. D. Mech (personal communication) studied a 31.8 kg (70-pound) wolf that had eaten 10 kg (22 pounds) of venison, 31.4% of its weight. This feed-move-feed regimen appears to be mediated by internal opioids and other neurotransmitters (Morley et al. 1983; Kreeger, Levine et al. 1991).

Environmental temperatures also affect the energy sources wolves use. Mammals increase metabolic heat production when environmental temperatures fall below their lower critical temperature. The principal type of fuel used under these conditions is fat. Animals in cold environments that mainly use fat for fuel should have more free fatty acids (FFA) in the blood than more temperate-dwelling species, and this is indeed the case among wolves. Plasma FFA concentrations in Arctic wolves fed fish and caribou (1,660 μEq/l) are 2–3 times higher than those of temperate-zone mongrel dogs (Ferguson and Folk 1970, 1971; Schultz and Ferguson 1974).

Wolves appear to utilize saturated FFA at a higher rate than unsaturated FFA, causing a deficit of saturated fats in the plasma (Schultz and Ferguson 1974; Ferguson and Schultz 1978). This preferential use of saturated FFA may be a life-saving strategy that has evolved in wolves. Saturated FFA in dogs at a low concentration (490 ± 170 μEq/l) cause sudden death by massive generalized thrombosis (Connor 1962). Unsaturated FFA do not seem to form thrombi as readily, so are better tolerated. Because saturated fat converts more rapidly to unsaturated fat in wolves (Ferguson and Shultz 1978) than in dogs (Havel and Fredrickson 1956), unsaturated fats accumulate in the blood, while the harmful saturated FFA are rapidly removed and metabolized.

Water is as critical for survival as food. For wolves, water may often be available only as ice or snow. Although water is an energy source, it is energetically expensive to extract. A wolf actually must expend more energy to melt snow than it receives. However, wolves do not conserve a significant amount of energy by suppressing snow intake (Philo 1985, 1986), and because water is so necessary, wolves continue to ingest snow whether fasting or gorging (Philo 1987; see also Peterson and Ciucci, chap. 4 in this volume).

Fasting in wolves is reflected by decreased serum urea nitrogen (SUN) and triiodothyronine (T_3) in the blood (DelGiudice et al. 1987). The decreased SUN probably indicates decreased hepatic gluconeogenesis and appears to be a sensitive indicator of nutritional status. The declining T_3 concentrations during fasting may be critical to the animal's survival because decreasing T_3 minimizes energy expenditures and fosters fuel conservation (DelGiudice et al. 1987). Conversely, fed wolves have increased concentrations of FFA, cholesterol, and insulin, all reflecting an increased intake of nutrients and the physiological responses needed to process them.

The nutritional status of wolves is also reflected in their urine, including urine deposits in snow (Mech, Seal, and DelGiudice 1987). The composition of urine is normally expressed as ratios to creatinine (C) so that samples taken under differing conditions can be compared. Urea/creatinine (U/C) ratios increase when wolves have fed and decrease during fasting, so they may indicate nutritional status (DelGiudice et al. 1987). However, U/C ratios also increase during extreme starvation. Sodium/creatinine (Na/C) and potassium/creatinine (K/C) ratios tend to parallel U/C ratios (DelGiudice et al. 1987).

When well fed, wolves store fat under the skin, around the heart, intestines, and kidneys, and in bone marrow. They tend to increase their fat stores during fall and winter (Seal and Mech 1983). During fasting, fat is mobilized for energy. As fasting progresses, fat from all subcutaneous and internal depots is used simultaneously, but subcutaneous fat is depleted first, then visceral fat, then kidney fat (LaJeunesse and Peterson 1993). Marrow fat is then rapidly metabolized. Foreleg bones are depleted before hind leg bones, and within those bones, proximal fat stores are depleted before distal stores (LaJeunese and Peterson 1993). During a prolonged fast (10 days), wolves lose 7–8% of their weight, about half in water, a fourth in fat, and a fourth in protein/ash. All this weight is regained after just 2 days of gorging (Kreeger et al. 1997).

Threats to Survival

Life is not easy for the wolf. The animal must locate and dispatch its prey with some regularity lest it starve. It must survive extreme climatic conditions and natural disasters. In addition, the wolf must survive infection by a host of parasitic, viral, and bacterial organisms that, if they don't kill it outright, can impair its prey-catching ability. This section deals with these threats.

Parasitic Diseases

Wolves host both internal parasites (endoparasites) and external parasites (ectoparasites). Most parasites live in and on the wolf in a way that does not kill their host, but allows both host and parasite to survive. Wolves tolerate a wide range of such parasites, usually with no harm. However, when factors such as malnutrition or viral or bacterial disease weaken the wolf, the effects of parasites may become more serious, or even fatal (Mech 1977b).

Protozoan Parasites

Protozoans are unicellular animals in which metabolism and locomotion are carried out by organelles within the cell. Few protozoan infections have been reported in wolves, but that may be only because no one has looked for them, since over 30,000 species have been described. Although most protozoans infecting wolves probably cause no harm, individual health and condition can predispose some wolves to morbidity.

Isospora. In 1997, three 4-month-old wolf pups from the Superior National Forest of Minnesota died, apparently of infection by *Isospora,* which causes coccidiosis (Mech and Kurtz 1999). The pups were from two adjacent packs that lived near humans and dogs and might have contracted the organism from dog feces (L. D. Mech, personal communication).

Toxoplasma. A protozoan called *Toxoplasma* has also been found in a wolf in the Superior National Forest. The animal was an adult male that was thought to have died from a combination of malnutrition, parasitism, and toxoplasmosis in 1971 (Mech 1977b).

Sarcocystis. *Sarcocystis* sp. have been reported in wolves from Russia and Minnesota (Zasukhin et al. 1979; Emnett 1986). Wolves experimentally fed venison containing infective *Sarcocystis* shed sporocysts 12–14 days later; however, only 3% of seventy-two scats examined in the wild contained sporocysts. Although wolves are susceptible to *Sarcocystis* infection, they may not be the definitive host (Emnett 1986). *Sarcocystis* sp. are not considered to be of any serious pathological significance (Soulsby 1968).

Babesia. Wolves in India have been infected with *Babesia gibsoni* (Howe 1971). *Babesia* invade and destroy red blood cells, causing anemia, icterus, and hemoglobinuria. However, no such signs have been reported in wolves.

Giardia. *Giardia* sp. are found in many parts of North America and infect a wide range of animals, including humans. Giardiasis has not been reported in wolves, but it is not uncommon in dogs, in which it causes diarrhea and dysentery (Soulsby 1968). The infective form is often found in water. Wolves probably harbor *Giardia* but suffer little or no morbidity as a result.

Other protozoans, such *Eimeria* and *Hepatozoon* (Pence and Custer 1981), have been found in coyotes and may be capable of infecting wolves.

Helminth Parasites

Three major groups of helminth parasites infect wolves: trematodes (flukes), cestodes (tapeworms), and nematodes (roundworms). Another small group, closely related to the nematodes, is the acanthocephalans (thorny-headed worms).

Trematodes. Several species of flukes parasitize wolves worldwide (table 7.5). Flukes have dorsoventrally flattened, one-piece bodies. Suckers, hooks, or clamps serve to attach them in or on internal organs. The life cycles of flukes that infect wolves involve multiple stages of development in intermediate hosts. In general, the adult fluke, which is hermaphroditic, sheds eggs in the feces of the wolf. A miracidium larva hatches under appropriate conditions and bores into a snail or slug. In the snail, the miracidium develops into a sporocyst, then (in some flukes) into a redia, and finally emerges from the snail as a cercaria. The cercaria may be consumed by or actively penetrates the wolf, or it may enter another intermediate host to become a metacercaria. When a cercaria or metacercaria is ingested by the wolf, it migrates to its target site.

Alaria sp. are some of the most common flukes found in North American wolves and other wild canids (Samuel, Ramalingam, and Carbyn 1978). Metacercariae develop in tadpoles and frogs, which are ingested by the wolf (Holmes and Podesta 1968). A further intermediate host may be a small rodent (Rausch and Williamson 1959). Within days of infection, mature *Alaria* migrate to the wolf's intestines. *Alaria* apparently harm wolves little, although they sometimes cause catarrhal duodenitis in dogs (Soulsby 1968).

TABLE 7.5. Trematodes found in wolves

Genus/species	Geographic location	Reference
Alaria sp.	Minnesota, Wisconsin	Byman et al. 1977; Archer et al. 1986
Alaria alata	Europe, Siberia, northern Asia	Stiles and Baker 1934; Morozov 1951; Zheleznov 1991
Alaria arisaemoides	Alberta, Manitoba, Yukon, Northwest Territories	Holmes and Podesta 1968; Samuel, Ramalingam, and Carbyn 1978
Alaria canis	Ontario, Alaska	Pearson 1956; Rausch and Williamson 1959
Alaria marcianae	Ontario, Alberta, Manitoba Yukon, Alaska	Law and Kennedy 1932; Holmes and Podesta 1968 Choquette et al. 1973; Samuel, Ramalingam, and Carbyn, 1978; Custer and Pence 1981b
Clonorchis sinensis	Northern Asia	Zheleznov 1991
Heterophyes persica	Europe, Asia	Stiles and Baker 1934
Metagonimus yokagawai	Northern Asia	Zheleznov 1991
Metorchis conjunctus	Alberta, Saskatchewan	Holmes and Podesta 1968; Wobeser et al. 1983
Nanophyetus schikhobalowi	Northern Asia	Zheleznov 1991
Opisthochis felineus	Siberia	Zheleznov 1991
Paragonimus westermanii	Korea	Stiles and Baker 1934
Pseudaphistomum sp.	Europe, Asia	Stiles and Baker 1934

Metorchis conjunctus larvae live in fish, and the organism ultimately resides in the wolf's liver or gall bladder (Wobeser et al. 1983). Seven Saskatchewan wolves with *M. conjunctus* infections showed liver pathology, and two of the wolves also had an inflamed and fibrous pancreas and were emaciated.

Little is known about most of the other flukes infecting wolves. Most apparently reside in the intestine, although *Paragonimus westermanii* lives in the lungs (Davis and Libke 1971).

Cestodes. Cestodes, or tapeworms, are also commonly found in wolves (table 7.6), primarily because these carnivores feed on ungulates, small mammals, and fish, which host their larvae. Tapeworms are hermaphroditic worms with a long, flat body. They range from a few millimeters to several meters long. For example, *Diphyllobothrium latum,* whose intermediate host is fish, can reach lengths of up to 20 m (22 yards) (Soulsby 1968).

The tapeworm body consists of a head, or scolex, which usually has suckers or hooks for attaching to the wolf's intestinal wall, and a strobila, which consists of a number of segments, or proglottids. Each proglottid contains both male and female sexual organs. Eggs in the proglottid are either self-fertilized or cross-fertilized. The gravid segments break off from the strobila, either singly or in groups, and are shed in the wolf's feces. The gravid segments then break down in the environment to release their eggs. An intermediate host ingests a tapeworm egg, the egg hatches, and the embryo penetrates the intestinal wall of the host in order to reach a suitable part of the body for further development. There, it grows into a cyst, composed of an outer cuticle, an inner germinal layer, and a cavity filled with fluid. When a wolf eats the intermediate host, the larval tapeworm attaches to the intestinal wall and develops into an adult.

Generally, tapeworms do little harm to wolves. When they do cause problems, their severity usually depends on the number and size of the worms and the susceptibility and sensitivity of the host. Their most common effects are intestinal blockage due to numerous large strobila; generalized toxic or allergic reactions; mechanical irritation of the intestinal mucosa; nutrient deprivation of the host; secondary bacterial infection of the intestine due to attachment of the scoleces to the mucosa; and absorption of proteins and vitamins from the intestinal mucosa (Leiby and Dyer 1971).

In many instances, larval infections of the intermediate host are much more serious than adult infections of the definitive host. For example, cysts of *Echinococcus multilocularis* (alveolar or hydatid cysts) produce neoplastic symptoms that can eventually destroy the host's liver, and germinal tissues of the liver cysts commonly metastasize to other organs (Leiby and Dyer 1971). Alveolar echinococcosis is often fatal to humans because no drugs are effective in killing the larvae; furthermore, sur-

TABLE 7.6. Cestodes found in wolves

Genus/species	Geographic location	Reference
Alveococcus multilocularis	Northern Asia	Zheleznov 1991
Diphyllobothrium latum	North America, northern Asia	Young and Goldman 1944; DeVos and Allin 1949; Archer et al. 1986; Zheleznov 1991
Dipylidium caninum	Russia	Morozov 1951; Zhelenznov 1991
Echinococcus granulosus	North America, Europe, Asia	DeVos and Allin 1949; Riley 1933; Cowan, 1947; Stiles and Baker 1934; Rausch and Williamson 1959; Freeman et al. 1961; Samuel, Rama-lingam, and Carbyn 1978; Messier et al. 1989; Guberti et al. 1991; Zheleznov 1991
E. multilocularis	Manitoba, Russia	Abuladze 1964; Samuel, Ramalingam, and Carbyn 1978
Joyeauxiella pasqualei	Europe, Asia	Stiles and Baker 1934
Mesocestoides kirbyi	Yukon	Choquette et al. 1973
M. lineatus	Europe, Russia, Asia	Stiles and Baker 1934; Morozov 1951; Zheleznov 1991
Spirometra erinaceieuropaei	Europe, Asia	Stiles and Baker 1934
S. janickii	Poland	Furmaga 1953
Taenia crassiceps	Ontario, Europe, Russia, Asia	Stiles and Baker 1934; Abuladze 1964; Freeman et al. 1961; Samuel, Ramalingam, and Carbyn 1978; Zheleznov 1991
T. hydatigena	North America, Europe, Asia	Morozov 1951; Cowan 1947; Erickson 1944; Rausch and Williamson 1959; Freeman et al. 1961; Samuel, Ramalingam, and Carbyn 1978; McNeill et al. 1984; Zheleznov 1991
T. krabbei	Alaska, Alberta, northern Asia	Morozov 1951; Freeman et al. 1961; Rausch and Williamson 1959; Holmes and Podesta 1968; Zheleznov 1991
T. laticollis	Ontario	Freeman et al. 1961
T. macrocystis	Northern Asia	Abuladze 1964; Zheleznov 1991
T. multiceps	North America, Europe, Asia	Stiles and Baker 1934; Erickson 1944; Bondareva 1955; Holmes and Podesta 1968; Zheleznov 1991
T. omissa	Alberta	Holmes and Podesta 1968
T. packii	North America, Russia	Erickson 1944; Abuladze 1964
T. parenchimatosa	Northern Asia	Zheleznov 1991
T. pisiformis	North America, Asia	Erickson 1944; Bondareva 1955; Law and Kennedy 1932; Freeman et al. 1961; Holmes and Podesta 1968; Samuel, Ramalingam, and Carbyn 1978; Zheleznov 1991
T. polyacantha	Russia	Abuladze 1964
T. serialis	Northern Asia	Abuladze 1964; Zheleznov 1991
T. skriabini	Russia	Abuladze 1964
T. taeniaeformis	Alberta	Holmes and Podesta 1968
Tetratirotaenia polyacantha	Northern Asia	Zheleznov 1991

gical removal of the cysts is fraught with difficulty, as un-intentional rupture of the fragile cysts simply spreads the infection further.

The debilitating effects of hydatid cysts on the inter-mediate host, however, can work to the wolf's advantage. There is indirect evidence that heavy *Echinococcus gra-nulosus* infections may predispose moose to wolf preda-tion. In moose, the majority of hydatid cysts develop in the lungs, which may reduce the animal's stamina during wolf pursuits. As many as fifty-seven cysts have been found in a single moose (Mech 1966b). The rate of

E. granulosus infection in moose also increases as wolf numbers increase (Messier et al. 1989). Higher wolf den-sity means a greater chance of exposure to the shed eggs for browsing moose, which subsequently develop cysts, become susceptible to predation, and are killed and eaten by wolves. This pattern ensures continuation of the tapeworm life cycle (Mech 1966b).

Nematodes. Nematodes, or roundworms, are a di-verse group of parasites with over thirty species found in wolves (table 7.7). Roundworms are elongate, unseg-mented, and vary greatly in size. They usually inhabit the

TABLE 7.7. Nematodes found in wolves

Genus/species	Geographic location	Reference
Ancylostoma sp.	Minnesota	Byman et al. 1977
A. braziliense	Asia	Stiles and Baker 1934
A. caninum	Asia, Russia, Minnesota	Stiles and Baker 1934; Panin and Lavrov 1962; Kreeger, Seal, Callahan, and Beckel 1990b; Zheleznov 1991
Capillaria sp.	Minnesota, Wisconsin,	Byman et al. 1977; Archer et al. 1986
C. aerophila	Russia, Alberta	Shaldybin 1957; Holmes and Podesta 1968
C. plica	Europe, Russia	Stiles and Baker 1934; Shaldybin 1957; Zheleznov 1991
Crenosoma vulpis	Europe, Russia	Stiles and Baker 1934; Shaldybin 1957; Zheleznov 1991
Dioctophyma renale	Minnesota, Quebec, Europe, Asia	Stiles and Baker 1934; Erickson 1944; McNeill et al. 1984; Zheleznov 1991
Dirofilaria immitis	North America, northern Asia	Coffin 1944; Mech and Fritts 1987; Kreeger, Seal, Callahan, and Beckel 1990b; Zheleznov 1991
D. repens	Northern Asia	Zheleznov 1991
Eucoleus aerophilus	Europe, Asia	Stiles and Baker 1934
Filaroides sp.	Minnesota	Byman et al. 1977
F. osleri	Minnesota	Erickson 1944
Mastophorus muris	Northern Asia	Zheleznov 1991
Oslerus osleri	Minnesota	Erickson 1944
Physaloptera papilloradiata	Persia	Stiles and Baker 1934
P. sibirica	Northern Asia	Zheleznov 1991
Riticularia affinis	Russia	Panin and Lavrov 1962; Zheleznov 1991
R. cahirensis	Russia	Panin and Lavrov 1962
R. lupi	Russia	Panin and Lavrov 1962
Sprirocerca lupi	Canada, Germany, Russia	Stiles and Baker 1934; Morozov 1951; Choquette et al. 1973; Zheleznov 1991
S. arctica	Canada, Northern Asia	Choquette et al. 1973; Zheleznov 1991
Strepropharagus sp.	Russia	Panin and Lavrov 1962
Thominx aerophilus	Northern Asia	Zheleznov 1991
Toxacaro sp.	Minnesota	Byman et al. 1977
T. canis	North America, Russia, Asia	Stiles and Baker 1934; Morozov 1951; Cowan 1947; Kreeger, Seal, Callahan, and Beckel 1990b; Zheleznov 1991
T. mystax	Europe, Asia	Stiles and Baker 1934
Toxascaris leonina	Russia, Alaska, Canada	Morozov 1951; Rausch and Williamson 1959; Fuller and Novakowski 1955; Holmes and Podesta 1968; Choquette et al. 1973; Samuel, Ramalingam, and Carbyn 1978; Zheleznov 1991
Trichinella sp.	North America, Europe, Russia	Machinskii 1966; Machinskii and Semov 1971; Gunson and Dies 1980; Pozio et al. 1989, 1996, 2001; Stancampiano et al. 1993; Zarnke et al. 1999.
T. nativa	Northern Asia	Zheleznov 1991
T. spiralis	North America, Europe	Rausch and Williamson 1959; Choquette et al. 1973; Fraga de Azevedo et al. 1974; Worley et al. 1990
Trichuris sp.	Minnesota	Byman et al. 1977
T. vulpis	Europe, Asia	Stiles and Baker 1934
Uncinaria sp.	Minnesota	Byman et al. 1977
U. stenocephala	North America, Russia	Erickson 1944; Rausch and Williamson 1959; Holmes and Podesta 1968; Choquette et al. 1973; Samuel, Ramalingam, and Carbyn 1978; Zheleznov 1991

intestine, but can be found in other organs, such as the heart *(Dirofilaria immitis)* or kidney *(Dioctophyma renale).* They may have simple life cycles, in which the shed eggs of one definitive host are ingested by another, or complex ones with one or more intermediate hosts. The larvae of some species penetrate the skin or mouth and migrate to the lungs, where they are coughed up and swallowed to become mature adult roundworms in the intestine. The sexes are usually separate and can be dimorphic.

Most roundworm infections of wolves appear to be benign, but large numbers of worms or infection of pups can cause pathologies. The dog hookworm, *Ancylostoma caninum,* which attaches to the intestinal wall to feed on blood, can cause hypochromic anemia (Kreeger, Seal, Callahan, and Beckel 1990b) and possibly emaciation, diarrhea, and death (Brand et al. 1995).

Toxocara canis is another hookworm commonly found in wolf pups; large numbers of these parasites can cause intestinal irritation, bloating, vomiting, and diarrhea (Kreeger, Seal, Callahan, and Beckel 1990b). Intestinal blockage and aspiration of the vomitus, resulting in pneumonia, can cause death in pups. Pups are infected in utero with *T. canis* by migrating larvae from their infected mother. The larvae continue to migrate in the pup prior to taking up residence in the intestine, and migration through the lungs can also cause pneumonia. No deaths of wild wolf pups due to hookworm infection have been reported. This does not mean that they do not happen, however. Because most biologists avoid studying newborn wolf pups in the wild for fear of jeopardizing them, little information is available on this stage of life.

The giant kidney worm, *Dioctophyma renale,* has been found in wolves, but the mink, *Mustela vison,* is considered the definitive host in North America. The giant kidney worm has a complex life cycle involving oligochaetes and perhaps fish as intermediate hosts. When infective larvae are eaten, the parasite develops in the kidney, reaching lengths up to 100 cm with a diameter of 1 cm in wolves (Fyvie 1971). The presence of *D. renale* in the kidney results in the complete destruction of the functional tissue, leaving only a distended and thickened capsule. The ureter remains functional and allows passage of eggs into the urine. To compensate for the loss of one kidney, the other kidney hypertrophies. If both kidneys are parasitized, the infection is fatal (Fyvie 1971).

The dog heartworm, *Dirofilaria immitis,* has been found in both captive and free-ranging gray wolves (Cof-

fin 1944; Pratt et al. 1981; Mech and Fritts 1987; Kreeger, Seal, Callahan, and Beckel 1990b) as well as in red wolves (Riley and McBride 1975). The adults usually live in the right ventricle of the heart, and sometimes in the larger pulmonary arteries as well. The eggs are shed into the blood, where they are ingested by mosquitoes, which in turn transmit the resulting infective larvae, or microfilaria, to another definitive host. Dogs, and presumably wolves, can tolerate low levels of heartworm infection for many years without noticeable clinical signs. However, large numbers of heartworms can cause cardiac enlargement, chronic passive congestion, and death (Fraser et al. 1991). Infection causing cardiopulmonary impairment may retard the wolf during heavy exertion, such as pursuing prey for long distances.

Trichinella spiralis is a small nematode that produces the disease trichinosis in humans and wild animals. Over 104 wild animal species have been infected with *T. spiralis* (Zimmerman 1971), but whereas infection is common, disease is rare. The major concern with trichinosis is the possible spread of the disease from wildlife reservoirs to humans; over 150 symptoms and signs have been related to trichinosis in humans (Gould 1945). Wolves commonly harbor *Trichinella,* with infection rates reaching 50% in Russian wolves (Machinskii and Semov 1971). No clinical signs attributable to trichinosis have been reported in wolves, but heavy infections in dogs can cause roughing of the fur, emaciation, increased salivation, and extreme muscle pain with inability to walk (Zimmerman 1971).

The *Trichinella* life cycle is unique for helminths in that the entire cycle transpires within a single host. When a host ingests meat containing viable trichina larvae, the larvae penetrate the intestine, pass through the lymph vessels to enter the blood circulation, and eventually penetrate the muscle fibers. Generally, the more active muscles, such as the diaphragm, masseter, and tongue, harbor the greatest concentration of parasites (Zimmerman 1971). Wolves probably become infected by feeding on the carcasses of other carnivores, including other wolves (Choquette et al. 1973; Gunson and Dies 1980).

Acanthocephalans. Acanthocephalans, or thornyheaded worms, are a group of evolutionary old organisms that rarely infect wolves. They are distinguished from other nematodes by their anterior attachment organ, the proboscis, which bears specific arrangements of articulated hooks that can be protruded or retracted

(DeGiusti 1971). The parasite attaches to the mucosal wall of the intestine, where mechanical damage can cause nodule formation. Secondary bacterial infection can result in perforation of the gut with consequent peritonitis. In light infections, there may be no outward clinical signs; heavy infections may result in loss of appetite, diarrhea, and emaciation (DeGiusti 1971). Thorny-headed worms require an arthropod intermediate host to complete their life cycle. Thorny-headed worms have been found in wolves in Russia, and the following species have been reported: *Onicola skrjabini* (Morozov 1951), *Macrocantorhynchus catulinus* (Zheleznov 1991), and *Moniliformis maniliformis* (Panin and Lavrov 1962).

Endoparasitic Diagnosis and Treatment

Most endoparasitic infections of wolves are diagnosed by microscopic examination of feces for helminth eggs, protozoan oocysts, or protozoa. Shed tapeworm proglottids can often be grossly seen in the feces. Heartworm infection can be diagnosed by microscopic examination of the blood for microfilaria or by serology tests.

There are several antihelmintics available, but few studies have been conducted on their efficacy in wolves. *Toxocara canis, Ancylostoma caninum,* and microfilaria of *Dirofilaria immitis* have been effectively eliminated in wolves given ivermectin (Kreeger, Seal, Callahan, and Beckel 1990b). Adult heartworms in wolves can be successfully treated with thiacetarsamide, but toxic reactions can occur, and wolves should be monitored both during and for several days after treatment (Kreeger,

Seal, Callahan, and Beckel 1990b). Tapeworms, including *Echinococcus,* have been effectively treated in wolves with praziquantel, and coccidiosis in foxes has been treated with sulfadimethoxine (T. J. Kreeger, unpublished data).

Ectoparasites

A variety of arthropod parasites—including fleas, ticks, lice, and mites—bite, suck, chew, and otherwise make life miserable for wolves (table 7.8). Most are irritating; some are potentially lethal. Other arthropods, such as biting flies, are only transitory parasites of wolves and are not included in this discussion.

Perhaps the most harmful ectoparasite is the mange mite, *Sarcoptes scabiei.* Mange mites are spread to new hosts by direct body contact or by transfer from common rubs used to relieve the pruritis caused by the mites. The life cycle of the mite includes egg, larva, nymph, and adult. All the active stages are able to burrow into the skin, but most burrows are made by the fertilized females, which lay their eggs in the burrow. The burrowing causes intense pruritis, depilation, scab formation, hyperkeratosis (thickening of the skin), and seborrhea, especially on the head and neck. In advanced stages, the entire body may be involved, causing emaciation, decreased flight distance, staggering, and death (Sweatman 1971).

Sarcoptic mange can be a serious threat to wolves. Cowan (1947) reported on several mangy wolves in Canada, one of which weighed only 17 kg (37 pounds) at death.

TABLE 7.8. Ectoparasites of wolves

Genus/species	Common name	Reference
Amblyomma americana	Lone Star tick	Cooley and Kohls 1944; Bishopp and Trembley 1945
A. maculatum	Gulf Coast tick	Bishopp and Trembley 1945
Dermacentor variabilis	American dog tick	Bishopp and Trembley 1945
D. albipictis	Winter tick	Gregson 1956
Haemaphysalis bispinosa	New Zealand cattle tick	Stiles and Baker 1934
Ixodes hexagonus	Hedgehog tick	Stiles and Baker 1934
I. kingi	Rotund tick	Bishopp and Trembley 1945
I. pacificus	California black-legged tick	Bishopp and Trembley 1945
I. ricinus	Castor-bean tick	Bishopp and Trembley 1945
Linguatula serrata	Tongue worm	Stiles and Baker 1934
Linognathous setosus	Louse	Hopkins 1949
Pulex irritans	Flea	Stiles and Baker 1934; Bishopp and Trembley 1945
Sacroptes scabei	Mange mite	Stiles and Baker 1934
Trichodectes canis	Louse	Stiles and Baker 1934; Hopkins 1949; Mech et al. 1985; Zarnke and Spraker 1985

Epizootics of mange can result in high mortality rates, especially among pups (Todd, Gunson, and Samuel et al. 1981), which may help control high populations (Pence and Custer 1981). In Wisconsin's and Michigan's recovering wolf populations, mange was known to be the cause of death of nine and six wolves respectively (Wisconsin Department of Natural Resources 1999; Michigan Department of Natural Resources 1997). In 1992 and 1993, 58% of the Wisconsin wolves examined had signs of mange, but this percentage then declined. Despite the mange, the population increased at high rates. Wolves in northeastern Minnesota were mangy during the late 1990s and in 2000 (L. D. Mech, unpublished data).

Lice, such as *Trichodectes canis,* also cause high morbidity in wolf packs, but they probably do not cause high mortality (Mech et al. 1985). A louse epizootic occurred in Alaska in 1981–1983, and the affected wolves suffered from alopecia and seborrhea, with up to 75% of the body involved (Schwartz et al. 1983; Taylor and Spraker 1983; Zarnke and Spraker, 1985). Self-inflicted trauma caused by the associated pruritis resulted in inflammation and infection. Although individual wolves were captured and effectively treated, the overall treatment program was not successful in eradicating lice from the wolf population (Taylor and Spraker 1983; Zarnke and Spraker 1985).

Ticks of the *Ixodes* genus can transmit a variety of diseases, such as Lyme disease and Rocky Mountain spotted fever. Lyme disease will be discussed under bacterial diseases.

Ectoparasitic Diagnosis and Treatment

Most ticks and lice can be seen with the unaided eye. Louse eggs may also be seen attached to hair. Mites can be observed by microscopic examination of scrapings of affected sites. Wolves infected with *Trichodectes canis* were successfully treated with ivermectin (Taylor and Spraker 1983). Ivermectin is also effective against *Sarcoptes* mites, although no reports of its use for mites on wolves are known. If necessary, anesthetized wolves could be dipped or bathed in shampoos containing pyrethrins to kill fleas, ticks, and lice (T. J. Kreeger, unpublished data).

Viral Diseases

Rabies

Rabies is an acute infectious disease of the central nervous system caused by a rhabdovirus that generally persists as a salivary gland infection of carnivores (Sikes 1981). Rabies is one of the oldest recorded infectious diseases, and it is enzootic on every continent except Australia (Tabel et al. 1974; Adamovich 1985; Butzeck 1987; Theberge et al. 1994). Although rabies has been reported in wolves, they are not considered the primary vector of rabies except in the eastern Mediterranean region and India (Cherkasskiy 1988). Domestic dogs and foxes are generally considered the primary vectors worldwide.

Rabies is almost always transmitted by the bite of an infected animal; however, transmission of rabies by aerosol routes has been demonstrated in a bat cave (Constantine 1962). The incubation period in wolves is 8–21 days (Cherkasskiy 1988); in dogs that contracted rabies from infected wolves, it was 17–39 days (Cowan 1949). In humans it is usually 20–60 days (Hattwick 1982), but can be as short as 9–11 days following severe injury inflicted by rabid wolves (Rausch 1973).

Signs of rabies in infected wolves are mostly anecdotal, as no controlled studies on wolves have been conducted. There are two forms of rabies: dumb and furious. Dumb rabies is characterized by early paralysis of the throat, loss of voice, excessive salivation, general paralysis, and death (Hattwick 1982). There is little inclination to bite. Furious rabies proceeds through three stages: a prodromal period, during which the animal exhibits abnormal behavior, modification of voice, and increased salivation; an excitement phase, during which the animal becomes severely agitated and attacks inanimate objects, wanders aimlessly, and bites other animals or people; and a paralytic phase, during which the voice is lost, the tongue protrudes from paralyzed jaws, and ascending paralysis soon leads to death (Rausch 1973). These signs, however, do not necessarily apply to wild animals under all conditions.

Rabid wolves certainly appear to lose their "fear" of humans because most documented wolf attacks on humans have been attributed to rabies. Such attacks have been more numerous in eastern Europe and Asia (Baltazard and Bahmanyar 1955; Cherkasskiy 1988) than in North America (Rausch 1958, 1973; Chapman 1978; Johnson 1995). Rabid wolves will attack dogs, livestock, and other wolves (Ballantyne and O'Donoghue 1954; Baltazard and Bahmanyar 1955; Tabel et al. 1974; Chapman 1978; Cherkasskiy 1988).

The effect of rabies on wolf populations may be severe. Infected wolves seem to become agitated, desert their packs, alter their den usage (Weiler et al. 1995), and run up to 80 km (48 mi)/day, which suggests that such animals could spread the disease to other wolf packs (Chapman 1978; Cherkasskiy 1988). A rabies epizootic among several packs in Alaska resulted in a population decline while unaffected wolf populations were increasing, thus suggesting that the disease may sometimes be a significant factor limiting population growth (Ballard and Krausman 1997). There is evidence, however, that the disease could be somewhat self-limiting in that it spreads within a pack and then disappears when all the pack members have either died or left the area prior to exposure (Chapman 1978; Ritter 1991; Brand et al. 1995).

Rabies can be diagnosed by direct microscopic examination of stained hippocampus, cerebral cortex, or cerebellum sections for Negri bodies; by fluorescent antibody detection of the rabies antigen; by mouse inoculation with a preparation of suspect brain tissue or skin punches from the muzzle; or by detection of rabies antibodies in serum by a rapid fluorescent focus inhibition test (Ritter 1981; Sikes 1981; Zarnke and Ballard 1987).

Clinical rabies in wolves is untreatable. Suspect animals should be euthanized in a manner that spares brain tissue. The head should be removed, stored in a leak-proof container, and sent to an appropriate diagnostic laboratory. Samples can be cooled or frozen if shipment will be delayed. Extreme care should be taken while removing and handling the infected tissue to avoid exposure to the virus; one must wear gloves and eye protection, and work carefully.

There are currently no U.S. Department of Agriculture (USDA)-approved rabies vaccines for wolves. Vaccinated captive wolves or wolf hybrids that have bitten a person are considered unvaccinated by departments of public health, and they are usually euthanized and the brain examined. Captive wolves have been routinely vaccinated for years using vaccines approved for dogs (Kreeger et al. 1987), and there is evidence that such vaccines result in seroconversion in wolves (Federoff 2001). There have been no reports, however, of vaccinated wolves being challenged by the rabies virus, so the efficacy of dog vaccines in wolves remains unknown.

Canine Distemper

Canine distemper is an acute or subacute febrile (fever-causing) disease of carnivores caused by a paramyxovirus (Budd 1981). Distemper has been known in Europe since the mid-sixteenth century and is currently a worldwide problem in domestic dogs. Reports of the disease in free-ranging wildlife, however, are few. Most information about the existence of distemper in wolves has come from serological evidence, which demonstrates exposure to the virus, but reveals little of its epidemiology.

The distemper virus is spread by aerosol or direct contact; nasal and conjunctival exudates, feces, and urine can all contain the virus (Budd 1981). Within days of exposure, signs of the disease appear, which include oral jaundice and ulceration, swollen feet, anorexia, ataxia, dyspnea, and neurological abnormalities (Monson and Stone 1976).

The seroprevalence of distemper in North American wolves is about 17% (Choquette and Kuyt 1974; Stephenson 1982; Zarnke and Ballard 1987; Bailey et el. 1995; Brand et al. 1995). There is disagreement as to whether distemper is enzootic within wolf populations or sporadically introduced by infected domestic dogs (Stephenson et al. 1982; Zarnke and Ballard 1987; Brand et al. 1995).

Although distemper is a highly lethal disease of many carnivores, mortality from distemper in wild wolves has been documented only in Canada (Carbyn 1982b) and Alaska (Peterson, Woolington, and Bailey 1984). It is generally thought that distemper is not a major mortality factor in wolves because despite evidence of exposure to the virus, affected wolf populations demonstrate good recruitment (Brand et al. 1995).

Distemper can be diagnosed by clinical signs combined with the identification of intranuclear inclusion bodies or fluorescent antibody detection of the viral antigen in circulating leukocytes or epithelial cells of the conjunctiva or tonsils. Preparations of lung, spleen, or other suitable tissues can be inoculated into ferrets to produce the disease. This technique is useful for frozen or partially decomposed samples that are unfit for histopathological examination. Clinical signs in inoculated ferrets may not appear for 30 days, however (Budd 1981).

There is no known cure for distemper in wild animals. Captive animals may be given supportive and symptomatic treatment such as antibiotics and fluids, but in many cases this is futile. Prevention through the use of modified live virus (MLV) vaccines approved for dogs may be effective in protecting wolves (Rausch 1967; cited in Mech 1970), although no clinical challenges have been conducted. Vaccination with modified live virus in a species other than the approved species may result in reversion of the virus to virulent form and actually induce the disease (Budd 1981). There have been no reports of reversion to virulent form of distemper vaccine in wolves, but MLV vaccines have caused distemper in black-footed ferrets and lesser pandas (Budd 1981).

Killed distemper vaccine protected one wolf pup, but failed its littermate (Mech 1970).

Canine Parvovirus

Canine parvovirus (CPV) is a relatively new infectious organism that caused epizootics of severe gastrointestinal and myocardial disease in domestic dogs in the late 1970s and early 1980s. Although CPV was not considered "discovered" in dogs until 1977 (Eugster and Nairn 1977), there was evidence of CPV exposure in wolves as early as 1975 (Goyal et al. 1986).

Contact with CPV is primarily through feces from an infected animal; the virus may also be shed in the saliva and vomitus during acute illness. Direct contact with the feces is not required, as CPV conceivably could be carried by vectors such as insects. The virus is quite hardy in the environment, and it is thought to survive for months in feces (Greene 1984a). Clinical signs are first noted 5–10 days post-inoculation. The course and severity of the disease depend on the age of the animal and such predisposing factors as the presence of maternal antibodies, concurrent disease, stress, *Giardia,* or other parasitic infections.

CPV is characterized by severe enteritis, diarrhea, and vomiting. Death can be caused by dehydration, electrolyte imbalance, and endotoxic shock or sepsis. In pups, myocardial disease following in utero or early neonatal infection can result in rapid death at less than 12 weeks of age. Cardiac damage caused by myofiber necrosis leads to myocarditis, which results in myocardial irritability and life-threatening cardiac arrhythmias (Greene 1984a).

There is only one published report of mortality or clinical illness due to canine parvovirus in a free-ranging wolf (Mech et al. 1997). Serological evidence of CPV infection has been reported in wild wolves from Minnesota (Goyal et al. 1986; Mech et al. 1986), Alaska (Zarnke and Ballard 1987; Bailey et al. 1995), Michigan (Peterson 1995b), Wisconsin (Brand et al. 1995), Montana and British Columbia (Johnson et al. 1994), and Italy (Baumgartner and Guberti 1996; Fico et al. 1996). Morbidity and mortality due to CPV infection have been reported in captive wolves (Mech et al. 1986).

There is growing evidence that CPV infection can affect wolf populations. A precipitous crash in the wolf population of Isle Royale National Park has been circumstantially linked to CPV. The crash was coincident with an outbreak of the disease among dogs in Michigan, and subsequent investigations demonstrated titers to CPV in Isle Royale wolves (Peterson 1995b). Mech and Goyal (1993, 1995) have demonstrated an inverse relationship between CPV seroprevalence and both the number of pups captured each year and the percent change in the wolf population in the following year. Although the wolf population did not decline during their 10-year study, there was evidence of a decline in pup recruitment. These findings suggested that CPV mortality compensated for mortality factors such as starvation and intraspecific strife that had been previously restricting this population prior to the CPV epizootic (Mech and Goyal 1995).

The serological evidence from wild wolves suggests that CPV infection is a survivable disease, despite the potential for high mortality in pups (Mech et al. 1986; Mech and Goyal 1995). In controlled studies on captive wolves, 30% of the challenged animals developed clinical illness, and it was estimated that 10% of the animals would have died without supportive care (Brand et al. 1995).

Antibodies against CPV have persisted in a wolf pup for more than a year (Mech et al. 1986), but how long antibodies persist in wolves without active infection or reexposure is unknown (Mech and Goyal 1995). Dogs and coyotes passed maternal antibodies derived from infection or vaccination on to some, but not all, of their pups. These antibodies persisted for up to 8 weeks (Green et al. 1984) after which pups that had maternal antibodies were no longer protected (Meunier et al. 1981).

Canine parvovirus can be diagnosed by a variety of serological tests, including hemagglutination inhibition, virus neutralization, and indirect immunofluorescence. Virus particles can also be detected in the feces via electron microscopy (Muneer et al. 1988). Clinically ill captive wolves can be successfully treated for CPV infection. Treatment generally consists of vigorous fluid therapy twice daily with lactated Ringer's solution to compensate for vomiting and diarrhea until signs abate (T. J. Kreeger et al., unpublished data). Although controversial, concurrent antibiotic therapy in severely ill wolves to prevent secondary bacterial infection has been recommended (Kreeger et al. 1983).

CPV infection can be prevented through the use of domestic dog vaccines. However, the normal dog regimen of vaccinating pups every 3–4 weeks from 8 to 20 weeks of age may be insufficient for wolves. I have had two cases of confirmed CPV infection in wolves that were vaccinated under the dog regimen, only to contract the disease at 4–5 months of age. These cases could be attributable either to persistent maternal antibodies negating the effects of the vaccine or to infection by a variant CPV strain (KF-11) that overrides high levels of maternal antibodies to cause disease (L. D. Munson, unpublished data). I currently recommend vaccinating wolf pups every 4 weeks until at least 24–28 weeks of age with a vaccine against the KF-11 strain.

Infectious Canine Hepatitis

Infectious canine hepatitis (ICH) is a worldwide infectious disease of dogs and foxes caused by canine adenovirus-1 (CAV-1). Transmission of the virus is usually through direct contact with contaminated respiratory tract discharge, saliva, urine, feces, fomites, or ectoparasites (Cabasso 1981). There are no reports of the course of ICH disease in wolves. In dogs, signs of the disease range from mild anorexia, depression, and fever to death within 12–24 hours of onset. After an incubation of 2–6 days in ranch foxes, ICH can cause rhinitis, ataxia, anorexia, bloody stool, and sometimes convulsions, flaccid and spastic paralysis, coma, and death. In nonfatal cases, a transitory interstitial keratitis in one or both eyes may be seen (Cabasso 1981).

Serological evidence of exposure to the ICH virus in wolves has been reported in Canada and Alaska (Choquette and Kuyt 1974; Stephenson et al. 1982; Zarnke and Ballard 1987). Seroprevalence ranged from 13% to 95%, with higher rates in Alaska than in Canada. Although the disease in wolves could have come from exposure to infected domestic dogs, there is no evidence of this transmission, and the disease is thought to be enzootic in wolf populations (Zarnke and Ballard 1987). Morbidity or mortality in wolves from ICH has not been confirmed.

Diagnosis of ICH can be accomplished through serological testing, virus isolation, or electron microscopic examination for inclusion bodies in hepatic cells. If ICH is suspected, supportive treatment with fluids and antibiotics may be helpful. Systemic glucocorticoid therapy, though controversial, may decrease the immunologically mediated ocular reaction to the virus.

Canine adenovirus-1 is antigenically distinct from canine adenovirus-2 (CAV-2), which produces respiratory disease in the dog, but either virus can be attenuated and used as a vaccine against ICH. A single vaccination with a MLV should provide lifelong immunity (Greene 1984b).

Papillomatosis

Oral papillomatosis has been reported only in two wolf pups found dead near a poisoned bait station in Alberta (Samuel, Chalmers, and Gunson 1978). The papillomas were few and small. The virus particles observed were indistinguishable from those seen in coyotes. Papillomatosis is not thought to cause serious morbidity or mortality in wolves, but may alter behavior or feeding. Spontaneous recovery with long-lasting immunity has been suggested in infected coyotes (Trainer et al. 1968).

Canine Coronavirus

Recently, prevalence of serum antibody to canine coronavirus in Alaskan wolves averaged 70% in spring and 25% in autumn, and indications were that transmission occurs primarily during winter (Zarnke et al. 2001).

Bacterial Diseases

Brucellosis

Brucellosis is a highly contagious, worldwide infection of both ungulates and carnivores. The disease is caused by *Brucella* bacteria, which are small, gram-negative, nonmotile, non-spore-forming rods. Although *Brucella canis* is a specific pathogen for domestic dogs, it has not been reported in wolves. The pathogens of interest in wolves are *Brucella suis* type 4, which infects wild caribou and domestic reindeer, and *Brucella abortus* type 1, which infects elk and bison in Wyoming, Montana, and Idaho (Cheville et al. 1998) and bison in Canada (Brand et al. 1995).

Wolves probably become infected with *Brucella* when they consume infected prey. The disease in ungulates generally causes bursitis-synovitis, metritis, mastitis, abortion, or orchitis (Witter 1981). In experimentally infected pregnant wolves, *Brucella suis* type 4 did not cause any apparent clinical signs in the adults (Neiland and Miller 1981). However, of the eight pups born in two litters, six were dead shortly after birth, and two died within 24 hours. The first six pups had signs consistent with mother-induced trauma. *B. suis* type 4 was cultured from several organs of the adults and from the liver, spleen, and blood of the pups. This experiment demonstrated that (1) wolves are susceptible to *B. suis* type 4 infection; (2) *Brucella* organisms are widely distributed at 24 days post-inoculation; (3) infection of the uterus and fetuses may lead to reproductive failure; (4) organisms may be shed in the urine, saliva, and milk; and

(5) serological responses to *B. suis* type 4 in wolves are similar to those in other host species (Neiland and Miller 1981). Although the deaths of the pups could not be directly attributed to *Brucella* infection, the organism could have indirectly caused their deaths (Carmichael and Kennedy 1970).

Brucella infection in wolves has been reported in Alaska (Neiland 1970, 1975; Zarnke and Ballard 1987), Canada (Brand et al. 1995), and Russia (Pinigin and Zabrodin 1970; Grekova and Gorban 1978). Wolf seroprevalences were 1% and 45% in Alaska (Neiland 1975; Zarnke and Ballard 1987), 31% in Canada (Brand et al. 1995), and 11% in Russia (Pinigin and Zabrodin 1970). The range in prevalence most likely reflects differing prevalences in the caribou herds on which the wolves prey (Zarnke and Ballard 1987).

The effect of *Brucella* infection on wolf populations is unknown. Ungulates infected with *Brucella* may initially abort or show signs of the disease, but recover and go on to reproduce normally later (Witter 1981). The same phenomenon may occur in wolves, since spontaneous recovery from *B. canis* infection in dogs can occur within 1–3 years (Greene and George 1984).

Brucella can be diagnosed by a variety of serological tests, including rapid slide agglutination, complement fixation, and tube agglutination as well as tissue, blood, urine, or semen culture. Infected animals can be treated with antibiotics such as tetracycline, streptomycin, or sulfonamides, but such treatment cannot be relied on to eliminate the infection (Witter 1981). High doses of minocycline (25 mg/kg) combined with streptomycin have been shown to eliminate infection in dogs (Greene and George 1984), although the costs are high ($500–1,000).

Lyme Disease

Lyme disease, which infects humans and animals, is caused by the spirochete *Borrelia burgdorferi*. The principal vertebrate reservoirs are thought to be the white-tailed deer and white-footed mouse, and the bacterium is primarily spread through the bite of infected deer ticks, *Ixodes dammini* (Bosler et al. 1984; Burgdorfer et al. 1982). These reservoir hosts do not show clinical signs of the disease. Transplacental transmission and contact are other possible routes of infection (Brand et al. 1995). Although Lyme disease has been reported in forty-three states and eastern Canada, niduses of the disease exist in the northeastern United States, Minnesota-Wisconsin,

and the Pacific coast. Of these areas, only Minnesota-Wisconsin has viable wolf populations.

Clinical Lyme disease has not been reported in wolves, but the disease can be debilitating in dogs, causing fever, lymphadenopathy, arthralgia, and arthritis (Kornblatt et al. 1985). Abortion and fetal mortality have also been reported in infected humans and horses. Seroprevalence in Minnesota and Wisconsin wolves is about 3% (Kazmierczak et al. 1988; Thieking et al. 1992), with a higher incidence in wolves found near Lake Superior (Thieking et al. 1992).

Kazmierczak et al. (1988) experimentally inoculated wolves with *B. burgdorferi* intravenously, subcutaneously, and orally. The wolf receiving the intravenous inoculation showed the highest and most prolonged titer, and both popliteal lymph nodes were enlarged at necropsy 75 days post-inoculation; no other clinical or pathological findings were seen. The other routes of inoculation did not result in infection.

Lyme disease may debilitate individual wolves, but is not thought to significantly affect wolf numbers or population health. Lyme disease can be diagnosed by detecting serum titers via an indirect fluorescent antibody test or direct immunofluorescent staining of tissue sections. Culture tests are usually unrewarding. Lyme disease can be treated with antibiotics such as oxytetracycline or ampicillin.

Leptospirosis

Leptospirosis is a worldwide zoonotic disease caused by *Leptospira* spirochetes. All pathogenic leptospires are classified as one species, *interrogans,* which contains more than 170 serovars based on differences in serological testing (Blood and Radostits 1989). Leptospirosis may cause morbidity and sometimes mortality in numerous species. In dogs, leptospirosis can cause fever, anorexia, vomiting, anemia, hematuria, icterus, and death if severe liver or kidney damage occurs (Greene 1984c).

Transmission of leptospirosis is usually by contact with infective urine or by feeding on infected prey (Reilly et al. 1970). Interspecies transmission in wolves may also occur by contact with urine scent marks. Seroprevalence of *Leptospira* infection in wolves in Alaska was 1% (Zarnke and Ballard 1987) and in Minnesota, 11% (Khan et al. 1991). Seroprevalence in Minnesota wolves near farms was 2.6 times greater than that in wilderness wolves, perhaps due to increased exposure to infected domestic animal waste (Khan et al. 1991).

Leptospirosis can be diagnosed by a microscopic ag-glutination test of serum. The disease can be treated with penicillin, streptomycin, or the tetracyclines, which are effective against all serotypes. Bivalent leptospira vac-cines approved for dogs have been given to wolf pups and adults with no adverse reactions (T. J. Kreeger et al., unpublished data), but the effectiveness of these bac-terins in wolves has not been demonstrated.

Other Bacterial Diseases

Tularemia is an acute febrile disease of wild lagomorphs and rodents caused by the bacterium *Francisella tularen-sis*. Zarnke and Ballard (1987) reported an overall 25% seroprevalence in Alaskan wolves, perhaps reflecting the high prevalence of the disease in the hare population. The effect of tularemia on wolves is unknown, but the authors hypothesized that most healthy wolves would re-cover from an uncomplicated bout with the disease.

Bovine tuberculosis, which primarily affects domestic and wild ungulates, is caused by *Mycobacterium bovis*. The only report of bovine tuberculosis in wolves comes from Canada, where two pups were found dead and *M. bovis* was found in one isolate (Carbyn 1982b). The source of the infection was not identified, but it could have come from infected carrion, an infected bitch, or contaminated soil. The bacterium can remain viable in the soil for many years. Although Carbyn (1982b) attrib-uted a decline in wolf numbers to disease, including tu-berculosis, this disease is not generally regarded as a ma-jor threat to wolf populations.

Other bacterial diseases that are not considered even minor threats to wolf populations include listeriosis (Rutter and Pimlott 1968), salmonella (Fox 1941), salmon poisoning (Young and Goldman 1944), ehrlichiosis (Harvey et al. 1979), Q fever (Zarnke and Ballard 1987), and *Heliobacter* infection (Jakob et al. 1997).

Fungal Diseases

Blastomycosis

Blastomycosis is a chronic pyogranulomatous pulmo-nary disease caused by the dimorphic fungus *Blasto-myces dermatitidis*. The disease has been reported mostly in humans and dogs in North America, but a few cases have appeared in Africa and Central America. The dis-ease is considered endemic in the Mississippi, Missouri,

Ohio, and St. Lawrence river drainages and in the mid-Atlantic states.

Blastomyces is thought to reside in the soil, but isola-tions have been few, and experiments have not been able to maintain the fungus for extended periods in soil. Transmission probably occurs by inhalation of spores. Incubation times may be up to 3 months in dogs. The lung is the primary focus of infection, but the organism can disseminate to other organs. General clinical signs of infection include weight loss, fever, anorexia, weak-ness, skin lesions, depression, dyspnea, lameness, ocular and nasal discharge, chronic cough, draining lymph nodes, swollen joints, blindness, lymphadenopathy, in-coordination, and enlarged testicles. Although the illness can last up to 3 years, most animals survive less than 4 months (Barsanti 1984).

Thiel et al. (1987) reported a case of blastomycosis in a wild wolf from Minnesota and serological evidence from another wolf in Wisconsin. In the confirmed case, the wolf was so weak and debilitated that people were able to capture it by hand. Its lungs contained many firm nodules and areas of granulomatous inflammation. *Blas-tomyces* was confirmed by immunofluorescent staining of yeast cells in the lung tissue. Presumptive blastomy-cosis was diagnosed in another wolf that had been cap-tured and radio-collared. The wolf appeared healthy at capture and was radio-tracked for the next 9 months with no indication of abnormal behavior (Thiel et al. 1987). The prevalence of blastomycosis in wolves appears to be low, and the disease probably does not significantly affect populations. Most recently, the disease was diag-nosed in three wolves from just northeast of Lake Supe-rior in Ontario (Krizan 2000; P. C. Paquet et al., personal communication).

Blastomycosis can be diagnosed by cytological ex-amination of exudates or skin lesions, fluorescent anti-body testing of tissue biopsies, or culture of exudates, as well as several serological tests. Blastomycosis in captive wolves could possibly be treated with amphotericin B administered over a prolonged period (Barsanti 1984).

Dermatomycosis

Dermatomycosis, or ringworm, is a mycotic infection of the skin caused by *Microsporum gypseum* or other der-matophytes. Although not usually debilitating, infection can cause intense inflammation, pruritis, and hair loss. The disease has not been reported in wild wolves, but it has occurred in captive wolves (Fischman et al. 1987).

TABLE 7.9. Miscellaneous pathological conditions of the wolf

Condition	Reference
Adenocarcinoma of the thyroid	Fox 1926
Adenomatous polyp of uterus	Fox 1926
Arthritis	Cross 1940; Fritts and Caywood 1980
Bladder stone and chronic nephritis	Hamerton 1945
Blindness	Laikre et al. 1993
Bone deformation	Ossent et al. 1984; Wobeser 1992
Carcinoma	Ryan and Nielsen 1979
Carcinoma of liver	Lucas 1923
Carcinoma of neck	Plimmer 1915
Carcinoma of tonsil	Scott 1928
Carcinoma of thyroid	Lucas 1923
Cataracts	Williams et al. 1977
Chronic diffuse nephritis	Fox 1926
Chronic intestinal nephritis	Hamerton 1932
Chronic parenchymatous nephritis	Hamerton 1931
Cretinism	Fox 1923
Degeneration of liver, kidney, and heart	Hamerton 1945
Duodenal ulcer	Fox 1927
Encephalitis	Young and Goldman 1944
Exopthalmic goiter	Fox 1923
Glanders	Blair 1919
Hemorrhagic cystitis	Plimmer 1916
Hip dysplasia	Douglass 1981
Hyperplasia of thyroid	Fox 1923
Jaundice	Fox 1927
Jaw, skull abnormalities	Haugen and Stephenson 1985; Mallory et al. 1994; Federoff 1996
Keratitis	Harwell et al. 1985
Multicentric hypernephroma	Fox 1926
Myocardial and arterial degeneration	Hamerton 1936
Pancreatitis	Fox 1923
Papillomatosis	Samuel, Chalmers, and Gunson 1978
Perianal gland tumor	Wheler and Pocknell 1996
Rickets	Blair 1908
Tooth abnormalities	Hell 1990, Vilà, Uríos, and Castroviejo 1993
Traumatic, degenerative lesions	Wobeser 1992
Tumor of lung, adrenal, kidney, thyroid, cerebellum	Fox 1926

Dermatomycosis probably is not a disease of concern in wolves. However, it can be transmitted from animals to humans, so care should be taken when handling suspect animals. The disease can be treated with griseofulvin.

Miscellaneous Diseases and Pathologies

Like most animals, wolves are subject to a variety of diseases, injuries, and accidents (Mech 1991b; Boyd et al. 1992) throughout life (table 7.9; see also Fuller et al., chap. 6 in this volume). Several such reports exist in the literature, but most concern captive animals. The majority of these pathologies affect individual animals, not wolf populations.

Health and Condition Assessment

Hematology and Serum Chemistries

Much of the above discussion on physiology and diseases was based on findings revealed by blood samples taken from captive or wild wolves. Blood can tell us something about the age, nutrition, reproductive status, and health of a wolf. Blood samples can easily be obtained from the cephalic (foreleg), femoral or saphenous (hind leg), or jugular (neck) veins. The cephalic or saphenous vein is the preferred route in adult wolves because jugular veins are difficult to locate in the thickly furred neck. Jugular veins, however, are readily accessible in pups. Blood samples should be analyzed as soon as possible; otherwise they should be cooled, but not frozen, with the exception of serum samples for serological tests, which can be frozen.

A variety of analyses can be performed on blood. Blood data are usually divided into hematological and serum chemical values. An explanation of each parameter is beyond the scope of this chapter, but can be found in veterinary or human clinical pathology texts. There have been several reports on wolf "baseline" blood values, but such reports are often limited in the number of wolves and the geographic location sampled (Dieterich 1970; Seal et al. 1975; Seal and Mech 1983; DelGiudice et al. 1987; DelGiudice, Mech, and Seal 1991; Pospisil et al. 1987; Drag 1991; Kreeger, Seal, Callahan, and Beckel 1990b; Constable et al. 1998b).

A composite of several hematological data reports is presented in table 7.10, and serum chemistry data are presented in table 7.11 (ISIS 1995). These data can serve as baselines with which other experimental or observational data can be compared. Blood data for wolf pups will differ from these adult values because pups have lower red blood cell counts until fetal erythrocytes are replaced by mature erythrocytes. Pup values for erythrocyte count, hematocrit, mean corpuscular volume, and hemoglobin increase slowly during the first year of life (Drag 1991).

Analysis of blood data over time can be used to assess the nutritional status and health of wolf populations (Seal et al. 1975; Seal and Mech 1983; DelGiudice, Mech, and Seal 1991), although such data must be interpreted with care due to the many variables of sample collection. Long-term monitoring of blood parameters in captive wolves under controlled conditions revealed a consistent circannual rhythm in erythrocyte count, hemoglobin, hematocrit, mean corpuscular hemoglobin concentration, and thyroxine level (Seal and Mech 1983). Knowledge of such rhythms is important in data interpretation because baseline values fluctuate. The use of a remote recapture collar (Mech and Gese 1992) allows repeated

TABLE 7.10. Hematological data from adult gray wolves worldwide

Parameter (units)	n	Mean	SD
Erythrocyte count ($\times 10^6/\mu l$)	76	6.97	0.77
Hematocrit (%)	76	48.22	4.63
Hemoglobin (g/dl)	64	16.50	1.09
Mean corpuscular hemoglobin (pgm)	45	22.87	1.40
Mean corpuscular hemoglobin concentration (g/dl)	64	35.12	2.91
Mean corpuscular volume (fl)	51	67.19	6.98
Leukocyte count ($\times 10^3/\mu l$)	76	12.25	1.97
Neutrophils ($\times 10^3/\mu l$)	75	8.51	2.12
Band neutrophils ($\times 10^3/\mu l$)	36	0.45	0.57
Lymphocytes ($\times 10^3/\mu l$)	76	1.64	0.19
Monocytes ($\times 10^3/\mu l$)	72	0.41	0.24
Eosinophils ($\times 10^3/\mu l$)	75	0.54	0.32
Basophils ($\times 10^3/\mu l$)	20	0.01	0.03

Source: Data from Pospisil et al. 1987; Drag 1991; Kreeger, Seal, Callahan, and Beckel 1990b; ISIS 1995.

data collection from the same free-ranging individuals, which should provide an even more accurate picture of the physiological responses of wolves to environmental and population changes.

Wolf Capture and Handling

Wolves must be captured and safely handled to obtain blood or other samples, to attach radio collars, or to treat diseases. Wild wolves are usually caught in modified steel foot traps (Mech 1974b; Kuehn et al. 1986) and chemically anesthetized, shot with a dart from a helicopter (Ballard et al. 1982; Ballard, Ayres, Roney, and Spraker 1991; Fuller and Keith 1981a), or restrained by a net gun fired from a helicopter. Although captive coyotes and wolves can be safely restrained in a large fish landing net and hand-injected with drugs (Kreeger and Seal 1986b), this technique must be carefully applied to adult wolves. Captive wolves can also be restrained against a cage wall with a forked stick or a snare pole while drugs are administered by pole syringe.

Trapped wolves can be given drugs with a pole syringe, blowgun, or tranquilizer pistol. Dart rifles powered by carbon dioxide or .22-caliber blanks are the only suitable devices for anesthetizing wolves from a helicopter. However, care must be taken with power settings and dart placement to avoid penetrating the thorax or

TABLE 7.11. Serum chemistry values for adult gray wolves

Parameter (units)	n	Mean	SD
Alanine aminotransferase (IU/l)	111	55.72	26.35
Albumin (g/dl)	90	3.35	0.43
Alkaline phosphatase (u/l)	108	34.65	27.18
Amylase (IU/l)	23	427.43	144.71
Aspartate aminotransferase (IU/l)	103	53.46	27.51
Bicarbonate (mEq/l)	13	20.15	1.75
Chloride (mEq/l)	97	117.08	4.17
Cholesterol (mg/dl)	106	194.50	51.28
Creatine kinase (IU/l)	34	143.59	88.74
Creatinine (mg/dl)	114	1.31	0.34
Direct bilirubin (mg/dl)	32	0.02	0.04
Gamma glutamyl transpeptidase (IU/l)	9	4.11	2.28
Globulin (g/dl)	91	2.89	0.77
Glucose (mg/dl)	111	6.97	0.77
Iron (μg/dl)	34	146.49	46.38
Lactate dehydrogenase (IU/l)	100	150.67	123.15
Magnesium (mEq/l)	8	1.96	0.13
Osmolarity (mOsm/kg)	4	163.75	134.28
Potassium (mEq/l)	103	4.64	0.47
Serum urea nitrogen (mg/dl)	116	22.89	10.11
Sodium (mEq/l)	98	150.07	3.46
Total bilirubin (mg/dl)	102	0.25	0.61
Total CO_2 (mEq/l)	11	19.45	2.27
Total protein (g/dl)	112	6.24	0.76
Triglycerides (mg/dl)	30	46.08	24.14
Uric acid (mg/dl)	27	0.53	0.49

Source: ISIS 1995.

abdomen (Kreeger et al. 1995). The best site for drug injection is the large muscles of the hindquarters; the shoulder muscles can also be used for injection by hand or pole syringes.

Wolves are generally easy to anesthetize, and complications from overdosing are rare. During 19 years of research on captive wolves, we have conducted upward of ten thousand wolf immobilizations (Seal et al. 1979; Seal and Mech 1983; Kreeger et al. 1986, 1987, 1988, 1989, 1992; Kreeger, Seal, Callahan, and Beckel 1990a; Kreeger, Seal et al. 1991), and I can recall only three deaths that could be attributed to the anesthesia process. The most common complication in anesthetizing wild wolves is hyperthermia. A struggling or running wolf generates significant heat. Anesthesia increases the heat load by decreasing the wolf's ability to dissipate heat through panting as well as by disrupting the wolf's thermoregu-

latory mechanism. Other complications include hypothermia, respiratory depression (often temporary), and dart wounds. Once anesthetized, wolves should be blindfolded to protect their eyes from the sun. Antibiotics should always be given to prevent infection from dart or needle wounds.

Wolves have been anesthetized with many different drugs and drug combinations (Seal and Kreeger 1987; Kreeger 1992, 1996), including neuromuscular blocking drugs (Dyson 1965), cyclohexanes (Ballard et al. 1982; Ballard, Ayres, Roney, and Spraker 1991; Fuller and Kuehn 1983; Kreeger et al. 1986, 1987; Kreeger, Seal, Callahan, and Beckel 1990a; Jalanka and Roeken 1990), and opioids (Fuller and Keith 1981a; Ballard et al. 1982; Kreeger et al. 1989; Kreeger and Seal 1990). The most commonly used and safest drugs are a combination of a cyclohexane (ketamine, tiletamine) and a tranquilizer

TABLE 7.12. Suggested sedative and anesthetic drug doses for captive or free-ranging wolves

Drug	Dose (mg/kg)	Comments
Ketamine	10.0	Good, safe combination for adult captive wolves.
+ Acepromazine	0.15	
Ketamine	10.0	Good dose for trapped wolves. Xylazine can be antagonized with 0.15 mg/kg yohimbine.
+ Xylazine	2.0	
Ketamine	4.0	Alternative to ketamine-xylazine; medetomidine can be antagonized with 0.4 mg/kg atipamezole.
+ Medetomidine	0.08	
Tiletamine	5.0	Tiletamine and zolazepam come premixed in equal amounts, so the dose actually becomes
+ Zolazepam	5.0	10.0 mg/kg of the mix. Good for trapped wolves, but recovery can be slow (2–3 hrs).
Tiletamine	5.0	Tiletamine and zolazepam come premixed in equal amounts, so the dose actually becomes
+ Zolazepam	5.0	10.0 mg/kg for the two. Good dose for helicopter darting. Generally dose for a 50 kg wolf; chased
+ Xylazine	1.5	wolves can be difficult to immobilize, so do not underdose. Propylene glycol can be added to retard freezing of drugs in winter. Xylazine can be antagonized with 0.15 mg/kg yohimbine.
Xylazine	1.0	Sedative dose only for captive wolves. Use for minor, non-painful procedures. Wolves can recover spontaneously, so use caution. Xylazine can be quickly antagonized with 0.15 mg/kg yohimbine.
Medetomidine	0.05	Sedative dose only for captive wolves. Use for minor, non-painful procedures. Wolves can recover spontaneously, so use caution. Medetomidine can be quickly antagonized with 0.25 mg/kg atipamezole.

Source: After Kreeger 1996.

(acepromazine, xylazine, zolazepam, medetomidine) (table 7.12).

Some of the tranquilizers (xylazine, medetomidine) can be antagonized with yohimbine or atipamezole, which decreases the total time the wolf remains anesthetized (Kreeger et al. 1986, 1987, 1988; Kreeger, Callahan, and Beckel 1996; Jalanka and Roeken 1990). There is no antagonist for cyclohexane drugs (Kreeger and Seal 1986a). I do not recommend the use of opioid (narcotic) drugs because they often cause complications such as respiratory depression or arrest; are legally controlled, which imposes additional responsibilities on the user; and are potentially lethal to humans.

Conclusion

Physiologically, we know a great deal about the wolf, although we still have much to learn. But why should we continue to study the wolf? Some people curse the animal; others deify it. As scientists study it, we may be able to blunt these extremes and place the wolf in proper perspective. Viewed from the inside, the wolf is a large, intelligent canid predator with a variety of interesting biochemical, neural, and hormonal adaptations. It is neither good nor evil.

Like all other creatures, the wolf strives daily to survive, reproduce, raise its young, and pass on its genes despite a variety of parasites, diseases, and other pathologies. When it is time, the animal dies from any of a variety of factors, including malnutrition, parasites, or disease.

From a physiological viewpoint, then, the wolf's life cycle is similar to that of other animals, including humans. Wolves, however, tend to roughen the edges of a world being smoothed by human hands. For many of us, that is good reason to learn what we can about them, inside and out, and certainly good reason to work for their conservation.

8

Molecular Genetic Studies of Wolves

Robert K. Wayne and
Carles Vilà

A WOLF STEPS OUT OF the forest into full view, and several human spectators gaze at it. They immediately recognize the animal as a wolf by its size, shape, and other physical features. The precise nature of these features, of course, is determined by the wolf's genetic makeup. Its seventy-eight chromosomes (Wurster-Hill and Centerwall 1982) carry the usual mammalian complement of genes, including those specific to wolves and to that individual wolf. The genes, complex strands of DNA composed of long chains of hydrogen-bonded nucleotide pairs, thus form potentially powerful sources of valuable information.

By applying modern molecular techniques that allow the specific sequence of nucleotides on a DNA strand to be read, the exact genetic makeup of individuals can be decoded. This chapter discusses how molecular genetic techniques have been applied to a variety of questions about wolves, ranging from individual patterns of kinship and dispersal to population genetics and systematics. Molecular genetic techniques can be used to quantify relationships among individuals, populations, and species. Consequently, molecular genetic data provide a framework for studying topics that range from mating system diversity to higher-order systematics (Avise 1994, 2000).

The basic approach of molecular genetics is to assay variation in DNA or in the protein sequences that DNA specifies. The assay may be indirect; for example, allozyme electrophoresis is a technique that allows the separation of protein variants (allozymes) according to their net electrical charge and structure. However, only one-half or less of amino acid changes can be detected by

this technique. Further, because of redundancy in the genetic code, many DNA sequence changes do not affect the amino acid composition of a protein. These "silent" changes cannot be detected by allozyme electrophoresis; detecting them requires DNA sequencing techniques. Therefore, direct sequencing of genomic regions is often preferred because it yields more information about the underlying genetic differences among genomes.

If we wish to understand the evolution of closely related species, or even of populations and individuals within the same species, the study of molecular traits offers several advantages over similar analyses based on morphology, physiology, or behavior. First, nucleotide changes in DNA, or amino acid changes in protein sequences, represent uncommon events, so individuals that share specific mutations often do so because of common ancestry. Convergent, parallel, and reverse mutations may occur (called homoplasies), but when the appropriate DNA region or protein is described through the process of chemical sequencing, the phylogenetic signal is often strong, as the changes in the sequence that can be used to support a phylogeny—can number in the hundreds or thousands. In contrast, the morphology of a single tooth can be influenced by several genes, but the number of phylogenetically independent characters that can be measured often is very limited.

Second, many DNA mutations that are assayed by molecular techniques are silent (have no phenotypic effect), so differences in natural selection among taxa do not confound phylogenetic reconstruction based on these techniques. Further, approximate mutation rate constancy among taxa offers a potential means to estimate divergence time among species or populations. Third, molecular genetic changes are, with a few exceptions, entirely heritable and are not affected by environmental factors. Consequently, in contrast to morphological studies involving many fewer characters, and which may be influenced by environmental differences, molecular genetic studies can potentially offer more precise insight into evolutionary history (Avise 1994, 2000).

Studies of wolves typify the varied applications of molecular genetic techniques (table 8.1). We will start by discussing mitochondrial DNA (mtDNA) nucleotide sequence data and microsatellite loci. We will then discuss how species, subspecies, and units for conservation have been defined based on molecular genetic data and examine problems with their application in wolflike canids. Thereafter, we will review studies spanning an evolutionary hierarchy beginning with the systematic relationships of wolves to other canids, followed by studies of the relationships among wolf populations at local, regional, and continental scales. Finally, we will discuss population studies that assess levels of genetic variation and address genetic aspects of mating systems. This section will be followed by one examining the role of hybridization in the evolution of North American canids and its conservation and taxonomic implications. We will conclude with a summary of molecular genetic studies and future research questions.

Molecular Genetic Approaches

The first studies of population genetic variability in wolves examined variation in allozymes (see table 8.1). Allozymes are the various forms of an enzyme that have similar activity but differ in their amino acid sequence, representing different alleles for the same DNA locus. These allelic forms can be identified by their mobility through a gel matrix in response to an electric field (electrophoresis). Wolflike canids have only low to moderate levels of allozyme polymorphism; consequently, systematic and population-level studies using this technique often are not definitive (Ferrell et al. 1980; Wayne and O'Brien 1987; Wayne et al. 1991; Kennedy et al. 1991; Randi et al. 1993; Lorenzini and Fico 1995).

More recent studies have utilized nucleotide sequence variation in mitochondrial DNA, assessed through indirect or direct techniques of sequencing (see table 8.1). Mitochondria are cytoplasmic organelles that are responsible for the production of energy in the cell. Mito-

chondria often contain multiple copies of a small circular DNA molecule, about 16,000 to 18,000 base pairs (nucleotides) in length, that codes for proteins that function in the organelle as well as specifying messenger (mRNA) and transfer RNA (tRNA). Hundreds of mitochondria may occur within a cell, and thus mitochondrial genes are severalfold more abundant than their nuclear counterparts. This trait facilitates their genetic characterization. Additionally, mtDNA sequences have a very high mutation rate. In mammals this rate is often three to five times faster than that of nuclear genes (Avise 2000). Consequently, closely related species and populations may have accumulated diagnostic mtDNA mutations in the absence of changes in a similar-sized fragment of the nuclear genome. Finally, with only a few exceptions, mtDNA is maternally rather than biparentally inherited, so there is no recombination. Therefore, phylogenetic analysis of mtDNA sequences within species provides a history of maternal lineages that can be represented as a simple branching phylogenetic tree (Avise 1994).

The first studies of mtDNA sequence variation were indirect and utilized a panel of restriction enzymes that cut mtDNA wherever specific four- or six-base-pair sequence patterns are located. The fragments cut by each restriction enzyme are then separated by size electrophoretically. Differences in the fragment pattern among individuals indicate nucleotide changes at the restriction site and allow the total number of nucleotide sequence differences between mitochondrial genomes (haplotypes: e.g., Lehman et al. 1991) to be estimated. Such restriction fragment length polymorphisms (RFLP; see table 8.1) provided the first estimates of nucleotide sequence variation within populations and allowed the relationships of populations to be determined (e.g., Avise et al. 1987; Slatkin and Madison 1989; Avise 1991, 1994).

Beginning in the late 1980s, the advent of the polymerase chain reaction (PCR) technique, which allowed the enzymatic amplification of DNA, in combination with new DNA sequencing methods, made population-level sequencing studies feasible. DNA sequence data allow a more precise reconstruction of historical demographic events, such as colonization and gene flow, than was ever possible before (Avise 1994, 2000). Both restriction fragment analysis (e.g., Lehman et al. 1991; Wayne et al. 1992; Randi et al. 1995; Pilgrim et al. 1998) and, more recently, mtDNA sequencing by PCR (Vilà, Amorim et al. 1999; Randi et al. 2000; Wilson et al. 2000) have

TABLE 8.1. Molecular genetic studies of gray wolves and related canids

Study	Species	No. individuals (populations)	Genetic marker	Subject
Kennedy et al. 1991	Gray wolves	188 (1)	Allozymes	Population differentiation in northwestern Canada
Lehman et al. 1991	Gray wolves	276 (25)	mtDNA RFLP	Hybridization
	Coyotes	240 (17)		
Wayne and Jenks 1991	Gray wolves	276 (25)	mtDNA RFLP	Origin of the red wolf
	Gray wolves	3 (3)	Cytochrome b	
	Coyotes	327 (17)	mtDNA RFLP	
	Coyotes	5 (5)	Cytochrome b	
	Red wolves (captive)	4 (1)	mtDNA RFLP and cytochrome b	
	Texas/Lousiana (historic canids)	77 (1)	mtDNA RFLP	
	Red wolves (museum skins)	6 (5)	Cytochrome b	
	Golden jackal	1 (1)	Cytochrome b	
Wayne et al. 1991	Gray wolves	171 (14)	Allozymes, mtDNA RFLP, and fingerprinting	Genetic variability in Isle Royale wolves
Lehman et al. 1992	Gray wolves	110 (3)	Fingerprinting and mtDNA RFLP	Pack structure
	Captive colonies	36 (2)		
Wayne et al. 1992	Gray wolves	350 (26)	mtDNA RFLP	Population differentiation
Randi et al. 1993	Gray wolves	38 (1)	Allozymes	Variability in Italian wolves
Roy, Geffen et al. 1994	Gray wolves	151 (7)	Microsatellites	Population differentiation and hybridization
	Coyotes	120 (6)		
	Captive red wolves	40 (1)		
	Golden jackals	20 (1)		
Lorenzini and Fico 1995	Gray wolves	46 (1)	Allozymes	Comparison between Italian wolves and dogs
	Dogs	53 (1)		
Randi et al. 1995	Gray wolves	46 (1)	mtDNA RFLP	Genetic variability in Italian wolves
Ellegren et al. 1996	Gray wolves	15 (1)	mtDNA Control region and microsatellites	Genetic variability in Swedish wolves
	Captive colony	20 (1)		
García-Moreno et al. 1996	Gray wolves	151 (7)	Microsatellites	Genetics of captive Mexican wolves
	Mexican wolves	39 (1)		
	Coyotes	140 (6)		
	Captive red wolves	40		
	Golden jackals	20 (1)		
	Dogs	42 (1)		
Roy et al. 1996	Captive red wolves	32 (1)	Cytochrome b and microsatellites	Hybridization and origin of red wolves
Taberlet, Gielly, and Bouvet 1996	Gray wolves	6[a]	mtDNA control region	Origin of wolves arriving in France
Forbes and Boyd 1997	Gray wolves	172 (6)	Microsatellites	Population differentiation and migration
Hedrick et al. 1997	Gray wolves	(review)	Allozymes, fingerprinting, mtDNA control region, and microsatellites	Genetics of captive Mexican wolves
D. Smith et al. 1997	Gray wolves	130 (2)	Microsatellites	Pack structure

TABLE 8.1 *(continued)*

Study	Species	No. individuals (populations)	Genetic marker	Subject
Tsuda et al. 1997	Gray wolves	19 (3)	mtDNA control region	Origin of the domestic dog
	Dogs	34 (24)		
Vilà et al. 1997	Gray wolves	162 (27)	mtDNA control region	Origin of the domestic dog
	Dogs	140 (67)		
Wayne et al. 1997	Canid species	24 (24)	mtDNA	Phylogeny of canids
Pilgrim et al. 1998	Gray wolves	90 (4)	mtDNA control region and mtDNA RFLP	Genetic tests for hybridization
	Coyotes	30 (1)		
Ellegren 1999	Gray wolves	13 (1)	mtDNA control region and microsatellites	Bottlenecks and loss of genetic variability
	Captive colony	29 (1)		
Vilà, Amorim et al. 1999	Gray wolves	259 (30)	mtDNA control region	Phylogeography and demographic history
	Coyotes	17 (1)		
Hedrick et al. 2000	Mexican wolves	36 (3)	MHC	Genetic diversity and evolution
	Gray wolves	15 (1)		
	Red wolves	12 (1)		
	Coyotes	10 (1)		
	Gray wolves	150 (10) [b]		
Randi et al. 2000	Dogs	192 (4)	mtDNA control region	Population differentiation, bottlenecks, and hybridization
Wilson et al. 2000	Algonquin wolves	68 (1)	mtDNA control region and microsatellites	Origin of the Algonquin wolf
	Captive red wolves	60		
	Coyotes	24 (1)		
	Gray wolves	67 (1)		
Sundqvist et al. 2001	Gray wolves	97 (4)	Y chromosome	Origin and variability of Scandinavian wolves
	Captive gray wolves	13 (1)	Microsatellites	
Carmichael et al. 2001	Gray wolves	491 (3)	Microsatellites	Population differentiation
Andersone et al. 2002	Gray wolves	31 (1)	mtDNA control region and microsatellites	Wolf-dog hybridization
Hedrick et al. 2002	Red wolves	48 (1)	MHC	Genetic diversity and evolution
	Coyotes	39 (2)		
Lucchini et al. 2002	Gray wolf scats	402 (2)	Microsatellites	Origin of the population
Randi and Lucchini 2002	Gray wolves	211 (1)	Microsatellites	Wolf-dog hybridization
	Dogs	30 (1)		
Vilà, Sundqvist et al. 2002	Gray wolves	187 (2)	Sex chromosomes, autosomal microsatellites, and mtDNA control region	Origin of Scandinavian wolf population
	Museum gray wolves	30 (1)		
	Dogs	66 (1)		
Vilà, Walker et al. 2003	Gray wolves	103 (5)	Y chromosome, autosomal microsatellites, and mtDNA control region	Wolf-dog hybridization
	Dogs	44 (1)		
Flagstad et al. 2003	Gray wolves	22 (1)	Autosomal microsatellites and mtDNA control region	Genetics of the current and historic Scandinavian wolf population
	Museum gray wolves	55 (1)		

[a] Some reference material from a variety of wolves and dogs were compared with hair samples.

[b] This study was centered on two populations.

been applied to wolflike canids (see table 8.1). A hypervariable region of the mitochondrial genome called the mtDNA control region, which is noncoding (the DNA is not transcribed), has been the focus of recent sequencing efforts.

However, phylogenetic trees based on mtDNA record the history of only a single linked set of genes (Avise et al. 1984; Pamilo and Nei 1988; Avise 1991, 1994). Additionally, because of the smaller effective population size (the number of breeding individuals in a population in which all individuals breed equivalently), levels of mtDNA variability are more severely affected by changes in population size than are those of nuclear loci (Avise 1994). Recently, nuclear loci with high mutation rates that can easily be surveyed in large population samples have been identified. Techniques that use these loci include multilocus DNA fingerprinting (Burke et al. 1996, 251–77), in which complex DNA banding patterns simultaneously represent approximately ten to twenty unspecified minisatellite loci.

A generally more desirable approach involves microsatellite loci (tandem repeats of two to six nucleotide sequences) (Bruford and Wayne 1993; Hancock 1999). Microsatellite loci are often preferable because simple sequence repeats can be amplified by PCR from minute or highly degraded samples of DNA, such as those from bones, hair, and feces (Roy, Girman, and Wayne 1994; Roy et al. 1996; Taberlet, Griffin et al. 1996; Foran et al. 1997; Kohn and Wayne 1997; Kohn et al. 1999; Lucchini et al. 2002). Additionally, each locus is scored separately, either through autoradiography or staining of acrylamide gels or by an automated sequencer (Bruford et al. 1996, 278–97). These methods allow for the identification of the two alleles inherited from the parents at each locus (both alleles are detected, as in codominant markers). Since half of the alleles are shared between parent and offspring, these methods also allow for a robust determination of family relationships. Consequently, microsatellite data can be analyzed by traditional population genetic approaches developed for codominant loci (Goldstein and Pollock 1997; Rousset and Raymond 1997; Bossart and Prowell 1998).

In contrast, other techniques, such as the minisatellite, RAPD (randomly amplified polymorphic DNA), and AFLP (amplified fragment length polymorphisms) approaches (Smith and Wayne 1996; Avise 1994, 2000), generally analyze many loci simultaneously. Furthermore, a heterozygote may not be distinguishable from a homozygote genotype (i.e., the loci are not codominant). Thus these procedures require additional assumptions for statistical analyses. A panel of ten or fewer microsatellite loci may be sufficient to quantify components of variation within and among populations accurately and to study individual relatedness within social groups (Bruford and Wayne 1993; Queller et al. 1993; Amos et al. 1993; Roy, Geffen et al. 1994; Forbes and Boyd 1996; D. Smith et al. 1997; Bossart and Prowell 1998).

Recently, researchers studying humans have identified single-nucleotide polymorphisms and microsatellites on the Y chromosome (Cooper et al. 1996; White et al. 1999; Jobling and Tyler-Smith 2000; Sundqvist et al. 2001). These discoveries have allowed a new paternal view of evolutionary patterns that complements studies on maternally inherited mtDNA and biparentally inherited nuclear genes (Jorde et al. 2000; Sundqvist et al. 2001). Consequently, Y-chromosome studies represent an independent test for hypotheses based on mitochondrial sequences or microsatellites (Pritchard et al. 1999; Seielstad et al. 1999; Thomson et al. 2000). They also permit estimation of sex-biased migration and dispersal (Seielstad et al. 1998). Y-chromosome studies are still uncommon in other mammal species (Boissinot and Boursot 1997; Hanotte et al. 2000), but with the development of canine-specific markers (Olivier and Lust 1998, 1999; Sundqvist et al. 2001), a new line of population genetic and phylogenetic studies will be possible in wolves.

Species, Subspecies, and Units for Conservation

The biological species concept (BSC) maintains that the unifying characteristic of species is reproductive independence or isolation from other species (Mayr 1963; O'Brien and Mayr 1991). However, reproductive isolation is difficult to assess for populations living in different areas, and hybrid zones may form between populations thought to represent distinct species. These problems with the BSC have been the subject of a long discussion (reviewed by Hull 1997).

Operationally, species are often defined as morphologically and behaviorally distinct entities (e.g., Nowak 1979); however, the level of morphological distinction separating various taxonomic units (species, subspecies, populations) may be somewhat arbitrary and dependent on the measurements taken by the researcher. Moreover, morphologically distinct populations may interbreed (reviewed by Barton and Hewitt 1985, Harrison

1990, and Arnold 1997). Consequently, purportedly more objective definitions have been developed, such as the phylogenetic species concept, which defines species according to "diagnosable" characteristics reflecting a common ancestry (e.g., McKitrick and Zink 1988; Vogler et al. 1993).

Combining elements of the biological and phylogenetic species concepts, Avise and Ball (1990) suggested that subspecies be defined as populations that are generally allopatric (live in different areas) and have a series of concordantly divergent traits, but may interbreed if barriers to dispersal are removed. In contrast, species are defined by a similar suite of concordantly divergent traits, but do not widely interbreed if barriers to dispersal are removed.

The notion that consequential units below the species levels should show evidence of common ancestry (i.e., should be monophyletic groups; fig. 8.1), as defined by the possession of uniquely shared or diagnosable traits, is exemplified by Moritz's definition of evolutionarily significant units (ESUs: Moritz 1994). Under this definition, ESUs have mtDNA sequences that define reciprocally monophyletic groups and are significantly differentiated with respect to nuclear loci (fig. 8.1). Populations lose mtDNA haplotypes through random genetic drift, with the rate of loss increasing inversely with population size (Hartl and Clark 1997). In populations isolated for more than $4N$ generations (where N is the effective population size for females), mitochondrial sequences found in each population will tend to have exclusive ancestors because haplotypes with closer ancestry to other populations will have been lost due to drift (Avise 1994, 2000). Because populations with exclusive ancestors have a long history of genetic isolation, their protection preserves potentially unique adaptations and evolutionary potential to form new species (Moritz 1994, 1995). Populations that differ in allele frequencies but do not have exclusive ancestors are considered management units (MUs) that should be managed separately if possible (Moritz 1994).

Two problems are apparent when applying these definitions to wolflike canids. The first is that wolves disperse over great distances and across topographic barriers to find mates and territories (see Mech and Boitani, chap. 1 in this volume). As a result, rates of gene flow are high, so wolf populations are rarely isolated long enough to produce reciprocal monophyly in their mitochondrial sequences. Even rapidly evolving microsatellite loci may not show much differentiation between popula-

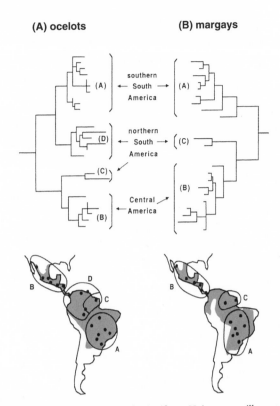

FIGURE 8.1. The Evolutionarily Significant Unit concept illustrated by an analysis of the mtDNA control region of (A) ocelots (*Felis pardalis*) and (B) margays (*F. wiedii*), two small South and Central American cats. Three reciprocally monophyletic clades (A, B, C) are common to the two phylogenies, each corresponding to cats from different areas. Therefore, three ESUs are suggested by the trees, plus a fourth one, D, for ocelots. (From Eizirik et al. 1998.)

tions. Rather than populations being discrete, a limited pattern of genetic differentiation with distance may be apparent (Forbes and Boyd 1996; see below).

For this reason, the division of wolves into discrete subspecies and other genetic units may be somewhat arbitrary and overly typological (conforming to a specific ideal type). In reality, wolves are better viewed as a series of intergrading populations having subtle or undetectable patterns of clinal genetic change (Lehman et al. 1991; Roy, Geffen et al. 1994; Forbes and Boyd 1997). Importantly, populations may differ in attributes important to fitness in spite of being connected by high rates of gene flow (e.g., T. B. Smith et al. 1997). Therefore, units for conservation should be based on fitness-related characters or their surrogates, rather than on largely neutral changes in mitochondrial sequences or microsatellite loci (Crandall et al. 2000).

A second problem stemming from high rates of gene

flow concerns the importance of hybridization. The width of a hybrid zone reflects dispersal distance and the degree of selection against hybrids (Barton and Hewitt 1985). Therefore, if selection against hybrids is weak and dispersal distances are large, interspecific hybridization can affect the genetic composition of a population over a wide geographic area. As discussed below, independent genetic studies suggest hybridization between coyotes and wolves and their hybrids over a wide area in southeastern Canada. As a consequence of hybridization, physically distinct populations may actually represent hybrids containing various proportions of genes from otherwise distinct species (see below). The pres-

ence of such introgressed populations greatly confounds taxonomic and conservation efforts (Jenks and Wayne 1992; Wayne and Brown 2001, 145–62).

Systematic Relationships

The evolutionary relationship of the gray wolf to other canids was reconstructed by phylogenetic analysis of DNA sequences from mitochondrial protein-coding genes (fig. 8.2A; Wayne et al. 1997). Further phylogenetic resolution among three closely related wolflike canids was achieved by sequencing part of the hypervariable mtDNA control region (fig. 8.2B; Gotelli et al. 1994; Vilà,

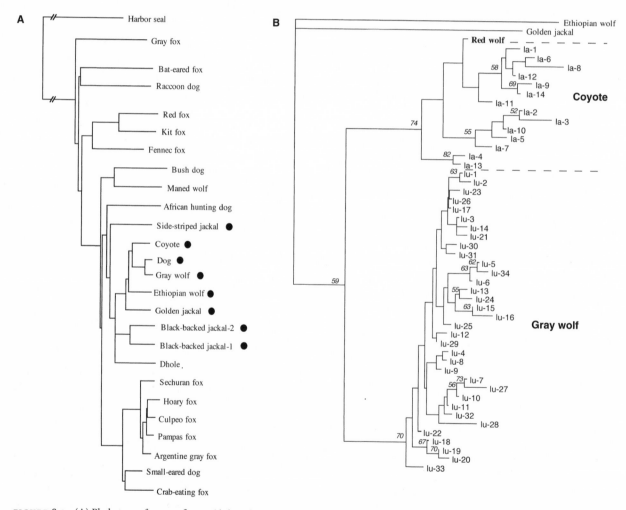

FIGURE 8.2. (A) Phylogeny of twenty-five canids based on 2001 base pairs of mtDNA sequence from three protein-coding genes. The solid circles indicate species belonging to the genus *Canis*. (B) Phylogeny of wolves (lu) and coyotes (la) based on mtDNA control region sequence. Bootstrap support values over 50% are indicated at nodes. Both phylogenies were constructed using a neighbor-joining algorithm. (A from Wayne et al. 1997; B from Vilà, Amorim et al. 1999.)

Maldonado, and Wayne 1999). The protein-coding gene phylogeny shows that the wolf genus *Canis* is a monophyletic group that also includes the dhole or Asian wild dog *(Cuon alpinus)*. The gray wolf, coyote *(C. latrans)*, and Simien jackal or Ethiopian wolf *(C. simensis)* form a monophyletic group or clade (i.e., a group with a single common ancestor), with the golden jackal *(C. aureus)* as the most likely sister taxon (fig. 8.2A). Basal to *Canis* and *Cuon* are the African hunting dog *(Lycaon pictus)* and a clade consisting of two South American canids, the bush dog *(Speothos venaticus)* and the maned wolf *(Chrysocyon brachyurus)*. Consequently, although the African hunting dog (or painted wolf) preys on large mammals as the gray wolf and dhole do, it is not closely related to either species.

The Domestic Dog

The species most closely related to the gray wolf is the domestic dog *(Canis lupus familiaris)* (Tsuda et al. 1997; Vilà et al. 1997; Vilà, Maldonado, and Wayne 1999; Leonard et al. 2002; Savolainen et al. 2002) (fig. 8.2B). A comprehensive survey of mitochondrial control region sequences in 140 dogs and 162 wolves showed that gray wolves were the exclusive ancestors of dogs (Vilà et al. 1997). Recently, this result was confirmed in a study of 654 dogs and 38 Eurasian wolves (Savolainen et al. 2002). These studies contradict previous theories suggesting that golden jackals also were involved in dog ancestry (Darwin 1871; Lorenz 1954; Coppinger and Schneider 1995). Coyotes and Ethiopian wolves are the closest wild relatives of gray wolves. However, they show sequence divergence values greater than about 4%, as opposed to an average of 1.8% between dogs and gray wolves (Wayne and Jenks 1991; Girman et al. 1993; Gotelli et al. 1994; Vilà et al. 1997; Vilà, Maldonado, and Wayne 1999). Nevertheless, dogs and wolves are not two monophyletic groups (see below).

A comparison of mitochondrial control region sequences of dogs from sixty-seven breeds and wolves from twenty-seven populations shows that domestic dogs possess four mtDNA lineages. This result suggests four independent domestication events or, alternatively, a single ancient domestication events and several backcrosses with female wolves (Vilà et al. 1997). Further studies on dogs (Tsuda et al. 1997) and wolves (Randi et al. 2000) have provided additional support for a multiple origin of dogs from wolves. A recent study comparing mtDNA sequences from native American dogs dating from before the arrival of Europeans has shown that these dogs did not derive from an independent domestication of North American wolves (Leonard et al. 2002). Instead, native American dogs derive from dogs that arrived in the New World with the first human immigrants.

About 80% of dog mitochondrial lineages are contained in one clade. The diversity in this clade was used to estimate the date of first domestication at over 100,000 years ago (Vilà et al. 1997). Although this estimation is based on many assumptions concerning mutation rate and has wide confidence intervals, it clearly points toward a domestication date older than that suggested by the archaeological record of about 12,000 to 14,000 years ago (Davis and Valla 1978; Nobis 1986; Clutton-Brock 1999). However, considerable debate still surrounds the date of the domestication of dogs from wolves (Federoff and Nowak 1997; Scott et al. 1997; Clutton-Brock 1999; Coppinger and Coppinger 2001). A recent genetic study showed that mtDNA sequence data can be consistent with a single origin from East Asia about 15,000 years ago (Savolainen et al. 2002).

Hybridization in *Canis*

Due to their close relationship, wolves and dogs easily hybridize (see below). So too do wolves and coyotes (Lehman et al. 1991), although hybridization with coyotes may be limited to a single group of wolves recently suggested to be a distinct species, *Canis lycaon* (Wilson et al. 2000). Additionally, Ethiopian wolves hybridize with feral dogs (Gotelli et al. 1994). In fact, all species in the genus *Canis*, as well as the dhole and the African hunting dog, possess identical numbers of chromosomes ($2n = 78$: Wurster-Hill and Centerwall 1982; Wayne, Nash, and O'Brien 1987a,b), and several *Canis* species can hybridize with each other (Gray 1954, Kolenosky 1971).

The mtDNA sequencing results and studies using other genetic markers, including microsatellites and Y-chromosome sequences, clearly establish the close relationship between wolves and the coyote and Ethiopian wolf (Roy, Geffen et al. 1994; Gotelli et al. 1994; Vilà et al. 1997; Vilà, Amorim et al. 1999; C. Vilà, unpublished data). These findings suggest a Pleistocene divergence among these taxa and the possibility of extensive hybridization caused by insufficient reproductive isolating barriers. In fact, as discussed below, hybridization may be causing the extinction of a distinct North American wolf.

Population Relationships

Gray wolf populations show evidence of genetic differentiation on regional and continental scales. Wolves in the Old and New World, with one exception, do not share mtDNA haplotypes, as defined by restriction fragment length polymorphisms (Wayne et al. 1992). Nor do Old and New World populations share mtDNA control region sequences (Vilà, Amorim et al. 1999). However, sequence divergence values and phylogenetic analysis of control region sequence data (fig. 8.3) imply that the New World was invaded multiple times by wolves representing distinct haplotype clades (e.g., haplotypes 28, 29, 30, 31, 32, and 33 in fig. 8.3). Alternatively, North America may have been invaded from the Old World by a genetically diverse founder population. Analysis of DNA from Arctic permafrost specimens accumulated over the last 50,000 years may resolve these alternatives (Wayne et al. 1999; Leonard et al. 2000). However, the similarity of sequences in some Alaskan and Russian wolves suggests that Siberian wolves may have been a recent source of migrants prior to the closing of the Bering land bridge about 10,000 years ago (Wayne et al. 1992).

Wolf populations of the Old and New World show varying degrees of genetic subdivision (Wayne et al. 1992; Roy, Geffen et al. 1994; Ellegren et al. 1996; Forbes and Boyd 1996, 1997; Ellegren 1999; Randi et al. 1993, 1995, 2000; Vilà, Amorim et al. 1999). In the Old World, mtDNA data suggest that most populations are genetically differentiated, with the exception of neighboring populations such as those in Spain and Portugal or in recently invaded areas such as France, where Italian wolves have migrated (Taberlet, Gielly, and Bouvet 1996; Vilà, Amorim et al. 1999; Randi et al. 2000). In western Europe, genetic subdivision may reflect habitat fragmentation that occurred over the past few hundred years with the loss of forests and, more importantly, the dramatic decrease in the size of all wolf populations due to human persecution (Wayne et al. 1992; Vilà, Amorim et al. 1999; Vilà, Sundqvist et al. 2002; Flagstad et al. 2003).

However, mtDNA control region sequence relationships do not show a consistent relationship with geography. For example, control region sequences from China (17, 22, and 23 in fig. 8.3) and Spain (1, 2, and 4) are more similar than are those from Spain and Italy (5). Moreover, some localities, such as Greece, contain highly di-

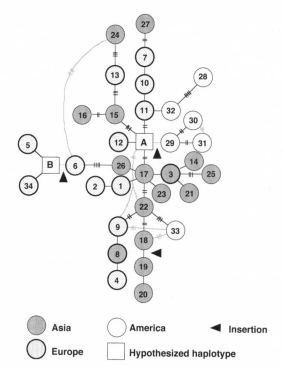

FIGURE 8.3. Minimum spanning network of Old and New World mtDNA control region haplotypes. Each circled number indicates a unique wolf haplotype. Squares indicate hypothetical bridging haplotypes. The number of sequence substitutions separating haplotypes is one, unless otherwise indicated by hatch marks (one per substitution) on the lines connecting haplotypes. Alternative connections between haplotypes are indicated as light lines. Note that haplotypes from the same geographic regions do not cluster on the same branches (they do not form reciprocally monophyletic clades). None of the haplotypes were shared between America and Europe or Asia, and only one was shared between Europe and Asia. (Note that the numbered control region haplotypes in this figure are not homologous to the numbered haplotypes in table 8.2, as they are defined with different kinds of data.) (Data from Vilà, Amorim et al. 1999.)

vergent control region sequences (3, 6, 10, 11). This pattern suggests the effect of multiple invasions following the many glaciations of the Pleistocene added to the extremely high mobility of wolves. With each glacial retreat and the advance of forest into formerly glaciated areas, new waves of immigrating wolves may have added diversity to refugial populations, resulting in very poorly defined patterns of genetic differentiation in mtDNA sequences across populations (Vilà, Amorim et al. 1999). The degree of genetic similarity between populations appears to depend more on the population-specific history of immigration and demography than on the geographic distance separating populations. Patterns of genetic differentiation arising after Ice Age glaciations may reflect

TABLE 8.2. Distribution of mtDNA RFLP haplotypes in North American and European populations of wolves

Population (sample size)	Old World haplotypes						
	W3	W5	W6	W15	W16	W17	W18
Portugal (3)						1.00	
Sweden (9)	1.00						
Estonia (3)	0.67						0.33
Italy (14)				1.00			
Israel (2)					1.00		
Iran (2)		1.00					
China (2)			1.00				

Population	New World haplotypes				
	W1	W2	W3	W4	W14
Alaska—Denali (22)			0.73	0.27	
Alaska—Anaktuvuk (14)			1.00		
Alaska—Nome (8)			1.00		
Alaska—Kenai (6)			0.17	0.83	
Yukon Territory (10)			1.00		
Northwest Territones—Inuvik (53)	0.75		0.23	0.02	
Northwest Territories—Yellowknife (8)	0.88		0.12		
British Columbia—Vancouver I (15)			1.00		
Alberta (4)	0.25		0.75		
Montana (5)	0.20		0.60	0.20	
Manitoba (4)		1.00			
Minnesota—Northeast (20)	0.95			0.05	
Minnesota—North (11)	1.00				
Central Ontario (12)	0.67			0.33	
Mexico (4)					1.00

Population	Coyote-like haplotypes[a]						
	W7	W8	W9	W10	W11	W12	W13
Manitoba (4)	1.00						
Minnesota—Northeast (46)	0.52		0.48				
Minnesota—North (9)	0.22		0.78				
Michigan—Isle Royale (7)		1.00					
Western Ontario (15)	0.53		0.47				
Central Ontario (13)	0.39	0.08	0.46				0.08
Eastern Ontario (3)				0.33	0.33	0.33	
Quebec (9)				0.07	0.07	0.43	0.43

Source: Wayne et al. 1992.

[a] These haplotypes are grouped phylogenetically with those from coyotes.

chance dispersal events rather than geographic distance or a long history of genetic isolation (Leonard et al. 2000).

The presence of genetic subdivision in Europe contrasts with the pattern in North America, where clinal changes in haplotype frequencies are apparent. For example, in the study by Wayne et al. (1992), mitochondrial haplotype W3 was common in Alaska and the Northwest Territories, but absent from populations in eastern Canada. Conversely, haplotype W1 was absent in Alaskan wolves, but common in eastern Canada (table 8.2). A similar pattern was observed for mitochondrial control region sequences by Vilà, Amorim et al. (1999). Perhaps these clinal patterns reflect to some extent prior isolation in southern and Alaskan Pleistocene refugia, followed by expansion and intergradation during interglacials (Hewitt 2000; see Nowak, chap. 9 in this volume), as is evident in North American brown bears (Talbot and Shields 1996; Waits et al. 1998).

Alternatively, the southern wolf population may have been a distinct red wolf–like canid, *Canis lycaon*, that is now interbreeding with gray wolves that migrated into eastern Canada after the last glaciation (Wilson et al. 2000). Regardless, North American gray wolf populations have proved not to be as dramatically structured as their Old World counterparts, as evidenced by the fact that variation among populations was high and several of the common mtDNA haplotypes were widely distributed (Wayne et al. 1992; see table 8.2).

The most highly differentiated North American gray wolf taxon is the Mexican wolf *(Canis lupus baileyi)*. Except for a reintroduced experimental population, this subspecies is thought to be extinct in the wild and exists only in three captive populations, each initiated by a small number of founders (Hedrick et al. 1997; Fritts et al., chap. 12, and Boitani, chap. 13 in this volume). Two of the captive Mexican wolf populations display a single divergent mtDNA haplotype, found nowhere else (33 in fig. 8.3), that is more closely related to a subset of Old World haplotypes than to any New World haplotype, suggesting that these Mexican wolves share a more recent ancestry with wolves from the Old World. The most similar haplotypes in the New World differ from theirs by five substitutions (2.2%) and one insertion-deletion event, whereas the most similar haplotypes found in Eurasian populations differ from theirs by only three substitutions (1.3%) (see fig. 8.3; Vilà, Amorim et al. 1999). Further, the basal position of the Mexican wolf haplotype in phylogenetic trees suggests that it is a relict form stemming from an early invasion of gray wolves from Asia (Wayne et al. 1992, Vilà, Amorim et al. 1999). However, an additional mtDNA haplotype similar to those of North American gray wolves has been found (S. Fain, unpublished data) in the third captive Mexican wolf population (Hedrick et al. 1997).

Analyses of microsatellite loci in North American wolves confirmed some of the patterns shown by the mtDNA data (Roy, Girman, and Wayne 1994; Roy, Geffen et al. 1994; Forbes and Boyd 1996). Genetic differentiation among populations in North America was low to moderate (table 8.3). A statistical analysis of allele fre-

quencies suggested that populations may often be exchanging a few individuals per generation, enough to prevent appreciable genetic differentiation by random drift (Slatkin 1987). Studies done on Rocky Mountain wolves sampled in Wyoming, Montana, Alberta, and British Columbia showed them to be slightly differentiated (Forbes and Boyd 1996, 1997). These studies, which combined genetic and field observations, supported long-distance dispersal in wolves (see Mech and Boitani, chap. 1 in this volume). Based on these results, the genetic diversity of naturally recolonized populations is likely to remain high (see below). Ecological factors are important as well; a recent study suggested that wolves specializing on different caribou herds in the Canadian Northwest, as well as populations on Banks and Victoria Islands, are differentiated (Carmichael et al. 2001). Additionally, when genetic differentiation was considered over a wide range of geographic scales, the microsatellite data revealed a pattern of differentiation with distance (Forbes and Boyd 1996). However, in an earlier study that primarily considered more widely spaced populations, the relationship between geographic and genetic distance was not significant (Roy, Geffen et al. 1994). Last, the most highly differentiated population was the Mexican wolf (García-Moreno et al. 1996; Hedrick et al. 1997), a population also shown to be distinct by mtDNA data (see above).

Several conservation implications are suggested by these genetic results. First, because the endangered Mexican gray wolf is genetically and physically distinct and historically isolated from other gray wolves (Nowak 1979), the breeding of pure Mexican wolves in captivity for reintroduction into the wild (Fritts et al., chap. 12, and Boitani, chap. 13 in this volume) is well justified. Second, because most wolf populations in North America are not strongly differentiated genetically and gene flow is high among populations, reintroduction need not include only the nearest extant populations as source material. Although the reintroduced Yellowstone wolves are slightly different from naturally recolonizing wolves in Montana, this minor difference is not a conservation concern (Forbes and Boyd 1997). However, reintroducing wolves from populations where hybridization with coyotes has occurred may not be advisable (see below). Finally, although contemporary wolf populations in Europe appear more genetically subdivided than their North American counterparts (Wayne et al. 1992; Vilà, Amorim et al. 1999; Randi et al. 2000; Vilà,

TABLE 8.3. Nei's unbiased genetic distance among populations of wolves and coyotes, based on a survey of ten microsatellite loci

	Wolves							Coyotes						Red wolf
	Vancouver	Kenai, AK	Alberta	*Minnesota	*S. Quebec	N. Quebec	Northwest Territories	Washington	Kenai	Alberta	Minnesota	Maine	California	
Wolves														
Vancouver														
Kenai	0.243													
Alberta	0.300	0.214												
*Minnesota	0.672	0.425	0.408											
*S. Quebec	0.519	0.272	0.295	0.135										
N. Quebec	0.418	0.208	0.374	0.296	0.281									
Northwest Territories	0.259	0.223	0.182	0.468	0.251	0.357								
Coyotes														
Washington	0.712	0.540	0.437	0.345	0.240	0.565	0.387							
Kenai	0.813	0.581	0.578	0.423	0.325	0.642	0.510	0.190						
Alberta	0.871	0.624	0.527	0.445	0.342	0.429	0.413	0.233	0.271					
Minnesota	0.728	0.559	0.491	0.402	0.393	0.427	0.425	0.221	0.212	0.123				
Maine	0.662	0.525	0.566	0.385	0.225	0.461	0.429	0.200	0.214	0.091	0.106			
California	0.672	0.535	0.522	0.448	0.261	0.480	0.393	0.116	0.190	0.228	0.248	0.181		
Red wolf	0.662	0.671	0.603	0.323	0.255	0.466	0.534	0.309	0.365	0.358	0.418	0.323	0.271	
Golden jackal	1.217	0.841	1.219	1.284	1.089	1.015	0.985	1.108	1.761	1.183	1.066	1.183	1.303	1.459

Note: Boxes enclose pairwise distances among gray wolf and among coyote populations. Values outside boxes involve comparisons among different species. Asterisks indicate gray wolf populations thought to hybridize with coyotes (Roy, Geffen et al. 1994b). Genetic divergence among populations increases with genetic distance values.

Sundqvist et al. 2002; Flagstad et al. 2003), the North American pattern might well reflect the ancestral condition in western Europe prior to habitat fragmentation and population decimation. Therefore, efforts to increase gene flow among European wolf populations to levels similar to that in North America could be defended.

Population Variability

Within large, interconnected wolf populations, mtDNA and microsatellite variability are often high (tables 8.4 and 8.5). For example, large populations in the Old and New World generally have several mtDNA control region or RFLP haplotypes (Wayne et al. 1992; Vilà, Amorim et al. 1999; Randi et al. 2000). Similarly, the average number of microsatellite alleles in North American wolf populations, except for Mexican wolves, is 5.0, and their average heterozygosity is 54% (Roy, Geffen et al. 1994). These values are similar to those found in other vertebrates (Avise 1994, 2000).

The mtDNA variation in wolves is less than that in coyotes (Lehman and Wayne 1991; Vilà, Amorim et al. 1999), as might be predicted from the current higher abundance of coyotes (Voigt and Berg 1987; Ginsberg and Macdonald 1990). Even though only a small number of coyotes were sampled, and even though their distribution is restricted to North America, the average sequence divergence observed among coyotes was higher than that among wolves sampled throughout the world (Vilà, Amorim et al. 1999). However, genealogical measures of nucleotide diversity suggest that wolves were more abundant than coyotes in the past and that both populations and species declined throughout the late Pleistocene (Vilà, Amorim et al. 1999). Nucleotide diversity data imply a decline in wolves from over 5 million breeding females (about 33 million wolves) worldwide in the late Pleistocene to about 173,000 breeding females (1.2 million wolves) in the recent past. Today about 300,000 wolves exist worldwide (see Boitani, chap. 13 in this volume). This pattern contrasts with a decline followed by a very recent increase in coyotes. As suggested by nucleotide diversity values, coyote numbers decreased from about 3.7 million breeding females (about 18 million coyotes) in the late Pleistocene to 460,000 breeding females (2.2 million coyotes) in the recent past. This drop was followed by an increase to about 7 million coyotes today (Vilà, Amorim et al. 1999 and references therein). The differences between the abundance estimates for the recent past and for today may reflect human-related changes in forested habitats and direct persecution that reduced wolf numbers but increased

TABLE 8.4. Number of haplotypes and levels of nucleotide diversity based on mtDNA control region sequences in gray wolf populations

Population (sample size)	No. haplotypes	Nucleotide diversity[a]	Reference[a]
Italy (101)	1	0.0000	Vilà, Amorim et al. 1999; Randi et al. 2000
France (7)	1	0.0000	Vilà, Amorim et al. 1999
Portugal (19)	2	0.0009	Vilà, Amorim et al. 1999
Spain (84)	3	0.0038	Vilà, Amorim et al. 1999
Bulgaria (29)	6	0.0145	Randi et al. 2000
Croatia (6)	2	0.0072	Vilà, Amorim et al. 1999
Yugoslavia (7)	2	0.0271	Vilà, Amorim et al. 1999
Greece (7)	4	0.0160	Vilà, Amorim et al. 1999
Israel (16)	1	0.0000	Vilà, Amorim et al. 1999
Saudi Arabia (7)	5	0.0201	Vilà, Amorim et al. 1999
Iran (6)	2	0.0015	Vilà, Amorim et al. 1999
Afghanistan (8)	2	0.0139	Vilà, Amorim et al. 1999
Mongolia (8)	4	0.0148	Vilà, Amorim et al. 1999
Mexico (captive) (6)	1	0.0000	Vilà, Amorim et al. 1999
Scandinavia (93)	1	0.0000	Ellegren 1999; Vilà, Sundqvist et al. 2002

[a]All studies used 29 microsatellite loci.

TABLE 8.5. Average number of alleles and expected heterozygosity based on microsatellite loci in gray wolf and coyote populations

Population (average sample size)	No. alleles[a]	Expected heterozygosity	Microsatellite loci	Reference
Gray wolf				
Vancouver (12.6)	3.4	0.566	10	Roy, Geffen et al. 1994
Kenai (18.9)	4.1	0.581	10	Roy, Geffen et al. 1994
Alberta (18.2)	4.5	0.668	10	Roy, Geffen et al. 1994
Minnesota (19.8)	6.3	0.686	10	Roy, Geffen et al. 1994
Southern Quebec (20.0)	6.4	0.741	10	Roy, Geffen et al. 1994
Northern Quebec (13.3)	4.1	0.565	10	Roy, Geffen et al. 1994
Northwest Territories (20.9)	6.4	0.721	10	Roy, Geffen et al. 1994
Mexico (captive) (38.9)	2.5	0.314	10	García-Moreno et al. 1996
Sweden (captive) (29)	2.9	0.510	10	Ellegren 1999
Scandinavia (93)	3.7	0.550	29	Ellegren 1999; Vilà, Sundqvist et al. 2002
Finland and northwestern Russia (93)	8.0	0.770	29	Ellegren 1999; Vilà, Sundqvist et al. 2002
Coyote				
Washington (15.9)	5.8	0.666	10	Roy, Geffen et al. 1994
Kenai (12.8)	4.9	0.627	10	Roy, Geffen et al. 1994
Alberta (16.8)	6.1	0.702	10	Roy, Geffen et al. 1994
Minnesota (18.4)	5.7	0.709	10	Roy, Geffen et al. 1994
Maine (16.2)	6.1	0.702	10	Roy, Geffen et al. 1994
California (22.1)	6.9	0.644	10	Roy, Geffen et al. 1994
Red wolf				
Captive colony (29.9)	5.3	0.548	10	Roy, Geffen et al. 1994
Golden jackal				
Kenya (16.4)	4.8	0.520	10	Roy, Geffen et al. 1994

[a]The number of alleles and sample size are correlated, hence caution should be used in interpreting differences in diversity among populations (see Roy, Geffen et al. 1994).

coyote abundance and distribution (see discussion in Lehman et al. 1991 and Vilà, Amorim et al. 1999). However, these changes have not been completely paralleled by the expected changes in genetic diversity.

Population Bottlenecks

Some wolf populations do show significantly low levels of genetic variation expected from recent population declines. For example, the Italian wolf population declined dramatically in the eighteenth and nineteenth centuries due to habitat loss and predator control programs (Randi et al. 1993, 1995, 2000). By the 1970s, only about a hundred wolves were left in Italy, mostly in the central and southern Apennine Mountains (Zimen and Boitani 1975). Extensive mtDNA RFLP and control re-

gion sequencing studies showed these wolves to have a single mitochondrial haplotype, a diversity generally lower than in other Old World populations (Wayne et al. 1992; Vilà, Amorim et al. 1999; Randi et al. 1995, 2000). Moreover, this haplotype is unique, being otherwise found only in French wolves, a population recently founded by wolves from Italy (Taberlet, Gielly, and Bouvet 1996; see also Scandura et al. 2001).

Scandinavian wolves have a similar history. Since the middle 1800s, the Scandinavian wolf population had decreased steadily to functional extinction during the 1960s (Wabakken et al. 2001). Genetic analyses of museum skulls collected during the nineteenth and twentieth centuries show that genetic diversity was lost during this time (except for some occasional immigrants from the neighboring wolf populations [Flagstad et al. 2003]).

However, a breeding pack appeared in southern Sweden in 1983, more than 900 km (540 mi) away from the known limits of the wolf distribution in Finland and Russia (Wabakken et al. 2001). A heated public debate arose about the origin of these wolves (Ellegren et al. 1996). The combination of maternal (mtDNA), paternal (Y chromosome microsatellites), and biparentally inherited genetic markers (autosomal and X chromosome microsatellites) showed that the population had been founded by one male and one female from Finland and Russia. An origin from the captive wolf population or from dogs could be excluded (Ellegren et al. 1996; Vilà, Sundqvist et al. 2002). These diverse genetic markers, combined with annual surveys of the wild population (Wabakken et al. 2001), also allowed reconstruction of its history. After the founding female was killed in 1985, some litters were born from incestuous mating, leading to a decrease in genetic diversity (Vilà, Sundqvist et al. 2002). During this time the population remained very small, with only one breeding pack, which failed to reproduce one year.

However, in 1991, a second pack was established by the arrival of one additional male immigrant. From that moment, the population started to grow exponentially at 29% per year (Wabakken et al. 2001), and by summer 2001 it was estimated to contain 120–140 individuals (see Boitani, chap. 13 in this volume). The arrival of just one immigrant male may have been crucial for the recovery of the population, and at least 68 of the 72 individuals born after 1993 for which samples were analyzed carry at least one of the alleles introduced by the immigrant (Vilà, Sundqvist et al. 2002).

Since the entire population may derive from as few as three founders, its genetic variability is much reduced. Only one mtDNA haplotype (Ellegren et al. 1996, Ellegren 1999) and two Y chromosome haplotypes (Sundqvist et al. 2001) are found, and variability is also reduced at autosomal and X chromosome microsatellites (Vilà, Sundqvist et al. 2002). The level of inbreeding observed in the Scandinavian wolves was similar to that of the Swedish captive population, in which inbreeding depression was detected (Laikre and Ryman 1991; Laikre et al. 1993). Since gene flow from neighboring wolf populations is low and genetic variability may still be lost (Vilà, Sundqvist et al. 2002), the risk of inbreeding depression remains important.

Finally, two captive Mexican wolf populations possess a very low number of alleles and a low level of heterozygosity (see table 8.5). Moreover, only two mtDNA haplotypes are found in the three existing captive populations. The total founding population numbered about seven. In the past, only the certified lineage, founded from three individuals of known Mexican wolf ancestry, was used in the captive breeding program (Fritts et al., chap. 12, and Boitani, chap. 13 in this volume). Recent genetic analyses established a close kinship among the three captive populations and found no evidence of dog, coyote, or northern gray wolf ancestry (García-Moreno et al. 1996; Hedrick et al. 1997). Although previous studies (Kalinowski et al. 1999) failed to find evidence of inbreeding depression in Mexican wolves, Fredrickson and Hedrick (2002) suggest that, as in the captive Scandinavian wolves, some inbreeding depression does exist in the Mexican wolf.

Inbreeding may have contributed to the temporary decline of the isolated Isle Royale wolf population in northern Lake Superior. The population was probably founded by a single pair of gray wolves that crossed an ice bridge from the Canadian mainland about 1949 (Mech 1966b). Thereafter, this well-monitored population increased to over fifty wolves by 1980, but then dramatically declined to a dozen or fewer individuals by 1990 (Peterson and Page 1988; Peterson 1995b, 1999). For several years, no new litters were raised. Disease and changes in food abundance were first suggested as causes for the decline, but both became increasingly improbable explanations when no evidence of disease was found in serological surveys and when wolf numbers did not increase as expected when their main prey, moose, increased (Peterson and Page 1988; Peterson et al. 1998; Peterson 1999).

Molecular genetic analysis showed that the Isle Royale wolf population possessed a single mtDNA haplotype and only half the level of allozyme heterozygosity observed in an adjacent mainland population. Furthermore, results of a multilocus DNA fingerprint survey suggested that the Isle Royale wolves were related about as closely as full siblings or parent-offspring pairs in captivity (Wayne et al. 1991). Inbreeding depression was thus suggested as the explanation for the population decline. In captive populations, heterozygosity is inversely related to the severity of inbreeding (Ellegren 1999; Hedrick et al. 2001). Although there is no direct evidence that inbreeding depression is affecting the survival of the wolves on Isle Royale, the effects of inbreeding are extremely difficult to detect in most natural populations because individual viability and fecundity cannot be closely monitored.

A possible alternative to inbreeding depression as a cause of lack of breeding on Isle Royale is that after the population crash to a single breeding pair, only wolves that recognized each other as siblings or parent-offspring were available for pair bonding (Wayne et al. 1991). Thus, behavioral avoidance of incest may have prevented the formation of additional breeding pairs until individuals from litters with no temporal overlap were produced. The same hypothesis has been suggested to explain the lack of population growth in the Scandinavian wolf population during the 1980s (Vilà, Sundqvist et al. 2002). Evidence against this explanation, however, is the fact that in captivity, such close relatives, even including littermates, will mate (Medjo and Mech 1976). In any case, the Isle Royale wolf population began increasing again in the mid-1990s, reaching 29 individuals in 2000, 19 in 2001, and 17 in 2002 (Peterson and Vucetich 2002). Conceivably, recessive deleterious alleles may have been purged from the population (see discussion in Templeton et al. 1987; Templeton and Read 1998; Kalinowski et al. 2000; Keller and Waller 2002).

Pack Structure and Mating Systems

Generally a wolf pack consists of a mated pair, their immediate offspring, and adult helper offspring from previous years (see Mech and Boitani, chap. 1 in this volume). In areas that have sufficient resources to support many wolves, packs develop well-defined territories and inter-pack aggression can be intense (Mech and Boitani, chap. 1, and Fuller et al., chap. 6 in this volume). Thus, within each pack, most members should be closely related, and they should be less closely related to wolves in neighboring packs.

Microsatellite analyses of wolf packs in Denali National Park, Alaska, and in northern Minnesota confirmed these relationships (fig. 8.4; Lehman et al. 1992; D. Smith et al. 1997). Additionally, genetic fingerprinting confirmed that wolf packs are not inbred islands (fig. 8.4); rather, offspring disperse into neighboring packs, despite high levels of inter-pack aggression, or form new nearby packs (see Mech and Boitani, chap. 1 in this volume). In Minnesota, for example, seven inter-pack genetic similarity values were as large as those between known siblings or parent-offspring pairs (fig. 8.4), and in Denali National Park, six such inter-pack connections were discovered. However, no such inter-pack genetic similarity was observed among wolf packs in the Inuvik region of Canada's Northwest Territories. The difference in inter-pack relatedness among the three populations may indicate higher genetic turnover in the Inuvik population, where wolves are heavily controlled. Instances of female-female relatedness between wolf packs were more common than male-male connections, suggesting a sex bias in gene flow, with males dispersing greater distances, or dispersing more frequently, than

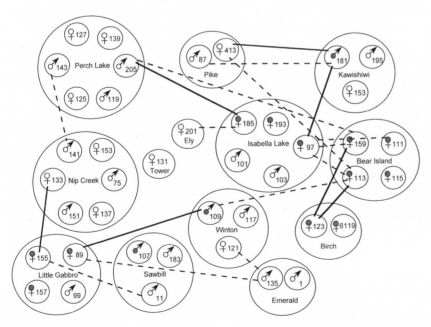

FIGURE 8.4. Pack membership, sex, and inter-pack kinship connections among individuals from Minnesota wolf packs, based on DNA fingerprinting. Large circles represent packs (the positions of packs in the figure do not indicate geographic proximity). Solid lines indicate genetic relatedness as great as that between parents and their offspring or siblings; dotted lines indicate a more distant but significant level of kinship. The mtDNA haplotype is "coyote-like" unless indicated by a shaded sex symbol. Note mixed-haplotype packs (see text and fig. 8.5). (From Lehman et al. 1992a.)

females. This finding contrasts with known dispersal behavior (see Mech and Boitani, chap. 1 in this volume). However, genetic data reflect dispersal that results in reproduction, whereas observational data on dispersal may not.

Hybridization

Wolf-Dog Hybridization

Since the origin of dogs from gray wolves, perhaps tens of thousands of years ago, dogs have interbred with wolves several times (Vilà et al. 1997). Dog breeders as well as many native American tribes have occasionally crossed their dogs with wolves to improve vigor (Schwartz 1997), and several hundred thousand captive wolf-dog hybrids may exist today in the United States (Hope 1994). In the wild, hybridization between gray wolves and dogs is likely to be most frequent near human settlements, where wolf density is low and feral and domestic dogs are common (see Nowak, chap. 9, Fritts et al., chap. 12, and Boitani, chap. 13 in this volume). The genetic integrity of wild wolf populations has been a concern among some conservationists (Blanco et al. 1992; Boitani 1993). However, analyses of mtDNA sequences have indicated that hybridization between wolves and dogs is rare (Vilà and Wayne 1999; Randi et al. 2000). Recently, methods have been developed that can detect hybridization using autosomal markers (Pritchard et al. 2000; Vilà, Walker et al. 2003).

A detailed genetic study was conducted on one apparent wolf-dog hybrid killed by a car in Norway (Vilà, Walker et al. 2003). Researchers examined mtDNA, Y-chromosome microsatellites, and eighteen autosomal microsatellites from the animal and from a large sample of Scandinavian, Finnish, and Russian wolves and domestic dogs. The suspected hybrid's haplotype was most consistent with that of a first-generation cross between a female Scandinavian wolf and a male domestic dog. Additionally, the study suggested that the Scandinavian wolf population is well differentiated from domestic dogs. Although this population consists of only 120–140 wolves living in a human-dominated landscape, hybridization with domestic dogs seems infrequent. Other studies have also shown rare hybridization between dogs and wolves in several areas across Europe (in Bulgaria: Randi et al. 2000; in Latvia: Andersone et al. 2002; in Italy: Randi and Lucchini 2002; in Spain: C. Vilà and L. Llaneza, personal observation). Limited data collected

in Latvia suggest that hybridization may be more frequent during wolf population declines (Andersone et al. 2002). Extensive wolf-dog hybridization also is not supported by morphological evidence (see Nowak, chap. 9 in this volume).

Wolf-Coyote Hybridization

Interbreeding between highly mobile species, such as wolves and coyotes, may result in the establishment of large hybrid zones. The gray wolf once ranged throughout most of North America and parts of Mexico, but over the past few hundred years, wolves have been eliminated from Mexico and much of the United States (see Boitani, chap. 13 in this volume). Similarly, the red wolf *(Canis rufus)* was exterminated by about 1975 from throughout its historic distribution, which included much of the southeastern United States, although it has since been reintroduced (see Phillips et al., chap. 11 in this volume).

Coyotes interbred extensively with red wolves as they approached extinction (Nowak 1979); consequently, genetic markers otherwise unique to coyotes are found in red wolves (Wayne and Jenks 1991; Roy, Geffen et al. 1994). However, an extensive genetic analysis characterizing microsatellite and mtDNA variation in coyotes, gray wolves, and historic and recent red wolves found no markers unique to red wolves, and instead found haplotypes and microsatellite alleles identical or very similar to those in gray wolves and coyotes (Roy, Geffen et al. 1994; Roy, Girman and Wayne 1994; Roy et al. 1996). Based on these findings, an origin of the red wolf through hybridization of gray wolves and coyotes in historic times or earlier was postulated (Wayne and Jenks 1991; Roy, Geffen et al. 1994; Roy et al. 1996; Reich et al. 1999). However, no single hypothesis for the origin of the red wolf is universally accepted (see Nowak, chap. 9, and Phillips et al., chap. 11 in this volume).

Evidence of hybridization between wolves and coyotes from Minnesota and eastern Canada was revealed by analysis of mtDNA and microsatellite loci (Lehman et al. 1991; Roy, Geffen et al. 1994). Coyotes invaded Minnesota about 100 years ago and then moved into eastern Canada and New England within the last 50 years (Nowak 1979; Moore and Parker 1992). Analyses of mtDNA from recent wolves from the Great Lakes region found a high proportion of haplotypes similar to those in coyotes. The frequency of "coyote" haplotypes in gray

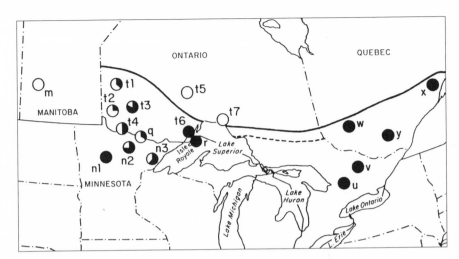

FIGURE 8.5. Frequency (solid portion of circles) of "coyote-like" RFLP haplotypes in gray wolves of the Great Lakes region (see table 8.2). Letters and numbers are used to identify sampling localities. The solid line describes the northern extent of coyotes in Ontario (Kolenosky and Standfield 1975) and in Quebec (Georges 1976). The dotted line describes the northern extent of *C. lupus lycaon* (boreal type) as determined by Kolenosky and Standfield (1975). (From Lehman et al. 1991.)

wolves increases to the east, from 50% in Minnesota to 100% in southern Quebec (fig. 8.5). The authors hypothesized that hybridization between coyotes and wolves had occurred in disturbed areas of eastern Canada where wolves had become rare through predator control efforts and habitat loss, but coyotes had become common.

Presumably, the recent arrival of coyotes, coupled with a decrease in the population of a small wolf subspecies in the Great Lakes area (*Canis lupus lycaon,* Brewster and Fritts 1995; Nowak 1995a), has led to interspecific hybridization. No sampled coyotes had wolflike haplotypes; since mtDNA is maternally inherited, this finding suggested that the predominant cross was between male wolves and female coyotes, followed by backcrosses to wolves.

This conclusion that wolves and coyotes had hybridized was also supported by microsatellite analysis, which showed that wolves from this area were genetically more similar to coyotes than to wolves elsewhere (fig. 8.6). The microsatellite data indicated that two or three successful hybridization events per generation could explain the allele frequency similarity found (Roy, Geffen et al. 1994). Thus, the genetic data imply both that significant hybridization has occurred between the two species and that introgression of coyote genes into the wolf population has occurred over a broad geographic region.

The directionality of hybridization between wolves and coyotes in the Great Lakes area is puzzling. Conceivably, male wolves might take advantage of their large size to impregnate female coyotes (Lehman et al. 1991). However, female wolves and male coyotes are more closely matched in size, and thus may also be likely to mate

(L. D. Mech, personal communication). The only published information on known wolf-coyote hybridization involved a female wolf and a male coyote (Kolenosky 1971; Schmitz and Kolenosky 1985a). This observation has been used to suggest the possibility of introgression

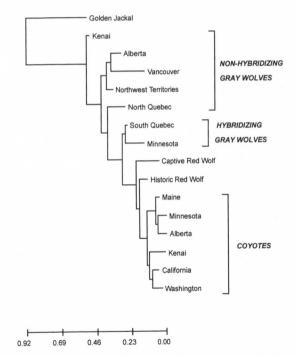

FIGURE 8.6. Neighbor-joining genetic distance tree of wolf and coyote populations. Scale is in Nei's genetic distance units (see table 8.3). The position of the historic red wolf population is based on the analysis of museum skins, generally obtained before 1940 (Roy et al. 1996). (Data from Roy, Geffen et al. 1994; Roy et al. 1996.)

of wolf DNA into coyote populations (Schmitz and Kolenosky 1985b). The possibility that coyote mtDNA haplotypes might have completely replaced wolf haplotypes in some populations is also puzzling (Lehman et al. 1991). This could possibly be due to hybridization if it took place when wolf numbers were reduced to a very low threshold in the presence of many more coyotes. However, definitive testing of these ideas will require field observations on mating between the two species in areas where wolf numbers are low and coyotes are abundant.

New genetic results question whether the red wolf and canids from the Great Lakes area are simply gray wolf-coyote hybrids (Wilson et al. 2000). Detailed genetic analysis of eastern Canadian wolflike canids and coyotes has found divergent mtDNA control region haplotypes with a distribution centered at Algonquin Provincial Park, Ontario. These divergent haplotypes appear to be phylogenetically similar to those of red wolves, which in turn are grouped with haplotypes of coyotes. However, the divergent genotypes are not reciprocally monophyletic with coyotes, as are those of long separated species such as gray wolves and coyotes (see above). Red wolves and Algonquin wolves also share similarity in microsatellite allele frequencies.

Consequently, Wilson et al. (2000) suggested that the smallish gray wolf that formerly inhabited the Great Lakes areas, *Canis lupus lycaon,* and the red wolf are the same species, designated *Canis lycaon.* Wilson et al. (2000) hypothesized that the Algonquin wolf is a native New World wolflike form that evolved independently from North American coyote-like ancestors (cf. Nowak, chap. 9 in this volume, and Nowak 2002). *Canis lycaon* can interbreed with coyotes, but Wilson et al. (2000) suggested that gray wolves *(Canis lupus)* cannot. *Canis lycaon,* on the other hand, can interbreed with the gray wolves that would have migrated into southern Canada from an Arctic refugium since the last glaciation. In Minnesota, some wolves have coyote (or *lycaon*)-like mtDNA sequences, whereas other members of the same packs have sequences like those in gray wolves (see fig. 8.4). Therefore, the hybrid zone in the Great Lakes area would represent a complex mix of *Canis lycaon* and hybrids with coyotes and gray wolves. "Gray wolves" of intermediate size and morphology have been described from two locations in southeastern Ontario (Kolenosky and Standfield 1975); however, past explanations have focused on differences in prey size or on hybridization between gray wolves and coyotes (Kolenosky and Standfield 1975; Hilton 1978; Nowak 1979; Schmitz and

Kolenosky 1985b; Thurber and Peterson 1991; Wayne 1992; Lariviere and Crête 1993).

The Wilson et al. (2000) interpretation of the new genetic data presents a novel paradigm that should be tested with additional genetic and morphological data from populations of coyotes and gray wolves (e.g., Mech and Federoff 2002; Nowak 2002; Hedrick et al. 2002). However, if wolves with coyote-like mtDNA haplotypes are actually a species genetically isolated from the gray wolf, then the common occurrence in Minnesota of wolf and coyote-like haplotypes in individuals belonging to the same wolf pack (Lehman et al. 1991) is problematic.

Conservation and Taxonomic Implications

The gray wolf has been divided into as many as thirty-two subspecies worldwide (Hall and Kelson 1959). Nowak (1995a) suggested that the twenty-four North American subspecies should be reduced to five. However, rates of gene flow among North American wolf populations are high, and differentiation by distance characterizes the genetic variation of wolves at some geographic scales. In this sense, typological species concepts may be inappropriate because geographic variation in the wolf is distributed along a continuum, rather than being partitioned into discrete geographic areas delineated by fixed boundaries (cf. Nowak, chap. 9 in this volume).

Physical differences are evident across the geographic range of wolves. These differences could arise in the presence of gene flow (T. B. Smith et al. 1997; Schneider et al. 1999) and might represent locality-specific adaptations to prey size or climate (e.g., Thurber and Peterson 1991) or size variation with latitude. However, the diversity of environments and prey found throughout the extensive geographic ranges of some subspecies (e.g., *C. lupus nubilus,* 30°–70° N and 60°–120° W: Nowak 1995a) argues against this view (L. D. Mech and L. Boitani, personal communication).

The possible presence of a hybrid zone between a native northeastern wolf species, *Canis lycaon,* and coyotes and gray wolves (see above) complicates taxonomic and conservation recommendations. If *Canis lycaon* is a distinct species, then captive breeding and conservation efforts in situ may be urgently needed. If *Canis lycaon* is a hybrid between gray wolves and coyotes or a hybrid between red wolves and coyotes (see Nowak, chap. 9, and Phillips et al., chap. 11 in this volume) and is a result of human-induced habitat changes and predator control efforts, then, from a conservation genetics viewpoint, *Ca-*

nis lycaon has no conservation merit (Jenks and Wayne 1992; Wayne and Brown 2001).

Furthermore, for the hybridization process to be of conservation concern—even hybridization between a unique North American wolf and other canids—it would have to be hybridization caused by human activities, not by natural hybridization. If the hybridization process is a natural one caused by migration of gray wolves to eastern Canada after the last glacial maximum, then the process should be allowed to continue (Wayne and Brown 2001). Additional genetic data from multiple mitochondrial, nuclear, and Y-chromosome markers are needed to resolve these issues. These markers could allow the identification of populations (and species) that have an independent evolutionary history and an estimate of their divergence times, as well as past and present effective population sizes and the degree of introgression in hybridizing populations. The genetic data collected to date are still too fragmentary to allow a definitive understanding of the evolutionary history of wolf populations. Field data on the environmental context of hybridization also are urgently needed. The specific demographic, ecological, and behavioral circumstances that encourage hybridization among coyotes, gray wolves, and *Canis lycaon* are critical information for the conservation of these species.

Other potential threats to the genetic integrity of the gray wolf include hybridization with dogs, as discussed above. However, this appears less a concern than originally reported for European wolves. There is no genetic evidence for hybridization between wolves and dogs in the wild in North America, except for one case in 2002 between a Mexican wolf and a dog (B. T. Kelly, unpublished data). Although a few episodes of hybridization have been reported in several European countries, extensive introgression of dog genes into wolf populations has not been detected.

Perhaps of greater concern is the loss of genetic variation in isolated wolf populations in the Old World (see Boitani, chap. 13 in this volume). Inbreeding depression has been documented in captivity (Laikre and Ryman 1991; Laikre et al. 1993; Federoff and Nowak 1998; Fredrickson and Hedrick 2002; but cf. Kalinowski et al. 1999). Italian, Scandinavian, and Isle Royale wolves have levels of average relatedness approaching those of inbred captive populations (see above), and could conceivably suffer a decrease in fitness that would eventually affect population persistence (Mace et al. 1996, Hedrick and Kalinowski 2000). High levels of gene flow probably

characterized Old World populations in the past, so there is reason to restore past levels of gene flow in parts of Europe, either through habitat restoration and protection along dispersal corridors or through translocation.

Future research should be aimed at monitoring and predicting genetic changes that will occur in wolf populations and trying to determine any possible population effects. With increasing habitat fragmentation and loss, wolf populations will become more isolated and inbred. Genetic data would allow identification of populations that are losing genetic variation at the greatest rate, so the data could provide guidance for conservation by locating suitable source populations for translocation. Estimates of gene flow would allow translocation efforts to mimic historic patterns of gene flow (e.g., Hedrick 1995).

Further, genetic changes in hybrid zones need careful scrutiny, as these zones are apparently dynamic and will change according to levels of habitat conversion and management. The coyote introgression found so far is currently limited to the Great Lakes region and the Alligator River Preserve in North Carolina, where the red wolf has been reintroduced (see Phillips et al., chap. 11 in this volume), but these hybrid zones may increase or decrease depending on management and habitat changes.

One outstanding question is the genetic effect of wolf harvesting. Preliminary fingerprint data suggested that one heavily controlled population had fewer kinship ties and more genetic turnover than two protected ones (Lehman et al. 1992). If inter-pack kinship affects social stability and pack persistence (Wayne 1996), then control plans that minimize the effects on genetic population structure may need to be considered (see McComb et al. 2001, Leader-Williams et al. 2001).

Another need is to assess genetic variation that is meaningful to survival and reproduction and hence to the persistence of populations (see Fuller et al., chap. 6 in this volume). The role of the major histocompatibility complex (MHC) in immunity to infectious disease is an important question in this regard (Hedrick and Kim 1999; Hedrick and Kalinowski 2000; Hedrick et al. 2000; Hedrick et al. 2002). Finally, new genetic techniques such as microsatellite analysis, Y-chromosome haplotyping, and fecal DNA typing promise a new understanding of wolf mating systems and of the role of kinship in behavior. Fecal DNA analysis is a noninvasive approach that can be used to census populations, document patterns of mating and kinship, and assess sex ratios and population differentiation (Kohn and Wayne

1997; Kohn et al. 1999). Lucchini et al. (2002) have used this approach to trace the origin of the wolves colonizing the Italian Alps. Using microsatellite genotypes obtained from feces, they define the areas used by packs and track their origins.

————

Although the field of molecular genetics is relatively new, it has provided numerous insights into the ecology, behavior, and conservation of the wolf. Furthermore, the rapid technological advances in molecular genetic techniques promise to add many more interesting and valuable contributions to our understanding of this species.

Adult wolves are very attentive to the pups. Both parents feed and care for them. Any older siblings similarly participate in pup care and feeding. Kin selection is probably the best explanation for the latter behavior. *Top:* Photograph by Isaac Babcock. *Bottom:* Photograph by L. David Mech.

Wolves sleep for much of the day. After a long hunt, it may take the pack 12 hours or even more before they are rested enough for another long trek and hunting spree. Photograph by L. David Mech.

Although the main prey of wolves worldwide are large hoofed mammals, wolves sometimes supplement their diets with smaller prey. In the High Arctic of Canada, wolves consume large numbers of arctic hares. Photograph by L. David Mech.

When an adult wolf returns to the pups after obtaining food, the pups rush the animal and excitedly lick at its mouth. The adult then regurgitates a load of meat from its stomach, and the pups frantically down it. Photographs by L. David Mech.

As they attack other large prey, wolves harass muskoxen to try to get them running. If the muskoxen flee, the wolves follow, dashing into the herd to try to grab a calf or other vulnerable member. Photograph by L. David Mech.

Once separated from a herd, a young muskox calf can be caught and subdued. Calves and old bulls constitute the highest proportion of muskoxen killed by wolves. Photograph by L. David Mech.

Wolves quickly open a carcass and sort through the viscera to find the most prized parts. Generally, liver, heart, and intestines are consumed first, followed by the flesh, bones, and hide. Photographs by L. David Mech.

Wolves learn to carry food in their mouths at an early age. They continue to do so throughout their lives to transport food to a peaceful spot to eat it, to bury it for later use, or to bring it to the breeding female or pups or both at the den. Photograph by Isaac Babcock.

Wolves seem to show great affection for pups. Regular touching and lying against one another may provide both comfort and bonding, or "affiliative ties." Usually only a tired father wolf just home from a long hunt disdains such overtures by pups. Photograph by L. David Mech.

Pups find their parents and older siblings handy playthings. Adult wolves, except some male parents, tend to tolerate all kinds of indignities by the pups, including not only tail-pulling but also ambushing, being pounced on, and constant pestering. Photograph by L. David Mech.

When a pup strays too far from the den, when danger threatens, or when adults want to move the pups, an adult may pick a pup up in its mouth and carry it a long distance. Most often it is the mother wolf that carries pups. Photograph by L. David Mech.

Bottom left: Like most young animals, wolf pups are extremely playful. Since litters are usually large, pups can play in numerous ways with littermates. Tag, hide-and-seek, ambush, and tug-of-war are all common games. Photograph by Isaac Babcock.

Top and bottom right: Wolf pups grow and develop rapidly. After going through pug-nosed and kitten-like stages, the pups soon begin taking on an adult conformation, as this 11-week-old pup shows. Photographs by Isaac Babcock.

Parent wolves dominate their offspring to control them, and the male breeder dominates the female breeder, although he routinely yields food to her. Dominance is shown by a raised tail and hackles, whereas submission is demonstrated by a lower body position and lowered tail and ears. Photograph by L. David Mech.

Although most wolves are a mottled gray, their color can vary from black to white and can include almost any shade in between. Their markings also vary, as illustrated by this collection of pelts taken during a wolf control program in Alaska. Photo by Alaska Department of Fish and Game.

Captive-bred red wolves were reintroduced to northeastern North Carolina in 1987. Photograph by Melissa McCaw. Photograph supplied by the North Carolina Wildlife Resources Commission.

9

Wolf Evolution and Taxonomy

Ronald M. Nowak

ABOUT 6 MILLION YEARS AGO, toward the end of the Miocene geological epoch, Earth's climate was slowly cooling, preliminary to the great glaciations that would follow in the Pliocene and Pleistocene (Ice Age). The forests and savannas that had dominated the landscape were being replaced in many regions by steppe or grassland (Webb 1984). It was a time of increasing challenge, in which only creatures that could adapt to the surrounding environmental changes would survive.

At opposite points of the globe, two very different lineages were demonstrating such an ability to adapt. The subsequent course of their evolution would produce species that were among the most widely distributed of mammals. It also would lead those species into intimate contact and, in varying circumstances, into conflict and cooperation. One lineage would arise in East Africa, where a critical split was taking place among large primates. Some primates, the precursors of modern chimpanzees, would remain largely associated with forests. Others, our own ancestors, would tend to move away from the trees, begin to walk upright on extended legs, and develop the enlarged brain associated with manipulability of the forelimbs and with the challenges of both avoiding predators and being a predator in open country.

In the other lineage, meanwhile, some of the small woodland foxes in the southern part of North America were growing larger and more cursorial. The Borophaginae, an unrelated group of much more massive but cumbersome dogs, were dying out (Kurtén and Anderson 1980). The conditions were right for the development of moderate-sized, swift, and intelligent canids. At some

point during the late Miocene, the first species of the genus *Canis* arose, the forerunner of all the coyotes, wolves, and domestic dogs that would follow.

It still would be a long time before true wolves appeared. Although the term "wolf" has been applied to various kinds of canids and to other animals as well, here it is restricted to *Canis lupus* and a few other living and extinct species of *Canis* that probably arose from a common ancestor. Two main areas of scientific study of this group of canids are the course of its evolution and the interrelationships of its extant members.

Both genetic (Wayne et al. 1995) and morphological (Nowak 1979) investigations suggest that wolves evolved during the Pliocene and early Pleistocene from an ancestral line that also led to coyotes (fig. 9.1). Other members of the genus *Canis*, particularly the jackals, had split off earlier. The modern coyote (*Canis latrans*) is closely related to the wolves. The domestic dog (*C. familiaris*) is most likely a descendant of *C. lupus* (see below).

Overview of Taxonomic Uncertainties

The so-called Simien jackal or Simien wolf (*C. simensis*) (also known as the Abyssinian or Ethiopian wolf) may also be part of the wolf-coyote group and is believed to have no close relationship to other animals called jackals. However, there is conflicting evidence as to how closely related it is to other animals called wolves. Genetic studies indicate, on one hand, that *C. simensis* is no more closely related to *C. lupus* than is *C. latrans*.

On the other hand, the genetic studies suggest that *C. simensis* diverged from *C. lupus* in the late Pleistocene

239

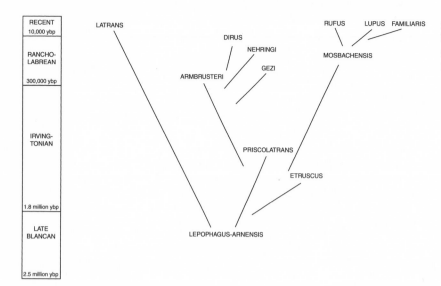

FIGURE 9.1. Hypothetical evolution and branching of the genus *Canis*. North American Land Mammal ages are shown to left (see also table 9.2).

(Wayne 1993; Gotelli et al. 1994; Roy, Geffen et al. 1994; Wayne et al. 1998). The late Pleistocene began just 130,000 years ago (Morgan and Hulbert 1995), and fossils show the wolf and coyote lines well separated by the early Pleistocene, about 1.5 million years ago (Kurtén 1974; Nowak 1979).

Moreover, the skull of *C. simensis* is the most distinctive of any living species of *Canis*, being far more different from that of *C. lupus* than is that of *C. latrans* (Clutton-Brock et al. 1976; Nowak 1979). While *C. simensis* warrants urgent attention from a conservation standpoint (Gotelli and Sillero-Zubiri 1992; Gotelli et al. 1994), it may be disregarded in a systematic discussion of wolf lineages.

Viewpoints also diverge concerning the development of wolves after their separation from the line leading to coyotes and other *Canis*. Perhaps nine species of wolves, as the term is used here, currently have general accep-

tance (table 9.1). However, some investigators have tended to consider most of those species, particularly some of North America *(priscolatrans, armbrusteri, rufus, lupus),* to be components of *C. lupus,* or at least to be in the direct line of transition between the coyote and *C. lupus* (Hoffmeister and Goodpaster 1954; Lawrence and Bossert 1967, 1975; Martin 1974; Webb 1974; Wayne and Jenks 1991). It even has been argued that the dire wolf *(C. dirus),* a massive animal best known from its abundant remains at the Rancho La Brea Tar Pits in Los Angeles, California, is nothing more than a large form of *C. lupus* that developed in response to ecological conditions of the late North American Pleistocene (Goulet 1993).

Disagreement intensifies with respect to the taxonomic status of extant species of *Canis*, in part because of its current management implications. Much recent argument has centered on whether the red wolf *(C. ru-*

TABLE 9.1. Fossil and modern species of wolves of the genus *Canis*

Species	Geological range	Geographic range	Reference
C. priscolatrans	Late Pliocene – early Pleistocene	North America	Nowak 1979
C. etruscus	Late Pliocene – early Pleistocene	Europe	Kurtén 1968
C. armbrusteri	Middle Pleistocene	North America	Nowak 1979
C. gezi	Middle Pleistocene	South America	Berta 1988
C. nehringi	Late Pleistocene	South America	Berta 1988
C. dirus	Late Pleistocene – early Recent	North and South America	Kurtén 1984
C. mosbachensis	Middle – late Pleistocene	Eurasia	Kurtén 1968
C. rufus	Late Pleistocene – Recent	Eastern North America	Nowak 1979
C. lupus	Late Pleistocene – Recent	Eurasia, North America	Nowak 1979

TABLE 9.2. Chronological development of wolves

Epoch	North American Land Mammal Age	Years ago	Major events
Holocene	Recent	0–10,000	Extinction of *dirus*, dominance of *lupus*, retreat of *rufus*
Late Pleistocene	Late Rancholabrean	10,000–130,000	Dominance of *dirus*, modern wolves enter New World
Middle Pleistocene	Early Rancholabrean	130,000–300,000	Emergence of modern wolves, extinction of *armbrusteri*
	Late Irvingtonian	300,000–700,000	Dominance of *armbrusteri* and *mosbachensis*
Early Pleistocene	Middle Irvingtonian	700,000–1 million	Emergence of larger wolves
	Early Irvingtonian	0–1.8 million	Dominance of *priscolatrans* and *etruscus*
Pliocene	Late Blancan	1.8–2.5 million	Divergence of wolf and coyote lineages
	Early Blancan	2.5–4.5 million	Emergence of small coyotelike forms
Miocene	Hemphillian	4.5–9 million	Origin of *Canis*

fus) of southeastern North America is a distinct species, a form of *C. lupus,* or the result of hybridization between *C. lupus* and *C. latrans* (Mech 1970; Brownlow 1996; Nowak and Federoff 1996, 1998; Roy et al. 1996; Wayne et al. 1998; Reich et al. 1999). Another point of contention is the domestic dog *(C. familiaris),* the recent trend being to follow Wozencraft (1993) in regarding it too as a form of *C. lupus.* A final major issue is subspeciation, especially in *C. lupus,* with its implications relative to reintroduction and other conservation measures. Each of those topics will be discussed in further detail below.

Evolution of Species

By the Pliocene, which in North America corresponds in large part with the Blancan Land Mammal Age (table 9.2), small kinds of *Canis* had become widespread in both the Old World and North America. A related branch of small canids had entered South America and begun an entirely separate evolutionary lineage (Kurtén 1968; Nowak 1979, 1992b; Kurtén and Anderson 1980; Anderson 1984; Berta 1988; Tedford et al. 1995; Tedford and Qiu 1996). There were other, larger canids, but they do not seem to have been involved in the ancestry of today's wolves.

Wolf Ancestry

Although the fossil record is incomplete, it is likely that all wolves arose from some population of those small, early canids (see fig. 9.1). According to Kurtén and Anderson (1980), the most probable ancestral candidate is *C. lepophagus* of North America. That species may have been derived from an earlier and smaller Miocene species of *Canis*. *C. lepophagus* seems to have spread to Eurasia, where a closely related or identical Pliocene species, *C. arnensis,* has been found (Kurtén 1974; Kurtén and Anderson 1980).

Specimens of the earlier populations of *C. lepophagus* possess small, delicate, narrowly proportioned skulls (Nowak 1979). They resemble small coyotes of today and probably are ancestral to *C. latrans.* Some of the later populations, especially those represented by specimens from Cita Canyon, in northern Texas, have larger, broader skulls. Johnston (1938) suggested that those populations were the ancestral stock of wolves. A few fragmentary specimens indicate further progression from *C. lepophagus* toward true wolves (Nowak 1979).

The Blancan Land Mammal Age ended about 1.8 million years ago and gave way to the Irvingtonian in North America (see table 9.2). The latter period corresponds to the late Villafranchian Age in Eurasia (Anderson 1984). The first definite wolves appeared from the late Blancan to the early Irvingtonian (Nowak 1979; Tedford and Qiu 1996).

In North America, the earliest named wolf species is *C. priscolatrans,* now known from Arizona, California, Colorado, Florida, Kansas, Oregon, Pennsylvania, Texas, and Mexico; a more familiar name, *C. edwardii,* is considered a synonym (Kurtén 1974; Nowak 1979; Kurtén and Anderson 1980; Anderson 1996; Albright 2000). Fossils indicate that *C. priscolatrans* resembled modern *C. rufus* in cranial size and proportion, but had more complex dental sculpturing, comparable to that of *C. latrans.* On the occlusal surface of the first upper molar of *C. priscolatrans,* there are broad and deep basins between the outer ridge formed by the paracone and metacone and a central cusp, the protocone, and between that cusp and the hypocone at the inner edge of the tooth. There also is a very prominent buccal cingulum on the outer edge and usually a pronounced anterior cingulum.

Kurtén (1974) thought *C. priscolatrans* represented a population of large coyotes, ancestral to modern *C. latrans*. It is more likely that *C. priscolatrans* is a small wolf and that it spread to Eurasia, where its counterpart is *C. etruscus* (Nowak 1979). Kurtén and Anderson (1980) recognized both forms as part of a group of primitive wolves, from which *C. lupus* arose. However, those authors were not certain whether those small wolves, in turn, had a common ancestry or evolved independently from among earlier populations of *C. lepophagus* and *C. arnensis*. The possibility also remains that the wolf line arose from Miocene-Pliocene canids that preceded *C. lepophagus* (Nowak 1979; Tedford and Qiu 1996).

The Red Wolf

Nowak's (1979, 1992b) suggestion that *C. rufus* also is a primitive wolf, closely related or even identical to *C. priscolatrans,* no longer seems tenable. Extensive material of the latter species, recently discovered in Florida (Berta 1995), shows the complexity of its dental sculpturing to considerably exceed that of *C. rufus*. The various Irvingtonian specimens of eastern North America identified as *C. rufus* by Nowak (1979) are more properly referred to *C. priscolatrans* (Nowak 2002). *C. rufus* thus is left without any fossil record prior to the late Rancholabrean.

C. rufus probably is not ancestral to *C. lupus,* as suggested by Nowak (1979). However, an oft-repeated claim that Nowak had argued that *C. rufus* is the progenitor of modern *C. latrans* as well is not accurate (Roy et al. 1996; Nowak and Federoff 1998; Wayne et al. 1998). Notwithstanding the above, evidence for a common North American origin of most wolves can be derived from the close morphological and genetic resemblance of early wolf populations (Nowak 1995a; Wayne et al. 1995). Although there is considerable question as to the status of *C. rufus,* it is similar systematically to *C. lupus lycaon* of southeastern Canada and *C. lupus pallipes* of southwestern Asia. The latter two subspecies are in turn systematically overlapped by adjoining populations of *C. lupus*.

Eurasian Wolves

While there is much individual variation, and some basis for subspecific distinction, within extant populations of the gray wolf (Sokolov and Rossolimo 1985; Nowak 1995a), there is considerable affinity between populations of wolves in North America and Eurasia. Certain subspecies on one continent seem more closely related to subspecies on the other continent than to other subspecies on the same continent (see below).

True wolves probably arose in North America, crossed into Eurasia, evolved there in the direction of *C. lupus,* and then reinvaded the New World. However, there was an intermediate stage in the Old World evolution of the lineage, between early Pleistocene *C. etruscus* and modern *C. lupus*. That stage is represented by *C. mosbachensis,* known from various sites in Eurasia.

I examined *C. mosbachensis* specimens from the Lake Baikal region of Siberia, on loan to Richard H. Tedford (American Museum of Natural History, personal communication, 24 March 1999), and found them smaller on average than those from most modern North American wolf populations, including even *C. rufus* of the southeastern United States. The Baikal specimens are thought to date from a period corresponding to the mid- to late Irvingtonian of North America.

Kurtén (1968) reported *mosbachensis* to be about the size of *C. lupus pallipes,* which is smaller than *C. rufus* (Nowak 1995a). Kurtén (1968) noted that *mosbachensis* was still present in Europe during a period corresponding to the early Rancholabrean of North America, and that increase in size did not occur until the late Rancholabrean. Since the evolution of wolves generally involves a progression from smaller to larger, *C. mosbachensis* seems a logical candidate for the ancestor of the modern North American (as well as Eurasian) populations. An initial invasion of North America by that ancestral *C. mosbachensis* may have become isolated by glaciation and developed into *C. rufus*. Eurasian *C. mosbachensis* then may have evolved into *C. lupus,* and subsequent invasions of North America may have led to the modern subspecific differentiation of the latter species (fig. 9.2).

There is still a question about what happened to the primitive stock of small wolves that remained from the outset in North America. It now seems likely that those canids gave rise, not to *C. rufus,* but to an entirely separate New World radiation of large wolves. Most of the many recently collected early Irvingtonian specimens from Florida are referable to *C. priscolatrans*. However, Berta (1995) assigned several, from the Leisey Shell Pit, Hillsborough County, and Haile 21A site, Alachua

County, to *C. armbrusteri*. The latter species is a relatively large wolf otherwise known mainly from mid- and late Irvingtonian sites in Florida and the early Rancholabrean Cumberland Cave in Maryland (Gidley and Gazin 1938; Martin 1974; Nowak 1979). Examination of specimens from all of these and other sites, as expressed in part by a bivariate analysis of the first upper molar (fig. 9.3), shows that *C. priscolatrans* grades into *C. armbrusteri* and suggests that it gave rise to the latter (Nowak 2002).

Fragmentary remains from Irvingtonian sites, including some previously assigned to *C. lupus* (Nowak 1979), hint that *C. armbrusteri* once inhabited much of North America, though it remains best known from Florida and Cumberland Cave. Its teeth are large and its molars well sculptured, though not so extensively as in *C. priscolatrans*. The skull also is distinguished by its huge tympanic bullae and relatively narrow rostrum (Nowak 1979).

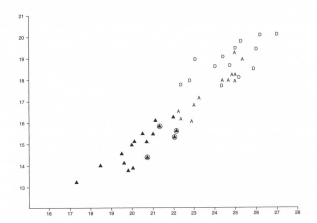

FIGURE 9.3 Bivariate analysis comparing measurements (in millimeters) of transverse diameter (horizontal axis) and anteroposterior length (vertical axis) of the first upper molar in individual specimens of *Canis priscolatrans* (triangles), specimens identified as *C. armbrusteri* by Berta (1995) but considered here to represent *C. priscolatrans* (circled triangles), *C. armbrusteri* (A), and *C. dirus* (D). (From Nowak 2002.)

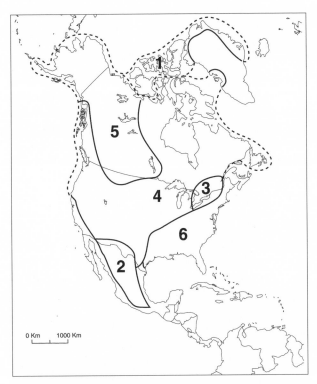

FIGURE 9.2 Original geographic distribution of wolves in North America, showing the two species and five subspecies of *Canis lupus* recognized by Nowak (1995a): 1, *C. l. arctos* (arctic wolf); 2, *C. l. baileyi* (Mexican wolf); 3, *C. l. lycaon* (eastern wolf); 4, *C. l. nubilus* (plains wolf); 5, *C. l. occidentalis* northwestern wolf; 6, *C. rufus* (red wolf).

While Martin (1974) suggested that *C. armbrusteri* is synonymous with *C. lupus*, he also pointed out that some specimens from Cumberland Cave approach *C. dirus* of the later Rancholabrean in size and other characters. There is also some overlap in skull measurements between *armbrusteri* and *dirus* (see fig. 9.3). Probably *armbrusteri* had no immediate relationship to *lupus* or any modern wolf, but was a central component of an archaic New World evolutionary sequence that progressed from *priscolatrans* to *dirus* (Nowak 2002).

The Dire Wolf

The rise of the dire wolf was the most dramatic event in the evolutionary history of *Canis*. The species appeared rather suddenly all across North America late in the Rancholabrean Land Mammal Age, which succeeded the Irvingtonian about 300,000 years ago. The dire wolf varied considerably in size, with some populations consisting of the largest members of the genus ever to exist, and others being only about the size of a large gray wolf. All individuals were characterized by massive heads, enormous teeth (fig. 9.4), and relatively short limbs.

Although Goulet (1993) and others have suggested that *C. dirus* may not be specifically distinct from *C. lupus*, there is otherwise a consensus that it is a highly evolved species (Berta 1988; Kurtén and Anderson 1980; Nowak 1979). Kurtén (1984) retained the subgeneric

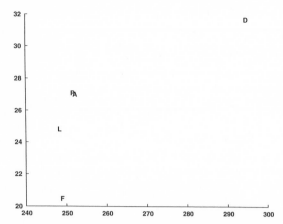

FIGURE 9.4. Bivariate analysis comparing the mean positions of measurements (in millimeters) of greatest length of skull (horizontal axis) and length of fourth upper premolar, or carnassial (vertical axis), of 94 *Canis lupus* from the mountainous region of the western contiguous United States (L), 29 *C. lupus arctos* from the Canadian High Arctic and Banks Island (A), 9 late Rancholabrean *C. lupus* from the Rancho La Brea and Maricopa Brea tar pits of southern California (P), 62 *C. dirus* from Rancho La Brea (D), and 20 *C. familiaris* with wolf-sized skulls (F).

designation *Aenocyon* for the dire wolf and recognized two subspecies: *C. dirus dirus,* from the eastern, central, and northwestern United States, and *C. d. guildayi,* from California and Mexico.

The dire wolf is the most common fossil canid in North America; its remains have been found from Pennsylvania and Florida to Oregon and California. Notwithstanding the data depicted in figure 9.3, none of those fossils clearly suggests transition from another species. At many sites, the dire wolf is found together with completely distinct specimens of gray wolves, red wolves, and coyotes (Nowak 1979). It is plausible that the Irvingtonian to early Rancholabrean *C. armbrusteri* was the progenitor, but *C. dirus* has several diagnostic cranial characters, involving the palate, inion, nasals, and foramina, that consistently distinguish it from all other North American *Canis.*

Possibly, *C. armbrusteri,* or an earlier species from the primitive wolf group, spread into South America and there developed into *C. dirus* (Nowak 1979, 1992b). *C. dirus* has been reported from various parts of South America, and in some of its critical characters it resembles two other, earlier wolves from that continent. Moreover, the sudden appearance of the dire wolf in North America, already fully evolved, suggests that it was an invader from elsewhere.

Kurtén (1968) and Berta (1988) hinted at affinity be-

tween *C. dirus* and *C. falconeri* from the early Pleistocene of Europe. However, the latter species is now thought to represent a separate genus, *Xenocyon,* and perhaps to belong to the lineage of the extant African hunting dog, *Lycaon pictus* (Rook 1994). Moreover, the dire wolf evidently was a warmth-adapted species that never moved into the northern part of North America. While many remains of large wolves have been collected from late Pleistocene deposits in Alaska and northwestern Canada, all are *C. lupus.*

Although Berta (1988) favored a North American origin for *C. dirus,* her analyses offer a reasonable basis for an evolutionary transition that includes South America. She indicated that from *C. armbrusteri,* or a related large, primitive wolf, there arose *C. gezi,* which is known with certainty only from the middle Pleistocene of South America. *C. gezi* in turn gave rise to *C. nehringi,* a large wolf of the late Pleistocene of South America, which closely resembles *C. dirus* and could have been its immediate ancestor.

Whatever its origin, the dire wolf disappeared about 8,000 years ago, in association with the extinction of other components of the North American late Pleistocene megafauna. Possibly its prey had been largely eliminated by the human hunters who were spreading across the continent at the time. It may also have lost in competition with gray and red wolves, which were better adapted to pursue the generally smaller and swifter prey species that had survived.

Intraspecific Variation in *Canis lupus*

With the extinction of *C. dirus,* the entire archaic New World line of wolves ended, and *C. lupus* was the only large and widespread species of wild *Canis* left. Its range was more extensive than that of any terrestrial mammal, other than *Homo sapiens.* As with humans, geographic barriers probably had resulted at times in restrictions or complete breaks in gene flow. When such a break occurs, as manifested in morphologically or genetically distinct specimens in a given region, it often is useful to designate the population involved with a subspecific (or racial) name. Various rules have been proposed as to how distinctive a population must be to warrant such designation, but none is followed uniformly. Some named subspecies would not withstand modern genetic or statistical analysis, and there has been a recent tendency to lump older subspecies and to name fewer new ones.

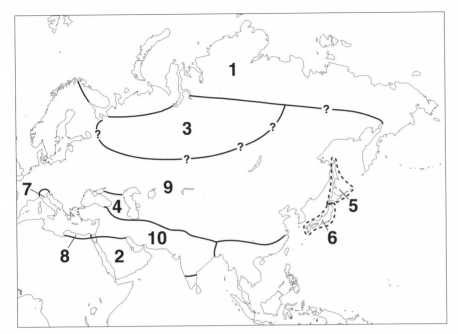

FIGURE 9.5. Original geographic distribution of gray wolves *(Canis lupus)* in the Old World: 1, *C. l. albus;* 2, *C. l. arabs;* 3, *C. l. communis;* 4, *C. l. cubanensis;* 5, *C. l. hodophilax;* 6, *C. l. hattai* (= *rex*); 7, *C. l. italicus;* 8, *C. l. lupaster;* 9, *C. l. lupus;* 10, *C. l. pallipes.* As indicated by the question marks (?), the extent of the range of *communis* is problematic, and there also is uncertainty as to whether northeastern Siberia is occupied by *communis, albus,* or *lupus.* (From Nowak 1995a.)

That trend is well illustrated by the systematics of wolves (table 9.3). Based largely on earlier work (Goldman 1944), E. R. Hall (1981) accepted twenty-four subspecies of *C. lupus* in North America. Sokolov and Rossolimo (1985) recognized nine additional subspecies in Eurasia, but reduced the number in the New World to seven. Other authorities, relying mainly on morphometric techniques, also have indicated that some of the subspecies listed by E. R. Hall (1981), particularly those in western parts of the United States and Canada, may be unwarranted (Rausch 1953; Jolicoeur 1959; Skeel and Carbyn 1977; Pedersen 1982; Walker and Frison 1982; Nowak 1983; Bogan and Mehlhop 1983; Friis 1985; Hoffmeister 1986; Mulders 1997).

Nowak (1995a) made the first attempt to systematically analyze statistical data on the morphology of all modern wolves. Computer-based multivariate analyses of ten measurements from each of 580 skulls of male wolves indicated the presence of five subspecies of *C. lupus* in North America (see fig. 9.2): *arctos,* a large-toothed Arctic wolf; *occidentalis,* a large animal of Alaska and western Canada; *nubilus,* a moderate-sized wolf, originally found from Oregon to Newfoundland and from Hudson Bay to Texas; *baileyi,* a usually smaller wolf of the Southwest; and *lycaon,* a small subspecies now restricted to southeastern Canada.

Eurasian subspecies (fig. 9.5) include an Arctic wolf *(albus),* a large north-central form *(communis),* and a widespread animal of moderate size *(lupus).* A small but broad-skulled subspecies *(cubanensis)* occurs in the Caucasus, and a narrow-skulled wolf *(pallipes)* inhabits most of southwestern Asia. Recent examination of specimens at the University of Rome and the Italian National Institute of Game indicates that *italicus,* the small wolf of the Italian peninsula, also is a distinct subspecies (Nowak and Federoff 2002). Four additional Old World subspecies may be recognizable—*arabs, hattai, hodophilax, lupaster*—but those four were not analyzed by Nowak (1995a).

The red wolf of southeastern North America was considered by Nowak (1995a) to have some affinity to *pallipes,* but to be best regarded as a species separate from *C. lupus.* Goldman (1944) recognized three subspecies of *C. rufus,* and recent reexamination (Nowak 2002) of all pertinent specimens dating prior to 1918 supports their retention (see table 9.3).

From Continent to Continent

Nowak's (1995a) interpretation of variation in North American *C. lupus* and *C. rufus* fits reasonably well with the hypothesis, discussed above, that the continent was reinvaded by modern wolves. The distribution of species and subspecies suggests that this reinvasion occurred in several waves, perhaps corresponding to the opening of passages across the Bering Sea or through the glacial ice.

TABLE 9.3. Species and subspecies of modern wolves

Canis rufus rufus Audubon and Bachman, 1851. Central and coastal Texas, southern Louisiana (evidently extirpated except in captivity and as a reintroduced population in eastern North Carolina and a few individuals released at other sites)

Canis rufus gregoryi Goldman, 1937. Southwestern Indiana, southern Missouri, and eastern Oklahoma to southern Mississippi, central Louisiana, and Big Thicket area of Texas (apparently extinct, though possibly represented by captive/reintroduced population, in part)

Canis rufus floridanus Miller, 1912. Maine and Ohio to Florida and Alabama (extinct)

Canis lupus pallipes Sykes, 1831. Israel to India

(?) *Canis lupus arabs* Pocock, 1934. Arabian Peninsula

(?) *Canis lupus lupaster* Hemprich and Ehrenberg, 1832. Egypt, Libya

(?) *Canis lupus hodophilax* Temminck, 1839. Japan except Hokkaido (extinct)

Canis lupus lycaon Schreber, 1775. Extreme southern Ontario, extreme southern Quebec, probably originally northern and western New York

Canis lupus baileyi Nelson and Goldman, 1929. Southern Arizona, southwestern New Mexico, extreme southwestern Texas, highlands of Mexico south to Oaxaca (extirpated except for a few possible survivors in northwestern Mexico, in captivity, and as a reintroduced population in southeastern Arizona)

Canis lupus cubanensis Ognev, 1923. Caucasus and adjacent parts of Turkey and Iran

Canis lupus italicus Altobello, 1921. Italian Peninsula

Canis lupus lupus Linnaeus, 1758 (*campestris, chanco, desertorum*). Europe east to an undetermined point in Russia, Central Asia, southern Siberia, China, Mongolia, Korea, Himalayan region (now greatly reduced in distribution, especially in western Europe)

Canis lupus nubilus Say, 1823 (*beothucus, crassodon, fuscus, hudsonicus, irremotus, labradorius, ligoni, manningi, mogollonensis, monstrabilis, youngi*). Southeastern Alaska, southern British Columbia, contiguous United States from Pacific to Great Lakes region and Texas, Ontario except southeastern, northern and central Quebec, Newfoundland, northern Manitoba, Keewatin, eastern Mackenzie, Baffin Island, occasionally west-central Greenland (now evidently extirpated in the western contiguous United States)

Canis lupus arctos Pocock, 1935 (*bernardi, orion*). Northern and eastern Greenland, Queen Elizabeth Islands (Ellesmere, Prince Patrick, Devon, etc.), Banks and Victoria Islands

Canis lupus albus Kerr, 1798. Extreme northern Eurasia

Canis lupus communis Dwigubski, 1804. Known with certainty only from the Ural Mountain region of north-central Russia, but probably once occurring over much of eastern Europe and Siberia

(?) *Canis lupus hattai* Kishida, 1931 (= *C. l. rex* Pocock, 1935). Sakhalin, Hokkaido (extinct)

Canis lupus occidentalis Richardson, 1829 (*alces, columbianus, griseoalbus, mackenzii, pambasileus, tundrarum*). Alaska, Yukon, Mackenzie, British Columbia, Alberta, Saskatchewan, southern Manitoba, northern Montana (now evidently expanding range into, and also reintroduced in, northwestern contiguous United States)

Source: Based primarily on Nowak 1995a and subsequent assessment of *Canis lupus italicus* (Nowak and Federoff 2002) and *Canis rufus* (Nowak 2002).

Note: Names in parentheses are additional subspecies recognized by Hall (1981) and Sokolov and Rossolimo (1985) that here would be regarded as synonyms. Question marks (?) indicate subspecies not subjected to statistical analysis. It must be emphasized that the above classification is tentative and based only on canonical discriminant analyses of ten measurements of the skulls of male wolves. Other systematic methodologies or assessment of additional material could produce different results.

The most distinct wolves are found at the periphery of the range (see fig. 9.2): *C. rufus* at the corner of the continent opposite the point of invasion, *C. lupus baileyi* far to the south, *C. lupus lycaon* in the eastern forests, and *C. lupus arctos* on the High Arctic islands. Those groups may represent survivors of the earliest migrations that had long been isolated. Indeed, *C. l. lycaon*, *C. l. baileyi*, and *C. rufus* possess certain primitive characteristics and show some systematic affinity to one another (Nowak 1979, 1995a; Wayne et al. 1992, 1995).

Goulet (1993) suggested that North American gray wolves of the late Rancholabrean generally were larger and had broader skulls than those of the Holocene and modern times. However, her analysis may be skewed by the predominance in the Pleistocene sample of specimens from a single area—Rancho La Brea and a few nearby sites in southern California. Most of the dozen or so skulls from that area are indeed massive, and they are characterized by relatively broad rostra and frontal shields and large teeth, especially the carnassials (Nowak 1979). One of the specimens actually was originally designated a separate species, *C. milleri*, and was placed together with the dire wolf in the genus *Aenocyon* (Merriam 1912, 1918). However, further analysis suggested that the resemblance of *C. milleri* and the other late Pleistocene specimens from California to *C. dirus* is strictly superficial and perhaps representative of convergence (Martin 1974; Nowak 1979).

Most specimens of late Pleistocene *C. lupus* from coastal southern California closely approach modern *C. l. arctos* of the High Arctic islands of Canada and *C. l. albus* of the extreme northern parts of Eurasia (see fig. 9.4). The geographic distribution of such wolves hints at the possibility that they once were widespread, and that some Arctic populations were driven southward by glaciation. Coastal southern California probably was much cooler in the late Pleistocene. No wolves of any kind are known to have occurred there in modern times (Goldman 1944), though possibly they were killed off by Spanish cattlemen in the early 1800s.

Not all specimens of Pleistocene *C. lupus* demonstrate the massiveness of Arctic wolves. A few specimens suggest that *C. l. baileyi* once extended as far as Kansas and southern California. That subspecies is known in modern times only from western Mexico and the adjacent fringe of the southwestern United States (Nowak 1979, 1995a). The smallest adult specimen of *C. lupus* ever collected in North America was found at the late Rancholabrean San Josecito Cave site in Nuevo Leon, northeastern Mexico. A number of large dire wolf skulls were found at the same site and form a remarkable contrast (Nowak 1979).

The distribution of fossil gray wolves during the late Pleistocene suggests a geographic fluctuation of populations, with large, Arctic forms moving far to the south during glacial advances and small, warmth-adapted subspecies expanding when the climate moderated. Nevertheless, there is little evidence of a long history of modern wolves in North America or a sufficient period for much fluctuation to occur. A few fragments suggesting Irvingtonian and early Rancholabrean occurrence of *C. lupus* and *C. rufus* in North America (Nowak 1979) are now thought to be *C. armbrusteri* and *C. priscolatrans* (Nowak 2002).

Phylogenetic Change or Adaptation?

Another critical question is whether any process of morphological fluctuation that took place involved phylogenetic change or merely adaptation to extrinsic (environmental, climatic, ecological) factors. Does the distribution of modern wolves represent many once-separated populations of subspecific or specific rank? Or were there only a few genetically homogeneous populations, the members of which varied because of intrinsic factors?

The second view is supported by some mtDNA evidence (Jenks and Wayne 1992; Wayne et al. 1992, 1995). Although several genotypes of *C. lupus* have been identified, they generally are not grouped geographically. A given locality may have more than one genotype, and one of those types may show more affinity to types in another part of the world than to types at the same locality. In any event, only one geographic population of *C. lupus* consistently demonstrates a unique genotype, *C. lupus baileyi*. In that regard, there is agreement between the molecular and morphological approaches to the systematics of wolves. However, it is important to note that the population sampled in the molecular study included only a few founders; conceivably a larger number might have included other genotypes.

Differences between geographic areas sometimes are held to result from responses to extrinsic factors. Such trends were suggested by Goulet (1993), who designated ecological races of *C. lupus* that adjusted morphologically in time and space to their environment. Further evidence is provided by recent studies indicating that the average size of wolves on Isle Royale has grown significantly in just a few decades, possibly in association with a change in prey from deer to moose (Nowak 1995b).

Kolenosky and Standfield (1975) recognized two kinds of *C. lupus* in Ontario, a larger animal in western Ontario and a smaller one in the southeast. They practically suggested specific separation, stating (p. 71) that "the ranges of the two types overlap throughout a broad band across east-central Ontario, but there is no conclusive evidence of their interbreeding." Subsequently, however, Schmitz and Kolenosky (1985b) presented a modified interpretation, concluding that both types are merely clinal variants of the same population. Although the wolf from western Ontario is nearly identical morphologically to the animal just across the border in northern Minnesota, Schmitz and Kolenosky (1985b) believed that the two are unrelated and that their phenotype is simply an expression of convergence in adapting to similar habitat and prey. Given that Minnesota wolves disperse extensively into Ontario (Mech 1987a; Gese and Mech 1991), this view seems untenable.

Hybridization with Coyotes

Hybridization of *C. lupus* and *C. latrans* in the Great Lakes region has been indicated by both morphological (Kolenosky and Standfield 1975; Nowak 1979; Schmitz and Kolenosky 1985b; Sears 1999) and mtDNA (Lehman et al. 1991) investigations. However, while all wolves of the eastern part of that region and on Isle Royale, and most wolves in Minnesota and western Ontario, reportedly have coyote-like mtDNA (Lehman et al. 1991), there is no other evidence — morphological, ecological, or behavioral — that the hybridization process has spread beyond southeastern Ontario and southern Quebec. The wolves of Minnesota and western Ontario are fully wolf-like in all those respects (Nowak 1991, 1992a, 1995a). Several earlier investigations (Mech and Frenzel 1971b; Skeel and Carbyn 1977; Van Ballenberghe 1977) actually had indicated that wolves of the western Great Lakes region have greater systematic affinity to populations farther to the west than to *C. l. lycaon* of the eastern Great Lakes region. New DNA analyses of Ontario wolves go even further (Wilson et al. 2000), supporting a return to the idea of specific distinction between eastern and western wolf populations suggested by Kolenosky and Standfield (1975).

If a species does vary gradually across a great geographic distance in a steady cline of size or other characters, it is reasonable to think that there has been continuous and uninterrupted gene flow, and that the variation represents ecological adaptation. Such a situation is indeed what is suggested by certain studies of mtDNA for most of the range of *C. lupus* (Jenks and Wayne 1992; Wayne et al. 1992, 1995). According to that rationale, wolves are such large and highly mobile animals that any divisions that may temporarily form are lost quickly as the various segments spread back and merge with one another. Moreover, across most of North America at least, there are no barriers that would stop the movement of wolves and reinforce any local distinctions that might develop. Wolves easily pass over mountain ranges, push through forests and swamps, and swim great distances. They routinely cross vast stretches of frozen sea to move among the Arctic islands.

Genetic Differentiation

The above notwithstanding, there remains considerable evidence for genetic differentiation of a number of populations and consequently for valid subspecific, even specific, designations. Just because there are no obvious barriers today does not mean that such was always the case. Indeed, we know with certainty that the world's populations of wolves were once divided by continent-wide stretches of glacial ice and, alternately, by sea levels higher than those of today. Moreover, the apparent natural distribution of certain species of *Canis*, such as *C. dirus, C. rufus,* and *C. latrans,* indicates adaptation only to certain habitats, prey, or climatic conditions. Changes in such conditions did cause fluctuations in their distributions and could have served to divide and differentiate populations.

Thus, it is reasonable to assume that at various times in geological history, great gaps appeared in the range of *C. lupus* and its progenitor species *C. mosbachensis,* because of either insurmountable physical barriers or regions of ecologically unsuitable habitat. Such gaps may have persisted for thousands of years. The populations on each side of the barriers may have undergone different genetic modification over time and developed different size, morphology, or behavior. When glaciers withdrew or other barriers moderated, the populations would have come back together, but physical or unseen genetic differences may have prevented the blending of groups into a single panmictic population. Those differences could have reinforced a reduction in gene flow that persists to the present.

The distributions of modern species and subspecies of wolves, examination of their morphology, and consideration of fossil material allows for a biogeographic hypothesis. *C. priscolatrans* or its immediate relatives evolved in North America by the early Irvingtonian (1.5 million years ago) and spread throughout the Northern Hemisphere during periods of conducive environmental conditions. The form *C. lupus pallipes* may be the most primitive Old World descendant of this early migration.

At some time, the lineage of *C. mosbachensis* and *C. lupus* developed in Eurasia and began moving back across the world in a series of waves, ebbing and flowing with development and withdrawal of glaciers and other barriers. When the first wave arrived in North America is not clear. Although the lineage is represented by late Irvingtonian fossils from ice-free portions of Alaska, there is no clear evidence of its presence farther south on the continent until the late Rancholabrean (about 15,000 years ago). It is pure speculation, but perhaps the first movement to the south occurred before the transition from *mosbachensis* to *lupus* and was by way of unglaciated coastal segments of Alaska and western Canada. Such a route may have opened before the development of the main inland passage during the terminal Pleistocene and may have been more conducive to a small wolf that could subsist along the continent's shore.

The Modern Wolf

In any event, the initial invasion of North America by the modern wolf lineage may be represented today by *C. rufus*. The small modern subspecies *C. lupus baileyi* may represent another early wave, or one that became isolated by desert barriers in the southwestern United States and Mexico. The much more massive *C. l. arctos* also could be a surviving early population, but one that had become isolated in the ice-free Pearyland refugium of northern Greenland and subsequently spread back across the Arctic. Other wolves pushed to the south when a route through the glaciers opened, and are today represented by *C. l. nubilus*. Finally, near the end of the Rancholabrean, a larger form developed, probably in the ice-free Beringian refugium in Alaska and adjacent areas (Nowak 1992b). When the last glaciers withdrew

about 10,000 years ago, that large wolf (*C. l. occidentalis*) moved out to occupy most of Alaska and western Canada, and even today is pushing into the western contiguous United States (Nowak 1995a).

The origin of *C. l. lycaon* of southeastern Canada is problematic. While it was sometimes thought to represent another surviving contingent of an early invasion, glaciers covered its range even longer than they did regions to the west. Moreover, new DNA analyses suggest that it has genetic affinity to *C. rufus* and may even be specifically distinct from *C. lupus* (Wilson et al. 2000), even though morphologically and ecologically it appears closer to the latter species. *Lycaon* may have resulted from natural hybridization as its range was simultaneously invaded by *C. lupus nubilus* from the west and *C. rufus* from the south when the most recent Pleistocene glaciation retreated (see below).

Meanwhile, in the Old World, a similar process was taking place, with smaller, more primitive wolves surviving on the fringes of the distribution of *C. lupus* (*arabs, cubanensis, hodophilax, italicus, lupaster, pallipes*). There is also a remnant Arctic population of a more massive form (*albus*) and a widespread group of moderate-sized wolves (*C. lupus lupus*). Apparently, the large late Rancholabrean-Holocene form moved westward from Beringia toward the Ural Mountains, where today it is represented by *communis*. A few specimens also suggest that it advanced southward to Sakhalin and Hokkaido islands (Nowak 1995a).

The overall intercontinental picture that emerges includes a few advanced subspecies occupying relatively large, central ranges while more primitive forms occupy smaller, peripheral areas (see figs. 9.2 and 9.5). Such a pattern is common in both zoological and botanical systematics and was designated "centrifugal evolution" by Groves (1993). Although the peripheral populations may be geographically far separated, some of them may be more closely related to one another than to adjoining central groups.

The idea of successive waves of wolves spreading across vast intercontinental distances in relatively brief geological intervals may seem far-fetched, but even in our own time canid populations have repeatedly demonstrated an ability to spread rapidly under suitable conditions. The coyote colonized the entire eastern half of North America in less than a century, and the raccoon dog (*Nyctereutes procyonoides*) spread through most of Europe in less time (Nowak 1999).

The above biogeographic hypothesis is based largely on morphological differences observed in modern populations and association of those observations with available fossil material and knowledge of geological history. An alternative explanation for the differences would be fairly recent and rapid adaptation to local ecological conditions. However, the degree of variation among certain of the recognized kinds of wolves suggests interruption of gene flow through isolation. Because there are no substantive isolating barriers among most of the kinds, especially in North America, the observed variation probably reflects populations that formerly were separated but that now have merged. Considering the size and mobility of wolves, it is unlikely that geographically proximal populations could have diverged so much morphologically.

The Red Wolf Enigma

Audubon and Bachman (1851) described the Texas red wolf as a subspecies, *Canis lupus rufus*. For nearly a century, *rufus* and the other named southeastern wolves (*ater, niger, floridanus*) were considered to be either subspecies or full species, depending on whether taxonomic lumping or splitting was in fashion, but to be no more or less distinct than were any of the other named kinds of wolves around the world. Not until the work of Goldman (1937, 1944) was a basic dichotomy set forth, with *C. rufus* (including *ater, niger,* and *floridanus*) regarded as a species of wolf separate from all others, which together were regarded as *C. lupus*. That view generally has been accepted for many years (E. R. Hall 1981; Corbet and Hill 1991; Wozencraft 1993; Jones et al. 1997; Nowak 1999).

As discussed in detail elsewhere (Nowak 1979, 1991, 1992a,b, 1995a; Kurtén and Anderson 1980), *C. rufus* once was believed to be a primitive species, closely related, or even identical, to *C. priscolatrans* (= *C. edwardii*) of the early Irvingtonian. As indicated above, new fossil evidence indicates that *C. priscolatrans* gave rise to a separate line of wolves and that *C. rufus* probably is more closely associated with the modern lineage of *C. mosbachensis* and *C. lupus*. However, *C. rufus* is similar to *C. latrans* in some characters and probably is more susceptible to hybridization with the latter species than are other modern wolves. Such hybridization seems to have begun in the late nineteenth century and eventually engulfed most surviving populations of *C. rufus* (McCarley 1962; Nowak 1979).

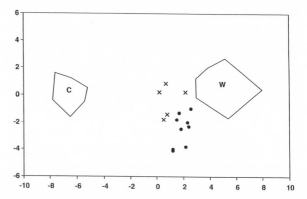

FIGURE 9.6. Statistical distribution of four groups of female *Canis*, based on the first (horizontal axis) and second (vertical axis) canonical variables, as produced by a canonical discriminant analysis of ten skull measurements. Solid lines represent the limits of eighteen western *C. latrans* (letter C shows mean) and twenty Minnesota *C. lupus* (letter W shows mean); solid circles, nine *C. rufus* taken 1919–1924 in southeastern Missouri; X's, five suspected hybrids of *C. latrans* and *C. lupus* from southeastern Canada. (Cf. Nowak 1995b and Jenks and Wayne 1992.)

A point of disagreement is whether the red wolf actually originated as a hybrid between *C. lupus* and *C. latrans*. Such an idea is not new, having been suggested by Mech (1970) and even by ranchers in late-nineteenth-century central Texas (Nowak 1979). One argument has been that because *C. rufus* seems morphologically intermediate to *C. lupus* and *C. latrans,* and indeed seems to be statistically close to specimens believed to be hybrids of the latter two species (fig. 9.6), the red wolf logically could represent a hybrid population (Jenks and Wayne 1992; Wayne 1992, 1995). However, the data used to arrive at that suggestion may not be fully appropriate (Nowak 1995b).

Furthermore, the red wolf's intermediate morphology could represent an evolutionary stage between the other two species. As indicated above, *C. mosbachensis,* a small wolf of the Eurasian Pleistocene, is thought to be intermediate to the primitive wolves, *C. priscolatrans* and *C. etruscus,* and modern *C. lupus,* and also may have given rise to *C. rufus* before the development of *C. lupus*. Interestingly, Kurtén (1974) regarded *C. priscolatrans* as an ancestral coyote, not a wolf. In any case, there would have to have been an evolutionary stage through which *Canis* needed to pass to change from coyote-like forms to modern Holarctic *C. lupus*. *C. rufus* could be a survivor of that stage. To paraphrase a popular saying, "if the red wolf did not exist, we would have to invent it."

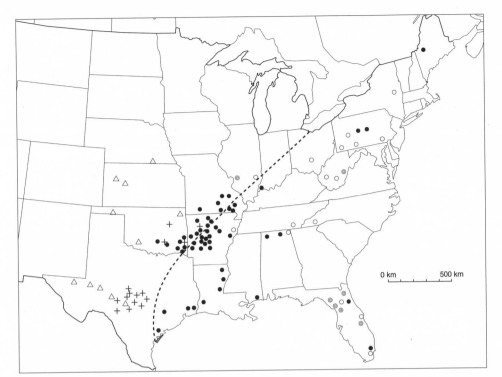

FIGURE 9.7. Localities of specimens dating prior to 1930 (except for the Mississippi record, which is 1931) in the south-central and southeastern United States. Triangles, *Canis lupus nubilus;* solid circles, complete adult specimens of *C. rufus;* open circles, archeological remains of *C. rufus* dating from 200 to 2,000 years ago; double circles, paleontological remains of *C. rufus* dating from about 10,000 years ago; crosses, specimens evidently representing hybridization of *C. rufus* and *C. latrans;* dashed line, approximate original eastern limits of range of *C. latrans* in historical time. The exact localities of the two complete Pennsylvania specimens are unknown. The complete specimens from northeastern Louisiana and east of the Mississippi (except for the ones from Indiana, Mississippi, and northern Florida) and the specimens of *C. lupus nubilus* are among the series tested by the analysis shown in figure 9.9. The hybrids from central Texas are those tested in the analysis shown in figure 9.12. (Based largely on Nowak 1979, Nowak and Federoff 1996, 1998.)

Coyote Invasion of the Southeastern United States

For the red wolf to have originated as a hybrid, its range in southeastern North America would have to have been inhabited by *C. lupus* and *C. latrans.* There has been repeated insistence that the putative hybrid population probably formed in response to agricultural development along the Atlantic coast by European settlers, starting about 250 years ago (Roy et al. 1996; Wayne et al. 1998; Reich et al. 1999). However, all available specimens—paleontological, archeological, and modern—

indicate that neither *C. lupus* nor *C. latrans* was present in the southeastern United States at a time that would have allowed formation of a hybrid population (fig. 9.7; Nowak and Federoff 1996, 1998).

The coyote did indeed invade the southeastern United States, partly as a result of human environmental disruption (and partly because extirpation of the red wolf by humans opened a niche for another canid). That invasion, which is well documented, began only in the twentieth century and did not cross the Mississippi River until about 1960 (Paradiso 1966; Nowak 1979; Hill et al. 1987).

It is true that *C. latrans* was present in the southeastern United States during parts of the Pleistocene. Apparently, as the archaic New World wolf line grew larger, passing from *C. priscolatrans* to the very large early Rancholabrean form of *C. armbrusteri,* a niche opened for the coyote. A few specimens indicate that it coexisted with *C. armbrusteri* at Cumberland Cave, Maryland, and subsequent fossils indicate widespread sympatry with the huge dire wolf. There are also late Rancholabrean records of *C. latrans* in Pennsylvania, West Virginia, Mississippi, Alabama, and Florida (Kurtén 1974; Nowak 1979; Kurtén and Kaye 1982; Morey 1994a; Graham and

Lundelius 1994). It has been suggested that the Florida specimens actually represent *C. familiaris* (Martin and Webb 1974), but recent reexamination confirms their identity as *C. latrans* (Nowak 2002).

In any case, that eastern coyote population did not persist later than about 10,000 years ago (Nowak 2002). It may not have been able to compete with *C. rufus*, which was becoming established in the East at about that time. Specimens from a few sites in Florida show a brief (less than 5,000 years) sympatric presence of *C. latrans*, *C. rufus*, and *C. dirus*, and that all were morphologically distinct. Of the three, only *C. rufus* seems to have survived the first human invasion of the region. Remarkably, when the red wolf succumbed to a second human invasion 10,000 years later (i.e., by the 1960s), the coyote returned.

The red wolf did interbreed with the coyote, though that had nothing to do with its origin. On the contrary, hybridization had much to do with its demise (McCarley 1962; Nowak 1979). All available evidence indicates that interbreeding began in the late nineteenth century on the western fringes of the range of the red wolf. Prior to 1930, the only specimens that showed morphological signs of hybridization are from one zone in central Texas and another in eastern Oklahoma and northeastern Arkansas (see fig. 9.7).

A poignant series of skulls was collected from 1919 to 1924 in the vicinity of the St. Francois Mountains in southeastern Missouri, where an isolated red wolf population then persisted. The specimens are unmistakably divisible into two groups, *C. rufus* and *C. latrans*, showing that the species still occurred together without hybridization (Nowak 1979).

Red Wolf Hybridization with Coyotes

All *Canis* specimens collected up to 1930 in most of Arkansas, and up to about 1950 in most of southeastern Texas, Louisiana, and states farther east, represent only a small wolf. The extirpation of that wolf by humans and the clearing of the forests for agriculture probably facilitated the eastward movement of the coyote and its interbreeding with remnant wolf populations (Nowak 1979). A substantial zone of hybridization apparently formed in the Ozark region of southern Missouri and Arkansas from about 1930 to 1950 (fig. 9.8).

By the 1960s, the hybridization process had engulfed most of southeastern Texas, and the coyote was spread-

FIGURE 9.8. Zones of hybridization between *Canis rufus* and *C. latrans* from 1930 to 1950 (1) and from 1960 to 1980 (2), and between *C. lupus lycaon* and *C. latrans* (3). In 1987, when *C. rufus* was reintroduced at Alligator River National Wildlife Refuge in eastern North Carolina (4), the eastern limits of the range of *C. latrans* were approximately as shown by the dashed line. Those limits now have reached the reintroduction site.

ing east of the Mississippi (see fig. 9.8). Animals that appeared to be like the original southeastern wolf still predominated in a stretch of coastal prairie from Galveston Bay in southeastern Texas to Vermilion Bay in southwestern Louisiana. A number of individuals were live-trapped in that area, and fourteen of them eventually served as the founding stock of the current captive and reintroduced *C. rufus* populations (Nowak et al. 1995). Subsequently, the hybrids evidently took over that area as well, and coyotes went on to occupy the entire Southeast (Nowak 1979; Bekoff 1999). This modern process of invasion by the coyote, hybridization, and decline of the red wolf is documented with specimens and other records and stands in marked contrast to the lack of specimens to support the hypothesis of an earlier hybrid origin as set forth by Roy, Girman and Wayne (1994) and Roy et al. (1996).

DNA and the Origin of the Red Wolf

We are still left with the second main line of evidence for hybrid origin, the reported presence of only *C. latrans* and *C. lupus* DNA genotypes in older samples of red wolves and of only *C. latrans* genotypes in recent samples (Wayne and Jenks 1991; Wayne et al. 1991, 1995, 1998; Jenks and Wayne 1992; Wayne 1992, 1993, 1995; Reich et al. 1999; Roy, Girman, and Wayne 1994; Roy, Geffen et al. 1994; Roy et al. 1996; Wayne and Gittleman 1995). The older samples were taken from skins collected before 1944, predominantly in the northwestern part of the original range of the red wolf. The recent samples include material derived from the current living population.

Although the reported presence of only coyote and gray wolf DNA in the red wolf could be viewed as indicative of hybrid origin for *C. rufus*, another interpretation has been applied by some of the same workers to a similar situation in the Great Lakes region (Lehman et al. 1991; Wayne et al. 1991). Those investigators reportedly found only DNA from the coyote in all *C. lupus* from southeastern Ontario and Isle Royale and in most *C. lupus* from Minnesota and western Ontario. While there is also morphological evidence of hybridization in southeastern Ontario, statistically, all cranial specimens from Minnesota, western Ontario, and Isle Royale are morphologically identical to those of *C. lupus* from farther west (Nowak 1992a, 1995a).

These findings have been interpreted to mean that limited hybridization with *C. latrans* did occur and did introduce a coyote genotype that somehow spread through most of the wolf population without otherwise affecting it. The situation in the southeastern United States could be interpreted similarly: hybridization, not between the coyote and gray wolf, but between the coyote and red wolf, and thus the spread of coyote genotypes. This explanation would mean that *C. rufus* is not a hybrid, but rather is just as valid a wolf as is *C. lupus* of Isle Royale, Minnesota, and other parts of the Great Lakes region.

A last problem pointed out by some workers is that, whereas the wolf component of gray wolf DNA from the Great Lakes region corresponds specifically to the DNA of *C. lupus* found elsewhere, the wolf component of southeastern red wolf DNA also corresponds specifically to that of *C. lupus*. No species-specific DNA has been found for *C. rufus;* only the DNA of *C. lupus* and *C. latrans* is present (Wayne and Jenks 1991; Roy, Geffen et al. 1994; Roy et al. 1996; Wayne and Gittleman 1995). One suggested explanation is that there were specific *C. rufus* DNA markers, which may even have been found in ear-

lier studies (Ferrell et al. 1980), but that they were lost with the near-total extirpation of the species (Nowak et al. 1995).

It must also be emphasized that the above suggestions of hybrid origin for *C. rufus* have been set forth by only some of the investigators who have utilized DNA analysis. Hybrid origin has not been supported by some other geneticists who have reviewed the issue (Dowling, DeMarais et al. 1992; Dowling, Minckley et al. 1992; Cronin 1993), by morphological studies (Nowak 1979, 1992a, 1995a,b), or by observation of living animals (Phillips and Henry 1992; Nowak et al. 1995).

A New Hypothesis

Indeed, some new DNA analyses have produced a very different assessment (Wilson et al. 2000). Essentially, those studies indicate that *C. rufus* and *C. lupus lycaon* have genetic sequences similarly divergent from those of *C. latrans* and distinct from those of other *C. lupus*. One interpretation is that *lycaon* and *rufus* together compose a separate species that is more closely related to *C. latrans* than to *C. lupus*. Such a species would take the name *Canis lycaon* (Wilson et al. 2000).

The close morphological similarity of *C. rufus* and *C. lupus lycaon* has long been recognized (Goldman 1944; Nowak 1979). Lawrence and Bossert (1967, 1975) thought *C. rufus* to be no more than subspecifically distinct from *C. lupus*. They pointed out that if initial taxonomic studies of the red wolf had concentrated on the eastern United States, rather than on Texas, it is doubtful whether *C. rufus* would have been set apart as a full species. The skull morphology of the population of *C. l. lycaon* in southeastern Ontario almost overlaps statistically with that of *C. rufus* (Nowak 1995a; Nowak and Federoff 1996). It might be supposed that specimens of gray wolves from the northeastern United States, if available, would overlap substantially with *C. rufus*.

Unfortunately, there are relatively few complete skulls of wild adult *Canis* dating prior to 1918 — that is, well before the modern coyote invasion — from the region east of the Mississippi River and south of the Prairie Peninsula, Lakes Erie and Ontario, and the St. Lawrence River. Those few specimens, two each from Pennsylvania, Alabama, and Florida and one each from Maine and Indiana (see fig. 9.7), had been assigned in the past either to *C. lupus* or *C. rufus*, sometimes with disagreement as to which (Goldman 1944; Lawrence and Bossert 1967;

Nowak 1979; Nowak and Federoff 1996). Of those specimens, six are known or judged to be males. In addition, there are seven males and two females collected just west of the Mississippi in northeastern Louisiana in 1898–1905 (see fig. 9.7), about 50 years before the coyote is known to have entered that state. These specimens were the basis of a morphometric analysis to assess the original relationships of southeastern wolves to *C. lupus, C. latrans,* and *lycaon.* Only males were sufficiently numerous to form meaningful series for analysis, though nothing about the females indicated any different kind of relationship.

The new analysis involved subjecting ten cranial and dental measurements to canonical discriminant analysis. In that operation, the measurements, weighted by their ability to distinguish designated groups, assign each specimen a total abstract numerical value—the first canonical variable. The next best distinguishing combination of measurements, uncorrelated with the first, provides a second variable, and so on. Commonly, a single graphical position is plotted based on the first two canonical variables arranged as perpendicular axes. (Illustrations and descriptions of the ten measurements, and a more detailed explanation of the statistical procedures, is provided by Nowak 1995a; see also Nowak 2002.)

In the analysis, the thirteen pre-1918 males from northeastern Louisiana and east of the Mississippi were compared as a group with ninety-six male western *C. latrans* and with six samples (totaling ninety-seven specimens) of male *C. lupus* from the western United States. Of those six samples, two were collected immediately west of the range of the red wolf: in central and western

Texas and in southern Nebraska, Kansas, and Oklahoma (see fig. 9.7). The other four were taken just north and west of the first two, in Minnesota and in the southern, central, and northern Rocky Mountains. Another series comprised ten male *lycaon,* taken in 1905–1933 in extreme southeastern Ontario, mainly in the vicinity of Algonquin Provincial Park, and in extreme southern Quebec. The coyote did not become established there until the 1940s (Nowak 1979).

The analysis shows all six series of western *C. lupus* to overlap extensively with one another; their mean positions group together closely (fig. 9.9). Those series are completely distinct from the thirteen old eastern wolves, which group together separately. There is no indication of an approach of the old eastern series to either *C. latrans* or western *C. lupus,* as would be expected if the eastern population had originated as a hybrid of the other two species. The series of old *lycaon* is statistically intermediate to *C. rufus* and *C. lupus* and closely approaches both (fig. 9.9).

A Distinct Species

Considering also the new DNA data, a reasonable interpretation of the morphometric analysis is that *C. rufus* is a species distinct from *C. latrans* and *C. lupus,* but that it hybridized with the latter in a limited area of the Northeast, producing the population known as *lycaon.*

While assessment of all available complete skulls from the eastern United States supports continued recognition of *C. rufus* as a distinct species, the specimens are rather few, and most date from well after human environmental disturbance had begun. There is some older

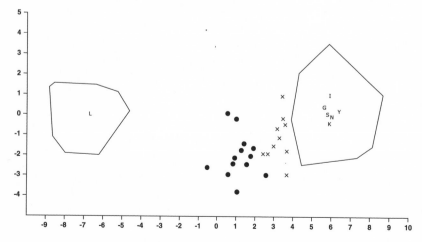

FIGURE 9.9. Statistical distribution of nine groups of North American *Canis,* based on the first (horizontal axis) and second (vertical axis) canonical variables, as produced by a canonical discriminant analysis of ten skull measurements. Solid lines represent the limits of 96 western *C. latrans* (letter L shows mean) and of 97 western *C. lupus* in six samples (means: G, 17 from southern Rocky Mountains; I, 14 from northern Rocky Mountains; K, 23 from Minnesota; N, 8 from Nebraska, Kansas, and Oklahoma; S, 7 from central and western Texas; Y, 28 from central Rocky Mountains); X's, 10 individual *C. lupus lycaon;* solid circles, 13 individual *C. rufus.*

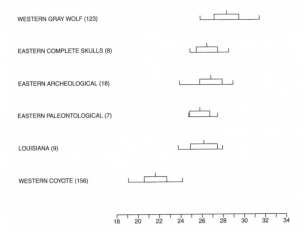

FIGURE 9.10. Length (in millimeters) of first lower molar tooth, or carnassial, in six samples (size in parentheses) of both sexes of *Canis.* The "western gray wolf" and "western coyote" series were taken in the mountainous region of the western contiguous United States. The "eastern complete skulls" date from before 1918 and were collected east of the Mississippi River and south of the Prairie Peninsula, Lakes Erie and Ontario, and the St. Lawrence River (see fig. 9.7). The "eastern archeological" specimens are from the same region and date from 200 to 2,000 years ago. The "eastern paleontological" specimens are from the same region and date from about 10,000 years ago. The "Louisiana" series was collected from 1898 to 1905. The slender horizontal line represents the range in size, the bar indicates one standard deviation on either side of the mean, and the vertical line above the bar shows the mean. (From Nowak 2002.)

material, though it consists mostly of fragmentary remains from paleontological and archeological sites. The most common elements recovered at such sites are the mandible and the large first lower molar (carnassial) tooth. The available material falls conveniently into two groups: paleontological specimens dating from about 10,000 years ago, and archeological fragments from about 200 to 2,000 years ago (Nowak 2002) (see fig. 9.7).

I made a comparison of carnassial length in these and other pertinent groups (fig. 9.10). The two groups of fragmentary material are closely similar to the two series of complete skulls from east of the Mississippi and northeastern Louisiana. Those four groups differ substantially from the series of western *C. lupus,* although that difference by itself is not necessarily of specific rank. Nonetheless, the analysis demonstrates a continuity between the mostly modern complete skulls, found to compose a statistically well defined species (see fig. 9.9), and much earlier specimens from the same region.

Like the multivariate study, the univariate analysis shows *C. latrans* to be well removed from the other

groups. Since both sexes are represented in the univariate analysis, the slight overlap shown could represent large male coyotes and small female wolves. It must be emphasized that no coyote specimens are known from the region that date from less than 10,000 years ago. There are older ones, as already noted, but they average smaller than the modern series of *C. latrans* in figure 9.10, and all are smaller than the smallest wolf depicted.

The close morphology of *C. rufus* and *C. lupus* in the Northeast is not seen where their ranges meet in the western United States (see fig. 9.7); indeed, the major evidence supporting separate species status has always come from that region. As Goldman (1944) observed, *C. rufus* seems to become smaller toward the western United States and to differ sharply from *C. lupus* of the Great Plains. Although Goldman, as he realized, may have been influenced by hybrids in that region, there is a striking statistical contrast (fig. 9.11) between large western *C. lupus* and the most westerly samples of unmodified *C. rufus* (Nowak 1995a; Nowak and Federoff 1996, 1998). Large specimens of the gray wolf were found as far east as central Texas and northeastern Oklahoma, very close to the most westerly specimens of the red wolf (see fig. 9.7). Variation across that front is far greater than among any samples of *C. lupus* in the analysis and about twice as great as that between *C. rufus* and *C. l. lycaon* from the Algonquin region.

The clearest indication of specific distinction between red and gray wolves comes from the analysis of a large series of specimens taken on the Edwards Plateau of central Texas from about 1890 to 1920 (Nowak 1979, 1995a).

FIGURE 9.11. Statistical distances (D^2 of Mahalanobis), as produced by a canonical discriminant analysis of ten skull measurements, between three series of male *Canis rufus* (Old Eastern, Louisiana, Missouri) and eight nearby populations of male *C. lupus.* (From Nowak and Federoff 1996.)

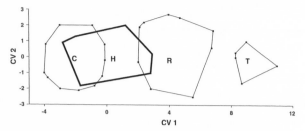

FIGURE 9.12. Statistical distribution of four groups of male *Canis*, based on the first (horizontal axis) and second (vertical axis) canonical variables, as produced by a canonical discriminant analysis of fifteen skull measurements. Light solid lines with centrally located letters indicate limits of known groups: C, 165 *C. latrans* from the western United States; R, 70 *C. rufus* collected before 1930 in the south-central United States; T, 7 *C. lupus* from western and central Texas. The heavy line around letter H indicates the statistical limits of 43 specimens taken before 1930 in central Texas (see fig. 9.7) and thought to represent a hybrid population of *C. latrans* and *C. rufus*. (Adapted from Nowak 1979; Nowak et al. 1995.)

That analysis (fig. 9.12) shows the presence of a hybrid population in the area (see fig. 9.7) that completely bridges the morphological gap between the populations of wolves then found farther east (that is, *C. rufus*) and the populations of coyotes farther west (*C. latrans*). The two species blend completely, with no place to draw a statistical or geographic line of division. It must be emphasized, though, that the above blending occurs only between the coyote and the animal that has been referred to as the red wolf. The gray wolf also was present in the area (see fig. 9.7), but was statistically far removed from the hybridization process. Based on more limited material, a similar situation apparently existed in eastern Oklahoma and northwestern Arkansas (see fig. 9.7).

The situation in central Texas is a major impediment both to the view that *C. rufus* and *C. lupus* are conspecific and to the idea that *C. rufus* originated through hybridization between *C. lupus* and *C. latrans*. If the gray wolf is the same as the red wolf, or indeed, if it crossed with the coyote to produce the red wolf, why did it not blend into the hybrid red wolf-coyote population of central Texas? How can the specimens of *C. lupus* in that area stand out so clearly from the other kinds of *Canis* present?

That distinction between *C. lupus* and *C. rufus*, compared with their morphological closeness in the Northeast, also long posed a dilemma for studies seeking to define the species limits of *C. rufus*. How can two large, highly mobile carnivores be so different from each other

along part of the line where their ranges meet, but resemble each other so closely at another part? The above suggestion as to the origin of the northeastern subspecies *C. l. lycaon* may finally offer a resolution. *C. rufus* and *C. lupus* once would have been fully distinct in all areas of contact, but hybridization in the Northeast produced an intermediate population.

Domestication of the Dog

Most authorities agree that the gray wolf (*C. lupus*) is the progenitor of the domestic dog (*C. familiaris*) (see also Wayne and Vilà, chap. 8 in this volume), and that domestication occurred in only one region, probably southern Asia (Lawrence 1966; Scott 1967, 1968; Clutton-Brock et al. 1976; Olsen and Olsen 1977; Nowak 1979). However, there have been suggestions that various dog breeds arose through multiple domestication events involving different subspecies of wolves, times, and places (Morey 1994b; Trumler 1990). Recent studies of mitochondrial DNA support the single origin hypothesis and suggest that initial domestication occurred in the eastern part of Asia (Savolainen et al. 2002). Perhaps the best-documented case for a separate origin was made by Olsen (1985), whose statistical analysis of cranial material suggests an early transition from wolf to Eskimo dog (malamute) in Alaska. However, Leonard et al. (2002) found mitochondrial DNA evidence to support an origin of New World dogs from multiple Old World lineages. In addition, based on extensive measurements of fossil and modern material in Europe, De Quoy (1993) argued that the Irish wolfhound originated independently, possibly as long ago as the late Pleistocene (ca. 10,000 years ago).

Populations of primitive dogs, known as pariahs, still survive in southern Asia. These animals, which may either be associated with people or live completely wild, appear to represent an evolutionary transition between the wolf and modern breeds of dogs (Fox 1978; Trumler 1990). They evidently developed on the mainland of southern Asia and then, about 3,500 to 4,000 years ago, were spread to other regions by seafarers and migrating peoples (Corbett 1995; Ginsberg and Macdonald 1990). The dingo of Australia probably resulted from an introduction there of some of those dogs at that time. Other populations became established in North Africa, the Balkan peninsula, New Guinea, and the East Indies. There is evidence that the animals were taken as far as southern

Africa, Japan, the Pacific Islands, and possibly North America. It may be appropriate to classify all of these dogs in the subspecies *dingo* (whether of the species *C. lupus* or *C. familiaris*). They are threatened because of human persecution, environmental disruption, and hybridization with true domestic dogs (Corbett 1995; Ginsberg and Macdonald 1990).

The oldest well-documented remains of *C. familiaris*, dating from about 11,000 and 12,000 years ago, respectively, were found in Idaho and Iraq. Other specimens, possibly representing domestication, date from as far back as 14,000 years in the Middle East (Davis and Valla 1978; Turnbull and Reed 1974; Walker and Frison 1982). Early remains of dogs also have been found in northern and central Europe (Mertens 1936; Degerbol 1961; Musil 1974; Nobis 1986; Benecke 1987). A report of a specimen from the northern Yukon, with a minimum age of 20,000 years (Beebe 1978), was considered doubtful by Olsen (1985). Based on an analysis of mitochondrial DNA, however, Vilà et al. (1997) suggested that the genetic lineage of the dog could have originated as long ago as 135,000 years, though it might not have become morphologically apparent until the more commonly accepted date for domestication. Nevertheless, a more recent DNA study showed that genetic results could be consistent with domestication starting 15,000 years ago (Savolainen et al. 2002).

The transition from wolf to "prototypic" wolf-dog to domestic dog was long and probably dependent on selective pressures and other related factors. However, Belyaev (1979) experimentally developed domestic traits in foxes in only twenty generations by selecting for foxes that were social toward humans. Early wolf-dogs probably were not selectively bred for any specific traits, but more likely bred freely, and associated with humans only for food (Olsen 1985). Imprinting on humans from an early age also would have facilitated the domestication process (Scott and Fuller 1965; Scott 1967; Mech 1970).

While there is general agreement that the gray wolf is the progenitor of the domestic dog, the question of whether the two should be treated as separate species may never be resolved. There has been a recent tendency to follow Wozencraft (1993) in including *familiaris* within *C. lupus*, as *C. lupus familiaris*. On the basis of genetic consistency, Wayne (1986) considered the dog, despite its diversity in size and proportion, nothing more than a gray wolf. In contrast, statistical analysis of skulls has repeatedly demonstrated total separation between wolf

and dog (Lawrence and Bossert 1967; Nowak 1979, 1994). The primary change in the skull during the domestication process was a general reduction in size, with shortening of the muzzle and, thus, a relatively broader palate, a steeper forehead, and more crowded teeth (Olsen 1985; Morey 1994b). These characters are comparable to those of a young wolf and suggest that an adult dog retains, to some extent, a juvenile morphology (paedomorphosis). Such retention is also seen with respect to behavior. It is as though some wolves, over time, never grew into adults.

Evolutionary forces independent of human control may have guided the original domestication process. Although people may initiate such a process, by doing so, they set into motion other associated and unintended events. Alterations in developmental rate and timing (heterochrony) and changes in genetic regulation may be the driving force of domestic reorganization (Wayne 1986; Morey 1992).

Coppinger and Coppinger (2001) argued that the dog is not a paedomorphic wolf, but has its own unique set of characters, both morphological and behavioral, that arose when humans created a new niche by providing a food source in association with villages. This hypothesis supports recognition of *C. familiaris* as a full species, but one that separated from *C. lupus* not more than 12,000–15,000 years ago. The initial development of the dog involved a process of natural selection resulting from voluntary action by the animal, not human domestication, according to this view.

Deliberate selection by humans probably is responsible for the characters of modern breeds, including the large size of some. In contrast to small dogs, those with large (wolf-sized) skulls have relatively tiny, widely spaced teeth (see fig. 9.4). An almost universal character of domestic dogs, whether large or small, is a broad and heavy frontal shield at the top of the skull (Nowak 1979). Could it have evolved selectively as a protection against human abusers, striking downward with clubs?

Wolf Hybridization with Dogs

Although many workers now consider the wolf and dog as the same species, one of the best arguments for recognizing them as separate species, in addition to their clear physical differences, is that the two generally do not voluntarily interbreed, even though they have shared vast parts of their ranges for thousands of years. Nevertheless, the two can produce viable offspring in captivity, and all subsequent generations are fertile (Iljin 1941). The deliberate breeding of wolf-dog hybrids and various backcross combinations recently has proliferated. An estimated 300,000 such animals now are present in the United States (Willems 1995).

The crossing of wolves with dogs by humans is not a recent phenomenon, but probably has occurred throughout history. Aristotle (1783) wrote of such crosses in the fourth century B.C., and Pliny (1771) (A.D. 23–79) reported that the Gauls tied their female dogs to trees so nearby wolves would have access to them for mating. The Eskimo peoples and the American Indians also were known to use wolves and hybrids as sled dogs (Olson 1985). Before snowmobiles or horses, native peoples had only their dogs on which to rely. The powerful draft animals used by the Plains Indians were produced by wolf-dog interbreeding, which resulted in dogs much like wolves in size and general appearance (Morey 1986). Specimens of those dogs, as well as wolves, have been recovered from archeological sites in North Dakota, dating from the late sixteenth century. Evidence of extensive early hybridization between wolves and dogs also has been found in Wyoming and the Dakotas (Walker and Frison 1982; Morey 1986). Numerous reports from early explorers substantiate this evidence (Audubon and Bachman 1851; Bailey 1926; Brackenridge 1816; Maximilian 1906).

There are a few documented cases of wolves, mostly females, soliciting and mating with dogs independent of human involvement (Maagaard and Graugaard 1994; Young 1944). There also have been reports of introgression from the domestic dog to wild populations of the gray wolf in Italy (Boitani 1984) and a few other areas (see Wayne and Vilà, chap. 8, Fritts et al., chap. 12, and Boitani, chap. 13 in this volume). However, Lorenzini and Fico (1995) and Randi et al. (1993) questioned whether such introgression is a substantive threat to the integrity of the wolf in Italy. Examination of a large series of wolf specimens recently collected in Italy at the University of Rome and the Italian National Institute of Game revealed no indication of modification (Nowak and Federoff 2002).

Based on a claimed progressive reduction in the size of the skull of the Arctic wolf *(C. lupus arctos)*, Clutton-Brock et al. (1994) suggested that there had been widespread introgression from escaped sled dogs. However, Nowak (1995b) found no such size reduction and doubted that hybridization had been extensive. According to Mengel (1971), hybrids of *C. familiaris* and *C. lupus* breed annually, like wild wolves, but the timing is shifted so that, on average, hybrids breed about 3 months earlier. Therefore, offspring of hybrids are born during winter and would have little chance of survival in the wild. The Arctic thus would be the last place where introgression from dogs into a wild population of wolves would be expected.

Nowak's (1979, 1995a,b) analysis of specimens also revealed no evidence of dog introgression into wolf populations of the western contiguous United States during the early twentieth century, when these populations were declining in response to environmental disruption and persecution, or into the populations of wolves now present on the fringes of agricultural zones in western Canada and expanding back into the western United States. Considering also other available information, most of the world's populations of wild wolves appear to be genetically intact.

10

Wolf Interactions with Non-prey

Warren B. Ballard, Ludwig N. Carbyn, and Douglas W. Smith

WOLVES SHARE THEIR ENVIRONMENT with many animals besides those that they prey on, and the nature of the interactions between wolves and these other creatures varies considerably. Some of these sympatric animals are fellow canids such as foxes, coyotes, and jackals. Others are large carnivores such as bears and cougars. In addition, ravens, eagles, wolverines, and a host of other birds and mammals interact with wolves, if only by feeding on the remains of their kills.

Wolves and Guilds

Ecological guilds are groups of species using common resources in a similar way (Root 1967), so wolves are members of a guild that includes other large carnivores, such as bears and cougars. In this chapter, we will also consider birds and mammals that are important scavengers on wolf prey as part of that guild.

Although wolves frequently interact with many other carnivore species and guilds, no studies have been conducted to determine the effects of these interactions on carnivore community structure and population dynamics. Consequently, the only available information concerning wolf-non-prey interactions consists largely of anecdotal observations. In this chapter we review the available literature and data and attempt to synthesize information about interactions between wolves and non-prey animals.

Except for the seminal works of Rosenzweig (1966), Johnson et al. (1996), and Palomares and Caro (1999), few researchers have addressed the subject of carnivore community dynamics, and none has dealt solely with wolves and non-prey species. The inherent genetic, behavioral, and morphological flexibility of wolves has allowed them to adapt to a wide range of habitats and environmental conditions in Europe, Asia, and North America. Therefore, the role of wolves varies considerably among specific ecosystems. To address the community role of wolves within different systems would require in-depth studies of sympatric wildlife populations. For now, we can only review and summarize information about wolf interactions with non-prey and interpret the relevance of these interactions to wolf populations and to the role of wolves within a carnivore and scavenger community or ecosystem.

Interactions among Guild Members

Interactions among members of the carnivore guild are ubiquitous, although opportunities to document such interactions are uncommon due to the elusive nature of most carnivores. The most common type of interaction is probably competition, which is generally most intense between the most similar species (Johnson et al. 1996; Palomares and Caro 1999). The principle of competitive exclusion holds that two competing species may coexist in a stable environment if they have adequate niche differentiation (Krebs 1994). If there is no such differentiation, then one species will exclude the other. The degree of competitive exclusion depends on the degree of niche overlap, the degree of spatial overlap, and the availability of limited resources (i.e., food and space).

Two basic types of competition are recognized: exploitation competition and interference competition.

Exploitation competition is indirect and is based on differential efficiency in accessing and using shared resources. Competition for food is a process in which carnivores interact with one another to access a shared prey base (Murie 1944; Haber 1977; Ballard 1980; Ballard 1982; Boertje et al. 1988) and is a form of exploitation competition. Evidence for such behavior is not always obvious, and it is harder to demonstrate than interference competition. Outcomes of exploitation competition are expressed slowly, through differential survival and reproduction, ultimately leading to extinction or evolutionary divergence (Krebs 1994). Exploitation competition is pervasive and has been predicted in earlier models (Hairston et al. 1960).

Interference competition is direct and is expressed through aggressive behavior. Interspecific killing, for example, is common among mammalian carnivores and may influence population and community structure (Palomares and Caro 1999). Interference competition causes the immediate exclusion of a competing individual or population from a resource (Krebs 1994). Among canids, interference competition is asymmetrical, with only the larger species benefiting from the interaction (Peterson 1995a) by excluding the smaller competitor from the resource. Interspecific competition can also influence spatial patterns in habitat selection and geographic distribution (Connor and Bowers 1987).

Competition, at both individual and population levels, may be influenced in subtle ways. Factors such as different seasonal movements, availability of alternative food resources, topography, snow cover, morphological differences, population characteristics, and reproductive histories may all be important in reducing niche overlap and increasing resource partitioning.

The availability of wolf-killed carcasses in winter tends to concentrate interspecific competition for some species (e.g., coyotes and foxes). Communal feeding at kills by wolves, bears, coyotes, foxes, and common ravens has been observed (Mech 1966b; Peterson 1977; Ballard 1982; Paquet 1992; Peterson 1995a). Such occurrences probably take place when the wolves are satiated and resting and may not truly represent tolerance, but rather the wolves' inability or lack of motivation to chase or catch the scavengers (Peterson 1995a).

As indicated above, degree of niche overlap, food availability, and species behavioral differences influence the intensity of competitive interactions among wolves and other carnivores. Some predators are more robust (i.e., more resilient to random events), can convert available energy into population numbers more quickly, can use food more efficiently, or may have quicker growth rates, than others. Some or all of these characteristics may confer a competitive advantage to one of the interacting species.

We found no case histories in the literature in which competition between wolves and other species resulted in a pronounced and long-lasting spatial partitioning of resources within the same area. There were, however, examples of elimination of some predators (e.g., coyotes) in the presence of wolves because of interference competition (Mech 1966b; Berg and Chesness 1978; Fuller and Keith 1981b; Johnson et al. 1996; Crabtree and Sheldon 1999a).

Most of the descriptions in this chapter discuss events that occur at the individual level and should not be confused with population-level concepts such as the species competition hypotheses used in ecological theory (e.g., ecological niche, principle of competitive exclusion). In addition, observations of individuals should not necessarily be considered important at the population level. Few opportunities have allowed the ecological consequences of competition with wolves to be quantified in terms of ecological theory.

The nature of wolf interactions with non-prey varies according to the size of the animal with which wolves are interacting. Thus we will discuss wolf interactions with different groups of species according to their size.

Interactions with Large Carnivores

The largest non-prey residents of wolf range, such as bears, cougars, and tigers, are competitors, and can even be adversaries, of wolves. Because their sizes, food habits, densities, and other relevant characteristics vary so much, the nature of their interactions with wolves also varies.

Brown Bears

Geographic overlap between wolves and brown (or grizzly) bears was once much more widespread than at present. In Yellowstone National Park (YNP), wolf and brown bear remains were found in the same cave deposits from 960 B.P. (Hadley 1989). Throughout most of their North American and Eurasian ranges, bear populations have experienced human-caused declines in recent years. Nevertheless, brown bears and wolves are still sympatric in significant portions of their former ranges, and interactions between them have been frequently observed. The most extensive observations come from Alaska and northern Canada.

One of the first biologists to report on interactions between wolves and brown bears was Adolph Murie (1944). He concluded that brown bears easily took ownership of wolf-killed carcasses. Murie did not record any fatal interactions between wolves and bears, although he did describe several harmless skirmishes. Since Murie's pioneering work, many other observations of wolf-bear interactions have been recorded.

We classified wolf-bear interactions into sixteen types of behavior that were modified from classifications orig-inally defined by T. J. Meier and M. D. Jimenez (personal communication). Wolves outnumbered brown bears during 54% of the interactions (table 10.1). Bears outnumbered wolves in only 19% of the interactions, and nearly all of these involved bears accompanied by cubs, yearlings, or 2-year-olds. Most (65%) wolf-bear interactions involved bears without young.

Of the 108 reported interactions between brown bears and wolves (excluding those in YNP), the most common types involved bears and wolves fighting and chasing each other (24%) and bears defending kill sites against wolf packs (see table 10.1). Feeding sites (i.e., kills made by either species) were the most common locations

TABLE 10.1. Summary of wolf-brown bear interactions in North America outside of Yellowstone National Park

| | Interaction type [a] | | | | | | | | | | | | | | | | | |
	1	2	3	4	5	6	7	8	9	10	11	12	13	14	15	16	Total	%
No. occurrences	6	7	3	26	11	7	8	1	4	23	2	2	1	2	3	2	108	
% occurrences	6	6	3	24	1	6	7	1	4	21	2	2	1	2	3	2	10	
Type of site																		
Feeding	6	7	3		7				4	23	2	2	1	2	2	2	62	57
Wolf den				13			2										15	14
Other				12	11	6	1								1		31	29
Numbers of each																		
Bears > wolves	1			8	8	1	1		1	1							21	19
Wolves > bears	4	2	1	9	2	6	3	1	3	16	2	2		2	3	2	58	54
Bears = wolves	1	5	1	9	1		3			5							25	23
Unknown			1				1			1			1				4	4
Outcome																		
Bear wins	5	2	1	1	3					22	2					2	38	35
Wolf wins				9	1				4			2		3			19	18
Neither wins	1	4	1	15	7	7	7	1									43	4
Both win		1	1										1				3	3
Unknown				1			1			1				2			5	5
Bears with young																		
Yes	1			10	9		1			3					3		27	25
NA			2	1		7								1			11	10
No	5	7	1	15	2		7	1	4	20	2	2	1	1		2	70	65

Sources: Data sources included Lent 1964; Ballard 1980, 1982; Peterson, Woolington, and Bailey 1984; Hornbeck and Horejsi 1986; Hayes and Mossop 1987; Hayes and Baer 1992; and MacNulty et al. 2001. Also included are previously unpublished data from Denali National Park and Preserve for 1970–1974, 1979–1989, and 1995 (J. W. Burch and T. J. Meier, personal communication), from northwestern Alaska during 1978 (D. James, personal communication), and from the Northwest Territories during 1988 (F. Messier and P. Clarkson, personal communication) and during 1996–1999 (D. Cluff, personal communication). A preliminary analysis of the Denali observations was presented in Servheen and Knight 1993, although locations of observations were not reported.

[a] 1, bear feeding, wolf in area, no mortalities; 2, bear and wolf feeding on same kill at same time, no mortalities; 3, bear feeding on wolf kill, wolves not present, no mortalities; 4, bear and wolves fight and chase each other, no mortalities; 5, wolf stalking bear, no mortalities; 6, wolf feeding on bear, believed to be scavenging, but could have been predation; 7, wolf and bear in same area, no mortalities; 8, other, information not specific; 9, wolf displaces bear from kill site, no mortalities; 10, bear defends kill from wolf, no mortalities; 11, bear displaces wolves from kill, no mortalities; 12, wolf defends kill from bear, no mortalities; 13, bear kills animal wounded by wolf, no other mortalities; 14, both bear and wolf sign at kill, cause of death unknown, no mortalities; 15, wolves kill bear; 16, bear kills wolf.

(57%) for all types of interactions. Interactions at a variety of different sites made up the second most common category (29%), followed by those near wolf dens (14%) (see table 10.1). The outcome of wolf-bear interactions varied depending on the type of interaction (see table 10.1). At feeding sites (i.e., kills that could have been made by either species), bears won all (22) of the encounters. Near wolf dens, wolves frequently won. In 3 of the 108 cases, wolves killed bears, and in 2 others, vice versa; most such mortal interactions occurred at feeding sites (see table 10.1).

Wolf-bear encounters can be quite aggressive and may last for several hours, as evidenced by Murie's (1944, 205) account:

> A female with three lusty 2-year-olds approached the den from down wind. They lifted their muzzles as they sniffed the enticing smell of meat, and advanced expectantly. They were not noticed until they were almost at the den, but then the four adult wolves that were at home dashed out at them, attacking from all sides. The darkest yearling seemed to enjoy the fight, for he would dash at the wolves with great vigor, and was sometimes off by himself, waging a lone battle. (On later occasions I noticed that this bear was particularly aggressive when attacked by wolves.) The four bears remained at the den for about an hour, feeding on meat scraps and uncovering meat the wolves had buried. During all of this time, the bears were under attack. When the pillaging was complete the bears moved up the slope.

In YNP, where wolves were recently reintroduced, we suspected that wolf-brown bear interactions might differ from those elsewhere in North America where these relationships have been long established. We found a significant difference in proportions of types of interactions between YNP and other areas in North America (see tables 10.1 and 10.2; $\chi^2 = 114$, $P < .0001$). Some of the differences we found may be rather arbitrary because of our classification system, but some noticeable differences did occur, as we point out below.

The most common interactions between wolves and brown bears in YNP involved wolves and bears simply being in the same area (34%), followed by bears defending kills from wolves (19%; probably wolf kills usurped by bears) and bears usurping wolf kills (19%) (table 10.2). Interactions most often occurred at kill sites (66%).

Most encounters at most sites were won by bears (40%), or the winner could not be determined (40%), even though wolves outnumbered bears during 76% of the interactions. Adult bears without cubs were involved in 88% of the encounters. Although wolves lost most disputed kills to bears, wolves were quite successful at defending their dens, and even wolf pups 6 to 7 months old chased bears away from wolf rendezvous sites (R. McIntyre, unpublished data). Two likely instances of wolves in YNP killing grizzly bear cubs have been recorded. One cub was found near an elk carcass and the other near a bison carcass. Necropsy of the cubs, and the circumstances around the carcasses, indicated death from wolves.

Much less is known about wolf-brown bear interactions in Eurasia, but wolves are known to have attacked young bears. Biologists have concluded that wolves and bears show neither spatial nor trophic influences on each other's distributions (Bromlei 1965; Portenko 1944, cited in Yudin 1992). As in North America, wolves successfully defended their dens against bears (Grachev and Fedosenko 1972).

In all areas, most wolf-brown bear interactions took place near ungulate kills that either predator could have made. Most adult ungulates are probably killed by wolves and then usurped by bears, although bears are also quite capable of killing both young and adult ungulates (Boertje et al. 1988; Ballard et al. 1990). Brown bears commonly usurped kills or defended them from wolves. Younger members of bear families were sometimes killed by wolves at such sites, and wolves were sometimes killed by bears. Wolves sometimes ate bears, but bears usually ate only young wolves.

Such interactions could have profound effects on predator-prey relationships because both wolves and bears can exert considerable pressure on the same prey species. Brown bears are often the greatest source of mortality to moose calves where brown bear densities exceed 16/1,000 km² (390 mi²), even though black bears and wolves may be equally abundant (Ballard 1992). Black bears are the greatest source of moose calf mortality where they are at least ten times more numerous than brown bears or wolves and their densities are greater than 200/1,000 km² (Ballard 1992). Where wolves lose kills to bears, their kill rates are probably higher than in systems without bears (Boertje et al. 1988).

The availability of ungulate carcasses to brown bears in systems occupied by wolves undoubtedly results in a higher protein intake for the bears. Reintroduction of

TABLE 10.2. Summary of wolf-brown bear interactions in Yellowstone National Park, Wyoming, during 1996–2001

	Interaction type[a]																Total	%
	1	2	3	4	5	6	7	8	9	10	11	12	13	14	15	16		
No. occurrences	0	3	0	7	4	0	20	0	0	11	11	2	0	0	0	0	58	
% occurrences	0	5	0	12	7	0	34	0	0	19	19	3	0	0	0	0	1	
Type of site																		
Feeding		3		2	1		8			11	11						38	66
Wolf den				3	1		6										10	17
Other				2	2		6										10	17
Numbers of each																		
Bears > wolves		1					1			1							3	5
Wolves > bears		2		7	3		13			7	10	2					44	76
Bears = wolves					1		6			3	1						11	19
Unknown																	0	0
Outcome																		
Bear wins				1			1			10	11						23	4
Wolf wins				3	2		1					2					8	14
Neither wins		2		2	2		16			1							23	4
Both win													1				0	0
Unknown		1		1			2							1			4	88
Bears with young																		
Yes		1		2	1		2				1						7	12
NA																	0	0
No		2		5	3		18			11	10	2					51	88

Source: D. W. Smith, unpublished data.

[a]Interaction types as in table 10.1.

wolves into areas such as YNP could provide benefits to rare and threatened bear populations (Servheen and Knight 1993). Such additional protein may aid bear reproduction. Black bear populations with access to relatively high densities of moose calves had higher productivity (Schwartz and Franzmann 1991); the same might be true in wolf-brown bear systems.

Carrion is an important food resource for brown bears upon their emergence from dens in late winter or spring (Servheen and Knight 1993). In YNP before wolf reintroduction, ungulate carrion was not available during mild winters (Houston 1978; Coughenour and Singer 1996). However, during such winters from 1998 to 2001, brown bears were able to usurp wolf kills.

The use of wolf-killed ungulate remains by bears in spring is particularly high in Pelican Valley, where most elk emigrate in winter, but some bison (in numbers dependent on snow depth) remain (Smith et al. 2000). Wolves that kill bison or early-returning elk routinely lose carcasses to brown bears. In one case near Pelican Valley, a brown bear emerged from a den and went directly to a wolf kill. This wolf-bear relationship may become even more important, for some predictions of the results of wolf reintroduction include reduced ungulate numbers (Boyce 1993). This would mean less late winter and early spring carrion, making wolf kills key food sources for brown bears.

In summary, wolf-brown bear interactions appear to involve both interference and exploitative competition. In such systems, ungulate carcasses are probably frequented by both predator species throughout the year. In addition, wolves may feed on bear carcasses, but bears usually eat only the wolf pups.

Black Bears

There are fewer observations of interactions between wolves and black bears (*n* = 26) than between wolves and brown bears, probably because of the different habitats occupied by the two bears. Brown bears live in open habitats, whereas black bears use dense closed-canopy habitats and are therefore less observable from aircraft. All reported wolf-black bear interactions occurred within the northern portions of black bear range. There are no reported Mexican wolf-black bear interactions, although little was known about Mexican wolves prior to their extermination and subsequent reintroduction in the United States (Ballard and Gipson 2000).

Of the five types of wolf-black bear interactions we classified, wolves killing black bears occurred most often (9 of 26 interactions) (table 10.3). Six of the nine mortalities involved wolves seeking out black bears in their dens, while single bear mortalities occurred at a feeding

TABLE 10.3. Summary of wolf-black bear interactions in North America

	Interaction type[a]					Total	%
	1	2	3	4	5		
No. occurrences	9	4	3	9	1	26	
% occurrences	35	15	12	35	4	1	
Type of site							
Feeding	3	2	1	1		7	27
Wolf den	5		1	1	1	8	31
Bear den				6		6	23
Other	1	2	1	1		5	19
Numbers of each							
Bears > wolves	1					1	4
Wolves > bears	5	4	3	9		21	81
Bears = wolves	3					3	15
Unknown					1	1	4
Outcome							
Bear wins	3				1	4	15
Wolf wins	5	2	2	9	0	18	69
Neither win	1	2				3	12
Both win							
Unknown			1			1	4
Bears with young							
Yes	3			6		9	35
NA						0	0
No	6	4	3	3	1	17	65

Sources: Data sources included Young and Goldman 1944; Joslin 1966; Theberge and Pimlott 1969; Rogers and Mech 1981; Horejsi et al. 1984; Paquet and Carbyn 1986; Gehring 1993; Veitch et al. 1993. Also included are unpublished data from Wood Buffalo National Park during 1995 (L. Carbyn, personal communication), Glacier National Park (D. Boyd, personal communication), and Yellowstone National Park during 1997–1999 (D.W. Smith, unpublished data).

[a]1, bears and wolves fight and chase each other, no mortalities; 2, wolf displaces bear from kill site, no mortalities; 3, wolves and bears in same area, no mortalities; 4, wolves kill bear; 5, bear kills wolf.

site, near a wolf den, and at an unclassified site. Only one observation of a black bear killing a wolf was reported, and this occurred near a wolf den.

In 81% of wolf-black bear interactions, wolves outnumbered bears, suggesting that wolves had an advantage in such interactions (see table 10.3). Wolves won 69% of the interactions, while black bears won only 15%. Young black bears were involved in 35% of encounters, which was much higher than the percentage of young bears reportedly involved in wolf-brown bear interactions.

In their review of interspecific killing among all mammalian carnivores, Palomares and Caro (1999) indicated that larger species generally kill both young and adults of smaller species. The outcomes of wolf interactions with brown bears, coyotes, and red and swift foxes fit this pattern, but killings of adult black bears by wolves (Rogers and Mech 1981; Paquet and Carbyn 1986) did not. Wolves apparently sought out black bears in their dens and killed them, only sometimes consuming them. Wolves usually outnumbered black bears in such interactions and won a high percentage of the encounters. Such interactions suggest interference competition between wolves and black bears. Even at kill sites, wolves usurped kills occupied by black bears. These types of interactions contrast sharply with those between wolves and brown bears.

Polar Bears

Wolves and polar bears probably come into contact only rarely (Ramsay and Stirling 1984). During 1980–1983, Ramsay and Stirling (1984) observed six interactions between wolves and polar bears. One interaction involved wolves killing and consuming a bear cub during the bears' spring migration, while another interaction occurred next to a caribou kill adjacent to a polar bear den, although there was no mortality. Both F. Messier (personal communication) and D. Cluff (personal communication) have observed wolves attacking sow polar bears with cubs of the year, but the attacks were unsuccessful. It is doubtful that such interactions are important to either species.

Cougars

Wolves and cougars share geographic ranges along portions of the Rocky Mountains and adjacent mountain ranges in North America. Both carnivores subsist on ungulates, but they use very different hunting techniques. Cougars are solitary predators and typically do not consume their kills quickly (Murphy 1998).

Since wolf packs are highly mobile, especially in winter, the potential exists for wolves to interact with cougars near kills. The degree of interaction between wolves and cougars probably varies temporally and spatially. In mountainous terrain, winter accumulation of snow often forces common prey species into valleys, which may increase spatial overlap between wolves and cougars (Hornocker and Ruth 1997). During summer, cougars follow prey species to higher elevations, whereas wolves tend to restrict their movements to denning areas in valley bottoms (Hornocker and Ruth 1997; Kunkel 1997).

Although relatively few wolf-cougar interactions have been observed, the animals occasionally do kill each other (Schmidt and Gunson 1985; White and Boyd 1989; Boyd and Neale 1992). Furthermore, usurping of cougar kills by wolves may cause cougars to increase their kill rates (Kunkel 1997; Hornocker and Ruth 1997; Murphy 1998). Based on current data and the paucity of available information, it is doubtful that either wolves or cougars are a significant mortality factor for the other species.

Tigers

Tigers, the largest living felids, overlap the ranges of wolves in Asia. Like other large cats, tigers depend on stealth to kill large prey. They are solitary "stalk and ambush" hunters, exploiting medium-sized to large prey.

In Siberia, prey are scattered, and tiger densities are low (Pikunov 1981; Yudakov and Nikolaev 1987, cited in Yudin 1992). Yudin (1992) suggested that tigers mounted no territorial defense against wolves and that at times, there was a degree of commensalism (Gromov and Matyushkin 1974; Yudin 1992). However, there were other cases in which their interactions appeared to reflect interference competition. Yudin (1992) indicated that there were no known cases of tigers pursuing or killing wolves, but that at least three wolf packs had been displaced by tigers. These observations appear to be an example of spatial partitioning by two carnivore species occupying similar ecological niches on the basis of trophic competition. Whether direct antagonism or predation was involved is unknown because it could well have occurred without being observed.

In fact, this is just what Makovkin (1999) found in the Lazovsky Reserve of the Russian Far East. He indicated that the relationship between wolves and tigers there was dependent on the density of each species, with the tigers outcompeting the wolves. He reported two instances of tigers killing wolves. In one case, the wolf had been wounded by a hunter; in the other, the wolf was killed at an ungulate carcass. In neither case did the tiger consume the wolf (Makovkin 1999). We found no other reports of observations concerning wolf-tiger interactions, but they do warrant further study.

Interactions with Mid-sized Carnivores

Wolf interactions with mid-sized carnivores are dominated by the wolves' superior predatory capacity. Thus the commonest type of interaction with members of this group of non-prey is to chase and attempt to kill them.

Lynx

The ranges of wolves and lynx overlap considerably, but we found only one North American record of a lynx interacting with wolves. In Jasper National Park, a warden watched a lynx feeding at an ungulate carcass for several days; a single wolf nearby did not get a chance to feed at the carcass at any time when the observer was watching (Dekker 1998).

Eurasian lynx are two or three times the size of North American lynx and thus are closer in size to wolves. In Russia, Yudin (1992) described wolf-lynx interactions as highly variable. Apparently there was evidence that wolves and lynx sometimes compete for prey, depending on the prey base. The common prey species for these two carnivores in Europe and parts of Asia is roe deer. In one location, lynx specialized on hares, with roe deer making up only 10–15% of their diet, while wolves specialized on roe deer. Thus, each species exerted a different degree of influence on different prey species. In eastern Poland, competition between wolves and lynx for roe deer was reportedly extensive at times (P. Suminski, personal communication).

Bobcats

We found no report of a wolf-bobcat interaction, although Stenlund (1955) suggested that bobcats benefited directly by scavenging on wolf kills. With the expansion of wolf range into the northwestern and southwestern

United States, opportunities for interaction between wolves and bobcats will probably increase, and may provide opportunities for additional study.

Wolverines

Interactions between wolves and wolverines have been described by a number of researchers (Freuchen 1935; Murie 1963; Burkholder 1962; Boles 1977; Bjärvall and Isakson 1982; Mech et al. 1998; White et al., in press; W. Ballard, personal communication; T. J. Meier, personal communication). Eight of the fourteen documented wolf-wolverine interactions resulted in death for the wolverines. Interestingly, the wolves did not consume the wolverines. Five accounts involved wolves chasing wolverines, but the wolverines reached escape habitats such as trees or caves. The interactions appeared opportunistic in that only three involved wolf kills, one was near a wolverine den, and the other ten occurred away from kills.

Wolf-wolverine interactions can be quite aggressive, as evidenced by an observation made by T. J. Meier during 1987 (Mech et al. 1998, 21):

> In January 1987, pilot Jim Cline and I were radio-tracking the East Fork pack when we spied seven wolves running up a creek bed near the Teklanika River. The wolves overtook and attacked a fleeing wolverine, forming a ring around the animal, lifting it off the ground and shaking it. Making a low pass, we saw that the wolverine was on its back with one wolf continuing the attack. On the next pass, some of the wolves were rolling on the ground, and the others were resting. Several ravens had also arrived. However, we could not find the wolverine.

Meier and Cline searched for 10 minutes.

> The seven wolves eventually arose and moved on up the creek. Finally, after another 20 minutes, we spotted the wolverine running rapidly down the creek the way it and the wolves had come. The creature appeared unhurt, and no blood was visible at the attack site. I visited the scene on the ground the next day. Approaching on the wolves' exit trail, I saw drops of blood in their tracks. At the attack site were a few drops of blood. It appeared that the wolverine had escaped under a shelf of ice until the wolves left. I saw no blood in the wolverine's exit trail, and I believe it escaped unharmed.

Interactions between wolves and wolverines may represent one of the better examples of interference competition. Although most documented interactions between the two species occurred away from wolf kills, we speculate that many of the interactions may have originally begun at kills where wolverines were attempting to scavenge.

Hyenas and Jackals

Although the distributions of wolves overlap with those of hyenas and jackals in Eurasia, there are few published reports of interactions between wolves and these two species. Mendelssohn (1982) reported that wolves and striped hyenas often met at garbage dumps, and that the wolves generally made way for the larger hyenas (weighing 25–40 kg, or 55–88 pounds), but there were occasions when wolf packs displaced hyenas. Wolf interactions with jackals may well be similar to those between wolves and coyotes.

Coyotes

Wolves and coyotes are close relatives (see Wayne and Vilà, chap. 8, and Nowak, chap. 9 in this volume). Although individuals of both species vary greatly in size, coyotes tend to weigh about a third as much as wolves. Many studies indicate that coyote and wolf population densities are inversely related (Berg and Chesness 1978; Paquet 1991b; Thurber et al. 1992; Peterson 1995a), suggesting interference competition. In other areas, the two species may coexist at low densities or remain spatially segregated. Range overlap and killings of coyotes by wolves have been most often documented in winter, when coyotes scavenge on ungulate carcasses (Crabtree and Sheldon 1999a).

The frequency of wolf-coyote encounters might be determined by the availability of food. Coyotes might be largely excluded by wolves where the main food is deer, since wolves are likely to consume all or most of a deer carcass after killing it. Moose and elk, on the other hand, are large enough to satiate a wolf pack and allow scavenging by other species.

Coyote sociality is flexible, and under suitable conditions, coyotes may form packs (Camenzind 1978; Bowen 1978; Gese 1995). Wolves, on the other hand, live in packs of up to forty-two animals (see Mech and Boitani, chap. 1 in this volume), but may become less social under some conditions and develop more coyote-like lifestyles (e.g., searching for food in pairs or in small packs) (Boitani 1982).

Reports of wolves killing coyotes are common (Seton 1929; Young and Goldman 1944; Munro 1947; Stenlund 1955; Carbyn 1982a; Paquet 1991b; Thurber et al. 1992). Generally, wolf-killed coyotes are not consumed, but rather are left with fatal wounds in the head, neck, rib cage, and back. These wounds often result in massive subcutaneous and internal hemorrhaging, muscle laceration, and trauma. By July 2001, at least twenty-seven coyotes had been killed by wolves in YNP, eighteen (67%) near wolf kills when coyotes approached to scavenge. There are no reported cases of coyotes killing wolves.

Crabtree and Sheldon (1999a) suggested that in YNP, coyote group size is an important factor in avoiding being killed by wolves. Most wolf-coyote interactions there occurred around wolf kills (122 of 145 encounters; 84%), and wolves typically "won" (121 of 145 encounters; table 10.4) these interactions, even when wolf numbers were equal to or less than the number of coyotes (D. W. Smith, unpublished data). Only four instances of coyotes chasing wolves were recorded, and in all four cases, there were at least as many coyotes as wolves. Also, three of these four interactions took place away from kill sites; one to three of them were near a coyote den. Coyotes apparently need special circumstances (e.g., motivation and at least equal numbers) before they will take on wolves. Other, more aggressive interactions have also been observed away from kills. On three occasions wolves attacked coyotes near coyote dens, digging into the dens and killing at least one pup.

It seems significant that only four wolf-coyote interactions took place where a single wolf had denned (not included in table 10.4). This wolf was a subordinate animal bred by the alpha male in the main pack, but she separated from the pack at whelping and denned alone. She produced three surviving pups. Her den was about 16 km (10 mi) from the main pack's den, and wolves from the main pack would occasionally visit her den. Three times in May 2001 coyotes were observed approaching her den. Once, when she was inside the den, a lone coyote carefully approached and raised-leg urinated at the entrance. The coyote left, and the wolf did not exit the den. On two other occasions, one and two coyotes approached her den, and both times she chased them away. In July 2001, when the lone female was not at her den, a coyote encountered one of her pups, chased it, and tackled it twice; however, it did not pursue the pup as it ran off. The pup did not appear injured.

The outcomes of wolf-coyote interactions appear to depend on three related factors: (1) coyotes benefit from scavenging on wolf carcasses; (2) wolves tend to kill coyotes, but do not usually consume them (i.e., killing appears to be opportunistic); and (3) coyotes may space themselves away from wolves (Berg and Chesness 1978; Fuller and Keith 1981b; Carbyn 1982a; Thurber et al. 1992).

Predator control programs in the early 1900s throughout North America greatly reduced or eliminated wolf populations (Young and Goldman 1944), allowing coyotes to expand their range. In addition, agricultural practices provided favorable habitats for coyotes and appeared to increase opportunities for hybridization between wolves and coyotes (Lehman et al. 1991; Roy, Geffen et al. 1994). This process has resulted in a unidirectional introgression of coyote mitochondrial DNA into wolf populations (see Wayne and Vilà, chap. 8, and Nowak, chap. 9 in this volume).

TABLE 10.4. Summary of wolf-coyote interactions in Yellowstone National Park, Wyoming, during 1995–2001

	Interaction type[a]					
	1	2	3	4	Total	%
No. occurrences	14	113	17	5	149	
% occurrences	9	76	11	3		1
Type of site						
Kill site	9	99	13	1	122	22
Coyote den	0	1	2	1	4	3
Wolf den	1	2	0	1	4	3
Other	4	11	2	2	19	13
Numbers of each						
Coyotes > wolves	2	13	0	3	18	12
Wolves > coyotes	8	88	17	0	113	76
Coyotes = wolves	3	12	0	2	17	11
Unknown	1	0	0	0	1	1
Outcome						
Wolf wins	3	103	17	0	123	83
Coyote wins	1	1	0	4	6	4
Neither wins	9	9	0	0	18	12
Unknown	1	0	0	1	2	1

Source: D. W. Smith, unpublished data.

[a]1, No chase, kill, or mortality; 2, wolf chases coyote, no mortality; 3, wolf kills coyote; 4, coyote chases wolf, no mortalities; 5, coyote kills wolf.

It seems reasonable to conclude that much of the observed wolf-coyote interaction on the individual level is of ecological consequence at the population level for both wolves and coyotes. The implication is that the more closely the interacting species are related, the more significant the long-term ecological consequences.

Can wolves and coyotes coexist in the same area? The answer is not simple. On Isle Royale, colonizing wolves apparently extirpated coyotes just a few years after the wolves arrived on the island (Mech 1966b; Krefting 1969). In other areas (e.g., Riding Mountain National Park), coyotes maintain high densities in the presence of moderate wolf densities (Paquet 1991b; Crabtree and Sheldon 1999a). In Alaska, the survival of coyotes living in wolf range was high (Thurber et al. 1992). Each situation appears to have its own set of dynamics.

Johnson et al. (1996) indicated that the general pattern of canid sympatry throughout most of North America, Eurasia, and Africa involved the occurrence of three sympatric canid species of differing size and forage requirements. In general, the pattern usually consisted of a large (i.e., > 20 kg, or 44-pound) canid, a medium-sized (i.e., 10–20 kg, or 22–44-pound) canid, and a small canid that was more omnivorous than the other species. In North America, this assemblage historically consisted of wolves, coyotes, and red foxes.

In the case of wolves, humans have changed these historical relationships. Wolves and red foxes were sympatric across North America, and coyotes probably occurred mostly along wolf territory boundaries (Fuller and Keith 1981b). In areas where this system was reduced to only two species of canids, as in much of North America when wolves were largely extirpated, several scenarios became possible between coyotes and foxes. These included exclusion, partial exclusion, scattered interspecies territories, or complete overlap (Johnson et al. 1996). Wolf recovery in Glacier and Yellowstone National Parks and on Alaska's Kenai Peninsula resulted in significant changes in coyote numbers, behavior, and distributions (Thurber et al. 1992; Arjo and Pletscher 1999; Crabtree and Sheldon 1999a,b), and probably in similar, but positive, changes in red fox populations.

The current changes within populations of carnivores in YNP as a result of wolf reintroductions present new scenarios that will probably result in long-term changes in the composition of the carnivore guild. There have been several short-term changes in coyote populations in the Lamar Valley of YNP since wolf reintroduction:

25–33% of the coyote population has been killed by wolves each winter; coyote numbers have decreased by 50%; and average coyote pack size has decreased from 6 to 3.8 (Crabtree and Sheldon 1999b). Coyotes in the Lamar Valley have also changed their behavior since wolf reintroduction by denning closer to roads and by reducing the frequency of their vocalizations, behaviors that probably reduce detection by wolves (R. Crabtree, personal communication).

On the other hand, the first record of a wolf and coyotes cooperating, or at least not attacking each other, during the killing of a prey animal was recorded in YNP recently. Four coyotes attacked a bison calf's hindquarters while a wolf bit the animal's neck (Smith et al. 2001). When the calf was dead, the wolf prevented the coyotes from feeding on it.

Wolves no doubt interacted with other canids and carnivores much more extensively in the past (pre-European times), since wolves were present across North America prior to European settlement (Young and Goldman 1944). Coyotes occurred in the more arid regions and open western plains and east to the midwestern states (Nowak 1978). Since the extirpation of wolves from much of their historic North American range, coyotes have greatly expanded their distribution and are now found in nearly every state and province north to Alaska. Possibly wolf elimination from the northern Great Plains influenced coyote densities there, which in turn may have influenced the decline in swift fox numbers (Carbyn 1994).

The question is often raised whether the principle of competitive exclusion is based on similarity of ecological niches. However, Schmidt (1986) pointed out that there is no evidence that aggression between wolves and coyotes is tied to niche overlap. Interference behavior is greatest between species closest in size, regardless of niche overlap, or between species that are most closely related taxonomically.

Interactions with Small Carnivores

Larger species are superior in interference competition, but not in exploitation competition (Persson 1985). This principle holds true for wolf-coyote relationships. Theoretically, smaller animals are likely to be more successful in exploitation competition because the advantage of being small would offset the evolutionary advantages leading to better interference abilities in larger competitors

(Palomares and Caro 1999). Smaller species tend to be more numerous, have smaller home ranges, and exploit resources more efficiently (Palomares and Caro 1999).

Red Foxes

Early explorers and ranchers knew long ago that wolves killed foxes (Young and Goldman 1944). Such behavior has subsequently been documented on Isle Royale in Michigan (Mech 1966b; Peterson 1977; Allen 1979), in Denali National Park and Preserve, Alaska (T. J. Meier, personal communication), and in Wood Buffalo National Park, Alberta (J. Turner, personal communication). The wolves may or may not consume the foxes (Mech 1966b).

Although wolves have killed foxes at a variety of sites and under a variety of conditions, most such mortalities apparently occur near wolf kills where foxes scavenge. T. J. Meier (personal communication) thought that all wolf kills in Denali National Park were ultimately visited by red foxes. Wolf kills undoubtedly provide an important source of food for the foxes. Carbyn et al. (1993) and Peterson (1995a) suggested that competition between red foxes and wolves was less pronounced than competition between coyotes and wolves, and our analyses suggest that wolves kill coyotes more often than they kill red foxes. In Wood Buffalo National Park, Alberta, there was evidence that, in the presence of wolves, fox populations increased (Carbyn et al. 1993). In a second area, Kenai Peninsula, Alaska, Peterson, Woolington, and Bailey (1984) predicted that red fox populations were likely to increase in the presence of wolves.

Arctic Foxes

Little is known about the interaction between wolves and arctic foxes, although there is no reason to believe that such interactions should be any different from those between wolves and red foxes. Wolves do chase arctic foxes whenever they are encountered, and arctic foxes do feed on wolf kills. In one instance, wolves spent considerable time and effort trying to fend off an arctic fox at a fresh muskox kill (Mech and Adams 1999). This was the case even though the wolves were full and were caching and there was still a considerable amount of food left on the carcass.

Interactions with Other Species

It is only natural that a carnivore with such a widespread distribution as the wolf would interact with a wide range of smaller mammalian and avian carnivores. Such encounters have been documented in a variety of anecdotal accounts. For example, Stenlund (1955) and Route and Peterson (1991) reported that river otters were occasionally killed by wolves. D. Boyd (personal communication) found a striped skunk killed by wolves; only the head was consumed. White et al. (in press) reported a single wolf killing an American marten, and L. D. Mech (personal communication) watched a pack of seven arctic wolves chase a weasel. On three occasions, D. Boyd (personal communication) found evidence of wolves killing golden eagles that were attempting to scavenge at ungulate carcasses; none of the eagles was consumed.

In the Great Indian Bustard Sanctuary in Maharashtra State, India, Kumar (1996) observed a pair of wolves with three pups feeding on a road-killed blackbuck near their den. An adult short-toed eagle swooped at the wolves five times. During each swoop, the adult male jumped at the eagle. The fifth time the eagle swooped much lower and was caught and killed by the wolf, but was not eaten. The wolves resumed eating the blackbuck carcass, and later the pack abandoned the carcass, ignoring the dead eagle. The short-toed eagle is not reported to be a scavenger, but feeds on a variety of small mammals up to the size of a hare, so perhaps it was aiming at the wolf pups rather than the carcass (Kumar 1996).

In Poland, Jedrzejewski et al. (1992) reported that wolves regularly inspected raccoon dog and European badger dens and occasionally killed and consumed raccoon dogs. In YNP, five wolves were observed attacking a lone badger (D. W. Smith, unpublished data). Two wolves successively attacked the badger individually, but quickly dropped it. Then all five wolves surrounded the badger, bit it, and violently shook it. The badger appeared to be dead, but two wolves continued to bite it, then carry it. One wolf carried it and dropped it five times before finally leaving it uneaten and joining the other wolves.

The species that probably interacts the most with wolves in North America is the common raven. The two species have a close association, from which the ravens benefit by scavenging wolf kills (Murie 1944; Mech 1966b; Peterson 1977; Carbyn et al. 1993). However, the benefits for the wolves are unclear, and at times wolves may kill ravens near carcass remains (D. Boyd, personal communication).

One result of wolf-raven interactions can be intense competition for food. Promberger (1992) studied wolf-raven interactions in the Yukon Territory. Sixteen ungulate carcasses were set out for scavengers during late winter, and the biomass of meat taken was measured every 24 hours. Ravens removed as much as 37 kg (81 pounds) of flesh per day. Based on his observations, Promberger estimated that up to 66% of ungulate kills made by single wolves might be consumed by ravens and other scavengers, but only 10% was taken from kills made by wolf packs with ten or more members. At these rates, lone wolves or wolf pairs would have to kill ungulates about twice as often as large packs in order to obtain the same amount of food. Ravens, therefore, when common, could have a considerable effect on wolf kill rates.

Wolf-raven interactions can also have a playful aspect, as indicated by the observations of Mech (1966b, 159) on Isle Royale, Michigan:

As the pack travelled across a harbor, a few wolves lingered to rest, and four or five accompanying ravens began to pester them. The birds would dive at a wolf's head or tail, and the wolf would duck and then leap at them. Sometimes the ravens chased the wolves, flying just above their heads, and once, a raven waddled to a resting wolf, pecked its tail, and jumped aside as the wolf snapped at it. When the wolf retaliated by stalking the raven, the bird allowed it within a foot before arising. Then it landed a few feet beyond the wolf and repeated the prank.

Recently, Stahler (2000) studied wolf-raven interactions in YNP to determine how much ravens associated with wolves at and away from wolf-killed carcasses. The birds usually stuck close to the wolves while these carnivores were traveling, resting, and hunting. In contrast, ravens did not associate with coyotes or elk or frequent areas that lacked wolves. In Yellowstone, ravens discovered 100% of wolf-killed ungulates in winter.

By associating with wolves, ravens appear to experience a socially facilitated reduction of their fear of large carcasses when first discovered. Stahler (2000) speculated that interactions between wolves and ravens may be important for experience-based modifications of behavior, perhaps built on innate responses, and may ultimately benefit both species throughout their lives. He concluded that these interactions reflected various forms of social symbiosis that hinted at a shared evolutionary history. It is clear that wolf-raven interactions are complex and important and warrant further study.

Wolves interact with a number of other smaller species, but with the exception of foxes and ravens, these interactions are probably opportunistic events that are likely to have an insignificant effect on the species involved.

———

This evaluation of wolf interactions with non-prey species has revealed a wide range of possibilities. The mechanisms of competition and coexistence between wolves and non-prey species range from interference and exploitative competition and avoidance behavior to tolerance and mutual acceptance. Avoidance behavior becomes important to survival strategies. Tolerance among species of different sizes can be influenced by factors such as food availability, use of different habitats, or temporal segregation in use of the same geographic areas. All of these processes may be of mutual benefit to select species and may help maintain the diversity of ecosystems. Recently, Berger, Stacey et al. (2001) indicated that extirpation of brown bears and wolves from the Greater Yellowstone Ecosystem had resulted in a moose population eruption that altered riparian habitats and caused a reduction in numbers of avian Neotropical migrants. They argued that restoration of bears and wolves provides a management option for restoring biological diversity.

The population characteristics of wolves and their associated non-prey are important in influencing the nature of interspecific competition (Sargeant et al. 1987). Population densities, the presence of adjacent pools of dispersers, reproductive rates, ages of females at first reproduction, and age-specific mortality rates are all important parameters in regulating the outcomes of competition (Sargeant et al. 1987).

During the twenty-first century, wolves will probably become more common in many areas where they once existed. Increasing wolf distributions and reintroductions into historical ranges, along with the development of advanced telemetry systems, better data collection and analytical methods, and more sophisticated research designs may result in a better understanding of the relationships between wolves and non-prey species.

In Yellowstone, cooperative efforts to examine carnivore-carnivore interactions are under way. One study located brown bears, cougars, and wolves before and after an elk hunting season on YNP's northern boundary. Preliminary data suggest that each carnivore had a different response to the hunting season: bears were drawn toward hunter activity, cougars moved away, and wolves had no response (D. W. Smith, unpublished data). The next phase of this study includes instrumenting each carnivore with Global Positioning System transmitters so that more locations per day and at night can be obtained.

Despite the competitive nature of the interactions between individuals of competing species, coexistence among carnivores of similar sizes or similar ecological niches does occur. The ranges of wolves, bears, coyotes, and foxes overlap in many areas where the species coexist in the same ecological systems. Wolves can exclude coyotes, and coyotes can exclude red foxes, at a number of scales ranging from individual encounters and territories to entire regions, yet they all coexist over many regions of North America (Crabtree and Sheldon 1999a). Wolves have been eliminated in many other parts of the world (see Fritts et al., chap. 12, and Boitani, chap. 13 in this volume), so the absence of this apex predator must also have created changes in the structures of ecosystems there. Conversely, in some parts of Europe, wolf populations are now extending their ranges into formerly occupied regions, probably causing more such changes in the opposite direction.

11

Restoration of the Red Wolf

Michael K. Phillips, V. Gary Henry, and Brian T. Kelly

"WOLFERS" IN NORTHEASTERN North Carolina were busy on February 5, 1768. Records from the Tyrrell County courthouse read:

> Giles Long and Thomas Wilkinson awarded one pound for a certified wolf scalp; Jeremiah Norman awarded two pounds for certified wolf and wild-cat scalps; Davenport Smithwick awarded one pound for a certified wolf-scalp.

Such was the nature of the war on the wolf: people killed them for money. The belief of the time held that the war was necessary because it was humankind's manifest destiny to tame the wilderness. And for the wilderness to be tame, the wolf had to be exterminated. The wolf was resourceful and hardy, but the wolfers persisted with increasingly sophisticated methods of killing. The war lasted 200 years, and the wolf lost.

History of the Red Wolf

In the late 1700s, naturalist William Bartram traveled throughout the southeastern United States. In his book *Travels* (Bartram 1791), he described the wolf he encountered in Florida:

> Observing a company of wolves (lupus niger) under a few trees, about a quarter of a mile from shore, we rode up towards them, they observing our approach, sat on their hinder parts until we came nearly within shot of them, when they trotted off towards the forests, but stopped again and looked at us, at about two hundred yards distance: we then whooped, and made a feint to pursue them; when they separated from each other, some stretching off into the plains, and others seeking covert in the groves on the shore: when

we got to the trees we observed they had been feeding on the carcase of a horse. The wolves of Florida are larger than a dog, and are perfectly black, except the females, which have a white spot on the breast; but they are not so large as the wolves of Canada and Pennsylvania, which are of a yellowish brown colour.

About 60 years later, researchers concluded that the Florida wolf inhabited other southeastern states and that it was structurally different from wolves inhabiting the rest of North America (Audubon and Bachman 1851). Goldman (1944) supported this conclusion after examining a large series of wolf specimens from the southeastern United States. He concluded that all the animals shared important cranial and dental characteristics and assigned them to one species, the red wolf (*Canis rufus*), which has both red and black phases.

Even though the red wolf was first described during the eighteenth century, the species' natural history remained poorly understood until the latter part of the twentieth century. This lack of understanding was largely due to a lack of interest in studying the species before the 1960s, and by then red wolves were endangered (McCarley 1962).

During the late 1960s and early 1970s, most efforts were directed toward determining the red wolf's status in the wild and identifying individuals to be placed in a captive breeding program. Because of this, our knowledge of red wolves prior to the restoration effort we describe in this chapter (Riley and McBride 1972; Shaw 1975) is based on relatively small samples from remnant and probably atypical red wolf populations. Phillips and Henry (1992) characterized the behavior and ecology of the red wolf using preliminary data from the restoration program. In this chapter, we present a more detailed analysis of these data and compare and contrast our findings, when possible, with the early information on the red wolf.

From the restored population, we know that the red wolf, like the gray wolf, is a monestrous species that typically becomes sexually mature by its second year. From historical data and the restoration to date, we know that litters average three pups (Riley and McBride 1972) and that red wolves live in family groups similar to those of gray wolves (Riley and McBride 1972; Shaw 1975). Data from the restored population indicate that the offspring of a breeding pair are tolerated in their natal home range until they disperse, and that dispersal is apparently related to social factors most typically associated with the onset of sexual maturity.

We have noted some fundamental differences in the prey consumed by the remnant populations of red wolves and the restored population. Principal prey prior to extinction included nutria, rabbits, and rodents (Riley and McBride 1972; Shaw 1975). In contrast, the restored wolves relied on white-tailed deer, raccoon, and rabbits, with resource partitioning evident within packs. Data from the restoration program indicate that dens can be located both above and below ground, and that mortality is due to a variety of factors, including vehicles, parasitism, and intraspecific aggression.

The demise of the red wolf was a result of many factors. Human persecution of wild canids and human settlement of most of the southeastern United States forced the last few red wolves to use marginal habitat in Louisiana and Texas, where they bred with coyotes and suffered heavy parasite infestation (Nowak 1972, 1979; Riley and McBride 1972; Carley 1975; Custer and Pence 1981a; Pence et. al. 1981).

The red wolf was listed by the United States as endangered in 1967, and a recovery program was initiated with passage of the Endangered Species Act (ESA) of 1973. The initial objective of the recovery program was to document the current distribution and abundance of red wolves in Texas and Louisiana. Fieldwork quickly revealed that free-ranging red wolves were rare, while coyotes were common (Riley and McBride 1972; Carley 1975). Red wolf-coyote hybrids were also common (Carley 1975). The U.S. Fish and Wildlife Service (USFWS) concluded that the red wolf could be recovered only through captive breeding and reintroductions (Carley 1975).

Captive Breeding of Red Wolves

In November 1973, a red wolf captive breeding program was established at the Point Defiance Zoological Gardens, Tacoma, Washington. To supply animals to the breeding program, the USFWS captured over 400 canids from southwestern Louisiana and southeastern Texas from 1973 to 1980 (Carley 1975; McCarley and Carley 1979; USFWS 1990). Measurements, vocalization analyses, and skull X rays were used to distinguish red wolves from coyotes and red wolf-coyote hybrids (Carley 1975; Paradiso and Nowak 1971, 1972; Riley and McBride 1972; Shaw 1975), although these criteria had their critics (Jordan 1979). Of the 400 animals captured, only 43 were believed to be red wolves and sent to the breeding facility. The first litters were produced in captivity in May 1977. Some of the pups were believed to be hybrids, so they and their parents were removed from the captive program. Of the original 43 animals, only 14 were considered pure red wolves and became the breeding stock for the captive program (USFWS 1990).

Although Bartram (1791) observed the black phase of the red wolf, he saw very few individuals. Had he viewed more, he would have realized that red wolves most often show a mixture of gray, black, and cinnamon-buff (Goldman 1944). Physically, the red wolf is intermediate to the coyote and gray wolf *(Canis lupus)* (Bekoff 1977a; Mech 1974a; Paradiso and Nowak 1972). The disproportionately long legs and large ears are two obvious features that separate red wolves from coyotes and gray wolves (Riley and McBride 1972).

It is difficult, however, to distinguish red wolves from red wolf-coyote hybrids (Carley 1975). This difficulty, combined with the intermediate morphology of red wolves and the commonness of hybrids, fueled a lasting debate over the taxonomic status of the red wolf (see Wayne and Vilà, chap. 8, and Nowak, chap. 9 in this volume). Some authorities consider the red wolf a full species (Paradiso 1968; Atkins and Dillon 1971; Paradiso and Nowak 1971; Elder and Hayden 1977; Ferrell et al. 1980; Gipson et al. 1974; Nowak 1979), while others consider it a subspecies of the gray wolf (Lawrence and Bossert 1967, 1975) or a hybrid resulting from interbreedings of gray wolves and coyotes (Mech 1970; Wayne and Jenks 1991).

In response, the USFWS conducted an exhaustive review of the issue and concluded that the red wolf is either a separate species or a subspecies of the gray wolf (Phillips and Henry 1992; Nowak 1992a; Nowak et al. 1995). Since then, molecular genetic data from wolves in southeastern Ontario have led Wilson et al. (2000) to contend that the red wolf and eastern timber wolf *(Canis lupus lycaon)* were closely related and shared a common lineage with the coyote until 150,000 to 300,000 years ago. However, Nowak (2002) presented morphological data countering this claim and supporting a taxonomic separation between the red wolf and gray wolf. Some genetic work provides similar evidence (Mech and Federoff 2002). Regardless of its true identity, the red wolf continues to be worthy of recovery efforts.

The Reintroduction Program

In 1984, the American Zoological Association (AZA) included the red wolf in its Species Survival Plan (SSP) program. This action helped intensify management of the species in captivity. A population viability assessment (PVA) conducted by the AZA estimated that recovering the red wolf and maintaining 85% of its genetic diversity for 150 years would require retaining at least 330 red wolves in captivity and restoring at least 220 wolves in the wild at three or more sites. This strategy would insure against random events that could wipe out a small population (USFWS 1990) (however, cf. Fuller et al., chap. 6 in this volume).

Long before the red wolf SSP was undertaken, the USFWS had been considering reintroduction. Indeed, the 1974 decision to place the last few wild red wolves in captivity was based on the belief that the animals or their offspring could eventually be reintroduced into the wild.

The red wolf reintroduction program was initiated in 1986. Warren Parker coordinated the effort, and M. K. Phillips was assigned to direct it. An excerpt from Phillips's field journal, written as he began his involvement with the red wolf recovery program, proved especially prophetic:

> I was mesmerized by the adult pair of red wolves racing about the large enclosure at the Point Defiance breeding facility. I knew what a red wolf looked like, but seeing live specimens was revealing in ways I had not anticipated. They acted wild, much more intolerant of people than I expected. And they moved silently as if floating inches above the ground. I was excited by these characteristics because they suggested that these wolves could survive in the wild.

Because previous attempts to translocate gray wolves to Isle Royale National Park (Mech 1966b; Allen 1979),

arctic Alaska (Henshaw et al. 1979), and Michigan (Weise et al. 1975) had failed, the USFWS had no protocol for successfully reintroducing wolves. Thus, during 1976 and 1977, the USFWS focused efforts on developing reintroduction methodology (e.g., acclimation, release, and recapture techniques). To assess the relative merits of various approaches to reintroduction, the USFWS released two groups of wild-caught red wolves onto Bulls Island, a 5,000 acre (2,000 ha) component of the Cape Romain National Wildlife Refuge in South Carolina (Carley 1979, 1981). These experiments demonstrated that red wolves acclimated at release sites for 6 months exhibited more restricted movements and higher persistence rates than red wolves released without being acclimated. This finding became the cornerstone of logic that supported the contention that it was feasible to reintroduce red wolves at select mainland sites.

After a failed proposal to use "Land Between the Lakes" in western Kentucky and Tennessee as the first mainland site for restoring red wolves (Carley and Mechler 1983), the USFWS chose the Alligator River National Wildlife Refuge (ARNWR) in northeastern North Carolina as the site for this landmark restoration project. ARNWR includes 120,000 acres (48,582 ha) of coastal plain habitats that are ideal for red wolves. ARNWR supports abundant prey, no coyotes, and few livestock; is bounded on three sides by large bodies of water; is sparsely settled by humans; and lies adjacent to 51,135 acres (20,702 ha) of undeveloped habitat owned by the Department of Defense (DOD) (Lee et al. 1982; Noffsinger et al. 1984; Phillips et al. 1995).

In 1990, the USFWS began adding Pocosin Lakes National Wildlife Refuge to the program to enlarge the restoration area. Pocosin Lakes was also ideal for red wolves because of its large size (110,000 acres or 44,534 ha), remoteness, abundant prey, small populations of coyotes and livestock, and proximity to ARNWR. While the restoration effort is still being carried out, this chapter presents specifics about the project from 1987 through 1994.

Preparations for Wolf Reintroduction

To promote reintroductions of endangered species, Congress amended the ESA in 1982 to allow reintroduced populations to be legally designated as "experimental/nonessential" rather than endangered. That designation allows the USFWS to relax restrictions of the ESA to encourage cooperation from those likely to be affected by the reintroduction (Bean 1983; Fitzgerald 1988; Parker and Phillips 1991).

Before the red wolf reintroduction program was initiated, the USFWS briefed representatives of environmental organizations in Washington, D.C., the North Carolina congressional delegation, the North Carolina Department of Agriculture, the governor's office, local county officials, and local landowners. The U.S. Air Force and Navy were briefed because they conduct training missions in the 40,000 acres (18,000 ha) adjacent to the refuge. Numerous personal contacts were made with local citizens, especially hunters and trappers, in preparation for four public meetings held during February 1986. At these meetings, the experimental/nonessential designation was explained clearly.

Comments resulting from the meetings were integrated into the proposed regulations (Parker et al. 1986). For example, the county government and local sportsmen supported the reintroduction on the condition that hunting and trapping still be permitted. In response, the USFWS decided to permit those activities even though they might result in the accidental "take" of a red wolf. The USFWS decided that the taking of a red wolf would not be prosecuted when it was unavoidable, unintentional, or did not result from negligent conduct, provided that the incident was reported immediately to the refuge manager or other authorized personnel. The USFWS further decided that wolves could be taken by citizens in defense of human life, but not to prevent or reduce depredations (e.g., of livestock or pets). In instances of depredations, citizens were required to contact USFWS or state conservation officers authorized to institute control measures. Without doubt, the flexibility of the experimental/nonessential designation was important in soliciting support for the proposed project.

The wolves we selected for release were taken from the USFWS's certified captive breeding stock. We considered each animal's age, health, genetics, reproductive history, behavior, and physical traits. Before release, we acclimated each wolf in a 225 m^2 (277-yard^2) pen at ARNWR. We acclimated the wolves to prepare them for life in the wild and to attenuate their possible tendency to travel widely after release. Acclimation periods were lengthy and averaged 19 months ($n = 42$, range 5 to 49 months), except for three adults, one yearling, and six pups that we acclimated for an average of one month (14 days to 2.5 months). The wolves were either released directly from the acclimation pens or transported to a remote location and released from a shipping container.

Because we were concerned that confinement would increase the wolves' tolerance of humans (a life-threatening trait for wolves about to be released in areas that might be used by some unsympathetic members of the public), we minimized human contact with them during acclimation, hoping to reduce their tolerance of humans. Additionally, we tried to provide the wolves with experiences they would encounter in the wild. For example, we varied the feeding regime to expose the animals to feast-or-famine conditions, and we weaned them from dog food and fed them an all-meat diet. We provided live prey to the first eight wolves we released to give them the opportunity to hone their predatory skills.

To keep the wolves in the area immediately after release and to facilitate their development of predatory skills and knowledge of prey habits, we provided the wolves with supplemental food in the form of deer carcasses placed near the release sites for a month or two after release. This approach was more cost-effective and practical in promoting the wolves' transition to the wild than providing live prey in the acclimation pens. Accordingly, that practice was halted after the first eight releases.

Just before release, we gave the wolves a final health check; administered various vaccines, vitamin supplements, and a parasiticide; took blood samples; determined weights; and fitted the wolves with motion-sensitive radio collars. Since pups were too small to wear radio collars, we implanted abdominal radio transmitters in them at about 10 weeks of age. Most of these animals were recaptured as adults and outfitted with radio collars. In addition, we captured 83% of the known wild-born offspring and outfitted them with radio collars.

Wolf Releases

Phillips's field journal described the first red wolf release:

Monday, 9/14/87: weather—clear, cool, and calm during morning; afternoon, light southeast breeze and temperatures in the upper 80's. At 0904 h Warren Parker, John Taylor, Chris Lucash and I departed the houseboat in the small Boston whaler to the South Lake pen. The calm weather made for a smooth ride but added to our anxiety because we knew that wolves 140M and 231F could hear us coming. At 0912 h we turned the engine off and floated the last

50 yards. Chris and I steadied the boat as Warren and John muscled the 110 lb. deer carcass out of the boat and began the wet walk through the sawgrass marsh to the pen. At 0924 h they returned breathless, anxious, and nervous. Both were unusually quiet. Taylor said nothing, but Parker uttered "we did it, we let them go."

From that rather humble beginning grew an aggressive restoration effort that eventually resulted in the release of 63 wolves on 76 occasions from October 1987 through December 1994 (tables 11.1–11.3). We released wolves directly from acclimation pens 46% of the time; for all other releases we transported wolves to remote sites and released them from shipping containers. Each wolf was released once, except for six adults that we released twice and three that we released three times. We defined a release as an initial release or a re-release of a wolf in a different area or with a different social group. Because the intent of the reintroduction was to restore a self-sustaining population, we considered a release successful if the animal eventually bred and raised pups in the wild.

Most initial releases involved adult pairs ($n = 14$) or families ($n = 8$), although additional releases included two siblings, an adult with a yearling, and an adult with a pup. We conducted most releases (71%) between August and October, when pups were 4–6 months old. We define adults as animals over 24 months of age, yearlings between 12 and 24 months of age, and pups less than 12 months of age. The adults we released ranged from 2 to 7 years.

Because wolves are wide-ranging and secretive, radio-tracking was our most important field technique. Thus, capturing wolves to attach or replace radio collars was a common field activity. Once a wolf was captured, we

could also implement management actions that had been specifically crafted for that particular wolf (e.g., return to captivity).

Radio-tracking greatly facilitated our determination of wolf movements, results of releases, and fates of wolves. The length of time we telemetrically monitored a wolf depended on the animal's fate and ranged from 0.1 months to 77.1 months ($\bar{x} = 15.4$, SE = 1.6). We monitored wolves frequently from the ground and the air. For example, from September 1987 through December 1994 we logged 1,453 hours in fixed-wing aircraft during 755 telemetry flights and recorded more than 10,000 wolf locations. The monitoring was so successful that we determined the outcome of 93% of the releases of captive-born wolves and the fates of 77% of the known wild-born wolves ($n = 66$). We also learned the cause of death for 94% of the wolves that died ($n = 51$). In addition, intensive monitoring allowed us to respond quickly to management issues that arose.

Only 21% of the releases with known outcomes were successful (table 11.1). The successful releases led to eleven adults and three pups establishing themselves and eventually producing pups in the wild (tables 11.1–11.3). One adult female was involved in two successful releases. Successfully restored adults persisted in the wild an average of 22 months, or about two reproductive cycles (table 11.2), whereas adults involved in unsuccessful releases persisted for an average of only about 3 months (table 11.2). Pups involved in successful releases persisted in the wild an average of 61 months, or about five reproductive cycles, whereas pups involved in unsuccessful releases persisted for an average of 7 months (table 11.3).

Success was not affected by the manner of release, as 29% and 25% of the releases from acclimation pens and shipping containers were successful, respectively.

TABLE 11.1. Outcomes of red wolf releases in northeastern North Carolina, September 1987–December 1994

	N		Outcomes			No. individuals involved in successes
Age, sex	Wolves	Releases	Success[a]	Failure	Unknown	
Adult males	16	22	6	16	0	6
Adult females	16	23	6	16	1	5
Male pups	16	16	2	12	2	2
Female pups	15	15	1	12	2	1
Totals	63	76	15	56	5	14

[a] A release was considered successful if the animal raised pups in the wild.

TABLE 11.2. Results of forty-four red wolf releases of known outcomes involving thirty-one adult red wolves

| | % of total releases ($n = 45$) | Families ($n = 8$) | Pairs ($n = 14$) | Outcomes[a] | | | | | Average ± SD persistence in the wild (months) |
				Death		Return to captivity		Free-ranging	
Success[b] ($n = 12$)	27%	33%	24%	58%	+	33%	+	9%	22 ± 18
Failure ($n = 32$)	73%	67%	76%	36%	+	57%	+	7%	3 ± 4

[a] Outcomes were determined through 31 December 1994.

[b] A release was considered successful if the wolf raised pups in the wild.

TABLE 11.3. Results of twenty-seven red wolf releases of known outcomes involving twenty-seven pups

| | % of total releases ($n = 27$) | Outcomes[a] | | | Average ± SD persistence in the wild (months) |
		Death	Return to captivity	Free-ranging	
Success[b] ($n = 3$)	11%	0	0	100%	62 ± 4
Failure ($n = 24$)	89%	68%	25%	7%	7 ± 7

[a] Outcomes were determined through 31 December 1994.

[b] A release was considered successful if the wolf raised pups in the wild.

Additionally, the type of social group (family versus adult pair) a wolf was released with did not appear to greatly affect the probability of success (see table 11.2).

The eventual fates of the released adults varied. Most adults involved in successful releases eventually died in the wild, whereas adults involved in failed releases were commonly returned to captivity within 3 months after release (see table 11.2). In contrast, the one female and two male pups that were involved in successful releases were free-ranging through December 1994. A higher proportion of pups than adults that failed died in the wild (see table 11.3).

Most successful adults (91%) and 60% of the pups—successful or not—established home ranges that included the release area. Establishment of home ranges began immediately following release; wide-ranging exploratory forays were not common. Only one adult and one pup that eventually bred in the wild did so after establishing home ranges that did not include their release sites. About 30% of the unsuccessful adults established home ranges that included their release sites, whereas the remaining 70% traveled widely immediately after being freed; on average these animals traveled a straight-line distance of 11 ± 4 SD miles (18.3 km ± 6.4 SD km) before dying or being returned to captivity. This trend

was much less pronounced for unsuccessful pups, as only 40% of these animals abandoned their release sites immediately after being freed; they traveled an average straight-line distance of 9 ± 3 SD miles (15.5 ± 4.7 SD km) before dying or being returned to captivity.

Because almost all the adults we released were acclimated for lengthy periods, there exists limited opportunity to clarify the effect of acclimation duration on the probability of success and post-release movements. However, some insight can be gained by examining the results of releases involving three adults and seven pups that, for various reasons, were acclimated for an average of only 0.9 months. Two of these pups and one of the adults established home ranges that included the release site and persisted for an average of 28.5 months (SD = 33.1 months). One of these pups survived to sexual maturity and bred.

Of the remaining seven wolves, one experienced an unknown fate, while six others persisted in the wild for only 1.0 month (SD = 0.7 months); none of these animals restricted their movements to the release area. Five of these seven wolves were members of a family that we acclimated and released on Durant Island. Immediately after release, the adult male drowned leaving the island; the adult female wandered widely and was returned to captivity. Following her departure, two of the three pups drowned, and the third disappeared. Possibly the behavior of these wolves resulted from the short acclimation period.

Success: Reproduction and Colonization

During the telemetry flight on May 5, 1988 I observed adults 211M and 196F. It was the first time in two weeks that 196F was away from what we hoped was her den. She seemed slimmer and spryer than two weeks earlier. As we circled for one last look, a small black ball of fur hurried to keep pace with the adults. The pudgy pup, known officially as 344F but affectionately referred to as "slick and steady," was the first red wolf born in the wild in North Carolina in many decades.

This observation from Phillips's field journal indicated that captive-born red wolves like 211M and 196F could make the transition from captivity to the wild and produce offspring. Indeed, fourteen captive-born wolves and twelve wild-born wolves bred in the wild.

From 1988 through 1994, thirteen adult pairs produced twenty-three litters that contained a minimum of sixty-six pups (table 11.4). The average litter contained three pups (range 1–5). Individual wolves contributed differentially to production. For example, two males (16% of the males that bred) and three females (21% of the females that bred) produced 36% and 42% of the known pups born, respectively. The wolves produced litters in the wild every year except 1989, when no wolves were paired during the breeding season. However, only 28% of the pups were produced during the first 4 years, whereas 65% were produced during the final 2 years (table 11.4).

We estimated parturition dates by noting when the adult pair began showing affinity to a particular area, indicating probable denning. Whelping extended from mid-April through early May, with most litters being produced during late April.

Wild-born wolf 670M was the youngest red wolf to breed; he sired a litter at about 10 months, much earlier than most wild gray wolves (see Mech and Boitani, chap. 1, Fuller et al., chap. 6, and Kreeger, chap. 7 in this volume). In contrast, breeding by yearling coyotes can be significant (Knowlton 1972; Kennelly 1978; Todd, Keith, and Fischer 1981). Male 442, who sired a litter at about 46 months (about 4 years), was the oldest wild-born male to breed, but we had no older wild-born males. The youngest recorded breeding for wild-born females was 22 months ($n = 3$); the oldest wild-born female bred at about 70 months (about 6 years), but we had no older wild-born females.

The limits of breeding age for captive-born wolves largely depended on when they were released. The earliest breeding for a captive-born male was 22 months, and for a female about 46 months. Captive-born male 184 bred at about 82 months (about 7 years) of age, and female 205 at about 106 months (about 9 years) of age.

Despite our best efforts at matchmaking by keeping unrelated adult males and females together in acclimation pens for several months, only four (28%) of the adult pairs that we released together stayed together and produced litters in the wild. Most reproduction resulted from nine pairs that formed naturally in the wild. In seven of these pairs, the adults began consorting about 4 months before the breeding season. The other two pairs were together for 8 and 17 months before successfully breeding. About 80% of the adult pairs that were together during a breeding season produced a litter the following spring ($n = 27$).

We learned little about the persistence of pairs of wild-born wolves because only three females (344F, 358F, and 496F) produced multiple litters. Female 344 had the same mate for all four of her litters, but female 508 had different mates in 1993 and 1994. Female 496 also gave birth to several litters, but we never determined the identity of her mate(s). Captive-born wolves provided more information about pair persistence. Six captive-born wolves produced multiple litters in the wild, including four animals that retained their original mates. The remaining two accepted new mates only after their original mates were returned to captivity or killed. Female 300 and male 319 produced litters in 1990, 1991, and

TABLE 11.4. History of red wolf production in the wild in northeastern North Carolina, 1988–1994

Year	No. of litters	Minimum no. of pups (M.F.?)
1988	2	00.02.00
1989	0	00.00.00
1990	1	01.02.00
1991	4	04.08.02
1992	2	02.02.00
1993	5	12.05.01
1994	9	11.08.06
Totals	23	30.27.09

1992 and remained together during 1993 and 1994 even though they did not produce pups during those years. Two pairs consisting of captive-born wolves that formed in the wild and one captive-born pair we released failed to breed during their first year together, but bred successfully the following year.

Biology of the Restored Wolves

Restoration of the red wolf population allowed us to study many aspects of red wolf biology, natural history, and behavior that had never been investigated before.

Den Characteristics

Three dens we inspected were aboveground nests (Mech 1993b) situated under dense vegetation, where the water table probably precluded underground dens. Through aerial radio-tracking, we learned the locations of twenty other dens that we did not inspect. Most were located along the sides or tops of brushy windrows in agricultural areas where the soil was friable and the water table low. Many were probably underground dens.

The three females that produced multiple litters showed varying patterns of annual den use. For example, 344F used the same den for 4 consecutive years, and 394F for 2 consecutive years. Both dens were burrows. In contrast, female 300 established a new den every year for 3 years, probably because her home range was dominated by swamps and her dens were aboveground nests. Using aboveground nests would make myriad sites available, which would increase the odds that she would den in a different location every year. In addition, she may have needed to do little to prepare the nests for pups, and that may have reduced her affinity for any particular site.

Red wolves routinely used den areas from mid-April until mid-July. For packs consisting of more than an adult pair, we documented all wolves frequenting dens, although we located breeding pairs there the most. By mid-July, wolves began moving more widely and seldom visited the dens.

Fates of Wild-Born Wolves

As of 31 December 1994, 36 (54%) of the 66 wolves conceived and born in the wild were free-ranging, 15 (23%) had unknown fates, 10 (15%) had died, and 5 (8%) had been placed in captivity. By December 1994, the oldest

TABLE 11.5. Persistence time of red wolves involved in the northeastern North Carolina restoration effort, 14 September 1987–31 December 1994

Origin	No. with known fates	Mean ± SD persistence (months)[a]
Captive-born adults[b]	51	8 ± 13
Captive-born pups[b]	27	13 ± 19
Wild-born[c]	44	22 ± 18

[a]Persistence times are minimums because some wolves were free-ranging through December 1994.

[b]There was no significant difference in persistence times between these two samples ($P = .14$, d.f. $= 45$, Kruskal-Wallis test statistic $= 55.15$).

[c]Persistence time was significantly different from each of the other two samples (for adults, $P = .01$, d.f. $= 57$, Kruskal-Wallis test statistic $= 87.18$; for pups, $P = .02$, d.f. $= 49$, Kruskal-Wallis test statistic $= 72.64$).

wild-born red wolf was 80 months of age. Wild-born pups persisted significantly longer than wolves we released ($P < .02$). There was no significant difference in average persistence times between captive-born pups and captive-born adults ($P = .14$) (table 11.5).

We placed four wild-born wolves in captivity at the behest of landowners who felt the wolves would eventually cause problems, and another that a farmer thought had been abandoned.

Red Wolf Dispersal

11/25/91, Monday: Flew today and located all wolves except 497M, despite a wide-ranging search. I suspect that he's dispersed as did his sister a few days ago.

This entry from Phillips's field journal was an important portent for the restoration program. We documented dispersal from natal ranges by eight male and ten female wolves born in the wild. The lack of a sex bias among red wolf dispersers ($P = .48$, $\chi^2 = 0.50$, d.f. $= 1$) is consistent with reports for gray wolves (Fritts and Mech 1981; Peterson, Woolington, and Bailey 1984; Ballard et al. 1987; Fuller 1989b; Gese and Mech 1991; Boyd et al. 1995). Five dispersing males and seven females were members of intact natal packs. On average, these males and females dispersed at about the same age, 27 ± 9 SD months and 23 ± 10 SD months, respectively ($P = .52$, $t = .66$, d.f. $= 10$). Similar ages have been reported for gray wolves in Minnesota (Mech 1987a), Montana (Boyd et al. 1995), and Alaska (Ballard et al. 1987).

Of the fifteen wolves born in the wild with fates unknown, only 502F dispersed (at 22.5 months of age) before we lost radio contact with her. Of the remaining fourteen, eleven were about 3.5 months old when we lost contact with them. The final three remained in their natal ranges for 13 to 20 months before disappearing.

Six pups dispersed after disruption of their natal pack's social cohesion. Four of these dispersed from their natal ranges within 3 months after we captured and returned their parents to captivity. Two other male pups dispersed within 2 months following the displacement of their father by an unrelated male. Apparently the disruption of social bonds between adults and offspring prompted these pups to disperse at the relatively young average age of 8 ± 1 SD months. Dispersal by small numbers of gray wolf pups has also been documented (Fuller 1989b; Gese and Mech 1991).

Our findings of high dispersal rates for yearlings are similar to those of Fritts and Mech (1981), Peterson, Woolington, and Bailey (1984), and Boyd et al. (1995). All of the wolf populations in these studies were at low density or increasing, intraspecific strife was uncommon, and all occupied areas of relatively high prey densities.

Dispersing red wolves settled new ranges in 1–44 days (average = 9 days, SD = 13 days, $n = 12$). Males and females dispersed similar mean distances of 36 ± 22 SD km and 45 ± 58 SD km, respectively, or 22 ± 13 miles and 27 ± 35 miles ($P = .74$, $t = -.34$, d.f. = 8). Similarly, gray wolves do not show a sex bias in dispersal distance (Ballard et al. 1987; Mech 1987a; Fuller 1989b; Gese and Mech 1991; Mech et al. 1998).

Almost 90% of red wolf dispersers traveled southward or westward to areas without wolf packs that contained good habitat and abundant prey. For most of these animals, established pack territories lay to the north and east. Only one wolf was killed while dispersing; she was hit by a vehicle. The other seventeen dispersing wolves settled new areas; 65% of them eventually paired and produced offspring (table 11.6).

All dispersals occurred between September and March, with 72% between November and February. Gray wolves show a similar peak in dispersal, although some gray wolves disperse at other seasons (Fritts and Mech 1981; Peterson, Woolington, and Bailey 1984; Ballard et al. 1987; Mech 1987a; Fuller 1989b; Gese and Mech 1991; Boyd et al. 1995; Mech et al. 1998).

Since wolves often dispersed at about the age of sexual maturity, it is likely that was a predisposing factor in

TABLE 11.6. Fates of seventeen red wolves that dispersed and settled in new areas in northeastern North Carolina, 14 September 1987–31 December 1994

Fate	No. of wolves (M.F.)	Litters produced	Minimum pups produced
Paired and bred	4.7	12	36
Paired but no pups	1.0		
Lived alone	1.2		
Consorted with coyotes	2.0	1	3

dispersal (see Mech and Boitani, chap. 1 in this volume). Dispersal seemed to be an effective means of maximizing genetic fitness, given that 76% of our wild-born animals that dispersed eventually consorted with other canids (usually with other wolves, but also with coyotes; see below), and 70% of the animals eventually produced pups (see table 11.6). Clearly, dispersal facilitates genetic exchange, thus reducing the frequency of inbreeding and associated problems (Mech 1987a; D. Smith et al. 1997; Mech et al. 1998). Boyd et al. (1995) pointed out that dispersal may help to ensure the genetic health of low-density, recolonizing wolf populations.

Dispersal also greatly affects the politics of wolf restoration. Through dispersal, a wolf population can spread out over a large area fairly quickly. This fact is tremendously important to acknowledge because many opponents of wolf restoration argue that wolves will not stay put, that they will wander widely and establish themselves well beyond the intended area. Regardless of where wolves are released, they are a "fluid" resource that will move about regardless of political boundaries. To be successful, restoration design must take this into account.

Red Wolf-Coyote Interactions

We observed one captive-born female and two wild-born male wolves consorting with coyotes. We returned the female to captivity before she achieved sexual maturity. One male was shot and probably did not sire a litter with a coyote. The other apparently did sire a litter of three hybrid pups during spring 1993. In July we captured two of these pups (both females) and observed the third. All were in poor health from sarcoptic mange (*Sarcoptes scabiei*). We believe the one pup died shortly after we observed it, and we placed the two captured animals in captivity, treated them, and studied their morphological development.

At about 8 months of age, the hybrid pups weighed an average of 12 kg (26 pounds), about the same size as four adult female coyotes we captured in the area (average weight = 13 kg, SD = 1 kg) but much smaller than female red wolves of comparable age (average weight = 18 kg, SD = 2 kg, n = 13). One female acted like the three coyotes we maintained in captivity: she was withdrawn and would often slink around the pen in our presence. The other female's behavior was wolflike: she was bold and ran excitedly around the pen in our presence. Both their physical appearance and their behavior suggested that they were the progeny of a male red wolf breeding a female coyote.

At the outset of the restoration effort we assumed that unmated red wolves would readily breed coyotes because historically they had done so in Texas and Louisiana. In those areas, red wolves were rare and coyotes were common, as discussed earlier. Historical hybridization between red wolves and coyotes could have been due to the fact that wolves encountered far more coyotes than conspecifics. In contrast, in northeastern North Carolina after restoration began, wolves were common and coyotes rare. Indeed, the scarcity of coyotes in northeastern North Carolina was one reason the ARNWR was selected for red wolf restoration. From 1987 through 1994 we captured 106 wolves, but only 4 coyotes. Although our trapping targeted wolves, coyotes would have been captured if they were present.

Even though hybridization between red wolves and coyotes was not a serious problem through 1994, it became so about then (Kelly et al. 1999). A comprehensive population and habitat viability assessment (Kelly et al. 1999) facilitated the development of an adaptive management plan to address the hybridization problem (Kelly 2000). The plan, implemented in April 1999, called for hybridization to be eliminated or reduced by euthanizing or sterilizing coyotes and hybrids and promoting the formation and maintenance of wolf breeding pairs. By 2002 the results were beginning to show that hybridization could potentially be reduced to an acceptable level. Even if this proves to be the case, there is little likelihood of restoring a red wolf population elsewhere without intensive management. There are no suitable restoration areas in the historic range of the red wolf that are not inhabited by coyotes.

As part of the adaptive management plan, intensive genetic, morphological, and ecological research is under way on red wolves and other canids in northeastern North Carolina. Such studies will improve our knowledge of certain aspects of wolf-coyote interactions, including the extent of introgression between the species and the parentage and identity of canids of unknown origin. Such knowledge will help determine whether it is possible to restore the red wolf as a unique taxon functioning as an important component of the southeastern landscape.

Home Range Characteristics

Location data from ninety-six wolves were obtained from aircraft and by triangulation from the ground. Locations per wolf ranged from 2 to 1,085 ($\bar{x}$ = 113, SD = 12). We chose thirteen wolves from three packs to represent the home range size of red wolves at ARNWR. The packs were chosen for the completeness of their data sets. The Milltail, Gator, and Airport packs had established themselves early in the restoration (more than a year before collection of the data we analyzed), were tracked intensively, and occupied significantly different habitats. To ensure more valid comparisons between packs and individuals, wolves with similar temporal distributions of location data were selected. For each wolf's location data, we calculated the 95% minimum convex polygon (Ackerman et. al. 1990). We used the habitat where scats were collected to represent the habitat used by a pack.

Home range sizes averaged 88.5 ± 18.3 SD km^2 (35 ± 7 SD mi^2) for individuals and 123.4 ± 53.5 SD km^2 (48 ± 21 SD mi^2) for packs (table 11.7). Range size differed significantly among packs (F = 17.5, P = .0005). The Gator pack used an area significantly larger than either the Milltail or Airport packs (table 11.7). Although home range size has been positively correlated with pack size in gray wolves (Ballard et al. 1987; Peterson, Woolington, and Bailey 1984; but cf. Mech and Boitani, chap. 1, and Fuller et al., chap. 6 in this volume), habitat type appears to interact with this relationship for the red wolves at ARNWR. The Airport pack, which had the fewest individuals, did have the smallest home range. However, the Gator pack established a home range that was two to three times larger than the home range used by the Milltail pack, even though the Gator and Milltail packs were similar in size (i.e., included four to five animals).

This disparity was probably a function of the produc-

TABLE 11.7. Home range estimates of free-ranging red wolves in northeastern North Carolina

Pack	n	km^2	mi^2	Dates tracked Begin	End
Milltail					
205F	105	35.1	13.7	10/04/90	09/30/91
331M	110	37.2	14.5	10/04/90	09/30/91
351F	42	76.9	30.0	10/05/90	01/24/91
394F	106	58.6	22.9	10/04/90	09/30/91
$\bar{x}$	4	52.0	20.3		
SE	4	9.9	3.9		
Composite	363	98.9	38.6	10/04/90	09/30/91
Gator					
300F	109	190.5	74.4	10/04/90	09/30/91
319M	103	199.9	78.1	10/19/90	09/30/91
442M	82	108.2	42.3	10/19/90	09/30/91
443F	89	197.5	77.1	10/19/90	09/30/91
444F	63	98.7	38.6	02/08/91	09/30/91
$\bar{x}$	5	159.0[a]	62.1		
SE	5	22.7	8.9		
Composite	446	225.8	88.2	10/04/90	09/30/91
Airport					
313F	115	39.4	15.4	10/04/90	09/17/91
328M	92	44.9	17.5	10/04/90	06/30/91
426M	93	29.4	11.5	10/04/90	07/15/91
430F	118	34.4	13.4	10/04/90	09/30/91
$\bar{x}$	4	44.9	17.5		
SE	4	3.3	1.3		
Composite	418	45.6	17.8	10/04/90	09/30/91
Overall					
Individual $\bar{x}$	13	88.5	34.6		
SE		18.3	7.1		
Composite $\bar{x}$	3	123.4	48.2		
SE		53.5	20.9		

[a]Different from the other packs ($P < .05$, Fisher's least significant difference test).

tivity of habitat. Of 893 scats attributed to the Gator pack, 99% ($n = 888$) were collected in pine-hardwood habitats where prey is relatively scarce (Lee et al. 1982; Noffsinger et al. 1984; M. K. Phillips, unpublished data). In contrast, 71% and 98% of the scats attributed to the Milltail and Airport packs, respectively, were collected in agricultural habitats where prey were abundant (Lee et al. 1982; Noffsinger et al. 1984; M. K. Phillips, unpublished data). Variation in home range size due to prey density has also been observed in gray wolves (Ballard et al. 1987; Wydeven et al. 1995; Fuller et al., chap. 6 in this volume), coyotes (Gese et al. 1988), and bobcats (Litvaitis et al. 1986).

Home range sizes for red wolves in Texas were similar to those in North Carolina, ranging from 25 km^2 to 130 km^2 (10–51 mi^2) (Riley and McBride 1972; Shaw 1975). Overall, red wolf home ranges appear to be intermediate to coyote ranges, which vary from 4 km^2 to 84 km^2 (1.5–34.6 mi^2) (Andelt 1985; Gese et al. 1988; Sargent et al. 1987), and gray wolf territories, which range beyond 2,600 km^2 (1,015 mi^2) (see Mech and Boitani, chap. 1 in this volume).

Food Habits

Between 27 November 1987 and 11 March 1993, we collected and analyzed 1,890 red wolf scats. When possible,

TABLE 11.8. Analysis of 1,890 red wolf scats collected in northeastern
North Carolina, November 1987–March 1993

Prey species	% of biomass[a]
White-tailed deer	43
Raccoon	31
Lagomorph	13
Rodent	11
Domestic ungulate	2

[a]Raw data converted as per Weaver 1993.

scats were assigned to individual wolves, or packs, via radioisotope marking (Crabtree et al. 1989) or intensive tracking of the wolf.

Scat content analyses based on the percentage of scats containing a given item, commonly referred to as frequency of occurrence or percent frequency, are biased (Kelly 1991). Accordingly, we used Weaver's (1993) model to refine our scat analysis and estimate the proportions of various prey red wolves consumed. Although Weaver's model was developed for gray wolves, its application to red wolves is tenable, with the caveat that prey smaller than snowshoe hares will probably be overestimated (Kelly 1991, 66).

White-tailed deer, raccoons, and marsh rabbits constituted 86% of the red wolves' diet (table 11.8). These results differ from previous reports about red wolf food habits. Nutria, rabbits, and cotton rats were the primary prey of red wolves in Texas (Shaw 1975; Riley and McBride 1972).

Differences in prey consumption by pack were evident at ARNWR. The Milltail pack consumed more small prey (rodents and rabbits) than the Gator pack, which consumed more large prey (deer and raccoons)

(table 11.9). This difference in food habits was related to the abundance and distribution of prey. While rabbits and rodents were abundant in the agricultural fields used by the Milltail pack, they were uncommon in the pine-hardwood swamps used by the Gator pack (Lee et al. 1982; Noffsinger et al. 1984; M. K. Phillips, unpublished data).

Rodents were consumed more by juvenile wolves than by adults, and analysis of the scats from the Milltail pack indicates a decrease in rodent consumption with age (table 11.10). A similar pattern of prey use was not evident for the Gator pack. However, resource partitioning similar to that manifested by the Milltail pack was documented among members of coyote packs in Yellowstone National Park (Gese et al. 1996).

The differential use of prey by the Milltail and Gator packs may have played a role in determining their home range sizes (see above). If the predominance of agricultural habitat in the Milltail pack's range provides enough prey variety to allow the pack to partition prey resources, their home range should be smaller than it would be otherwise (Harestad and Bunnell 1979). Additionally, the relatively abundant and diverse prey in the Milltail pack range may explain why this pack was able to produce and raise an average of 4.0 pups per litter ($n = 3$), whereas the Gator pack produced and raised an average of only 2.3 pups per litter ($n = 3$).

Mortality

Of the 135 red wolves involved in the restoration effort, 51 (38%) died while free-ranging, most during the first year after release or birth (table 11.11). The first wolf to die was female 231, whose death prompted this entry in Phillips's field journal:

TABLE 11.9. Analysis of 494 and 831 scats from two red wolf packs in northeastern
North Carolina, November 1987–March 1993

Prey species	Milltail pack		Gator pack	
	% of biomass[a]	% of scats	% of biomass[a]	% of scats
White-tailed deer	25	29	40	43
Raccoon	38	52	46	59
Lagomorph	14	25	8	15
Rodent	19	38	2	8
Domestic ungulate	—	—	1	—

[a]Raw data converted as per Weaver 1993.

TABLE 11.10. Mammalian prey consumed (% of biomass[a] and % of scats below) by wolves of different ages[b] as determined from analysis of scats collected from two red wolf packs in northeastern North Carolina, November 1987–March 1993

| | Milltail pack: | | Gator pack: | | Milltail pack: | | | |
| | Age of wolf | | Age of wolf | | Age of wolf (months) | | | |
Prey species	Juvenile $n = 17$	Adult $n = 191$	Juvenile $n = 46$	Adult $n = 390$	≤ 24 $n = 23$	24–48 $n = 29$	48–72 $n = 88$	>72 $n = 208$
White-tailed deer	14	21	42	39	17	17	23	21
	18	26	56	46	22	24	30	26
Raccoon	15	59	31	51	11	16	42	56
	24	74	46	69	26	35	61	70
Lagomorph	3	9	13	9	2	7	9	8
	12	20	20	15	9	21	19	19
Rodent	62	9	1	1	66	57	25	14
	71	19	9	6	74	69	40	24

[a]Raw data converted as per Weaver 1993.

[b]Ages were based on isotope labeling of scats from known individuals or from intensive tracking of known individuals (see text).

TABLE 11.11. Number (percentage) of thirty-six captive-born red wolves released in northeastern North Carolina dying, and causes of death, 14 September 1987–31 December 1994

| | No. months after release | | | | | |
Cause	1	2	6	12	>12	Totals
Vehicle	5 (14)	3 (8)	1 (3)	3 (8)		12 (33)
Intraspecific aggression	5 (14)					5 (14)
Malnutrition and parasitism				4 (11)	3 (8)	7 (19)
Drowning	3 (8)		1 (3)			4 (11)
Shot	1 (3)					2 (6)
Miscellaneous causes[a]	1 (3)			1 (3)	4 (11)	6 (17)
Totals	15 (42)	3 (8)	2 (5)	9 (25)	7 (20)	36 (100)

[a]Includes uterine infection (1), suffocation (1), pleural effusion and internal bleeding from unknown causes (1), handling accident (1), and unknown (2).

12/18/88, Friday: It was cold, clear, and windy all day. At 1530 we found 231F dead on the beach about 1 mile south of Long Shoal point. We had last located her on December 11 about 2 miles west, but weather had prevented monitoring since then. We found her laying on her side. She had obviously been dead for some time as the tides had nearly covered her with sand.

Female 231 died because of internal bleeding and fluid in her chest from an unknown cause. Most other deaths were caused by vehicles (30%), malnutrition and parasitism (27%), or intraspecific aggression (12%) (see Mech and Boitani, chap. 1, and Fuller et al., chap. 6 in this volume). In addition, four wolves drowned, four

were shot, one died of complications from a uterine infection, one choked on a raccoon kidney, one was poisoned, and one died during a handling accident. The causes of death of three wolves were unknown.

Because two paved highways bisect ARNWR, we expected vehicles to be an important source of mortality. To reduce vehicle strikes, the North Carolina Department of Transportation erected red wolf road-crossing signs. In addition, we produced public service announcements on local radio to alert motorists to the presence of wolves.

Despite the fact that the captive-born wolves had little or no experience hunting, none died solely from an inability to feed itself. Those that were malnourished were either very old or also suffered heavy parasite infestations (see Kreeger, chap. 7 in this volume). For example, wolves 300F and 319M, who had been together for almost 5 years and were 8 and 7 years old, respectively, died from malnutrition within 4 months of each other. Both possessed heavily worn teeth, and we supposed that they had grown too old to hunt successfully (but see Mech 1997).

Four other wolves, all from one pack, died from malnutrition and parasitism by ticks (*Dermacentor variabilis* and *Amblyomma americanum*) and intestinal worms (*Ancylostoma caninum* and *Dioctophyme renale*). Another wolf, adult 358M, succumbed to sarcoptic mange. We also captured three pups, sired by 358M, that harbored large numbers of *Sarcoptes scabiei* and were in marginal condition. We treated them with parasiticides for 18 to 21 days and released them; they survived at least through December 1994. Two wild-born pups died at 10 and 11 months, respectively, from complications of demodectic mange *(Demodex canis)*, which has not been reported before for red or gray wolves (see Kreeger, chap. 7 in this volume). Mange, ticks, and intestinal worms were known causes of mortality for naturally occurring red wolves in Texas and Louisiana (Riley and McBride 1972; Carley 1975; Custer and Pence 1981a; Pence et. al. 1981).

Five wild-born wolves that presumably died from malnutrition were littermates whose only parent (383F) was killed by a vehicle when they were about 40 days old. Despite extensive searches, we were unable to locate the litter after 383F's death. Because of their young age, we presumed that they all died.

Intraspecific aggression led to the deaths of five recently released wolves that entered the territories of established wolves. Possibly these inexperienced captive-born wolves were unaware of the grave consequences that sometimes accompany trespass (Mech 1994a; Mech et al. 1998). The other death from intraspecific aggression involved a 33-month-old, wild-born female killed by her pack, apparently in competition over the only breeding-age male in the area. This is one of the few records of a wild red or gray wolf killed by close relatives (see Mech and Boitani, chap. 1 in this volume).

Four wolves drowned, including a female pup accidentally captured in a foothold trap set for a bobcat. The other three were from a pack we acclimated on Durant Island, as mentioned earlier.

The four wolves that were illegally shot included two captive-born and two wild-born animals. Two of them had been mistaken for coyotes, which can be legally harvested in North Carolina. Over 90% of the red wolf deaths were accidental or natural.

Management during Restoration

From the outset of the restoration program, intensive management of the wolves was necessary to ensure quick establishment of a breeding population and adequate resolution of wolf-human conflicts. Most management required capture of wolves for reasons discussed below. We made 110 captures of 45 (71%) of the 63 captive-born wolves and 125 captures of 59 (83%) of the 71 wild-born wolves. We accomplished 195 of these captures (83%) using foothold traps (Mech 1974b). We also modified acclimation pens to act as traps for 27 captures (11%) of 12 captive-born wolves and 1 wild-born wolf. The remaining 13 captures involved a variety of techniques, including dart guns, box traps, and nets.

On 42 occasions the solution we adopted for the management problem at hand was to return a wolf to captivity or translocate the animal to another area before re-release. We returned one pup to captivity because of concern for its welfare; a farmer had found it and believed it had been abandoned. We placed two wild-born wolves in captivity because they were malnourished and harbored significant parasite infestations. Intraspecific aggression prompted four captive-born wolves to wander widely, which forced us to return them to captivity. Decisions to recapture these wolves were based on judgments that their future movements would continue to be wide-ranging and that it was likely that they would be involved in negative encounters with humans.

We returned six wolves to captivity on seven occasions for breeding because their mates had died or, in one case, had been returned to captivity. We removed another wolf from the wild to breed her so as to improve the representation of a rare genetic lineage.

Conflicts with people led to twenty-eight (70%) of the incidents that prompted us to return wolves to captivity or translocate them. Eighteen of these incidents involved captive-born wolves that, for mostly unknown reasons, frequented small areas inhabited by people. Although these animals rarely caused actual problems, their mere presence was unacceptable to the residents. In contrast, the ten incidents involving wild-born wolves resulted from the animals colonizing uninhabited private land. Even though they did not cause problems, their presence was unacceptable to the landowners, who requested their removal.

These management issues and a few others that did not involve the public were resolved without significantly injuring the wolves or inconveniencing residents. We were able to manage the wolves successfully because radio collars allowed us to determine their whereabouts almost at will. Knowledge of a wolf's location simplified all aspects of management.

The importance of managing wolves successfully during restoration cannot be overstated. For wolf restoration to succeed, the public must support, or at least tolerate, the program, and managing wolves successfully is one way to generate and maintain support and tolerance. Because successful management is so important, all or most wolves involved in a restoration effort should be radio-collared during the first several years of the program.

Capturing wolves was not the only intensive management strategy we employed to ensure establishment of the red wolf population. For example, during the first 2 years of the project, when the population consisted of just a few wolves, we implemented a parasite control program that prevented or ameliorated parasitism in selected wolves (Phillips and Scheck 1991). As the population grew, however, it become extremely laborious to continue this effort, and the importance of individual wolves decreased, so we terminated the parasite control program.

Conclusions

The red wolf restoration program progressed considerably from 1987 to 1994. As of June 2002, approximately a hundred red wolves (all wild-born animals), distributed in twenty packs, inhabited a restoration area that had grown to encompass about 680,000 ha (1.7 million acres). From the project's inception through June 2002, free-ranging wolves had given birth to 281 pups over four generations (USFWS, unpublished data).

The restoration area is now composed of 60% private land and 40% public land, which includes three national wildlife refuges. Since 1988 we have officially integrated about 78,800 ha (197,000 acres) of private land into the restoration area through cooperative agreements, at a total cost of $3,951 per year for 5 years (Phillips et al. 1995).

The red wolf restoration program has generated benefits that extend beyond the immediate preservation of red wolves, positively affecting local citizens and communities, larger conservation efforts, and other imperiled species (Phillips 1990). Indeed, the program is an effective model for restoring other controversial endangered carnivores, such as gray wolves, African wild dogs, and black-footed ferrets.

The red wolf program also illustrates that the designation of a population as "experimental/nonessential" can be beneficial for wide-ranging species introduced into areas not designated critical habitat, or where an introduced population may expand into nonpublic land not designated critical habitat. The experience gained by reintroducing red wolves suggests that such a designation would help other introduction programs succeed.

However, the red wolf program also serves as an example of a potential overrelaxation of regulations under the experimental designation. Despite the utility of the original final rule that resulted from the experimental designation, local opposition to the red wolf program during the early 1990s prompted the USFWS to modify it (Henry 1995). The revised rule requires the USFWS to remove wolves from private land at the behest of the landowner if possible, even if the only problem is the mere presence of the animal(s). A similar rule has been adopted by the Mexican wolf recovery program (Parsons 1997). The revised red wolf rule also contains a provision that allows issuance of a permit for landowners to take red wolves (for simply being present) after USFWS efforts to remove the animals have concluded.

Regulations that provide landowners such flexibility are potentially inappropriate for at least two reasons: first, because they are nearly impossible to implement effectively as the wolf population grows because of the difficulties of responding simultaneously to a large number of landowners, and second, because they might establish a precedent that could be used to argue for the removal of individuals from other populations of endangered species (both reintroduced and naturally occurring) inhabiting private land. However, given that traditional wildlife management concepts and attendant regulations assume that wildlife is public property and not subject to removal from private property in the absence of a problem, concern that the red wolf rule might establish a precedent may be moot except for specific situations involving reintroduced predatory species that are perceived to conflict with private interests.

Certainly local opposition to the red wolf and Mexican wolf reintroduction programs greatly affected the regulations governing management of the wolves. Indeed, the recovery program coordinators and Phillips (for the red wolf project) assumed from personal knowledge of local politics and sentiments that more restrictive rules would have significantly hindered and possibly caused the termination of the project (V. G. Henry, personal communication, 1994; D. R. Parsons, personal communication, 1996). Additionally, there was a need to clarify regulations for the red wolf program so that they accurately reflected long-standing commitments made by the USFWS that wolves that inhabited private land would be removed if so desired by the landowner. The revised regulations published in 1995 may have contributed to the widespread local support for red wolf recovery (Quintal 1995; Mangun et al. 1996).

Nonetheless, it has been argued that the 1995 regulations were excessively relaxed (Phillips and Smith 1998) and may have contributed to the current level of hybridization by allowing wolves to be managed in a manner that continually disrupted their social affinities (Kelly and Phillips 2000). Phillips and Smith (1998) believed that the argument that relatively relaxed regulations were necessary to ensure successful restoration of red wolves is contrary to experiences from reintroduction of gray wolves to central Idaho and Yellowstone National Park (YNP). Local opposition to these programs was substantial, but the authors of the regulations did not provide landowners a level of flexibility similar to that afforded landowners affected by red wolf reintroductions.

It is true that the central Idaho and YNP projects were much less dependent on private land than the red wolf project. However, throughout the planning period for the gray wolf projects, landowners expressed grave concern over problems that would arise if wolves came to inhabit private property. And during the first two years of the YNP project several contentious management incidents arose involving wolves and private land (Phillips and Smith 1998). Nonetheless, the relatively restrictive regulations in no way hindered resolution of those incidents nor the maturation of the two projects; both are viewed as unqualified successes (Bangs and Fritts 1996; Phillips and Smith 1996).

During the restoration program several important points became apparent:

1. Acclimating and releasing captive-born adults in a manner that predisposed them to remain near the release site and establish a home range there seemed to increase the chances that the wolves would breed in the restoration area. Furthermore, it simplified the task of initial telemetric monitoring and management.

2. Given that the manner of release (i.e., directly from an acclimation pen versus transport to a distant site and release from a shipping container) did not affect success, we concluded that it was most cost-effective to use a central facility for acclimation rather than a multitude of remote sites.

3. Most releases failed to result in the wolf breeding in the wild, so numerous releases over an extended period were required. This fact and the differential pup production by a few individuals emphasize the importance of individual wolves early in the program. Accordingly, it was appropriate during the first few years of the project to monitor and manage the wolves intensively to ensure their survival. Similar results have been reported for other restoration projects (Griffith et al. 1989).

4. Even though most captive-born wolves did not contribute to population growth, a large enough number (at least 18% of the total number released) did to serve as the catalyst for population formation. Indeed, fourteen captive-born wolves were involved in the production of at least 50% of the pups born from 1987 through 1994. Clearly captive-born red wolves were appropriate "seed stock" for restoring a free-ranging population.

5. Our matchmaking of captive pairs was not very effective. Of the fourteen adult pairs we released, only 28% remained together and produced pups in the wild. Most reproduction during the first 7 years resulted from nine pairs that formed naturally in the wild.

6. Maintaining radio contact with free-ranging wolves was essential to determining the fates of individual animals and for resolving management issues.

7. The management flexibility afforded by the experimental/nonessential designation was critical in soliciting and maintaining support for the restoration effort from local citizens and state and federal agencies. This flexibility also provided field biologists with the latitude necessary to resolve conflicts in innovative and cost-effective ways.

8. Because red wolves traveled long distances, dispersal greatly affected the politics of restoration. It is critical when designing a wolf restoration program to realize that the wolf population will occupy a large area, regardless of political boundaries.

9. Significant land use restrictions were not necessary for wolves to survive. Indeed, the rather lenient hunting and trapping regulations for the refuge remained unchanged or were further relaxed during the experiment. The lack of land use restrictions facilitated the integration of private land into the program, which greatly increased the area wolves could inhabit, which facilitated population growth. The prognosis for landowners and red wolves to coexist is good, since the wolves do not fit their stereotypical image and are not a threat to personal safety and landowner rights.

10. It will be necessary to study the extent of introgression between red wolf and coyote populations and to actively manage both to prevent hybridization. Intensive management seems to be the only way to ensure the coyotes will not again genetically "swamp" red wolves.

11. Most management issues that arose resulted in extensive press coverage, which promoted the perception that wolves are less manageable and more difficult to live with than other wildlife. This perception may subside as local residents become accustomed to living with wolves and as the species becomes less "newsworthy." However, we feel that wolf conservation will continue to be controversial as discussions shift from whether to restore the species to how best to manage free-ranging populations. A similar trend has been predicted for conservation of gray wolves in the northern Rocky Mountains (Bangs and Fritts 1996).

12. A well-trained and dedicated field crew with appropriate expertise was crucial to program success. Administrative continuity also facilitated success. The importance of these two aspects of the program should not be overlooked. Reintroduction programs using captive-born animals are especially sensitive to staff changes and administrative inefficiencies because they are long-lived, because they require that many difficult decisions be made in crisis situations, and because mistakes with small populations can be hard to reverse (Miller et al. 1996).

13. Red wolves can flourish in a wide variety of habitats, and there is sufficient habitat available in the southeastern United States to meet the population objectives of the Red Wolf Recovery Plan (USFWS 1990), assuming that the problem of hybridization between red wolves and coyotes can be resolved. Much of that area, however, is privately owned. Consequently, recovery of the red wolf is not dependent on setting aside undisturbed habitat, but rather on overcoming hybridization with coyotes and the political, logistical, and emotional obstacles to human coexistence with wild wolves.

12

Wolves and Humans

Steven H. Fritts, Robert O. Stephenson,
Robert D. Hayes, and Luigi Boitani

TRY TO IMAGINE a small group of wolves sitting at a table engaged in vigorous debate. These wolves are from various parts of the globe and are perhaps a bit more scholarly than most. In fact, they are especially knowledgeable about the biology of that notorious two-legged species, *Homo sapiens*. They have been brought together to document their relationship with humans over the last several millennia. Pause for a few moments and consider what they might say . . .

Perhaps the wolves' discussion would chronicle the evils of the human species, including details of atrocities committed against lupine ancestors down through the centuries. They might discuss the bizarre workings of the human imagination and the hopeless confusion of fact and fiction about wolf relationships with humans. The discussion might also express admiration for the way early humans respected wolves and imitated their living in family bands, maintaining pair bonds for years at a time, communicating in complex ways, and hunting cooperatively. The effects of advances in human technology might be detailed. The recent and long-awaited legal protection for wolves and the soaring popularity of wolves among some humans would certainly deserve mention. After an exhaustive review of the wolf-human relationship, the wolves might finally conclude that it has taken so many forms, depending on time and place, that generalizations are impossible.

We begin this chapter with the incredibly broad range of relationships between wolves and humans in mind. Our focus will be on the following topics: past and current human perceptions of wolves, wolf behavior toward humans, depredations on domestic animals, and the economic impacts of wolves, especially their predation on big game animals of importance to hunters.

Human Attitudes toward Wolves

Wolves have been of special interest to many human cultures around the Northern Hemisphere from prehistoric times to the present. Attitudes toward the animal range from reverence to hatred. Neither the historical record of humanity nor of wolves (if wolves could write one) would be complete without something being said about the other species.

Humans often determine where wolves can exist and influence their ecology and behavior in various ways (Young and Goldman 1944; Mech 1970; Boitani 1995; Stephenson et al. 1995; Thiel and Ream 1995; Hayes and Gunson 1995; Bangs and Fritts 1996). The wolf's range has waxed and waned during the past 2,000 years as humans alternately turned up and then turned down the heat of persecution (Okarma 1993). Humans are a major cause of wolf mortality in much of the wolf's current range. We tend to think of wolves as creatures of wilderness (Theberge 1975), yet they often exploit niches in which they are intimately intertwined with human communities (Thiel et al. 1998). In Romania, some travel city streets at night in search of food (Promberger et al. 1997). In Italy, some depend on village garbage dumps for food (Boitani 1982).

Many aspects of the wolf-human relationship are based on sometimes irrational cultural perceptions. Persecution of the wolf has often been out of proportion to the threat it actually posed to people. Consider the destruction of Scotland's forests to rid the country of the last wolf (Boitani 1995), the relentless pursuit of the last wolves in the American West during the 1920s and 1930s (Young and Goldman 1944; Young 1970; Lopez 1978; Brown 1983; McIntyre 1995; Hampton 1997), and the continuing fear among some people of wolf attacks despite the overwhelming odds against such attacks (Kellert 1999).

Similarly, public reaction to contemporary wolf management programs is often extreme, as occurred when Alaska proposed wolf control in a small part of that state's vast area (Stephenson et al. 1995). These diverse reactions are fairly typical of the historical relationship between human cultures and wolves.

The negative image of the wolf in the psyche of many people may be deeply ingrained, and not only because of the last ten to twenty centuries of history. In advancing his proposal for the coevolution of a gene-culture system, Wilson (1984, 1993) pointed to the almost universal human fear of snakes and suggested it is related to genetically prepared learning and retention of negative experiences. Ulrich (1993) provided convincing evidence that humans are biologically prepared to acquire and retain adaptive biophobic responses to certain natural situations and stimuli that contained some kind of risk in former times.

Predators probably posed an important risk to humans for much of our history, and wolves, though not as widespread as snakes, have flanked the development of culture from the time early humans colonized Eurasia. Conservation efforts around the world must contend with these long-standing fears. Negative perceptions of the wolf make it difficult to find a compromise between human interests and wolf conservation. Additional concern about wolves comes from the negative effects wolf predation can have on livestock producers, rural communities, and local economies, as discussed below.

Ultimately, the wolf exists in the eye of the beholder. There is the wolf as science can describe it, but there is also the wolf that is a product of the human mind, a cultural construct—sometimes called the "symbolic wolf"—colored by our individual, cultural, or social conditioning (Lawrence 1993). This wolf is the sum total of what we believe about the animal, what we think it represents, and what we want and need it to be. To many humans, this animal is the ultimate symbol of wilderness and environmental completeness. To others—for example, a Wyoming rancher or an Italian shepherd—it represents nature out of control, a world in which the rights and needs of rural people are subjugated by city-dwelling animal lovers intent on imposing their conservation values on others.

The symbolic status of the wolf, or shall we say "wolf mythology," is so strong that biological facts about the animal are often irrelevant—a situation especially vexing to biologists (Mech 2000b,c). For example, when biologists brief public officials about the actual numbers of livestock that wolves kill, the officials often focus instead on their constituents' perception of the problem and perhaps on their own prospects for re-election, not facts and figures about wolf depredations.

What people *choose* to believe about wolves can be more important than the objective truth, or at least those beliefs can have a greater effect. Whether looking at the past, the present, or the future, it is beliefs and perceptions that primarily affect the survival of wolves. For example, the battle of wills that happened over restoration of the wolf to Yellowstone National Park and central Idaho had more to do with what wolves symbolize than with the animal itself (Fritts et al. 1995).

Why does the wolf arouse diverse passions in humans that are not kindled to the same degree by most other animals, such as the bears and the large cats? How do we explain the pervasiveness of the wolf in folklore, and why have we relentlessly exterminated wolf populations in the past? And what might the recent popularity of wolf-dog hybrids as pets tell us about humans and wolves? Why, as we enter the twenty-first century, does wolf recovery and management attract such strong public interest (Mech 2000b)? The answers to these questions are elusive and complex, but might tell us much about our own species. We hope that the following discussion will provide some insight into our relationship with wolves.

Early Humans

Ethnographic accounts from early historic times, as well as evidence from archaeological sites, provide clues about attitudes toward wolves among prehistoric peoples. Cave paintings and associated artifacts in France show that humans have had a close relationship with animals for at least 100,000 years (Pfeiffer 1982). Rituals, ceremony, and art associated with animals increased about 30,000 years ago, at the beginning of the Upper Paleolithic period. This development is thought to represent an effort by early hunters to increase their likelihood of success.

A complex spiritual relationship apparently existed

between early hunters and the prey on which they depended for food and clothing. Like many more recent societies, Upper Paleolithic people may have believed in "a master of the hunt or keeper of the animals, an exalted being that provided game, established rules for the chase and punishments if the rules were broken, and who had to be obeyed and appeased when angry" (Pfeiffer 1982).

Early humans and wolves occupied similar ecological niches. Both were broadly adapted predators of large herbivores and hunted in family groups (Schaller and Lowther 1969; Mech 1970; Peters and Mech 1975a; Hall and Sharp 1978). People and wolves lived in loosely analogous societies that shared such characteristics as pair bonding, staying together year-round (not just for a breeding season), extended family clans, group cooperation, communal care and training of young by both males and females, group ceremonies, leadership hierarchies, and the sharing of food with kin. Like early humans, wolves often defended their hunting territory from other packs (see Mech and Boitani, chap. 1 in this volume). Although wolves and humans probably scavenged from each other's kills, we do not know whether Paleolithic people saw wolves as competitors.

Some authors suggest that wolves may be models for understanding early humans (Hultkrantz 1965; Schaller and Lowther 1969; Hall and Sharp 1978). We can be sure that early people observed wolves at length on the open plains, steppes (Kumar and Rahmani 2001), and tundra (Mech 1998a) and came to be familiar with their behavior and some of its similarities to their own (Stephenson and Ahgook 1975; Stephenson 1982). This sort of relationship between humans and wolves as fellow predators persisted in North America longer than in Europe because of the later transition from hunting to predominantly agricultural economies.

Native Americans

Most of North America's indigenous people were familiar with wolves and often regarded them as spiritually powerful and intelligent animals. Wolves were "medicine" animals and were sometimes identified with a particular individual, tribe, or clan (Lopez 1978). Some tribes believed that wearing the skin of the wolf brought about a supernatural union of human and wolf (fig. 12.1). Unlike elements of contemporary society, however, native cultures did not elevate wolves above other animals. Wolves were hunted and trapped by many Native Amer-

FIGURE 12.1. Many indigenous peoples in North America regarded wolves as spiritually powerful animals, and some used their pelts to symbolize a wolf-human relationship.

ican tribes, often with rituals and apologies to the spirit of wolves, but rarely with rancor or guilt.

The Nunamiut Inupiat in Alaska's central Brooks Range have had a long association with wolves. Like many other Eskimo peoples, they relied historically on wolves and other furbearers as an important part of their economy (E. S. Hall 1981). Furs provided clothing and were traded with other natives and later with Europeans. During the 1970s, not long after the Nunamiut had settled in their village of Anaktuvuk Pass after centuries of semi-nomadic life, R. O. Stephenson was able to work with them as an apprentice. He gained an understanding of their view of wolves, which may be indicative of those of other North American hunting societies (Stephenson and Ahgook 1975). Through their long experience observing and hunting wolves in the open foothills and mountain valleys of the Brooks Range, the Nunamiut acquired a refined understanding of wolves. They regard wolves as very smart animals and skilled hunters, possessing keen senses.

Like other northern peoples who hunted and trapped wolves, the Nunamiut often howled to call wolves, stalked sleeping wolves, or used deadfalls, traps, and snares to capture them (Boas 1888; Stephenson 1982; Nelson 1983; Mary-Rousseliere 1984). Hunters often expressed appreciation for the wolf's abilities and social complexity. They never spoke harshly about wolves, bragged about their ability to capture wolves, or announced their intention to hunt them. To do so could offend wolves or other animals and bring bad luck. The wolf did not evoke fear, although the Nunamiut were less sanguine than is our society about the possibility of attacks on people. This caution stemmed from a few attacks on people by hungry wolves prior to the advent of firearms and several incidents involving rabid wolves.

Relations between wolves and North American Indians were in many respects similar to those between wolves and Eskimos. Most tribes took precautions to avoid offending wolves and engendering bad luck or other consequences. However, historic accounts suggest considerable diversity in attitudes. In Alaska, the Tanaina people believed that wolves were once men (Osgood 1936) and viewed wolves as brothers. It was said that if a man was hungry and lost, he need only ask his brother the wolf for help (Townsend 1981). In contrast, the wolf was generally feared by the Chilicotin of British Columbia, and contact with the animal was thought to cause nervous illness and possibly death (Lane 1981).

Among Indians of the U.S. western plains, the wolf personified craft in war. Scouts often wore wolf skins, and the sign for "scout" and for "wolf" was the same in sign language. It was believed that wolves sometimes talked to people and warned them of the presence of enemies. Boys were told to imitate the wolf's habit of pausing to look back at its trail, even when running for its life, and to acquire its ability to endure severe conditions (Mails 1995).

Ethnographic studies do not suggest that there was widespread concern among Native Americans about the effect of wolf predation on important game populations. However, the oral history of several northern Athabascan groups includes descriptions of efforts to reduce wolf predation by killing pups at dens (Peter John, First Traditional Chief, Tanaina Chiefs Region,

AK; Tom Denny, Tanaina Village, AK; Ron Chambers, Champagne-Aishihik First Nation, Yukon, Canada; and Nick Bobbie, Tanaina Athabascan, AK, personal communications). Those efforts may have been prompted by the often limited and unpredictable supply of game typical of the northern interior of the continent (Burch 1972).

Eurasians

In Eurasian cultures, the socioeconomic relationship between early human societies and their environment largely determined their perception of the wolf. During much of history, economies were based on hunting and making war. Nomadic and sedentary shepherding came later, followed by crop and farm animal production (Boitani 1995). Like Native Americans, the early Eurasian cultures admired the wolf and, in some ways, tried to emulate it. However, societies that made their living as nomadic shepherds were vulnerable to wolf depredations and came to hate the animal (Boitani 1995).

The wolf appears in the earliest stories about European gods and was credited with involvement in human ancestry (Boitani 1995). Early Germanic warriors regarded the wolf as a totem. Anglo-Saxon nobles and kings, like American Indians, named themselves after wolves, attempting to associate themselves with admirable characteristics of the animal. According to Romanian biologist O. Ionescu (personal communication), ancient inhabitants of what is now his country portrayed the wolf on their battle flag.

The wolf was also viewed positively in the mythology of the Celts and the Greeks (Boitani 1995). Apollo, the god of light and order, was associated with the wolf in a predominantly positive way. Building on an earlier Greek legend, a well-known story describes the founding of Rome by the twins Romulus and Remus, who were raised by a nurturing female wolf. The Sabines regarded the wolf as a totem animal and had religious practices that centered around it. The positive view of the wolf among the Greeks and Romans survived for several centuries despite an influx of negative attitudes from northern Europe. The resulting ambivalent attitude in parts of Europe, especially the Mediterranean area, helped prevent the complete extermination of the wolf on that continent (see Boitani, chap. 13 in this volume).

The changes in Western thought about the environment and the wolf that were brought about by Christianity were second in importance only to those accompanying the domestication of animals. They were felt first in Europe (Ortalli 1973; Boitani 1995) as man switched from considering himself part of the natural world to master of it. The Bible does not seem to judge animals as "good" or "bad." All were created by God and declared "good" by God in the beginning (Genesis 1:25), and God intentionally saved all "kinds" during the great flood (Genesis 6:19–20). The wolf is mentioned in the Old and New Testament only as a *symbol* of rapacity, wantonness, cunning, and deceit, in reference to human characteristics. Nonetheless, the animal itself came to be viewed as evil, symbolizing threats to the Roman Catholic Church (Boitani 1995). During the early Middle Ages the wolf was viewed as evil and was a major character in the legends of the saints (Ortalli 1973). For over a thousand years books influenced primarily by the Catholic Church, such as the *Physiologi,* presented animals, including wolves, in highly fanciful ways by way of teaching moral lessons.

Science and natural history writings before the mid-twentieth century typically portrayed the wolf in a negative light. In the early nineteenth century, one could turn to *The Natural History of Quadrupeds* and read:

> Wolves are such ferocious and useless creatures that all other animals detest them, yea they even hate each other, and therefore scarcely ever live together, each one in its own separate hole. . . . Perhaps of all other animals, wolves are the most hateful while living and the most useless when dead. . . . The continual agitations of this restless animal renders him so furious, that he frequently ends his life in madness. (Robinson 1828)

Danger from wolves was a common theme in early literature and folklore. European and Russian literature abounded with fables, legends, references to werewolves, and tales about children raised by wolves. Werewolves were feared even more than real wolves because they added the supernatural power of the devil to the strength, ferocity, and cunning attributed to wolves (Lopez 1978; Stekert 1986; Fogleman 1988; Slupecki 1987). Folk tales such as "Little Red Riding Hood" and "The Three Little Pigs" taught carefulness and a work ethic. Though intended to be symbolic or metaphorical, they had a profound effect on how wolves were viewed in Western culture (Levin 1986; Greenleaf 1989). The nega-

tive view was so persuasive that it was not until the mid-twentieth century that Western culture considered the wolf worthy of scientific inquiry (see Boitani, chap. 13 in this volume).

In Japan, the relationship between religion and wolf conservation was quite different. The Japanese word for wolf, *ookami,* translates as "great god." During the era of the Shoguns (710–1867 A.D.), damage to agricultural crops by deer and other wildlife was a common problem. Farmers regarded wolves as beneficial because they killed wildlife that damaged crops. In the 1600s, people prayed to wolves at shrines throughout Japan, asking them to kill the crop-eating wildlife. One shrine reportedly bred wolves and rented them to villagers to combat wildlife pests (N. Maruyama, Tokyo Noko University, personal communication). This era ended in 1868 when the Shoguns lost power and Western advisors were brought to Japan to modernize agriculture (McIntyre 1996; N. Maruyama, Tokyo Noko University, personal communication). The Japanese were advised to poison their wolves, thus ending their reverence and tolerance for the animal. That farmers and ranchers in different parts of the world were simultaneously praying to wolves and finding new ways to kill them attests to the diversity in wolf-human relations.

Post-settlement Americans

European colonists brought to America a fear and hatred of the wolf based largely on Old World myth and folklore. Attitudes were strongly negative even in the earliest settlements (Young and Goldman 1944; Young 1970; Nash 1967; Lopez 1978; Fogleman 1988; McIntyre 1995; Hampton 1997). There were rational reasons to impugn the wolf, as its depredations on livestock posed a real threat to early settlements (see references in Fogleman 1988 and McIntyre 1995). The wolf ultimately became a metaphor for the environmental challenges the new North Americans had to contend with and felt a moral obligation to subdue. The goals of subjugating wolves and wilderness became synonymous.

This decidedly negative view of wolves prevailed during their eradication from most of the United States and large portions of Canada. The fervor with which European settlers and pioneers killed wolves (Young 1944; Lopez 1978; Brown 1983; Fogleman 1988; Thiel 1993; McIntyre 1995) far exceeded the intensity of persecution in Europe, where campaigns were more localized and short-lived (see Boitani, chap. 13 in this volume). Hampton (1997) called it "the longest, most relentless, and most ruthless persecution one species has waged against another."

Native ungulate populations were decimated by settlers and market hunters during the late 1800s, and large numbers of sheep and cattle were introduced into open range in the American West. Wolves and other large predators turned increasingly to livestock to survive, and the human determination to kill these carnivores increased.

The fate of the wolf in the American West was sealed when Congress established the federal Bureau of Biological Survey and its Division of Predator and Rodent Control (PARC) in 1915, with the mission of eliminating wolves and other large predators from all federal lands (Dunlap 1988). The threat to livestock became the strongest argument for killing every last wolf at taxpayer expense, even in areas remote from livestock range (Young and Goldman 1944; Curnow 1969; Weaver 1978; Lopez 1978; Brown 1983; McIntyre 1995; Hampton 1997).

During 1890–1930, the perception of the wolf by the U.S. public and Congress was strongly influenced by accounts of outlaw wolves that allegedly killed stock in large numbers. Many of these accounts were embellished and were developed, at least in part, by members of the U.S. Biological Survey to generate and maintain funding for their programs (Gipson et al. 1998). However, they continue to influence the perception of wolves among ranchers. Kellert et al. (1996) suggested that wolf destruction in the United States and Canada reflected an urge to rid the world of an unwanted and feared element of nature, including, perhaps, the possibility that settlers might succumb to the attractions of wildness and the absence of civilization.

Prior to the mid-twentieth century, most American biologists denigrated the wolf (Dunlap 1988). E. A. Goldman defended PARC and its poisoning at the 1924 meeting of the American Society of Mammalogists: "Large predatory mammals, destructive to livestock and game, no longer have a place in our advancing civilization" (Dunlap 1988, 51). However, when PARC nearly exterminated wolves in the American West, several biologists in the American Society of Mammalogists did object.

Aldo Leopold (1949) was one of the first Americans to speak in defense of the wolf. In his essay "Thinking like a Mountain," he related how the experience of killing a wolf and watching the "fierce green light" fade from her eyes helped change his opinion on the need to eradicate wolves, although he continued to push for wolf bounties (Flader 1974).

Contemporary Views

In the early 1940s, Leopold (1944) proposed restoring wolves to Yellowstone National Park, where they had been eradicated by the government only a decade earlier (Jones 2002). The first detailed field studies of wolves were launched in North America in the late 1930s (Olson 1938; Murie 1944) and 1940s (Cowan 1947; Stenlund 1955). By the 1960s, researchers such as Durward Allen, Douglas Pimlott, David Mech, and others were presenting more objective and balanced information about wolves and arguing for their conservation.

The prevailing attitude toward the wolf in Europe remained negative long after the animal was exterminated from most of the continent. This was true even in countries where no wolves remained. The Mediterranean countries, where an ambivalent attitude persisted, were an exception (Boitani 1995). The first wolf conservation programs in Italy and Spain began during the 1970s. Able European spokespersons emerged, including Erkki Pulliainen (Finland), Dimitry Bibikov (USSR), Anders Bjärvall (Sweden), Luigi Boitani (Italy), Eric Zimen (Germany), and others. However, negative attitudes toward wolves have generally persisted in eastern Europe and in the former Soviet Union.

The book *Never Cry Wolf* (Mowat 1963), a mostly fictional work (Banfield 1964; Pimlott 1966; Mech 1970; Goddard 1996), was the first positive presentation of wolves in the popular literature, with over a million copies sold. Despite its depiction of fiction as fact, this widely read book probably played a greater role than any other in creating support for wolves. A Disney movie based on the book reached millions of Americans and Canadians. Other early books that touched the public and biologists alike were *The Wolves of Mount McKinley* (Murie 1944), *Arctic Wild* (Crisler 1958), *The Custer Wolf* (Caras 1966), *The World of the Wolf* (Rutter and Pimlott 1968), and *The Wolf: The Ecology and Behavior of an Endangered Species* (Mech 1970), still in print with over 100,000 copies in circulation. In 1978, Barry Lopez's *Of Wolves and Men* provided a lucid and poignant exploration of the human relationship with wolves during recorded history, including the following provocative observation:

Throughout history man has externalized his bestial nature, finding a scapegoat upon which he could heap his sins and whose sacrificial death would be his atonement. He has put his sins of greed, lust, and deception on the wolf and put the wolf to death—in literature, in folklore, and in real life." (Lopez 1978, 226)

Increasingly favorable attitudes toward the wolf reflected a general change in outlook on wildlife and the environment. Legal protection of game animals was finally extended to various predators, and bounties were gradually eliminated (Dunlap 1988; Keiter and Holscher 1990). Objections to the extensive government wolf control programs in Alaska and Canada were raised (Theberge 1973). By the late 1960s, there were more calls to restore wolves to Yellowstone National Park (Mech 1991a).

During the 1970s, organizations with the sole mission of wolf conservation were formed. Key among them was the Wolf Specialist Group (Pimlott 1975; Mech 1982b) of the International Union for the Conservation of Nature and Natural Resources (IUCN), recently renamed the World Conservation Union. In 1973, D. H. Pimlott formed the Wolf Specialist Group at a meeting in Stockholm. The group then developed a "Manifesto on Wolf Conservation" (Pimlott 1975) as a guide for countries wishing to recover and conserve wolves, and this manifesto has been updated twice and approved by the IUCN. Globally, IUCN (2000) classified the wolf in its "Vulnerable" category in 2000.

Mainstream public conservation organizations in the United States such as the National Wildlife Federation, Audubon Society, and Defenders of Wildlife also became involved in wolf conservation (Tilt et al. 1987), as did the World Wildlife Fund in both the United States and Eurasia. In 1974, wolves were classified under the Endangered Species Act of 1973 as "endangered" in the contiguous United States. That action triggered an intense debate over whether U.S. wolves actually needed legal protection (Van Ballenberghe 1974; Llewellyn 1978; Thiel 1993).

As concern about human effects on the natural world increased, much of the public feared that wolves would soon be extinct. This fear was fostered by the failure of the U.S. government's Endangered Species List to distinguish between species that were endangered globally, such as the California condor and Kirtland's warbler, and those that were endangered only locally (Mech 2000c). In truth, tens of thousands of wolves survived in Canada and Alaska and hundreds in Minnesota, and the former Soviet Union supported 50,000 (Bibikov 1975). A small management program in the Yukon in the early 1980s (involving 2% of the Yukon wolf population) was incorrectly reported in Germany as an indiscriminate program to kill 5,000 wolves and to protect people from attacks (R. D. Hayes, personal observation). Across the United States, privately owned colonies of captive wolves were established with the expectation that those wolves would be used to reestablish the species in the wild (Mech 1995a). Several people appointed themselves "wolf educators," propaganda and inaccurate information were disseminated (Blanco 1998; Mech 2000b), and opposition to any form of wolf control broadened.

Numerous studies of human attitudes toward wolves in the United States in recent decades have documented strong public support for wolves (Kellert 1986, 1991; McNaught 1985; Lenihan 1987; Biggs 1988; Tucker and Pletscher 1989; Bath and Phillips 1990; Johnson 1990; Bath 1991a,b; Thompson and Gasson 1991; Duda and Young 1995; Bright and Manfredo 1996; Kellert et al. 1996; Pate et al. 1996; Wilson 1999). Most have focused on areas in the Upper Midwest where wolves were present and on western states where reintroduction was being planned or discussed. Residents of western states predominantly favored wolves and preferred they be restored. Studies of attitudes toward red wolves and their restoration have revealed even stronger regional support (Quintal 1995; Mangun et al. 1996; Rosen 1997). Except for Alaskans, who are generally positive and knowledgeable about wolves, residents living close to wolves are less positive about them than those living farther from wolf habitat (Williams et al. 2002).

Farmers and ranchers hold the most negative view of wolves in the United States, and probably elsewhere, with surveys showing up to 90% disapproval (Buys 1975; Kellert 1985, 1986; Nelson and Franson 1988; Bath and Buchanan 1989). This is true regardless of whether the farmers live close to a wolf population or have had any experience with wolves. However, Minnesota farmers regarded wolves far more positively in 1998 that in 1985 (Kellert 1999). The most positive and protectionist views of wolves are held by urban people and members of environmental organizations (Kellert 1987, 1999; Quintal 1995; Bath and Buchanan 1989; Duda et al. 1998). In general, more negative views are found among older, less educated, and lower-income people (Kellert 1996).

Most Americans, however, know little about wolves.

Some studies indicate that greater knowledge of wolves is related to a more positive attitude about them. However, many urbanites with little knowledge of wolves are highly positive about the animals (Kellert 1999).

The origins of current American attitudes about wolves are complex and are linked to the symbolic and economic value of wolves. People favorable toward wolves and their restoration often cite values related to ecosystem completeness, the right of the wolf to exist, and recreational value. Reasons for disliking wolves or opposing wolf restoration include the expectation of attacks on livestock, pets, and humans; cost; declines in big game populations; loss of self-determination; erosion of private property rights; and fear of more restrictions on the private use of federal land (Bright and Manfredo 1996; Wilson 1997; Scarce 1998).

In the western United States, wolf restoration is inextricably linked to a long-standing debate over how federal land is used—an issue that often pits local and regional views against national perspectives. Government is widely distrusted, perhaps especially by rural people. There are fundamental differences in the way urban and rural people in the West view nature (Wicker 1996). Various surveys show that although most Americans value wolves, they do not do so to the exclusion of important human needs (Kellert 1986, 1987, 1999; Tucker and Pletscher 1989; Thompson and Gasson 1991; Wolstenholme 1996).

Attitudes toward wolves in Canada are similar to those in the United States (Murray 1975). In British Columbia, viewpoints vary on wolf management and control and on the effects of wolves on the livestock and ungulate populations (Hoffos 1987). Attitudes toward wolves and wolf restoration in New Brunswick are strongly influenced by anticipated effects on deer hunting and are related to gender (females are more favorable to restoration), education level, knowledge of wolves, size of community, level of fear of wolves, and big game hunting experience (Lohr et al. 1996). Similar factors determined the willingness of Manitoban residents to maintain wolves in Riding Mountain National Park (Kellert et al. 1996; Ponech 1997).

The views of contemporary Native Americans toward wolves appear to vary depending on how "traditional" a person is (Vest 1988; Segal 1994). The reintroduction of wolves to Idaho was of great significance to the Nez Percé tribe, restoring pride and spiritual power and providing an opportunity for economic revitalization (Robbins 1997). However, in the southwestern United States,

Apaches attribute no special significance to the wolf and opposed its reintroduction (D. Parsons, USFWS, personal communication). Young Native Americans are often concerned that the return of the wolf would upset or restrict their modern lifestyle. Younger and middle-aged Kalispell Indians in Washington were more likely than older individuals to fear wolves (Segal 1994). First Nations in Canada and Alaska often have polarized views of wolves, which depend on the status of wildlife around their communities (R. D. Hayes, personal observation); attitudes toward wolves reflect a balance between their sometimes negative economic impacts on other wildlife uses and their cultural and spiritual importance to First Nations societies (Chambers 1995). Some First Nations in Alaska and Canada are involved in developing wolf control programs to help maintain ungulate numbers (Dekker 1994; Hayes and Gunson 1995).

Modern European attitudes about wolves have generally improved during the past two decades, especially in urban areas. Resentment toward the wolf is still strong in many rural areas (Promberger and Schröder 1993). The only attitude study in Italy, carried out in 1975–1976 in the Abruzzo region, revealed that fears and prejudices were strongly correlated with ignorance about the wolf (Serracchiani 1976). Attitudes have gradually improved in Finland (Pulliainen 1993). In Sweden and Norway, most people, even in rural areas, want the wolf to survive (Andersson et al. 1977; Bjärvall 1983; Bjerke et al. 1998). However, over 70% of reindeer owners and farmers in Sweden are against protective measures (Andersson et al. 1977).

In Scotland, 44% of the general public and 58% of local people are against wolf reintroduction to the Highlands; 17% of local residents and 36% of the general public are in favor (D. MacMillan, Macaulay Land Use Research Institute, personal communication). Fifty-three percent of Spanish gamekeepers say wolves should be eradicated, and 38% favor some control in areas adjacent to their operations (Blanco et al. 1992). In several European countries, rural law enforcement personnel often sympathize with poachers and fail to arrest and prosecute those who illegally kill wolves (Francisci and Guberti 1993; Boitani and Ciucci 1993). A recent expansion in European wolf range (Promberger and Schröder 1993) is partly the result of greater tolerance, but protective laws probably have played a more important role.

Attitudes about wolves in Croatia have also improved recently, corresponding to a decline in both wolf numbers and the number of livestock killed (Gyorgy 1984;

Huber, Berislav et al. 1993; Huber, Mitevski, and Kuhar 1993). The treatment of the wolf in Croatia changed from persecution to protection during the height of the Serb and Muslim war; 1994 was declared the "Year of the Wolf" and a commemorative stamp was issued (Gibson 1996). However, respondents in a small survey in Macedonia in 1992 unanimously favored maintaining the bounty for killing wolves (Huber, Mitevski, and Kuhar 1993). A belief in werewolves lingers in some Slavic countries, as well as in Poland and Bulgaria (S. Tolstoy, Russian Academy of Science Institute of Slavic Studies, personal communication; Slupecki 1987; Tolstoy 1995).

The potential for natural or human-assisted recovery of the wolf in Asia is limited. Though studies of attitudes are lacking, the prevailing view of the wolf is negative throughout most of Asia (Shahi 1983; Bibikov 1988; Fox and Chundawat 1995). A 1987–1988 study in Kazakhstan, where 60,000–62,000 wolves remain, indicated that 59% of people preferred elimination of the wolf using any method; only 3% favored protection (Stepanov and Pole, presentation at the 1994 Large Carnivore Conference in Bieszczady, Poland). Surveys in 1993 and 1996 in Japan revealed only moderate interest in wolves and their possible restoration (Koganezawa et al. 1996).

Perspectives of Biologists

Because wolves and wolf management are so controversial, wolf biologists face a variety of challenges in different parts of the world. In some countries, biologists may be among the few people working toward conservation of this predator (Zimen and Boitani 1979). In Western countries, they usually function in a complex environment in which supporters of wolves are many, but so are their views and demands (Mech 1995a, 2000b). Wolf managers must find a balance between idealism and pragmatism and between their focus on populations and animal rights activists' emphasis on individual animals. North American wolf biologists often disagree about the extent to which wolves regulate prey populations and about the need for, and effects of, wolf control (see Mech and Peterson, chap. 5, and Boitani, chap. 13 in this volume).

Strong public interest and the clash of human values often result in unusual demands on biologists who work with wolves (Bass 1992; Bangs 1995; Steinhart 1995; McNamee 1997). The bureaucratic working environment can be complex. For example, the wolf recovery program in the northwestern United States involved five federal agencies, three state wildlife departments, at least seven Native American groups, and land management agencies in at least four levels of government (Fritts et al. 1995). Whether researchers or managers, biologists often find themselves in the media spotlight and in the midst of controversy. Criticism from anti-wolf groups has been common historically, but criticism from pro-wolf organizations has intensified recently (Mech 1995a, 2000b; Blanco 1998).

Educating the Public about Wolves

Worldwide professional efforts to educate the public about wolves began in the early 1970s. The IUCN Manifesto on Wolf Conservation (Pimlott 1975) and all four U.S. Wolf Recovery Plans recommended public education to promote wolf conservation. Volunteers and conservation organizations took up the challenge of combating the wolf's negative image in both North America and Eurasia with varying degrees of accuracy and effectiveness (Mech 2000b). Prior to the reintroduction of wolves to Yellowstone, project biologists spent about 60% of their time on some form of public outreach (Fritts et al. 1995). Similarly, in all areas of Europe where wolves remained by 1970, wolf biologists promoted and conducted public education about wolves.

Although an informed public is essential to wolf conservation, defining what public education should consist of is problematic. There are important and critical differences between objective wolf education and wolf advocacy or activism. An unbiased portrayal of wolf and wolf management issues may not be possible, in part because ethical and other subjective values are involved (Gilbert 1995). If not carefully tempered, wolf "education" can reflect personal values (Haber 1996). Most wolf biologists believe that an objective portrayal of the wolf is needed to sustain wolf recovery. This means that the conflicts caused by wolves must be fairly expressed along with the solutions and compromises necessary to resolve those conflicts (Fritts et al. 1995; Mech 1995a,e; Blanco 1998).

Many different approaches have been used to inform and educate the public about wolves: one-on-one visits with key landowners and opinion leaders, wolf education kits in schools, wolf howling excursions, traveling and permanent wolf exhibits, public lectures, and tame "ambassador" wolves.

Dozens of nonfiction books and magazines about wolves are now available for all ages and levels of biological expertise. In 1990, the International Wolf Center

(IWC) launched *International Wolf* magazine, which includes wolf conservation news from around the world. Technical literature about wolves proliferated in the 1980s and 1990s. One of at least four wolf bibliographies contains 420 pages (Mech 1995e). Numerous Internet sites offer information (and misinformation) on wolves. For example, the International Wolf Center's home page receives over a million hits and 60,000 unique visits per month (V. Du Vernet, IWC, personal communication).

Wolves and the News Media

Television and newspapers are the public's primary sources of information about wolves. Several accurate and well-balanced documentaries about wolves and wolf recovery have been produced. However, news media are attracted to controversy, and wolf recovery, depredations, control programs, and most any other wolf-related topics seem irresistible. The Yellowstone wolf reintroduction was intensively covered by sixty international media. Popular information about wolves is often biased or inaccurate (Blanco 1998; Mech 2000b). When wolf stories appear, the extreme views of opponents and supporters of wolves are often highlighted, further polarizing the issue. The way the media covers wolves leaves the impression that they are more of a problem than other animals (Bangs and Fritts 1996).

Wolf-Related Organizations

About forty nongovernmental organizations (NGOs) in North America and at least a dozen in Europe exist to promote wolf conservation (M. Ortiz, IWC, personal communication; J. Warzinik, Timber Wolf Information Network, personal communication). Reintroduction of wolves to Yellowstone and Idaho might not have happened without the Wolf Fund, Defenders of Wildlife, the Wolf Education and Research Center, the National Wildlife Federation, and other organizations that continually lobbied both the U.S. Congress and federal agencies (Fischer 1995).

In addition to advocating for wolf recovery, a few NGOs, including Defenders of Wildlife in the northern Rockies and southwestern United States (Fischer 1989) and the World Wildlife Fund in Italy, have even sponsored livestock depredation compensation funds to assist wolf recovery. Private foundations and public contributions augmented government funding for the recovery program in Yellowstone and the red wolf program in the eastern United States. Conservation organizations have also furthered wolf conservation by holding numerous meetings worldwide that bring together biologists, managers, educators, and the public. For example, the International Wolf Center sponsored international wolf symposia in 1990, 1995, and 2000, and plans to continue this endeavor.

Various factors motivate pro-wolf organizations (Boitani 1995), and these groups often differ in approach. Some pro-wolf groups appear to be most concerned with the ethics of wolf management. In 1996, a group called "Friends of the Wolf" opposed the capture of wolves in British Columbia for reintroduction into the United States, offering a $5,000 reward for release of captured wolves. The Sierra Club attempted to prevent reintroduction of wolves into Idaho as an "experimental/nonessential" population (see Boitani, chap. 13 in this volume), preferring that colonization happen naturally.

Although most wolf-related groups are pro-wolf, some are anti-wolf. When wolves from Italy recolonized France's southern Alps in 1992, local shepherds joined with hunters for the first time in Europe to form a league for wolf eradication. In the United States, organized opposition to wolf restoration emerged during the 1990s. Preventing wolf reintroduction in the U.S. northern Rockies was the objective of the No Wolf Option Committee, the Abundant Wildlife Society, and the American Farm Bureau's Wyoming chapter. The "Wise Use" movement also opposed wolf recovery based on anticipated restrictions on use of public land and other resources by local residents. Some people suspect that wolf recovery is part of a conspiracy by the government and environmentalists to prohibit grazing, mineral extraction, and recreational use of public land (Fischer 1995; Wicker 1996).

Economic Value of the Wolf

Wolves have intangible values to many people, such as the important role some think they play in an ecosystem (but cf. Mech 1996) and the enrichment of nature (Pimlott 1975; Kellert and Wilson 1993). But wolves also have a complex economic value, which is hard to measure and overlain with emotional issues. In the past, the wolf was believed by most of society to have a mainly negative economic impact because it killed livestock and game animals. Economic benefit came to the few who sold furs or collected some form of payment for killing wolves (Thiel 1993), and economic loss was one of the most common arguments for wolf eradication.

Untold amounts of private and public money were spent to eradicate or control wolves (Dunlap 1988; Thiel

1993; McIntyre 1995; Hampton 1997). By one estimate, over three centuries of wolf bounties in North America cost governments, stock associations, and private individuals about $100 million (Hampton 1997, 136). During the Soviet period (1917–1991), Russia spent over $300 million on wolf bounties, stock insurance, and other payments related to wolf damage (D. I. Bibikov, interview by *Russian Conservation News,* Managing Editor Anya Menner, reprinted in *Natural Area News* 1[2]:5–7).

Economics is often brought into arguments about the desirability of wolf recovery and conservation. The cost of wolf recovery in the U.S. northern Rockies was projected to be $12 million over a 30-year period. Although this is only 5 cents for each American citizen (Bangs and Fritts 1996), cost was the main reason people gave for opposition.

The annual regional economic losses from the Yellowstone and Idaho wolf reintroductions were predicted to be $187,000–$465,000 in lost hunter benefits, $207,000–$414,000 in potential reduced hunter expenditures, and $1,888–$30,470 in livestock losses. However, the yearly gain would be $23 million per year in increased tourist expenditures (Duffield 1992; USFWS 1994b; Bangs and Fritts 1996).

Wolf management often requires substantial resources (Archibald et al. 1991; Mech 1998b). Wolf control to enhance deer hunting on Vancouver Island produced $5.90 of resident deer hunter benefits for every dollar spent (Reid and Janz 1995). Wolf reductions in interior Alaska and southern Yukon cost $500–$1,500 per wolf, but returns were high in terms of additional ungulate harvest (Boertje et al. 1995). The least expensive management methods (poisoning and aerial shooting by the public) are currently the least acceptable to the public (Fritts 1993; Boertje et al. 1995; Cluff and Murray 1995).

Tourism associated with wolves has recently emerged as a significant economic benefit. Wolf-related tourism helps fund wolf research in Poland and Romania (C. Promberger, personal communication). Such opportunities are limited by the elusive habits of the wolf, terrain, the need for a well-developed tourism infrastructure (technology, guides, accessibility), and cost (Wilson and Heberlein 1996). Opportunities to see wolves without professional assistance are rare and limited to some areas of open terrain (Mech 1995b). For example, from 1995 through 2000, some 70,000 visitors observed wolves in a nonforested part of Yellowstone National Park (R. McIntyre, U.S. National Park Service,

personal communication). Denali National Park, Jasper and Banff National Parks in Alberta, and several areas of Alaska and Canada outside parks also provide opportunities to observe wolves.

Fairly expensive expeditions to see or hear wolves and their signs are available in Idaho, Minnesota, Alaska, and Canada, as described in magazines devoted to wolves. Businesses on the outskirts of Yellowstone National Park quickly profited from interest in the newly established wolf population. There is growing concern about the effect of tourists on wolves and wilderness environments. Wolf education centers can also be an economic boost to local communities. The International Wolf Center in Ely, Minnesota, brings an estimated $3 million benefit to the local economy each year and stimulates the equivalent of sixty-six full-time jobs (Schaller 1996).

Wolves also have a certain consumptive value, although that value generally was more important in the past. Sales of pelts in the United States and Canada fluctuate widely because of market demand, ranging from about 21,000 in 1927–1928 to about a thousand in 1956 (Obbard et al. 1987). The number of wolves sold for fur in Canada declined by 40%, from 3,738 in 1983 to 2,285 in 1990, reflecting a general decline in the fur market (Hayes and Gunson 1995). Wolf pelts are still valued for parka trim, fur coats, and rugs and are an important component in the local manufacture of clothing in virtually all Arctic communities in Canada and Alaska, where they provide a significant part of winter income. In Alaska and in Canada's Northwest Territories and Nunavut, wolf harvests remain fairly stable because of this strong local demand for their fur.

Some economic values of wolves are more elusive. Economists have recently developed ways to assess the potential value of nonconsumptive uses of wildlife, such as viewing, and to define preservation or existence value (Krutilla 1967). Many people value simply knowing that wolves exist in the wild, without ever expecting to see or hear one. This type of value can be economically evaluated by asking individuals their willingness to pay, contingent on a hypothetical situation. Using this approach, the existence value of wolves was estimated at $8,300,000 per year in Yellowstone and $8,400,000 per year in Idaho (Duffield 1992). Similarly, the benefit from red wolf restoration was estimated to be at least $18,270,000 per year to the nation and $3,240,000 per year in eight southeastern states nearest the two reintroduced populations (Rosen 1997). We must note, however, that no attempt was made to similarly assess the negative value wolves have to other people, which would tend to offset these hypothetical positive values.

Currently the wolf is riding a wave of marketing popularity. Books, magazine articles, conferences, T-shirts, jewelry, paintings, photographs, sculptures, coffee mugs, and audio- and videotapes are all part of the economic activity associated with wolves. The charisma of the wolf has been used—and sometimes abused—to raise funds for conservation and advocacy organizations (Mech 2000b). Appeals for financial support from organizations purporting to be "saving the wolf" have proliferated.

In 1995, a direct mail solicitation from a major pro-wolf organization informed readers that "a war on wolves, begun a hundred years ago, still rages today." Citing atrocities against wolves committed more than a century ago, the letter went on to convince readers that money is urgently needed "at this critical time . . . in the fight to save America's wolves." Such appeals tap the guilt, vague environmental concern, and resources of people, especially in cities, who wish to do something for wildlife and "the environment." The widespread use of such techniques by environmental organizations to raise funds ($3.5 billion in 1999) were explored in a recent newspaper series (Knudson 2001). This approach has created problems for wolf recovery and the long-term coexistence of wolves and people and fostered a growing resistance to some environmental causes (Mech 2000b).

Wolf Behavior toward Humans

Overall Reactions to Humans

How wolves react to humans depends on their experience with people. Wolves with little negative experience with people, or wolves that are positively conditioned by feeding, including in parts of the High Arctic, may exhibit little fear of humans (Parmelee 1964; Grace 1976; Miller 1978, 1995; Mech 1988a, 1998a). Perhaps prehistoric humans and wolves feared each other less in open habitat because each species could watch the other from a distance, thus removing some of the mutual apprehension (Stephenson and Ahgook 1975; Hampton 1997). Wolves on the American Great Plains often seemed to be unafraid of humans. Explorer Meriwether Lewis once killed a wolf in present-day Montana with a bayonet (Hampton 1997). Forest-dwelling wolves, however, were rarely observed, thus remaining mysterious and feared, and they themselves were generally afraid of people

(Fogleman 1988; McIntyre 1995). After wolves on the open prairie encountered firearms, they became secretive and elusive.

Denizens of Wilderness?

Society has come to believe that wolves are incompatible with civilization, and to many people, the wolf symbolizes wilderness (Theberge 1975). Mech (1995a) argued that equating wolves with wilderness is an artifact of wolves being exterminated in most areas except wilderness, creating a misconception that they *require* habitat free of human influences to survive. Whereas wolves in some areas of Canada, Alaska, and Russia might never see, smell, or hear a human, most of the world's wolves live somewhere near people. They encounter the sights, sounds, and scents of civilization in their daily travels.

Human population density in areas occupied by wolves ranges from less than 1 to at least 200/km^2 (Shahi 1983; Mech 1988a; Promberger and Schröder 1993; Marquard-Petersen 1995). Living near people requires caution about where and when to travel. Behavioral adaptations to humans are most evident in parts of Europe where wolves survived in heavily populated areas. For example, wolves in Italy and Spain avoid activity during daylight (except during foggy or hazy weather) to minimize contact with people (Zimen and Boitani 1979; Boitani 1982, 1986; Vilà et al. 1995; Ciucci et al. 1997). In remote Lapland, wolves are said to be afraid to cross a ski track, while those near the large cities of Finland have learned to move around houses and cross highways while still avoiding contact with humans (Pulliainen 1993). Romanian wolves have entered towns at night, totally unbeknownst to residents (Promberger et al. 1995, 1997). In one area, the animals travel into town in search of food, crossing a large industrial area, a highway, and a busy railroad several times during a night. Italian wolves also enter mountain villages at night in search of food; one pack even denned in an abandoned house (Boitani 1982). In India, wolves regularly live around people, and one pack denned in a concrete pipe (Kumar and Rahmani 2001). In most forested areas of wolf range around the world, however, wolves are rarely seen except in winter by researchers aided by radio-telemetry and aircraft.

Within any wolf population, individuals vary in their caution toward humans and human modifications of the environment (Fox 1972b). Bold individuals may occur in any population. Less cautious wolves are probably the first to be killed by hunting or trapping, but they can survive when protected. Wolves in protected populations generally are less fearful of humans than those in exploited populations (McNay 2002a,b). Several individuals in protected colonizing populations have demonstrated very little fear. Recolonizing wolves have passed within a few meters of houses and vehicles on many occasions in the Ninemile Valley and Glacier National Park areas of Montana (M. D. Jimenez and D. K. Boyd, University of Montana, personal communication).

Wolves recolonizing Varmland, Sweden, were unusually bold, setting off a debate about whether they had been released by wolf advocates (Promberger, Dahlstrom et al. 1993). Minnesota citizens claim that wolves there are more bold around people after 25 years of legal protection. Nevertheless, wolves on Isle Royale, Michigan, still avoid humans after being protected for over 50 years (Thurber et al. 1994), although they encounter people only during 5 months of the year. Elsewhere, chance encounters between humans and wolves increase during autumn big game hunting seasons, when the number of people in wolf habitat soars and less cautious pups are about.

Wolves enjoy a high degree of protection in North American national parks and often show unusual tolerance of humans in these environments. Wolves in places such as Denali and Yellowstone are often watched at short distances by park visitors. Yellowstone's Druid and Rose Creek packs are regularly observed along the main road as the animals sleep, travel, howl, hunt, and feed.

Wolves show a surprising willingness to live near humans after legal protection. Italian wolves have colonized habitat near the outskirts of Rome. Minnesota wolves have dispersed into open agricultural areas, even though they were raised in a forested environment (Licht and Fritts 1994; Merrill and Mech 2000). Wolves live near a military training facility at Fort Ripley, Minnesota, where they encounter explosions, low-flying aircraft, human shouts, troop movements, and noisy vehicles (Merrill 1996; Thiel et al. 1998). In parts of Spain wolves live primarily in sunflower and wheat fields (Vilà, Castroviejo, and Urios 1993). Clearly, wolves are not wilderness dependent, but their survival depends on the availability of cover that allows them to avoid humans, and on human attitudes that are relatively positive, or at least benign.

Wolves and Roads

One-third of all documented mortality among wolves east of the central Rockies in Canada was related to roads (Paquet 1993), and 75% of human-caused wolf mortality in the U.S. northern Rockies and adjacent Canada occurred within 250 meters of a road (Boyd-Heger 1997). Roads that follow narrow mountain valleys may increase the chance of human-related mortality or substantially alter wolf movement patterns (Paquet and Callaghan 1996). The Trans-Canada Highway and Railroad through Banff National Park, Alberta, accounted for over 90% of local wolf mortality (P. Paquet, World Wildlife Fund-Canada, personal communication).

Thus, roadways can have a strong effect on the way wolves perceive and move about the landscape, and are both a blessing and a curse to wolves. Abandoned roads become travel routes and make travel easier. Secondary roads are often scent-marked in Minnesota (Peters and Mech 1975b) and, like lakes and streams, often represent boundaries between territories. Primary or secondary highways defined 25–90% of the boundaries of the territories of seven of eight packs in Wisconsin (Frair et al. 1996). On the Kenai Peninsula, Alaska, wolves selected or avoided roads depending on human use, and roads influenced the spatial organization of packs (Thurber et al. 1994). Closed roads were preferred winter routes for wolves near Glacier National Park, Montana (Singleton 1995). Wolves commonly use roads in Denali National Park, Alaska (Mech et al. 1998).

Roads that provide access to remote areas can result in vehicle strikes and increased harvest, poaching, or disturbance of wolves. As wolves were just starting to recolonize various areas, they were absent from areas where road density exceeded about 0.6 km/km^2 (Thiel 1985; Jensen et al. 1986; Mech, Fritts, Radde, and Paul 1988; Fuller et al. 1992; Boyd-Heger 1997). Most recolonizing packs in Wisconsin selected areas with a road density of less than 0.45 km/km^2 (Mladenoff et al. 1995). However, as recolonization continued, wolves occupied areas where human populations were relatively high and road density was much higher than 0.6 km/km^2 (Mech, Fritts, Radde, and Paul 1988; Berg and Benson 1999; Merrill 2000; Corsi et al. 1999).

Trains and snowmobiles are also a factor in the wolf's environment. Train tracks often parallel highways, as do pipelines and power lines, thus widening the corridor and increasing the risk for wolves that try to cross or travel along them. Snowmobile trails are commonly used by wolves because the packed snow allows easy travel. Most use is at night when snowmobile traffic is lowest, but wolves have been seen leaving trails to let snowmobiles pass and then going back to them.

Reactions to Humans Near Pup-Rearing Sites

Wolves vary in their tolerance of human activity around pup-rearing sites. Those not often exposed to humans tend to avoid denning near human activity. However, several den and rendezvous sites have been found within 1–2 km (0.6–1.2 mi) of roads in North America (Jimenez 1992; Mattson 1992; Thiel et al. 1998; Mech et al. 1998) and Italy (Boitani 1986). In remote tundra areas, wolves abandoned dens after people established temporary camps within 1.0 km (0.6 mi) (Chapman 1977, 1979), while some denning wolves in Denali National Park were more tolerant of disturbance (Mech et al. 1998).

Three wolf dens in Yellowstone National Park were located near paved roads (one within 0.4 km), and two became visitor attractions (Smith 1998). Wolves twice moved litters as a result of disturbance by park visitors; the second move resulted in loss of the pups. A Montana pack maintained a rendezvous site at the edge of an active timber cutting operation despite regular low-level helicopter flights directly over the site (Jimenez 1995). Another Montana pack reused its traditional den the year after the area was clear-cut (J. Till, USFWS, personal communication). Thiel et al. (1998) documented active dens and rendezvous sites near active gravel pits, peat mining operations, and military firing ranges, and researchers in Romania found a pack of wolves denning near the city of Brasov (Promberger et al. 1997).

Most countries, states, and provinces provide no special protection for wolf dens or rendezvous sites. However, when wolves were reintroduced to Yellowstone and Idaho, the U.S. Fish and Wildlife Service (USFWS) established the option of closing to humans a 1.6 km area around their active dens and rendezvous sites on public land during the denning season (Fritts et al. 1995). The measure, implemented only once, was intended to protect wolves from disturbance that might cause adults to move pups to another site at too vulnerable an age. In contrast, Denali National Park maintains closures around some dens and rendezvous sites that have been inactive for many years (Mech et al. 1998).

Attacks on Humans

Do wolves attack humans? As already indicated, fear of wolves has been pervasive in human societies. At one time in the 1980s, armed parents escorted their children to school in Whitehorse, Yukon, because they feared wolf attacks, and children in Norway were being bused short distances to school for the same reason (R. D. Hayes, personal observation). A few years later, Montana's U.S. Senator Conrad Burns, opposing wolf reintroduction to Yellowstone, predicted "a dead child within a year" (Fischer 1995). Fear of wolves was an important reason for wolf persecution in both the Old and New World (Young and Goldman 1944; Rutter and Pimlott 1968; Mech 1970; Lopez 1978) and still influences current attitudes about wolves (Kellert et al. 1996).

Cultures that had regular contact with wolves (e.g., Eskimos, American Indians) did not generally regard them as dangerous (Ingstad 1954; Stepehenson and Ahgook 1975), although wolves have killed some Eskimos and Indians (Lopez 1978; Hampton 1997). Biblical references (Matthew 10:16) to wolves allude to their ferocity and threat to sheep, but do not describe them as dangerous to humans. Written accounts of wolves attacking humans are far more common in Europe and Russia than anywhere in North America (Mech 1970).

Clarke (1971) reviewed historic reports of wolf attacks in Europe and central Asia and concluded that nearly all incidents involved wolf-dog hybrids or rabid wolves. Rutter and Pimlott (1968) concurred, although Pimlott (1975) believed reports of wolf attacks on children in Spain. Nevertheless, most North American biologists have been skeptical about reported wolf attacks in the United States and Canada and have downplayed wolf danger to humans.

Records of wolf attacks on humans in Europe and Asia are numerous. In 1994, Ilmar Rootsi, of the Estonian Naturalists Society, presented a paper at a conference in Poland entitled "Man-Eater Wolves in 19th-Century Estonia." The report was based on a study of folklore archives, annual reports of clergy, court records, government correspondence, and other press reports and literature. These sources suggest that 108 children and 3 adults were killed by non-rabid wolves in Estonia from 1804 to 1853, but that tame wolves and wolf-dog hybrids were involved in these attacks. Rootsi also found records of 82 registered cases of attacks by rabid wolves from 1763 to 1891, with most occurring in winter and spring.

Cagnolaro et al. (1996) analyzed state and communal archives from the fifteenth to nineteenth centuries in northern Italy and found at least 440 accounts of humans killed by non-rabid wolves. Most were children less than 12 years old. The percentage of children killed was highest in rural areas, while adults were more often attacked near villages or towns. According to these records, 67 persons, including 58 "youths," were killed by wolves in the Po Valley of Northern Italy between 1801 and 1825.

Mivart (1890; cited in Mech 1970) reported that 161 people were killed by wolves in Russia in 1875 alone. Wolves allegedly attacked people in several regions in Russia during the nineteenth century and earlier, and also in 1944–1953 (Pavlov 1990; Bibikov and Rootsi 1993). Bibikov (1994) suggested that these incidents occurred "during and after [human] hostilities when wolves became accustomed to corpses, or some individuals were to blame that were raised in captivity and became feral." There are a variety of other reports of wolves scavenging from human corpses (Young and Goldman 1944; Lopez 1978; Shahi 1983; Fogleman 1988; Hampton 1997).

A few attacks on people were reported in Kazakhstan in 1995 and 1996 (Sergei B. Pole to L. D. Mech, personal communication, 12 March 1996), and several fatal attacks were reported in Poland prior to the mid-nineteenth century (Krawczak 1969). Pulliainen (1984, citing Godenhjelm 1981) described a "fairly well documented" case based on church records in which 23 children were killed by a "wolf-like" canid in southwestern Finland during 1878–1881. However, there were no subsequent reports of wolf attacks there (Pulliainen 1993).

Haken Eles reviewed kyrkbocker ("church books") kept by parish priests in twenty-five Varmland (Sweden) parishes from 1749 to 1859 (Eles 1986; H. Eles, personal communication, 1995). In one parish he found records of a 4-year-old boy "clawed to death" and "mainly consumed" by a wolf in 1727 and a 9-year-old boy killed 25 days later. Church records also indicate that a wolf killed another child during the 1700s. Eles nonetheless concluded that such events were "something very, very rare."

The most compelling evidence of wolves killing humans recently comes from India. Both Shahi (1983) and Jhala and Sharma (1997) investigated reports of wolves carrying away and eating small children ("child-lifting") and concluded that some were true. In 1996, the latter

biologists investigated fatal and nonfatal attacks on 76 children, aged 4 months to 9 years, in rural villages of eastern Uttar Pradesh. Over 7 months, attacks occurred about every third day, and children were killed every fifth day on average. Several partly consumed bodies were examined. Evidence pointed to a single bold wolf. The general poverty of the area was thought to contribute to the attacks. Small children were allowed to roam untended. They outnumbered unguarded livestock, and wild prey were scarce. High government compensation payments for the children may have fostered this situation (Jhala and Sharma 1997).

The dearth of fatal wolf attacks on humans in North America following European settlement contrasts with the situation in Europe and Asia. Virtually no early explorer or trapper in the United States and Canada regarded wolves as dangerous (Hampton 1997). Many observers on the Western frontier were astonished that wolves did not kill humans, in view of the stories they had heard (Casey and Clark 1996; Hampton 1997). However, Young and Goldman (1944) described a number of instances of aggression by wolves toward people in various parts of the United States during the nineteenth and early twentieth centuries.

In recent decades, incidents of aggressive behavior in wolves toward humans seem to have increased in North America. McNay (2002b) analyzed eighty cases in which wolves exhibited fearless behavior toward humans between 1900 and 2001, and elsewhere (McNay 2002a) provided detailed accounts of these incidents, which included incidents detailed in earlier studies (Young and Goldman 1944; Mech 1990b; Munthe and Hutchison 1978; Jenness 1985; Scott et al. 1985; The Raven 1997, 1999; Aho 2000; National Post 2000) as well as more recent incidents. Aggression by wolves was evident in fifty-one cases. Most incidents were attributed to self-defense, defense of other wolves, rabies, or aggression toward people who were accompanied by dogs. However, nineteen cases of apparently unprovoked aggression involved displays, charges, or bites associated with agonistic or predatory behaviors; eighteen of those occurred after 1968. Among the thirteen biting incidents recorded in cases of unprovoked aggression, eleven involved wolves that were habituated to humans. The apparent increase in aggressive encounters after 1970 was thought to be the result of greater protection for wolves and increased wolf numbers, combined with increased visitor use of parks and other remote areas. These factors have created increased opportunities for wolf habituation and food conditioning.

Two of the most serious attacks occurred in Algonquin Provincial Park in Ontario, where five people have been bitten in the last 25 years (The Raven 1997, 1999). In 1996, a 12-year-old boy sleeping outdoors was bitten in the face and dragged about 2 meters before the wolf was driven away. In 1998, a wolf grabbed a 19-month-old boy as he played alongside his parents in a campground. The wolf tossed the boy in the air, leaving several puncture wounds on his chest and back before being driven away.

Other wolf-human incidents in North America involved rabid wolves (Chapman 1977; McNay 2002a,b), or were thought to (Peterson 1947). Rabid wolves have rarely killed people in North America, but Native Americans were aware of the danger from rabid animals, including wolves (Young and Goldman 1944; Lopez 1978; Hampton 1997). Currently there is little concern about rabid wolves in Canadian communities in the Arctic, despite epizootic outbreaks in arctic foxes (P. L. Clarkson, Gwitch'in Renewable Resources Board, Inuvik, N.W.T., personal communication). In parts of Eurasia and the Middle East, however, attacks by rabid wolves have been more common (Baltazard and Bahmanyar 1955; Cherkasskiy 1988; Linnell et al. 2002).

Hampton (1997) suggested that the subject of non-rabid wolves preying on humans is "veiled in a hopeless tangle of fact, fear, myth, and folklore passed down through the generations." However, even allowing for exaggerations and fertile imaginations, it is now clear that even non-rabid wolves sometimes attack humans. What is puzzling is why such incidents have been so rare in Europe and Asia in recent years in view of the historical accounts (Linnell et al. 2002). We suspect that a number of factors are responsible, including changes in animal husbandry practices in Europe, where children once herded livestock; the decline of wolves in many parts of Eurasia; and the advent of firearms and consequent selection against wolves that are aggressive toward people. Wolves may have learned that modern humans are especially dangerous and changed their behavior accordingly.

Wolves may perceive humans as being unique in their environment. A human walking upright and wearing clothes is unlike anything else in the wolf's world, and upright humans do evoke strong fear in captive wolves (Joachim 2000). Perhaps the best way to put the safety issue in perspective is to realize that each day millions of people live, work, and recreate in areas occupied by wolves. Attacks by wild wolves are nonetheless rare, and fatal attacks are ever rarer and hard to document (note especially Linnell et al. 2002 and McNay 2002a,b).

Wolves and Hybrids as Pets

The popularity of wolves and wolf-dog crosses (hybrids) as pets is one manifestation of the modern fascination with wolves (Hope 1994). Ironically, ownership and commercial trade in these animals is yet another form of human exploitation. Some figure that there are more than 100,000 captive wolves and 400,000 hybrids in the United States alone (Hope 1994); others estimate the number of privately owned wolves or hybrids at 8,000 to 2 million (Kramek 1992). However, accurate information about the numbers of these animals and the problems they cause is difficult to obtain.

Keeping wolves as pets has become popular despite the danger and other problems that usually result, and despite recommendations discouraging private ownership (IUCN Wolf Specialist Group Resolution, 24 April 1990). The U.S. Endangered Species Act of 1973, as amended, forbids ownership of pure wolves, but hybrids are subject to little, if any, regulation in all but a few states. Such animals are offered for sale in newspapers for $250–$1,500 each. States that try to regulate their ownership encounter complex problems relating to identification; no genetic or other test can consistently distinguish pure wolves from hybrids (see Wayne and Vilà, chap. 8 in this volume).

Hybrids and tame wolves have little fear of humans, are less predictable and manageable than dogs (Mech 1970), and are considerably more dangerous to people (R. Lockwood, American Humane Society, personal communication). Pet wolves and wolf-dog hybrids killed at least nine children in the United States from 1986 to 1994 (Hope 1994), and many children have been maimed. An unknown number of tame wolves and hybrids are released to the wild in the United States (Wisconsin Department of Natural Resources 1999), and distinguishing these animals from wild wolves that are abnormally bold can be difficult or even impossible (Bangs et al. 1998; Boyd et al. 2001).

Wolves Nurturing Humans

Can wolves adopt and rear human infants? The notion that wolves can nurture children occurred in both Eurasia and North America and dates back at least as far as

Romulus and Remus, but this idea has now been debunked (Mech 1970).

Depredations on Domestic Animals

The domestication of animals that began some 12,000–13,000 years ago brought profound changes in the human view of wolves (Boitani 1995). The Sami people (formerly called Lapps) of northern Sweden, for example, changed from respecting the wolf to disdaining it after they began herding reindeer (Turi 1931; cited in Boitani 1995). Over millennia, selective breeding reduced the natural defenses of domesticated animals. Meanwhile, human societies developed more effective means of killing wolves (e.g., the Sami now use snowmobiles and modern rifles).

Depredations on livestock became the primary reason for attempts to exterminate the wolf, first in the Old World and later in North America (Young and Goldman 1944; Bibikov 1982). Wolves preyed on the livestock of European colonists in New England beginning in the 1600s. As settlers advanced westward, so did the wolf-livestock problem. This conflict, along with a host of secondary factors, fueled an outright war on wolves in America for 300 years (Young and Goldman 1944; Lopez 1978).

Depredations on livestock continue to be a major problem in wolf conservation. Wolves prey on domestic animals in every country where the two coexist, killing cattle in Minnesota, reindeer in northern Scandinavia, sheep and goats in India, and horses in Mongolia (Ginsberg and Macdonald 1990). Aside from the economic losses, the very threat of depredation creates stress for livestock producers. Human ingenuity and technology have so far been unable to resolve this conflict, short of eradicating wolves in areas near livestock (Fritts 1982; Mech 1995a). The public and the media are intensely interested in these controversies; human values clash, emotions run high, and misinformation abounds (Blanco 1998; Mech 2000b).

Human tolerance for wolf depredations and ability to combat the problem vary among cultures. Native Americans lost horses to wolves, but did not react with the hostility shown by northern Europeans and Euro-Americans (Hampton 1997). There seems to be greater tolerance for wolf depredations on livestock in the parts of southern Europe and Asia where wolves were never completely eradicated and agricultural societies have adjusted to their presence (Boitani 1995).

Nature and Extent of Depredations

Wolves kill every kind of livestock available to them. Sheep are the most common domestic prey in Europe because of their vulnerability and relative abundance in wolf areas. Aside from turkeys, cattle are the most frequent domestic prey of wolves in North America and greatly outnumber sheep, which have declined sharply in recent decades.

As populations of wild prey were depleted in much of Europe and Asia, livestock became more important to surviving wolves. In the American West, losses of livestock increased following the depletion of bison, elk, deer, and other ungulates and the replacement of those species with cattle and sheep (Young and Goldman 1944). Healthy populations of wild prey have been restored in most parts of North America where livestock are raised, and the proportion of livestock lost to wolves now is generally low (Dorrance 1982; Fritts 1982; Gunson 1983; Tompa 1983a; Fritts et al. 1992; Mack et al. 1992; Bangs et al. 1995, 1998; Treves et al. 2002). We know of no place in North America where livestock compose a major portion of wolf prey, or where wolves rely mainly on livestock to survive.

In Europe and Asia, however, livestock make up a larger part of the wolf's diet, although the proportion varies among regions. In western and southern Europe and the Middle East, wolves have survived in areas with highly degraded natural habitat by eating livestock, livestock carrion, and human refuse. In Gujarat and Rajasthan, India, wolves subsist mainly on sheep and goats because wild prey is scarce outside of preserves (Shahi 1983; Jhala and Giles 1991). In the Hustain Nuruu Reserve of Mongolia, wolves feed mostly on livestock, with over half the diet composed of horses and sheep (Hovens et al. 2002). In an area almost devoid of wild ungulates in northern Portugal, wolves appeared to feed exclusively on livestock, especially goats (Vos 2000). On the other hand, there is relatively little livestock depredation in areas where populations of wild ungulates are healthy (Promberger and Schröder 1993). In Poland (Okarma 1993; Bobek 1995), Romania (Almasan et al. 1970; Ionescu 1993), and Finland (Pulliainen 1965, 1993), depredations on livestock declined after populations of native ungulates were restored. Improved animal husbandry is also thought to be partly responsible for the decline.

Wolves kill dogs wherever the two canids occur, and dogs are an important food for wolves in some areas

(Boitani 1982; Brtek and Voskar 1987; Bibikov 1988; Fritts and Paul 1989; Pulliainen 1993; Bangs et al. 1998; Kojola and Kuittinen 2002). A survey in Croatia indicated that dogs were the most frequent domestic prey of wolves, outranking even sheep (Huber, Mitevski, and Kuhar et al. 1993a). At least twenty-five dogs were killed in Minnesota in 1998 (Mech 1998b), and wolves appear to limit the number of stray dogs in Russia (Bibikov 1988). More compensation has been paid for dogs than for livestock in Wisconsin (Wisconsin DNR 1999; Treves et al. 2002). Attacks on pets in the United States and Finland often occur near human dwellings. Wolves that attacked dogs near homes in Minnesota seemed to focus on them so intently that they temporarily lost their fear of humans (Fritts and Paul 1989; Mech 1990b). If a dog happens to be a beloved companion, the owner experiences an emotional loss and a grieving process (Anderson et al. 1984). In Europe and Wisconsin, wolves often kill hunting dogs, perhaps because they are more likely to be in wolf habitat.

Numbers of Livestock Killed

It is difficult to determine the number of livestock injured or killed by wolves. In the past there was less scrutiny of alleged losses to wolves, so older records should be viewed with caution. For example, Bibikov (1982) cited early Russian reports of about 1 million cattle (0.5% of all cattle available) being killed in the Soviet Union in 1924–1925. As recently as 1987, some 150,000 domestic animals (mainly sheep) were claimed to have been killed by wolves in Kazakhstan, based on a survey of local people (Stephanov and Pole, presentation at 1994 Large Carnivore Conference, Bieszczady, Poland).

In North America, reliable long-term data on livestock losses to wolves are available for Alberta, British Columbia, Minnesota (fig. 12.2), and Montana. (Records are also accumulating from Wisconsin, Idaho, and Wyoming.) Although they are increasing in some of those areas, wolf depredations involve less than 1% of the available livestock (table 12.1), and less than 1% of producers within wolf range experience losses to wolves each year. Information from other states and Canadian provinces suggests a similar pattern (Gunson 1983). However, these figures are all from places where wolves were long ago exterminated from most of the main livestock-producing areas.

The extent of livestock killing by wolves varies greatly by area and by year and is difficult to predict. In Min-

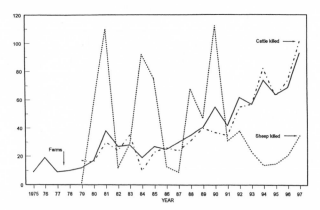

FIGURE 12.2. Numbers of Minnesota farms on which wolves killed livestock and numbers of cattle and sheep verified as killed by wolves.

nesota, there has been less livestock depredation following the most severe winters, apparently because winter conditions increased the vulnerability of white-tailed deer fawns to wolf predation (Mech, Fritts, and Paul 1988; but cf. Fritts et al. 1992). In contrast, the worst year for depredations in Montana (1997) followed an extremely severe winter. The resulting sharp decline in deer in northwestern Montana was believed to be responsible (Bangs et al. 1998).

Number Killed per Attack

The number of livestock wolves kill during an attack is related to the size and abundance of the prey. Most attacks on cattle or horses result in one animal being killed or wounded, whereas more than one sheep is usually killed in one attack. Losses in individual incidents in Minnesota averaged 1.2 animals for cattle, 4.4 for sheep, and 53.5 for turkeys (Fritts et al. 1992). In the Abruzzo region of Italy, the average was 5.9 for sheep and goats, 1.1 for cattle, and 1.1 for horses (Fico et al. 1993). Wolves killed 3 sheep per attack in Tuscany (range 1–18), excluding incidents in which some sheep were killed but not eaten (Ciucci and Boitani 1998b), and 7.6 sheep were killed per attack in Spain (Telleria and Saez-Royuela 1989).

Wolves often kill far more domestic prey than they can eat, especially sheep (Pulliainen 1965; Zimen 1981; Boitani 1982), reindeer (Bjärvall and Nilsson 1976), and turkeys (Fritts et al. 1992). Wolves killed or injured 34 sheep and 200 turkeys in a single night in Minnesota (Paul and Gipson 1994). Turkeys often panic and concentrate in corners of their pens, where hundreds may

TABLE 12.1. Annual rates of depredations on cattle and sheep in selected areas

Area	Period	$\bar{x}$ no. of cattle killed	Cattle killed per 10,000	$\bar{x}$ no. of sheep killed	Sheep killed per 10,000	Reference
Alberta	1974–90	235	8.9	31	31	Dorrance 1982; Gunson 1983; Mack et al. 1992
British Columbia	1978–80	137	2.3	26	5.4	Tompa 1983a; Mack et al. 1992
Minnesota	1979–97	41	2	42	26	Fritts et al. 1992; W. J. Paul, un-published data
Montana	1987–97	4.6	0.2	3.8	1	Niemeyer et al. 1994; USFWS 1998[a]
Wisconsin	1990–97	3	0.3	1.1	3	Wisconsin Department of Natural Resources 1999
Spain (mountainous area)	—	—	—	—	13	Telleria and Saez-Royuela 1989
Tuscany, Italy	1991–95	30	2	2549	35	Ciucci and Boitani 1998b
North Karelia, Finland	1959–63	11	3	103	32	Pulliainen 1965[b]
Mongolia[c]	1993–97	24	120	121	87	Hovens et al. 2000
India	—	—	—	—	250–670[d]	Shahi 1983

[a]See also Bangs et al. 1998.

[b]Adapted from that publication.

[c]In and around the Hustain Nuruu Steppe Reserve.

[d]Sheep and goats combined.

die of suffocation. Surplus killing (see Mech and Peterson, chap. 5 in this volume) resulted in 21–113 sheep being killed per attack in Tuscany (Ciucci and Boitani 1998b) and up to 80 in Czechoslovakia (Hell 1993). Excess killing leaves the impression that wolves kill "for fun" and are wasteful, thus enhancing the negative attitude of livestock producers.

Selection of Domestic Prey

Wolves killed more sheep than cattle where both were available in Finland (Pulliainen 1963), and more goats than sheep in India (Kumar and Rahmani 2001) and Portugal (Vos 2000). Depredation rates on sheep (loss/availability) in Minnesota, Alberta, and British Columbia were about 5–10 times higher than on cattle (Mack et al. 1992). In the Carpathian Mountains of Poland (Bobek 1995) and in Tuscany, Italy, 97% of the livestock killed in recent years were sheep (Ciucci and Boitani 1998b).

With cattle, horses, and reindeer, wolves usually attack the young. Calves constituted 67–85% of all cattle killed by wolves in Minnesota, Alberta, British Columbia, and the U.S. northern Rockies (Dorrance 1982; Gunson 1983; Tompa 1983a; Fritts et al. 1992; Mack et al. 1992; C. C. Niemeyer, USDA Wildlife Services, personal communication) and 100% in Wisconsin (Treves et al. 2002). In contrast, wolves appear to select adult sheep and goats

rather than lambs and kids (Gunson 1983; Fico et al. 1993).

Seasonality of Losses

Most livestock are killed during the summer grazing season, which is fairly short in northern areas. Because livestock tend to be on open range longer in more southerly areas, the depredation season there is not as sharply defined. About 83% of all verified losses in Minnesota occur from May through September, when cattle, sheep, and turkeys are on summer range. Depredations on cattle in Minnesota peak during the calving season in May and June; sheep losses peak in July and August; and most turkeys are killed in August and September (Fritts et al. 1992). In western Canada, wolves kill more calves in mid-to late summer than in other seasons (Dorrance 1982; Carbyn 1983a; Gunson 1983; Tompa 1983a; Mack et al. 1992).

In Italy and Spain, wolves attack sheep and goats mainly during August and September, when flocks are on pasture (Brangi et al. 1992; Fico et al. 1993; Telleria and Saez-Royuela 1989; Ciucci and Boitani 1998b). Wolf attacks on cattle in the Abruzzo Mountains of Italy occur mainly during the May calving season (Fico et al. 1993), but attacks on calves continue through September. In Spain's Cantabrian Mountains, wolves concentrate on cattle, horses, and sheep in summer (Vignon 1995). Most attacks on horses occur during the foaling season (Lampe 1997). The increasing amount of food required by growing wolf pups probably explains the relatively high losses of sheep in August and September.

Wolf Behavior and Livestock Depredations

Considering the availability of relatively vulnerable livestock, why don't wolves kill more of them? Wolves often spend considerable time near livestock without showing much interest in them. Hundreds of wolves in North America surely pass near livestock in their daily travels, especially in summer, yet rarely take advantage of what would seem to be an easy meal. The territories of some recolonizing Minnesota packs bordered farms with livestock, but the wolves were not known to kill livestock or even to venture into open pastures (Fritts and Mech 1981). Since 1980, wolves have occupied the North Fork of the Flathead River in northwestern Montana, where residents raise cattle and horses, but wolves have killed none to date (D. Pletscher, University of Montana, personal communication). A pack territory in Montana's Ninemile Valley includes both private land and grazing leases with hundreds of cattle, yet wolves killed only two during 9 years (M. Jimenez, University of Montana, personal communication). A newly formed pair of wolves denned on Montana's East Front in the middle of an open pasture used by dozens of cows and calves, but walked past the cattle to hunt elk and deer instead (Diamond 1994) and did not kill any cattle for about a year (J. Fontaine, USFWS, personal communication). Biologist Jim Till watched as a radio-collared Montana wolf sighted a calf and immediately charged toward it, only to come to a stop within 2 meters of the startled animal and then casually walk away. In Wisconsin, R. P. Thiel (personal communication) watched a pack walking single file through a herd of cattle, with no apparent reaction by either predator or prey.

These observations and many others tell us that wolves often react to livestock differently than to wild prey. The difference may have something to do with exposure to livestock. Because livestock often inhabit the wolf's environment for only part of the year, wolves may not become sufficiently familiar with them to react as they would to wild prey.

Typically, when North American wolves do prey on cattle, they kill only a few and then resume hunting wild prey. Cattle may not be attacked again for several weeks, if at all. More vulnerable animals such as sheep, goats, pigs, and turkeys seem to be taken more regularly. Wolf packs in Minnesota sometimes move their pups close to flocks of turkeys in August and September, with the apparent intention of preying on them for an extended period (Fritts et al. 1992).

Few attacks on livestock are actually witnessed, partly because most occur at night (Lampe 1997; Ciucci and Boitani 1998b; Vos 2000). Determining the age, breeding status, or number of wolves involved is rarely possible (Fritts et al. 1992). There is little evidence that wolves that kill livestock are old, injured, or otherwise less able to kill wild prey (Fritts et al. 1992). Pups apparently do not kill livestock in their first summer, except perhaps poultry and small lambs (W. J. Paul, USDA/ WS, personal communication).

Husbandry and Depredations

Higher levels of depredations are associated with certain husbandry practices. Untended livestock in remote pastures sustain the highest losses from wolf depredations in both North America and Europe (Fritts 1982; Dorrance 1982; Bjorge and Gunson 1983, 1985; Stardom 1983; Tompa 1983a; Blanco et al. 1992; Paul and Gipson 1994; Bangs et al. 1995; Okarma 1995; Ciucci and Boitani 1998b; Vos 2000). In Alberta, Canada, wolves killed three times more cattle on heavily forested, less managed grazing leases than on pastures where most trees had been removed and cattle were managed intensively (Bjorge 1983, but cf. Mech et al. 2000).

Newborn livestock in remote locations are also much more likely to be killed by wolves (Hatler 1981; Fritts 1982; Tompa 1983a; Paul and Gipson 1994, but cf. Mech et al. 2000). Therefore, delaying the release of newborns onto spring pastures is one way farmers can sometimes reduce losses (Fico et al. 1993; Paul and Gipson 1994).

Poor surveillance of livestock is the most important factor associated with wolf depredations in Italy (Ciucci and Boitani 1998b), Spain (Blanco et al. 1992; Vilà et al. 1995); Karelia (Pulliainen 1963, 1993), Romania (Ionescu 1993), and Russia (Bibikov 1982, 1994). Untended livestock do not always suffer heavy losses, however, even in areas with high wolf populations (Mech et al. 2000). For example, only about 50 sheep are lost each year in the Bieszczady Mountains in Poland, even though large flocks of sheep and other livestock graze untended each summer (Perzanowski 1993).

A third factor increasing the risk of wolf depredations

may be the presence of livestock carcasses (Hatler 1981; Fritts 1982; Tompa 1983a; Bjorge and Gunson 1985; Fritts et al. 1992, but cf. Mech et al. 2000). Carcasses or other edible refuse can attract wolves. In Minnesota, there were several instances in which wolves killed young calves near cattle carcasses close to farmyards (Fritts 1982). Wolves conditioned to livestock in this manner often subsequently kill livestock on neighboring farms (Fritts 1982; Tompa 1983a). Robel et al. (1981) found that sheep producers who buried carcasses or had them hauled away lost fewer sheep to coyotes. However, a study in Minnesota produced equivocal evidence about the importance of carcass disposal in reducing wolf depredations (Mech et al. 2000).

Misperceptions about the Depredation Problem

Agriculturists generally view wolves as relentless killers of livestock. When a few wolves recolonized Scandinavia in the 1970s, there was an uproar (Bjärvall 1983). Wolves are often blamed for depredations even when evidence points to other predators, including coyotes (Fritts and Mech 1981; Thiel 1993), dholes (Fox and Chundawat 1995), and, especially in Europe, dogs (Salvador and Abad 1987; Magalhaes and Fonseca 1982; Boitani and Fabbri 1983).

Wolf involvement was confirmed in 36% of the complaints of wolf depredation in Alberta (Gunson 1983), 49% in Wisconsin (Treves et al. 2002), 25% in northwestern Montana (E. E. Bangs, USFWS, personal communication), 55% in Minnesota (Fritts et al. 1992; W. J. Paul, USDA/WS, unpublished data), and less than 50% in Italy (Zimen and Boitani 1979; Boitani 1982). Cattle producers in eighteen western U.S. states reported losses of 1,400 cattle to wolves in 1991 (National Agricultural Statistics Board, USDA, 1992), 1,200 of which were reported in states where wolves did not exist (Bangs et al. 1995).

In some newly colonized areas, however, wolves have lived up to their reputation as the archenemy of agriculture. When they kill excessively in reoccupied range, their exploits draw strong attention. For example, a wolf killed 80–100 reindeer in Sweden during one month in 1977 (Bjärvall 1983). The first pack to colonize France's Mercantour National Park killed 36 sheep in the first year (Lequette et al. 1995).

Even experienced investigators cannot always identify wolf depredation from evidence at a kill site. Clues used to help identify predators include tooth marks,

placement of bites, pattern and extent of feeding, and tracks, scats, and hair left near the carcass (Roy and Dorrance 1976; Wade and Brown 1982; Acorn and Dorrance 1990; Paul and Gipson 1994). Wolves usually bite large cattle and horses on the hindquarters, flanks, and upper shoulders. Young calves and sheep are usually bitten on the throat, head, neck, back, or hind legs (Acorn and Dorrance 1990; Paul and Gipson 1994).

Economic Impacts and Compensation Programs

Carbyn (1987) estimated that wolves in North America cause livestock damage of $280,000 to $320,000 annually, equivalent to about $6 per wolf. The annual market value of losses in Poland is estimated to be U.S.$32,900 (Bobek 1995). Bibikov (1994) estimated that the value of livestock losses in ten Russian regions in January–May 1986 totaled 2,438 million rubles.

Published estimates of damage on a per wolf basis vary widely, with the highest levels reported in Spain (U.S.$2,773/wolf/year) and Italy (U.S.$1,200–3,200/wolf/year) (Blanco et al. 1992). A few farmers and ranchers usually experience a disproportionate share of the losses in a given area.

Compensation programs (Fritts 1982; Fischer 1989) or state insurance (Lampe 1997) help offset economic losses in some areas. From 1977 through 1997, $658,260 was paid for wolf damage to livestock in Minnesota, Wisconsin, Montana, Wyoming, and Idaho, of which $81,270 was for 1997 losses. This amounts to about $30 per wolf per year in the contiguous United States, assuming a total of 2,700 wolves in 1997. In 1989, U.S. livestock and poultry producers reported losing $138 million to predators (Wywialowski 1994), suggesting that wolves account for about 6/100 of 1% of the total losses to predation.

Defenders of Wildlife, a nongovernmental organization, established a compensation program to help lower resistance to wolf recovery in the U.S. northern Rockies (Fischer 1989; Fischer et al. 1994). From 1987 through 2000, it paid 134 ranchers $149,415 for the loss of 173 cattle, 385 sheep, 5 equids, 10 guarding dogs, and 8 herding dogs (H. Fischer, Defenders of Wildlife, personal communication). This program also reimbursed ranchers in the northern Rockies for hay to lure cattle away from a wolf den and for an electric fence, and paid two landowners $5,000 each for allowing wolves to den and raise pups on their property (H. Fischer, Defenders of Wildlife, personal communication). Compensation pay-

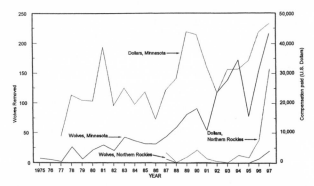

FIGURE 12.3. Numbers of wolves destroyed or placed in captivity by government programs in Minnesota and the northern Rockies because of depredation on livestock and compensation paid to livestock producers for wolf depredations. Additional wolves were captured, but were released on site or translocated. Most problem wolves in the northern Rockies were translocated or released on site.

ments will continue to grow as wolf populations increase (fig. 12.3). Defenders had also paid $6,008 in compensation for livestock and dogs killed by Mexican wolves reintroduced in the southwestern United States as of December 2000.

Compensation payments are high in Europe. Spain expends U.S.$1–$1.5 million annually for damage caused by a population of about 2,000 wolves (Vilà, Castroviejo, and Urios 1993). In Tuscany, Italy, annual compensation for wolf and dog depredations averaged U.S.$345,000 during 1991–1995; this figure includes damage caused by 80–100 wolves (Ciucci and Boitani 1998b). Payments are highest in Greece, where the government paid full compensation for 2,729 cows and 21,000 sheep and goats from April 1989 through June 1991 (Papageorgiou et al. 1994).

Despite its inherent problems, compensation does play a role in wolf conservation, especially in Europe, where wolf control is not legal (Promberger and Schröder 1993). Fair and timely compensation can help reduce animosity toward wolves. Without it, wolves probably would not survive in some places, but it is not a long-term solution (Wagner et al. 1997). The cost may increase to the point at which the public will demand reduction in payments or in numbers of wolves (Mech, Fritts, and Nelson et al. 1996). Most western European programs do little to ameliorate wolf-human conflict (Lampe 1997), so compensation is viewed as offering only temporary relief rather than an enduring solution (Cozza et al. 1996). In fact, compensation programs

could actually "encourage a state of permanent conflict" (Ciucci and Boitani 1998b) and could result in subsidizing wolf populations that then increase, making the problem worse.

An innovative compensation program has been implemented in the northern half of Sweden, where the Sami have their traditional reindeer-herding areas; the yearly loss to all large predators there is as high as 20,000 reindeer, although most of the damage is done by wolverines and bears. Since 1996, compensation for damages caused by large predators has been paid on the basis of verified reproduction or confirmed presence of predators in community grazing areas, and compensation is paid to the local Sami community rather than to individual reindeer owners (Berg and Bjärvall 2000). During 2000, the total cost for this system was 35 million SEK (U.S.$3.3 million in 2001). Compensation for losses of livestock other than domestic reindeer is paid on the basis of animals killed, but the county administrative boards also contribute funds for measures to prevent damage by predators.

Depredation Control

When wolves prey on livestock, some form of wolf management is usually inevitable, whether lethal or nonlethal, legal or illegal. If the government does not act, livestock owners often try to resolve problems themselves, which can mean indiscriminate killing of wolves. Many biologists believe that government removal of problem wolves is in the best interest of wolf conservation (Mech 1995a). Wolf management in response to depredations on livestock can take several forms.

Lethal Control Methods

It is important to remember that the low rates of livestock losses in the recent past in North America generally happened while there was some degree of lethal wolf control. Depredations would certainly be much higher if not for the removal of problem wolves. Killing wolves to reduce livestock depredation is generally tolerated by the American public (Kellert 1985, 1999), but is subject to increased scrutiny, and the public would prefer nonlethal methods if any were effective (Gilbert 1995; Kellert 1999). However, no consistently effective nonlethal method is anticipated soon (Mech, Fritts, and Nelson 1996).

In the contiguous United States, where the wolf is listed as either "endangered" or "threatened," only government agents can legally kill or translocate wolves. An exception is that members of "experimental/nonessential" populations in the northern Rockies, Arizona, and New Mexico can be shot by livestock owners if found in the act of killing livestock.

In Canada, wolf control is conducted by government agents and the public. Lethal control by government agents can be either general or specific. General control attempts to prevent losses by removing wolves from certain areas, whereas site-specific actions target only problem wolves. Site-specific control has little effect on wolf populations, and its results are often short-lived. In 2000, all government wolf livestock depredation control in the United States and Canada was reactive and site-specific.

The number of problem wolves removed in several Canadian provinces was less than 100 per year from 1987 to 1991, totaling less than 1% of the population (Hayes and Gunson 1995). The number of wolves euthanized in Minnesota has increased steadily during the past 20 years, with an average of 152 killed annually from 1995 to 1999 (W. J. Paul, USDA/WS, unpublished data). Thus about 5% of Minnesota's wolf population is killed each year to keep livestock depredations in check, at an annual cost of $255,000 in 1998 (Mech 1998b). In Montana, about 6% of the wolf population is removed annually, at a cost of $19,000 (Bangs et al. 1995, 1998). The cost of wolf management in the western United States will undoubtedly increase now that wolves also inhabit Wyoming, Idaho, Arizona, and New Mexico. During fiscal year 2000, Wildlife Services in Idaho spent $135,880 on wolf control.

Steel-jawed foot traps (Mech 1974b) are used to capture essentially all problem wolves in Minnesota and Wisconsin, but are illegal in Europe. In Montana, 42% of wolves taken for control were taken with traps, and 58% were captured by helicopter. Helicopters can be an extremely effective tool, either to dart and drug wolves or to kill them. This technique, in combination with trapping, has satisfied ranchers' doubts that wolves can be controlled in the western United States (Niemeyer et al. 1994).

Poisons can be effective, inexpensive, and highly selective in removing problem wolves, although they are poorly regarded by the public. Poison (strychnine and compound 1080) for predator management was banned in the United States in 1972 (Dunlap 1988), but along with traps and snares, is still used on a limited basis in Alberta (Gunson 1992; Hayes and Gunson 1995). Poison

is also used legally in many parts of Russia, the Middle East (including Saudi Arabia, but not Israel), and India and illegally in many parts of Europe, including Spain, Portugal, Italy, and Greece.

Cyanide and strychnine are hard to obtain in Europe, but livestock owners can buy several anticoagulants used to kill rodents. These poisons provide an easy substitute for traps, which are more conspicuous and difficult to use. Before the collapse of the Soviet Union, aerial shooting from helicopters was widely used, but this method has been discontinued because of its high cost. Poison is now preferred and is applied even in protected areas (D. Bibikov, Institute of Animal Evolutionary Morphology and Ecology, personal communication). Poisoning is on the increase in parts of India, and stone pits or deadfall traps are also commonly used there (Fox and Chundawat 1995). Wolves caught in pit traps are killed with stones.

Wildlife managers are sometimes pressured by livestock producers to exercise more lethal control than needed or allowed by law. Clear guidelines governing how wolf control actions can be conducted make the jobs of field personnel easier. Legal actions against the USFWS in Minnesota helped clarify the circumstances under which management of a "threatened" wolf population can occur (Fritts 1982; O'Neill 1988). USFWS regulations required that wolves be killed only *after* they had committed "significant depredations on lawfully present domestic animals" (USFWS 1978).

Nonlethal Methods of Preventing Losses

Several nonlethal methods have been tried for alleviating livestock losses, but none has proved consistently effective (Fritts 1982; Mech, Fritts, and Nelson 1996; Bangs and Shivik 2001). Translocating wolves is an option where lethal methods are illegal or a wolf population is so low that every wolf needs to be saved. However, most translocated wolves move extensively after being released (Fritts et al. 1984, 1985), and the USFWS has recommended that translocations be discontinued in the northern Rocky Mountains (Bangs et al. 1998). Bringing wolves into permanent captivity has also been suggested, but wild wolves adjust poorly to confinement; euthanasia is probably more humane.

One of the oldest nonlethal methods of preventing depredations involves guard dogs. They have been used in Eurasia for centuries, and can be quite effective as long as they are used by trained shepherds (Coppinger and Coppinger 1982; Adamakopoulos and Adamako-poulos 1993; Hell 1993; Vilà, Castroviejo, and Urios 1993; Promberger et al. 1997; L. Boitani, personal observation). However, in the U.S. northern Rockies, where shepherds are rarely present, wolves have killed several guard dogs (Bangs and Shivik 2001). Promberger et al. (1997) cited inadequate numbers of dogs per herd, inadequate training, proximity of bedding ground to forest, and absence of shepherds as important factors limiting the effectiveness of guard dogs.

Lampe (1997) concluded that guarding of livestock, when done correctly, was effective in reducing losses in Europe. Interestingly, losses appear to be lower in parts of Europe where wolves were never extirpated. There, livestock producers never lost the "know-how" to protect their herds nor developed the attitude that the government should assist them in dealing with wolves.

Many other nonlethal techniques have also been tried. Taste aversion (Gustavson 1982; Gustavson and Nicolaus 1987) did not appear to be effective in Minnesota, and its application elsewhere has declined (Conover and Kessler 1994). The Minnesota program experimented with blinking highway lights, light-siren devices, and surveyors' flagging on fences to simulate "fladre" used in Europe for funneling wolves during hunting (Fritts 1982; Fritts et al. 1992). Some methods appeared useful in some instances, especially in small pastures, but none was consistently effective. Recently, however, closely spaced and well-maintained fladre seem to have succeeded in preventing livestock losses to wolves on a few ranches (Musiani et al., in press). Fencing, propane exploders, cracker shells, pyrotechnics, diversionary feeding, and other techniques (Cluff and Murray 1995; Bangs and Shivik 2001) have met with only limited success because wolves habituate to them. Fertility control might be useful to limit pup production and wolf density in disjunct wolf populations near livestock (Mech, Fritts, and Nelson 1996; Haight and Mech 1997).

Future Outlook
Controlled experiments to test the effectiveness of different control methods are sorely needed (Fritts et al. 1979). The number of spatial and temporal variables involved make these tests difficult to design. Experience in British Columbia (Tompa 1983a,b), Alberta (Bjorge and Gunson 1985), Minnesota (Fritts et al. 1992), and Montana (Bangs et al. 1995) indicates that the reactive, site-specific wolf removal currently being used usually reduces future livestock depredation problems.

Despite short-term success with a site-specific approach in Minnesota, biologists recommended preventive control where several turkeys and sheep were killed almost every year (Fritts et al. 1992). A zoning system in which the level of control is based on the depredation potential is probably the most effective way to limit losses (Mech 1995a). This approach includes preventing wolves from colonizing areas where the potential for depredation is high, as recommended by the USFWS Eastern Timber Wolf Recovery Team (USFWS 1992). The cost of wolf presence in agricultural areas and the resulting ill will could be substantial and could undermine wolf conservation in the long run (see Boitani, chap. 13 in this volume).

A combination of zoning for wolf population control, indemnity payments, lethal and nonlethal control methods, animal husbandry modifications, and research offers the best hope of balancing wolf conservation with livestock production. At the extreme, some livestock producers may be able to bring livestock into shelters or remain with them overnight. The willingness of farmers and ranchers to make such changes, however, ultimately depends on the cost, the potential for future losses, the feasibility of changing husbandry practices, and the availability of compensation (see Boitani, chap. 13 in this volume).

Wolf Politics and Conflicts among Humans

The conservation and management of wildlife is a complex endeavor in which the biology of animals interacts with human values (Nie 2003). Whether an animal population is lost, restored, or ignored usually reflects human decisions and actions. Wolf management is especially challenging, not only because wolves cause socioeconomic problems, but also because of the universally contrasting viewpoints about wolves. The wolf is one of the most studied mammals there is, and we have most of the information needed to manage it (Mech 1995e). Our understanding of the human aspects of wolf management, however, is more limited, and the application of policy development, mediation, and conflict resolution has only begun (Clark 1993; Haggstrom et al. 1995).

Canis lupus politicus
Wolves have been the subject of political attention since the first bounty was established by Solon of Greece in the sixth century B.C. Predation on livestock has prob-

ably generated more furor than any other facet of wolf-human relations, with wolf control to maintain wild ungulate populations running a close second. Political debate that, on the surface, is about wolves often involves underlying issues that reflect conflict within human societies, especially rural-urban differences.

In some instances, lawmakers recognize exaggerated claims by the livestock industry but ignore scientific data. On the other hand, some wolf advocacy groups minimize existing and potential problems and misinform their members and the public (Blanco 1998; Mech 2000b). Legislators from urban areas, and their constituents, may not sympathize with farmers or hunters in distant parts of the nation, or understand the need to manage wolves.

Throughout most of history, finding consensus and taking action against wolves was easy because most people either supported reducing wolves or didn't care. However, wolf management has become increasingly complex and contentious in recent years. The difficulty in simply defining the degree of protection for wolves in various parts of the world is a good example. Application of the IUCN (2000) classification for threatened species considers only biological criteria. However, individual countries necessarily operate on a national or regional scale and encompass different levels of governmental authority. As the scale becomes smaller, local opinion becomes a greater factor. This is evident in Europe, where local attitudes toward wolves are predominately negative. In North America, national pro-wolf interests now usually dominate local interests. This creates regional hardships and animosity and works against wolf conservation in places such as Alaska, Minnesota, and western Canada where wolf populations are secure and thriving.

Local versus National Interests

Regional and local interests continually compete with national or biological considerations. "State's rights" issues can also come into play. For example, Wyoming legislators tried to reinstate a wolf bounty in 1995 in response to the reintroduction of wolves to Yellowstone National Park. At the national (federal) level, the wolf's legal status reflects the status of wolves nationwide, as well as a national view that wolves should be protected and restored. At the local level (e.g., Montana, Wyoming), the livestock industry and other interests influence state governments, although national laws supersede local laws. A similar conflict is evident in most European nations where small numbers of wolves are present; national law is often resented locally. Ideally, a global conservation strategy would be based on population biology and implemented regionally according to local priorities.

One fundamental change in the roles and responsibilities of governments and individuals in wolf management should be noted, however. The payment of compensation for wolf damages is a fairly new development, and the change happened at about the same time as the introduction of economic incentives and subsidies for agricultural products. Although having society share the costs of wolves seems socially appropriate (livestock producers cannot be asked to bear the costs alone, especially when laws prevent them from protecting their interests), this policy leads to a philosophical dilemma. Currently, any damage from natural calamities can be the object of a compensation claim in Europe. This policy increases the separation of humans from the natural environment on both ideological and practical levels.

Local Economies, Conservation, and Wolf Management

Here we explore the relationship between large populations of wolves and the way in which local values are considered in wolf management decisions. Earlier we recognized the importance of depredation control and compensation in making wolf recovery possible where livestock occur. This principle also applies where extensive populations of wolves and wild ungulates coexist with people, and where big game are as important to local economies as livestock are in other areas. Nevertheless, there is a wide range of opinions on and reactions to wolf management in wild systems (Gasaway et al. 1992; Haber 1996; Theberge 1998).

Most of the 60,000 or so wolves in North America inhabit Alaska and Canada, preying primarily on wild ungulates. Although human density is low, hundreds of small communities and dozens of cities are scattered throughout this vast area. Both Native and non-Native people depend on local wildlife resources for economic, material, and spiritual sustenance. Agricultural potential in these northern communities is almost nonexistent, and harvesting local fish and wildlife is a long-standing tradition (Weeden 1985).

The concerted efforts to eradicate wolves that peaked during the late nineteenth century continued during the 1930s–1960s in western and northern Canada and Alaska (Pimlott 1961; Harbo and Dean 1983; Carbyn 1987; Hayes and Gunson 1995). Since then, attitudes toward wolves have improved, as discussed earlier. In recent decades, only small-scale, temporary wolf control programs have been implemented, and they have adhered appropriately to Principle 7 of the IUCN Wolf Specialist Group's Manifesto on Wolf Conservation, which sets out new, rigid scientific guidelines and conditions for wolf population management. Several such programs in western and northern Canada (Hayes and Gunson 1995) and Alaska (Stephenson et al. 1995) were conducted to allow low or declining ungulate populations to recover.

The Manifesto on Wolf Conservation was of paramount importance to these control programs because its Principle 7 set out clear guidelines for wolf management developed by international conservation authorities. Nevertheless, these efforts generated intense controversies, reflecting fundamental differences in values between rural and urban people. These differences should be evaluated in light of the emerging understanding of the role of local economies and sustainable use of local resources, including wildlife, in long-term conservation strategies.

Biodiversity, Wolf Management and Traditional Uses

The World Conservation Union's (IUCN's) mission statement, called the World Conservation Strategy, includes as a primary objective "to ensure the sustainable utilization of species and ecosystems." The IUCN Specialist Group for the Sustainable Use of Wild Species was formed in the 1990s to promote sustainable local use of wildlife as a primary goal. Sustainable-use principles have been incorporated into conservation biology because classic preservation (i.e., parks and reserves) often failed to protect wildlife and ecosystems, perhaps most notably in Africa (Leader-Williams 1990; Lewis et al. 1990; Saether and Jonsson 1991; deBie 1990).

Conservation models recognizing economic uses by local people have benefited the conservation of African elephants (Leader-Williams and Albon 1988; Leader-Williams 1990; de Meneghi and Kaweche 1990) and of threatened wildlife in South America (Robinson and Redford 1991). Human use of nature can even play an integral role in maintaining biodiversity (Wilson 1992;

Berry 1977, 1987, 1992). However, this approach has not been widely considered in North America.

The definition of conservation embodied in the World Conservation Strategy contrasts with predominant environmental attitudes in developed countries, where the use of renewable resources, particularly large mammals, tends to be regarded as unnecessary and undesirable. The alternative view holds that maintaining healthy economic relationships between human societies and wildlife provides both incentives for conservation and an environmentally sound alternative to the conversion of wild systems to other uses—including domestic food production. The effectiveness of such a conservation model in the North has not yet been objectively evaluated (Herscovici 1985).

These viewpoints conflict with the widely held view of lay environmental organizations that wildlife conservation is best served by minimizing or eliminating consumptive use. Gilbert (1995) concluded that rural people tend to value wildlife for their own use and see wolf control as "utilitarian" because it reduces competition for wild food. Urban people value wolves for "naturalistic" reasons, assuming that "natural" systems are better than managed ones. Wolf predation, however, can hold prey densities down for extended periods, during which little or no harvest is available to people (Hayes and Gunson 1995; Stephenson et al. 1995; see Mech and Peterson, chap. 5 in this volume).

When does wolf control conflict with wolf conservation? According to IUCN principles (Pimlott 1975), such conflict occurs when control is protracted, indiscriminate, and not biologically justified. However, indiscriminate wolf control ended in North America in about 1950, and it is unlikely to recur as long as the current affluence, the resulting availability of alternative resources, and contemporary environmental attitudes prevail.

Mech (1995a), Bangs et al. (1995), and each American wolf recovery plan recognized that some wolf control is needed to provide balance and thus foster local support for recovering wolf populations in the United States. Likewise, wolf recovery in Europe may ultimately depend on the removal of dispersing wolves from densely populated agricultural areas (Boitani 2000). In recent decades, however, environmental groups have campaigned and litigated against wolf control and have usually prevailed over the interests of rural communities.

Wolves and Local Wildlife Values

One of the important results of opposing wolf control relates to the environmental, social, and economic costs of reducing the amount of renewable sustenance obtainable from natural systems by local people (Weeden 1985). Reducing use of natural systems increases reliance on energy-intensive domestic food production and distribution systems that carry a high environmental cost, are not sustainable, and diminish or eliminate wildlife habitat. In other words, when wolves are not controlled, local residents must rely more on domestic than on wild foods, which will affect the environment and natural systems elsewhere.

A major concern of wolf biologists is that demands for complete protection for wolves increase the resistance to wolves and wolf recovery in additional areas (Mech 1995a). Wolf recovery plans recognized the importance of local support early on, and control programs were designed to respond to public demand for livestock protection (Fritts 1982; Bangs et al. 1995; Mech 1995a). However, the economic and political dynamics are the same whether wolves affect privately owned livestock or publicly owned wildlife that humans depend on.

A paradox in the modern debate about wolf management is that the recent expansion in wolf numbers and range in North America and Europe, and the policies that fostered it, were possible because of the affluence of these areas. However, these economies depend on the consumption of tremendous amounts of finite resources, intensive agriculture, and an elaborate transportation network. This affluence has allowed relative tolerance of the predators that sometimes compete directly with humans elsewhere. In less affluent countries, such as Russia, wolves and other predators have been jeopardized because conservation, and especially maintaining predators, is an unaffordable luxury. The future of wolves in affluent countries in many respects depends on how well the elaborate system of production and distribution endures.

In affluent countries, opposing wolf control can be morally gratifying, creating the illusion of doing something positive for the environment. However, few people today, including wildlife managers, support eliminating wolves to maximize ungulate harvest. Rather, the issue is one of balance, of providing a reasonable share of wildlife for both wolves and people. In addition, organized opposition to wolf control diverts attention from the more important and challenging issue of long-term maintenance of wild lands where wolves can live.

Alaska and Yukon Wolf Management Plans

Earlier we stated that intensive government control of wolves is declining, at least in North America and Europe. Hummel (1995) predicted that some current wolf control methods would end early in the twenty-first century. This may already be happening. Aerial killing of wolves by government agents is the most efficient, effective, and humane method, but is highly controversial (Boertje et al. 1995; Cluff and Murray 1995) and is not practiced anywhere in North America at present.

Recent control programs in Alaska and the Yukon involved killing 60% or more of the wolves in local areas for 3–7 years. While these efforts usually resulted in substantial increases in ungulates (Gasaway et al. 1992) and eventually benefited wolf populations, they also provoked controversy among people who may otherwise share the same long-term objectives for wildlife conservation. Moving beyond such impasses requires negotiation and compromise among diverse interests, as well as a shift from position-based arguments to interest-based negotiations about where, when, and how wolves should be managed.

Are local people capable of caring for northern wildlife—including wolves—or do they simply have a utilitarian view of wolves as unwanted competitors for wild food (Gilbert 1995)? Both Native and non-Native people living in wolf range are only beginning to articulate a rationale for sustainably using wolves and other wildlife, and part of the responsibility for making regional management decisions in the North is shifting from central governments to local communities.

As part of recent land claim settlements in Alaska and northern Canada, wildlife management policies are now based on principles of co-management. Governments, local resource councils, and First Nations share responsibility for wildlife management. Rather than ignoring local values and engendering further opposition from northern communities, some urban conservation interests are pursuing a dialogue with local communities.

When allowed to influence the direction of wolf conservation, northern people have developed progressive and balanced plans. During the 1990s, public planning teams in Alaska and the Yukon Territory produced wolf management plans that were initially well received (Yukon Wolf Management Planning Team 1992; Haggstrom et al. 1995). Both plans established rigid guidelines for ensuring long-term wolf conservation and guarded against unnecessary wolf control by limiting its scale and duration. The plans also recommended ways to increase human respect for wolves through education, more conservative wolf hunting laws, and recognition of nonconsumptive values. However, both plans were eventually opposed by environmental groups that wanted no wolf control at all.

Another example of successful interest-based negotiation is the Fortymile caribou recovery plan in Alaska and Yukon (Todd 1995; Fortymile Caribou Herd: Management Plan 1995). The goal of restoring this once abundant herd was supported by local residents, conservation groups, and Alaskan and Canadian wildlife agencies. A planning team of government officials and members of the public recommended some new approaches to ungulate recovery. First, the plan was based on principles of fairness and respect for differing views. Rather than attempting to effect a short-term, large-scale increase in caribou, the plan outlined a more moderate approach that required less intrusive methods. Intensive aerial control of wolves was rejected in favor of a combination of public trapping, experimental fertility control, and translocation of wolves. Wolf population control was limited to the caribou herd's post-calving range, where reduced predation was most likely to increase calf survival and herd recruitment. Caribou harvest was reduced to the level required to meet minimum subsistence needs. Public support for the plan was widespread, although some U.S. environmental groups opposed it.

The Fortymile planning process illustrates how consensus on some issues can be found. Compromises and concessions were derived by establishing a fair balance of urban and rural values. A key compromise by rural people involved substituting the experimental technique of fertility control for more effective aerial control. Local people also agreed to reduce the scale of wolf control and to further restrict caribou hunting to show respect for wolves. Environmental groups compromised by recognizing that local people have legitimate concerns and an interest in participating in wildlife conservation, and by accepting wolf reduction by local trappers.

Conclusion

Many factors, both historical and current, are involved in humans' perceptions of wolves. The status of wolf populations in much of the world has improved in recent decades, largely because human societies have become more urban and affluent and more tolerant of the species. However, attitudes toward wolves continue to be diverse, and the wolf-human relationship is often strained. Wolves are revered as a symbol of wilderness and ecological harmony by some, while others regard them primarily as a threat to human interests. In many parts of the world, especially where livestock are a means of economic survival, people continue to have an antagonistic relationship with wolves that is not likely to change in the foreseeable future.

The wolf's future depends to a large degree on how the values and economic interests of people that live in wolf range are incorporated into wolf management. It also depends on the future status of human economies in North America and Eurasia, and on the degree to which wolf populations can be managed in a way that will maintain predominantly positive, or at least neutral, attitudes toward them. The long-term coexistence of wolves and people will benefit if depredations on livestock and pets can be minimized, and if predation on populations of wild ungulates can be managed to allow a fair share of wildlife for both people and wolves. Ultimately, the survival and well-being of wolf populations will require negotiated compromises that balance the needs, values, and desires of different interest groups with the biological needs of the wolf.

The wolves we imagined at the beginning of this chapter would probably agree that their present circumstances are better than they have experienced for a long, long time. They would also be optimistic about their prospects for coexisting with the human race far into the new millennium. However, they would remain keenly aware that the two-legged species largely controls their destiny.

13

Wolf Conservation and Recovery

Luigi Boitani

ON THE NIGHT OF 26 June 1976, as on other nights, our radio-collared wolves were out roaming the roads, towns, fields, and woods of the Maiella, a mountainous area east of Rome in central Italy. I was following them with an antenna on my Land Rover at a distance of often no more than 50 meters. In the absence of large prey, the wolves made the rounds to alternative food sources, such as village dumps or the folds where sheep were kept at night.

Normally, the sheepdogs and the rickety fences were enough to discourage the wolves from attempting a raid: the wolves would approach the fold upwind, watch and wait for an opportunity for hours, and then simply slink away—without having caused any damage. Often the sheepdogs didn't even notice.

But there was something in the air that night. It was midsummer, and the fold had been left in the dogs' care while the shepherds went to town. I had the impression that the two wolves were determined in the way they approached the fold. In the darkness, I could only guess their movements from the signals and the sounds from about 250 meters away.

The dogs began barking, meaning they had discovered the wolves. But suddenly the barking became furious, and then it moved away from the flock. This, and the radio signals indicating full activity by the two wolves in the middle of the fold, meant that our wolves were busy. They left the site a half hour later, and I followed them to their rendezvous. The next day I discovered they had killed forty sheep.

I distinctly remember my mixed feelings during that half hour of silence broken only by the radio's beeps.

First, identifying with the wolves, I was excited by the success of the raid. But then, putting myself in the shepherds' shoes, I worried about the consequences. Sooner or later, a wolf would die because of this or some other raid. It is easy to be completely on the side of either the wolves or the shepherds. Going beyond that, to try to find a compromise, is much harder.

But that is often what conservation is all about: an attempt to reconcile the needs and requirements of wildlife with the needs, expectations, and desires of human beings. Even that is relatively easy when a species does not require large areas or special resources; it is hard, on the other hand, when a species conflicts directly with humans. With the wolf, a preconceived negative image in the culture and psychology of many people further complicates the problem (see Fritts et al., chap. 12 in this volume).

In general, one is most concerned about a species that is threatened with extinction, but about 200,000 wolves remain today. Why, then, is wolf conservation a problem? Why is there a large public movement committed to wolf conservation and recovery?

I will answer these questions in the following pages, but suffice it to say here that the wolf is irregularly distributed across a variety of diverse ecological, cultural, political, and economic settings, in which local pressure is often far stronger than any holistic planning. If wolf management could be planned globally, wolf conservation would be assured. However, the wolf's range is broken into a myriad of patches, each bounded by political borders imposing different management. No world authority for natural resources management has yet been set up. Thus wolf management is still a national or local prerogative, subject to the wide variety of biological, philosophical, and political approaches dominating local conservation policies. This fragmentation is one of the threats that justifies continued concern about wolf conservation.

Extermination of the Wolf

The wolf originally ranged over the entire Northern Hemisphere north of 13°–20° north latitude, including central Mexico, the Arabian peninsula, and southern India (Kumar and Rahmani 2001). This enormous range, the most extensive of any land mammal, was cut back considerably over the centuries by humans.

The first reports of planned action against the wolf date to the sixth century B.C., when Solone of Athens introduced a 5-drachma bounty for every male wolf killed and 1 drachma for every female. Among various human epochs and cultures, the motivations, techniques, and effects of the anti-wolf campaign have varied. In some cultures the war was purely defensive, while in others it lacked rational foundation. Persecution was carried out by diverse means, but most commonly by traps and poisons, of which there are innumerable variations and adaptations (Young and Goldman 1944; Cluff and Murray 1995).

A wide variety of traps have been used throughout the centuries, from the spiral corrals illustrated in the medieval book on hunting by Gaston Phebous to snares, stomach piercers, pitfalls, and the classic steel foot traps (Young and Goldman 1944). The shooting party, with or without nets, involving a large number of people, is still used in Poland (Okarma 1992) and has recently been used to capture wolves for research (Okarma and Jedrzejewski 1997). Hunting dogs were used mainly in open terrain, and the ancestors of the modern Irish wolfhound were bred for that purpose by the Celts as early as the third century B.C.

It was in the early Middle Ages that persecution of the wolf became an organized effort aimed at exterminating the largest number of wolves—an effort that continued until the late 1800s (Mallinson 1978; Zimen 1978). The fact that economies were based on nomadic sheep herding in central Europe and sedentary farming and livestock raising in southern Europe generated different attitudes toward the wolf (Boitani 1995). Periods of economic wealth and positive attitudes alternated with

more difficult times, when nature and wildlife were perceived as hostile forces. Movements of people across the continent with their traditions and attitudes contributed to the mix of viewpoints about the wolf.

The British Isles

The history of wolf extermination in Great Britain is ancient (Fiennes 1976). It was a very effective process driven by the constant and strong competition between humans and wolves. There were few alternatives to domestic animals as a source of wealth and sustenance, and the wolf was a constant threat. Wolf persecution was enforced by legislation, and England's last wolf was killed in the early sixteenth century under Henry VII (Harmer and Shipley 1902).

In Scotland, despite intense efforts to kill wolves, the immense Scottish forests offered safe retreats (Fiennes 1976). Scotland's final solution was to burn the forests (McIntyre 1993). A few wolves reportedly persisted in the forests of Braemar and Sutherland until 1684, the probable date of the species' extinction there. In Ireland, the story was similar, leading to the extinction of the wolf around 1770 (Harmer and Shipley 1902).

Central Europe

The wolves of central Europe were dramatically reduced during the eighteenth century, and the last animals were seen in the early nineteenth century. Large ungulates were also reduced to very low levels by organized hunts. Both persecuted and deprived of their prey, wolves eventually disappeared. In Denmark, the last wolf was killed in 1772. In Switzerland, wolves disappeared before the end of the nineteenth century, although some animals dispersing from other countries were seen as recently as 1946 in Sion, near Lake Geneva. The last Bavarian wolf was slain in 1847, and by 1899, the wolf had disappeared from the Rhine regions (Zimen 1978; Boitani 1986).

In France, Charlemagne issued two laws between 800 and 813 A.D. (*Capitolare de Villis* and *Capitolare Aquisgranensis*) establishing a special corps of wolf hunters, the "louveterie" (Hainard 1961). When a member of the corps killed a wolf, he had the right to payment from residents within two leagues (about 7 km, or 4 mi) around. Famous events, such as the famine and plague of 1033 and the incursion of wolves in Montmartre, near the center of Paris, on 17 December 1438, gave further impetus to the drive to exterminate the wolf.

The French Revolution in 1789 ended the louveterie, but the corps was revived in 1814. The number of wolves killed was as high as 1,386 in 1883. Many more were eliminated by poison (Victor and Lariviere 1980). The last wolves from the original population in France (and they may not even have been wolves) were spotted in March 1934 in the Chantal Forest near Saone-et-Loire (Beaufort 1987).

Scandinavia

In Sweden, the first wolf bounty was introduced in 1647. The wolf's main prey in Scandinavia, the moose and the reindeer, were soon exterminated, killed either for meat or because of competition with livestock. There, too, the wolf was left without wild prey and had to resort to domestic animals, further incurring human wrath. In Sweden's northern mountains, the Sami (formerly called Lapps) extirpated wolves using organized drives involving hundreds of people (Turi 1931; Promberger and Schröder 1993).

In the twentieth century, the use of snowmobiles left the wolf no chance; by 1960 only a few remained, although the moose population had in the meantime recovered to a peak of 70,000. The last wolf in Sweden was slain in 1966, after which the species was declared legally protected and eventually recolonized the country. The situation in Norway followed the same pattern, and its last wolf was killed in 1973; legal protection was then declared, followed by a parallel increase in wolves (Promberger and Schröder 1993).

Although wolves dispersing from Russia regularly enter Finland, its wolf population was also decimated during the last century. By 1900, the wolf was present only in the eastern and northern parts of the country (Palmén 1913). Following a population increase after World War II, wolf numbers fluctuated depending on the status of the wolf population in neighboring Soviet Karelia (Pulliainen 1965, 1985, 1993).

Eastern Europe

In eastern Europe, human population density and social and political organization never reached the levels they did in central Europe. Furthermore, the vastness and diversity of the region, along with its contiguity with Asia, prevented wolf extermination. Nevertheless, wolf populations in eastern Europe were reduced to very low numbers by the end of the nineteenth century. By the middle of the twentieth century, wolves were limited to a few forested areas in eastern Poland, with numbers probably at an all-time low (Okarma 1992). The wolf disappeared altogether in Slovakia in the first decade of the twentieth century, but later became relatively common in the eastern regions (Hell 1993).

In the eastern Balkan areas, wolves benefited from their proximity to the vast populations in the former Soviet Union, and the mosaic of plains, mountains, and farmlands allowed some populations to persist. In Hungary, the wolf's range covered less than half the country at the turn of the century, confined as it was to the Carpathian basin (Farago 1993).

In Romania, on the other hand, the wolf population remained relatively substantial (Ionescu 1993). Data before World War II are lacking, but from 1955 to 1965, an average of 2,800 wolves per year were killed in a population that averaged about 4,600. The all-time low was reached in 1967, when there were an estimated 1,550 wolves in Romania (Ionescu 1993). In Bulgaria, estimates ranged from about 1,000 animals in 1954 to 100–200 in 1964 (Bibikov 1988), confirming that extermination occurred relatively recently.

The small wolf populations in northern Greece are still connected to the larger Balkan population, but the southernmost Peloponnesian wolves disappeared in 1930 (Zimen and Boitani 1979; Adamakopoulos and Adamakopoulos 1993). In the rest of the western Balkans, from Albania to the regions of the former Yugoslavia, wolves never disappeared, despite periods of intense hunting during the eighteenth century. Indeed, the wolf could be found in almost all forested regions until the early 1900s (Adamic 1993). Organized persecution began only in 1923 with the setting up of a Wolf Extermination Committee in Kocevje, Slovenia. This campaign was remarkably successful in reducing the wolf population in the central part of the Dinaric Alps (Frkovic et al. 1988; Frkovic and Huber 1993).

Southern Europe

Attitudes toward wolves in southern Europe are notably different from those in the rest of Europe. Historical, geographic, and above all, cultural factors linked to the ecologies of the Mediterranean human populations have fostered relative tolerance toward wolves (Boitani 1995). As a result, persecution of wolves on the Iberian peninsula and in Italy was not nearly as complete as in central and northern Europe. On the Iberian peninsula, there are reports of hunts similar to those in France organized by Carlo VI in 1404. The wolf's range there, however, did not start to shrink until early in the 1800s, although by 1900 it was already only half its previous size (Blanco et al. 1990; Grande del Brio 1984).

In Italy, wolf bounties were paid regularly in the twelfth and thirteenth centuries and as recently as 1950. As in France, wolf hunting was financed by voluntary contributions from farmers where wolves were killed, and wolves were rarely hunted in uninhabited areas. But Italian wolf hunters had neither the organization nor the persistence of their French counterparts: they did not form a team with a special "esprit du corps." Most wolf hunters were farmers who, during winter, hunted wolves for their pelts (Boitani 1986).

Exterminated in the Alps by the late 1800s, wolves in Italy numbered no more than 100 individuals in 1973 and inhabited only 3–5% of their original range, primarily in rugged mountains (Zimen and Boitani 1975). Nevertheless, the wolf never became extinct in Italy, and the small surviving population made recovery possible (Boitani 1992; Boitani and Ciucci 1993).

Asia

Over most of the immense territories of the former Soviet Union, wolves were never exterminated, despite major fluctuations in their numbers and distribution associated with changes in human populations and the status of wild prey (Bibikov 1988).

Little historical information is available about wolf populations in south-central Asia. However, we can infer the pattern of extermination from the vast areas in which the species is no longer present, including 80% of China and India (Ginsberg and Macdonald 1990). In northern Inner Mongolia, both wolf numbers and range have been greatly reduced since 1940, primarily because of livestock depredations, but also because of the poaching of gazelles, the wolf's main wild prey there (Maruyama et al. 1996).

The Contiguous United States

The American Indian's relationship with the wolf was a fortunate one for the wolf; it was hunted, but also appreciated and respected (Lopez 1978). The campaign to exterminate wolves began in North America as soon as the Pilgrim fathers arrived from England with all the preju-

dices, beliefs, laws, and devices that had just eradicated the wolf at home (Young and Goldman 1944; Matthiessen 1959; Lopez 1978; Dunlap 1988; McIntyre 1995; Busch 1995). In-depth accounts have been provided of wolf extermination in the Southwest (Brown 1983), in Wisconsin and Michigan (Thiel 1993), and in the northern Rockies (Curnow 1969; Fritts et al. 1994).

The war against the wolf began officially in 1609, when the first livestock arrived in Jamestown, Virginia. Plymouth Colony enacted a wolf bounty in 1630, and bounties were soon established by all the other settlements along the eastern seaboard. By 1700, the wolf had disappeared from New England.

After 1750, new and more effective hunting devices, including Newhouse traps (1843) and poison (primarily strychnine), were introduced (Young and Goldman 1944; Cluff and Murray 1995). When the frontier moved west, hunters were temporarily diverted by the immense herds of bison. But the huge number of carcasses left by the hunters may have fostered an increase in the wolf population (Curnow 1969; McIntyre 1995). This, and a growing trade in wolf furs, soon revived interest in killing wolves.

Around 1870, the westward expansion of the livestock industry coincided with the disappearance of the huge buffalo herds. In Montana, the state perhaps most symbolic of this process, the last bison was killed in 1884, while cattle increased from 67,000 to 1.1 million and sheep from 300,000 to 2.2 million between 1867 and 1890 (Montana Agricultural Statistics Service 1992). The scarcity of natural prey caused an increase in wolf predation on livestock, and the extermination of all predators became one of the main concerns of landowners and local authorities (Rutter and Pimlott 1968; Dunlap 1988). Wolf hunting was even permitted in protected areas such as parks. Between 1870 and 1877, an estimated 100,000 wolves were supposedly killed in Montana alone (Curnow 1969; Fritts et al. 1994), although L. D. Mech (personal communication) thinks this unlikely, for often coyotes were not distinguished from wolves (Stenlund 1955).

Finally, in 1915, the wolf war became the responsibility of the U.S. government with the establishment of the Division of Predator and Rodent Control (PARC) within the Biological Survey (Young and Goldman 1944; Dunlap 1988). Official hunters were paid full-time to kill the last wolves, and wolf persecution became an irrational obsession with no objective relationship to the actual threat. This is the stuff of which legends of elusive beasts of imaginary cunning and resistance were made: Old Whitey of Bear and Rags the Digger in Colorado, the Truxton Wolf in Arizona, and Three Toes and the Custer Wolf in South Dakota (Young and Goldman 1944; Brown 1983; Busch 1995; Gipson et. al. 1998).

By 1930, the wolf had disappeared from almost all the forty-eight contiguous states, even including Yellowstone National Park (Jones 2002): the last wolves were killed in Arkansas in 1928, in Washington in 1940, and in Colorado and Wyoming in 1943 (Busch 1995). Only the wolves of the Lake Superior region survived a bit longer: the last wolves in Wisconsin were slain between 1950 and 1970, although bounties in Wisconsin and Michigan were repealed in 1957 and 1960, respectively (Thiel 1993). A few wolves may have remained in Michigan after 1970 (Hendrickson et al. 1975).

Despite the continuation of bounties in Minnesota until 1965, several hundred wolves survived in the northern part of the state, thanks to the remoteness of the area and its proximity to the vast Canadian wilderness (Mech 1970; Van Ballenberghe 1974), and in Isle Royale National Park in Lake Superior (Mech 1966b).

Wolves were finally protected in Minnesota, Wisconsin, and Michigan in 1974, and throughout the forty-eight contiguous states in 1978, by the federal Endangered Species Act (ESA) of 1973. In the Southwest, the last Mexican wolves were killed in Texas in 1970, although unconfirmed reports of wolves in Arizona and New Mexico continued through the early 1980s (USFWS 1982).

Canada and Alaska

In Canada and Alaska, wolves fared much better, surviving in 90% of their original Canadian range (Theberge 1973) and in almost all of their Alaskan range (Stephenson et al. 1995). Elsewhere in Canada, the wolf was already rare in southern Quebec and Ontario by 1870; it disappeared from New Brunswick in 1880, from Nova Scotia in 1900, and from insular Newfoundland before 1930 (Hayes and Gunson 1995). By the first decade of the twentieth century, persecution had been very successful in eliminating many of southern Canada's populations.

Following the formidable campaigns of the early 1900s, wolf populations were greatly reduced even in western Canada, and in 1930 they reached their all-time low (Hayes and Gunson 1995). However, they quickly increased again thereafter (Nowak 1983). Government wolf control programs using poison still continue in Canada's western provinces (Hayes and Gunson 1995; G. Court, personal communication). Ontario repealed its wolf bounty in 1972, but retains a year-round open season on wolves.

Current Status of the Wolf

With the exceptions of those in Alaska, Canada, and northern Asia, most wolf populations continued declining until the late 1960s, when most were classified as endangered or lingering at dangerously low numbers (Mech 1982b). It appeared inevitable that the wolf's range would continue to shrink, especially in the case of the small, isolated populations of Europe.

However, this realization brought strong and effective responses by governments, and by the early 1980s wolf populations had begun to increase (Ginsberg and Macdonald 1990). The viability of many small wolf populations and the tenacity of the wolf, which allowed it to survive in even the most degraded habitats, had been underestimated. Legal protection and a ban on poison was all it took in most areas. In western Europe, ecological conditions were substantially changing as large numbers of people from rural communities moved to cities. Mountainous areas, in particular, quickly reverted to a more natural state.

The expansion of the wolf's range and the improved status of many populations during the last two decades are relatively well documented (Carbyn et al. 1995; Schröder and Promberger 1992), although there is still considerable uncertainty about many Asian areas. Today, *Canis lupus* is considered "vulnerable" globally according to the new criteria of the World Conservation Union's threat categories (IUCN 2000), with several small, isolated populations endangered locally (table 13.1).

North America

Wolves once ranged over almost all of North America north of Mexico City, except possibly parts of California. The current status and management of wolves in North America varies by country, state, and province.

Canada and Alaska

Alaskan wolves, numbering 6,000–7,000, may be legally taken during hunting and trapping seasons with bag limits and other restrictions; an estimated 15% are harvested annually (Stephenson et al. 1995).

Wolf status in Canada has been periodically reviewed

TABLE 13.1. Numbers, trends, and legal status of gray wolf populations in 2000

Country	No. wolves	Population trend	Legal protection	Damage compensation
United States	9,000	Incr.		
Alaska	6,000–7,000	Stable/incr.	Game species	No
Minnesota	2,500	Incr.	Yes but depredation control	Yes
Montana	70	Incr.	Yes (endangered)	No
Idaho	185	Incr.	Yes (experimental/nonessential)	No
Wyoming	165	Incr.	Yes (experimental/nonessential)	No
Washington	?	?	Yes	No
Michigan	200	Incr.	Yes	No
Wisconsin	200	Incr.	Yes	No
Canada	52,000–60,000	Stable/Incr.	Game species/protected in 3% of Canada	No
Northwest Territories	5,000	Stable	Game species	No
Nunavut	5,000	Stable	Game species	No
Yukon	5,000	Stable	Game species	No
British Columbia	8,000	Incr.	Game species	No
Alberta	4,200	Incr.	Game species	No
Saskatchewan	4,300	Stable	Game species	No
Manitoba	4,000–6,000	Stable	Game species	No
Ontario	9,000	Incr.	Game species	No
Quebec	5,000	Incr.	Game species	No
Labrador	2,000	Stable	Game species	No
Greenland (Denmark)	50–100	?	Yes (in 90% of range)	No
Portugal	200–300	Stable	Yes	Yes
Spain	2,000	Incr.	Game species (protected in the south)	Yes, but varies with regional laws
France	40	Incr.	Yes	Yes
Italy	400–500	Incr.	Yes	Yes, by regional governments
Switzerland	1–2 ?	—	Yes	Yes, by cantons
Germany	5 ?	Stable	Yes	No
Norway	10–15	Incr./Stable	Yes	Yes
Sweden	70–80	Incr.	Yes	Yes
Finland	100	Incr./Stable	Hunted only in reindeer areas	Yes, by the state and insurance companies
Poland	600–700	Incr.	Yes, except Bieszczady	No
Estonia	< 500	Decr./Stable	No (the only outlawed species)	No (insurance too expensive)
Lithuania	600	Incr.	No	No (only if animals were insured)
Latvia	900	Stable	No	No
Belarus	2,000–2,500	Incr./Stable	No	No
Ukraine	2,000	Stable	No	?
Czech Republic	< 20	Stable	Yes	No
Slovakia	350–400	Stable	Yes (exceptions)	No
Slovenia	20–40	Incr.	Yes	No
Croatia	100–150	Incr.	Yes, since May 1995	Yes
Bosnia-Herzegovina	400 ?	Decr.	No	No
Yugoslav Federation	500	Stable	?	No
Hungary	< 50	Stable	Yes (exceptions)	No
Romania	2,500	Incr.	Yes	No
Bulgaria	800–1,000	Stable	No	No
Greece	200–300	Stable	Yes	Yes, 80% paid by insurance
Macedonia	> 1,000	Incr.	No	No

TABLE 13.1 (continued)

Country	No. wolves	Population trend	Legal protection	Damage compensation
Albania	250	Incr.	Yes	No
Turkey	1,000 ?	?	No	No
Syria	200 ?	?	No	No
Lebanon	< 50	?	No	No
Israel	150	Stable	Yes	No
Jordan	200 ?	?	No	No
Egypt (Sinai)	< 50	Stable	No	No
Saudi Arabia	300 – 600	Stable	No	No
India	1,000	Decr.	Yes	No
China				
Cheiludjiang	500 ?	Decr.	No	No
Xinjiang	10,000	Decr.	No	No
Tibet	2,000	Decr.	No	No
Mongolia	10,000 – 20,000	Stable ?	No	No
Russia	25,000 – 30,000	Incr. / stable	No	No
Kazakhstan	30,000	Stable	No	No
Turkmenistan	1,000	Stable	No	No
Uzbekistan	2,000	Stable	No	No
Kirgizstan	4,000	Stable	No	No
Tadjikistan	3,000	Stable	No	No

Note: This information was obtained by assembling data from available bibliographic sources and the informed and subjective estimates provided by the experts of the IUCN/*SSC* Wolf Specialist Group and the Large Carnivore Initiative for Europe (Boitani 2000). Except for a few local situations that are well known, most of the numerical estimates should be considered no more than indicative of the general status of the populations. There are no recent reliable estimates for Iraq, Iran, Afghanistan, Nepal, and Bhutan.

(Pimlott 1961; Theberge 1973, 1991; Carbyn 1983a, 1987; Hayes and Gunson 1995). Generally, the Canadian wolf population is in excellent biological condition, numbering about 52,000–60,000 (Hayes and Gunson 1995), and has been stable except for local changes related to fluctuations in prey biomass.

The legal status of wolves in Canada varies from province to province and territory to territory. Natives can hunt without restriction, whereas other residents require licenses for different hunting and trapping seasons depending on local regulations. An average of 4,000 wolves are killed annually in Canada, accounting for 4–11% of the population (Hayes and Gunson 1995), but such harvests probably affect only wolf populations at the edge of the range (Theberge 1991).

The Contiguous United States

During the 1970s, wolves in southwestern Canada increased and expanded into the northern Rocky Mountains of the United States, specifically Montana and possibly Washington (Ream et al. 1991). Recolonization of northwestern Montana began during the late 1970s (Ream and Mattson 1982), and for the first time in 50 years, wolves denned in Montana (in Glacier National Park) in 1986 (Ream et al. 1989). The population increased to about 84 individuals and thirteen packs or pairs in 2001 (U.S. Fish and Wildlife Service 2002) and is expanding back into Canada (Boyd et al. 1995) and dispersing southward into other parts of Montana, Idaho, and Wyoming (Fritts et al. 1995).

As part of a wolf recovery effort for the northern Rocky Mountain region, the federal government proposed reintroducing wolves into Yellowstone National Park, where they had been deliberately exterminated by 1930, and into central Idaho (USFWS 1987). These reintroductions were highly controversial and required an Environmental Impact Statement (EIS). Over a 32-month period, some 130 public meetings were held,

750,000 informational documents distributed, and 170,000 comments received from the public. The USFWS proposed that the animals used in both reintroductions be designated "experimental/nonessential" populations (USFWS 1994a). That designation allowed more management options to minimize conflicts with local people than would otherwise be possible under the ESA. A major concern about the reintroductions was that the wolves would separate and not remain in the areas intended for them.

Wolves were captured in Alberta and British Columbia, and dispersal-aged individuals were released immediately into Idaho ("hard release"), whereas small family groups were held in Yellowstone National Park in acclimation pens for several weeks before release ("soft release") (Fritts et al. 1997). In winter 1995, 14 wolves in three packs were released into Yellowstone, and 15 were released into Idaho. In 1996, 17 wolves in four packs were released into Yellowstone and 20 into Idaho (Bangs and Fritts 1996).

The released wolves survived well, most remained in the intended areas, and reproduction began during the first year (Bangs and Fritts 1996). The pace of population establishment exceeded projections, so additional reintroductions were canceled. As of early 2002, 250 wolves in twenty-eight packs lived in the Yellowstone area and 260 wolves in twenty-five packs in Idaho (D. W. Smith, personal communication; C. Mack, personal communication). The reintroduced wolves and their offspring had killed some 97 cattle and 426 sheep up to that time, a fraction of the projected amount; 59 wolves were killed because of repeated depredations (E. Bangs, personal communication).

The Mexican wolf was also protected under the ESA in 1976 (USFWS 1982). All efforts at finding evidence of the Mexican wolf, even in remote areas of Mexico such as Chihuahua and Durango, have so far failed: no reliable sighting of a wild Mexican wolf had been reported since the last 5 individuals were trapped in 1980 to establish a captive breeding program in Arizona (Parsons and Nicholopoulos 1995). In March 1998, 11 Mexican wolves from captive breeding programs were released in different areas of the Apache National Forest in Arizona. From then until March 2001, another 45 wolves were released. By mid-November 2001, some 23 wolves in six packs inhabited Arizona and New Mexico (Phillips et al. in press), and by late 2002, there were 28 wolves in eight packs (USFWS 2002).

These wolf restorations have shown that reintroduction, although controversial, is a viable option for reestablishing wolves in suitable parts of their former range (Bangs et al. 1998). However, it is essential that procedures be in place before reintroduction in order to deal with wolves that cause problems for people in the area (Mech 1979b; Phillips and Smith 1998).

As already mentioned, the wolf received protection in Minnesota, Wisconsin, and Michigan in 1974 under the ESA. The species was already increasing in Minnesota and responded quickly to protection (Fuller et al. 1992), soon spreading to Wisconsin (Wydeven et al. 1995) and Michigan (Michigan Department of Natural Resources 1997). Minnesota's wolf population was downlisted to "threatened" in 1978, and Wisconsin's and Michigan's were proposed for reclassification to "threatened" in 2001. In 2003, the proposal was finalized.

In North America today, there is essentially one single continent-wide wolf population that extends over most of Alaska and Canada and southward into Minnesota, Wisconsin, Michigan, Montana, Wyoming, and Idaho (fig. 13.1B). From Minnesota and Manitoba, wolves have also begun dispersing into the Dakotas, where at least 10 wolves were killed between 1981 and 1992 (Licht and Fritts 1994). The only disjunct populations are those of the Mexican wolf in Arizona and New Mexico and the red wolf in North Carolina (see Phillips et al., chap. 11 in this volume).

Europe

In Europe, wolf status became much more dynamic after 1980 (fig. 13.2). Small, isolated populations expanded into habitats that became suitable after human density decreased in many rural areas and wild prey recovered (Delibes 1990; Promberger and Schröder 1993; Boitani 2000). The traditional pastoral and rural economies that lasted until the 1950s changed deeply: the number of livestock dropped dramatically, animal husbandry techniques changed, and cultivation of poor soils and mountain areas became uneconomical. These changes eliminated the need to persecute the wolf in most of Europe, while greater appreciation of nature increased the number of pro-wolf Europeans.

Although the quality and quantity of data on wolf populations vary, there is nevertheless sufficient information to provide a realistic picture for each country (see table 13.1). In Scandinavia, after the slaying of the

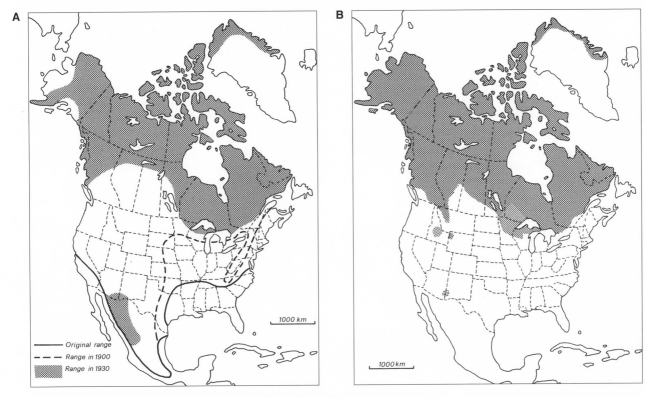

FIGURE 13.1. Past range (A) and current range (B) of the wolf in North America.

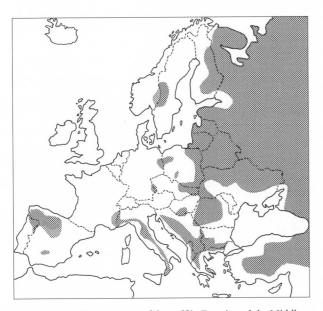

FIGURE 13.2. Current range of the wolf in Eurasia and the Middle East.

last wolves in Sweden (1966) and Norway (1973) and the simultaneous introduction of protection, the first new pack was observed in central Sweden in 1978 (Wabakken et al. 2001). The total number of wolves today is estimated to be at least a hundred, with eleven breeding pairs in the forested areas of central Sweden and southern Norway (A. Bjärvall, personal communication). Vilà, Sundqvist et al. (2002) have shown that the viability of the new population increased when a single immigrant boosted the productivity of the original founding pair. Wolves are fully protected in Sweden and partially controlled in Norway.

In Finland, the powerful lobby of northern reindeer farmers imposed zonal protection of the wolf. Wolves are fully protected only in the southern third of the country, while they may be killed during specific seasons in other areas (Pulliainen 1993; Council for Environment and Natural Resources 1996). Thus the Scandinavian wolf owes its perpetuation to Finland's contiguity with the Republic of Karelia, which has a large wolf population (Danilov et al. 1978).

In central Europe, Poland plays a fundamental role in

providing routes for the wolf's expansion. The population reached an all-time low of 100 in the 1950s, but the wolf was classified as game in 1976 and began to increase to its current level of 800–900 (Okarma 1989; Bobek et al. 1993). In the north, the population's range overlaps with that of the Lithuanian, Belarusian, Ukrainian, and Slovakian populations. A small new population in western Poland (Bobek et al. 1993) sends dispersers to Germany. Since 1945, 23 wolves have been killed in eastern Germany (Promberger, Vogel, and von Loeper 1993), and the possibility of a reproducing pack has recently been recorded.

In Slovakia between 1968 and 1990, an average of 60 wolves were killed annually, with peaks of 131, 102, and 115 in 1988, 1989, and 1990, respectively (Hell 1993). These figures suggest either a substantial population of a few hundred or significant immigration from Poland. The wolf has been protected in Slovakia since 1995, but animals guilty of substantial depredations on livestock may be killed. A few animals disperse into the Czech Republic, where they are completely protected.

Hungary receives dispersing wolves from Slovakia, Romania, and Croatia (Farago 1993). Hungary's relative lack of cover has hindered the buildup of an autonomous population. The wolf can be hunted there, even though it has a special status, and offending animals can be killed year-round by permit.

Romania is important to the European wolf population. After an all-time low of 1,550 animals in 1967, Romanian wolves increased to about 2,500 individuals (Ionescu 1993). The wolf could be hunted year-round until 1996. It is now protected, although the law is not enforced (C. Promberger, personal communication).

Only general estimates are available for wolf numbers in Macedonia and Albania. Both are vitally important in linking the wolves of Greece with those of Bosnia and Croatia. Official estimates for Greece are approximate and possibly inaccurate (Adamakopoulos and Adamakopoulos 1993; Y. Mertzanis, personal communication). Wolves south of 39° N are protected by the Bern Convention and European Union laws, but dozens of wolves are killed each year, and the animal's future in Greece seems poor (Hatzirvassanis 1991).

The latest wolf population estimates for Bosnia (400 wolves in 36,000 km², or 14,000 mi²) are from 1986; however, all reports point to a sharp decline since then (Promberger and Schröder 1993). The small remnant populations in Slovenia and Croatia are fully protected and are increasing (Huber 1999).

The situations of wolves on the Italian and Iberian peninsulas are different, since both populations have been isolated for at least a century. In Italy, the wolf was partly protected in 1971 and fully protected in 1976 (Ciucci and Boitani 1998a). The population and range grew constantly beginning in the late 1970s, and the population now stands at 450–500 individuals (Boitani 1992; Boitani and Ciucci 1993; Corsi et al. 1999). Illegal hunting takes 15–20% of the population annually (Boitani 1986), but that has not been enough to prevent a 6% average annual increase of the population.

Italian wolves have now recolonized the entire Apennine Mountain range into France. In 1993, wolves denned in the Mercantour National Park area of France (Poulle et al. 1999), and by 2000 at least 50 roamed the western Alps across the Italian and French border. One wolf originating from the Italian population was found in the Eastern Pyrenees in 2000 (B. Lequette, personal communication). In 1996 a wolf from the Italian population killed several sheep in Switzerland near Italy, but disappeared after being wounded (Breitenmoser 1998). Since then, at least 6 other wolves have been killed in southern Switzerland.

On the Iberian peninsula, the wolf inhabits northwestern Spain (Galicia, Leon, and Asturia) and northeastern Portugal. Only wolves south of the river Duero are protected by European Union legislation. The total population is estimated at 2,000, of which approximately 150 inhabit Portugal (Vilà, Castroviejo, and Urios 1993; J. C. Blanco, personal communication). Hundreds are killed each year, and only a few are taken legally (Vilà, Castroviejo, and Urios 1993). Still, the Iberian population has expanded south across the river Duero and eastward in the Asturias and Pyrenees Mountains (Blanco and Cortés 2002), and may reach France soon.

Because of the dynamic nature of all the European wolf populations, wolf distributions in southern and eastern Europe will probably change considerably in the coming years, above all in the Alps as far as Switzerland and in southeastern France. Furthermore, the return of the wolf to areas from which it has been absent for more than a century will undoubtedly cause serious management problems.

The Middle East

Except for Israel and Saudi Arabia, few reliable data on wolf status are available from the Middle East. Some anecdotal information has been provided by members of the IUCN/SSC Wolf Specialist Group and others who have visited the region or had contacts with conservationists there. The resulting population estimates are often purely indicative and should not be used in management planning.

Three Middle Eastern countries probably play a central role in maintaining wolves in the region: Turkey, with its natural contiguity with the vast areas of central Asia; Israel, whose conservation policies and their efficient enforcement maintain a moderate wolf population that radiates into neighboring countries; and Saudi Arabia, with its vast and relatively undisturbed deserts. The mountains of Turkey have served as a reservoir for the few wolves surviving in Syria.

The Golan Heights, the hills lying between Syria and Israel, support a small nucleus of wolves that is well protected by the military activities in the area (T. Ron, Nature Reserve Authority, Israel, personal communication). In 1980, wolves still occupied a large part of eastern and southern Israel (Mendelssohn 1982). The wolves in the southern Negev desert are contiguous with those of the Egyptian Sinai and Jordan (Hefner and Geffen 1999).

While wolves have been protected in Israel since 1954, they are not protected in neighboring countries. Indeed, the wolf is hunted year-round by Bedouins defending their flocks (Spalton 2002). However, the fact that sheep raising flourishes has probably promoted wolf survival, as many wolves raid the Bedouins' livestock and retreat safely into the desert (P. Ciucci, personal communication).

Wolves on the Arabian peninsula are estimated at 300–600 (I. A. Nader, personal communication). Although hunted year-round, the Arabian wolf is naturally protected by the inaccessibility of the western mountain areas and the central and northern deserts.

On the whole, wolf numbers seem to be stable in the Middle East, and probably will remain so.

Southern Asia

From Syria all across the Middle East and southern Asia, there are no data on wolf numbers except for India. In Iran, the wolf was distributed over much of the country at low densities in the mid-1970s (Joslin 1982). However, much has changed in Iran in the last 20 years, and this situation may also have changed.

Afghanistan and Pakistan are certainly important strongholds of the wolf, especially in the north. Fox and Chundawat (1995) discussed the low densities of wolves in India's trans-Himalayan region and suggested that their densities and conservation status are similar in northern Pakistan and adjacent China. They estimated that there are about 300 wolves in approximately 60,000 km² (23,000 mi²) of Jammu and Kashmir in northern India and 50 more in the similar habitat of Himachal Pradesh.

Altogether, India supports an estimated 800 (Shahi 1983) to 2,000–3,000 (Jhala 2000) wolves, scattered among several remnant populations. Although the wolf was fully protected in India in 1972, it is considered endangered, and many populations linger at low densities (Jhala and Giles 1991; Fox and Chundawat 1995) or are found in areas increasingly used by humans. Wolves inhabit Nepal and Bhutan, but no data from those countries are available.

Northern and Central Asia

The wolf in most of Asia has been homogeneously distributed over time and space. Indeed, the immensity of Asia makes any precise assessment of wolf populations at the continental level meaningless. Estimates are approximate to the hundreds or thousands and have been based on indicators such as harvest levels (fig. 13.3).

Wolf numbers in the Soviet Union (Bibikov 1975, 1982, 1985, 1994; Promberger and Schröder 1993) apparently peaked twice in the twentieth century, following the two world wars. In the late 1940s the population was estimated at 200,000 animals. About 30,000 wolves were harvested each year, with 40,000–50,000 taken during peak years. Around 1970, the population reached its

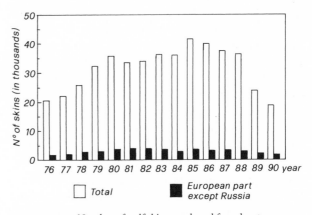

FIGURE 13.3. Number of wolf skins purchased from hunters by government authorities in the former Soviet Union. (From Promberger and Schröder 1993, 24.)

lowest level (as in most of Europe and North America), and the wolf disappeared from large areas of the European Soviet Union. By 1980 the population had increased again to about 75,000 wolves, with 32,000 killed in 1979 alone (Bibikov 1982).

The Soviet wolf population peaked at 80,000–120,000 in the early 1980s (Promberger and Schröder 1993), followed by control that probably limited the population. Then, with the collapse of the Soviet system, continent-wide control also collapsed. The number of wolves taken has dropped to about 20,000 per year, and populations appear to be increasing again. Some 25,000–30,000 wolves were estimated to inhabit the former Soviet Union in 2000 (N. G. Ovsyanikov, personal communication).

Wolves are not protected in Mongolia or China except in reserves. The Mongolian population is estimated at 10,000 (D. I. Bibikov, unpublished data) to 30,000 (K. Tungalagtuya, personal communication). Information from China is more fragmentary. Gao Zhong Xin (personal communication) estimated the population in the north at about 400 wolves, scattered over a vast area and declining from persecution. D. I. Bibikov (unpublished data) estimated 10,000 wolves in Xinjiang and another 2,000 in Tibet. These estimates greatly exceed other reports (Schaller 1988; Schaller et al. 1987; Achuff and Petocz 1988; Fox and Chundawat 1995).

Threats to Wolf Conservation

A conservation threat is any factor likely to wipe out or continually depress a wolf population. To evaluate the significance of a conservation threat, however, it must be placed in the appropriate temporal and spatial perspective. For example, annual human-caused mortality of 35–50% of wolves older than 5 months may cause a population decline (see Fuller et al., chap. 6 in this volume). Thus a higher rate could be a reason for concern; however, in a large population, mortality of this magnitude for only a year would cause only a slight fluctuation without further consequences. On the other hand, the same high mortality in a small population, in which other environmental factors or low reproduction or recruitment rates could randomly play a negative role, could endanger the population.

While scientists recognize the need to analyze conservation threats in the appropriate contexts, the public often does not. Distinctions between a truly endangered population and one the public might believe is endangered are not easy to make, since animal populations follow ecological patterns, while conservation threats usually result from human actions constrained by political borders and national laws. In the following brief analysis of conservation threats, I will deal only with the biological significance of some factors that potentially threaten local wolf populations.

The wolf is singled out here as if its conservation could be planned and implemented without considering the intricacies of the ecological relationships of a predator with its environment. This is obviously a simplification. Conservation goals can vary from the mere persistence of a species at low densities to the maintenance or restoration of the full array of species and functions of an ecosystem. This range of variation is the battleground of conservation, where ideologies, philosophies, lobbies, and politics attempt to find their compromises.

Human Persecution

The dominant factor in the extirpation of the wolf from its historical range has been persecution by humans (see above). This persecution, however, had varying intensities and different motivations at various times and places. From the establishment of the "louveterie" by Charlemagne and the extensive poisoning of wolves to the more sophisticated and selective removal of offending individuals (Fritts 1982; Fritts et al. 1992), the objective has usually been to reduce wolf predation on domestic animals. However, this local need was often extended to the extermination of wolves over large areas.

The most important difference between defense of livestock and attempted extermination was the intensity of the effort. In the former Soviet Union from 1925 to 1992, an estimated 1.6 million wolves were killed, ten times the population of that vast territory (Bibikov 1985; D. I. Bibikov, unpublished data). However, this massacre was not uniform over time and space, allowing the population to recover each time. In the United States, on the other hand, wolf extermination continued until the last wolf was killed, except in Minnesota and Alaska (Young and Goldman 1944). Although the scales of the kill were comparable, the effects were very different. Human persecution never really threatened the wolf in the Soviet Union, except locally. The effect of an extermination effort is much greater with smaller populations.

Wolf Harvesting

Harvesting of wolves by hunting or trapping has rarely been a conservation threat, primarily because it is so difficult, inefficient, and time-consuming. Hunting certainly was not a threat when conducted as it was in Europe until 1900, as a form of aristocratic activity. That wolves menaced livestock sometimes provided additional rationale for sport hunting of wolves (Lopez 1978). Hunting by airplane and snowmobile became popular in Alaska from the late 1940s through the 1970s (Rausch and Hinman 1977), in Minnesota from the late 1940s to the early 1950s (Stenlund 1955), and in parts of Canada (Cluff and Murray 1995). These methods made wolf hunting easier and may have affected local populations in Alaska and Canada, although there is no evidence that it ever threatened regional populations.

Wolf hunting and trapping peaked in North America in the 1920s and 1930s, when 21,000 wolves were taken each year (Busch 1995). More recently, that figure has stabilized at around 3,000 from Canada and Alaska. In the former Soviet Union between 1976 and 1988, about 30,000 pelts were produced annually (Promberger and Schröder 1993; D. I. Bibikov, unpublished data; see fig. 13.3). The most recent statistics from the CITES organization indicate that 6,000–7,000 wolf skins are traded internationally each year, with Canada, the former Soviet Union, Mongolia, and China the largest exporters and the United States and Great Britain the largest importers (Ginsberg and Macdonald 1990).

The harvesting of wolves for fur is probably not a threat to wolf populations, since only furs from northern countries are of commercial value and the northern populations are the most secure. China is the only area that may be of concern, as there are insufficient data to evaluate the effects of wolf harvesting there.

Habitat Destruction

The greatest long-term threat to the wolf is habitat destruction and reduction. In the IUCN/SSC publications on wolf status (Mech 1982b; Ginsberg and Macdonald 1990), habitat destruction ranked second as a cause of decline in almost all countries. As human populations continue to rise and cultivation and development of land increase, less and less land suitable for wolves is left (Mech 1996).

Habitat suitable for wolves means, first, habitat that can provide them with food. Thus, in the habitat currently available for wolves, habitat destruction means the destruction of the wolf's prey base or the prey's habitat. Wolves can thrive at relatively high densities on a wide variety of foods besides their main prey, hoofed mammals; alternative foods include rodents, garbage, and carrion (see Peterson and Ciucci, chap. 4 in this volume). Thus it is difficult to destroy all potential wolf food in a given habitat, except perhaps where cultivated fields predominate. However, it is possible to destroy the wolf's main prey base by reducing the number, availability, and quality of primary prey. In North America, some 72% of the variation in wolf density is accounted for by ungulate biomass (Keith 1983; Fuller 1989b, 21).

Depletion of the prey base, however, has not actually seriously threatened any wolf population. Even in Italy and Israel, where wolves often rely on small animals or garbage, they are usually in good physical condition, and their populations persist (Boitani and Ciucci 1993; Ciucci and Boitani 1998a; Mendelssohn 1982). In Saudi Arabia, medium and large prey have been almost completely exterminated, yet the wolf persists on small animals and livestock. On the other hand, careful management of food supplies in relation to desired wolf densities is a key factor in any wolf management plan.

The second important characteristic of wolf habitat is that it is habitat where humans do not kill wolves faster than they can reproduce. In Spain (Vilà, Castroviejo, and Urios 1993), Italy (Boitani 1982), Israel (Mendelssohn 1982), Romania (Christoph Promberger, personal communication), and the United States (Thiel et al. 1998), some wolves live and den near human activities as long as they are not killed. However, until about 1975, the only places where they were not killed were inaccessible areas such as mountains in Italy or wilderness in the United States. This led people to believe that wolves required wilderness habitat (Theberge 1975; Mech 1995a).

Some habitat features, such as road density (Thiel 1985; Mech, Fritts, Radde, and Paul 1988; Mladenoff et al. 1995), human population density (Weise et al. 1975; Fuller et al. 1992), forest cover (Hell 1993; Boitani and Fabbri 1983), or a combination of such features (Corsi et al. 1999), have sometimes defined wolf distributions. These factors are really indexes of human effects on wolves (see Fuller et al., chap. 6 in this volume), and their values are discussed in wolf habitat management recommendations (Wydeven and Schultz 1993; Fuller 1995b).

It is not clear to what extent human infrastructure affects wolf presence. Highways have been suggested as a major factor, assuming that wolves have difficulty crossing them (Paquet and Callaghan 1996). However, preliminary data from dedicated research in Spain showed that even fenced highways apparently do not greatly hinder wolves' movements and territories (J. C. Blanco, personal communication). Highways might negatively affect local wolf populations under some conditions, but there is considerable evidence of wolves crossing highways, railways, and intensively used areas (Boitani 1986; Licht and Fritts 1994; Mech, Fritts, and Wagner 1995; Merrill and Mech 2000).

Small Population Risks

Isolated or semi-isolated small populations are at risk of losing their genetic diversity and lowering their fitness (Allendorf and Leary 1986). Genetic diversity can be lost through several mechanisms related to population size: founder effects, demographic bottlenecks, genetic drift, and inbreeding. Several wolf populations are so small (i.e., fewer than 200 individuals) that theoretically these mechanisms could have an effect, making small population size a potential conservation threat (Theberge 1983).

However, there is no evidence that any wild wolf population has been threatened by loss of genetic diversity (Fritts and Carbyn 1995), although the growth of the Scandinavian population may have been retarded by inbreeding problems (Vilà, Sundqvist et al. 2002a). Nor has inbreeding been found to cause the decline of any wild species (Caro and Laurenson 1994). I do tend to agree with Ballou (1997), however, that we may lack such evidence because all cases supporting the theory went extinct before we could get insight into the decline processes. Small captive populations of wolves show some signs of inbreeding depression (Laikre and Ryman 1991; Fredrickson and Hedrick 2002; but cf. Kalinowski et al. 1999).

Searching for evidence of inbreeding depression in small populations of wild wolves has produced little. In one of the best-studied small populations, that of Isle Royale National Park, Michigan, Wayne et al. (1991) found that after a population decline to about a dozen individuals, only 50% of the heterozygosity in mainland wolves had been retained. However, the Isle Royale wolves have survived inbreeding for 50 years: from the single pair that colonized the island in 1949 (Mech 1966b), the population increased to a maximum of 50

before crashing to about a dozen in 1993, and by 2000 had recovered to about 29, close to its long-term average (Peterson 2000). Whether inbreeding has had or will have an effect on the Isle Royale population is still unknown. Many other wolf populations have undergone conditions subjecting them to reductions in genetic diversity, but we have not yet been able to obtain a convincing indication of minimum viable population (MVP) size; that is, the threshold for persistence of the population (Fritts and Carbyn 1995).

Among other small wolf populations, the Italian population is also instructive. That population has been isolated for at least 100–150 years and suffered a bottleneck of about 100 individuals in the early 1970s (Zimen and Boitani 1975). This population has a characteristic monomorphic mtDNA type, confirming that it has been a small, isolated population for a significant period (Wayne et al. 1992). However, the population still shows an average polymorphism level for nuclear and enzyme-coding genes and has apparently not suffered any loss of variability (Randi et al. 1993, 1995, 2000; Lorenzini and Fico 1995). Moreover, its recovery during the last 20 years (see above) is an indication of its viability.

Genetic considerations are primarily of academic importance, for genetic stochasticity probably has the least effect on the viability of small populations; demographic and environmental stochasticity have the greatest effect (Lande 1988; see also Wayne and Vilà, chap. 8 in this volume).

One aspect of genetics that has been little explored in wolves is the possible evolutionary significance, and thus the conservation value, of locally adapted populations (Shields 1983; Theberge 1983). Because wolves occupy a wide variety of habitats over a large distribution, they might have local adaptations that enhance the fitness of local populations. If so, then careless management of these genetic aspects of local populations would pose an important conservation threat, for example, through reintroductions and translocations. Nevertheless, the wolf's long dispersal distances (see Mech and Boitani, chap. 1 in this volume) and the success of transplanted and colonizing populations are evidence against this concern. Caution is probably the best approach until we learn more.

Hybridization and Genetic Swamping

Wolves sometimes hybridize with coyotes and dogs (see Wayne and Vilà, chap. 8, Nowak, chap. 9, and Fritts et al., chap. 12 in this volume), so hybridization is a possible conservation threat. The most significant problem resulted from the "swamping" of the red wolf by hybridization with coyotes (see Phillips et al. chap. 11 in this volume).

In the 1970s, wolf-dog competition, including hybridization, appeared to be a major threat to the wolf in Italy (Boitani 1983). The competition appeared particularly dangerous in three spheres: competition for food, mainly at garbage dumps; competition for space for territories; and hybridization. While the existence of the first two forms of competition has been partly confirmed (Boitani et al. 1995), hybridization has been documented in Italy in only one case (Zimen 1978; Boitani 1982). After years of negative results (Randi et al. 1993, 1995, 2000), fast-improving techniques have recently shown evidence of some introgression of domestic dog genes into wild Italian wolf populations (Randi and Lucchini 2002), and at least six wolf-dog hybrids have been identified (E. Randi and V. Lucchini, personal communication). Wolf-dog hybrids were documented in Latvia using microsatellite DNA markers (Andersone et al. 2002).

Wolf-dog hybridization was also proposed as an important conservation issue in Israel (Mendelssohn 1982) and the former Soviet Union. Bibikov (1982, 1988) described extensive proliferation of hybrids in several regions. The hybrids were bolder and more aggressive than wolves and had a major effect on both wild prey and livestock. Rjabov (1973, 1979) documented the high variation of morphological traits among first- and second-generation hybrids there.

In Estremadura (Spain), Blanco et al. (1990) described hybrids and discussed hybridization in the early 1970s as the possible cause of the low density of wolves there. Teruelo and Valverde (1992, 261) reported several cases of wolf-dog hybrids in various parts of Spain. Nevertheless, wolf-dog hybridization is an extremely difficult phenomenon to assess and quantify. Genetic analysis has begun only recently (see Wayne and Vilà, chap. 8 in this volume). As a result, concern is justified until more definite information is available.

Diseases

The role of diseases and parasites in wolf conservation has received little attention so far. Mech (1970), Brand et al. (1995), and Kreeger (chap. 7 in this volume) have synthesized the literature on the infectious and parasitic diseases of the wolf, discussing their implications for management and for reintroduction programs. Although the role of diseases in limiting wolf populations remains unknown (Brand et al. 1995), they are certainly a concern, as is any mortality factor, when populations are small and threatened. Rabies, canine distemper, sarcoptic mange, and canine parvovirus have been shown to be potential mortality factors that can have substantial effects on wolf populations (see Fuller et al., chap. 6, and Kreeger, chap. 7 in this volume).

Assessing Conservation Status and Needs

Estimating Wolf Numbers and Densities

In areas with dense cover or little snow, the wolf is one of the most difficult animals to census. Hayes and Gunson (1995) described the diverse methods used in Canada, such as trapper questionnaires, field observations, past estimates, trends in prey, long-term radiotelemetry and ground studies, localized aerial censuses, snow tracking, extrapolations, and correlations. These techniques can provide the order of magnitude of the regional population, but there is no way to evaluate their statistical error.

The most common method of estimating wolf numbers over large areas remains extrapolation of densities known for small areas. One technique that can yield accurate counts, though it is expensive and time-consuming, is aerial radiotelemetry, which has been widely used in North America (Kolenosky and Johnston 1967; Mech and Frenzel 1971a; Mech 1974b, 1979a, 1982a; Peterson, Woolington, and Bailey 1984; Ballard et al. 1987, 1997; Fuller 1989b; Mech et al. 1998; Hayes and Harestad 2000a,b).

In the semi-forested or open terrain that characterizes much of the northern wolf areas, aerial snow tracking with small aircraft and counting of wolves and their tracks can provide relatively accurate census data (Stenlund 1955; Burkholder 1959; Mech 1966b; Parker 1973; Fuller 1982; Ballard et al. 1995). Areas of several thousand square kilometers can be accurately censused in a few days (Ballard et al. 1995). Aerial strip censuses (Fuller 1982) can also be used, which involve flying low transects in search of wolf tracks and following them until the animals can be counted directly. The method is suited only for open and snow-covered areas, and its reliability under certain conditions has been questioned (Miller and Russel 1977).

Tracks, scats, and interviews with outdoorspeople have often been used to evaluate wolf presence; although they can only index abundance (Fuller and Sampson 1988; Mech, Fritts, Radde, and Paul 1988; Fuller et al. 1992; Wydeven et al. 1995), they can inexpensively monitor population trends even in large areas (Crête and Messier 1987).

Following tracks on the ground in fresh snow until all individuals were counted was the method used to census wolves in about 120 km² in Italy (Zimen and Boitani 1975). In Finland, the Border Patrol Establishment has monitored the wolf immigration/emigration balance by daily counting all wolf tracks crossing the border with Russia since 1968 (Pulliainen 1985, 1993). In central and eastern Europe, hunters count tracks to estimate the numbers of game animals in their management units (Okarma 1989; Bobek et al. 1993). The method's reliability in estimating wolf numbers has never been tested, and Becker et al. (1998) have proposed a more robust technique based on network sampling of tracks in the snow.

By counting wolf responses to human (or recorded) howling, the number and location of packs can be inferred (Pimlott 1960; Joslin 1967; Theberge and Strickland 1978). However, extrapolations of this index to true wolf numbers (Harrington and Mech 1982a) may prove inaccurate (Fuller and Sampson 1988). Moreover, its applicability remains untested where dogs are also present (Boitani and Ciucci 1993).

While accurate estimates are necessary in scientific studies, the same accuracy is not always required for management and conservation planning. The order of magnitude of a population, or even its trend, is often sufficient to decide on conservation actions.

Assessing Genetic Identity

The genetic identity of a population is also important to wolf conservation decisions. Investigation methods include both classic taxonomy and conservation genetics (see Wayne and Vilà, chap. 8, and Nowak, chap. 9 in this volume). It is important to clarify whether the diversity of a population has a natural basis or springs from a fragmentation of its range caused by human activity. Molecular genetics can provide decisive information.

Considering Tradition

A relatively new approach to conservation has begun to consider the importance of tradition in animal popula-

tions. This approach is particularly interesting in the case of the wolf, since the animal's sociality gives it considerable potential for cultural transmission. A few workers believe that tradition is very important to wolf conservation and that issues of tradition should be carefully considered (J. B. Theberge, personal communication; Haber 1996). On the other hand, natural wolf populations experience a high turnover (Mech et al. 1998), which minimizes the potential for traditions, and no evidence of the importance of traditions has been documented for wolves. Furthermore, the success of transplanted and colonizing populations suggests that tradition in wolves is not of critical importance (see Mech and Boitani, chap. 1 in this volume).

Assessing Wolf Distributions

Another important assessment in wolf conservation is that of distribution patterns. Often relationships between wolf density and environmental features such as road density (Thiel 1985; Mech, Fritts, Radde, and Paul 1988; Fuller et al. 1992; Mladenoff et al. 1995) or prey biomass (Keith 1983; Fuller 1989b; Messier 1995b) are straightforward and strong. However, sometimes no single environmental factor can account for a significant part of the observed variation in wolf density. Corsi et al. (1999), using a Geographic Information System (GIS) and multivariate statistics, proposed that in Italy, a probabilistic distribution map more appropriately describes wolf distribution. GIS-based techniques such as gap analysis (Scott et al. 1993; Boyce and McDonald 1999) appear to be the most promising for analysis of distribution patterns, metapopulation fragmentation, and habitat and corridor suitability (Haight et al. 1998; Mladenoff and Sickley 1998; Wydeven et al. 1998; Harrison and Chapin 1998). However, it must be stressed that such models primarily describe current wolf distributions, which do not necessarily predict future distributions (L. D. Mech and L. Boitani, personal communication).

Population Viability

Two pieces of information that would be useful in evaluating the conservation status of a small population are the minimum viable population (MVP), or the minimum number of individuals necessary for population survival (Shaffer 1987), and the minimum area required (MAR) to sustain that number of individuals. Theoretical models (population viability analyses, or PVAs) using these measures have been proposed and applied (Soulé 1980, 1987; Ciucci and Boitani 1991), but they have proved unsatisfactory (Boitani 1984; Boyce 1992; Fritts and Carbyn 1995).

Fritts and Carbyn (1995) challenged the reliability of the present PVA models and MVP estimates on the grounds that most theoretical analyses do not give the proper weight to the resilience of wolf populations and to the complexity of local conditions. These authors went even further in claiming that previous theoretical treatments of wolf population viability have contributed little to wolf conservation and have created unnecessary dilemmas for recovery programs. More recently, White (2000) challenged existing PVAs as theoretically flawed because they do not account for individual variation, a particularly important issue in wolf populations. Nevertheless, the PVA process may be useful as one more analytical tool for evaluating different variables and indicating research and conservation priorities.

Assessing Wolf-Prey Relationships

Many models have been constructed to analyze wolves' functional and numerical relations with prey. The general relationship between wolf density and ungulate biomass developed by Keith (1983) and Fuller (1989b), while having wide confidence limits, is robust and appears to be valid for North American environments (see Fuller et al., chap. 6 in this volume).

A more complex stochastic model (Boyce 1990) simulates wolf recovery in Yellowstone. This model allows exploration of the dynamics of wolf and prey populations in different contexts of management and geography. However, although the model predicts that the wolf population in Yellowstone will reach 50–120 within a century (Boyce 1990, 3–34), wolf numbers had already reached 131 within 6 years after reintroduction (U.S. Fish and Wildlife Service et al. 2002). Other, more sophisticated wolf population models also include social considerations (Vucetich et al. 1997; Haight and Mech 1997; Haight et al. 1998; and Cochrane 2000) (see also Fuller et al., chap. 6 in this volume).

Conservation Strategies and Management Plans

Modern wolf management requires precise objectives and consistency of action; this can best be achieved by developing management plans that address specific areas and conservation needs. Such plans include the Wolf Recovery Plans in the United States (see below); the IUCN-World Conservation Union's Canid Action Plan (Ginsberg and Macdonald 1990); the Wolf Conservation Strategy for Europe (Schröder and Promberger 1992) and its Management Action Plan (Boitani 1993) and Research Action Plan (Promberger 1993); the Action Plan for the Conservation of Wolves in Europe (Boitani 2000); and the management plans developed on a national scale in Italy (Boitani and Fabbri 1983; Boitani and Ciucci 2000) and on a local scale in Germany (Promberger and Hofer 1994).

Wisconsin, Michigan, and Minnesota, after extensive consultation and public participation, have approved management plans for their recovering wolves (Michigan DNR 1997; Wisconsin DNR 1999; Minnesota DNR 2001). These plans aim to manage the recovery of wolf populations after delisting from the federal Endangered Species List by identifying clear objectives, approaches needed, time frame, costs, and means required.

Wolf Management and Conservation

Management and conservation are human constructs dictated by diverse ethical, economic, political, and social motivations, and wolf management has included everything from eradication to complete protection. The wolf has been the object of continuous political attention during the last several millennia, a privilege and a curse that few other species ever had to the same degree.

Until recently, consensus on wolf management was automatic in that it consisted essentially of killing wolves or, at best, ignoring them. But recently, with the increased urbanization, interest in the environment, and prominence of the media discussed above (and by Fritts et al. in chap. 12 in this volume), wolf management has become far more complex and controversial. Following is a brief review of the major current approaches, policies, and techniques used to manage the wolf throughout its range.

Problems in Wolf Management

Despite the enormous biological, social, and economic diversity in which the wolf is managed today, certain management problems are common. The first springs from the clash between the wolf's biology and the human legal structure (Sax 1991). Wolf biology includes dynamic natural spatial and temporal dimensions, whereas human society is underlain by artificial political and legal structures. If the wolf has overcome this conflict, it is thanks to the animal's enormous flexibility and the buffering in its natural history.

The second problem derives from the wolf's complex biology and ecology. Understanding this complexity is difficult for specialists, let alone the public. But political decisions, as we have seen, are based more on emotions than on data, and the public does not master complex or confusing situations. Rather, everything becomes either black or white, simplifying situations that should, instead, retain their nuances.

The third problem arises from the fact that conservation is a multidisciplinary process. As such, it might benefit from multidisciplinary teams, including biologists, sociologists, land use planners, representatives of "stakeholder" groups, economists, and other specialists as required, although there is not unanimous agreement on this point (Mech 1996, 2000b; L. D. Mech, personal communication). However, management is usually in the hands of specialists in one sector or another (Clark 1993).

The fourth problem is that prejudice, ignorance, and superficial knowledge of the wolf is still widespread; this is true among both the wolf's adversaries and its supporters. Ultimately, the wolf's survival depends on the public's attitude toward it, but informed decisions must be based on actual facts undistorted by prejudices and impossible expectations.

Managing Small Populations

Legislation

Except for the Mexican wolf, most of the smaller, more endangered wolf populations are located in western Europe, the Middle East, and India (see above and table 13.1). All European nations have subscribed to the Bern Convention (Convention on the Conservation of European Wildlife and Natural Habitat, 19 September 1979), which provides for full legal protection of both wolf populations and their habitat, but individual countries do not all obey their obligations.

Spain, Slovakia, Poland, and Finland have made an exception for wolf protection, shifting the wolf from the Convention's Appendix II (totally protected species) to Appendix III (protected species that can be subjected to controlled hunting). In Spain, each Autonomous Region has the right to decide its own policies, so the Spanish wolf is hunted in six of nine Regions (Blanco et al. 1992). European wolves are also listed in Appendix II (species in need of habitat conservation) and Appendix IV (fully protected) of the European Habitat Directive (92/43 of 21 May 1992), except for the Greek populations north of 39° and the Spanish population north of the river Duero (the primary population).

Outside Europe, concern for their small wolf populations prompted Israel to declare full protection in 1954 (Mendelssohn 1982), India in 1972 (Shahi 1983), and Croatia in 1995, although Croatia is reconsidering (Huber 1999). The Convention on International Trade in Endangered Species of Wild Fauna and Flora (CITES, 3 March 1973) lists the wolf in its Appendix II (potentially endangered), except in Bhutan, Pakistan, India, and Nepal, where it is listed in Appendix I (endangered).

Law enforcement

Despite the apparently robust umbrella of legal protection for the wolf, the degree of enforcement varies considerably. In Eurasia, the most usual approach is a policy of benign neglect. In the last 25 years in Europe, not a single wolf poacher has ever been apprehended or legally prosecuted. This policy is unwritten, but has been adopted not only in Italy, Greece, Portugal, and Spain, but also in various Eastern European countries and India. Thus wolf management in all of the above countries relies on poaching, which tends to slow expansion of the relatively small populations. In Italy an estimated 15–20% of the population is illegally taken each year (Boitani and Ciucci 1993), while in Spain the total legal and illegal (especially in Asturias) annual human-caused mortality is about 40% (Blanco et al. 1992). Because poaching cannot be directed spatially or temporally, it could cause serious damage to populations locally. In other nations, such as France, Israel, and to some extent, Poland and Slovakia, citizens are more likely to obey the law.

National and International Coordination

Another problem plaguing wolf management is that comprehensive management strategies are still lacking. Not only is wolf management fragmented into systems of varying effectiveness, but international and interstate collaboration is almost nonexistent, even between jurisdictions such as Spain and Portugal, or Minnesota and Wisconsin, which share common wolf populations. In

India, the fragmentation of wolf management into different state jurisdictions is making conservation difficult (Shahi 1983; Jhala and Giles 1991). On the other hand, significant collaboration is taking place in Sweden and Norway, which have been monitoring the conservation of their wolf packs (Bjärvall 1983; Promberger and Schröder 1993).

Although some individual U.S. states have their own wolf management plans, only a few European countries have national plans. Germany has adopted a plan for the few wolves that have reappeared in the Brandenburg area (Promberger and Hofer 1994), and France recently (2000) proposed a draft plan to mitigate wolf conflict with livestock in the Alps. Italy is still working on its first plan. If management remains restricted to individual populations, it can offer only scant assurance of viable conservation in the long term.

Coordination of national strategies by a European plan is essential for conserving Europe's small, fragmented populations. The "Large Carnivore Initiative for Europe" (LCIE), recently launched by WWF International in collaboration with a subgroup within the IUCN/SSC Wolf Specialist Group called the European Wolf Network, has produced such a plan. This "Action Plan for the Conservation of Wolves in Europe" (Boitani 2000), which was adopted by the Bern Convention, is based on the concept of a network of areas and connecting corridors where the wolf can survive with the least conflict with human activities. A zoning system of modular wolf management is a key part of the plan, providing for wolf removal in the areas of greatest conflict with livestock. To date, the only European countries that have adopted such a zoning system are Finland (Pulliainen 1993) and Norway, and they include total protection in some zones.

Protected areas in Europe are generally much smaller than the "several thousand km²" suggested by Mech (1979b), Soulé (1980), and Fritts and Carbyn (1995) as the minimum range for viable wolf populations. Nevertheless, these protected areas could be integrated into a network of larger areas earmarked for wolves and, in extreme cases, could function as temporary retreats for a threatened population.

Although national parks and wilderness areas can serve as refuges and reservoirs for wolves in North America, conservation of the small residual wolf populations in Eurasia cannot depend on the preservation of large wild areas, for there are few such areas left there. Thus wolf conservation in Eurasia must be based on in-

tegrating the wolf with human activities to the maximum extent permitted by local economies. This philosophy is foreign to wolf management in North America, but it may be the only feasible approach there in the future (Mech 1995a, 1996) (see below).

Wolf-Human Conflicts

The small wolf populations in the Old World were able to persist as long as they did, even in an environment dominated by human activity, for two main reasons: first, local people tolerated a certain level of depredations, and wolves' attacks were limited by traditional methods of livestock defense; and second, relatively inaccessible mountain ranges provided wolves some refuge. Today, economic damage due to wolves is sometimes very high (see Fritts et al., chap. 12 in this volume).

The most rational and effective approach to mitigating wolf depredations on livestock has rarely been implemented in a coordinated way. Such an approach would involve three stages: (1) prevention, by providing incentives to improve protection of livestock (fences, guard dogs, shepherds, etc.); (2) compensation of farmers for damage; and (3) elimination of individual wolves that cause damage.

In Europe, the first two components of this approach are being implemented where endangered wolf populations survive, but the third action has never been implemented, primarily because of fierce opposition by animal protection groups. Nevertheless, direct wolf control is becoming harder to avoid, and it is being incorporated into national management policies. (The French Action Plan is based on this premise.)

The first two kinds of action are expensive and insufficient. Fences, guard dogs, and shepherds can be used only where local methods of husbandry and traditions allow for them. And incentives and compensation are often partial and involve considerable red tape (Blanco et al. 1990; Fritts et al. 1992; Boitani and Ciucci 1993). Nevertheless, these measures are generally accepted because wolf populations are small and threatened. As populations recover, however, this approach will probably have to be altered significantly. In North America, greater emphasis already is placed on killing wolves and less on improving livestock protection, although compensation is also paid in some areas (Fritts et al. 1992; see Fritts et al., chap. 12 in this volume).

In Europe, the management of livestock depredations is further complicated by the substantial subsidies paid by the European Union and national governments to farmers. Each sheep in the European Union countries is subsidized for 60% of its value; in Norway and Switzerland, the supports reach 82% (Savelli et al. 1998). The paradox, therefore, is that when wolves kill a sheep, they kill an animal that in great part belongs to the public. It is obvious, then, that management of livestock depredations must be planned and implemented within the context of the entire rural economic policies of these countries.

A different type of wolf conflict with humans that can have a considerable effect on wolf conservation involves direct wolf-human interactions (Linnell et al. 2002). In India, for example, wolves were eating human bodies at burial sites, and other wolves attacked several children in the early 1980s (Shahi 1983) and in 1995–1996 (see Fritts et al., chap. 12 in this volume). These reports seriously curbed wolf conservation efforts in India. Wolves were officially killed in those areas, and the attacks ceased, at least temporarily. However, wolf attacks on humans, especially children, have also begun recently in North America (see Fritts et al., chap. 12 in this volume), and what their effect on wolf conservation there will be is not yet known (McNay 2002b).

Habitat and Prey Restoration

Conservation planning for small wolf populations in Europe has also had to address the key problem of environmental recovery; in Italy and Portugal, this has meant reintroducing wolf prey such as red deer and roe deer where these species had been exterminated (Boitani 1992). Another important habitat restoration action would be the control of feral and stray dogs, which compete with wolves for food and space. However, all attempts at dog control seem destined to fail, at least in the Mediterranean countries, because of the magnitude of the problem and its cultural underpinnings (Boitani et al. 1995).

Research and Monitoring

An essential element in the conservation of small wolf populations is the monitoring of their status. Almost all small wolf populations are being monitored, albeit by very different methods and means (Promberger and Schröder 1993; Mech, Pletscher, and Martinka 1995; Boitani 2000). Scientific research on small populations cannot produce the amount of information provided by large populations. Nonetheless, the results are extremely useful and important for understanding the needs of

small populations and the adaptations of wolves living close to human beings (Vilà, Castroviejo, and Urios 1993; Boitani and Ciucci 1993; Mendelssohn 1982).

Managing Wolf Recovery

The reasons for promoting wolf recovery are ethical (Pimlott 1975), biological, genetic, economic (Schaller 1996; Promberger et al. 1998), and cultural. However, there are also reasons why wolf recovery must be controlled. As wolves expand into agricultural and urbanized areas, conflicts with humans will undoubtedly increase (Mech 1995a, 1998b), and all means of protecting livestock from wolves over large areas are largely ineffective and expensive (Fritts et al. 1992; Mech et al. 2000; see Fritts et al., chap. 12 in this volume). High levels of conflict are obviously not sustainable, and they risk threatening wolf conservation in its entirety.

After discussing possible alternatives, Mech (1995a, 2001a) and Mech, Fritts, and Nelson (1996) suggested that, where wolf populations are expanding, lethal control will remain the ultimate means of curbing wolf damage to livestock and pets. Without constant control, wolf populations can recover quickly, as shown by experiments (Ballard et al. 1987; Hayes and Harestad 2000a) and by the repopulations following different levels of persecution in the former Soviet Union (Bibikov 1985; Pulliainen 1965) and in Poland (Okarma 1992).

Wolf control, however, must be based on a system of zoning to allow for the persistence of viable protected populations (Mech 1979b, 1995a; Boitani 1982, 2000; Clarkson 1995). North American national parks are better suited to protect wolves than are European parks. Their large size and the absence of human economic activity allow these protected areas to play an essential role in the conservation of wolves, especially as reservoirs for recovery. Nevertheless, if fully protected inside a park, wolf populations naturally overflow the protected area and disperse into neighboring areas, presenting the same problems and demanding the same compromises.

Much public opposition to wolf control has been stirred by animal protectionist groups. Whereas some in Europe correctly claim a lack of balanced wolf management by governments, more radical groups have often failed to understand or consider wolf biology (the wolf's rate of increase, dispersal, ecological adaptability, etc.) and the animal's potential effect on local economies (see Fritts et al., chap. 12 in this volume). Mech (1998b, 2001a) pointed out the folly of resisting limited wolf control today and delaying it until control will necessarily have to be much greater. Thus, in the end, many more wolves will have to be killed.

The three-pronged wolf depredation management approach mentioned above appears to be the only feasible way to allow wolves to spread acceptably and sustainably over larger areas of Europe and the Middle East, where wolf survival must be coordinated with human activities with minimal conflict (Boitani and Fabbri 1983). However this is also true of wolf recovery across the contiguous United States, where conditions are evolving—albeit with some difference in scale—into a similar scenario (Fritts and Carbyn 1995; Mech 1995a).

United States Recovery Plans

With the enactment of the Endangered Species Act, American federal agencies were charged with recovering threatened or endangered species to the point at which they can be "delisted," or removed from the Endangered Species List. Recovery teams were appointed by the U.S. Fish and Wildlife Service for four regions and for certain wolf subspecies recognized before 1995 (Nowak 1995a), including the eastern timber wolf, the northern Rocky Mountain wolf, the Mexican wolf, and the red wolf (see Phillips et al., chap. 11 in this volume). Originally the plans pertained to these wolf subspecies, but by 1978, when the wolf was listed not by subspecies but by geographic area—the forty-eight contiguous states—the plans applied to whatever wolves inhabited the geographic area with which each plan dealt.

Four plans with similar objectives were developed, each suited to markedly different environmental and sociopolitical situations: two relied on captive-bred animals for reintroductions and two on recovery of existing wild populations. The four plans together form an excellent example of a coordinated, extended effort with a strong base of technical, organizational, and policy tools. Recovery plans were also formulated at state levels, as in Wisconsin (Thiel and Valen 1995; Wisconsin DNR 1999). Although these plans emphasize single species in a specific region, the actions they call for are also of great value for conservation of the entire environment.

The Eastern Timber Wolf Recovery Plan. The first wolf recovery plan was finalized for the eastern timber wolf in 1975, and was revised in 1990 and in 1992 (USFWS 1992). It was based on the original range of this subspecies that had been recognized until 1995, when that range was revised (Nowak 1995a). The plan includes increasing the Minnesota population to a minimum of 1,251 wolves and reestablishment of a second population of at least 100 wolves for at least 5 years in Wisconsin and Michigan. The plan zoned Minnesota and suggested different recovery targets for each zone. It also identified four critical factors for long-term recovery: large wild areas with low human densities and accessibility, ecologically sound management, availability of adequate wild prey, and adequate understanding of wolf ecology and management (USFWS 1992).

The recovery plan succeeded, with populations in Minnesota, Wisconsin, and Michigan currently more than double the recovery minimums (see table 13.1), and, as mentioned earlier, the process of delisting the wolf is under way. This success marked a significant advance in wolf management by setting standards, targets, and methods that proved attainable and reliable. The recovery plan, however, did not stop debate about wolf conservation (see www.wolf.org), nor did it stop poaching (Peterson 1986).

The Northern Rocky Mountain Wolf Recovery Plan. The recovery plan for the northern Rocky Mountain wolf *(Canis lupus irremotus)* was approved in 1980 and revised in 1987: it defined recovery as at least ten breeding pairs of wolves inhabiting northwestern Montana, Yellowstone National Park, and central Idaho for 3 successive years (USFWS 1987). Nowak's (1995a) reclassification of wolf subspecies mentioned above did not recognize *C. l. irremotus* and indicated that originally *C. l. nubilus* and *C. l. occidentalis* occupied the Northern Rockies. The plan, therefore, pertained to those two subspecies.

In Montana, natural recolonization from Canada was considered the best method for recovery, and that has occurred (Ream et al. 1991). Reintroduction was prescribed for Idaho and Yellowstone. As indicated earlier, those reintroductions have been done and have been highly successful.

The Mexican Wolf Recovery Plan. The Mexican wolf *(Canis lupus baileyi)* has been considered extinct in the United States since the 1970s (see above), although occasional and unverified sightings and sign have been reported in Mexico (J. Carrera, personal communication; J. Servín-Martínez, personal communication). The subspecies was listed as endangered in 1976. Between 1977 and 1980, four males and one pregnant female were captured in Durango and Chihuahua and moved to the

United States to establish a captive breeding program (Parsons and Nicholopoulos 1995).

A Mexican Wolf Recovery Team was appointed in 1979, and a recovery plan was approved in 1982. Its main objective was to maintain a captive breeding program and reestablish a self-sustaining wild population of at least 100 Mexican wolves within the Mexican wolf's historical range in the United States (USFWS 1982). As indicated above, this restoration is now under way, with a growing wolf population inhabiting Arizona and New Mexico. Restoration of the Mexican wolf to the southern Rockies is also being proposed by the Turner Endangered Species Fund, a nongovernmental organization that hopes to release a few animals on a large private property in northern New Mexico (M. K. Phillips, personal communication). Like the reintroduced Yellowstone and Idaho wolves, the reestablished Mexican wolves have been designated "experimental/nonessential," in accordance with Section 10(j) of the ESA, to allow for more flexible management.

Managing Abundant and Unknown-Sized Populations

Uncontrolled Harvest

Most wolf populations have been managed without any planning of harvesting times, areas, or methods and without any monitoring of the effects of harvesting. Russia's vast wolf populations are legally hunted in any season, in any place, and by any means (Bibikov 1985; N. G. Ovsyanikov, personal communication), as are most of Canada's (Hayes and Gunson 1995). The Russian system of bounties and unregulated harvesting, which for many years controlled the number of wolves, has broken down with the collapse of the old Soviet regime, and now works in a fragmented way. This approach does not seem to have damaged the wolf population.

The same absence of policies and regulation characterizes wolf management in many of the former Soviet republics and in Romania, where wolf hunting (even poisoning) was permitted year-round until a few years ago (Ionescu 1993). The situation is similar in Mongolia, Saudi Arabia, and the region lying between Syria and Pakistan. The wolf populations in these countries live in deserts, and their density would not appear to support indiscriminate harvests; nevertheless, they survive.

Game Status

In a few countries and states, such as Alaska (see Fritts et al., chap. 12 in this volume), the wolf is simply consid-

ered a game species. This was the case in Poland for many years, as it is today in part of Spain. With an estimated total of 900 wolves in Poland, the annual bag quota was approximately 110 wolves (Bobek et al. 1993); since poaching was very limited, this take did not prevent the Polish wolf population from expanding. Today wolves in Poland are officially protected, and a population increase is expected. In Spain, the average annual bag quota in the six communities of northern Spain is about 19% of the population, plus poaching (Blanco et al. 1992). Still, the population continues to spread.

Management of the wolf as a game animal is strongly opposed by the more radical protectionists, who call for a reduction in the harvest and better law enforcement. However, full protection of the wolf in Spain cannot be rationally proposed. The population of approximately 2,000 wolves would soon grow to unsustainable numbers for the rural economy. Each Spanish wolf causes an average U.S.$500 in damage per year, with peaks of $2,083 (Blanco et al. 1992), and this damage occurs even where wild prey is abundant; rural public opinion appears to be prevalently negative toward the wolf (Blanco et al. 1992). Widespread regulated hunting in Spain, which seems inevitable, ideally should be conducted as part of a national management plan.

Active Management Policies and Plans

The wolf management plans being implemented in Alaska and Canada have been widely sensationalized and exploited by the media (see Fritts et al., chap. 12 in this volume; for reviews see Carbyn 1983c, 1987; Theberge 1991; Hayes and Gunson 1995; Harbo and Dean 1983; Stephenson et al. 1995). Controversy increased when a greater protectionist sentiment led to public questioning of classic methods of wolf management, which until the 1960s and early 1970s basically meant wolf control sponsored by the state (Carbyn 1983c). Public opinion was split between acceptance and nonacceptance of wolf populations being controlled by government action for the benefit of a particular social group—namely, hunters.

In Alaska (6,000–7,000 wolves) and Canada (50,000–60,000 wolves), the wolf is hunted as big game, with long hunting and trapping seasons and often without any bag limits (Stephenson et al. 1995; Hayes and Gunson 1995). The most extreme situation may be in Alberta and Ontario, where unlicensed landowners can kill wolves on their land at any time.

Aerial hunting has been prohibited in Alaska since

1972, but has been replaced by "land-and-shoot" hunting, which allows sighting and harassment of wolves from the air, but shooting only after landing. This technique is difficult to control and, associated with liberal bag limits, can reduce local wolf densities significantly in areas where sparse tree cover, lakes, and rivers are common (Van Ballenberghe 1991). It has also been strongly contested on ethical grounds, such as the ease with which it can be extended to remote areas (Stephenson et al. 1995).

Alaska does not control wolves for livestock depredations, as they rarely occur given that few livestock raisers live in high-density wolf country, but the Alaska Department of Fish and Game (ADF&G) has carried out at least five programs for wolf control with the aim of increasing the moose and caribou populations for the benefit of other wildlife and hunters. However, while hunters took about 1,400 wolves per year in the years 1976–1986, only about 100–150 per year were killed during this period by control programs. Neither the wolf nor the ungulates have ever risked extinction because of these numerical fluctuations (ADF&G 1992). Today Alaska is still far from having found a solution to the problem of wolf management that satisfies everyone. Litigation and controversy are likely to continue.

In Canada, the estimated percentage of the wolf population taken annually (mostly trapped) varies regionally from 4% to 11%, and declined by 40% between 1983 and 1990. Wolves are not limited by these takings except perhaps along the southern range of their distribution, where most of the harvest takes place (Theberge 1991; Hayes and Gunson 1995). Most provinces control wolves for livestock depredations; the annual average take for this purpose for the whole country in the years 1986–1991 was about 240 wolves. In addition, British Columbia took about 125 wolves per year to help increase ungulate populations, while the Yukon cropped about 40 per year (Hayes and Gunson 1995). However, in spite of the wolf management plans in several provinces and significant participation by the public in the creation of those plans, fierce opposition has been raised against these control schemes, mainly for ethical reasons.

Intensive government control of wolves is now declining. New techniques of wolf density control, such as the surgical sterilization (see Kreeger et al., chap. 7 in this volume) of free-ranging wolves (Mech, Fritts, and Nelson 1996), are promising, but far from providing a viable alternative to classic methods (Spence et al. 1999). Solutions to the impasse will require concessions on both sides and a shift away from position-based arguments to interest-based negotiations about where and when wolves should be managed.

Managing the Public

Because wolves and their management are highly controversial, managing human attitudes toward wolves is of paramount importance in wolf management (Boitani and Zimen 1979; USFWS 1982, 1987, 1989; Ginsberg and Macdonald 1990; Jhala and Giles 1991; Blanco et al. 1992; Boitani 1992; Promberger and Schröder 1993; Stephenson et al. 1995; Hayes and Gunson 1995; Fritts et al. 1995; Clarkson 1995; Thiel and Valen 1995; Mech 1995a, 2000b,c; Haggstrom et al. 1995).

Some professionals believe that considering public attitudes is crucial to wolf conservation, while others believe that managing the public is an opportunity that managers cannot afford to miss (Nie 2003). In any case, the issue is seldom ignored. There are two stages in the process of public involvement: the first is to analyze human attitudes and expectations, and the second is to manage them to help attain management objectives.

The study of human attitudes toward wildlife was pioneered in the United States, and the wolf has been one of its most frequent subjects (see Fritts et al., chap. 12 in this volume). The results of such studies reveal how different human attitudes toward the wolf can be, depending on region, age, education, profession, closeness to natural habitat, and other social and economic parameters. Knowledge of attitudes has been used in promoting conservation projects; for example, proposals for the Yellowstone wolf reintroduction were encouraged by the finding that most state resident and nonresident park visitors favored it (McNaught 1987).

Still, there is no standard approach to incorporating public opinion into wolf management decisions. Although the role of social sciences in wildlife management was emphasized long ago (Williamson and Teague 1971), public involvement is sometimes ignored and sometimes, at the other extreme, given priority over all other input (Thiel and Valen 1995; Mech 1996, 2000c).

There are several degrees of possible public involvement in wolf management. The first consists of informing the public of activities under way. Gilbert (1964), Boitani and Zimen (1979), Mech (1979b), and Boitani (1992) have stressed the importance of an informed public to wolf conservation.

However, decision making must also include public

participation, or at least participation of the main stakeholders. This approach, named and implemented in various ways in different contexts (e.g., collaborative management) (Borrini-Feyerabend 1996; McNeeley 1995), appears to be the only framework in which biological and social conflicts can be reconciled. The approach is based on two main considerations. The first postulates that democracy requires public participation; ignoring this principle can lead to management by public referendum, a potential wildlife management disaster (Mech 1996). The second consideration is more pragmatic and seeks to use public involvement to reconcile controversy and generate the support of the public (Thiel and Valen 1995). It is no coincidence that wolf management was among the first conservation issues to involve the public, since opinions regarding the wolf are so polarized. Several attempts to use this process have been made, with varying results (see Fritts et al., chap. 12 in this volume).

Whereas the task of government agencies is to facilitate consensus among interest groups by informing them of basic technical data, the role of nongovernmental organizations is to organize diverse public opinions. The role of organizations of wolf advocates and wolf opponents is both legitimate and fundamental, even when their proposals are extreme. In particular, the role of watchdog over natural resource programs is basic to the health of a democratic system and to maintaining ecologically and socially sound wolf management programs.

Conclusion

Wolves and wolf conservation mean different things to different people, but rarely mean indifference (Mech 1970). The range of public attitudes runs from considering the wolf vermin to making it a symbol of nature conservation. The wolf itself—that is, the real animal—

shows such great ecological flexibility and complex behavior that scientists, in spite of some of the most expensive and long-lasting research programs ever undertaken, still have much to learn about wolf biology. Thus there cannot be a single recipe for wolf conservation that can be applied in all ecological and social contexts. Rather, there are several diverse solutions depending on the needs of both humans and wolves at the local level.

In the early 1970s, Douglas Pimlott (1975), a wolf conservation pioneer, claimed that wolf conservation policy needed to reflect a balance among the plurality of viewpoints held about wolves. Thirty years later, and after much research and discussion by many biologists, sociologists, and activists around the world, we suggest that finding that balance is possible only at the local level. International treaties and governments can and must provide the overall institutional and legislative framework, but effective conservation can be implemented only with the participation of all stakeholders.

What history teaches us is the relative ease with which wolves were wiped out of large areas, but also the formidable resilience of this species and its ability to recover wherever it is given a chance. Thus, humans have the technical power to decide the fate of the wolf.

Currently these are good times for the wolf, and probably those good times will continue in the near future. However, the key to long-term wolf conservation is the degree of tolerance and rationality that humans will be able to muster. Tolerance may mean accepting that wolves are totally protected in some areas, forbidden in others, and controlled in still others; that wolves remain free to kill their large prey and even livestock in some areas; and that some wolves live in marginal habitats and others in wilderness.

The wolf as a species knows how to handle all this; humans still have to learn.

Conclusion

L. David Mech and
Luigi Boitani

WOLVES CAN LIVE almost anywhere in the Northern Hemisphere, and almost everywhere they do, they are an issue. In the vast emptiness of the northern tundra or the Arabian desert, on the outskirts of a European town or in the safety of an American national park, in meager agricultural lands in India or mountains in rich Norway or Switzerland, wolves always attract people's attention. Wolves form a key part of many ecosystems, and they are considered charismatic creatures by most human cultures. Thus they polarize public opinion and make headlines year after year.

If we look back 60 years to the first landmark monograph by Young and Goldman (1944), or just 30 years to Mech's (1970) volume, we can see that both scientific knowledge of wolf biology and human attitudes toward the wolf have improved tremendously. The wolf has benefited from, and has often been a protagonist and a symbol of, the remarkable changes in the way Western societies regard conservation. However, much of this improvement paralleled the increasing distance between urban and rural cultures, and most of the changes occurred in urban populations.

These changes were useful in reversing some of the negative trends in conservation, such as the decline of some small wolf populations, but they also resulted in large portions of our societies having an increasingly idealized and possibly biased perception of nature and its dynamics. In short and crude terms, the number of people who love the wolf has increased, but the number of those who understand its ecological context has probably decreased. From the excesses of indiscriminate wolf killing we often moved to excesses of wolf protection.

We are now facing the difficult challenge of redirecting the vast support for wolf conservation toward more rational and contextual reasoning in which not only the wolf, but also the whole environment, including the legitimate interests of humans, is considered. After decades of advocacy for wolf conservation using all possible means to sell the goal of wolf recovery, it is now necessary to start advocating for compromise between wolf and human interests.

Scientific research plays a special role in this process, as it provides the basis for rational common ground. However, research efforts within the wolf's range have been diverse, with the majority of data pertaining to North America. So too have the ways in which scientific data have been used for management and conservation. Too often, particularly in Europe, we have seen management action taken without appropriate consideration for existing data, missing a precious opportunity to move conservation away from uninformed confrontation of opposing lobbies. We need to find more efficient ways for policymakers to use the available data or conduct management-oriented research. Society at large would benefit from increased use of and familiarity with scientific data, especially about the wolf, which has been idealized and misunderstood as few other animal species.

In the preceding chapters, we covered the historical reasons for the continuing battle over wolf conservation and management as well as the wolf's extraordinary biological adaptability, which makes it one of the most resilient animals in the world. Despite a remarkable amount of available scientific data and many excellent accounts of wolf management issues, it is hard to find general conclusions on how to manage wolf-human conflicts. If any conclusion can be drawn, it is that every case is unique.

We and many of our colleagues around the world have been involved in wolf management for years, and each case is a different story, a unique blend of the attitudes and laws of the local people and the ecology of the local wolves. Therefore, there is no single solution to wolf-human conflicts—there must be many, one for every context. Nor is there any recipe for crafting solutions. The wolf has proved to be a particularly tough challenge for policymakers everywhere, the main reason being not so much the amount of conflict involved as the high level of emotion and prejudice pervading all confrontations. The difficulties of navigating through the many positions of stakeholders, lobbyists, public opinion, and politicians have been discussed elsewhere (Mech 2000b,c).

However, as we end a successful period of wolf management in which many small wolf populations have been restored to safe levels and new populations established, we can perhaps build on these experiences to look for common ground for future wolf management. If North America can claim the best data bases on wolf biology and ecology, Europe and Asia offer several living examples of the extent to which wolves can thrive in areas with high human density. There is an emerging need for a revised conservation philosophy to guide us into the next decades, based on wise management of the current positive trends of many wolf populations. If trench warfare was justified in the past, when we had to reverse the negative trends in wolf conservation, in the future we should adopt a strategy to suit the rapid emergence of new patterns of wolf-human coexistence.

The first point of this strategy will have to be the abandonment of the old prejudice that wolves are denizens of the wilderness and that they need wilderness to survive. Of course, in pristine areas wolves will be exposed to the full range of natural conditions, and they will have a life free from human influence. These areas should remain essential components of a broader conservation strategy, but the concept that wolves can or should be saved only in human-free areas is passé. Wolves appear to cope well with extreme wilderness, but they also inhabit crowded agricultural lands at the outskirts of towns and villages. The concept that wolves living near human settlements have a "degraded" life is strongly anthropocentric and the product of a stereotyped view of nature. This concept is often used to jus-

tify removing wolves from human-inhabited areas, as if to save them from a degenerate life, but it thus prevents wolves from exploiting another niche. We must forget about wolves being only beasts of the wilderness and focus on the wolf-human interface: this is the real challenge for conservation and is where wolf conservation most benefits overall biodiversity.

Second, we need to fully accept that wolves and humans can live an integrated coexistence in the same area, rather than having to be segregated forever in separate districts (nature reserves vs. human-dominated lands). Many good examples of wolves inhabiting multi-use landscapes can be found throughout most wolf range in Europe, the Middle East, and Asia, and increasingly in North America, with wolves now regularly visiting the outskirts of large cities in Minnesota, Montana, and Wisconsin. Appropriate local tactics for keeping the integration within sustainable limits must be found, but the overall strategy should be maintained, at least in areas lacking wilderness. Besides preserving existing wilderness against expanding human encroachment, it may well be that this is the only option we have for the future of wolves and many other large carnivores in increasingly human-dominated landscapes.

Third, we need a shift in our long-standing conservation paradigm, from measuring success in terms of wolf numbers toward new goals in which success means expanding wolf ranges rather than numbers. Demanding that wolf populations be allowed to continue to increase is not only a false conservation goal, but also a counterproductive tactic that is bound for short-term failure. It is strategically preferable to promote wolf range expansion and to accept reduction of unacceptable levels of conflict through scientifically planned and managed culling rather than through uncontrolled poaching. Full protection of wolf populations living near, or interspersed with, human settlements leads sooner or later to surplus wolves being killed, legally or illegally. Opposing wolf killing altogether implies accepting that all wolves will eventually be removed from these areas, whereas accepting some wolf control will allow wolves over much larger ranges (Mech 1995a). This vision requires a fundamental shift in the way wolves are perceived by folks who consider every wolf a symbol of the conservation battle or an animal with special rights among all other species. In the end, this approach probably will yield many more wolves than we could afford to keep in a few fully protected areas, no matter how large.

Fourth, we should make an extra effort at all levels of

management to keep the objectivity of scientific data separate from our legitimate emotional bonds with wolves. Far too often confrontations on wolf issues mix scientific data with emotion. Both are important, but they belong to two different stages of the negotiating process that leads to the final political decisions. Scientists are particularly touchy on this issue, as they often feel they could lose credibility if they also act as conservation advocates. On the contrary, scientists are morally obliged to be advocates for the conservation of the species they are working on (Bekoff 2001); their knowledge of ecology and their training in the use of criticism make them an irreplaceable force to inform and facilitate the decisions of all other stakeholders. However, scientists advocating conservation must strive continually to separate their feelings from their research and their objective knowledge.

Finally, a fifth point of the revised strategy is that methods of wolf management should be independent of a society's wealth. The outcome of a conservation strategy cannot depend on the amount of money a country is able to pay to sustain wolves, but must be the result of a philosophical acceptance of wolf-human coexistence. The recent recovery of several wolf populations in Europe and North America has brought a great variety of responses at local levels, depending on old and new attitudes toward wolves. Each society has its own body of cultural and technical means to achieve rational wolf management and will rely on traditional and modern methods to prevent wolf damage to livestock, to increase the level of tolerance toward damage, and to control wolf populations. Whatever the outcome of this strategy, there will be countless variations of possible compromises between the wolf's needs and people's expectations, depending more on social and political factors than on technical means.

Wolf conservation tends to focus discussions on the management of the animal, often with little regard for the rest of the environment in which a wolf population lives, but wolves are just one of the many elements of the environment, and their conservation is often best accomplished by managing several other components of the ecosystem in a holistic approach. Wolves should be saved and managed as part of the whole context, not because they are singled out as special species.

A central challenge that we will have to face as conservation proceeds into the coming decades is to revise the ways we sell conservation efforts. In the recent past, wolves were labeled a flagship species or an umbrella, in-

dicator, or keystone species, depending on what conservation market one was trying to penetrate. Some of the authors of the foregoing chapters may not agree, but we think arguments can be made that wolves do not necessarily deserve any of these labels (Linnell et al. 2000).

A flagship species is an attraction to nearly all society's strata, but wolves are not welcomed by all factions of society. With a few rare exceptions, the rural world opposes wolves, so the animal's flagship role is restricted primarily to urbanites or to local areas. Wolves are certainly a powerful flagship species for the conservation movement, particularly that of affluent societies with strong lobbies in large cities, but a true flagship species should be able to move an entire society toward a goal.

Neither are wolves a good umbrella species (i.e., a species, usually high in the ecological pyramid, whose conservation necessarily fosters that of the rest of the chain) in that they can live well on a variety of food resources and in areas with an impoverished prey base. Wolves are not a keystone species (sensu Simberloff 1998) either, in that they are not essential for the presence of many other species (e.g., herbivores flourish in areas devoid of wolves). And wolves are not necessarily indicators of habitat quality or integrity because they are too generalist to be good indicators of the presence of a pristine trophic chain.

The above labels have been very useful in many circumstances and have contributed significantly to wolf recovery. They may still be useful in the future, but we should be aware that they are shortcuts to "sell a product" rather than good scientific grounds on which to build conservation. In the near future, when hopefully the primary concern of wolf conservation will be the management of recovered populations, we will need to abandon the use of inappropriate labels and turn to more substantial concepts and solutions for conservation.

Such an approach will be particularly important as we attempt to address the difficult issues of expanding and increasing wolf populations using such unpopular tools as zoning, delisting, and population control. Labels have been of tremendous help in engaging emotions and obtaining quick support for wolf recovery, but managing expanding wolf populations will require solid and consistent arguments rather than emotional pressures. We will need to change the values, strategies, and tactics of wolf conservation, as well as using different mechanisms for conflict resolution and decision making. The temporal and spatial scales on which we have considered conservation actions in the past 30 years need now to be expanded to incorporate longer-term strategies: the fast responses needed to reverse negative trends at local levels should be replaced by more thoughtful and concerted efforts that expand across national boundaries.

If we give up using the old labels, we are left with the true core of wolf conservation, which is the understanding of the animal's biology and the acceptance of the creature for its intrinsic aesthetic and ethical values, even though it means tolerance for some inevitable conflict.

We hope this book will help shape this new attitude toward the wolf.

Appendix: Species Names
Used in the Text

COMMON NAME	LATIN BINOMIAL	COMMON NAME	LATIN BINOMIAL
African hunting dog	*Lycaon pictus*	Deer, mule	*Odocoileus hemionus*
African lion	*Panthera leo*	Deer, musk	*Moschus moschiferus*
African wild dog	*Lycaon pictus*	Deer, red	*Cervus elaphus*
Arctic fox	*Alopex lagopus*	Deer, roe	*Capreolus capreolus*
Arctic hare	*Lepus arcticus*	Deer, sika	*Cervus nippon*
Badger, American	*Taidea taxus*	Deer, white-tailed	*Odocoileus virginianus*
Badger, European	*Meles meles*	Deermouse	*Peromyscus leucopus*
Beaver	*Castor canadensis*	Demodectic mange	*Demodex canis*
Bison, American	*Bison bison*	Dhole or Asian wild dog	*Cuon alpinus*
Bison, European	*Bison bonasus*	Dingo	*Canis familiaris dingo* or *Canis lupus* (in dispute)
Black bear	*Ursus americanus*		
Blackbuck	*Antilope cervicapra*	Dog (domestic)	*Canis lupus familiaris*
Black-footed ferret	*Mustela nigripes*	Elephant	*Loxodonta africana*
Bobcat	*Lynx rufus*	Elk	*Cervus elaphus*
Brown bear	*Ursus arctos*	Fox	*Vulpes* spp.
Bush dog	*Speothos venaticus*	Goat (domestic)	*Capra hircus*
California condor	*Gymnogyps californianus*	Golden eagle	*Aquila chrysaetos*
Caribou	*Rangifer tarandus*	Golden jackal	*Canis aureus*
Cat (domestic)	*Felis domesticus*	Gray jay	*Perisoreus canadensis*
Cattle (domestic)	*Bos taurus*	Guinea pig	*Cavia porcellus*
Chamois	*Rupicapra rupicapra*	Hare	*Lepus* spp.
Chimpanzee	*Pan troglodytes*	Horse	*Equus caballus*
Cotton rat	*Signodon hispidus*	Human	*Homo sapiens*
Cougar	*Felis concolor*	Hyena	Hyaenidae
Coyote, American	*Canis latrans*	Ibex	*Capra ibex*
Coyote, Old World (extinct[b])	*Canis arnensis*	Kirtland's warbler	*Dendroica kirtlandii*
		Lesser panda	*Ailurus fulgens*
Coyote, rabbit-eating (extinct[b])	*Canis lepophagus*	Lynx, Canadian	*Lynx canadensis*
		Lynx, Eurasian	*Lynx lynx*
Deer, black-tailed	*Odocoileus hemionus*	Maned wolf	*Chrysocyon brachyurus*
Deer, fallow	*Dama dama*	Margay	*Felis wiedii*

COMMON NAME	LATIN BINOMIAL
Marsh rabbit	*Sylvilagus palustris*
Marten, American	*Martes americana*
Mink	*Mustela vison*
Moose	*Alces alces*
Mouflon	*Ovis musimon*
Mountain goat (tur)	*Capra caucasica* (Europe)
Mountain goat	*Oreamnos americanus* (North America)
Mountain sheep	*Ovis canadensis*
Muskox	*Ovibos moschatus*
Nutria	*Myocastor coypus*
Ocelot	*Felis pardalis*
Pig (domestic)	*Sus scrofa*
Polar bear	*Ursus maritimus*
Pronghorn	*Antilocapra americana*
Rabbit	*Sylvilagus* spp.
Raccoon	*Procyon lotor*
Raccoon dog	*Nyctereutes procyonoides*
Rat	*Rattus* spp.
Raven	*Corvus corax*
Red fox	*Vulpes vulpes*
Reindeer	*Rangifer tarandus*
River otter	*Lutra canadensis*
Saiga	*Saiga tatarica*
Sarcoptic mange	*Sarcoptes scabiei*
Sheep, bighorn	*Ovis canadensis*
Sheep, Dall	*Ovis dalli*
Sheep (domestic)	*Ovis aries*
Sheep, stone	*Ovis dalli*
Short-toed eagle	*Circactus gallicus*
Simien jackal (Ethiopian wolf)	*Canis simensis*
Snowshoe hare	*Lepus americanus*
Striped hyena	*Hyaena hyaena*
Striped skunk	*Mephitis mephitis*
Swift fox	*Vulpes velox*
Ticks	*Dermacentor variabilis, Amblyomma americanum*
Tiger	*Panthera tigris*
Weasel	*Mustela* spp.
White-footed mouse	*Peromyscus leucopus*
Wild boar	*Sus scrofa*
Wolf, Algonquin	*Canis lycaon*[a]
Wolf, Armbruster's (extinct[b])	*Canis armbrusteri*
Wolf, dire (extinct[b])	*Canis dirus*
Wolf, eastern timber	*Canis lupus lycaon*

COMMON NAME	LATIN BINOMIAL
Wolf, Ethiopian (Simien jackal)	*Canis simensis*
Wolf, Etruscan (extinct[b])	*Canis etruscus*
Wolf, Falconer's (extinct[b])	*Xenocyon falconeri*
Wolf, Geza's (extinct[b])	*Canis gezi*
Wolf, gray	*Canis lupus*
Wolf, Irvingtonian (extinct[b])	*Canis priscolatrans*
Wolf, Mexican	*Canis lupus baileyi*
Wolf, Miller's (extinct[b])	*Canis milleri*
Wolf, Mosbachen (extinct[b])	*Canis mosbachensis*
Wolf, Nehring's (extinct[b])	*Canis nehringi*
Wolf, red	*Canis rufus*
Wolverine	*Gulo gulo*

[a] Proposed by Wilson et al. 2000.

[b] Known from fossils.

Contributors

Cheryl S. Asa
Research Department
St. Louis Zoo
1 Government Drive
St. Louis, MO 63110
asa@slu.edu

Warren B. Ballard
Department of Range, Wildlife, and Fisheries Management
Texas Tech University
Box 42125
Lubbock, TX 79409
wballard@ttacs.ttu.edu

Luigi Boitani
Department of Animal and Human Biology
University of Rome "La Sapienza"
Viale Universita, 32
Rome, Italy 00185
luigi.boitani@uniroma1.it

Ludwig N. Carbyn
Canadian Wildlife Service
4999, 98th Ave., 2nd Floor
Edmonton, Alberta
CANADA T6B 2X3
lu.carbyn@ec.gc.ca

Paolo Ciucci
Department of Animal and Human Biology
University of Rome
Viale Universita, 32
Rome, Italy 00185
paolo.ciucci@uniroma1.it

Jean Fitts Cochrane
U.S. Fish and Wildlife Service
P.O. Box 668
Grand Marais, MN 55604
jean_cochrane@fws.gov

Steven H. Fritts
U.S. Fish and Wildlife Service
P. O. Box 25486
Denver, CO 80225
steve_fritts@fws.gov

Todd K. Fuller
Department of Natural Resources Conservation
University of Massachusetts
Amherst, MA 01003-4210
tkfuller@forwild.umass.edu

Fred H. Harrington
Psychology Department
Mt. St. Vincent University
Halifax, Nova Scotia
CANADA B3M 2J6
fred.harrington@msvu.ca

Robert D. Hayes
Box 5499
Haines Junction
Yukon, CANADA Y0B 1L0
hayes@yknet.ca

V. Gary Henry
206 Arrowhead Lane
Asheville, NC 28806

Brian T. Kelly
USDA-APHIS
National Wildlife Research Center
Utah State University
Logan, UT 84322-5295
brian_t_kelly@fws.gov

CURRENT ADDRESS:
U.S. Fish and Wildlife Service
P.O. Box 1306
Albuquerque, NM 87103-1306

Terry J. Kreeger
Wyoming Game & Fish Department
Sybille Wildlife Research Unit
2362 Hwy. 34
Wheatland, WY 82201
tekreege@wyoming.com

L. David Mech
U.S. Geological Survey
8711 37th Street, SE
Northern Prairie Wildlife Research Center
Jamestown, ND 58402-7317
mechx002@tc.umn.edu

MAILING ADDRESS:
Department of Fisheries, Wildlife, and Conservation Biology
The Raptor Center
University of Minnesota
1920 Fitch Avenue
St. Paul, MN 55108

Ronald M. Nowak
2101 Greenwich Street
Falls Church, VA 22043
ron4nowak@cs.com

Jane M. Packard
Department of Wildlife and Fisheries Sciences
Texas A & M University
College Station, TX 77843-2258
ethology@tamu.edu

Rolf O. Peterson
School of Forest Resources and Environmental Science
Michigan Technological University
Houghton, MI 49931
ropeters@mtu.edu

Michael K. Phillips
Turner Endangered Species Fund
1123 Research Drive
Bozeman, MT 59718
tesf@montana.net

George B. Rabb
Brookfield Zoo
3300 Golf Road
Brookfield, IL 60513
bzadmin@brookfieldzoo.org

Douglas W. Smith
U.S. Park Service
Yellowstone Center for Resources
P.O. Box 168
YNP, WY 82190
doug_smith@nps.gov

Robert O. Stephenson
Alaska Department of Fish & Game
1300 College Road
Fairbanks, AK 99701
bob_stephenson@fishgame.state.ak.us

Carles Vilà
Department of Evolutionary Biology
Uppsala University
Norbyvägen 18D
S-752 36 Uppsala (Sweden)
carles.vila@ebc.uu.se

Robert K. Wayne
Biology Department
University of California
621 Circle Drive South
Los Angeles, CA 90024
rwayne@biology.lifesci.ucla.edu

References

Abrams, P. A. 2000. The evolution of predator-prey interactions. *Annu. Rev. Ecol. Syst.* 31:79–105.

Abuladze, K. I. 1964. *Osnovy Tsestodologii.* Vol. IV. *Teniaty–lentochnye gel'minty zhivotnykh i cheloveka i vyzyvaevaniia.* Nauka, Moscow. 530 pp.

Achuff, P. L., and R. Petocz. 1988. Preliminary resource inventory of the Arjin Mountains Nature Reserve, Xinjiang, People's Republic of China. World Wide Fund for Nature, Gland, Switzerland. 78 pp.

Ackerman, B. B., F. A. Leban, M. D. Samuel, and E. O. Garton. 1990. User's manual for program Home Range. 2d ed. Technical Report no. 15. Forestry, Wildlife, and Range Experiment Station, University of Idaho, Moscow.

Acorn, R. C., and M. J. Dorrance. 1990. Methods of investigating predation of livestock. Alberta Agriculture, Edmonton. 36 pp.

Adamakopoulos, P., and T. Adamakopoulos. 1993. Wolves in Greece: Current status and prospects. Pp. 56–61 in C. Promberger and W. Schröder, eds., *Wolves in Europe: Status and perspectives.* Munich Wildlife Society, Ettal, Germany.

Adamczewski, J. Z., P. F. Flood, and A. Gunn. 1995. Body composition of muskoxen *(Ovibos moschatus)* and its estimation from condition index and mass measurements. *Can. J. Zool.* 73:2021–34.

Adamic, M. 1993. Status of the wolf in Slovenijia. Pp. 70–73 in C. Promberger and W. Schröder, eds., *Wolves in Europe: Status and Perspectives.* Munich Wildlife Society, Ettal, Germany.

Adamovich, V. L. 1985. Rabies in wolves in the central part of the Russian valley. *Zool. Zh.* 64:590–99.

Adams, D. R., and M. D. Wiekamp. 1984. The canine vomeronasal organ. *J. Anat.* 138:771–87.

Adams, L. G., and B. W. Dale. 1998a. Reproductive performance of female Alaskan caribou. *J. Wildl. Mgmt.* 62:1184–95.

———. 1998b. Timing and synchrony of parturition in Alaskan caribou. *J. Mammal.* 79:287–94.

Adams, L. G., B. W. Dale, and L. D. Mech. 1995. Wolf predation on caribou calves in Denali National Park, Alaska. Pp. 245–60 in L. N. Carbyn, S. H. Fritts, and D. R. Seip, eds., *Ecology and conservation of wolves in a changing world.* Canadian Circumpolar Institute, Edmonton, Alberta.

Adams, L. G., F. G. Singer, and B. W. Dale. 1995. Caribou calf mortality in Denali National Park, Alaska. *J. Wildl. Mgmt.* 59:584–94.

Adams, L. G., and R. O. Stephenson. 1986. Wolf Survey, Gates of the Arctic National Park and Preserve—1986. U.S. National Park Service, Anchorage, AK.

Adams, M. G. 1980. Odour-producing organs of mammals. *Symp. Zool. Soc. Lond.* 45:57–86.

Adorjan, A. S., and G. Kolenosky. 1969. A manual for the identification of hairs of selected Ontario mammals. Research Report (Wildlife), no. 90. Ontario Department of Lands and Forests, Toronto.

Afik, D., and B. Pinshow. 1993. Temperature regulation and water economy in desert wolves. *J. Arid Environ.* 24:197–209.

Aho, K. 2000. Wolf attacks 6-year-old near Yukatat. *Anchorage Daily News,* 27 April.

Alaska Department of Fish and Game. 1992. Strategic wolf management plan for Alaska. *Alaska's Wildlife,* Jan./Feb., S7–S14.

Al-Bagdadi, F., and J. Lovell. 1979. The integument. Pp. 78–101 in H. E. Evans and G. C. Christensen, eds., *Miller's anatomy of the dog,* 2d ed. W. B. Saunders, Philadelphia.

Albone, E. S. 1984. *Mammalian semiochemistry: The investigation of chemical signals between mammals.* John Wiley and Sons, New York.

Albone, E. S., and G. C. Perry. 1976. Anal sac secretions of the red fox, *Vulpes vulpes;* volatile fatty acids and diamines: Implications for a fermentative hypothesis of chemical recognition. *J. Chem. Ecol.* 2:101–11.

Albright, L. B., III. 2000. *Biostratigraphy and vertebrate paleontology of the San Timoteo Badlands, southern California.* University of California Publications in Geological Sciences, 144. University of California Press, Berkeley. 121 pp.

Allen, D. L. 1979. *Wolves of Minong: Their vital role in a wild community.* Houghton Mifflin, Boston.

Allen, W. F. 1937. Olfactory and trigeminal conditioned reflexes in dogs. *Am. J. Physiol.* 118:532–40.

Allendorf, F. W., and R. F. Leary. 1986. Heterozygosity and fitness in natural populations of animals. Pp. 57–76 in M. E. Soulé, ed., *Conservation biology: The science of scarcity and diversity.* Sinauer Associates, Sunderland, MA.

Almasan, H., G. Scarlatescu, V. Nesterov, and L. Manolache. 1970. Contribution a la connaissance du regine de nourriture du loup (*Canis lupus* L.) dans les Carpathes roumaines. *Trans. Int. Congr. Game Biol.* 9:523–29.

Altmann, D. 1974. Beziehungen zwischen socialer rangordnung und jungenautzucht bei *Canis lupus. Zool. Gart. N.F. Jena* 44:235–36.

———. 1987. Social behavior patterns in three wolf packs at Tierpark Berlin. Pp. 415–24 in H. Frank, ed., *Man and wolf: Advances, issues and problems in captive wolf research.* Dr. W. Junk Publishers, Dordrecht, The Netherlands.

Álvares, F. J. 1995. Aspectos da distribuição e ecologia do lobo no noroeste de Portugal. Master's thesis, Universidade de Lisboa, Portugal.

Amos, B., C. Schlötterer, and D. Tautz. 1993. Social structure of pilot whales revealed by analytical DNA profiling. *Science* 260:670–72.

Andelt, W. F. 1985. Behavioral ecology of coyotes in south Texas. Wildlife Monographs, no. 94. The Wildlife Society, Bethesda, MD. 45 pp.

Anderson, A. C. 1970. *The beagle as an experimental dog.* Iowa State University Press, Ames.

Anderson, E. 1984. Who's who in the Pleistocene: A mammalian bestiary. Pp. 40–89 in P. S. Martin and R. G. Klein, eds., *Quaternary extinctions: A prehistoric revolution.* University of Arizona Press, Tucson.

———. 1996. Preliminary report on the Carnivora of Porcupine Cave, Park County, Colorado. Pp. 259–82 in K. M. Stewart and K. L. Seymour, eds., *Palaeoecology and palaeoenvironments of late Cenozoic mammals: Tributes to the career of C. S. Churcher.* University of Toronto Press, Toronto.

Anderson, R. K., B. L. Hart, and L. A. Hart, eds. 1984. *The pet connection.* Proceedings of the Minnesota-California Conferences on the human-animal bond. Center to Study Human-Animal Relationships and Environments, University of Minnesota, Minneapolis.

Andersone, Z., V. Lucchini, E. Randi, and J. Ozolins. 2002. Hybridisation between wolves and dogs in Latvia as documented using mitochondrial and microsatellite DNA markers. *Mammal. Biol.* 67:79–90.

Andersson, T., A. Bjärvall, and M. Bromberg. 1977. Attitudes to the wolf in Sweden: An interview study. The Swedish Environment Protection Board PM 850, 65 pp. General results given in discussion of paper by Bjärvall. 1983. Scandinavia's response to a natural repopulation of wolves. *Acta Zool. Fenn.* 174:273–75.

Andreyev, F. V. 1985. Organ of sight. [In Russian; trans. T. Neklioudova.] Pp. 267–77 in D. I. Bibikov, ed., *The wolf: History,* systematics, morphology, ecology. [In Russian.] Izdatetvo Nauka, Moskva.

Anisko, J. 1976. Communication by chemical signals in Canidae. Pp. 283–93 in R. L. Doty, ed., *Mammalian olfaction, reproductive processes and behavior.* Academic Press, New York.

Aoki, T., and M. Wada. 1951. Functional activity of the sweat glands in the hairy skin of the dog. *Science* 114:123–24.

Appelberg, B. 1958. Species differences in the taste qualities mediated through the glossopharyngeal nerve. *Acta Physiol. Scand.* 44:129–37.

Appleby, M. C. 1993. How animals perceive a hierarchy: Reactions to Freeman et al. *Anim. Behav.* 46:1232–33.

Archer, J., S. J. Taft, and R. P. Thiel. 1986. Parasites of wolves, *Canis lupus,* in Wisconsin, as determined from fecal examinations. *Proc. Helminthol. Soc. Wash.* 53:290–91.

Archibald, W. R., D. Janz, and K. Atkinson. 1991. Wolf control: A management dilemma. *Trans. N. Am. Wildl. Nat. Resour. Conf.* 56:497–511.

Aristotle. 1783. *Histoire des animaux d'Aristote.*

Arjo, W. M., and D. H. Pletscher. 1999. Behavioral responses of coyotes to wolf recolonization in northwestern Montana. *Can. J. Zool.* 77:1919–27.

Arnold, M. L. 1997. *Natural hybridization and evolution.* Oxford University Press, New York.

Asa, C. S. 1995. Physiological and social aspects of reproduction of the wolf and their implications for contraception. Pp. 283–91 in L. N. Carbyn, S. H. Fritts, and D. R. Seip, eds, *Ecology and conservation of wolves in a changing world.* Canadian Circumpolar Institute, Edmonton, Alberta.

———. 1997. Hormonal and experiential factors in the expression of social and parental behavior in canids. Pp. 129–49 in N. G. Solomon and J. A. French, eds., *Cooperative breeding in mammals.* Cambridge University Press, Cambridge.

Asa, C., and L. D. Mech. 1995. A review of the sensory organs in wolves and their importance to life history. Pp. 287–91 in L. N. Carbyn, S. H. Fritts and D. R. Seip, eds., *Ecology and conservation of wolves in a changing world.* Canadian Circumpolar Institute, Edmonton, Alberta.

Asa, C. S., L. D. Mech, and U. S. Seal. 1985. The use of urine, faeces, and anal-gland secretions in scent-marking by a captive wolf (*Canis lupus*) pack. *Anim. Behav.* 33:1034–36.

Asa, C. S., L. D. Mech, U. S. Seal, and E. D. Plotka. 1990. The influence of social and endocrine factors on urine-marking by captive wolves (*Canis lupus*). *Horm. Behav.* 24:497–509.

Asa, C. S., E. K. Peterson, U. S. Seal, and L. D. Mech. 1985. Deposition of anal-sac secretions by captive wolves *Canis lupus. J. Mammal.* 66:89–93.

Asa, C. S., U. S. Seal, M. Letellier, E. D. Plotka, and E. K. Peterson. 1987. Pinealectomy or superior cervical ganglionectomy do not alter reproduction in the wolf (*Canis lupus*). *Biol. Reprod.* 37:14–21.

Asa, C. S., U. S. Seal, E. D. Plotka, M. A. Letellier, and L. D. Mech. 1986. Effect of anosmia on reproduction in male and female wolves (*Canis lupus*). *Behav. Neural Biol.* 46:272–84.

Asa, C. S., and C. Valdespino. 1998. Canid reproductive biology: An integration of proximate mechanisms and ultimate causes. *Am. Zool.* 38:251–59.

Ashmead, D. H., R. K. Clifton, and E. P. Reese. 1986. Development of auditory localization in dogs: Single source and precedence effect sounds. *Dev. Psychobiol.* 19:91–103.

Atkins, D. L., and L. S. Dillon. 1971. Evolution of the cerebellum in the genus *Canis. J. Mammal.* 52:96–107.

Audubon, J. J., and J. Bachman. 1851. *The viviparous quadrupeds of North America.* Volume 2. New York.

Avise, J. C. 1991. Ten unorthodox perspectives on evolution prompted by comparative genetic findings on mitochondrial DNA. *Annu. Rev. Genet.* 25:45–69.

———. 1994. *Molecular markers, natural history and evolution.* Chapman and Hall, New York.

———. 2000. *Phylogeography: The history and formation of species.* Harvard University Press, Cambridge, MA.

Avise, J. C., J. Arnold, R. M. Ball, E. Bermingham, T. Lamb, J. E. Neigel, C. A. Reeb, and N. C. Saunders. 1987. Intraspecific phylogeography: The mitochondrial DNA bridge between population genetics and systematics. *Annu. Rev. Ecol. Syst.* 18:489–522.

Avise, J. C., and R. M. Ball, Jr. 1990. Principles of genealogical concordance in species concepts and biological taxonomy. *Oxford Surv. Evol. Biol.* 7:45–67.

Avise, J. C., J. E. Neigel, and J. Arnold. 1984. Demographic influences on mtDNA lineage survivorship in animal populations. *J. Mol. Evol.* 20:99–105.

Baenniger, R. 1987. Some comparative aspects of yawning in *Betta spendens, Homo sapiens, Panthera leo,* and *Papio sphinx. J. Comp. Psychol.* 101:349–54.

Bailey, T. N., E. E. Bangs, and R. O. Peterson. 1995. Exposure of wolves to canine parvovirus and distemper on the Kenai National Wildlife Refuge, Kenai Peninsula, Alaska, 1976–1988. Pp. 441–46 in L. N. Carbyn, S. H. Fritts, and D. R. Seip, eds., *Ecology and conservation of wolves in a changing world.* Canadian Circumpolar Institute, Edmonton, Alberta.

Bailey, V. 1926. A biological survey of North Dakota. *N. Am. Fauna* 49:1–226.

Bakarich, A. C. 1979. Comparative development in wolves *(Canis lupus)* and dogs *(Canis familiaris).* Master's thesis, Trinity University, San Antonio, TX.

Bakker, R. T. 1983. The deer flees, the wolf pursues: Incongruencies in predator-prey coevolution. Pp. 350–82 in D. J. Futuyma and M. Slatkin, eds., *Coevolution.* Sinaur Associates, Sunderland, MA.

Ballantyne, E. E., and J. D. O'Donoghue. 1954. Rabies control in Alberta. *J. Am. Vet. Med. Assoc.* 125:316–26.

Ballard, W. B. 1980. Brown bear kills grey wolf. *Can. Field Nat.* 94:91.

———. 1982. Gray wolf-brown bear relationships in the Nelchina Basin of south-central Alaska. Pp. 71–80 in F. H. Harrington and P. C. Paquet, eds., *Wolves of the world: Perspectives of behavior, ecology, and conservation.* Noyes Publications, Park Ridge, NJ.

———. 1992. Bear predation on moose: A review of recent North American studies and their management implications. *Alces,* suppl. 1:162–76.

Ballard, W. B., L. A. Ayres, C. L. Gardner, and J. W. Foster. 1991. Den site activity patterns of gray wolves, *Canis lupus,* in south-central Alaska. *Can. Field Nat.* 105:497–504.

Ballard, W. B., L. A. Ayres, P. R. Krausman, D. J. Reed, and S. G. Fancy. 1997. Ecology of wolves in relation to a migratory caribou herd in northwest Alaska. Wildlife Monographs, no. 135. The Wildlife Society, Bethesda, MD. 47 pp.

Ballard, W. B., L. A. Ayres, K. E. Roney, and T. H. Spraker. 1991. Immobilization of gray wolves with a combination of tiletamine hydrochloride and zolazepam hydrochloride. *J. Wildl. Mgmt.* 55:71–74.

Ballard, W. B., and J. R. Dau. 1983. Characteristics of gray wolf *(Canis lupus)* den and rendezvous sites in south-central Alaska. *Can. Field Nat.* 97:299–302.

Ballard, W. B., R. Farnell, and R. O. Stephenson. 1983. Long distance movement by gray wolves *(Canis lupus). Can. Field Nat.* 97:333.

Ballard, W. B., A. W. Franzmann, and C. L. Gardner. 1982. Comparison and assessment of drugs used to immobilize Alaskan gray wolves *(Canis lupus)* and wolverines *(Gulo gulo)* from a helicopter. *J. Wildl. Dis.* 18:339–42.

Ballard, W. B., A. W. Franzmann, K. P. Taylor, T. Spraker, C. C. Schwartz, and R. O. Peterson. 1979. Comparison of techniques utilized to determine moose calf mortality in Alaska. *Proc. N. Am. Moose Conf.* 15:362–87.

Ballard, W. B., and P. S. Gipson. 2000. Wolf. Pp. 321–46 in S. Demarais and P. R. Krausman, eds., *Ecology and management of large mammals in North America.* Prentice Hall, Upper Saddle River, NJ.

Ballard, W. B., and P. R. Krausman. 1997. Occurrence of rabies in wolves of Alaska. *J. Wildl. Dis.* 33:242–45.

Ballard, W. B., M. E. McNay, C. L. Gardner, and D. J. Reed. 1995. Use of line-intercept track sampling for estimating wolf densities. Pp. 469–80 in L. N. Carbyn, S. H. Fritts, and D. R. Seip, eds., *Ecology and conservation of wolves in a changing world.* Canadian Circumpolar Institute, Edmonton, Alberta.

Ballard, W. B., S. D. Miller, and J. S. Whitman. 1990. Brown and black bear predation on moose in south-central Alaska. *Alces* 26:1–8.

Ballard, W. B., T. H. Spraker, and K. P. Taylor. 1981. Causes of neonatal moose calf mortality in south-central Alaska. *J. Wildl. Mgmt.* 45:335–42.

Ballard, W. B., and V. Van Ballenberghe. 1998. Moose-predator relationships: Research and management needs. *Alces* 34:91–105.

Ballard, W. B., J. S. Whitman, and C. L. Gardner. 1987. Ecology of an exploited wolf population in south-central Alaska. Wildlife Monographs, no. 98. The Wildlife Society, Bethesda, MD. 54 pp.

Ballou, J. D. 1997. Genetic and demographic aspects of animal reintroductions. Istituto Nazionale per la Fauna Selvatica—Supplemento alle Richerche di Biologia della Selvaggina

17:75–96. Atti del III Convegno Nazionale dei Biologia della Selvaggina, Bologna.

Baltazard, M., and M. Bahmanyar. 1955. Practical trial of antirabies serum in those bitten by a rabid wolf. *Bull. World Health Org.* 13:742–72.

Banfield, A. W. F. 1954. Preliminary investigation of the barren-ground caribou. Part I. Former and present distribution, migrations, and status. Wildlife Management Bulletin Series 1, no. 10A. Canadian Wildlife Service. 79 pp.

———. 1964. Review of F. Mowat's *Never cry wolf. Can. Field Nat.* 78:52–54.

Bangs, E. E. 1995. Wolf hysteria: Reintroducing wolves to the West. Pp. 397–410 in R. McIntyre, ed., *War against the wolf: America's campaign to exterminate the wolf.* Voyageur Press, Stillwater, MN.

Bangs, E. E., and S. H. Fritts. 1996. Reintroducing the gray wolf to central Idaho and Yellowstone National Park. *Wildl. Soc. Bull.* 24:402–13.

Bangs, E. E., S. H. Fritts, J. A. Fontaine, D. W. Smith, K. M. Murphy, C. M. Mack, and C. C. Niemeyer. 1998. Status of gray wolf restoration in Montana, Idaho, and Wyoming. *Wildl. Soc. Bull.* 26:785–98.

Bangs, E. E., S. H. Fritts, D. R. Harms, J. A. Fontaine, M. D. Jimenez, W. G. Brewster, and C. C. Niemeyer. 1995. Control of endangered gray wolves in Montana. Pp. 127–34 in L. N. Carbyn, S. H. Fritts, and D. R. Seip, eds., *Ecology and conservation of wolves in a changing world.* Canadian Circumpolar Institute, Edmonton, Alberta.

Bangs, E., and J. Shivik. 2001. Managing wolf conflict with livestock in the Northwestern United States. *Carnivore Damage Prevention News* 3:2–5.

Bannikov, A. G., L. V. Zhirnov, L. S. Lebedeva, and A. A. Fandeev. 1967. *Biology of the saiga.* [Translated from the Russian by M. Fleischmann.] U.S. Department of the Interior and National Science Foundation, Washington, D.C. 152 pp.

Barrette, C. 1993. The "inheritance of dominance," or of an aptitude to dominate? *Anim. Behav.* 46:591–93.

Barrette, C., and F. Messier. 1980. Scent-marking in free-ranging coyotes, *Canis latrans. Anim. Behav.* 28:814–19.

Barrientos, L. M. 1993. Evolution of the Iberian wolf (*Canis lupus signatus*) in highly humanized areas in Castilla y Leon (Spain). Pp. 42–44 in C. Vilà and J. Castroviejo, eds., Simposio Internacional sobre el Lobo, 19–23 October, Leon, Spain.

Barsanti, J. A. 1984. Blastomycosis. Pp. 675–86 in C. E. Greene, ed., *Clinical microbiology and infectious diseases of the dog and cat.* W. B. Saunders, Philadelphia.

Barton, N. H., and G. M. Hewitt. 1985. Analysis of hybrid zones. *Annu. Rev. Ecol. Syst.* 16:113–48.

Bartram, W. 1791. *Travels.* New Haven University Press, Philadelphia.

Baru, A. V. 1971. Absolute thresholds and frequency difference limens as a function of sound duration in dogs deprived of the auditory cortex. Pp. 265–85 in G. V. Gersuni, ed., *Sensory processes at the neuronal and behavioral levels* [trans. J. Rose]. Academic Press, New York.

Bass, R. 1992. *The Ninemile wolves.* Clarke City Press, Livingston, MT.

Bath, A. J. 1991a. Public attitudes about wolf restoration in Yellowstone National Park. Pp. 367–76 in R. Keiter and M. S. Boyce, eds., *The Greater Yellowstone ecosystem: Redefining America's wilderness heritage.* Yale University Press, New Haven, CT.

———. 1991b. Public attitudes in Wyoming, Montana and Idaho toward wolf restoration in Yellowstone National Park. *Trans. N. Am. Wildl. Nat. Resour. Conf.* 56:91–95.

Bath, A. J., and T. Buchanan. 1989. Attitudes of interest groups in Wyoming toward wolf restoration in Yellowstone National Park. *Wildl. Soc. Bull.* 17:519–25.

Bath, A. J., and C. Phillips. 1990. Statewide surveys of Montana and Idaho: Resident attitudes toward wolf reintroduction in Yellowstone. Report submitted to Friends of Animals, National Wildlife Federation, Fish and Wildlife Service, and National Park Service.

Baumgartner, K., and V. Guberti. 1996. Parvovirus CPV-2 in free-ranging wolves (*Canis lupus*) in Italy. Proceedings of the First Scientific Meeting of the European Association of Zoo and Wildlife Veterinarians, Rostock, Germany, 171–80.

Baun, M. M., N. Bergstrom, N. F. Langston, and L. Thoma. 1983. Physiological effects of human/companion animal bonding. *Nurs. Res.* 33:126–29.

Baxendale, P. M., H. S. Jacobs, and V. H. T. James. 1982. Salivary testosterone: Relationship to unbound plasma testosterone in normal and hyperandrogenic women. *Clin. Endocrinol.* 16:595–603.

Beach, F. A. 1974. Effects of gonadal hormones on urinary behavior in dogs. *Physiol. Behav.* 12:1005–13.

———. 1976. Sexual attractivity, proceptivity and receptivity in female mammals. *Horm. Behav.* 7:105–38.

Beach, F. A., and A. Merari. 1970. Coital behavior in dogs. V. Effects of estrogen and progesterone on mating and other forms of social behavior in the bitch. *J. Comp. Physiol. Psychol.* 70:1–22.

Bean, M. J. 1983. *The evolution of national wildlife law.* Praeger, New York.

Beaufort, F. G. de 1987. Ecologie Historique du Loup en France. These de Doctorate, Unversit de Rennes I. Rennes (France). 4 vol., 1104 pp.

Becker, E. F., M. A. Spindler, and T. O. Osborne. 1998. A population estimator based on network sampling of tracks in the snow. *J. Wildl. Mgmt.* 62:968–77.

Beebe, B. F. 1978. Two new Pleistocene mammal species from Beringia. American Quaternary Association, Abstracts, 5th Biennial Meeting, p. 159.

Bekoff, M. 1972. The development of social interaction, play, and metacommunication in mammals: An ethological perspective. *Q. Rev. Biol.* 47:413–34.

———. 1974a. Social play and play-soliciting by infant canids. *Am. Zool.* 14:323–40.

———. 1974b. Social play in coyotes, wolves and dogs. *BioScience* 24:225–30.

———. 1977a. *Canis latrans.* Mammalian Species, 79. American Society of Mammalogists, New York. 90 pp.

———. 1977b. Mammalian dispersal and the ontogeny of individual behavioral phenotypes. *Am. Nat.* 3:715–32.

———. 1977c. Social communication in canids: Evidence for evolution of a stereotyped mammalian display. *Science* 197: 1097–99.

———. 1979a. Behavioral acts: Description, classification, ethogram analysis, and measurement. Pp. 67–80 in R. B. Cairns, ed., *The analysis of social interactions: Methods, issues, and illustrations.* Lawrence Erlbaum Associates, Hillsdale, NJ.

———. 1979b. Scent-marking by free-ranging domestic dogs: Olfactory and visual components. *Biol. Behav.* 4:123–39.

———. 1981. Mammalian sibling interactions: Genes, facilitative environments and the coefficient of familiarity. Pp. 307–46 in D. J. Gubernick and P. H. Klopfer, eds., *Parental care in mammals.* Plenum Press, New York.

———. 1984. Social play behavior. *BioScience* 34:228–33.

———. 1989. Behavioral development of terrestrial carnivores. Pp. 89–124 in J. L. Gittleman, ed., *Carnivore behavior, ecology and evolution.* 2 volumes. Cornell University Press, Ithaca, NY.

———. 1995. Play signals as punctuation: The structure of social play in canids. *Behaviour* 132:419–29.

———. 1999. Coyote/*Canis latrans.* Pp. 139–41 in D. E. Wilson and S. Ruff, eds., *The Smithsonian book of North American mammals.* Smithsonian Institution Press, Washington, D.C.

———. 2001. Human-carnivore interactions: Adopting proactive strategies for complex problems. Pp. 179–95 in J. L. Gittleman, S. M. Funk, D. W. Macdonald, and R. K. Wayne, eds., *Carnivore Conservation.* Cambridge University Press, Cambridge.

Bekoff, M., H. L. Hill, and J. B. Mitton. 1975. Behavioral taxonomy in canids by discriminant function analyses. *Science* 190: 1123–25.

Bekoff, M., and L. D. Mech. 1984. Simulation analyses of space use: Home range estimates, variability, and sample size. *Behav. Res. Meth. Instrum.* 16:32–37.

Bekoff, M., and M. C. Wells. 1980. The social ecology of coyotes. *Sci. Am.* 4:130–48.

Belyaev, D. K. 1979. Destabilizing selection as a factor in domestication. *J. Hered.* 70:301–8.

Benecke, N. 1987. Studies on early dog remains from northern Europe. *J. Archaeol. Sci.* 14:31–49.

Berg, L., and A. Bjärvall. 2000. The situation of large carnivores in Sweden. Pp. 88–89 in Report on the meeting of the "Group of Experts on Conservation of Large Carnivores," Oslo, 22–24 June 2000. Council of Europe, T-PVS (2000) 33. 101 pp.

Berg, W. E., and S. Benson. 1999. Updated wolf population estimate for Minnesota, 1997–1998. Minnesota Department of Natural Resources, Grand Rapids, MN.

Berg, W. E., and R. A. Chesness. 1978. Ecology of coyotes in northern Minnesota. Pp. 229–47 in M. Bekoff, ed., *Coyotes: Biology, behavior, and management.* Academic Press, New York.

Berg, W. E., and D. W. Kuehn. 1980. A study of the timber wolf population on the Chippewa National Forest, Minnesota. *Minn. Wildl. Res. Q.* 40:1–16.

———. 1982. Ecology of wolves in north-central Minnesota. Pp. 4–11 in F. H. Harrington and P. C. Paquet, eds., *Wolves of the world: Perspectives of behavior, ecology, and conservation.* Noyes Publications, Park Ridge, NJ.

Berger, J. 1978. Group size, foraging, and antipredator ploys: An analysis of bighorn sheep decisions. *Behav. Ecol. Sociobiol.* 4:91–99.

Berger, J., and C. Cunningham. 1988. Size related effects on search times in North American grassland female ungulates. *Ecology* 69:177–83.

Berger, J., P. B. Stacey, L. Bellis, and M. P. Johnson. 2001. A mammalian predator-prey imbalance: Grizzly bear and wolf extinction affect avian Neotropical migrants. *Ecol. Appl.* 11:947–60.

Berger, J., J. E. Swenson, and I. L. Persson. 2001. Recolonizing carnivores and naive prey: Conservation lessons from Pleistocene extinctions. *Science* 291:1036–39.

Bergerud, A. T. 1974. Decline of caribou in North America following settlement. *J. Wildl. Mgmt.* 38:757–70.

———. 1985. Antipredator strategies of caribou: Dispersion along shorelines. *Can. J. Zool.* 63:1324–29.

———. 1992. Rareness as an antipredator strategy to reduce predation risk for moose and caribou. Pp. 1008–21 in D. R. McCullough and R. H. Barrett, eds., *Wildlife 2001: Populations.* Elsevier Applied Science, London.

Bergerud, A. T., H. E. Butler, and D. R. Miller. 1984. Antipredator tactics of calving caribou: Dispersion in mountains. *Can. J. Zool.* 62:1566–75.

Bergerud, A. T., and J. P. Elliott. 1998. Wolf predation in a multiple-ungulate system in northern British Columbia. *Can. J. Zool.* 76:1551–69.

Bergerud, A. T., R. Ferguson, and H. E. Butler. 1990. Spring migration and dispersion of woodland caribou at calving. *Anim. Behav.* 39:360–68.

Bergerud, A. T., and R. E. Page. 1987. Displacement and dispersion of parturient caribou at calving as an antipredator tactic. *Can. J. Zool.* 65:1597–1606.

Bergerud, A. T., W. Wyett, and J. B. Snider. 1983. Role of wolf predation in limiting a moose population. *J. Wildl. Mgmt.* 47:977–88.

Bernal, J. F., and J. M. Packard. 1997. Differences in winter activity, courtship and social behavior of two captive family groups of Mexican wolves *Canis lupus baileyi. Zoo Biol.* 16:435–43.

Berry, W. 1977. *The unsettling of America: Culture and agriculture.* Sierra Club Books, San Francisco, CA.

———. 1987. *Home economics.* North Point Press, San Francisco, CA.

———. 1992. *Sex, economy, freedom and community.* Pantheon Books, New York.

Berta, A. 1988. Quaternary evolution and biogeography of the large South American Canidae (Mammalia: Carnivora). University of California Publications in Geological Sciences, 132. University of California Press, Berkeley. 149 pp.

———. 1995. Fossil carnivores from the Leisey Shell Pits, Hillsborough County, Florida. *Bull. Fla. Mus. Nat. Hist.* 37 Pt. II(14): 463–99.

Bibikov, D. I. 1975. The wolf in the USSR. Pp. 29–36 in D. H. Pimlott, ed., Wolves: Proceedings of the First Working Meeting of Wolf Specialists and of the First International Conference on the Conservation of the Wolf. Supplementary Paper no. 43. IUCN, Morges, Switzerland. 145 pp.

———. 1982. Wolf ecology and management in the USSR. Pp. 120–33 in F. H. Harrington and P. C. Paquet, eds., *Wolves of the world: Perspectives of behavior, ecology, and conservation.* Noyes Publications, Park Ridge, NJ.

——— (ed.). 1985. *The wolf: History, Systematics, Morphology, Ecology.* [In Russian.] Izdatetvo Nauka, Muskva. 606 pp.

———. 1988. *Der Wolf. Dei Neue Brehm-Bucherci.* A. Ziemsen Verlag, Wittenberg Lutherstadt, Germany. 198 pp.

———. 1994. Wolf problem in Russia. *Lutreola* 3:10–14.

Bibikov, D. I., A. N. Filimonov, and A. N. Kudaktin. 1983. Territoriality and migration of the wolf in the USSR. *Acta Zool. Fenn.* 174:267–68.

Bibikov, D. I., and K. P. Filonov. 1980. The wolf in preserves of the USSR. *Priroda* 2.

Bibikov, D. I., A. N. Kudaktin, and L. S. Ryabov. 1985. Synanthropic wolves: Distribution, ecology. [In Russian.] *Zool. Zh.* 64(3): 429–41.

Bibikov, D. I., and I. Kh. Rootsi. 1993. Is the Russian wolf dangerous to man? *Priroda I Okhota* 4:43–45.

Biggs, J. R. 1988. Reintroduction of the Mexican wolf into New Mexico: An attitude survey. Master's thesis, New Mexico State University, Las Cruces. 66 pp.

Biknevicius, A. R., and B. Van Valkenburgh. 1996. Design for killing: Craniodental adaptations of predators. Pp. 393–428 in J. L. Gittleman, ed., *Carnivore behavior, ecology and evolution,* vol. 2. Cornell University Press, Ithaca, NY.

Bildstein, K. L. 1983. Why white-tailed deer flag their tails. *Am. Nat.* 121:709–15.

Bishopp, F. C., and H. L. Trembley. 1945. Distribution and hosts of certain North American ticks. *J. Parasitol.* 31:1–54.

Bjärvall, A. 1983. Scandinavia's response to a natural repopulation of wolves. *Acta Zool. Fenn.* 174:273–75.

Bjärvall, A., and E. Isakson. 1981. Algen favoritbytet for Varmlandsvargen. *Svansk Jakt* 119:762–67.

———. 1982. Winter ecology of a pack of three wolves in Northern Sweden. Pp. 146–57 in F. H. Harrington and P. C. Paquet, eds., *Wolves of the world: Perspectives of behavior, ecology, and conservation.* Noyes Publications, Park Ridge, NJ.

Bjärvall, A., and E. Nilsson. 1976. Surplus-killing of reindeer by wolves. *J. Mammal.* 57:585.

Bjerke, T., O. Reitan, and S. R. Kellert. 1998. Attitudes toward wolves in southeastern Norway. *Soc. Nat. Resour.* 11:169–78.

Bjorge, R. R. 1983. Mortality of cattle on two types of grazing areas in northwestern Alberta. *J. Range Mgmt.* 36:20–21.

Bjorge, R. R., and J. R. Gunson. 1983. Wolf predation of cattle on the Simonette River pastures in west-central Alberta. Pp. 106–11 in L. N. Carbyn, ed., *Wolves in Canada and Alaska: Their status, biology, and management.* Report Series, no. 45. Canadian Wildlife Service, Edmonton, Alberta.

———. 1985. Evaluation of wolf control to reduce cattle predation in Alberta. *J. Range Mgmt.* 35:483–87.

———. 1989. Wolf *(Canis lupus)* population characteristics and prey relationships near Simonette River, Alberta. *Can. Field Nat.* 103:327–34.

Blair, W. R. 1908. Report of the veterinary pathologist. 13th Annual Report of the New York Zoological Society, 137–42.

———. 1919. Report of the veterinarian. 24th Annual Report of the New York Zoological Society, 82–87.

Blanco, J. C. 1998. The extinction of the wolf in Spain: Account of a scientific fraud. *Biologica* 26:56–59.

Blanco, J. C., and Y. Cortés. 2002. *Ecología, censos, percepcion y evolucion del lobo en España: Analisis de un conflicto.* Sociedad Española para la Conservación y Estudio de los Mamíferos, Malaga. 176 pp.

Blanco, J. C., L. Cuesta, and S. Reig. 1990. El lobo en España: Una vision global. Pp. 69–93 in J. C. Blanco, L. Cuesta, and S. Reig, eds., *El lobo* (Canis lupus) *en España: Situation, problematica y apuntes sobre su ecología.* ICONA, Madrid. 118 pp.

Blanco, J. C., S. Reig, and Cuesta. 1992. Distribution, status, and conservation problems of the wolf *Canis lupus* in Spain. *Biol. Conserv.* 60:73–80.

Blaza, S. E. 1982. Energy balance, water balance, and the physiology of digestion and absorption. Pp. 1–12 in A. T. B. Edney, ed., *Dog and cat nutrition.* Pergamon Press, Oxford.

Block, M. L., L. C. Volpe, and M. J. Hayes. 1981. Saliva as a chemical cue in the development of social behavior. *Science* 211: 1062–64.

Blood, D. C., and O. M. Radostits. 1989. *Veterinary medicine.* Bailliere Tindall, London.

Boas, F. 1888. The Central Eskimo. Pp. 399–669 in 6th Bureau of Ethnology Annual Report. Smithsonian Institution, Washington, D.C.

Bobek, B. 1995. Status, distribution, and management of the wolf in Poland. (Abstract.) p. 23 in *Wolves and humans 2000: A global perspective for managing conflict.* International Wolf Center, Ely, MN.

Bobek, B., M. Kosobucka, K. Perzanowski, and K. Plodzien. 1993. Distribution and wolf numbers in Poland. Pp. 27–29 in C. Promberger and W. Schröder, eds., *Wolves in Europe: Status and perspectives.* Munich Wildlife Society, Ettal, Germany.

Boertje, R. D, W. C. Gasaway, D. W. Grangaard, and D. Kellyhouse. 1988. Predation of moose and caribou by radio-collared grizzly bears in east central Alaska. *Can. J. Zool.* 66:2492–99.

Boertje, R. D., D. G. Kelleyhouse, and R. D. Hayes. 1995. Methods for reducing natural predation on moose in Alaska and Yukon: An evaluation. Pp. 505–13 in L. N. Carbyn, S. H. Fritts, and D. R. Seip, eds., *Ecology and conservation of wolves in a changing world.* Canadian Circumpolar Institute, Edmonton, Alberta.

Boertje, R. D., and R. O. Stephenson. 1992. Effects of ungulate availability on wolf reproductive potential in Alaska. *Can. J. Zool.* 70:2441–43.

Boertje, R. D., P. Valkenburg, and M. E. McNay. 1996. Increases in moose, caribou, and wolves following wolf control in Alaska. *J. Wildl. Mgmt.* 60:474–89.

Bogan, M. A., and P. Mehlhop. 1983. Systematic relationships of gray wolves *(Canis lupus)* in southwestern North America. Occasional papers, University of New Mexico, Museum of Southwestern Biology, no. 1. 21 pp.

Boissinot, S., and P. Boursot. 1997. Discordant phylogeographic patterns between Y chromosomes and mitochondrial DNA in the house mouse: Selection on the Y chromosome? *Genetics* 146:1019–34.

Boitani, L. 1982. Wolf management in intensively used areas of Italy. Pp. 158–72 in F. H. Harrington and P. C. Paquet, eds., *Wolves of the world: Perspectives of behavior, ecology, and conservation.* Noyes Publications, Park Ridge, NJ.

———. 1983. Wolf and dog competition in Italy. *Acta Zool. Fenn.* 174:259–64.

———. 1984. Genetic consideration on wolf conservation in Italy. *Boll. Zool.* 51(3): 367–73.

———. 1986. *Dalla parte del lupo.* Milano: L'airone di G. Mondadore Associati Spa.

———. 1992. Wolf research and conservation in Italy. *Biol. Conserv.* 61:125–32.

———. 1993. Wolf management action required for conservation. Pp. 114–18 in C. Promberger and W. Schröder, eds., *Wolves in Europe: Status and perspectives.* Munich Wildlife Society, Ettal, Germany.

———. 1995. Ecological and cultural diversities in the evolution of wolf-human relationships. Pp. 3–11 in L. N. Carbyn, S. H. Fritts, and D. R. Seip, eds., *Ecology and conservation of wolves in a changing world.* Canadian Circumpolar Institute, Edmonton, Alberta.

———. 2000. Action Plan for the conservation of wolves in Europe *(Canis lupus)*. Nature and Environment Series, no. 113: Convention on the Conservation of European Wildlife and Natural Habitats. Council of Europe, Strasbourg. 81 pp.

Boitani, L., and P. Ciucci. 1993. Wolves in Italy: Critical issues for their conservation. Pp. 75–90 in C. Promberger and W. Schröder, eds., *Wolves in Europe: Current status and perspectives.* Munich Wildlife Society, Ettal, Germany.

———. 2000. Action plan for the conservation and management of the wolf in the Alps. Worldwide Fund for Nature, Rome, Italy. 58 pp.

Boitani, L., and M. L. Fabbri. 1983. Strategia nazionale di conservationi per il lupo *(Canis lupus)*. *Ricerche di Biologia della Selvaggina* 72:1–31.

Boitani, L., F. Francisci, P. Ciucci, and G. Andreoli. 1995. Population biology and ecology of feral dogs in central Italy. Pp. 217–44 in J. Serpell, ed., *The domestic dog, its evolution, behaviour and interactions with people.* Cambridge University Press, Cambridge.

Boitani, L., and E. Zimen. 1979. The role of public opinion in wolf management. Pp. 471–77 in E. Klinghammer, ed., *The behavior and ecology of wolves.* Garland STPM Press, New York.

Boles, B. K. 1977. Predation by wolves on wolverines. *Can. Field Nat.* 91:68–69.

Bondareva, V. I. 1955. Rol' domashnikh i dikikh plotoiadnykh v epidemiologii i epizootologii larval'nykh tsestodozov. Part 2. Fauna tsestod volkov. *Trudy Inst. Zool., Akad. Nauk Kazakh. SSR* 3:101–4.

Booth, W. D. 1972. The occurrence of testosterone and 5α-dihydrotestosterone in the submaxillary salivary gland in the boar. *J. Endocrinol.* 55:119–25.

Booth, W. D., M. F. Hay, and H. M. Dott. 1973. Sexual dimorphism in the submaxillary gland of the pig. *J. Reprod. Fertil.* 33: 163–66.

Borrini-Feyerabend, G. 1996. Collaborative management of protected areas: Tailoring the approach to the context. Issues in Social Policy. IUCN, Gland, Switzerland.

Bosler, E. M., B. G. Ormiston, J. L. Coleman, J. P. Hanrahan, and J. L. Benach. 1984. Prevalance of the Lyme disease spirochete in populations of white-tailed deer and white-footed mice. *Yale J. Biol. Med.* 57:651–59.

Bossart, J. L., and D. P. Prowell. 1998. Genetic estimates of population structure and gene flow: Limitations, lessons and new directions. *Trends Ecol. Evol.* 13:202–6.

Botkin, D. B. 1990. *Discordant harmonies: A new ecology for the twenty-first century.* Oxford University Press, Oxford.

Boutin, S. 1992. Predation and moose population dynamics: A critique. *J. Wildl. Mgmt.* 56:116–27.

Boving, P. S., and E. Post. 1997. Vigilance and foraging behaviour of female caribou in relation to predation risk. *Rangifer* 17(2): 55–63.

Bowen, W. D. 1978. Social organization of the coyote in relation of prey size. Ph.D. dissertation, University of British Columbia, Vancouver.

———. 1981. Variation in coyote social organization: The influence of prey size. *Can. J. Zool.* 59:639–52.

Bowen, W. D., and I. M. Cowan. 1980. Scent marking in coyotes. *Can. J. Zool.* 58:473–80.

Boyce, M. S. 1990. Wolf recovery for Yellowstone National Park: A simulation model. Pp. 3-4–3-58 in *Wolves for Yellowstone? A report to the United States Congress,* Vol. 2, *Research and analysis.* U.S. National Park Service, Yellowstone National Park, WY.

———. 1992. Population viability analysis. *Annu. Rev. Ecol. Syst.* 23:481–506.

———. 1993. Predicting the consequences of wolf recovery to ungulates in Yellowstone National Park. Pp. 234–69 in R. S. Cook, ed., *Ecological Issues on Reintroducing Wolves into Yellowstone National Park.* U.S. National Park Service Scientific Monograph Series NPS/NRYELL/NRSM-93/22.

———. 1998. Ecological-process management and ungulates: Yellowstone's conservation paradigm. *Wildl. Soc. Bull.* 26: 391–98.

Boyce, M. S., and E. M. Anderson. 1999. Evaluating the role of carnivores in the Greater Yellowstone Ecosystem. Pp. 265–83 in T. W. Clark, A. P. Curlee, S. C. Minta, and P. M. Kareiva, *Carnivores in ecosystems: The Yellowstone experience.* Yale University Press, New Haven, CT.

Boyce, M. S., and L. L. McDonald. 1999. Relating populations to habitats using resource selection functions. *Trends Ecol. Evol.* 14(7): 268–72.

Boyd, D. K., S. H. Forbes, D. H. Pletscher, and F. H. Allendorf. 2001. Identification of Rocky Mountain gray wolves. *Wildl. Soc. Bull.* 29:78–85.

Boyd, D. K., and M. D. Jimenez. 1994. Successful rearing of young by wild wolves without mates. *J. Mammal.* 75:14–17.

Boyd, D. K., and G. K. Neale. 1992. An adult cougar, *Felis concolor,* killed by gray wolves, *Canis lupus,* in Glacier National Park, Montana. *Can. Field Nat.* 106:524–25.

Boyd, D. K., P. C. Paquet, S. Donelon, R. R. Ream, D. H. Pletsher, and C. C. White. 1995. Transboundary movements of a colonizing wolf population in the Rocky Mountains. Pp. 135–40 in L. N. Carbyn, S. H. Fritts, and D. R. Seip, eds., *Ecology and conservation of wolves in a changing world.* Canadian Circumpolar Institute, Edmonton, Alberta.

Boyd, D. K., and D. H. Pletscher. 1999. Characteristics of dispersal in a colonizing wolf population in the central Rocky Mountains. *J. Wildl. Mgmt.* 63:1094–1108.

Boyd, D. K., D. H. Pletscher, and W. G. Brewster. 1993. Evidence of wolves burying dead wolf pups. *Can. Field Nat.* 107:230–31.

Boyd, D. K., R. R. Ream, D. H. Pletscher, and M. W. Fairchild. 1993. Variation in denning and parturition dates of a wild gray wolf *(Canis lupus)* in the Rocky Mountains. *Can. Field Nat.* 107:359–60.

———. 1994. Prey taken by colonizing wolves and hunters in the Glacier National Park area. *J. Wildl. Mgmt.* 58:289–95.

Boyd, D. K., L. B. Secrest, and D. H. Pletscher. 1992. A wolf, *Canis lupus,* killed in an avalanche in southwestern Alberta. *Can. Field Nat.* 106:526.

Boyd, R. J. 1978. American elk. Pp. 11–29 in J. L. Schmidt and D. L. Gilbert, eds., *Big game of North America: Ecology and management.* Stackpole, Harrisburg, PA.

Boyd-Heger, D. K. 1997. Dispersal, genetic relationships, and landscape use by colonizing wolves in the central Rocky Mountains. Ph.D. dissertation, University of Montana, Missoula.

Brackenridge, H. M. 1816. *Journal of a voyage up the River Missouri, performed in eighteen hundred and eleven.* Coale & Maxwell, Baltimore.

Bradshaw, J. W. S., and C. J. Thorne. 1992. Feeding behaviour. Pp. 115–30 in C. Thorne, ed., *The Waltham book of dog and cat behaviour.* Pergamon Press, Oxford and New York.

Brand, C. J., M. J. Pybus, W. B. Ballard, and R. O. Peterson. 1995. Infectious and parasitic diseases of the gray wolf and their potential effects on wolf populations in North America. Pp. 419–29 in L. N. Carbyn, S. H. Fritts, and D. R. Seip, eds., *Ecology and conservation of wolves in a changing world.* Canadian Circumpolar Institute, Edmonton, Alberta.

Brangi, A., P. Rosa, and A. Meriggi. 1992. Predation by wolves *(Canis lupus* l.) on wild and domestic ungulates in northern Italy. *Ongules/Ungulates* 91:541–43.

Breazile, J. 1978. Neurologic and behavioural development in the puppy. *Vet. Clin. N. Am.* 8:31–45.

Breitenmoser, U. 1998. Large predators in the Alps: The fall and rise of man's competitors. *Biol. Conserv.* 83(3): 279–89.

Bresler, D. E., G. Ellison, and S. Zamenhof. 1975. Learning deficits in rats with malnourished grandmothers. *Dev. Psychol.* 8:315–23.

Brewster, W. G., and S. H. Fritts. 1995. Taxonomy and genetics of the gray wolf in western North America: A review. Pp. 353–73 in L. N. Carbyn, S. H. Fritts, and D. R. Seip, eds., *Ecology and conservation of wolves in a changing world.* Canadian Circumpolar Institute, Edmonton, Alberta.

Bridges, R. S., R. DiBiase, D. D. Loundes, and P. C. Doherty. 1985. Prolactin stimulation of maternal behavior in female rats. *Science* 227:782–84.

Bright, A. D., and M. J. Manfredo. 1996. A conceptual model of attitudes toward natural resource issues: A case study of wolf reintroduction. *Human Dimensions of Wildlife* 1:1–21.

Brisbin, I. L., Jr., and S. N. Austad. 1991. Testing the individual odour theory of canine olfaction. *Anim. Behav.* 42:63–69.

———. 1993. The use of trained dogs to discriminate human scent: A reply. *Anim. Behav.* 46:191–92.

Briscoe, B. K., M. A. Lewis, and S. E. Parrish. 2002. Home range formation in wolves due to scent marking. *Bull. Math. Biol.* 64:261–84.

British Broadcasting Co. 1996. *Wolves and bison in Wood Buffalo National Park.* BBC, London. (Video.)

Brock, V. E., and R. H. Riffenburgh. 1960. Fish schooling: A possible factor in reducing predation. *Journal du Conseil International pour l'Exploration de la Mer* 25:307–17.

Bromlei, G. F. 1965. *Bears of the south of the Far East of the U.S.S.R.* Nauka, Moscow and Leningrad.

Bronson, F. H. 1968. Pheromonal influences on mammalian reproduction. Pp. 341–61 in M. Diamond, ed., *Perspectives in reproduction and sexual behavior.* Indiana University Press, Bloomington.

Bronson, F. H., and D. Caroom. 1971. Preputial gland of the male mouse: Attractant function. *J. Reprod. Fertil.* 25:279–82.

Brouwer, E., and H. J. Nijkamp. 1953. Occurrence of two valeric acids (β-methylbutyric acid and α-methylbutyric acid) in the hair grease of the dog. *Biochem. J.* 55:444–47.

Brown, D. E., ed. 1983. *The wolf in the Southwest: The making of an endangered species.* University of Arizona Press, Tucson.

Brown, D. S., and R. E. Johnston. 1983. Individual discrimination on the basis of urine in dogs and wolves. Pp. 343–46 in D. Müller-Schwarze and R. M. Silverstein, eds., *Chemical signals in vertebrates,* 3rd ed. Plenum Press, New York.

Brown, J. L. 1964. The evolution of diversity in avian territorial systems. *Wilson Bull.* 76:160–69.

———. 1982. Optimal group size in territorial animals. *J. Theor. Biol.* 95:793–810.

Brown, J. S., J. W. Laundré, and M. Gurung. 1999. The ecology of fear: Optimal foraging, game theory, and trophic interactions. *J. Mammal.* 80:385–99.

Brownlow, C. A. 1996. Molecular taxonomy and the conservation of the red wolf and other endangered carnivores. *Conserv. Biol.* 10:390–96.

Brtek, L., and J. Voskar. 1987. Food biology of the wolf in Slovak Carpathians. *Biologia* (Bratislava) 42:985–90.

Bruford, M. W., D. J. Chessman, T. Coote, H. A. A. Green, S. A. Haines, C. O'Ryan, and T. R. Williams. 1996. Microsatellites and their application to conservation genetics. Pp. 278–97 in T. B. Smith and R. K. Wayne, eds., *Molecular genetic approaches in conservation.* Oxford University Press, New York.

Bruford, M. W., and R. K. Wayne. 1993. Microsatellites and their application to population genetic studies. *Curr. Biol.* 3:939–43.

Bubenik, A. B. 1972. North American moose management in light of European experiences. *Proc. N. Am. Moose Conf.* 8: 279–95.

Bubenik, G. A., A. B. Bubenik, D. Schams, and J. F. Leatherland. 1983. Circadian and circannual rhythms of LH, FSH, testosterone (T), prolactin, cortisol, T_3, T_4 in plasma of mature, male white-tailed deer. *Comp. Biochem. Physiol. A. Comp Physiol.* 76:37–45.

Bubenik, G. A., P. S. Smith, and D. Schams. 1986. The effect of orally administered melatonin on the seasonality of deer pelage exchange, antler development, LH, FSH, prolactin, testosterone, T_3, T_4, cortisol, and alkaline phosphatase. *J. Pineal Res.* 3:331–49.

Budd, J. 1981. Distemper. Pp. 31–44 in J. W. Davis, L. H. Karstad, and D. O. Trainer, eds., *Infectious diseases of wild mammals.* Iowa State University Press, Ames.

Budgett, H. M. 1933. *Hunting by scent.* Eyre and Spottiswoode, London.

Burch, E. S. 1972. The caribou/wild reindeer as a human resource. *Am. Antiquity* 37(3): 339–68.

Burch, J. W. 2001. Evaluation of wolf density estimation from radiotelemetry data. Master's thesis, University of Alaska, Fairbanks. 55 pp.

Burgdorfer, W., A. G. Barbour, S. F. Hayes, J. L. Benach, E. Grunwaldt, and J. P. Davis. 1982. Lyme disease—A tick-borne spirochetosis? *Science* 216:1317–19.

Burke, T., O. Hanotte, and I. V. Pijlen. 1996. Microsatellite analysis in conservation genetics. Pp. 251–77 in T. B. Smith and R. K. Wayne, eds., *Molecular genetic approaches in conservation.* Oxford University Press, New York.

Burkholder, B. L. 1959. Movements and behavior of a wolf pack in Alaska. *J. Wildl. Mgmt.* 23:1–11.

———. 1962. Observations concerning wolverine. *J. Mammal.* 43: 263–64.

Burt, W. H. 1943. Territoriality and home range concepts as applied to mammals. *J. Mammal.* 24:346–52.

Busch, R. H. 1995. *The wolf almanac.* Lyons & Burford, New York. 226 pp.

Buskirk, S. W. 1999. Mesocarnivores of Yellowstone. Pp. 165–87 in T. W. Clark, A. P. Curlee, S. C. Minta, and P. M. Kareiva, *Carnivores in ecosystems: The Yellowstone experience.* Yale University Press, New Haven, CT.

Butzeck, S. 1987. The wolf, *Canis lupus* L., as a rabies vector in the 16th and 17th centuries. *Z. Gesamte Hyg.* 33:666–69.

Buys, C. 1975. Predator control and ranchers' attitudes. *Environ. Behav.* 7:81–98.

Byman, D., V. Van Ballenberghe, J. C. Schlotthauer, and A. W. Erickson. 1977. Parasites of wolves (*Canis lupus* L.) in northeastern Minnesota, as indicated by analysis of fecal samples. *Can. J. Zool.* 55:376–80.

Byrne, R. 1995. *The thinking ape.* Oxford University Press, Oxford.

Cabasso, V. J. 1981. Infectious canine hepatitis. Pp. 191–95 in J. W. Davis, L. H. Karstad, and D. O. Trainer, eds., *Infectious diseases of wild mammals.* Iowa State University Press, Ames.

Cagnolaro, L., M. Comincini, A. Martinoli, and A. Oriani. 1996. Dati storici sulla presenza e su casi di antropofagia del lupo nella Padania centrale. In F. Cecere, ed. Atti del Convegno "Della Parte del Lupo." Atti e Studi del WWF Italia, 10: 83–99.

Calder, W. A. III. 1984. *Size, function, and life history.* Harvard University Press, Cambridge, MA.

Camenzind, F. J. 1978. Behavioral ecology of coyotes on the National Elk Refuge, Jackson, Wyoming. Pp. 267–94 in M. Bekoff, ed., *Coyotes: Biology, behavior, and management.* Academic Press, New York.

Caraco, T., and L. L. Wolf. 1975. Ecological determinants of group sizes of foraging lions. *Am. Nat.* 109:343–52.

Caras, R. 1966. *The Custer wolf: Biography of an American regegade.* Boston: Little, Brown.

Carbyn, L. N. 1974. *Wolf predation and behavioral interactions with elk and other ungulates in an area of high prey diversity.* Canadian Wildlife Service, Edmonton, Alberta. 233 pp.

———. 1980. Ecology and management of wolves in Riding Mountain National Park, Manitoba. Canadian Wildlife Service, Edmonton, Alberta. 184 pp.

———. 1981. Territory displacement in a wolf population with abundant prey. *J. Mammal.* 62:193–95.

———. 1982a. Coyote population fluctuations and spatial distribution in relation to wolf territories in Riding Mountain National Park, Manitoba. *Can. Field Nat.* 96:176–83.

———. 1982b. Incidence of disease and its potential role in the population dynamics of wolves in Riding Mountain National Park, Manitoba. Pp. 106–16 in F. H. Harrington and P. C. Paquet, eds., *Wolves of the world: Perspectives of behavior, ecology, and conservation.* Noyes Publications, Park Ridge, NJ.

———. 1983a. Management of non-endangered wolf populations in Canada. *Acta Zool. Fenn.* 174:239–43.

———. 1983b. Wolf predation on elk in Riding Mountain National Park, Manitoba. *J. Wildl. Mgmt.* 47:963–76.

———, ed. 1983c. *Wolves in Canada and Alaska: Their status, biology, and management.* Report Series, no. 45. Canadian Wildlife Service, Edmonton, Alberta.

———. 1987. Gray wolf and red wolf. Pp. 378–93 in M. Novak, J. A. Baker, M. E. Obbard, and B. Malloch, eds., *Wild furbearer management and conservation in North America.* Ontario Ministry of Natural Resources, Toronto.

———. 1994. Swift Fox reintroduction program in Canada from 1983 to 1992. Pp. 247–71 in M. L. Bowles and C. J. Whelan, eds., *Restoration of endangered species.* Cambridge University Press, London.

———. 1997. Unusual movement by bison, *Bison bison*, in response to wolf, *Canis lupus*, predation. *Can. Field Nat.* 111:461–62.

Carbyn, L. N., S. H. Fritts, and D. R. Seip. 1995. *Ecology and conservation of wolves in a changing world.* Canadian Circumpolar Institute, Edmonton, Alberta.

Carbyn, L. N., and M. C. S. Kingsley. 1979. Summer food habits of wolves with emphasis on moose in Riding Mountain National Park. *Proc. N. Am. Moose Conf.* 15:349–61.

Carbyn, L. N., S. M. Oosenbrug, and D. W. Anions. 1993. Wolves, bison and the dynamics related to the Peace Athabaska

Delta in Canada's Wood Buffalo National Park. Circumpolar Research Series, no. 4. Canadian Circumpolar Institute, University of Alberta, Edmonton.

Carbyn, L. N., and T. Trottier. 1987. Responses of bison on their calving grounds to predation by wolves in Wood Buffalo National Park. *Can. J. Zool.* 65:2072–78.

———. 1988. Descriptions of wolf attacks on bison calves in Wood Buffalo National Park, Canada. *Arctic* 41:297–302.

Carl, G., and C. T. Robbins. 1988. The energetic cost of predator avoidance in neonatal ungulates: Hiding versus following. *Can. J. Zool.* 66:239–46.

Carley, C. J. 1975. Activities and findings of the red wolf recovery program from late 1973 to July 1, 1975. U.S. Fish and Wildlife Service, Albuquerque, NM.

———. 1979. Report on successful translocation experiment of red wolves *(Canis rufus)* to Bulls Island, South Carolina. Proceedings of the Portland Wolf Symposium, Lewis and Clark College, Portland, OR.

———. 1981. Red wolf experimental translocation summarized. Wild Canid Survival and Research Center Bulletin, Part I, Winter 1980, pp. 4,5,7; Part II, Spring 1981, pp. 8–9.

Carley, C. J., and J. L. Mechler. 1983. An experimental reestablishment of red wolves *(Canis rufus)* on the Tennessee Valley Authority's Land Between the Lakes. U.S. Fish and Wildlife Service, Asheville, NC.

Carmichael, L. E., J. A. Nagy, N. C. Larter, and C. Strobeck. 2001. Prey specialization may influence patterns of gene flow in wolves of the Canadian Northwest. *Mol. Ecol.* 10:2787–98.

Carmichael, L. E., and R. M. Kennedy. 1970. Canine brucellosis: The clinical disease, pathogenesis and immune response. *J. Am. Vet. Med. Assoc.* 156:1726–36.

Caro, T. 1994. *Cheetahs of the Serengeti Plains.* University of Chicago Press, Chicago.

Caro, T. M., and M. K. Laurenson. 1994. Ecological and genetic factors in conservation: A cautionary tale. *Science* 263:485–86.

Caro, T. M., L. Lombardo, A. W. Goldizen, and M. Kelly. 1995. Tail-flagging and other antipredator signals in white-tailed deer: New data and synthesis. *Behav. Ecol.* 6:442–50.

Casey, D., and T. W. Clark. 1996. *Tales of the wolf.* Homestead Publishing, Moose, WY.

Castroviejo, J., F. Palacios, J. Garzon, and L. Cuesta. 1975. Sobre la alimentazion de los Canidos ibericos. *Trans. Int. Congr. Game Biol.* 12:39–46.

Caswell, H. 1989. *Matrix population models.* Sinauer Associates, Sunderland, MA.

Caughley, G. 1977. *Analysis of vertebrate populations.* John Wiley and Sons, New York.

Chambers, R. 1995. Yukon First Nations, wolves, and wildlife: Past and present. (Abstract.) P. 3 in *Wolves and humans 2000: A global perspective for managing conflict.* International Wolf Center, Ely, MN.

Chandra, R. K. 1975. Antibody formation in first and second generation offspring of nutritionally deprived rats. *Science* 190:289–90.

Chapman, R. C. 1977. The effects of human disturbance on wolves *(Canis lupus* L.) Master's thesis, University of Alaska, Fairbanks.

———. 1978. Rabies: Decimation of a wolf pack in arctic Alaska. *Science* 201:365–67.

———. 1979. Human disturbance of wolf dens: A management problem. In R. M. Linn, ed., Proceedings of the First Conference on Scientific Research in the National Parks. Proceedings Series no. 5, Vol. 1. U.S. National Park Service, New Orleans, LA.

Chase, I. 1974. Models of hierarchy formation in animal societies. *Behav. Sci.* 19:374–82.

Cheney, C. D. 1982. Probability learning in captive wolves. Pp. 272–81 in F. H. Harrington and P. C. Paquet, eds., *Wolves of the world: Perspectives of behavior, ecology, and conservation.* Noyes Publications, Park Ridge, NJ.

Cherkasskiy, B. L. 1988. Roles of the wolf and raccoon dog in the ecology and epidemiology of rabies in the U.S.S.R. *Rev. Infect. Dis.* 10:634–36.

Cheville, N. F., D. R. McCullough, and L. R. Paulsen. 1998. Brucellosis in the Greater Yellowstone Area. National Research Council, Washington, D.C. 186 pp.

Chien, Y. W. 1985. *Transnasal systemic medications.* Elsevier Science Publishers, Amsterdam.

Child, K. N., K. K. Fujino, and M. W. Warren. 1978. A gray wolf *(Canis lupus columbianus)* and Stone sheep *(Ovis dalli stonei)* fatal predator-prey encounter. *Can. Field Nat.* 92:399–401.

Choquette, L. P., G. G. Gibson, E. Kuyt, and A. M. Pearson. 1973. Helminths of wolves, *Canis lupus* L., in the Yukon and Northwest Territories. *Can. J. Zool.* 51:1087–91.

Choquette, L. P., and E. Kuyt. 1974. Serological indication of canine distemper and of infectious canine hepatitis in wolves *(Canis lupus* L.) in northern Canada. *J. Wildl. Dis.* 10:321–24.

Christian, J. J. 1971. Population density and reproductive efficiency. *Biol. Reprod.* 4:248–94.

Christie, D. W., and E. T. Bell. 1971a. Endocrinology of the oestrous cycle of the bitch. *J. Small Anim. Pract.* 12:383–89.

Christie, D. W., and E. T. Bell. 1971b. Some observations on the seasonal incidence and frequency of oestrus in breeding bitches in Britain. *J. Small Anim. Pract.* 12:159–67.

Ciucci, P. 1994. Movimenti, attività e risorse del lupo *(Canis lupus)* in due aree dell'Appennino centro-settentrionale. [In Italian.] Ph.D. dissertation, Università di Roma "La Sapienza," Roma, Italy. 117 pp.

Ciucci, P., and L. Boitani. 1991. Viability assessment of the Italian wolf and guidelines for the management of the wild and a captive population. *Ricerche Biologia della Selvaggina* 89. 58 pp.

———. 1998a. Elementi di biologia, gestione, ricerca. Documenti Tecnici, 23. Istituto Nazionale per la Fauna Selvatica, Bologna. 114 pp.

———. 1998b. Wolf and dog depredation on livestock in central Italy. *Wildl. Soc. Bull.* 26:504–14.

Ciucci, P., L. Boitani, F. Francisci, and G. Andreoli. 1997. Home range, activity and movements of a wolf pack in central Italy. *J. Zool.* 243:803–19.

Ciucci, P., L. Boitani, E. Raganella Pelliccioni, M. Rocco, and I. Guj. 1996. A comparison of scat analysis techniques to assess the diet of the wolf *(Canis lupus)*. *Wildl. Biol.* 2: 267–78.

Ciucci, P., and L. D. Mech. 1992. Selection of wolf dens in relation to winter territories in northeastern Minnesota. *J. Mammal.* 73:899–905.

Clark, K. R. F. 1971. Food habits and behavior of the tundra wolf on central Baffin Island. Ph.D. dissertation, University of Toronto, Ontario.

Clark, T. W. 1993. Creating and using knowledge for species and ecosystem conservation: Science, organization, and policy. *Perspect. Biol. Med.* 36:497–525.

Clarke, C. D. H. 1971. The beast of Gevauden. *Nat. Hist.* 80:44–51, 66–73.

Clarkson, P. L. 1995. Recommendations for more effective wolf management. Pp. 527–34 in L. N. Carbyn, S. H. Fritts and D. R. Seip, eds., *Ecology and conservation of wolves in a changing world.* Canadian Circumpolar Institute, Edmonton, Alberta.

Cloninger, R. 1986. A unified biosocial theory of personality and its role in the development of anxiety states. *Psychol. Dev.* 3:167–226.

Cluff, H. D., and D. L. Murray. 1995. Review of wolf control methods in North America. Pp. 491–504 in L. N. Carbyn, S. H. Fritts, and D. R. Seip, eds., *Ecology and conservation of wolves in a changing world.* Canadian Circumpolar Institute, Edmonton, Alberta.

Clutton-Brock, J. 1999. *A natural history of domesticated mammals.* 2d ed. Cambridge University Press, Cambridge.

Clutton-Brock, J., G. B. Corbet, and M. Hills. 1976. A review of the family Canidae, with a classification by numerical methods. *Bull. Brit. Mus. (Nat. Hist.), Zool.* 29:119–99.

Clutton-Brock, J., A. C. Kitchener, and J. M. Lynch. 1994. Changes in the skull morphology of the Arctic wolf, *Canis lupus arctos,* during the twentieth century. *J. Zool.* 233:19–36.

Clutton-Brock, T. H., P. N. M. Brotherton, M. J. O'Riain, A. S. Griffin, D. Gaynor, R. Kansky, L. Sharpe, and G. M. Mc-Ilrath. 2001. Contributions to cooperative rearing in meerkats. *Anim. Behav.* 61:705–10.

Coady, J. W. 1974. Influence of snow on behavior of moose. *Nat. Can.* 101:417–36.

Cochrane, J. F. 2000. Gray wolves in a small park: Analyzing cumulative effects through simulation. Ph.D. dissertation, University of Minnesota, Minneapolis.

Coffin, D. L. 1944. A case of *Dirofilaria immitis* infection in a captive-bred timber wolf *(Canis occidentalis* Richardson). *N. Am. Vet.* 25:611.

Cohen, J. A., and M. W. Fox. 1976. Vocalizations of wild canids and possible effects of domestication. *Behav. Processes* 1: 77–92.

Coile, D., C. Pollitz, and J. Smith. 1989. Behavioral determination of critical flicker fusion in dogs. *Physiol. Behav.* 45: 1087–92.

Colmenares Gil, R. 1979. Analisis automatico de la estructura social de una manada de lobos ibericos *(Canis lupus signatus)* en cautividad. Dissertation, Universidad Autónoma de Madrid, Madrid. 103 pp.

———. 1983. Social structure of an Iberian wolf pack *(Canis lupus signatus)* in captivity: Social strategies and social niches. [In French.] *Biol. Behav.* 8:27–47.

Concannon, P. W. 1986a. Canine physiology of reproduction. Pp. 23–77 in T. Burke, ed., *Small animal reproduction and fertility.* Lea and Febiger, Philadelphia.

———. 1986b. Canine pregnancy and parturition. *Vet. Clin. N. Am., Sm. Anim. Prac.* 16:453–75.

Concannon, P. W., W. Hansel, and K. McEntee. 1977. Changes in LH, progesterone and sexual behavior associated with preovulatory luteinization in the bitch. *Biol. Reprod.* 17:604–13.

Concannon, P., W. Hansel, and W. Visek. 1975. The ovarian cycle of the bitch: Plasma estrogen, LH and progesterone. *Biol. Reprod.* 13:112–21.

Concannon, P. W., J. P. McCann, and M. Temple. 1989. Biology and endocrinology of ovulation, pregnancy and parturition in the dog. *J. Reprod. Fertil.* (Suppl.) 39:3–25.

Concannon, P. W., S. Whaley, D. Lein, and R. Wissler. 1983. Canine gestation length: Variation related to time of mating and fertile life of sperm. *Am. J. Vet. Res.* 44:1819–21.

Concannon, P. W., R. Weinstein, S. Whaley, and D. Frank. 1987. Suppression of luteal function in dogs by luteinizing hormone anti-serum and bromocriptine. *J. Reprod. Fertil.* 81:175–80.

Connor, E. F, and M. A. Bowers. 1987. The spatial consequences of interspecific competition. *Ann. Zool. Fenn.* 24:213–26.

Connor, W. E. 1962. The acceleration of thrombus formation by certain fatty acids. *J. Clin. Invest.* 41:1199–1205.

Conover, M. R., and K. K. Kessler. 1994. Diminished producer participation in an aversive conditioning program to reduce coyote predation on sheep. *Wildl. Soc. Bull.* 22:229–33.

Constable, P., K. Hinchcliff, N. Demma, M. Callahan, B. Dale, K. Fox, L Adams, R. Wack, and L. Kramer. 1998a. Electrocardiographic consequences of a peripatetic lifestyle in gray wolves *(Canis lupus). Comp. Biochem. Physiol. B. Biochem. Mol. Biol.* 120:557–63.

———. 1998b. Serum biochemistry of captive and free-ranging gray wolves *(Canis lupus). J. Zoo Wildl. Med.* 29:435–40.

Constantine, D. G. 1962. Rabies transmission by nonbite route. *Public Health Rep.* 77:287.

Cooley, R. A., and G. M. Kohls. 1944. The genus *Amblyomma* (Ixodidae) in the United States. *J. Parasitol.* 39:77–111.

Cooper, G., W. Amos, D. Hoffman, and D. C. Rubinsztein. 1996. Network analysis of human Y microsatellite haplotypes. *Hum. Mol. Genet.* 5:1759–66.

Coppinger, L., and R. Coppinger. 1982. Livestock guarding dogs that wear sheep's clothing. *Smithsonian,* April, 65–69.

Coppinger, R., and L. Coppinger. 1995. Interactions between livestock guarding dogs and wolves. Pp. 523–26 in L. N. Carbyn, S. H. Fritts, and D. R. Seip, eds., *Ecology and conservation of wolves in a changing world.* Canadian Circumpolar Institute, Edmonton, Alberta.

———. 2001. *Dogs: A startling new understanding of canine origin, behavior, and evolution.* Scribner, New York.

Coppinger, R., and R. Schneider. 1995. Evolution of working dogs. Pp. 21–47 in J. Serpell, ed., *The domestic dog: Its evolution, behaviour, and interactions with people.* Cambridge University Press, Cambridge.

Corbet, G. B., and J. E. Hill. 1991. *A world list of mammalian species.* London: Natural History Museum Publications/Oxford University Press, New York.

Corbett, L. K. 1995. *The dingo in Australia and Asia.* Comstock/Cornell University Press, Ithaca, NY.

Corsi, F., E. Dupre, and L. Boitani. 1999. A large-scale model of wolf distribution in Italy for conservation planning. *Conserv. Biol.* 13(1): 150–59.

Coscia, E. M. 1989. Development of vocalizations in timber wolves *(Canis lupus).* Master's thesis, Dalhousie University, Halifax, NS.

———. 1993. Swimming and aquatic play by timber wolf, *Canis lupus,* pups. *Can. Field Nat.* 107:361–62.

———. 1995. Ontogeny of timber wolf vocalizations: Acoustic properties and behavioral contexts. Ph.D. dissertation, Dalhousie University, Halifax, NS.

Coscia, E. M., D. P. Phillips, and J. C. Fentress. 1990. Den vocalizations of wolf pups *(Canis lupus)* during the first five postnatal weeks. *Am. Zool.* 30:103A. (Abstract.)

———. 1991. Spectral analysis of neonatal wolf *Canis lupus* vocalizations. *Bioacoustics* 3:275–93.

Coughenour, M. B., and F. J. Singer. 1996. Elk population processes in Yellowstone National Park under the policy of natural regulation. *Ecol. Appl.* 6:573–93.

Council for Environment and Natural Resources. 1996. Management of bear, wolf, wolverine and lynx in Finland. Ministry of Agriculture and Forestry 6a/1996, Helsinki.

Covey, D. S. G., and W. S. Greaves. 1994. Jaw dimensions and torsion resistance during canine biting in the Carnivora. *Can. J. Zool.* 72:1055–60.

Cowan, I. M. 1947. The timber wolf in the Rocky Mountain national parks of Canada. *Can. J. Res.* D. 25:139–74.

———. 1949. Rabies as a possible population control of Arctic Canidae. *J. Mammal.* 30:396–98.

Cozza, K., R. Fico, M. Battistini, and E. Rogers. 1996. The damage conservation interface illustrated by predation on domestic livestock in central Italy. *Biol. Conserv.* 78:329–36.

Crabtree, R. L. 1988. Sociodemography of an unexploited coyote population. Ph.D. dissertation, University of Idaho, Moscow.

Crabtree, R. L., F. G. Burton, T. R. Garland, D. A. Catalodo, and W. H. Rickard. 1989. Slow-release radioisotope implants as individual markers for carnivores. *J. Wildl. Mgmt.* 53:949–54.

Crabtree, R. L., and J. W. Sheldon. 1999a. Coyotes and canid coexistence in Yellowstone. Pp. 127–63 in T. W. Clark, A. P. Curlee, S. C. Minta, and P. M. Kareiva, *Carnivores in ecosystems: The Yellowstone experience.* Yale University Press, New Haven, CT.

———. 1999b. The ecological role of coyotes on Yellowstone's Northern Range. *Yellowstone Sci.* 7:15–23.

Crandall, K. A., O. R. P. Bininda-Emonds, G. M. Mace, and R. K. Wayne. 2000. Considering evolutionary processes in conservation biology. *Trends Ecol. Evol.* 15:290–95.

Crawley, M. J., and J. R. Krebs. 1992. Foraging theory. Pp. 90–114 in M. J. Crawley, ed., *Natural enemies: The population biology of predators, parasites and diseases.* Blackwell Scientific Publications, Oxford.

Creel, S. 2001. Social dominance and stress hormones. *Trends Ecol. Evol.* 16:491–97.

Creel, S., N. M. Creel, M. G. L. Mills, and S. L. Monfort. 1997. Rank and reproduction in cooperatively breeding African wild dogs: Behavioral and endocrine correlates. *Behav. Ecol.* 8:298–306.

Creel, S., J. E. Fox, A. Hardy, J. Sands, B. Garrott, and R. O. Peterson. 2002. Snowmobile activity and glucocorticoid stress responses in wolves and elk. *Conserv. Biol.* 16:809–14.

Creel, S. R., and P. M. Waser. 1997. Variation in reproductive suppression among dwarf mongooses: Interplay between mechanisms and evolution. Pp. 150–70 in N. G. Solomon and J. A. French eds., *Cooperative breeding in mammals.* Cambridge University Press, Cambridge.

Crête, M. 1987. The impact of sport hunting on North American moose. *Swed. Wildl. Res.* (Suppl. 1): 553–63.

Crête, M., and F. Messier. 1987. Evaluation of indices of gray wolf, *Canis lupus,* density in hardwood-conifer forests of southwestern Quebec. *Can. Field Nat.* 101:147–52.

Crête, M., J. Ouellet, and L. Lesage. 2001. Comparative effects on plants of caribou/reindeer, moose and white-tailed deer herbivory. *Arctic* 54:407–17.

Criddle, S. 1947. Timber wolf den and pups. *Can. Field Nat.* 115.

Crisler, L. 1956. Observations of wolves hunting caribou. *J. Mammal.* 37:337–46.

———. 1958. *Arctic wild.* Harper and Row, New York.

Cronin, M. A. 1993. Mitochondrial DNA in wildlife taxonomy and conservation biology: Cautionary notes. *Wildl. Soc. Bull.* 21:339–48.

Cross, E. C. 1940. Arthritis among wolves. *Can. Field Nat.* 54:2–4.

Cuesta, L., F. Barcena, F. Palacios, and S. Reig. 1991. The trophic ecology of the Iberian Wolf *(Canis lupus signatus* Cabrera, 1907). A new analysis of stomach's data. *Mammalia* 55(2): 239–54.

Curnow, E. 1969. The history of the eradication of the wolf in Montana. Master's thesis, University of Montana, Missoula. 99 pp.

Custer, J. W., and D. B. Pence. 1981a. Ecological analyses of helminth populations of wild canids from the Gulf coastal prairies of Texas and Louisiana. *J. Parasitol.* 67:289–307.

———. 1981b. Host-parasite relationships in wild Canidae of North America. I. Ecology of helminth infections in the genus *Canis.* Pp. 730–59 in J. A. Chapman and D. Pursley, eds., Worldwide Furbearer Conference Proceedings, vol. 2.

Dale, B. W., L. G. Adams, and R. T. Bowyer. 1994. Functional response of wolves preying on barren-ground caribou in a multiple-prey ecosystem. *J. Anim. Ecol.* 63:644–52.

———. 1995. Winter wolf predation in a multiple ungulate prey system: Gates of the Arctic National Park, Alaska. Pp. 223–30 in L. N. Carbyn, S. H. Fritts, and D. R. Seip, eds., *Ecology and conservation of wolves in a changing world.* Canadian Circumpolar Institute, Edmonton, Alberta.

Danilov, P. I., E. V. Ivanter, V. V. Belkin, and A. A. Nicolaevskij. 1978. Izmenenia cislennosti ohotnicih zverej Karelii po materialam zimnih marsrutnyh ucetov. Pp. 128–59 in V. V. Ivanter, ed., Fauna i ekologia Ptic i Mlekopitauscih Taeznogo Severo-Zapada SSSR, Petrozavodsk.

Darimont, C. T., and T. E. Reimchen. 2002. Intra-hair stable isotope analysis implies seasonal shift to salmon in gray wolf diet. *Can. J. Zool.* 80:1638–42.

Darling, F. F. 1937. *A herd of red deer.* Oxford University Press, Oxford.

Darwin, C. 1871. *The descent of man and selection in relation to sex.* Murray, London.

———. 1872. *The expression of the emotions in man and animals.* University of Chicago Press, Chicago.

———. 1896. *On the origin of species.* 6th ed. D. Appleton, New York.

Davis, J. L. 1978. History and current status of Alaska caribou herds. Pp. 1–8 in D. R. Klein and R. G. White, eds., Parameters of caribou population ecology in Alaska. Biological Papers of the University of Alaska. Special Report, no. 3. University of Alaska, Fairbanks.

Davis, J. W., and K. G. Libke. 1971. Trematodes. Pp. 235–57 in J. W. Davis and R. C. Anderson, eds., *Parasitic diseases of wild mammals.* Iowa State University Press, Ames.

Davis, S., and F. Valla. 1978. Evidence for domestication of the dog 12,000 years ago in the Natufian of Israel. *Nature* 276: 608–10.

Dawkins, R., and J. R. Krebs. 1978. Animal signals: Information or manipulation. Pp. 282–309 in J. R. Krebs and N. B. Davies, eds., *Behavioral ecology: An evolutionary approach.* Sinauer Associates, Sunderland MA.

———. 1979. Arms races between and within species. *Proc. R. Soc. Lond. B* 205:489–511.

DeBie, S. 1990. Is there a future for wildlife in West Africa? *Trans. Int. Congr. Game Biol.* 19:655–63.

Debrot, S., G. Fivaz, C. Mermod, and J.-M. Weber. 1982. *Atlas des poils de Mammifères d'Europe.* Institut de Zoologie, Université de Neuchâtel, Suisse. 208 pp.

Degerbol, M. 1961. On a find of preboreal domestic dog (*Canis familiaris* L.) from Starr Carr, Yorkshire, with remarks on other Mesolithic dogs. *Proc. Prehist. Soc.* 27:35–55.

DeGiusti, D. L. 1971. Acanthocephala. Pp. 140–57 in J. W. Davis and R. C. Anderson, eds., *Parasitic diseases of wild mammals.* Iowa State University Press, Ames.

Dehn, M. M. 1990. Vigilance for predators: Detection and dilution effects. *Behav. Ecol. Sociobiol.* 26:337–42.

Dekker, D. 1985. Responses of wolves (*Canis lupus*) to simulated howling during fall and winter on a homesite in Jasper National Park, Alberta. *Can. Field Nat.* 99:90–93.

———. 1994. *Wolf story: From varmint to favourite.* Edmonton, Alberta: BST Publications.

———. 1998. *Wolves of the Rocky Mountains from Jasper to Yellowstone.* Hancock House Publishing, Surrey, BC.

DelGiudice, G. D. 1998. Surplus killing of white-tailed deer by wolves in north-central Minnesota. *J. Mammal.* 79:227–35.

DelGiudice, G. D., L. D. Mech, and U. S. Seal. 1991. Gray wolf density and its association with weights and hematology of pups from 1970 to 1988. *J. Wildl. Dis.* 27:630–36.

DelGiudice, G. D., R. O. Peterson, and U. S. Seal. 1991. Differences in urinary chemistry profiles of moose on Isle Royale during winter. *J. Wildl. Dis.* 27:407–16.

DelGiudice, G. D., R. O. Peterson, and W. M. Samuel. 1997. Trends of winter nutritional restriction, ticks, and numbers of moose on Isle Royale. *J. Wildl. Mgmt.* 61:895–903.

DelGiudice, G. D., U. S. Seal, and L. D. Mech. 1987. Effects of feeding and fasting on wolf blood and urine parameters. *J. Wildl. Mgmt.* 51:1–10.

Delibes, M., ed. 1990. Status and conservation needs of the wolf (*Canis lupus*) in the Council of Europe member states. Nature and Environment Series no. 47. Council of Europe, Strasbourg. 46 pp.

de Meneghi, D., and G. B. Kaweche. 1990. Wildlife conservation and game management in Zambia. *Trans. Int. Congr. Game Biol.* 19:643–54.

Deputte, B. L. 1994. Ethological study of yawning in primates. I. Quantitative analysis and study of causation in two species of Old World monkeys (*Cercocebus albigena* and *Macaca fascicularis*). *Ethology* 98:221–45.

DeQuoy, A. 1993. *Irish wolfhound saga: A triology.* Volume 1. Alfred DeQuoy, McLeary, Virginia. xxix + 432 pages.

Derix, R. R. W. M., ed. 1994. The social organisation of wolves and African wild dogs: An empirical and model-theoretical approach. Universiteit Utrecht, Utrecht, Utrecht, The Netherlands.

Derix, R. R. W. M., H. de Vries, and J. A. R. A. M. van Hooff. 1994. Relationships in wolves (*Canis lupus*) and African wild dogs (*Lycaon pictus*) in captivity. Pp. 23–56 in R. R. W. M. Derix, ed., *The social organization of wolves and African wild dogs: An empirical and model-theoretical approach.* Universiteit Utrecht, Utrecht, The Netherlands.

Derix, R. R. W. M., J. van Hooff, H. de Vries, and J. Wensing. 1993. Male and female mating competition in wolves— female suppression vs. male intervention. *Behaviour* 127: 141–74.

Dethier, V. G. 1954. The physiology of olfaction in insects. *Ann. N.Y. Acad. Sci.* 58:139–57.

De Vos, A. 1949. Timber wolves killed by cars on Ontario highways. *J. Mammal.* 30:197.

De Vos, A., and A. E. Allin. 1949. Some notes on moose parasites. *J. Mammal.* 30:430–31.

Dewsbury, D. A. 1972. Patterns of copulatory behavior in male mammals. *Q. Rev. Biol.* 47:1–33.

Diamond, S. 1994. The prairie wolf returns. *Int. Wolf* 4:3–7.

Dieterich, R. A. 1970. Hematologic values of some Arctic mammals. *J. Am. Vet. Med. Assoc.* 157:604–6.

Dieulafé, L. 1906. Morphology and embryology of the nasal fossae of vertebrates. *Ann. Otol. Rhinol. Laryngol.* 15:267–349.

Dimond, S., and J. Lazarus. 1974. The problem of vigilance in animal life. *Brain Behav. Evol.* 9:60–79.

Dixson, A. F., and L. George. 1982. Prolactin and paternal behaviour in a male New World primate. *Nature* 299:551–53.

Dodman, N., R. Donnelly, L. Shuster, P. Mertens, W. Rand, and

K. Miczek. 1996. Use of fluoxetine to treat dominance aggression in dogs. *J. Am. Vet. Med. Assoc.* 209:1585–87.

Dolf, G., J. Schlapfer, C. Gaillard, E. Randi, V. Lucchini, U. Breitenmoser, and N. Stahlberger-Saitbekova. 2000. Differentiation of the Italian wolf and the domestic dog based on microsatellite analysis. *Genet. Sel. Evol.* 32:533–41.

Donovan, C. A. 1969. Canine anal glands and chemical signals (pheromones). *J. Am. Vet. Med. Assoc.* 155:1995–96.

Dorrance, M. J. 1982. Predation losses of cattle in Alberta. *J. Range Mgmt.* 35:690–92.

Doty, R. L. 1986. Odor-guided behavior in mammals. *Experientia* 42:257–71.

Doty, R. L., and I. Dunbar. 1974. Attraction of beagles to conspecific urine, vaginal and anal sac secretion odors. *Physiol. Behav.* 12:825–33.

Doty, R. L., P. A. Green, C. Ram, and S. L. Yankell. 1982. Communication of gender from human breath odors: Relationship to perceived intensity and pleasantness. *Horm. Behav.* 16:13–22.

Douglass, E. M. 1981. Hip dysplasia in a timber wolf. *Vet. Med. Sm. Anim. Clin.* 76:401–3.

Dowling, T. E., B. D. DeMarais, W. L. Minckley, M. E. Douglas, and P. C. Marsh. 1992. Use of genetic characters in conservation biology. *Conserv. Biol.* 6:7–8.

Dowling, T. E., W. L. Minckley, M. E. Douglas, P. C. Marsh, and B. D. DeMarais. 1992. Response to Wayne, Nowak, and Phillips and Henry: Use of molecular characters in conservation biology. *Conserv. Biol.* 6:600–603.

Drag, M. D. 1991. Hematologic values of captive Mexican wolves. *Am. J. Vet. Res.* 52:1891–92.

Duda, M. D., S. J. Bissell, and K. C. Young. 1998. *Wildlife and the American mind: Public opinion on and attitudes toward fish and wildlife management.* Harrisonburg, VA: Responsive Management.

Duda, M. D., and K. C. Young. 1995. New Mexico residents' opinions toward wolf reintroduction. (Responsive Management Report.) League of Women Voters, Harrisonburg, VA.

Duffield, J. W. 1992. An economic analysis of wolf recovery in Yellowstone: Park visitor attitudes and values. Pp. 2-31–2-87 in J. D. Varley and W. G. Brewster, eds., *Wolves for Yellowstone? A Report to the United States Congress,* Vol. 4, *Research and analysis.* U.S. National Park Service, Yellowstone National Park, WY.

Duke-Elder, S. 1958. *System of ophthalmology.* Vol. 1. Henry Kimpton, London.

Duke-Elder, S., and K. C. Wybar. 1961. *System of ophthalmology.* Vol. 2. Henry Kimpton, London.

Dunbar, I. 1977. Olfactory preferences in dogs: The response of male and female beagles to conspecific odors. *Behav. Biol.* 20:471–81.

Dunbar, I., and M. Buehler. 1980. A masking effect of urine from male dogs. *Appl. Anim. Ethol.* 6:297–301.

Dunbar, I., E. Ranson, and M. Buehler. 1981. Pup retrieval and maternal attraction to canine amniotic fluids. *Behav. Processes* 6:249–60.

Dunlap, T. R. 1988. *Saving America's wildlife.* Princeton University Press, Princeton, NJ.

Dutt, R. H., E. C. Simpson, J. C. Christian, and C. E. Barnhart. 1959. Identification of preputial glands as the site of production of sexual odor in the boar. *J. Anim. Sci.* 18:1557.

Dworkin, S., J. Katzman, G. A. Hutchison, and J. R. McCabe. 1940. Hearing acuity of animals as measured by conditioning methods. *J. Exp. Psychol.* 6:281–98.

Dyson, R. F. 1965. Experience with succinylcholine chloride in zoo animals. *Int. Zoo Yrbk.* 5:205–6.

Eaton, R. L. 1970. The predatory sequence, with emphasis on killing behavior and its ontogeny, in the cheetah (*Acinonyx jubatus* Schreber). *Z. Tierpsychol.* 27:492–504.

Eberhardt, L. L. 1997. Is wolf predation ratio-dependent? *Can. J. Zool.* 75:1940–44.

———. 1998. Applying difference equations to wolf predation. *Can. J. Zool.* 76:380–86.

———. 2000. Predator-prey ratio dependence and regulation of moose populations. *Can. J. Zool.* 78:511–13.

Eberhardt, L. L., R. A. Garrott, D. W. Smith, P. J. White, and R. O. Peterson. 2002. Assessing the impact of wolves on ungulate prey. *Ecol. Appl.* In press.

Eberhardt, L. L., and R. O. Peterson. 1999. Predicting the wolf-prey equilibrium point. *Can. J. Zool.* 77:494–98.

Eckels, R. P. 1937. Greek wolf-lore. Ph.D. dissertation, University of Pennsylvania, Philadelphia.

Edmonds, E. J. 1988. Population status, distribution, and movements of woodland caribou in west central Alberta. *Can. J. Zool.* 66:817–26.

Edwards, J. 1983. Diet shifts in moose due to predator avoidance. *Oecologia* 60:185–89.

Eisenberg, J. F., and D. G. Kleiman. 1972. Olfactory communication in mammals. *Annu. Rev. Ecol. Syst.* 3:1–32.

Eizirik, E., S. L. Bonato, W. E. Johnson, P. G. Crawshaw, J. C. Vie, D. M. Brousset, S. J. O'Brien, and F. M. Salzano. 1998. Phylogeographic patterns and evolution of the mitochondrial DNA control region in two Neotropical cats (Mammalia, Felidae). *J. Mol. Evol.* 47:613–24.

Elder, W. H., and C. M. Hayden. 1977. Use of discriminant function in taxonomic determination of canids from Missouri. *J. Mammal.* 58:17–24.

Eles, H. 1986. Vargen. Arsbok fran Varmlands Museum, Argang 84. Andra upplagan AB Ystads Centraltryckeri 74013.

Ellegren, H. 1999. Inbreeding and relatedness in Scandinavian grey wolves *Canis lupus. Hereditas* 130:239–44.

Ellegren, H., P. Savolainen, and B. Rosen. 1996. The genetical history of an isolated population of the endangered grey wolf *Canis lupus:* A study of nuclear and mitochondrial polymorphisms. *Phil. Trans. R. Soc. Lond. B* 351:1661–69.

Elliott, D., L. Stein, and M. Harrison. 1960. Determination of absolute intensity thresholds and frequency difference thresholds in cats. *J. Acoust. Soc. Am.* 32:380–84.

Emnett, C. W. 1986. Prevalence of sarcocystis in wolves and white-tailed deer in northeastern Minnesota. *J. Wildl. Dis.* 22:193–95.

Erickson, A. B. 1944. Helminths of Minnesota Canidae in relation to food habits and a host list and key to the species reported from North America. *Am. Midl. Nat.* 32:358–72.

Errington, P. 1967. *On predation and life.* Iowa State University Press, Ames.

Estes, R. D. 1966. Behaviour and life history of the wildebeest (*Connochaetes taurinus* Burchell). *Nature* 212:999–1000.

———. 1972. The role of the vomeronasal organ in mammalian reproduction. *Mammalia* 36:315–41.

Eugster, A. K., and C. Nairn. 1977. Diarrhea in puppies: Parvovirus-like particles demonstrated in their feces. *Southwest Vet.* 30:59.

Evans, E. 1933. The transport of spermatozoa in the dog. *Am. J. Physiol.* 105:287–93.

Ewer, R. F. 1968. *Ethology of mammals.* London: Elek Scientific Books.

Fancy, S. G., and W. B. Ballard. 1995. Monitoring wolf activity by satellite. Pp. 329–33 in L. N. Carbyn, S. H. Fritts, and D. R. Seip, eds., *Ecology and conservation of wolves in a changing world.* Canadian Circumpolar Institute, Edmonton, Alberta.

Farago, S. 1993. Current status of the wolf (*Canis lupus* L.) in Hungary. Pp. 44–49 in C. Promberger and W. Schröder, eds., *Wolves in Europe: Status and perspectives.* Munich Wildlife Society, Ettal, Germany.

Fau, J. F., and I. R. Tempany. 1976. Wolf observations from aerial bison surveys, 1972–76. Pp. 11–14 in J. G. Stelfox, ed., Wood Buffalo National Park Research 1972–76. Annual Report. Canadian Wildlife Service and Parks Canada, Edmonton, Alberta.

Fay, R. R. 1988. *Hearing in vertebrates: A psychophysics data book.* Hill-Fay Associates, Winnetka, Illinois.

———. 1992. Structure and function in sound discrimination among vertebrates. Pp. 229–63 in D. B. Webster, R. R. Fay, and A. N. Popper, eds., *The evolutionary biology of hearing.* Springer-Verlag, New York.

Federoff, N. E. 1996. Malocclusion in the jaws of captive bred Arctic wolves, *Canis lupus arctos. Can. Field Nat.* 110:683–87.

———. 2001. Antibody response to rabies vaccination in captive and free-ranging wolves (*Canis lupus*). *J. Zoo Wildl. Med.* 32:127–29.

Federoff, N. E., and R. M. Nowak. 1997. Man and his dog. *Science* 278:205.

Federoff, N. E., and R. M. Nowak. 1998. Cranial and dental abnormalities of the endangered red wolf *Canis rufus. Acta Theriol.* 43:293–300.

Fentress, J. C. 1967. Observations on the behavioral development of a hand-reared male timber wolf. *Am. Zool.* 7:339–51.

———. 1982. Conflict and context in sexual behaviour. Pp. 579–613 in J. Hutchinson, ed., *Biological determinants of sexual behaviour.* J. Wiley & Sons, New York.

———. 1983. A view of ontogeny. Pp. 24–64 in J. F. Eisenberg, and D. G. Kleiman, eds., *Advances in the study of mammalian behavior.* Special Publication no. 7. American Society of Mammalogists, Stillwater, OK.

———. 1992. The covalent animal: On bonds and their boundaries in behavioral research. Pp. 44–71 in H. Davis and D. Balfour, eds., *The inevitable bond: Examining scientist-animal interactions.* Cambridge University Press, Cambridge.

Fentress, J. C., R. Field, and H. Parr. 1978. Social dynamics and communication. Pp. 67–106 in H. Markowitz and V. Stevens, eds., *Behavior of captive wild animals.* Nelson-Hall, Chicago.

Fentress, J. C., and J. Ryon. 1982. A long-term study of distributed pup feeding in captive wolves. Pp. 238–61 in F. H. Harrington and P. C. Paquet, eds., *Wolves of the world: Perspectives of behavior, ecology, and conservation.* Noyes Publications, Park Ridge, NJ.

Fentress, J. C., J. Ryon, P. J. McLeod, and G. Z. Havkin. 1987. A multidimensional approach to agonistic behavior in wolves. Pp. 253–74 in H. Frank, ed., *Man and wolf: Advances, issues and problems in captive wolf research.* Dr. W. Junk Publishers, Dordrecht, The Netherlands.

Ferguson, J. H., and G. E. Folk, Jr. 1970. Free fatty acid levels of arctic mammals. *Fed. Proc.* 29:659.

———. 1971. Free fatty acid levels in several species of arctic carnivores. *Comp. Biochem. Physiol. B. Comp. Biochem.* 40:309–12.

Ferguson, J. H., and T. D. Schultz. 1978. Lipid turnover in the tundra wolf *(Canis lupus tundrarum). Comp. Biochem. Physiol. A. Comp. Physiol.* 61:439–40.

Ferguson, S. H., A. T. Bergerud, and R. Ferguson. 1988. Predation risk and habitat selection in the persistence of a remnant caribou population. *Oecologia* 76:236–45.

Fernandez, A., J. M. Fernandez, and G. Palomero. 1990. El lobo en Cantabria. Pp. 32–44 in J. C. Blanco, L. Cuesta, and S. Reig (eds.), *El lobo* (Canis lupus) *en España: Situacion, problematica y apuntes sobre su ecología.* Istituto Nacional para la Conservacion de la Naturaleza, Madrid.

Ferrell, R. E., D. C. Morizot, J. Horn, and C. J. Carley. 1980. Biochemical markers in species endangered by introgression: The red wolf. *Biochem. Genet.* 18:39–49.

Fico, R., F. Marsilio, and P. G. Tiscar. 1996. Antibodies against canine parvovirus, distemper, infectious canine hepititis, canine coronavirus, and *Ehrlichia canis* in wolves (*Canis lupus*) in central Italy. Istituto Nazionale per la Fauna Selvatica—Supplemento all Richerche di Biologia della Selvaggina 24:137–43.

Fico, R., G. Morosetti, and A. Giovannini. 1993. The impact of predators on livestock in the Abruzzo region of Italy. *Rev. Sci. Tech. O.I.E.* 12:39–50.

Field, R. 1978. Vocal behavior of wolves: Variability in structure, context, annual/diurnal patterns, and ontogeny. Ph.D. dissertation, Johns Hopkins University, Baltimore.

———. 1979. A perspective on syntactics of wolf vocalizations. Pp. 182–205 in E. Klinghammer, ed., *The behavior and ecology of wolves.* Garland STPM Press, New York.

Fiennes, R. 1976. *The order of the wolves.* Bobbs-Merrill, Indianapolis, IN.

Filibeck, U., M. Nicoli, P. Rossi, and G. Boscagli. 1982. Detection by frequency analyzer of individual wolves howling in a chorus: A preliminary report. *Boll. Zool.* 49:151–54.

Filonov, C. 1980. Predator-prey problems in nature reserves of the European part of the RSFSR. *J. Wildl. Mgmt.* 44: 389–96.

Fischer, H. 1989. Restoring the wolf—Defenders launches a compensation fund. *Defenders* 64 (Jan.–Feb.), 9, 36.

———. 1995. *Wolf wars.* Falcon Press, Helena and Billings, MT.

Fischer, H., B. Snape, and W. Hudson. 1994. Building economic incentive into the Endangered Species Act. *Endangered Species Tech. Bull.* 19:4–5.

Fischman, O., P. A. Siqueira, and G. Baptista. 1987. *Microsporum gypseum* infection in a gray wolf *(Canis lupus)* and a camel *(Camelus bactrianus)* in a zoo. *Mykosen* 30:295–97.

Fitzgerald, J. M. 1988. Withering wildlife: Whither the Endangered Species Act? A review of amendments to the Act. *Endangered Species Update* 5:27–34.

Flader, S. L. 1974. *Thinking like a mountain: Aldo Leopold and the evolution of an ecological attitude toward deer, wolves and forests.* University of Wisconsin Press, Madison.

Flagstad, Ø., C. W. Walker, C. Vilà, A.-K. Sundqvist, B. Fernholm, A. K. Hufthammar, Ø. Wiig, I. Kojola, and H. Ellegren. 2003. The historical grey wolf population of the Scandinavian Peninsula: Patterns of genetic variability and migration during an era of dramatic decline. *Mol. Ecol.* In press.

Floyd, T. J., L. D. Mech, and P. A. Jordan. 1978. Relating wolf scat content to prey consumed. *J. Wildl. Mgmt.* 42:528–32.

Fogleman, V. M. 1988. American attitudes toward wolves: A history of misperception. *Environ. Ethics* 10:63–94.

Folk, G. E., M. Fox, and M. Folk. 1970. Physiological differences between alpha and subordinate wolves in a captive sibling pack. *Am. Zool.* 10:487.

Folk, G. E., and R. S. Hedge. 1964. Comparative physiology of heart rate in unrestrained mammals. *Am. Zool.* 4:111. (Abstract.)

Fonseca, F. P. 1990. O lobo *(Canis lupus signatus* Cabrera, 1907) em Portugal: Problemática da sua conservação. Ph.D. dissertation, Universidade de Lisboa, Portugal.

Foran, D. R., S. C. Minta, and K. S. Heinemeyer. 1997. DNA-based analysis of hair to identify species and individuals for population research and monitoring. *Wildl. Soc. Bull.* 25:840–47.

Forbes, G. J., and J. B. Theberge. 1995. Influences of a migratory deer herd on wolf movements and mortality in and near Algonquin Park, Ontario. Pp. 303–13 in L. N. Carbyn, S. H. Fritts, and D. R. Seip, eds., *Ecology and conservation of wolves in a changing world.* Canadian Circumpolar Institute, Edmonton, Alberta.

Forbes, S. H., and D. K. Boyd. 1996. Genetic variation of naturally colonizing wolves in the Central Rocky Mountains. *Conserv. Biol.* 10:1082–90.

———. 1997. Genetic structure and migration in native and reintroduced Rocky Mountain wolf populations. *Conserv. Biol.* 11:1226–34.

Foromozov, A. N. 1946. The snow cover as an environment factor and its importance in the life of mammals and birds. (Moskovskoe obshchestvo ispytatelei priroda) Materialy k poznaniyu fauny i flory SSSR, Otdel. Zool. n. 5 (xx). [Translation from Russian published by Boreal Institute, University of Alberta, Edmonton.]

Fortymile Caribou herd: Management Plan. 1995. Unpublished report. Sponsored by U.S. Bureau of Land Management, U.S. Fish and Wildlife Service, U.S. National Park Service, and Alaska Department of Fish and Game. 21 pp.

Foss, I., and G. Flottorp. 1974. A comparative study of the development of hearing and vision in various species commonly used in experiments. *Acta Otolaryngol.* 77:202–14.

Fox, H. 1923. *Disease in captive wild mammals and birds.* J. B. Lippincott, Philadelphia.

———. 1926. Report of the Laboratory of Comparative Pathology, Philadelphia.

———. 1927. Report of the Laboratory of Comparative Pathology, Philadelphia.

———. 1941. Report of the Penrose Research Laboratory of the Zoological Society of Philadelphia.

Fox, J. 2001. Stress physiology and movement behavior of gray wolves in Voyageurs and Isle Royale national parks. Master's thesis, Michigan Technological University, Houghton. 140 pp.

Fox. J. L., and R. S. Chundawat. 1995. Wolves in the Transhimalayan region of India: The continued survival of a low density population. Pp. 95–103 in L. N. Carbyn, S. H. Fritts, and D. R. Seip, eds., *Ecology and conservation of wolves in a changing world.* Canadian Circumpolar Institute, Edmonton, Alberta.

Fox, J. L., and G. P. Streveler. 1986. Wolf predation on mountain goats in southeastern Alaska. *J. Mammal.* 67:192–95.

Fox, M. W. 1964. The ontogeny of behavior and neurologic responses of the dog. *Anim. Behav.* 12:301–10.

———. 1969. The anatomy of aggression and its ritualization in Canidae: A developmental and comparative study. *Behaviour* 35:242–58.

———. 1970. A comparative study of the development of facial expressions in canids: Wolf, coyote and foxes. *Behaviour* 36:49–73.

———. 1971a. *The behavior of wolves, dogs and related canids.* Jonathan Cape, London.

———. 1971b. *Integrative development of brain and behavior in the dog.* University of Chicago Press, Chicago.

———. 1971c. Ontogeny of prey-killing behavior in Canidae. *Behaviour* 35:259–72.

———. 1971d. Possible examples of high order behavior in wolves. *J. Mammal.* 52:640–41.

———. 1971e. Socio-infantile and socio-sexual signals in canids: A comparative and ontogenetic study. *Z. Tierpsychol.* 28: 185–210.

———. 1972a. The social significance of genital licking in the wolf, *Canis lupus. J. Mammal.* 53:637–40.

———. 1972b. Socio-ecological implications of individual differences in wolf litters: A developmental and evolutionary perspective. *Behaviour* 41:298–313.

———. 1973. Social dynamics of three captive wolf packs. *Behaviour* 47:290–301.

————. 1975. Evolution of social behavior in canids. Pp. 429–60 in M. W. Fox, ed., *The wild canids: Their systematics, behavioral ecology, and evolution.* Van Nostrand Reinhold, New York.

————. 1976. Inter-species interaction differences in play actions in canids. *Appl. Anim. Ethol.* 2:181–85.

————. 1978. *The dog: Its domestication and behavior.* Garland STPM Press, New York.

————. 1980. *The soul of the wolf.* Little, Brown, Boston.

Fox, M. W., and R. V. Andrews. 1973. Physiological and biochemical correlates of individual differences in behavior of wolf cubs. *Behaviour* 46:129–40.

Fox, M. W., and J. A. Cohen. 1977. Canid communication. Pp. 728–48 in T. A. Sebeok, ed., *How animals communicate.* Indiana University Press, Bloomington.

Fox, M. W., R. Lockwood, and R. Schideler. 1974. Introduction studies in captive wolf packs. *Z. Tierpsychol.* 35:39–48.

Fraga de Azevedo, J., J. M. Palmeiro, and P. Rombert. 1974. Aspects of trichinelliasis in Portugal: Apropos of a parasitic case in *Canis lupus* L. *Ann. Inst. Hyg. Med. Trop.* 2:349–56.

Frair, J. L., E. M. Anderson, B. E. Kohn, D. P. Shelley, T. E. Gehring, and D. Unger. 1996. Impacts of Highway 53 expansion on gray wolves: Preliminary results. Pp. 123–31 in N. Fascione and M. Cecil, eds., Proceedings of the Defenders of Wildlife's Wolves of America Conference, Albany, NY. Defenders of Wildlife, Washington, D.C.

Frame, L. H., J. R. Malcolm, G. W. Frame, and H. van Lawick. 1979. Social organization of African wild dogs (*Lycaon pictus*) on the Serengeti Plains, Tanzania 1967–1978. *Z. Tierpsychol.* 50:225–49.

Francisci, F., and V. Guberti. 1993. Recent trends of wolves in Italy as apparent from kill figures and specimens. Pp. 91–102 in C. Promberger and W. Schröder, eds., *Wolves in Europe: Status and perspectives.* Munich Wildlife Society, Ettal, Germany.

Frank, H., ed. 1987. *Man and Wolf: Advances, Issues, and Problems in Captive Wolf Research.* Dr. W. Junk Publishers, Dordrecht, The Netherlands. 439 pp.

Frank, H., and M. G. Frank. 1982. Comparison of problem-solving performance in six-week-old wolves and dogs. *Anim. Behav.* 30:95–98.

————. 1983. Inhibition training in wolves and dogs. *Behav. Processes* 8:363–77.

————. 1984. Information processing in wolves and dogs. *Acta Zool. Fenn.* 171:225–28.

————. 1985. Comparative manipulation-test performance in ten-week-old wolves (*Canis lupus*) and Alaskan malamutes (*Canis familiaris*): A Piagetian interpretation. *J. Comp. Psychol.* 99:266–74.

————. 1987. The University of Michigan canine information-processing project (1979–1981). Pp. 143–67 in H. Frank, ed., *Man and wolf: Advances, issues, and problems in captive wolf research.* Dr. W. Junk Publishers, Dordrecht, The Netherlands.

Frank, H., M. G. Frank, L. M. Hasselbach, and D. M. Littleton. 1989. Motivation and insight in wolf (*Canis lupus*) and Alaskan malamute (*Canis familiaris*): Visual discrimination learning. *Bull. Psychon. Soc.* 27:455–58.

Frank, H., L. M. Hasselbach, and D. M. Littleton. 1986. Socialized vs. unsocialized wolves (*Canis lupus*) and Alaskan malamutes (*Canis familiaris*): A Piagetian interpretation. Pp. 33–49 in M. W. Fox and L. D. Mickley, eds., *Advances in animal welfare science 1986/87.* The Humane Society of the United States, Washington, D.C.

Frank, M. G., and H. Frank. 1988. Food reinforcement versus social reinforcement in timber wolf pups. *Psychon. Soc. Bull.* 26:467–68.

Fraser, C. M., J. A. Bergeron, A. Mays, and S. E. Aiello, eds., 1991. *The Merck veterinary manual.* Merck & Co., Rahway, NJ.

Fredrickson, R., and P. Hedrick. 2002. Body size in endangered Mexican wolves: Effects of inbreeding and cross-lineage matings. *Anim. Conserv.* 5: 39–43.

Freeman, R. S., A. Adorjan, and D. H. Pimlott. 1961. Cestodes of wolves, coyotes, and coyote-dog hybrids in Ontario. *Can. J. Zool.* 39:527–32.

Frenzel, L. D., Jr. 1974. Occurrence of moose in food of wolves as revealed by scat analyses: A review of North American studies. *Nat. Can.* 101:467–79.

Freuchen, P. 1935. Report on the fifth Thule expedition, 1921–1924, the Danish expedition to Arctic North America in charge of Knud Rasmussen 2 (4 and 5). Mammals, Part 2. Field notes and biological observations. Copenhagen, Denmark.

Friis, L. K. 1985. An investigation of subspecific relationships of the grey wolf, *Canis lupus*, in British Columbia. Master's thesis, University of Victoria.

Frijlink, J. H. 1977. Patterns of wolf pack movements prior to kills as read from tracks in Algonquin Provincial Park, Ontario. *Bijdragen tot de Dierkunde* 47(1): 131–37.

Fritts, S. H. 1982. Wolf depredation on livestock in Minnesota. Resource Publ. 145. U.S. Department of the Interior, Fish and Wildlife Service, Washington, D.C. 11 pp.

————. 1983. Record dispersal by a wolf from Minnesota. *J. Mammal.* 64:166–67.

————. 1993. Controlling wolves in the greater Yellowstone area. Pp. 173–233 in R. S. Cook, ed., *Ecological issues on reintroducing wolves into Yellowstone National Park.* U.S. National Park Service Scientific Monograph Series NPS/NRYELL/NRSM-93/22.

Fritts, S. H., E. E. Bangs, J. A. Fontaine, W. G. Brewster, and J. F. Gore. 1995. Restoring wolves to the northern Rocky Mountains of the United States. Pp. 107–25 in L. N. Carbyn, S. H. Fritts, and D. R. Seip, eds., *Ecology and conservation of wolves in a changing world.* Canadian Circumpolar Institute, Edmonton, Alberta.

Fritts, S. H., E. E. Bangs, J. A. Fontaine, M. R. Johnson, M. K. Phillips, E. D. Koch, and J. R. Gunson. 1997. Planning and implementing a reintroduction of wolves to Yellowstone National Park and Central Idaho. *Restor. Ecol.* 5:7–27.

Fritts, S. H., E. E. Bangs, and J. F. Gore. 1994. The relationship of wolf recovery to habitat conservation and biodiversity in the northwestern United States. *Landsc. Urb. Plann.* 28:23–32.

Fritts, S. H., and L. N. Carbyn. 1995. Population viability, nature reserves, and the outlook for gray wolf conservation in North America. *Restor. Ecol.* 3:26–38.

Fritts, S. H., and D. D. Caywood. 1980. Osteoarthrosis in a wolf (*Canis lupus*) radio-tracked in Minnesota. *J. Wildl. Dis.* 16:413–17.

Fritts, S. H., and L. D. Mech. 1981. Dynamics, movements, and feeding ecology of a newly protected wolf population in northwestern Minnesota. Wildlife Monographs, no. 80. The Wildlife Society, Bethesda, MD. 79 pp.

Fritts, S. H., and W. J. Paul. 1989. Interactions of wolves and dogs in Minnesota. *Wildl. Soc. Bull.* 17:121–23.

Fritts, S. H., W. J. Paul, and L. D. Mech. 1979. Evaluation of methods for alleviating wolf depredations on livestock. Unpublished report. U.S. Fish and Wildlife Service, St. Paul, MN.

———. 1984. Movements of translocated wolves in Minnesota. *J. Wildl. Mgmt.* 48:709–21.

———. 1985. Can relocated wolves survive? *Wildl. Soc. Bull.* 13: 459–63.

Fritts, S. H., W. J. Paul, L. D. Mech, and D. P. Scott. 1992. Trends and management of wolf-livestock conflicts in Minnesota. Resource Publ. 181. U.S. Fish and Wildlife Service, Washington, D.C. 27 pp.

Frkovic, A., and D. Huber. 1993. Wolves in Croatia: Baseline data. Pp. 66–69 in C. Promberger and W. Schröder, eds., *Wolves in Europe: Status and perspectives.* Munich Wildlife Society, Ettal, Germany.

Frkovic, A., L. R. Ruff, L. Cicnjak, and D. Huber. 1988. Ulov vuka u Gorskom kotaru u razdoblju od 1945 do 1986 godine. *Sumarski list.* 62:519–30.

Frommolt, K. H., M. I. Kaal, N. M. Paschina, and A. A. Nikol'skii. 1988. Die Entwicklung der Lautgebung bein Wolf (*Canis lupus* L., Canidae L.) wahrend der postnatalen Ontogenese [Sound development of the wolf (*Canis lupus* L., Canidae L.) during the postnatal ontogeny]. *Zoologische Jahrbucher Abteilung fur Allegemeine Zoologie Physiologie der Tiere* 92:105–15.

Fryxell, J. M., J. Greever, and A. R. E. Sinclair. 1988. Why are migratory ungulates so abundant? *Ecol. Soc. Am. Bull.* 69:140.

Fuller, J. L., and E. M. DuBuis. 1962. The behavior of dogs. Pp. 415–52 in E. Hafez, ed., *The behavior of domestic animals.* Bailliere, Tindall & Cox, London.

Fuller, T. K. 1982. Wolves. Pp. 225–26 in D. E. Davis, ed., *CRC handbook of census methods for terrestrial vertebrates.* CRC Press, Boca Raton, FL.

———. 1989a. Denning behavior of wolves in north-central Minnesota. *Am. Midl. Nat.* 121:184–88.

———. 1989b. Population dynamics of wolves in north-central Minnesota. Wildlife Monographs, no. 105. The Wildlife Society, Bethesda, MD. 41 pp.

———. 1990. Dynamics of a declining white-tailed deer population in north-central Minnesota. Wildlife Monographs, no. 110. The Wildlife Society, Bethesda, MD. 37 pp.

———. 1991. Effect of snow depth on wolf activity and prey selection in north central Minnesota. *Can. J. Zool.* 69:283–87.

———. 1995a. Comparative population dynamics of North American wolves and African wild dogs. Pp. 325–28 in L. N. Car-

byn, S. H. Fritts, and D. R. Seip, eds., *Ecology and conservation of wolves in a changing world.* Canadian Circumpolar Institute, Edmonton, Alberta.

———. 1995b. Guidelines for gray wolf management in the Northern Great Lakes Region. International Wolf Center Technical Publication no. 271. Ely, Minnesota. 19 pp.

———. 1995c. An international review of large carnivore conservation status. Pp. 410–12 in J. A. Bissonette and P. R. Krauseman, eds., *Integrating people and wildlife for a sustainable future.* The Wildlife Society, Bethesda, MD.

Fuller, T. K., W. E. Berg, G. L. Radde, M. S. Lenarz, and G. B. Joselyn 1992. A history and current estimate of wolf distribution and numbers in Minnesota. *Wildl. Soc. Bull.* 20:42–55.

Fuller, T. K., and L. B. Keith. 1980a. Wolf population dynamics and prey relationships in northeastern Alberta. *J. Wildl. Mgmt.* 44:583–602.

———. 1980b. Woodland caribou population dynamics in northeastern Alberta. AOSERP Report no. 101. Alberta Oil Sands Environmental Resource Program, Edmonton, Alberta. 63 pp.

———. 1981a. Immobilization of wolves in winter with etorphine. *J. Wildl. Mgmt.* 45:271–73.

———. 1981b. Non-overlapping ranges of coyotes and wolves in northeastern Alberta. *J. Mammal.* 62:403–5.

Fuller, T. K., and D. B. Kittredge. 1996. Conservation of large forest carnivores. Pp. 137–64 in R. M. DeGraaf and R. I. Miller, eds., *Conservation of wildlife diversity in forested landscapes.* Chapman and Hall, London.

Fuller, T. K., and D. W. Kuehn. 1983. Immobilization of wolves using ketamine in combination with xylazine or promazine. *J. Wildl. Dis.* 19:69–72.

Fuller, T. K., and B. A. Sampson. 1988. Evaluation of a simulated howling survey for wolves. *J. Wildl. Mgmt.* 52:60–63.

Fuller, W. A., and N. S. Novakowski. 1955. Wolf control operations, Wood Buffalo National Park, 1951–1952. Wildlife Management Bulletin Series 1, no. 11. Canadian Wildlife Service.

Furmaga, S. 1953. *Spirometra janicki* sp. n. (Diphyllobothriidae). *Acta Parasitol. Polon.* 1(2): 29–59.

Fyvie, A. 1971. *Dioctophyma renale.* Pp. 258–62 in J. W. Davis and R. C. Anderson, eds., *Parasitic diseases of wild mammals.* Iowa State University Press, Ames.

Gadbois, S. 2002. The socioendocrinology of aggression-mediated stress in timber wolves (*Canis lupus*). Ph.D. dissertation, Dalhousie University, Halifax, NS.

Galton, F. 1871. Gregariousness in cattle and men. *Macmillan's Magazine* 23, 353.

Gao, Z. 1990. Feeding habits of the wolf in Inner Mongolia and Heilongjiang Provinces, China. *Trans. Int. Congr. Game Biol.* 19(2): 563–65.

García-Moreno, J., M. D. Matocq, M. S. Roy, E. Geffen, and R. K. Wayne. 1996. Relationships and genetic purity of the endangered Mexican wolf based on analysis of microsatellite loci. *Conserv. Biol.* 10:376–89.

Gasaway, W. C., R. D. Boertje, D. V. Grangaard, D. G. Kelleyhouse, R. O. Stephenson, and D. G. Larsen. 1992. The role of predation in limiting moose at low densities in Alaska and

Yukon and implications for conservation. Wildlife Monographs, no. 120. The Wildlife Society, Bethesda, MD. 59 pp.

Gasaway, W. C., R. O. Stephenson, J. L. Davis, P. E. K. Shepherd, and O. E. Burris. 1983. Interrelationships of wolves, prey, and man in interior Alaska. Wildlife Monographs, no. 84. The Wildlife Society, Bethesda, MD. 50 pp.

Gavrin, W. F., and S. S. Donaurov. 1954. Der Wolf in Wilde von Bielowieza. *Zool. Zh.* 33:904–24.

Geffen, E., M. Gompper, J. Gittleman, L. Hang-Kwang, D. Macdonald, and R. Wayne. 1996. Size, life-history traits, and social oranization in the Canidae: A reevaluation. *Am. Nat.* 147:140–60.

Gehring, T. M. 1993. Adult black bear, *Ursus americanus,* displaced from a kill by a wolf, *Canis lupus,* pack. *Can. Field Nat.* 107:373–74.

Geist, V. 1998. *Deer of the world: Their evolution, behavior, and ecology.* Stackpole Books, Mechanicsburg, PA.

Gensch, W. 1968. Notes on breeding timber wolves, *Canis lupus occidentalis,* at Dresden Zoo. *Int. Zoo Yrbk.* 8:15–16.

Georges, S. 1976. A range expansion of the coyote in Quebec. *Can. Field Nat.* 90:78–79.

Gese, E. M. 1995. Foraging ecology of coyotes in Yellowstone National Park. Ph.D. dissertation, University of Wisconsin, Madison.

———. 2001. Territorial defense by coyotes (*Canis latrans*) in Yellowstone National Park, Wyoming: Who, how, where, when, and why. *Can J. Zool.* 79:980–87.

Gese, E. M., and L. D. Mech. 1991. Dispersal of wolves (*Canis lupus*) in northeastern Minnesota, 1969–1989. *Can. J. Zool.* 69(12): 2946–55.

Gese, E. M., O. J. Rongstad, and W. R. Mytton. 1988. Home range and habitat use of coyotes in southeastern Colorado. *J. Wildl. Mgmt.* 52:640–46.

Gese, E. M., R. L. Ruff, and R. L. Crabtree. 1996. Foraging ecology of coyotes (*Canis latrans*): The influence of extrinsic factors and a dominance hierarchy. *Can. J. Zool.* 74:769–83.

Gibson, N. 1996. *Wolves.* Voyageur Press, Stillwater, MN. 72 pp.

Gidley, J. W., and C. L. Gazin. 1938. The Pleistocene vertebrate fauna from Cumberland Cave, Maryland. Bulletin (U.S. National Museum), no. 171.

Gilbert, D. L. 1964. *Public relations in natural resources management.* Burgess Publ. Co., Minneapolis, MN. 227 pp.

Gilbert, F. K. 1995. Historical perspectives on wolf management in North America with special reference to humane treatments in capture methods. Pp. 13–17 in L. N. Carbyn, S. H. Fritts, and D. R. Seip, eds., *Ecology and conservation of wolves in a changing world.* Canadian Circumpolar Institute, Edmonton, Alberta.

Gilmour, I., M. A. Johnston, C. T. Pillinger, C. M. Pond, C. A. Mattacks, and P. Prestrud. 1995. The carbon isotopic composition of individual fatty acids as indicators of dietary history in arctic foxes on Svalbard. *Phil. Trans. R. Soc. Lond. B* 349:135–42.

Ginsberg, J. R., and D. W. Macdonald. 1990. *Foxes, wolves, jackals, and dogs: An action plan for the conservation of canids.* IUCN World Conservation Union, Gland, Switzerland. 117 pp.

Ginsburg, B. E. 1987. The wolf pack as a socio-genetic unit. Pp. 401–13 in H. Frank, ed., *Man and wolf: Advances, issues and problems in captive wolf research.* Dr. W. Junk Publishers, Dordrecht, The Netherlands.

Ginsburg, B. E., and L. Hiestand. 1992. Humanity's "best friend": The origins of our inevitable bond with dogs. Pp. 93–108 in H. Davis and D. Balfour, eds., *The inevitable bond: Examining scientist-animal interactions.* Cambridge University Press, Cambridge.

Gipson, P. S., W. B. Ballard, and R. M. Nowak. 1998. Famous North American wolves and the credibility of early wildlife literature. *Wildl. Soc. Bull.* 26:808–16.

Gipson, P. S., W. B. Ballard, R. M. Nowak, and L. D. Mech. 2000. Accuracy and precision of estimating age of gray wolves by tooth wear. *J. Wildl. Mgmt.* 64:752–58.

Gipson, P. S., E. E. Bangs, T. N. Bailey, D. K. Boyd, H. D. Cluff, D. W. Smith, and M. D. Jiminez. 2002. Color patterns among wolves in western North America. *Wild. Soc. Bull.* 30:821–30.

Gipson, P. S., J. A. Sealander, and J. E. Dunn. 1974. The taxonomic status of wild *Canis* in Arkansas. *Syst. Zool.* 23:1–11.

Girman, D. J., P. W. Kat, M. G. L. Mills, J. R. Ginsberg, M. Borner, V. Wilson, J. H. Fanshawe, C. Fitzgibbon, L. M. Lau and R. K. Wayne. 1993. Molecular genetic and morphological analyses of the African wild dog (*Lycaon pictus*). *J. Hered.* 84:450–59.

Glowacinski, Z., and P. Profus. 1997. Potential impact of wolves (*Canis lupus*) on prey populations in eastern Poland. *Biol. Conserv.* 80:99–106.

Goddard, J. 1996. A real whopper. *Saturday Night* III(4), 46–50, 52, 54, 64.

Gogan, P. J. P., E. M. Olexa, N. Thomas, D. Kuehn, and K. M. Podruzny. 2000. Ecological status of gray wolves in and adjacent to Voyageurs National Park, Minnesota. Draft technical report, U.S. Geological Survey, Northern Rocky Mountain Science Center, Bozeman, MT.

Golani, I., and G. Moran. 1983. A motility-immobility gradient in the behavior of the "inferior wolf" during "ritualized fighting." Pp. 65–94 in J. R. Eisenberg and D. G. Kleiman, eds., *Advances in the study of mammalian behavior.* Special Publication no. 7. American Society of Mammalogists, Stillwater, OK.

Goldman, E. A. 1937. The wolves of North America. *J. Mammal.* 18:37–45.

———. 1944. Classification of wolves. Pp. 389–636 in S. P. Young, and E. A. Goldman, eds., *The wolves of North America.* American Wildlife Institution, Washington, D.C.

Goldman, J. A. 1993. An acoustic basis for maternal recognition in wolf pups (*Canis lupus*). Master's thesis, Dalhousie University, Halifax, NS. 101 pp.

Goldman, J. A., D. P. Phillips, and J. C. Fentress. 1995. An acoustic basis for maternal recognition in timber wolves (*Canis lupus*)? *J. Acoust. Soc. Am.* 97:1970–73.

Goldstein, D. B., and D. D. Pollock. 1997. Launching microsatellites: A review of mutation processes and methods of phylogenetic inference. *J. Hered.* 88:335–42.

Goodman, P. 1978. Scent rolling in wolves *(Canis lupus)*. Master's thesis, Purdue University, Lafayette, IN.

Goodmann, P. A., and E. Klinghammer. 1990. Wolf ethogram. Ethology Series, no. 3. North American Wildlife Park, Battle Ground, IN.

Goodrowe, K. L., M. A. Hay, C. C. Platz, S. K. Behrns, M. H. Jones, and W. T. Waddell. 1998. Characteristics of fresh and frozen-thawed red wolf spermatozoa. *Anim. Reprod. Sci.* 53:299–308.

Goodwin, M., K. M. Gooding, and F. Regnier. 1979. Sex pheromone in the dog. *Science* 203:559–61.

Gorman, M. L., M. G. Mills, J. P. Raath, and J. R. Speakman. 1998. High hunting costs make African wild dogs vulnerable to kleptoparasitism by hyenas. *Nature* 391:479–81.

Gosling, L. M. 1982. A reassessment of the function of scent marking in territories. *Z. Tierpsychol.* 60:89–118.

Gotelli, D., and C. Sillero-Zubiri. 1992. The Ethiopian wolf—an endangered endemic canid. *Oryx* 26:205–14.

Gotelli, D., C. Sillero-Zubiri, G. D. Applebaum, M. S. Roy, D. J. Girman, J. García-Moreno, E. A. Ostrander, and R. K. Wayne. 1994. Molecular genetics of the most endangered canid: The Ethiopian wolf *Canis simensis. Mol. Ecol.* 3: 301–12.

Gould, S. E. 1945. *Trichinosis.* Charles C. Thomas, Springfield, IL.

Goulet, G. D. 1993. Comparison of temporal and geographical skull variation among Nearctic, modern, Holocene, and late Pleistocene gray wolves *(Canis lupus)* and selected *Canis.* Master's thesis, University of Manitoba, Winnipeg. 116 pp.

Goyal, S. M., L. D. Mech, R. A. Rademacher, M. A. Khan, and U. S. Seal. 1986. Antibodies against canine parvovirus in wolves of Minnesota: A serologic study from 1975 through 1985. *Am. Vet. Med. Assoc. J.* 189:1092–94.

Grace, E. S. 1976. Interactions between men and wolves at an Arctic outpost on Ellesmere Island. *Can. Field Nat.* 90:149–56.

Grachev, Y. A., and A. K. Fedosenko. 1972. Interrelationships of bears and wolves in Dxhungarskij Alatay: Ecology, morphology, protection and use of bears. *Nauka:* 36–39.

Graham, R. W., and E. L. Lundelius, Jr. 1994. *FAUNMAP: A database documenting late Quaternary distributions of mammal species in the United States.* Illinois State Museum, Springfield, IL.

Grande del Brio, R. 1984. *El lobo iberico, biologia y mitologia.* H. Blume, Madrid. 175 pp.

Grant, P. R. 1968. Polyhedral territories of animals. *Am. Nat.* 102: 75–80.

Gray, A. P. 1954. Mammalian hybrids—A check-list with bibliography. Technical Communications, no. 10. Commonwealth Animal Breeding and Genetics, Edinburgh. 144 pp.

Gray, D. R. 1970. The killing of a bull muskox by a single wolf. *Arctic* 23:197–99.

———. 1983. Interactions between wolves and muskoxen on Bathurst Island, Northwest Territories, Canada. *Acta Zool. Fenn.* 174:255–57.

———. 1987. *The muskoxen of Polar Bear Pass.* Fitzhenry and Whiteside, Markham, Ontario. 191 pp.

———. 1993. The use of muskox kill sites as temporary rendezvous sites by arctic wolves with pups in early winter. *Arctic* 46:324–30.

Greaves, W. S. 1974. Functional implications of mammalian jaw joint position. *Forma et Functio* 74:363–76. (Cited in Biknevicius and Van Valkenburgh 1996.)

Green, D. M. 1973. Minimum integration time. In A. R. Møller, ed., *Basic mechanisms in hearing.* Academic Press, New York.

Green, J. S., M. L. Bruss, J. F. Evermann, and P. K. Bergstrom. 1984. Serologic response of captive coyotes *(Canis latrans* Say) to canine parvovirus and accompanying profiles of canine coronavirus titers. *J. Wildl. Dis.* 20:6–11.

Greene, C. E. 1984a. Canine viral enteritis. Pp. 437–60 in C. E. Greene, ed., *Clinical microbiology and infectious diseases of the dog and cat.* W. B. Saunders, Philadelphia.

———. 1984b. Infectious canine hepatitis. Pp. 406–18 in C. E. Greene, ed., *Clinical microbiology and infectious diseases of the dog and cat.* W. B. Saunders, Philadelphia.

———. 1984c. Leptospirosis. Pp. 588–98 in C. E. Greene, ed., *Clinical microbiology and infectious diseases of the dog and cat.* W. B. Saunders, Philadelphia.

Greene, C. E., and L. W. George. 1984. Canine brucellosis. Pp. 646–62 in C. E. Greene, ed., *Clinical microbiology and infectious diseases of the dog and cat.* W. B. Saunders, Philadelphia.

Greenleaf, S. 1989. The evolution of the wolf in children's books. *Appraisal, Science Books for Young People* 22:1–6.

Gregson, J. D. 1956. The Ixodidea of Canada. Can. Dept. Agric. Publ. 930. Ottawa. 92 pp.

Grekova, N. A., and L. V. Gorban. 1978. Pathogenicity of brucellae isolated from wild and game animals of the extreme north of the USSR. [In Russian]. *Zh. Mikrobiologii, Epidemiologii I Immunobiologii* 197:46–48.

Griffith, G., J. M. Scott, J. W. Carpenter, and C. Reed. 1989. Translocation as a species conservation tool: Status and strategy. *Science* 245:477–80.

Grinnell, G. B., ed. 1904. *American big game in its haunts.* A book of the Boone and Crockett Club. Forest and Stream Publishing Co., New York. 497 pp.

Gromov, E. I, and E. N. Matyushkin. 1974. Towards an analysis of competitive relationships of tigers in Sokhote-Aline. *Biol. Sci.* 2:20–25.

Groves, C. P. 1993. *A theory of human and primate evolution.* Clarendon Press, Oxford.

Grundlach, H. 1968. Brutvorsorge, Brutpflege, Verhaltensontogenese and Tagesperiodik beim europaischen Wildschwein *(Sus scrofa* L.) *Z. Tierpsychol.* 25:955–95.

Guberti, V., F. Francisci, U. Andreotta, and A. A. Andreoni. 1991. *Echinococcus granulosus* in wolves *(Canis lupus)* in Italy. 1st Naz. Biol. Selvaggina no. 18. 3 p.

Guitian, J. R., A. L. de Castro, S. L. Bas, and J. L. Sanchez. 1979. Nota sobre la dieta del lobo *(Canis lupus* L.) an Galicia. *Trabajos Compostelanos de Biologia* 8:95–104.

Gunson, J. R. 1983. Wolf predation of livestock in western Canada. Pp. 102–5 in *Wolves in Canada and Alaska: Their status, biology, and management.* Report Series, no. 45. Canadian Wildlife Service, Edmonton, Alberta.

———. 1992. Historical and present management of wolves in Alberta. *Wildl. Soc. Bull.* 20:330–39.

———. 1995. Wolves in Alberta: Their characteristics, history, prey relationships and management. Alberta Environmental Protection, Edmonton. 24 pp.

Gunson, J. R., and K. H. Dies. 1980. Sylvatic trichinosis in Alberta. *J. Wildl. Dis.* 16:525–28.

Gustavson, C. R. 1982. An evaluation of taste aversion control of wolf *(Canis lupus)* predation in northern Minnesota. *Appl. Anim. Ethol.* 9:63–71.

Gustavson, C. R., and L. K. Nicolaus. 1987. Taste aversion conditioning in wolves, coyotes and other canids: Retrospect and prospect. Pp. 169–200 in H. Frank, ed., *Man and wolf: Advances, issues and problems in captive wolf research.* Dr. W. Junk Publishers, Dordrecht, The Netherlands.

Guthrie, R. D. 1971. A new theory of mammalian rump patch evolution. *Behaviour* 38:132–45.

Gyorgy, J. 1984. Istra'ivanje javnog miljenja o vukovima u Hrvatskoj. Drugi kongres biologa hrvatske, Zadar, pp. 116–17.

Haase, E. 2000. Comparison of reproductive biological parameters in male wolves and domestic dogs. *Z. Saugetierk.* 65:257–70.

Haber, G. C. 1968. The social structure and behavior of an Alaskan wolf population. Master's thesis, Northern Michigan University, Marquette.

———. 1977. Socio-ecological dynamics of wolves and prey in a subarctic ecosystem. Ph.D. dissertation, University of British Columbia, Vancouver. 824 pp.

———. 1996. Biological, conservation, and ethical implications of exploiting and controlling wolves. *Conserv. Biol.* 10:1068–81.

Hadley, E. A. 1989. Holocene mammalian fauna of Lamar Cave, Yellowstone National Park, and its implications for ecosystem dynamics. Pp. 10–12 in F. J. Singer, ed., *Grazing influences on Yellowstone's northern range.* U.S. National Park Service, Yellowstone National Park, Wyoming.

Haggstrom, D. A., A. K. Ruggles, C. M. Harms, and R. O. Stephenson. 1995. Citizen participation in developing a wolf management plan for Alaska: An attempt to resolve conflicting human values and perceptions. Pp. 481–87 in L. N. Carbyn, S. H. Fritts, and D. R. Seip, eds., *Ecology and conservation of wolves in a changing world.* Canadian Circumpolar Institute, Edmonton, Alberta.

Haight, R. G, and L. D. Mech. 1997. Computer simulation of vasectomy for wolf control. *J. Wildl. Mgmt.* 61:1023–31.

Haight, R. G., D. J. Mladenoff, and A. P. Wydeven. 1998. Modeling disjunct gray wolf populations in semi-wild landscapes. *Conserv. Biol.* 12(4): 879–88.

Haight, R. G., L. E. Travis, K. Nimerfro, and L. D. Mech. 2002. Computer simulation of wolf-removal strategies for animal damage control. *Wildl. Soc. Bull.* 30:844–52.

Hainard, R. 1961. *Mammiferes sauvage d'Europe.* I. Delachaux et Niestle, Neuchatel (Switzerland). 350 pp.

Hairston, N. G., Jr., and N. G. Hairston. 1993. Cause-effect relationships in energy flow, trophic structure, and interspecific interactions. *Am. Nat.* 142:379–411.

Hairston, N. G., F. E. Smith, and L. B. Slobodkin. 1960. Community structure, population control and competition. *Am. Nat.* 44:421–25.

Hall, E. R. 1981. *The mammals of North America.* 2 vols. John Wiley and Sons, New York.

Hall, E. R., and K. R. Kelson. 1959. *The mammals of North America.* Vol 2. Ronald Press, New York.

Hall, E. S. 1981. Interior North Alaska Eskimo. Pp. 338–46 in W. C. Sturtevant, ed., *Handbook of North American Indians,* Vol. 6, *Subarctic.* Smithsonian Institution, Washington, D.C.

Hall, R. L., and H. S. Sharp, eds. 1978. *Wolf and man: Evolution in parallel.* Academic Press, New York.

Halpin, Z. T. 1980. Individual odors and individual recognition: Review and commentary. *Biol. Behav.* 5:233–48.

Hamerton, A. E. 1931. Report on the deaths occurring in the Society's gardens during 1930. *Proc. Zool. Soc. Lond.* 1931:527.

———. 1932. Report on the deaths occurring in the Society's gardens during 1931. *Proc. Zool. Soc. Lond.* 1932:613.

———. 1936. Report on the deaths occurring in the Society's gardens during 1935. *Proc. Zool. Soc. Lond.* 1936:659.

———. 1945. Report on the deaths occurring in the Society's gardens during 1944. *Proc. Zool. Soc. Lond.* 115:371.

Hamilton, W. D. 1971. Geometry for a selfish herd. *J. Theor. Biol.* 31:295–311.

Hammond, K. A., and J. Diamond. 1997. Maximal sustained energy budgets in humans and animals. *Nature* 386:457–62.

Hampton, B. 1997. *The great American wolf.* Henry Holt, New York.

Hancock, J. M. 1999. Microsatellites and other simple sequences: Genomic context and mutational mechanisms. Pp. 1–9 in D. B. Golstein and C. Schlötterer, *Microsatellites: Evolution and applications.* Oxford University Press, Oxford.

Hanotte, O., C. L. Tawah, D. G. Bradley, M. Okomo, Y. Verjee, J. Ochieng, and J. E. O. Rege. 2000. Geographic distribution and frequency of a taurine *Bos taurus* and an indicine *Bos indicus* Y specific allele amongst sub-Saharan African cattle breeds. *Mol. Ecol.* 9:387–96.

Harbo, S. J., and F. C. Dean. 1983. Historical and current perspectives on wolf management in Alaska. Pp. 51–64 in L. N. Carbyn, ed., *Wolves in Canada and Alaska: Their status, biology, and management.* Report Series, no. 45. Canadian Wildlife Service, Edmonton, Alberta.

Hare, B., M. Brown, C. Williamson, and M. Tomasello. 2002. The domestication of social cognition in dogs. *Science* 298:1634–36.

Hare, W. C. D. 1975. Carnivore respiratory system. Pp. 1559–75 in R. Getty, ed., *Sisson and Grossman's The anatomy of the domestic animals,* 5th ed. W. B. Saunders, Philadelphia.

Harestad, A. S., and F. L. Bunnell. 1979. Home range and body weight—a reevaluation. *Ecology* 60:389–402.

Harmer, S. F., and A. E. Shipley, eds. 1902. *Cambridge natural history: Mammalia.* (F. E. Beddard) MacMillan, London.

Harrington, F. H. 1975. Response parameters of elicited wolf howling. Ph.D. dissertation, State University of New York, Stony Brook. 385 pp.

———. 1981. Urine-marking and caching behavior in the wolf. *Behaviour* 76:280–88.

———. 1982a. Pseudo-urination by coyotes. *J. Mammal.* 63:501–3.

———. 1982b. Urine marking at food and caches in captive coyotes. *Can. J. Zool.* 60:776–82.

———. 1986. Timber wolf howling playback studies: Discrimination of pup from adult howls. *Anim. Behav.* 34:1575–77.

———. 1987. Aggressive howling in wolves. *Anim. Behav.* 35:7–12.

———. 1989. Chorus howling by wolves: Acoustic structure, pack size and the Beau Geste effect. *Bioacoustics* 2:117–36.

Harrington, F. H., and L. D. Mech. 1978a. Howling at two Minnesota wolf pack summer homesites. *Can. J. Zool.* 56:2024–28.

———. 1978b. Wolf vocalization. Pp. 109–32 in R. L. Hall and H. S. Sharp, eds., *Wolf and man: Evolution in parallel.* Academic Press, New York.

———. 1979. Wolf howling and its role in territory maintenance. *Behaviour* 68:207–49.

———. 1982a. An analysis of howling response parameters useful for wolf pack censusing. *J. Wildl. Mgmt.* 46:686–93.

———. 1982b. Fall and winter homesite use by wolves in northeastern Minnesota. *Can. Field Nat.* 96:79–84.

———. 1982c. Patterns of homesite attendance in two Minnesota wolf packs. Pp. 81–105 in F. H. Harrington and P. C. Paquet, eds., *Wolves of the world: Perspectives of behavior, ecology, and conservation.* Noyes Publications, Park Ridge, NJ.

———. 1983. Wolf pack spacing: Howling as a territory-independent spacing mechanism in a territorial population. *Behav. Ecol. Sociobiol.* 12:161–68.

Harrington, F. H., L. D. Mech, and S. H. Fritts. 1983. Pack size and wolf pup survival: Their relationship under varying ecological conditions. *Behav. Ecol. Socbiol.* 13:19–26.

Harrington, F. H., P. C. Paquet, J. Ryon, and J. C. Fentress. 1982. Monogamy in wolves: A review of the evidence. Pp. 209–22 in F. H. Harrington and P. C. Paquet, eds., *Wolves of the world: Perspectives of behavior, ecology, and conservation.* Noyes Publications, Park Ridge, NJ.

Harris, S., and P. C. L. White. 1992. Is reduced affiliative rather than increased agonistic behaviour associated with dispersal in red foxes? *Anim. Behav.* 44:1085–89.

Harrison, D. J., and T. G. Chapin. 1998. Extent and connectivity of habitat for wolves in eastern North America. *Wildl. Soc. Bull.* 26(4):767–75.

Harrison, R. G. 1990. Hybrid zones: Windows on evolutionary process. *Oxford Surv. Evol. Biol.* 7:69–128.

Hart, B. L. 1968. Alteration of quantitative aspects of sexual reflexes in spinal male dogs by testosterone. *J. Comp. Phys. Psychol.* 66:726–30.

———. 1970. Mating behavior in the female dog and the effects of estrogen on sexual reflexes. *Hormon. Behav.* 1:93–104.

———. 1974a. Environmental and hormonal influences on urine marking behavior in the adult male dog. *Behav. Biol.* 11:167–76.

———. 1974b. Gonadal androgen and sociosexual behavior of male mammals: A comparative study. *Psychol. Bull.* 81:383–400.

Hart, B. L., and C. Haugen. 1971. Scent marking and sexual behavior maintained in anosmic male dogs. *Commun. Behav. Biol.* 6:131–35.

Hartl, D. L., and A. G. Clark. 1997. *Principles of population genetics,* 3d ed. Sinauer Associates, Sunderland, MA.

Harvey, J. W., C. F. Simpson, J. M. Gaskin, and J. H. Sameck. 1979. Ehrlichiosis in wolves, dogs, and wolf-dog crosses. *Am. Vet. Med. Assoc. J.* 175:901–5.

Harwell, G. M., J. A. Angell, R. E. Merideth, and C. Carley. 1985. Chronic superficial keratitis in a Mexican wolf. *Am. Vet. Med. Assoc. J.* 187:1268.

Hatler, D. F. 1981. Analysis of livestock predation and predator control effectiveness in northwest British Columbia. Report prepared for the Ministry of the Environment, Fish and Wildlife Branch, Victoria, BC. 98 pp.

Hatter, I. W. 1984. Effects of wolf predation on recruitment of black-tailed deer on northeastern Vancouver Island. Master's thesis, University of Idaho, Moscow. 156 pp.

Hattwick, M. A. W. 1982. Rabies. Pp. 2097–2100 in J. B. Wyngaarden and L. H. Smith, eds., *Textbook of medicine.* W. B. Saunders, Philadelphia.

Hatzirvassanis, V., 1991. Observations sur l'etat des populations du loup en Grece. *Biol. Gallo-Hellenica* 18(1): 13–20.

Haufler, J. B., and F. A. Servello. 1994. Techniques for wildlife nutritional analyses. Pp. 307–23 in T. A. Bookout, ed., *Research and management techniques for wildlife and habitats,* 5th ed. The Wildlife Society, Bethesda, MD.

Haugen, H. S. 1987. Den site behavior, summer diet, and skull injuries of wolves in Alaska. Master's thesis, University of Alaska, Fairbanks. 205 pp.

Haugen, H. S., and R. O. Stephenson. 1985. Anomalies on Alaskan wolf skulls. *Proc. Alaska Sci. Conf.* 36:110.

Hauser, M. 2000. *Wild minds.* Henry Holt, New York.

Havel, R. J., and D. S. Fredrickson. 1956. The metabolism of chylomicra I. The removal of palmitic acid $1\text{-}C^{14}$ labelled chylomicra from dog plasma. *J. Clin. Invest.* 35:1025.

Havkin, Z. 1977. Symmetry shifts in the development of interactive behavior of two wolf pups (*Canis lupus*). Master's thesis, Dalhousie University, Halifax, NS.

———. 1981. Form and strategy of combative interactions between wolf pups (*Canis lupus*). Ph.D. dissertation, Dalhousie University, Halifax, NS.

Havkin, Z., and J. C. Fentress. 1985. The form of combative strategy in interactions among wolf pups (*Canis lupus*). *Z. Tierpsychol.* 68:177–200.

Hayes, B. 1993. Hunters of moose. *Int. Wolf* 3(2): 6–9.

Hayes, R. D. 1995. Numerical and functional responses of wolves and regulation of moose in the Yukon. Master's thesis, Simon Fraser University, Burnaby, BC. 144 pp.

Hayes, R. D., and A. M. Baer. 1992. Brown bear, *Ursus arctos,* preying upon gray wolf, *Canis lupus,* pups at a wolf den. *Can. Field Nat.* 106:381–82.

Hayes, R. D., A. M. Baer, and D. G. Larsen. 1991. Population dynamics and prey relationships of an exploited and recovering wolf population in the southern Yukon. Yukon Territory, Fish and Wildlife Branch, Department of Renewable Resources, Final Report TR-91-1. 67 pp.

Hayes, R. D., A. M. Baer, U. Wotschikowsky, and A. S. Harestad. 2000. Kill rate by wolves on moose in the Yukon. *Can. J. Zool.* 78:49–59.

Hayes, R. D., and J. R. Gunson. 1995. Status and management of wolves in Canada. Pp. 21–33 in L. N. Carbyn, S. H. Fritts, and D. R. Seip, eds., *Ecology and conservation of wolves in a changing world.* Canadian Circumpolar Institute, Edmonton, Alberta.

Hayes, R. D., and A. S. Harestad. 2000a. Demography of a recovering wolf population in the Yukon. *Can. J. Zool.* 78:36–48.

———. 2000b. Wolf functional response and regulation of moose in the Yukon. *Can. J. Zool.* 78:60–66.

Hayes, R. D., and D. H. Mossop. 1987. Interactions of wolves, *Canis lupus,* and brown bears, *Ursus arctos,* at a wolf den in the northern Yukon. *Can. Field Nat.* 101:603–4.

Hayssen, V., and R. C. Lacy. 1985. Basal metabolic rate in mammals: Taxonomic differences in the allometry of BMR and body mass. *Comp. Biochem. Physiol. A. Comp. Physiol.* 81:741–54.

Hayssen, V., A. van Tienhoven, and A. van Tienhoven. 1993. *Asdell's patterns of mammalian reproduction.* Comstock Publishing Associates, Ithaca, NY.

Heard, D. C. 1992. The effect of wolf predation and snow cover on musk-ox group size. *Am. Nat.* 139:190–204.

Heard, D. C., and T. M. Williams. 1992. Distribution of wolf dens on migratory caribou ranges in the Northwest Territories, Canada. *Can. J. Zool.* 70:1504–10.

Hebblewhite, M., and D. H. Pletscher. 2002. Effects of elk group size on predation by wolves. *Can. J. Zool.* 80:800–809.

Hebblewhite, M., D. H. Pletscher, and P. C. Paquet. 2002. Elk population dynamics in areas with and without predation by recolonizing wolves in Banff National Park, Alberta. *Can. J. Zool.* 80:789–99.

Hebert, D. M., J. Youds, R. Davies, H. Langin, D. Janz, and G. W. Smith. 1982. Preliminary investigations of the Vancouver Island wolf (*Canis lupus crassodon*) prey relationships. Pp. 54–70 in F. H. Harrington and P. C. Paquet, eds., *Wolves of the world: Perspectives of behavior, ecology, and conservation.* Noyes Publications, Park Ridge, NJ.

Hedrick, P. W. 1995. Gene flow and genetic restoration: The Florida panther as a case study. *Conserv. Biol.* 9:996–1007.

Hedrick, P. W., R. Fredrickson, and H. Ellegren. 2001. Evaluation of d^2, a microsatellite measure of inbreeding and outbreeding, in wolves with a known pedigree. *Evolution* 55:1256–60.

Hedrick, P. W., and S. T. Kalinowski. 2000. Inbreeding depression in conservation biology. *Annu. Rev. Ecol. Syst.* 31:139–62.

Hedrick, P. W., and T. J. Kim. 1999. Genetics of complex polymorphisms: Parasites and maintenance of the major histocompatibility complex variation. Pp. 204–34 in R. S. Singh and C. B. Krimbas, eds., *Evolutionary genetics: From molecules to morphology.* Cambridge University Press, Cambridge.

Hedrick, P. W., R. N. Lee, and D. Garrigan. 2002. Major histocompatability complex variation in red wolves: Evidence for common ancestry with coyotes and balancing selection. *Mol. Ecol.* 11: 1905–13.

Hedrick, P. W., R. N. Lee, and K. M. Parker. 2000. Major histo-compatibility (MHC) variation in the endangered Mexican wolf and related canids. *Heredity* 85:617–24.

Hedrick, P. W., P. S. Miller, E. Geffen, and R. K. Wayne. 1997. Genetic evaluation of the three captive Mexican wolf lineages. *Zoo Biol.* 16:47–69.

Heffner, H. E. 1983. Hearing in large and small dogs: Absolute thresholds and size of the tympanic membrane. *Behav. Neurosci.* 97:310–18.

Heffner, R. S., and H. E. Heffner. 1985. Hearing range of the domestic cat. *Hearing Res.* 19:85–88.

———. 1992a. Evolution of sound localization in mammals. Pp. 691–715 in D. B. Webster, R. R. Fay, and A. N. Popper, eds., *The evolutionary biology of hearing.* Springer-Verlag, New York.

———. 1992b. Hearing in large mammals: Sound-localization acuity in cattle (*Bos taurus*) and goats (*Capra hircus*). *J. Comp. Psychol.* 106:107–13.

———. 1992c. Visual factors in sound localization in mammals. *J. Comp. Neurol.* 317:219–32.

Hefner, R., and E. Geffen. 1999. Group size and home range of the Arabian wolf (*Canis lupus*) in southern Israel. *J. Mammal.* 80(2): 611–19.

Heinrich, B. 1989. *Ravens in winter.* Summit Books, New York.

Hell, P. 1990. Dental anomalies among west Carpathian wolves and their significance in dog breeding. *Z. Jagdwiss.* 36:266–69.

———. 1993. Current situation and perspectives of the wolf in Czechoslovakia. Pp. 36–42 in C. Promberger and W. Schröder, eds., *Wolves in Europe: Status and perspectives.* Munich Wildlife Society, Ettal, Germany.

Hendrickson, J., W. L. Robinson, and L. D. Mech. 1975. Status of the wolf in Michigan, 1973. *Am. Midl. Nat.* 94:226–32.

Henry, J. D. 1977. The use of urine marking in the scavenging behavior of the red fox (*Vulpes vulpes*). *Behaviour* 61:82–106.

Henry, V. G. 1995. Revision of special rule for nonessential experimental populations of red wolves in North Carolina and Tennessee: B: Final rule. *Fed. Reg.* 60:18940–48.

Henshaw, R. E., R. Lockwood, R. Shideler, and R. O. Stephenson. 1979. Experimental release of captive wolves. Pp. 319–45 in E. Klinghammer, ed., *The behavior and ecology of wolves.* Garland STPM Press, New York.

Henshaw, R. E., and R. O. Stephenson. 1974. Homing in the gray wolf (*Canis lupus*). *J. Mammal.* 55:234–37.

Henshaw, R. E., L. S. Underwood, and T. M. Casey. 1972. Peripheral thermoregulation in two arctic canines. *Science* 175:988–90.

Herscovici, A. 1985. *Second nature: The animal-rights controversy.* CBC Enterprises, Radio-Canada, Montreal. 254 pp.

Hertel, H. H. 1984. The role of social niche in juvenile wolves: Puppyhood to independence. Master's thesis, University of Minnesota, Duluth.

Hewitt, G. 2000. The genetic legacy of the Quaternary ice ages. *Nature* 405:907–13.

Hickerson, H. 1965. The Virginia deer and intertribal buffer zones in the upper Mississippi Valley. Pp. 43–66 in A. Leeds and A. P. Vayda, eds., *Man's culture and animals.* Publication 8. American Association for the Advancement of Science, Washington, D.C.

————. 1970. *The Chippewa and their neighbors: A study in ethno-history.* Holt, Rinehart and Winston, New York.

Hiestand, N. L. 1989. A comparison of problem-solving and spatial orientation in the wolf *(Canis lupus)* and dog *(Canis familiaris).* Ph.D. dissertation, University of Connecticut, Storrs.

Hildebrand, M. 1952. The integument of the Canidae. *J. Mammal.* 33:419–28.

Hill, E. P., P. W. Sumner, and J. B. Wooding. 1987. Human influences on range expansion of coyotes in the southeast. *Wildl. Soc. Bull.* 15:521–24.

Hillis, T. L. 1990. Demography and ecology of the tundra wolf in the Keewatin District. Master's thesis, Laurentian University, Sudbury, Ontario.

Hillis, T. L., and F. F. Mallory. 1996a. Fetal development in wolves, *Canis lupus,* of the Keewatin District, Northwest Territories, Canada. *Can. J. Zool.* 74:2211–18.

Hillis, T. L., and F. F. Mallory. 1996b. Sexual dimorphism in wolves *(Canis lupus)* of the Keewatin District, Northwest Territories, Canada. *Can. J. Zool.* 74:721–25.

Hilton, H. H. 1978. Systematics and ecology of the eastern coyote. Pp. 209–28 in M. Bekoff, ed., *Coyotes: Biology, behavior, and management.* Academic Press, New York.

Hinchcliff, K. W., G. A. Reinhart, J. R. Burr, C. J. Schreier, and R. A. Swenson. 1997. Metabolizable energy intake and sustained energy expenditure of Alaskan sled dogs during heavy exertion in the cold. *Am. J. Vet. Res.* 58:1457–62.

Hirth, D. H., and D. R. McCullough. 1977. Evolution of alarm signals in ungulates with special reference to white-tailed deer. *Am. Nat.* 111:31–42.

Hoefs, M., and M. Bayer. 1983. Demographic characteristics of an unhunted dall sheep *(Ovis dalli)* population in southwest Yukon, Canada. *Can. J. Zool.* 61:1346–57.

Hoefs, M., and I. M. Cowan. 1979. Ecological investigation of a population of Dall sheep *(Ovis dalli dalli* Nelson). *Syesis* 12(suppl. 1). 81 pp.

Hoffmeister, D. F. 1986. *Mammals of Arizona.* University of Arizona Press, Tucson.

Hoffmeister, D. F., and W. W. Goodpaster. 1954. The mammals of the Huachuca Mountains, southeastern Arizona. Illinois Biological Monographs, vol. 24, no. 1. University of Illinois Press, Urbana.

Hoffos, R. 1987. Wolf management in British Columbia: The public controversy. British Columbia Ministry of Environment and Parks, Wildl. Branch, Bull. no. B-52.

Holleman, D. F., and R. O. Stephenson. 1981. Prey selection and consumption by Alaskan wolves in winter. *J. Wildl. Mgmt.* 45:620–28.

Holling, C. S. 1959. The components of predation as revealed by a study of small-mammal predation on the European pine sawfly. *Can. Entomol.* 91:293–320.

Holmes, J. C., and R. Podesta. 1968. The helminths of wolves and coyotes from the forested regions of Alberta. *Can. J. Zool.* 46:1193–1204.

Holroyd, J. D. 1967. Observations of Rocky Mountain goats on Mount Wardle, Kootenay National Park, British Columbia. *Can. Field Nat.* 81:1–22.

Holst, P. A., and R. D. Phemister. 1975. Temporal sequence of events in the estrous cycle of the bitch. *Am. J. Vet. Res.* 36:705–6.

Holt, R. 1977. Predation, apparent competition, and the structure of prey communities. *Theor. Pop. Ecol.* 12:197–229.

Holt, T. D. 1998. A structural description and reclassification of the wolf, *Canis lupus,* chorus howl. Master's thesis, Dalhousie University, Halifax, NS.

Hone, E. 1934. The present status of the muskox in Arctic North America and Greenland. Special Publication no. 5. American Committee for International Wild Life Protection, Cambridge, MA.

Honeycutt, R. 1992. Naked mole-rats. *Am. Sci.* 80:43–53.

Hoogland, J. L. 1979. The effect of colony size on individual alertness of prairie dogs (Sciuridae: *Cynomys* spp.). *Anim. Behav.* 27:394–407.

Hope, J. 1994. Wolves and wolf hybrids as pets are big business—but a bad idea. *Smithsonian,* 25, 34–44.

Hopkins, G. H. E. 1949. The host associations of lice of mammals. *Proc. Zool. Soc. Lond.* 119:387–604.

Horejsi, B., G. E. Hornbeck, and R. M. Raine. 1984. Wolves, *Canis lupus,* kill female Black Bear, *Ursus americanus,* in Alberta. *Can. Field Nat.* 98:368–69.

Hornbeck, G. E., and B. L. Horejsi. 1986. Grizzly bear, *Ursus arctos,* usurps wolf, *Canis lupus,* kill. *Can. Field Nat.* 100:259–60.

Hornocker, M. G., and T. K. Ruth. 1997. Cougar-wolf interaction in the North Fork of the Flathead River, Montana. Report, Hornocker Wildlife Institute, Moscow, ID.

Hoskinson, R. L., and L. D. Mech. 1976. White-tailed deer migration and its role in wolf predation. *J. Wildl. Mgmt.* 40:429–41.

Houston, D. B. 1978. Elk as winter-spring food for carnivores in northern Yellowstone National Park. *J. Appl. Ecol.* 15:653–61.

Hovens, J. P. M., K. H. Tungalaktuja, T. Todgeril, and D. Batdor. 2000. The impact of wolves *(Canis lupus* L., 1758) on wild ungulates and nomadic livestock in and around the Hustain Nuruu Steppe Reserve (Mongolia). *Lutra* 43:39–50.

Howard, W. E. 1960. Innate and environmental dispersal of individual vertebrates. *Am. Midl. Nat.* 63:152–61.

Howe, D. L. 1971. Babesiosis. Pp. 335–42 in J. W. Davis and R. C. Anderson, eds., *Parasitic diseases of wild mammals.* Iowa State University Press, Ames.

Hradecky, P. 1985. Possible pheromonal regulation of reproduction in wild carnivores. *J. Chem. Ecol.* 11:241–50.

Huber, D. 1999. Wolf legal protection useless without management plan in Croatia. *Int. Wolf* 9(2): 14–15.

Huber, D., R. Berislav, D. Novosel, and A. Frkovic. 1993. Public attitude towards wolves in Croatia: Positive change with the wolf population drop. Paper submitted for the proceedings of the wolf meeting in Leon (Spain).

Huber, D., S. Mitevski, and D. Kuhar. 1993. Questionnaire on wolves in Croatia and Macedonia: Comparison of public attitudes. Pp. 124–25 in C. Promberger and W. Schröder, eds., *Wolves in Europe: Status and perspectives.* Munich Wildlife Society, Ettal, Germany.

Huggard, D. J. 1993a. Effect of snow depth on predation and scavenging by gray wolves. *J. Wildl. Mgmt.* 57:382–88.

———. 1993b. Prey selectivity of wolves in Banff National Park. I. Prey species. *Can. J. Zool.* 71:130–39.

Huggins, C. 1946. The prostatic secretion. *Harvey Lect.* 42:148–93.

Hughes, A. 1977. The topography of vision in mammals of contrasting life style: Comparative optics and retinal organization. Pp. 613–756 in F. Crescitelli, ed., *Handbook of sensory physiology*, Vol. VII/5. Springer-Verlag, New York.

Hull, D. L. 1997. The ideal species concept—and why we can't get it. Pp. 357–80 in M. F. Claridge, H. A. Dawah, and M. R. Wilson, eds., *Species: The units of biodiversity.* Chapman & Hall, New York.

Hultkrantz, A. 1965. *Type of religion in Arctic hunting cultures: A religion-ecological approach.* Lulea, Stockholm.

Hummel, M. 1995. A personal view on wolf conservation and threatened carnivores in North America. Pp. 549–51 in L. N. Carbyn, S. H. Fritts, and D. R. Seip, eds., *Ecology and conservation of wolves in a changing world.* Canadian Circumpolar Institute, Edmonton, Alberta.

Hutt, N. 2002. Wolves of the world. *Int. Wolf* 12(1): 16–21.

Huxley, J. S. 1934. A natural experiment on the territorial instinct. *Brit. Birds* 27:270–77.

Iljin, N. A. 1941. Wolf-dog genetics. *J. Genet.* 42:359–414.

Ims, R. A. 1990. The ecology and evolution of reproductive synchrony. *Trends Ecol. Evol.* 5:135–40.

Ingstad, H. 1954. *Nunamiut.* George Allen and Unwin, London.

International Species Information System (ISIS). 1995. Hematology/Chemistry/Serology Records Report—Reference Values. Apple Valley, Minnesota.

Ionescu, O. 1993. Current status and prospects for the wolf in Romania. Pp. 50–55 in C. Promberger and W. Schröder, eds., *Wolves in Europe: Status and perspectives.* Munich Wildlife Society, Ettal, Germany.

IUCN Species Survival Commission. 2000. 2000 IUCN Red List of Threatened Species. Compiled by Craig Hilton-Taylor. IUCN, Gland, Switzerland and Cambridge, UK. Xviii + 61 pp and CD.

Ivanov, V. K. 1988. Observation on the behaviour and biology of wolves (*Canis lupus* L.) in the Ithiman Sredna Gora Mountains. *Bulgarian Academy of Sciences, Ecology* 21:25–33.

Ivlev, V. S. 1961. *Experimental ecology of the feeding of fishes.* Yale University Press, New Haven, CT.

Jacobs, G. H. 1993. The distribution and nature of colour vision among the mammals. *Biol. Rev.* 68:413–71.

Jacobs, G. H., J. F. Deegan II, M. A. Crognale, and J. A. Fenwick. 1993. Photopigments of dogs and foxes and their implications for canid vision. *Vis. Neurosci.* 10:173–80.

Jacobson, S. G., K. B. J. Franklin, and W. I. McDonald. 1976. Visual acuity of the cat. *Vision Res.* 16:1141–43.

Jaeger, M. M., R. K. Pandit, and E. Haque. 1996. Seasonal differences in territorial behavior by golden jackals in Bangladesh: Howling versus confrontation. *J. Mammal.* 77:768–75.

Jakob, W., M. Stolte, A. Valentin, and H.-D. Schroder. 1997. Demonstration of *Heliobacter pylori*-like organisms in the gastric mucosa of captive exotic carnivores. *J. Comp. Pathol.* 116:21–33.

Jalanka, H. H., and B. O. Roeken. 1990. The use of medetomidine, medetomidine-ketamine combinations, and atipamezole in nondomestic animals: A review. *J. Zoo Wildl. Med.* 21:259–82.

Jedrzejewska, B., W. Jedrzejewski, A. N. Bunevich, L. Milkowski, and H. Okarma. 1996. Population dynamics of wolves, *Canis lupus*, in Bialowieza Primeval Forest (Poland and Belarus) in relation to hunting by humans, 1847–1993. *Mammal Rev.* 26:103–26.

Jedrzejewska, B., H. Okarma, W. Jedrzejewski, and L. Milkowski. 1994. Effects of exploitation and protection on forest structure, ungulate density and wolf predation in Bialowieza Primeval Forest, Poland. *J. Appl. Ecol.* 31:664–79.

Jedrzejewski, W., B. Jedrzejewska, H. Okarma, and A. L. Ruprecht. 1992. Wolf predation and snow cover as mortality factors in the ungulate community of the Bialowieza National Park, Poland. *Oecologia* 90:27–36.

Jedrzejewski, W., B. Jedrzejewska, H. Okarma, K. Schmidt, K. Zub, and M. Musiani. 2000. Prey selection and predation by wolves in Bialowieza Primeval Forest, Poland. *J. Mammal.* 81:197–212.

Jedrzejewski, W., K. Schmidt, J. Theuerkauf, B. Jedrzejewska, and H. Okarma. 2001. Daily movements and territory use by radio-collared wolves, *Canis lupus*, in Bialowieza Primeval Forest in Poland. *Can. J. Zool.* 79:1993–2004.

Jedrzejewski, W., K. Schmidt, J. Theuerkauf, B. Jedrzejewska, N. Selva, K. Zub, and L. Szymura. 2002. Kill rates and predation by wolves on ungulate populations in Bialowieza Primeval Forest (Poland). *Ecology* 83:1341–56.

Jenks, S. M. 1988. Behavioral regulation of social organization and mating in a captive wolf pack. Ph.D. dissertation, University of Connecticut, Storrs.

Jenks, S. M., and B. E. Ginsburg. 1987. Socio-sexual dynamics in a captive wolf pack. Pp. 375–99 in H. Frank, ed., *Man and wolf: Advances, issues and problems in captive wolf research.* Dr. W. Junk Publishers, Dordrecht, The Netherlands.

Jenks, S. M., and R. K. Wayne. 1992. Problems and policy for species threatened by hybridization: The red wolf as a case study. Pp. 237–51 in D. R. McCullough and R. H. Barrett, eds., *Wildlife 2001: Populations.* Elsevier Applied Science, London.

Jenness, S. E. 1985. Arctic wolf attacks scientist: A unique Canadian incident. *Arctic* 38:129–32.

Jensen, K. 1993. Development of triadic and dyadic interactions in wolves (*Canis lupus*) and red foxes (*Vulpes vulpes*). Master's thesis, Dalhousie University, Halifax, NS.

Jensen, W. F., T. K. Fuller, and W. L. Robinson. 1986. Wolf, *Canis lupus*, distribution on the Ontario-Michigan border near Sault Ste. Marie, Ontario. *Can. Field Nat.* 100:363–66.

Jhala, Y. V. 1993. Predation on blackbuck by wolves in Velavadar National Park, Gujarat, India. *Conserv. Biol.* 7:874–81.

———. 2000. Human-wolf conflict in India. Beyond 2000: Realities of global wolf restoration. (Abstract.) International Wolf Center, Ely, MN: 26.

Jhala, Y. V., and R. H. Giles. 1991. The status and conservation of the wolf in Gujarat and Rajasthan, India. *Conserv. Biol.* 5:476–83.

Jhala, Y. V., and D. K. Sharma. 1997. Child-lifting by wolves in eastern Uttar Pradesh, India. *J. Wildl. Res.* 2:94–101.

Jimenez, M. D. 1992. Establishment of and prey selection by a new wolf pack in the Wigwam River drainage. Master's thesis, University of Montana, Missoula.

———. 1995. Tolerance and respect help Nine-Mile wolves recover. *Int. Wolf* 5(3): 18–19.

Joachim, A. 2000. An observation of the influence of posture of humans on the flight distance of captive wolves, *Canis lycaon* [sic]. Poster. American Society of Mammalogists, Durham, NH.

Jobling, M. A., and C. Tyler-Smith. 2000. New uses for new haplotypes: The human Y chromosome, disease and fuction. *Trends Genet.* 16:356–62.

Johnson, M. R. 1995. Rabies in wolves and its potential role in a Yellowstone wolf population. Pp. 431–39 in L. N. Carbyn, S. H. Fritts, and D. R. Seip, eds., *Ecology and conservation of wolves in a changing world.* Canadian Circumpolar Institute, Edmonton, Alberta.

Johnson, M. R., D. K. Boyd, and D. H. Pletscher. 1994. Serologic investigations of canine parvovirus and canine distemper in relation to wolf (*Canis lupus*) pup mortalities. *J. Wildl. Dis.* 30:270–73.

Johnson, R. P. 1973. Scent marking in mammals. *Anim. Behav.* 21: 521–35.

Johnson, T. B. 1990. Preliminary results of a public opinion survey of Arizona residents and interest groups about the Mexican wolf. Arizona Game and Fish Department, Phoenix.

Johnson, W. E., T. K. Fuller, and W. L. Franklin. 1996. Sympatry in canids: A review and assessment. Pp. 189–218 in J. Gittleman, ed., *Carnivore behavior, ecology, and evolution,* vol. 2. Cornell University Press, Ithaca, NY.

Johnston, C. S. 1938. Preliminary report on the vertebrate type locality of Cita Canyon, and the description of an ancestral coyote. *Am. J. Sci.,* ser. 5, 35:383–90.

Johnston, S. D. 1986. Pseudopregnancy in the bitch. Pp. 490–91 in D. A. Morrow, ed., *Current theory in theriogeriology.* W. B. Saunders, Philadelphia.

Jolicoeur, P. 1959. Multivariate geographic variation in the wolf *Canis lupus* L. *Evolution* 13:283–99.

Jones, C., R. S. Hoffmann, D. W. Rice, M. D. Engstrom, R. D. Bradley, D. J. Schmidly, C. A. Jones, and R. J. Baker. 1997. Revised checklist of North American mammals north of Mexico, 1997. Occasional Papers, Texas Tech University Museum, no. 173. Museum of Texas Tech University, Lubbock. 19 pp.

Jones, K. R. 2002. *Wolf Mountains: A History of Wolves Along the Great Divide.* University of Calgary Press, Calgary, Alberta. 336 pp.

Jones, R. B., and N. W. Nowell. 1973. The effects of preputial and coagulating gland secretions upon aggressive behaviour in male mice: A confirmation. *J. Endocrinol.* 59: 203–4.

Jordan, P. A. 1979. The red wolf, an endangered canid. Pp. 525–57 in E. Klinghammer, ed., *The behavior and ecology of wolves.* Garland STPM Press, New York.

Jordan, P. A., P. C. Shelton, and D. L. Allen. 1967. Numbers, turnover, and social structure of the Isle Royale wolf population. *Am. Zool.* 7:233–52.

Jorde, L. B., W. S. Watkins, M. L. Bamshad, M. E. Dixon, C. E. Ricker, M. T. Seielstad, and M. A. Batzer. 2000. The distribution of human genetic diversity: A comparison of mitochondrial, autosomal, and Y-chromosome data. *Am. J. Hum. Genet.* 66:979–88.

Joslin, P. W. B. 1966. Summer activities of two timber wolf (*Canis lupus*) packs in Algonquin Park. Master's thesis, University of Toronto, Ontario. 99 pp.

———. 1967. Movements and home sites of timber wolves in Algonquin Park. *Am. Zool.* 7:279–88.

———. 1982. Status, growth and other facets of the Iranian Wolf. Pp. 196–203 in F. H. Harrington and P. C. Paquet, eds., *Wolves of the world: Perspectives of behavior, ecology, and conservation.* Noyes Publications, Park Ridge, NJ.

Kalinowski, S. T., P. W. Hedrick, and P. S. Miller. 1999. No inbreeding depression observed in Mexican and Red wolf captive breeding programs. *Conserv. Biol.* 13(6): 1371–77.

———. 2000. Inbreeding depression in the Speke's gazelle captive breeding program. *Conserv. Biol.* 14:1375–84.

Kalmus, H. 1955. The discrimination by the nose of dog of individual human odours and in particular of the odours of twins. *Brit. J. Anim. Behav.* 3:25.

Kay, C. E. 1998. Are ecosystems structured from the top-down or bottom-up: A new look at an old debate. *Wildl. Soc. Bull.* 26:484–98.

Kazmierczak, J. J., E. C. Burgess, and T. E. Amundson. 1988. Susceptibility of the gray wolf (*Canis lupus*) to infection with the Lyme disease agent, *Borrelia burgdorferi. J. Wildl. Dis.* 24:522–27.

Keiter, R. B., and P. K. Holscher. 1990. Wolf recovery under the Endangered Species Act: A study in contemporary federalism. *Public Land Law Rev.* 11:19–52.

Keith, L. B. 1974. Some features of population dynamics in mammals. *Trans. Int. Congr. Game Biol.* 11:17–58.

———. 1983. Population dynamics of wolves. Pp. 66–77 in L. N. Carbyn, ed., *Wolves in Canada and Alaska: Their status, biology, and management.* Report Series, no. 45. Canadian Wildlife Service, Edmonton, Alberta.

Keller, L. F., and D. M. Waller. 2002. Inbreeding effects in wild populations. *Trends Ecol. Evol.* 17: 230–41.

Kellert, S. R. 1985. Public perceptions of predators, particularly the wolf and coyote. *Biol. Conserv.* 31:167–89.

———. 1986. The public and the timber wolf in Minnesota. *Trans. N. Am. Wildl. Nat. Resour. Conf.* 51:193–200.

———. 1987. The public and the timber wolf in Minnesota. *Anthrozoos* 1:100–109.

———. 1991. Public views of wolf restoration in Michigan. *Trans. N. Am. Wildl. Nat. Resour. Conf.* 56:152–61.

———. 1996. *The value of life: Biological diversity and human society.* Island Press, Washington, D.C.

———. 1999. The public and the wolf in Minnesota. Report to the International Wolf Center, Minneapolis, MN.

Kellert, S. R., M. Black, C. R. Rush, and A. J. Bath. 1996. Human culture and large carnivore conservation in North America. *Conserv. Biol.* 10:977–90.

Kellert, S. R., and E. O. Wilson. 1993. *The biophilia hypothesis.* Island Press, Washington, D.C.

Kelly, B. T. 1991. Carnivore scat analysis: An evaluation of existing techniques and the development of predictive models of prey consumed. Master's thesis, University of Idaho. 200 pp.

———. 2000. Red wolf recovery program adaptive work plan FY00–FY02. Unpublished USFWS document. Alligator River National Wildlife Refuge, Manteo, NC. 15 pp.

Kelly, B. T., P. S. Miller, and U. S. Seal, eds. 1999. Population and habitat viability assessment workshop for the red wolf *(Canis rufus).* Conservation Breeding Specialist Group (SSC/IUCN), Apple Valley, MN. 88 pp.

Kelly, B. T., and M. K. Phillips. 2000. Red wolf. Pp. 247–52 in R. P. Reading and B. Miller, eds., *Endangered animals: A reference guide to conflicting issues.* Greenwood Press, Westport, CT.

Kelsall, J. P. 1957. Continued barren-ground caribou studies. Canadian Wildlife Service, Wildlife Management Bulletin Series, no. 12. 148 pp.

———. 1968. The migratory barren-ground caribou of Canada. Canadian Wildlife Service, Monograph no. 3. Queens Printer, Ottawa. 340 pp.

———. 1969. Structural adaptations of moose and deer for snow. *J. Mammal.* 50:302–10.

Kennedy, P. K., M. L. Kennedy, P. L. Clarkson, and I. S. Liepins. 1991. Genetic variability in natural populations of the gray wolf. *Can. J. Zool.* 69:1183–88.

Kennelly, J. J. 1978. Coyote reproduction. Pp. 73–92 in M. Bekoff, ed., *Coyotes: Biology, behavior, and management.* Academic Press, New York.

Khan, M. A., S. M. Goyal, S. L. Diesch, L. D. Mech, and S. H. Fritts. 1991. Seroepidemiology of leptospirosis in Minnesota wolves. *J. Wildl. Dis.* 27:248–53.

Kirk, R. W., and S. I. Bistner. 1981. *Handbook of veterinary procedures and emergency treatment.* W. B. Saunders, Philadelphia.

Kitchen, D. W. 1974. Social behavior and ecology of the pronghorn. Wildlife Monographs, no. 38. The Wildlife Society, Washington, D.C. 96 pp.

Kleiber, M. 1961. *The fire of life.* John Wiley & Sons, New York.

Kleiman, D. 1966. Scent marking in the canidae. *Symp. Zool. Soc. Lond.* 18:167–77.

Kleiman, D. G. 1977. Monogamy in mammals. *Q. Rev. Biol.* 52: 39–69.

Kleiman, D. G., and J. F. Eisenberg. 1973. Comparisons of canid and felid social systems from an evolutionary perspective. *Anim. Behav.* 21:637–59.

Klein, D. R. 1995. The introduction, increase, and demise of wolves on Coronation Island, Alaska. Pp. 275–80 in L. N. Carbyn, S. H. Fritts, and D. R. Seip, eds., *Ecology and conservation of wolves in a changing world.* Canadian Circumpolar Institute, Edmonton, Alberta.

Klinghammer, E., and P. A. Goodmann. 1987. Socialization and management of wolves in captivity. Pp. 31–59 in H. Frank, ed., *Man and wolf: Advances, issues and problems in captive wolf research.* Dr. W. Junk Publishers, Dordrecht, The Netherlands.

Klinghammer, E., and L. Laidlaw. 1979. Analysis of 23 months of daily howl records in a captive grey wolf pack *(Canis lupus).* Pp. 153–81 in E. Klinghammer, ed., *The behavior and ecology of wolves.* Garland STPM Press, New York.

Knick, S. T., and L. D. Mech. 1980. Sleeping distance in wild wolf packs. *Behav. Neural Biol.* 28:507–11.

Knowlton, F. F. 1972. Preliminary interpretations of coyote population mechanics with some management implications. *J. Wildl. Mgmt.* 36:369–82.

Knudson, T. 2001. Environment Inc. Four-part series. *Sacramento (California) Bee,* July 1–4, 2001.

Koch, S., and L. Rubin. 1972. Distribution of cones in retina of the normal dog. *Am. J. Vet. Res.* 33:361–63.

Kochetkov, V. V. 1988. Biology of the wolf in Verkhnevolzhe, a case of the Central-Forest State Reserve. Ph.D. dissertation, Gosudarstvennyj Agropromyshlennyj Komitat SSSR.

Koehler, J. K., C. C. Platz, Jr., W. Waddell, M. H. Jones, and S. Behrns. 1998. Semen parameters and electron microscope observations of spermatozoa of the red wolf, *Canis rufus.* *J. Reprod. Fertil.* 114:95–101.

Koganezawa, M., N. Maruyama, M. Takahashi, S. Chinen, and C. Angeli. 1996. Japanese peoples' attitudes toward wolves and their reintroduction into Japan. Pp. 298–300 in N. Fascione and M. Cecil, eds., Proceedings of Defenders of Wildlife's Wolves of America Conference, Albany, NY. Defenders of Wildlife, Washington, D.C.

Kohn, M. H., and R. K. Wayne. 1997. Facts from feces revisited. *Trends Ecol. Evol.* 12:223–27.

Kohn, M. H., E. C. York, D. A. Kanradt, G. Haught, R. M. Sauvajot, and R. K. Wayne. 1999. Estimating population size by genotyping feces. *Proc. R. Soc. Lond. B* 266:657–63.

Kojola, I., and J. Kuittinen. 2002. Wolf attacks on dogs in Finland. *Wildl. Soc. Bull.* 30:498–501.

Kolenosky, G. B. 1971. Hybridization between wolf and coyote. *J. Mammal.* 52:446–49.

———. 1972. Wolf predation on wintering deer in east-central Ontario. *J. Wildl. Mgmt.* 36:357–69.

Kolenosky, G. B., and D. Johnston. 1967. Radio-tracking timber wolves in Ontario. *Am. Zool.* 7:289–303.

Kolenosky, G. B., and R. O. Standfield. 1975. Morphological and ecological variation among gray wolves *(Canis lupus)* of Ontario, Canada. Pp. 62–72 in M. W. Fox, ed., *The wild canids: Their systematics, behavioral ecology, and evolution.* Van Nostrand Reinhold, New York.

Kolpashchikov, L. 1995. Wolf influence on the Temyr wild reindeer population. Page 203 in J. Gurnall, ed., 2nd European Congress of Mammalogy, 27 March–1 April 1995, Southampton, U.K.

Konishi, M. 1973. Locatable and nonlocatable acoustic signals for barn owls. *Am. Nat.* 107:775–85.

Kornblatt, A. N., P. H. Urband, and A. C. Steere. 1985. Arthritis caused by *Borrelia burgdorferi* in dogs. *J. Am. Vet. Med. Assoc.* 186:960–64.

Korschgen, L. J. 1980. Procedures for food-habits analyses (Wildlife). Pp. 113–27 in S. D. Schemnitz, ed., *Wildlife management techniques manual*, 4th ed. The Wildlife Society, Washington, D.C.

Kramek, B. J. 1992. The hybrid howl: Legislators listen—these animals aren't crying wolf. *Rutgers Law J.* 23:633–56.

Krawczak, C. 1969. Polowanie na wilki w Wielkim Ksiestwie Poznanskim (Wolf hunting in the Great Duke of Poznan). [In Polish.] Lowiec Polski, 4, 12.

Krebs, C. J. 1994. *Ecology: The experimental analysis of distribution and abundance.* Harper Collins College Publishers, New York.

Krebs, C. J., S. Boutin, and R. Boonstra, eds. 2001. *Ecosystem Dynamics of the Boreal Forest: The Kluane Project.* Oxford University Press, New York. 511 pp.

Krebs, C. J., A. R. E. Sinclair, R. Boonstra, S. Boutin, K. Martin, and J. N. M. Smith. 1999. Community dynamics of vertebrate herbivores: How can we untangle the web? Pp. 447–73 in H. Olff, V. K. Brown, and R. H. Drent, eds., *Herbivores: Between plants and predators: 38th Symposium of the British Ecological Society.* Blackwell Science, Oxford.

Krebs, J. R. 1977. The significance of song repertories: The Beau Geste hypothesis. *Anim. Behav.* 25:475–78.

Kreeger, T. J. 1992. A review of chemical immobilization of wild canids. Pp. 271–83 in Annual Proceedings of the Joint Conference of the American Association of Zoo Veterinarians and the American Association of Wildlife Veterinarians.

———. 1996. *Handbook of wildlife chemical immobilization.* International Wildlife Veterinary Services, Laramie, WY. 340 pp.

Kreeger, T. J., M. Callahan, and M. Beckel. 1996. Use of medetomidine for chemical restraint of captive gray wolves. *J. Zoo Wildl. Med.* 27:507–12.

Kreeger, T. J., G. DelGiudice, and L. D. Mech. 1997. Effects of fasting and feeding on body composition of gray wolves (*Canis lupus*). *Can. J. Zool.* 75:1549–52.

Kreeger, T. J., A. M. Faggella, U. S. Seal, L. D. Mech, M. Callahan, and B. Hall. 1987. Cardiovascular and behavioral responses of gray wolves to ketamine-xylazine immobilization and antagonism by yohimbine. *J. Wildl. Dis.* 23:463–70.

Kreeger, T. J., D. L. Hunter, and M. R. Johnson. 1995. Immobilization protocol for free-ranging gray wolves (*Canis lupus*) translocated to Yellowstone National Park and Central Idaho. Proc. Joint Mtg. of Wildl. Dis. Assoc., Am. Assoc. Wildl. Vet., Am. Assoc. Zoo. Vet.

Kreeger, T. J., K. P. Jeraj, and P. J. Manning. 1983. Bacteremia concomitant with parvovirus infection in a pup. *J. Am. Vet. Med. Assoc.* 184:196–97.

Kreeger, T. J., V. B. Kuechle, L. D. Mech, J. R. Tester, and U. S. Seal. 1990. Physiological monitoring of gray wolves (*Canis lupus*) by radiotelemetry. *J. Mammal.* 71:258–61.

Kreeger, T. J., A. S. Levine, U. S. Seal, M. Callahan, and M. Beckel. 1991. Diazepam-induced feeding in captive gray wolves (*Canis lupus*). *Pharmacol. Biochem. Behav.* 39:559–61.

Kreeger, T. J., R. E. Mandsager, U. S. Seal, M. Callahan, and

M. Beckel. 1989. Physiological response of gray wolves to butorphanol-xylazine immobilization and antagonism by naloxone and yohimbine. *J. Wildl. Dis.* 25:89–94.

Kreeger, T. J., D. L. Pereira, M. Callahan, and M. Beckel. 1996. Activity patterns of gray wolves housed in small vs. large enclosures. *Zoo Biol.* 15:395–401.

Kreeger, T. J., and U. S. Seal. 1986a. Failure of yohimbine hydrochloride to antagonize ketamine hydrochloride immobilization of gray wolves. *J. Wildl. Dis.* 22:600–603.

———. 1986b. Immobilization of coyotes with xylazine hydrochloride-ketamine hydrochloride and antagonism by yohimbine hydrochloride. *J. Wildl. Dis.* 22:604–6.

———. 1990. Immobilization of gray wolves (*Canis lupus*) with sufentanil citrate. *J. Wildl. Dis.* 26:561–63.

———. 1992. Circannual prolactin rhythm in intact dogs housed outdoors. *Chronobiologia* 19:1–8.

Kreeger, T. J., U. S. Seal, M. Callahan, and M. Beckel. 1988. Use of xylazine sedation with yohimbine antagonism in captive gray wolves. *J. Wildl. Dis.* 24:688–90.

———. 1990a. Physiological and behavioral responses of gray wolves to immobilization with tiletamine and zolazepam (Telazol). *J. Wildl. Dis.* 26:90–94.

———. 1990b. Treatment and prevention with ivermectin of dirofilariasis and ancylostomiasis in captive gray wolves (*Canis lupus*). *J. Zoo Wildl. Med.* 21:310–17.

Kreeger, T. J., U. S. Seal, Y. Cohen, E. D. Plotka, and C. S. Asa. 1991. Characterization of prolactin secretion in gray wolves. *Can. J. Zool.* 69:1366–74.

Kreeger, T. J., U. S. Seal, and A. M. Faggella. 1986. Xylazine hydrochloride-ketamine hydrochloride immobilization of wolves and its antagonism by tolazoline hydrochloride. *J. Wildl. Dis.* 22:397–402.

Kreeger, T. J., U. S. Seal, and E. D. Plotka. 1992. Influence of hypothalamic-pituitary-adrenocortical hormones on reproductive hormones in gray wolves (*Canis lupus*). *J. Exp. Zool.* 264:32–41.

Kreeger, T. J., F. O. Smith, and U. S. Seal. 1993. Method and apparatus for detection of pregnancy in canids. U. S. Patent no. 5,270,220 issued December 14, 1993.

Krefting, L. W. 1969. The rise and fall of the coyote on Isle Royale. *Naturalist* 20:24–31.

Krizan, P. 2000. Blastomycosis in a free ranging lone wolf, *Canis lupus*, on the north shore of Lake Superior, Ontario. *Can. Field Nat.* 114:491–93.

Krutilla, J. 1967. Conservation reconsidered. *Am. Econ. Rev.* 57:77–86.

Kruuk, H. 1972. Surplus killing by carnivores. *J. Zool.* 166:233–44.

Kudaktin, A. N. 1978. Selective wolf predation on ungulates in the Caucasian Nature Reserve. *Moskovskoe Obshchestvo Ispytatelei Priroda, Otdel Biologicheski Byulleten* 83(3):19–28.

———. 1979. Territorial distribution and structure of the wolf population in the Caucasian Nature Reserve. *Obshchestvo Ispytatelei Priroda Otdel Biologicheski Byulleten* 84:56–65.

Kuehn, D. W., T. K. Fuller, L. D. Mech, W. J. Paul, S. H. Fritts, and W. E. Berg. 1986. Trap related injuries to wolves in Minnesota. *J. Wildl. Mgmt.* 50:90–91.

Kumar, S. 1993. Bonelli's eagle *(Hieraatus fasciatus)* killing a Black-buck *(Antilope cervicapra)* fawn. *J. Raptor Res.* 27(4): 218–19.

———. 1996. Unusual interaction between wolf and short-toed eagle. *J. Raptor Res.* 30:41–42.

Kumar, S., and A. R. Rahmani. 2001. Livestock depredation by wolves in the Great Indian Bustard Sanctuary, Nannaj, Maharashtra, India. *J. Bombay Nat. Hist Soc.* 97:340–48.

Kunkel, K. E. 1997. Predation by wolves and other large carnivores in northwestern Montana and southeastern British Columbia. Ph.D. dissertation, University of Montana, Missoula.

Kunkel, K. E., and L. D. Mech. 1994. Wolf and bear predation on white-tailed deer fawns in northeastern Minnesota. *Can. J. Zool.* 72:1557–65.

Kunkel, K. E., and D. H. Pletscher. 1999. Species-specific population dynamics of cervids in a multipredator ecosystem. *J. Wildl. Mgmt.* 63:1082–93.

———. 2000. Habitat factors affecting vulnerability of moose to predation by wolves in southeastern British Columbia. *Can. J. Zool.* 78:150–57.

———. 2001. Winter hunting patterns of wolves in and near Glacier National Park, Montana. *J. Wildl. Mgmt.* 65:520–30.

Kunkel, K. E., D. H. Pletscher, D. K. Boyd, R. R. Ream, and M. Fairchild. In press. Kill rate of wolves in or near Glacier National Park, Montana. *J. Wildl. Mgmt.*

Kunkel, K. E., T. K. Ruth, D. H. Pletscher, and M. G. Hornocker. 1999. Winter prey selection by wolves and cougars in and near Glacier National Park, Montana. *J. Wildl. Mgmt.* 63:901–10.

Kurtén, B. 1968. *Pleistocene mammals of Europe.* Chicago: Aldine.

———. 1974. A history of coyote-like dogs (Canidae, Mammalia). *Acta Zool. Fenn.* 140. 38 pp.

———. 1984. Geographic differentiation in the Rancholabrean dire wolf *(Canis dirus* Leidy) in North America. Pp. 218–27 in H. H. Genoways and M. R. Dawson, eds., *Contributions in Quaternary vertebrate paleontology: A volume in memorial to John E. Guilday.* Special Publication no. 8. Carnegie Museum of Natural History, Pittsburgh, PA.

Kurtén, B., and E. Anderson. 1980. *Pleistocene mammals of North America.* Columbia University Press, New York.

Kurtén, B., and J. M. Kaye. 1982. Late Quaternary Carnivora from the Black Belt, Mississippi. *Boreas* 11:47–52.

Kuyt, E. 1972. Food habits and ecology of wolves on barren-ground caribou range in the Northwest Territories. Report Series, no. 21. Canadian Wildlife Service. 36 pp.

Labutin, V. 1972. Geograficeskie osobennosti pitanija volka i lisicy. Zoologiceskie problemy Sibiri, Martial IV sovesc. Pp. 413–15 in Zoologov Sibiri. Nowosibirsk. (Cited in Bibikov 1985.)

Lacey, E. A., and P. W. Sherman. 1997. Cooperative breeding in naked mole-rats: Implications for vertebrate and invertebrate sociality. Pp. 267–99 in N. G. Solomon and J. A. French, eds., *Cooperative breeding in mammals.* Cambridge University Press, Cambridge.

LaGory, K. E. 1986. Habitat, group size, and the behaviour of white-tailed deer. *Behaviour* 98:168–79.

———. 1987. The influence of habitat and group characteristics on the alarm and flight response of white-tailed deer. *Anim. Behav.* 35:20–25.

Laikre, L., and N. Ryman. 1991. Inbreeding depression in a captive wolf *(Canis lupus)* population. *Conserv. Biol.* 5:33–40.

Laikre, L., N. Ryman, and E. A. Thompson. 1993. Hereditary blindness in a captive wolf *(Canis lupus)* population: Frequency reduction of a deleterious allele in relation to gene conservation. *Conserv. Biol.* 7:592–601.

LaJeunesse, T. A., and R. O. Peterson. 1993. Marrow and kidney fat as condition indices in gray wolves. *Wildl. Soc. Bull.* 21:87–90.

Lampe, A. 1997. Der wolf-nutztier-konflikt in Europa: Moglichkeiten zu dessen Management. Diplomarbeit des Forstwissenschaftlichen Fachbereichs der Georg-August-Universitat Gottingen. 128 pp. [The wolf-livestock conflict in Europe: Possibilities of its management. Master's thesis of the Department of Forest Science of the Georg-August-University, Gottingen.]

Lamprecht, J. 1981. The function of social hunting in larger terrestrial carnivores. *Mammal Rev.* 11:169–79.

Lande, R. 1988. Genetics and demography in biological conservation. *Science* 241:1455–60.

Landis, R. 1998. *Yellowstone wolves—predation.* Trailwood-Landis films, Gardiner, MT.

Lane, R. B. 1981. Chilocotin. Pp. 404–12 in W. C. Sturtevant, ed., *Handbook of North American Indians,* Vol. 6, *Subarctic.* Smithsonian Institution, Washington, D.C.

Lariviere, S., and M. Crête. 1993. The size of eastern coyotes *(Canis latrans):* A comment. *J. Mammal.* 74:1072–74.

Lariviere, S., H. Jollicoeur, and M. Crête. 2000. Status and conservation of the gray wolf *(Canis lupus)* in wildlife reserves in Quebec. *Biol. Conserv.* 94:143–51.

Lauer, B. H., E. Kuyt, and B. E. Baker. 1969. Wolf milk. I. Arctic wolf *(Canis lupus arctos)* and husky milk: Gross composition and fatty acid constitution. *Can. J. Zool.* 47:99–102.

Laundré, J. W., L. Hernández, and K. B. Altendorf. 2001. Wolves, elk, and bison: Reestablishing the "landscape of fear" in Yellowstone National Park, U.S.A. *Can. J. Zool.* 79:1401–9.

Law, R. G., and A. H. Kennedy. 1932. Parasites of fur-bearing animals. Bulletin 4. Department of Game and Fisheries, Ontario. 30 pp.

Lawhead, B. 1983. Wolf den site characteristics in the Nelchina Basin, Alaska. Master's thesis, University of Alaska, Fairbanks. 65 pp.

Lawrence, B. 1966. Early domestic dogs. *Z. Saugetierk.* 32:44–59.

Lawrence, B., and W. H. Bossert. 1967. Multiple character analysis of *Canis lupus, latrans,* and *familiaris* with a discussion of the relationship of *Canis niger. Am. Zool.* 7:223–32.

———. 1975. Relationships of North American *Canis* shown by a multiple character analysis of selected populations. Pp. 73–86 in M. W. Fox, ed., *The wild canids: Their systematics, behavioral ecology, and evolution.* Van Nostrand Reinhold, New York.

Lawrence, E. A. 1993. The sacred bee, the filthy pig, and the bat out of hell: Animal symbolism as cognitive biophilia. Pp. 301–41 in S. R. Kellert and E. O. Wilson, eds., *The biophilia hypothesis.* Island Press, Washington, D.C.

Lawrence, R. D. 1990. *In praise of wolves.* NorthWord Press, Inc., Minocqua, WI. (Video.)

———. 1993. *Trail of the wolf.* Rodale Press, Toronto.

Leader-Williams, N. 1990. Allocation of resources for conserving African pachyderms. *Trans. Int. Congr. Game Biol.* 19: 633–39.

Leader-Williams, N., and S. D. Albon. 1988. Allocation of resources for conservation. *Nature* 336:533–35.

Leader-Williams, N., R. J. Smith, M. J. Walpole, K. McComb, C. Moss, and S. Durant. 2001. Elephant hunting and conservation. *Science* 293:2203–4.

Lechner, A. J. 1978. The scaling of maximal oxygen consumption and pulmonary dimensions in small mammals. *Respir. Physiol.* 34:29–44.

Lee, D. S., M. K. Clark, and J. B. Funderburg. 1982. A preliminary survey of the mammals of mainland Dare County, North Carolina. Pp. 20–61 in E. F. Potter, ed., *A survey of the vertebrate fauna of mainland Dare County, North Carolina.* U.S. Fish and Wildlife Service, Raleigh, NC.

Lehman, N. E., P. Clarkson, L. D. Mech, T. J. Meier, and R. K. Wayne. 1992. A study of the genetic relationships within and among wolf packs using DNA fingerprinting and mitochondrial DNA. *Behav. Ecol. Sociobiol.* 30:83–94.

Lehman, N. E., A. Eisenhawer, K. Hansen, L. D. Mech, R. O. Peterson, P. J. P. Gogan, and R. K. Wayne. 1991. Introgression of coyote mitochondrial DNA into sympatric North American gray wolf populations. *Evolution* 45:104–19.

Lehman, N. E., and R. K. Wayne. 1991. Analysis of coyote mitochondrial DNA genotype frequencies: Estimation of effective number of alleles. *Genetics* 128:405–16.

Lehner, P. N. 1978a. Coyote communication. Pp. 128–59 in M. Bekoff, ed., *Coyotes: Biology, behavior, and management.* Academic Press, New York.

———. 1978b. Coyote vocalizations: A lexicon and comparisons with other canids. *Anim. Behav.* 26:712–22.

———. 1982. Differential vocal response of coyotes to "group howl" and "group yip-howl" playbacks. *J. Mammal.* 63:675–79.

Leiby, P. D., and W. G. Dyer. 1971. Cyclophyllidean tapeworms of wild carnivora. Pp. 174–234 in J. W. Davis and R. C. Anderson, eds., *Parasitic diseases of wild mammals.* Iowa State University Press, Ames.

Lenihan, M. L. 1987. Montanans ambivalent on wolves. The Montana Poll. Unpublished report, Bureau of Business and Economic Research, University of Montana; cosponsored by Great Falls Tribune, Great Falls, MT. 6 pp.

Lent, P. C. 1964. Tolerance between grizzlies and wolves. *J. Mammal.* 45:304–5.

———. 1974. Mother-infant relationships in ungulates. Pp. 14–55 in V. Geist and F. Walther, eds., *The behaviour of ungulates and its relation to management.* IUCN Publ., New Ser. no. 24, vol. 1.

Lentfer, J. W., and D. K. Sanders. 1973. Notes on the captive wolf (*Canis lupus*) colony, Barrow, Alaska. *Can. J. Zool.* 51: 623–27.

Leonard, J. A., R. K. Wayne, and A. Cooper. 2000. Population genetics of Ice Age brown bears. *Proc. Nat. Acad. Sci. U.S.A.* 97:1651–54.

Leonard, J. A., R. K. Wayne, J. Wheeler, R. Valadez, S. Guillen, and C. Vilà. 2002. Ancient DNA evidence for Old World origin of New World dogs. *Science* 298:1613–16.

Leopold, A. 1944. Review of *The wolves of North America* by S. P. Young and E. H. Goldman. *J. Forestry* 42(12): 928–29.

———. 1949. *A Sand County almanac, and sketches here and there.* Oxford University Press, New York.

Lequette, B., T. Houard, M. L. Poulle, and T. Dahier. 1995. The wolf returns to France. *Int. Wolf* 5(4): 8–9.

Lesniewicz, K., and K. Perzanowski. 1989. The winter diet of wolves in Bieszczady Mountains. *Acta Theriol.* 34(27): 373–80.

Levin, J. 1986. Children's literature. Pp. 29–37 in Wolves in American Culture Committee, *Wolf! A modern look.* Northword, Ashland, WI.

Lewis, D., G. B. Kaweche, and A. Mwenga. 1990. Wildlife conservation outside protected areas: Lessons from an experiment in Zambia. *Conserv. Biol.* 4:171–80.

Lewis, M. A., and J. D. Murray. 1993. Modelling territoriality and wolf-deer interactions. *Nature* 366:738–40.

Lewis, S. E., and A. E. Pusey. 1997. Factors influencing the occurrence of communal care in plural breeding mammals. Pp. 335–63 in N. G. Solomon and J. A. French, eds., *Cooperative breeding in mammals.* Cambridge University Press, Cambridge.

Licht, D. S., and S. H. Fritts. 1994. Gray wolf (*Canis lupus*) occurrences in the Dakotas. *Am. Midl. Nat.* 132:74–81.

Lima, S. L., and L. M. Dill. 1990. Behavioral decisions made under the risk of predation: A review and prospectus. *Can. J. Zool.* 68:619–40.

Lindstrom, E. 1986. Territory inheritance and the evolution of group-living in carnivores. *Anim. Behav.* 34:1825–35.

Ling, G. V., and A. L. Ruby. 1978. Aerobic bacterial flora of the prepuce, urethra, and vagina of normal adult dogs. *Am. J. Vet. Res.* 39:695–98.

Linnell, J. D. C., R. Andersen, Z. Andersone, L. Balciauskas, J. C. Blanco, L. Boitani, S. Brainerd, U. Breitenmoser, I. Kojola, O. Liberg, J. Loe, H. Okarma, H. C. Pedersen, C. Promberger, H. Sand, E. J. Solberg, H. Valdmann, and P. Wabakken. 2002. The fear of wolves: A review of wolf attacks on humans. Norsk Institute for Naturforskning, Trondheim, Norway. 65 pp.

Linnell, J. D. C., J. E. Swenson, and R. Andersen. 2000. Conservation of biodiversity in Scandinavian boreal forests: Large carnivores as flagships, umbrellas, indicators, or keystones? *Biodiversity Conserv.* 9(7): 857–68.

Lipetz, V. E., and M. Bekoff. 1982. Group size and vigilance in pronghorns. *Z. Tierpsychol.* 58:203–16.

Lipman, E. A., and J. R. Grassi. 1942. Comparative auditory sensitivity of man and dog. *Am. J. Psychol.* 55:84–89.

Litvaitis, J. A., J. A. Sherburne, and J. A. Bissonette. 1986. Bobcat habitat use and home range size in relation to prey density. *J. Wildl. Mgmt.* 50:110–17.

Litvinov, V. P. 1981. The wolf *(Canis lupus)* and wild boar *(Sus scrofa)* in the Kyzyl-Agach State Reservation. *Zool. Zh.* 60:1588–91.

Llaneza, L., A. Fernandez, and C. Nores. 1996. Dieta del lobo en dos zonas de Asturias (España) que difieren en carga ganadera. *Donana Acta Vertebrata* 23(2):201–13.

Llewellyn, L. G. 1978. Who speaks for the timber wolf? *Trans. N. Am. Wildl. Nat. Resour. Conf.* 43:442–52.

Lockwood, R. 1976. An ethological analysis of social structure and affiliation in captive wolves. Ph.D. dissertation, Washington University, St. Louis, MO.

———. 1979. Dominance in wolves: Useful construct or bad habit? Pp. 225–44 in E. Klinghammer, ed., *The behavior and ecology of wolves.* Garland STPM Press, New York.

Lohr, C., W. B. Ballard, and A. Bath. 1996. Attitudes toward gray wolf reintroduction to New Brunswick. *Wildl. Soc. Bull.* 25:414–20.

Loop, M. S., and M. M. Martin. 1991. Dogs and cats: They may look different but they see the same. *Soc. Neurosci. Abstr.* 17:848.

Lopez, B. H. 1978. *Of wolves and men.* Charles Scribner's Sons, New York.

Lorenz, K. Z. 1954. *Man meets dog.* London: Methuen.

———. 1966. *On aggression.* Harcourt, Brace and World, New York.

Lorenzini, R., and R. Fico. 1995. A genetic investigation of enzyme polymorphisms of wolf and dog: Suggestions for wolf conservation in Italy. *Acta Theriol.* 3:101–10.

Lotka, A. J. 1925. *Elements of physical biology.* Williams and Wilkins Co., Baltimore.

Lovell, J. E., and R. Getty. 1957. The hair follicle, epidermis, dermis, and skin glands of the dog. *Am. J. Vet. Res.* 18:873–85.

Lucas, J. R., S. R. Creel, and P. M. Waser. 1997. Dynamic optimization and cooperative breeding: An evaluation of future fitness effects. Pp. 171–98 in N. G. Solomon and J. A. French, eds., *Cooperative breeding in mammals.* Cambridge University Press, Cambridge.

Lucas, N. S. 1923. Report on the deaths occurring in the Society's gardens during 1922. *Proc. Zool. Soc. Lond.* 1923:125.

Lucas, P. W., and D. A. Luke. 1984. Chewing it over: Basic principles of food breakdown. Pp. 283–301 in D. J. Chivers, B. A. Wood, and A. Bilsborough, eds., *Food acquisition and processing in primates.* Plenum Publishing, New York.

Lucchini, V., E. Fabbri, F. Marucco, S. Ricci, L. Boitani, and E. Randi. 2002. Noninvasive molecular tracking of colonizing wolf *(Canis lupus)* packs in the western Italian Alps. *Mol. Ecol.* 11:857–68.

Luisi, M., F. Franchi, P. M. Kicovic, D. Silvestri, G. Cossu, A. L. Catarsi, D. Barletta, and M. Gasperi. 1981. Radioimmunoassay for progesterone in human saliva during the menstrual cycle. *J. Steroid Biochem.* 14:1069–73.

Lynch, J. J. 1974. Psychophysiology and development of social attachment. *J. Nerv. Ment. Dis.* 151:231–44.

Lyons, C. A., P. M. Ghezzi, and C. D. Cheney. 1982. Reinforcement of cooperative behavior in captive wolves. Pp. 262–71 in F. H. Harrington and P. C. Paquet, eds., *Wolves of the world:*

Perspectives of behavior, ecology, and conservation. Noyes Publications, Park Ridge, NJ.

Maagaard, L., and J. Graugaard. 1994. Female Arctic wolf, *Canis lupus arctos,* mating with domestic dogs, *Canis familiaris,* in Northeast Greenland. *Can. Field Nat.* 108:374–75.

Macdonald, D. W. 1979. Some observations and field experiments on the urine marking behaviour of the red fox, *Vulpes vulpes* L. *Z. Tierpsychol.* 51:1–22.

———. 1983. The ecology of carnivore social behaviour. *Nature* 301:379–84.

Macdonald, D. W., L. Boitani, and P. Barrasso. 1980. Foxes, wolves and conservation in the Abruzzo mountains. *Biogeographica* 18:223–35.

Macdonald, D. W., and P. D. Moehlman. 1983. Cooperation, altruism, and restraint in the reproduction of carnivores. Pp. 433–67 in P. Bateson and P. Klopfer, eds., *Perspectives in ethology.* Plenum Press, New York.

MacDonald, K. B. 1980. Activity patterns in a captive wolf pack. *Carnivore* 3:62–64.

———. 1983. Stability of individual differences in behavior in a litter of wolf cubs *(Canis lupus). J. Comp. Psychol.* 97:99–106.

———. 1987. Development and stability of personality characteristics in pre-pubertal wolves: Implications for pack organization and behavior. Pp. 293–312 in H. Frank, ed., *Man and wolf: Advances, issues and problems in captive wolf research.* Dr. W. Junk Publishers, Dordrecht, The Netherlands.

Mace, G. M., T. B. Smith, M. W. Bruford, and R. K. Wayne. 1996. Molecular genetic approaches in conservation—an overview of the issues. Pp. 1–12 in T. B. Smith and R. K. Wayne, eds., *Molecular genetic approaches in conservation.* Oxford University Press, Oxford.

MacFarlane, R. R. 1905. Notes on mammals collected and observed in the northern Mackenzie River district, Northwest Territories of Canada. *Proc. U. S. Nat. Mus.* 28:673–764.

Machinskii, A. P. 1966. Epidemiology of trichinellosis in the Mordovian ASSR. [In Russian.] *Meditsinskaia Parazitologiia I Parazitarnye Bolezni* 35:373–74.

Machinskii, A. P., and V. N. Semov. 1971. Trichinosis of wild animals in Mordovia. [In Russian.] *Meditsinskaia Parazitologiia I Parazitarnye Bolezni* 40:532–34.

Mack, J. A., W. G. Brewster, and S. H. Fritts. 1992. A review of wolf depredation on livestock and implications for the Yellowstone area. Pp. 5-21–5-44 in J. D. Varley and W. G. Brewster, eds., *Wolves for Yellowstone? A report to the United States Congress,* Vol. 4, *Research and analysis.* U.S. National Park Service, Yellowstone National Park, WY.

MacNulty, D. R. 2002. The predatory sequence and the influence of injury risk on hunting behavior in the wolf. Master's thesis, University of Minnesota, St. Paul. 71 pp.

MacNulty, D. R., N. Varley, and D. W. Smith. 2001. Grizzly bear, *Ursus arctos,* usurps bison, *Bison bison,* captured by wolves, *Canis lupus,* in Yellowstone National Park, Wyoming. *Can. Field Nat.* 115:495–98.

Magalhaes, C. P., and F. P. Fonseca. 1982. The wolf in Braganca county: Impact on cattle and game. *Trans. Int. Congr. Game Biol.* 14:281–86.

Magoun, A. 1976. Summer scavenging activity in northeastern Alaska. Master's thesis, University of Alaska, Fairbanks. 168 pp.

Magrane, W. G. 1977. *Canine opthalmology*, 3rd ed. Lea and Febiger, Philadelphia.

Mails, T. E. 1995. *The mystic warriors of the plains.* Marlowe and Co., New York.

Makovkin, L. I. 1999. The sika deer of Lazovsky Reserve and surrounding areas of the Russian Far East. Almanac, Russki Ostrov, Vladivostok, Russia.

Makridin, V. P. 1962. The wolf in the Yamal north. *Zool. Zh.* 41: 1413–17.

Makridin, V. P., N. K. Zheleznov, E. I. Gromov, and G. I. Chuvashov. 1985. Kranjnij Sever. Pp. 467–76 in D. I. Bibikov, ed., *The wolf: History, systematics, morphology, ecology.* [In Russian.] Izdatetvo Nauka, Muskva.

Malcolm, J., and K. Marten. 1982. Natural selection and the communal rearing of pups in African wild dogs *(Lycaon pictus)*. *Behav. Ecol. Sociobiol.* 10:1–13.

Mallinson, J. 1978. *The shadow of extinction: Europe's threatened wild mammals.* Macmillan, London.

Mallory, F. F., T. L. Hillis, C. G. Blomme, and W. G. Hurst. 1994. Skeletal injuries of an adult timber wolf, *Canis lupus,* in northern Ontario. *Can. Field Nat.* 108:230–32.

Malm, K., and P. Jensen. 1996. Weaning in dogs: Within- and between-litter variation in milk and solid food intake. *Appl. Anim. Behav. Sci.* 49:223–35.

Mangun, W. R., J. N. Lucas, J. C. Whitehead, and J. C. Mangun. 1996. Valuing red wolf recovery efforts at Alligator River NWR: Measuring citizen support. Pp. 165–71 in N. Fascione and M. Cecil, eds., Proceedings of the Defenders of Wildlife's Wolves of America Conference, Albany, NY. Defenders of Wildlife, Washington, D.C.

Marhenke, P. 1971. An observation of four wolves killing another wolf. *J. Mammal.* 52:630–31.

Marler, P. 1955. Characteristics of some animal calls. *Nature* 176: 6–7.

———. 1965. Communication in monkeys and apes. Pp. 544–84 in I. DeVore, ed., *Primate behavior.* Holt, Rinehart and Winston, New York.

Marler, P., and W. J. Hamilton, III. 1966. *Mechanisms of animal behavior.* John Wiley and Sons, New York.

Marples, M. J. 1969. Life on the human skin. *Sci. Am.* 220:108–16.

Marquard-Petersen, U. 1995. Status of wolves in Greenland. Pp. 55–57 in L. N. Carbyn, S. H. Fritts, and D. R. Seip, eds., *Ecology and conservation of wolves in a changing world.* Canadian Circumpolar Institute, Edmonton, Alberta.

———. 1998. Food habits of Arctic wolves in Greenland. *J. Mammal.* 79:236–44.

Marshal, J. P., and S. Boutin. 1999. Power analysis of wolf-moose functional responses. *J. Wildl. Mgmt.* 63:396–402.

Marshall, D. A., and D. G. Moulton. 1981. Olfactory sensitivity to alpha-ionone in humans and dogs. *Chem. Senses* 6: 53–61.

Marten, K., and P. Marler. 1977. Sound transmission and its significance for animal vocalization. I. Temperate habitats. *Behav. Ecol. Sociobiol.* 2:271–90.

Marten, K., D. Quine, and P. Marler. 1977. Sound transmission and its significance for animal vocalization. II. Tropical forest habitats. *Behav. Ecol. Sociobiol.* 2:291–302.

Martin, P. S., and C. R. Szuter. 1999. War zones and game sinks in Lewis and Clark's West. *Conserv. Biol.* 13:36–45.

Martin, R. A. 1974. Fossil mammals from the Coleman IIA fauna, Sumter County. Pp. 35–99 in S. D. Webb, ed., *Pleistocene mammals of Florida.* University Presses of Florida, Gainesville.

Martin, R. A., and S. D. Webb. 1974. Late Pleistocene mammals from the Devil's Den fauna. Pp. 114–45 in S. D. Webb, ed., *Pleistocene mammals of Florida.* University Presses of Florida, Gainesville.

Martinet, L., D. Allen, and C. Weiner. 1984. Role of prolactin in the photoperiodic control of moulting in the mink *(Mustela vison). J. Endocrinol.* 103:9–15.

Maruyama, N., Z. Gao, and K. Shi. 1996. Decline of gray wolves in Xinhaerhuyougi and Ewenkegi Districts, Northern Inner Mongolia. Abstract. Pp. 90, Proceedings of the 2nd International Symposium on Coexistence of Large Carnivores with Man. Ecosystem Conservation Society, Saitawa, Japan.

Mary-Rousseliere, G. 1984. Iglulik. Pp. 431–46 in W. C. Sturtevant, ed., *Handbook of North American Indians,* vol. 5, *Arctic.* Smithsonian Institution, Washington, D.C.

Masterton, B., H. Heffner, and R. Ravizza. 1969. The evolution of human hearing. *J. Acoust. Soc. Am.* 45:966–85.

Matthiessen, P. 1959. *Wildlife in America.* Viking, New York.

Mattioli, L., M. Apollonio, V. Mazzarone, and E. Centofanti. 1995. Wolf food habits and wild ungulate availability in the Foreste Casentinesi National Park, Italy. *Acta Theriol.* 40(4): 387–402.

Mattson, M. Y. 1992. Denning ecology of wolves in northwest Montana and southern Canadian Rockies. Master's thesis, University of Montana, Missoula.

Maximilian, Prince of Wied. 1906. Travels in the interior of North America 1830. Pp. 22–24 in R. G. Thwaites, ed., *Early Western travels.* Clark, Cleveland.

Mayr, E. 1963. *Animal species and evolution.* Harvard University Press, Cambridge, MA.

McCarley, H. 1962. The taxonomic status of wild *Canis* (Canidae) in the south central United States. *Southwest Nat.* 7:227–35.

———. 1975. Long distance vocalizations of coyotes *(Canis latrans). J. Mammal.* 56:847–56.

———. 1978. Vocalizations of red wolves *(Canis rufus). J. Mammal.* 59:27–35.

McCarley, H., and Carley, C. J. 1979. Recent changes in distribution and status of wild red wolves *(Canis rufus).* Endangered Species Report no. 4. U.S. Fish and Wildlife Service, Albuquerque, NM.

McCay, C. M. 1949. *Nutrition of the dog.* Comstock Publishing, Ithaca, NY.

McComb, K., C. Moss, S. Durant, M. Sarah, L. Baker, and S. Sayialel. 2001. Matriarchs as repositories of social knowledge in African elephants. *Science* 292:491–94.

McCullough, D. R. 1969. The tule elk: Its history, behavior, and ecology. University of California Publications in Zoology, no. 88. Berkeley, CA.

McIntyre, R. 1993. *A society of wolves: National parks and the battle for wolves.* Voyageur Press, Stillwater, MN.

———, ed. 1995. *War against the wolf: America's campaign to exterminate the wolf.* Voyageur Press, Stillwater, MN.

———. 1996. Wolf reintroduction in Japan. *Int. Wolf* 6:13–15, 27.

McIntyre, R. and D. Smith. 2000. The death of a queen: Yellowstone mutiny ends tyrannical rule over Druid Pack. *Int. Wolf* 10(4): 8–11.

McKitrick, M. C., and R. M. Zink. 1988. Species concepts in ornithology. *Condor* 90:1–14.

McLaren, B. E., and R. O. Peterson. 1994. Wolves, moose, and tree rings on Isle Royale. *Science* 266:1555–58.

McLeod, P. J. 1987. Aspects of the early social development of timber wolves *(Canis lupus).* Ph.D. dissertation, Dalhousie University, Halifax, NS.

———. 1990. Infanticide by female wolves. *Can. J. Zool.* 68:402–4.

———. 1996. Developmental changes in associations among timber wolf *(Canis lupus)* postures. *Behav. Processes* 38:105–18.

McLeod, P. J., and J. C. Fentress. 1997. Developmental changes in the sequential behavior of interacting timber wolf pups. *Behav. Processes* 39:127–36.

McLeod, P. J., J. C. Fentress and, J. Ryon. 1991. Patterns of aggression within a captive timber wolf pack. *Aggress. Behav.* 17:84.

McLeod, P. J., W. H. Moger, J. Ryon, S. Gadbois, and J. C. Fentress. 1996. The relation between urinary cortisol levels and social behaviour in captive timber wolves. *Can. J. Zool.* 74: 209–16.

McNab, B. K. 1963. Bioenergetics and the determination of home range size. *Am. Nat.* 97:133–40.

McNamee, T. 1997. *The return of the wolf to Yellowstone.* Henry Holt and Co., New York.

McNaught, D. A. 1985. Park visitor attitudes toward wolf recovery in Yellowstone National Park. Master's thesis, University of Montana, Missoula.

———. 1987. Wolves in Yellowstone? Park visitors respond. *Wildl. Soc. Bull.* 15:518–21.

McNay, M. E. 2002a. A case history of wolf-human encounters in Alaska and Canada. Wildlife Technical Bulletin 13. Alaska Department of Fish and Game. 45 pp.

———. 2002b. Wolf-human interactions in Alaska and Canada: A review of the case history. *Wildl. Soc. Bull.* 30:831–43.

McNeeley, J. A., ed. 1995. *Expanding partnership in conservation.* Island Press, Washington, D.C.

McNeill, M. A., M. E. Rau, and F. Messier. 1984. Helminths of wolves *(Canis lupus* L.) from Southwestern Quebec. *Can. J. Zool.* 62:1659–60.

McRoberts, R. E., L. D. Mech, and R. O. Peterson. 1995. The cumulative effect of consecutive winters' snow depth on moose and deer populations: A defence. *J. Anim. Ecol.* 64:131–35.

Mech, L. D. 1966a. Hunting behavior of timber wolves in Minnesota. *J. Mammal.* 47:347–48.

———. 1966b. The wolves of Isle Royale. U.S. National Park Service Fauna Series, no. 7. U.S. Govt. Printing Office. 210 pp.

———. 1970. *The wolf: The ecology and behavior of an endangered species.* Natural History Press, Garden City, NY.

———. 1971. Wolves, coyotes, and dogs. Pp. 19–22 in M. Nelson, ed., Symposium on the white-tailed deer in Minnesota. Minnesota Department of Natural Resources, St. Paul. 88 pp.

———. 1973. Wolf numbers in the Superior National Forest of Minnesota. USDA Forest Service Research Paper NC-97. North Central Forest Experiment Station, St. Paul, MN. 10 pp.

———. 1974a. *Canis lupus.* Mammalian Species, no. 37. American Society of Mammalogists. 6 pp.

———. 1974b. Current techniques in the study of elusive wilderness carnivores. *Trans. Int. Congr. Game Biol.* 11:315–22.

———. 1975. Disproportionate sex ratios of wolf pups. *J. Wildl. Mgmt.* 39:737–40.

———. 1977a. Population trend and winter deer consumption in a Minnesota wolf pack. Pp. 55–83 in R. L. Phillips and C. Jonkel, eds., Proceedings of the 1975 predator symposium. Montana Forest and Conservation Experiment Station, University of Montana, Missoula. 268 pp.

———. 1977b. Productivity, mortality and population trend of wolves in northeastern Minnesota. *J. Mammal.* 58:559–74.

———. 1977c. Where can the wolf survive? *Natl. Geogr.* 152:518–37.

———. 1977d. Wolf-pack buffer zones as prey reservoirs. *Science* 198:320–21.

———. 1979a. Making the most of radio-tracking. Pp. 85–95 in C. J. Amlaner and D. W. Macdonald, eds., *A handbook on biotelemetry and radio-tracking.* Pergamon Press, Oxford.

———. 1979b. Some considerations in re-establishing wolves in the wild. Pp. 445–57 in E. Klinghammer, ed., *The behavior and ecology of wolves.* Garland STPM Press, New York.

———. 1982a. Censusing wolves (radio-tracking). Pp. 227–28 in D. W. Davis, ed., *CRC handbook of census methods for terrestrial vertebrates.* CRC Press, Boca Raton, FL.

———. 1982b. The IUCN-SSC Wolf Specialist Group. Pp. 327–33 in F. H. Harrington and P. C. Paquet, eds., *Wolves of the world: Perspectives of behavior, ecology, and conservation.* Noyes Publications, Park Ridge, NJ.

———. 1984. Predators and predation. Pp. 189–200 in L. K. Halls, ed., *White-tailed deer ecology and management.* Stackpole, Harrisburg, PA.

———. 1986. Wolf population in the central Superior National Forest, 1967–1985. USDA Forest Service Research Paper NC-270. North Central Forest Experiment Station, St. Paul, MN. 6 pp.

———. 1987a. Age, season, distance, direction and social aspects of wolf dispersal from a Minnesota pack. Pp. 55–74 in B. D. Chepko-Sade and Z. Tang Halpin, eds., *Mammalian dispersal patterns.* University of Chicago Press, Chicago.

———. 1987b. At home with the arctic wolf. *Natl. Geogr.* 171(5): 562–93.

———. 1988a. *The Arctic wolf: Living with the pack.* Voyageur Press, Stillwater, MN.

———. 1988b. Life in the High Arctic. *Natl. Geogr.* 173:750–67.

———. 1988c. Longevity in wild wolves. *J. Mammal.* 69:197–98.

———. 1989. Wolf population survival in an area of high road density. *Am. Midl. Nat.* 121:387–89.

———. 1990a. Snow as a driving force in a wolf-deer system. *Trans. Int. Congr. Game Biol.* 19:562. (Abstract.)

———. 1990b. Who's afraid of the big, bad wolf? *Audubon* 92: 82–85.

———. 1991a. Returning the wolf to Yellowstone. Pp. 309–22 in R. B. Keiter and M. S. Boyce, eds., *The Greater Yellowstone ecosystem: Redefining America's wilderness heritage.* Yale University Press, New Haven, CT.

———. 1991b. *The way of the wolf.* Voyageur Press, Stillwater, MN.

———. 1992. Daytime activity of wolves during winter in northeastern Minnesota. *J. Mammal.* 73:570–71.

———. 1993a. Details of a confrontation between two wild wolves. *Can. J. Zool.* 71:1900–1903.

———. 1993b. Resistance of young wolf pups to inclement weather. *J. Mammal.* 74:485–86.

———. 1994a. Buffer zones of territories of gray wolves as regions of intraspecific strife. *J. Mammal.* 75:199–202.

———. 1994b. Regular and homeward travel speeds of wolves. *J. Mammal.* 75:741–42.

———. 1995a. The challenge and opportunity of recovering wolf populations. *Conserv. Biol.* 9:270–78.

———. 1995b. How can I see a wolf? *Int. Wolf* 5(1): 8–11.

———. 1995c. Summer movements and behavior of an arctic wolf, *Canis lupus,* pack without pups. *Can. Field Nat.* 109: 473–75.

———. 1995d. A ten-year history of the demography and productivity of an arctic wolf pack. *Arctic* 48:329–32.

———. 1995e. What do we know about wolves and what more do we need to learn? Pp. 537–45 in L. N. Carbyn, S. H. Fritts, and D. R. Seip, eds., *Ecology and conservation of wolves in a changing world.* Canadian Circumpolar Institute, Edmonton, Alberta.

———. 1996. A new era for carnivore conservation. *Wildl. Soc. Bull.* 24:397–401.

———. 1997. An example of endurance in an old wolf, *Canis lupus. Can. Field Nat.* 111:654–55.

———. 1998a. *The arctic wolf: Ten years with the pack.* Voyageur Press, Stillwater, MN.

———. 1998b. Estimated costs of maintaining a recovered wolf population in agricultural regions of Minnesota. *Wildl. Soc. Bull.* 26:817–22.

———. 1999. Alpha status, dominance, and division of labor in wolf packs. *Can. J. Zool.* 77:1196–1203.

———. 2000a. Leadership in wolf, *Canis lupus,* packs. *Can. Field Nat.* 114:259–63.

———. 2000b. Wolf restoration to the Adirondacks and advantages and disadvantages of public participation in the decision. Pp. 13–22 in V. A. Sharpe, B. Norton, and S. Donnelley, eds., *Wolves and human communities: Biology, politics, and ethics.* Island Press, Washington, D.C.

———. 2000c. *The wolves of Minnesota: Howl in the heartland.* Voyageur Press, Stillwater, MN.

———. 2000d. Lack of reproduction in muskoxen and arctic hares caused by early winter? *Arctic* 53:69–71.

———. 2001a. Managing Minnesota's recovered wolves. *Wildl. Soc. Bull.* 29:70–77.

———. 2001b. "Standing over" and "Hugging" in wild wolves, *Canis lupus. Can. Field Nat.* 115:179–81.

———. 2002. Breeding season of wolves, *Canis lupus,* in relation to latitude. *Can. Field Nat.* 115:139–40.

Mech, L. D., and L. G. Adams. 1999. Killing of a muskox, *Ovibos mochatus,* by two wolves, *Canis lupus,* and subsequent caching. *Can. Field Nat.* 113:673–75.

Mech, L. D., L. G. Adams, T. J. Meier, J. W. Burch, and B. W. Dale. 1998. *The wolves of Denali.* University of Minnesota Press, Minneapolis.

Mech, L. D., and G. D. DelGuidice. 1985. Limitations of the marrow-fat technique as an indicator of body condition. *Wildl. Soc. Bull.* 13:204–6.

Mech, L. D., and N. E. Federoff. 2002. Alpha$_1$-Antitrypsin polymorphism and systematics of eastern North American wolves. *Can. J. Zool.* 80:961–63.

Mech, L. D., and L. D. Frenzel, Jr., eds. 1971a. *Ecological studies of the timber wolf in northeastern Minnesota.* USDA Forest Service Research Paper NC-52. North Central Forest Experiment Station, St. Paul, MN. 62 pp.

———. 1971b. The possible occurrence of the Great Plains wolf in northeastern Minnesota. Pp. 60–62 in L. D. Mech and L. D. Frenzel, Jr., eds., *Ecological studies of the timber wolf in northeastern Minnesota.* USDA Forest Service Research Paper NC-52. North Central Forest Experiment Station, St. Paul, MN. 62 pp.

Mech, L. D., L. D. Frenzel, Jr., and P. D. Karns. 1971. The effect of snow conditions on the ability of wolves to capture deer. Pp. 51–59 in L. D. Mech and L. D. Frenzel, Jr., eds., *Ecological studies of the timber wolf in Minnesota.* USDA Forest Service Research Paper NC-52. North Central Forest Experiment Station, St. Paul, MN.

Mech, L. D., and S. H. Fritts. 1987. Parvovirus and heartworm found in Minnesota wolves. *Endangered Species Tech. Bull.* 12:5–6.

Mech, L. D., S. H. Fritts, and M. E. Nelson. 1996. Wolf management in the 21st century: From public input to sterilization. *J. Wildl. Res.* 1:195–98.

Mech, L. D., S. H. Fritts, and W. J. Paul. 1988. Relationship between winter severity and wolf depredations on domestic animals in Minnesota. *Wildl. Soc. Bull.* 16:269–72.

Mech, L. D., S. H. Fritts, G. L. Radde, and W. J. Paul. 1988. Wolf distribution and road density in Minnesota. *Wildl. Soc. Bull.* 16:85–87.

Mech, L. D., S. H. Fritts, and D. Wagner. 1995. Minnesota wolf dispersal to Wisconsin and Michigan. *Am. Midl. Nat.* 133: 368–70.

Mech, L. D., and E. M. Gese. 1992. Field testing the Wildlink capture collar on wolves. *Wildl. Soc. Bull.* 20:221–23.

Mech, L. D., and S. M. Goyal. 1993. Canine parvovirus effect on wolf population change and pup survival. *J. Wildl. Dis.* 29:330–33.

———. 1995. Effects of canine parvovirus on gray wolves in Minnesota. *J. Wildl. Mgmt.* 59:565–70.

Mech, L. D., S. M. Goyal, C. N. Bota, and U. S. Seal. 1986. Canine parvovirus infection in wolves *(Canis lupus)* from Minnesota. *J. Wildl. Dis.* 22:104–6.

Mech, L. D., E. K. Harper, T. J. Meier, and W. J. Paul. 2000. Assessing factors that predispose Minnesota farms to wolf depredations on cattle. *Wildl. Soc. Bull.* 28:623–29.

Mech, L. D., and H. H. Hertel. 1983. An eight year demography of a Minnesota wolf pack. *Acta Zool. Fenn.* 174:249–50.

Mech, L. D., and P. D. Karns. 1977. Role of the wolf in a deer decline in the Superior National Forest. USDA Forest Service Research Paper NC-148. North Central Forest Experiment Station, St. Paul, MN. 23 pp.

Mech, L. D., and S. T. Knick. 1978. Sleeping distance in wolf pairs in relation to the breeding season. *Behav. Biol.* 23:521–25.

Mech, L. D., and M. Korb. 1978. An unusually long pursuit of a deer by a wolf. *J. Mammal.* 59:860–61.

Mech, L. D., and H. J. Kurtz. 1999. First record of coccidiosis in wolves, *Canis lupus. Can. Field Nat.* 113:305–6.

Mech, L. D., H. J. Kurtz, and S. Goyal. 1997. Death of a wild wolf from canine parvoviral enteritis. *J. Wildl. Dis.* 33:321–22.

Mech, L. D., and R. E. McRoberts. 1990. Survival of white-tailed deer fawns in relation to maternal age. *J. Mammal.* 71:465–67.

Mech, L. D., R. E. McRoberts, R. O. Peterson, and R. E. Page. 1987. Relationship of deer and moose populations to previous winters' snow. *J. Anim. Ecol.* 56:615–27.

Mech, L. D., T. J. Meier, and J. W. Burch. 1991. Denali Park wolf studies: Implications for Yellowstone. *Trans. N. Am. Wildl. Nat. Resour. Conf.* 56:86–90.

Mech, L. D., T. J. Meier, J. W. Burch, and L. G. Adams. 1995. Patterns of prey selection by wolves in Denali National Park, Alaska. Pp. 231–43 in L. N. Carbyn, S. H. Fritts, and D. R. Seip, eds., *Ecology and conservation of wolves in a changing world.* Canadian Circumpolar Institute, Edmonton, Alberta.

Mech, L. D., T. J. Meier, and U. S. Seal. 1993. Wolf nipple measurements as indices of age and breeding status. *Am. Midl. Nat.* 129:266–71.

Mech, L. D., and S. B. Merrill. 1998. Daily departure and return patterns of wolves, *Canis lupus,* from a den at 80° latitude. *Can. Field Nat.* 112:515–17.

Mech, L. D., and M. E. Nelson. 1990a. Evidence of prey-caused mortality in three wolves. *Am. Midl. Nat.* 123:207–8.

———. 1990b. Non-family wolf (*Canis lupus*) packs. *Can. Field Nat.* 104:482–83.

———. 2000. Do wolves affect white-tailed buck harvest in northeastern Minnesota? *J. Wildl. Mgmt.* 64:129–36.

Mech, L. D., M. E. Nelson, and R. E. McRoberts. 1991. Effects of maternal and grandmaternal nutrition on deer mass and vulnerability to wolf predation. *J. Mammal.* 72:146–51.

Mech, L. D., and R. M. Nowak. 1981. Return of the gray wolf to Wisconsin. *Am. Midl. Nat.* 105:408–9.

Mech, L. D., and J. M. Packard. 1990. Possible use of wolf (*Canis lupus*) den over several centuries. *Can. Field Nat.* 104:484–85.

Mech, L. D., M. K. Phillips, D. W. Smith, and T. J. Kreeger. 1996. Denning behaviour of non-gravid wolves, *Canis lupus. Can. Field Nat.* 110:343–45.

Mech, L. D., D. H. Pletscher, and C. J. Martinka. 1995. Gray wolf status and trends in the contiguous United States. Pp. 98–100 in E. T. LaRoe, G. S. Farris, C. E. Puckett, and P. D. Doran, eds., *Our living resources: A report to the nation on distribution, abundance and health of U.S. plants, animals, and ecosystems.* USDI National Biological Service, Washington, D.C. 530 pp.

Mech, L. D., and U. S. Seal. 1987. Premature reproductive activity in wild wolves. *J. Mammal.* 68:871–73.

Mech, L. D., U. S. Seal, and S. M. Arthur. 1984. Recuperation of a severely debilitated wolf. *J. Wildl. Dis.* 20:166–68.

Mech, L. D., U. S. Seal, and G. D. DelGiudice. 1987. Use of urine in snow to indicate condition of wolves. *J. Wildl. Mgmt.* 51:10–13.

Mech, L. D., D. W. Smith, K. M. Murphy, and D. R. MacNulty. 2001. Winter severity and wolf predation on a formerly wolf-free elk herd. *J. Wildl. Mgmt.* 64:998–1003.

Mech, L. D., R. P. Thiel, S. H. Fritts, and W. E. Berg. 1985. Presence and effects of the dog louse *Trichdectes canis* (Mallophaga, Trichdectidae) on wolves and coyotes from Minnesota and Wisconsin. *Am. Midl. Nat.* 114:404–5.

Mech, L. D., P. Wolfe, and J. M. Packard. 1999. Regurgitative food transfer among wild wolves. *Can. J. Zool.* 77:1192–95.

Medjo, D., C., and L. D. Mech. 1976. Reproductive activity in nine- and ten-month-old wolves. *J. Mammal.* 57(2): 406–8.

Meier, T. J., J. W. Burch, L. D. Mech, and L. G. Adams. 1995. Pack structure dynamics and genetic relatedness among wolf packs in a naturally regulated population. Pp. 293–302 in L. N. Carbyn, S. H. Fritts, and D. R. Seip, eds., *Ecology and conservation of wolves in a changing world.* Canadian Circumpolar Institute, Edmonton, Alberta.

Mellett, J. S. 1981. Mammalian carnassial function and the "Every effect." *J. Mammal.* 62:164–66.

Mendelssohn, H. 1982. Wolves of Israel. Pp. 173–95 in F. H. Harrington and P. C. Paquet, eds., *Wolves of the world: Perspectives of behavior, ecology, and conservation.* Noyes Publications, Park Ridge, NJ.

Mengel, R. M. 1971. A study of dog-coyote hybrids and implications concerning hybridization in *Canis. J. Mammal.* 52:316–36.

Meriggi, A., A. Brangi, C. Matteucci, and O. Sacchi. 1996. The feeding habits of wolves in relation to large prey availability in northern Italy. *Ecography* 19(3): 287–95.

Meriggi, A., P. Rosa, A. Brangi, and C. Matteucci. 1991. Habitat use and diet of the wolf in northern Italy. *Acta Theriol.* 36:141–51.

Meriwether, D., and M. K. Johnson. 1980. Mammalian prey digestibility by coyotes. *J. Mammal.* 61(4): 774–75.

Merriam, J. C. 1912. The fauna of Rancho La Brea. Part II. Canidae. Memoirs of the University of California 1:217–72.

———. 1918. Note on the systematic position of the wolves of the *Canis dirus* group. University of California Publ., Bull. Dept. Geol. 10:531–33.

Merrill, S. B. 1996. The wolves of Camp Ripley. *Int. Wolf* 6(1):7–8, 22.

———. 2000. Road densities and wolf, *Canis lupus,* habitat suitability: An exception. *Can. Field Nat.* 114:312–13.

———. 2002. An evaluation of the use of global positioning system telemetry in studying wolf biology. Ph.D. dissertation, University of Minnesota, St. Paul. 135 pp.

Merrill, S. B., L. G. Adams, M. E. Nelson, and L. D. Mech. 1998. Testing releasable GPS collars on wolves and white-tailed deer. *Wildl. Soc. Bull.* 26(4): 830–35.

Merrill, S. B., and L. D. Mech. 2000. Details of extensive movements by Minnesota wolves. *Am. Midl. Nat.* 144:428–33.

Mertens, H. 1936. Der hund aus dem Senkenberg-Moor, ein Begleiter des Ur's. *Natur und Volk* 66:506–10.

Mertl-Millhollen, A. S., P. A. Goodmann, and E. Klinghammer. 1986. Wolf scent marking with raised-leg urination. *Zoo Biol.* 5:7–20.

Messier, F. 1985a. Social organization, spatial distribution and population density of wolves in relation to moose density. *Can. J. Zool.* 63:1068–77.

———. 1985b. Solitary living and extra-territorial movements of wolves in relation to social status and prey abundance. *Can. J. Zool.* 63:239–45.

———. 1991. The significance of limiting and regulating factors on the demography of moose and white-tailed deer. *J. Anim. Ecol.* 60:377–93.

———. 1994. Ungulate population models with predation: A case study with the North American moose. *Ecology* 75:478–88.

———. 1995a. Is there evidence for cumulative effect of snow on moose and deer populations? *J. Anim. Ecol.* 64:136–40.

———. 1995b. On the functional and numerical responses of wolves to changing prey density. Pp. 187–97 in L. N. Carbyn, S. H. Fritts, and D. R. Seip, eds., *Ecology and conservation of wolves in a changing world.* Canadian Circumpolar Institute, Edmonton, Alberta.

Messier, F., and M. Crête. 1985. Moose *(Alces alces)* and wolf *(Canis lupus)* dynamics and the natural regulation of moose populations. *Oecologia* 65:503–12.

Messier, F., J. Huot, D. Le Henaff, and S. Luttich. 1988. Demography of the George River Caribou Herd: Evidence of population regulation by forage exploitation and range expansion. *Arctic* 41:279–87.

Messier, F., and D. Joly. 2000. The regulation of moose populations by wolf predation: A comment. *Can. J. Zool.* 78: 506–10.

Messier, F., M. E. Rau, and M. A. McNeill. 1989. *Echinococcus granulosus* (Cestoda: Taeniidae) infections and moose–wolf population dynamics in southwestern Quebec. *Can. J. Zool.* 67:216–19.

Meunier, P. C., L. T. Glickman, M. J. G. Appel, and S. J. Shin. 1981. Canine parvovirus in a commercial kennel: Epidemiologic and pathologic findings. *Cornell Vet.* 71:96–110.

Meyer, H., W. Drochner, M. Schmidt, R. Riklin, and A. Thomee. 1980. On the pathogenesis of alimentary disorders in dogs. Pp. 189–99 in R. S. Anderson, ed., *Nutrition of the dog and cat.* Pergamon Press, Oxford.

Meyer, H., and G. Stadtfeld. 1980. Investigations on the body and organ structure of dogs. Pp. 15–30 in R. S. Anderson, ed., *Nutrition of the dog and cat.* Pergamon Press, Oxford.

Michigan Department of Natural Resources. 1997. Michigan gray

wolf recovery and management plan. Wildlife Division, Michigan Department of Natural Resources, Lansing.

Miller, B., R. P. Reading, and S. Forrest. 1996. *Prairie night: Black-footed ferrets and the recovery of endangered species.* Smithsonian Institution Press, Washington, D.C.

Miller, D. R. 1975. Observations of wolf predation on barren ground caribou in winter. Pp. 209–20 in J. R. Luick, P. C. Lent, D. R. Klein, and R. G. White, eds., Proceedings of the 1st International Reindeer and Caribou Symposium. University of Alaska, Fairbanks.

Miller, F. L. 1978. Interactions between men, dogs, and wolves on western Queen Elizabeth Island, Northwest Territories, Canada. *Musk-ox* 22:70–72.

———. 1983. Wolf-related caribou mortality on a calving ground in north-central Canada. Pp. 100–101 in L. N. Carbyn, ed., *Wolves in Canada and Alaska: Their status, biology, and management.* Report Series, no. 45. Canadian Wildlife Service, Edmonton, Alberta.

———. 1995. Status of wolves in the Canadian Arctic Islands. Pp. 35–42 in L. N. Carbyn, S. H. Fritts, and D. R. Seip, eds. *Ecology and conservation of wolves in a changing world.* Canadian Circumpolar Institute, Edmonton, Alberta.

Miller, F. L., and E. Broughton. 1974. Calf mortality on the calving ground of Kaminuriak caribou. Report Series, no. 26. Canadian Wildlife Service.

Miller, F. L., A. Gunn, and E. Broughton. 1985. Surplus killing as exemplified by wolf predation on newborn caribou. *Can. J. Zool.* 63:295–300.

Miller, F. L., and R. H. Russel. 1977. Unreliability of strip aerial surveys for estimating number of wolves on Western Queen Elizabeth Islands, Northwest Territories. *Can. Field Nat.* 91:77–81.

Miller, P. E., and C. J. Murphy. 1995. Vision in dogs. *J. Am. Vet. Med. Assoc.* 207:1623–34.

Minnesota Department of Natural Resources. 2001. Minnesota wolf management plan. Minnesota Department of Natural Resources, Division of Wildlife, St. Paul. 36 pp.

Mirachi, F. E., B. E. Howland, P. F. Scanlon, R. L. Kirkpatrick, and L. M. Stanford. 1978. Seasonal variation in plasma LH, FSH, prolactin, and testosterone concentrations in adult male white-tailed deer. *Can. J. Zool.* 56:121–27.

Mitsuzuka, M. 1987. Collection, evaluation and freezing of wolf semen. Pp. 127–41 in H. Frank, ed., *Man and wolf: Advances, issues and problems in captive wolf research.* Dr. W. Junk Publishers, Dordrecht, The Netherlands.

Mladenoff, D. J., and T. A. Sickley. 1998. Assessing potential Gray wolf restoration in the northeastern United States: A spatial prediction of favorable habitat and potential population levels. *J. Wildl. Mgmt.* 62:1–10.

Mladenoff, D. J., T. A. Sickley, R. G. Haight, and A. P. Wydeven. 1995. A regional landscape analysis and prediction of favorable gray wolf habitat in the northern Great Lakes region. *Conserv. Biol.* 9:279–94.

Mladenoff, D. J., T. A. Sickley, and A. D. Wydeven. 1999. Predicting gray wolf landscape recolonization: Logistic regression models vs. new field data. *Ecol. Appl.* 9:37–44.

Moehlman, P. D. 1986. Ecology of cooperation in canids. Pp. 64–86 in D. I. Rubenstein and R. W. Wrangham, eds., *Ecological aspects of social evolution.* Princeton University press, Princeton, NJ.

———. 1987. Social organization in jackals. *Am. Sci.* 75:366–75.

———. 1989. Intraspecific variation in canid social systems. Pp. 143–63 in J. L. Gittleman, ed., *Carnivore behavior, ecology and evolution.* Cornell University Press, Ithaca, NY.

Moehlman, P. D., and H. Hofer. 1997. Cooperative breeding, reproductive suppression, and body mass in canids. Pp. 76–128 in N. G. Solomon and J. A. French, eds., *Cooperative breeding in mammals.* Cambridge University Press, Cambridge.

Moger, W. H., L. E. Ferns, J. R. Wright, S. Gadbois, and P. J. McLeod. 1998. Elevated urinary cortisol in a timber wolf *(Canis lupus):* A result of social behaviour or adrenal pathology? *Can. J. Zool.* 76:1957–59.

Molvar, E. M., and R. T. Bowyer. 1994. Costs and benefits of group living in a recently social ungulate: The Alaskan moose. *J. Mammal.* 75:621–30.

Monson, R. A., and W. B. Stone. 1976. Canine distemper in wild carnivores in New York. *Fish Game J.* 23:149–54.

Montagna, W., and H. F. Parks. 1948. A histochemical study of the glands of the anal sac of the dog. *Anat. Rec.* 100:297–318.

Montana Agricultural Statistics Service. 1992. Montana agricultural statistics: Historic state series. Montana Department of Agriculture and U.S. Department of Agriculture National Agricultural Statistics Service, Helena, Montana. 128 pp.

Moore, A. J. 1993. Towards an evolutionary view of social dominance. *Anim. Behav.* 46:594–96.

Moore, G. C., and G. R. Parker. 1992. Colonization by the eastern coyote *(Canis latrans).* Pp. 23–28 in A. H. Boer, ed., *The ecology and management of the eastern coyote.* University of New Brunswick Press, Frederickton.

Moran, G. 1978. The structure of movement in supplanting interactions in the wolf. Ph.D. dissertation, Dalhousie University, Halifax, NS.

———. 1982. Long-term patterns of agonistic interactions in a captive group of wolves. *Anim. Behav.* 30:75–83.

———. 1987. Dispensing with the fashionable fallacy of dispensing with description in the study of wolf social behavior. Pp. 205–18 in H. Frank, ed., *Man and wolf: Advances, issues and problems in captive wolf research.* Dr. W. Junk Publishers, Dordrecht, The Netherlands.

Moran, G., and J. C. Fentress. 1979. A search for order in wolf social behavior. Pp. 245–83 in E. Klinghammer, ed., *The behavior and ecology of wolves.* Garland STPM Press, New York.

Moran, G., J. C. Fentress, and I. Golani. 1981. A description of relational patterns during "ritualized fighting" in wolves. *Anim. Behav.* 29:1146–65.

Morey, D. F. 1986. Studies on Amerindian dogs: Taxonomic analysis of canid crania from the northern plains. *J. Archaeol. Sci.* 13:119–45.

———. 1992. Size, shape, and development in the evolution of the domestic dog. *J. Archaeol. Sci.* 119:181–204.

———. 1994a. *Canis* remains from Dust Cave. *J. Alabama Archaeol.* 40:163–72.

———. 1994b. The early evolution of the domestic dog. *Am. Sci.* 82:336–47.

Morgan, G. S., and R. C. Hulbert. 1995. Overview of the geology and vertebrate biochronology of the Leisey Shell Pit Local Fauna, Hillsborough County, Florida. *Bull. Fla. Mus. Nat. Hist.* 37 Pt. I(1): 1–92.

Moritz, C. C. 1994. Defining "evolutionarily significant units" for conservation. *Trends Ecol. Evol.* 9:373–75.

———. 1995. Uses of molecular phylogenies for conservation. *Phil. Trans. R. Soc. Lond. B* 349:113–18.

Morley, J. E., A. S. Levine, E. D. Plotka, and U. S. Seal. 1983. The effect of naloxone on feeding and spontaneous locomotion in the wolf. *Physiol. Behav.* 30:331–34.

Morozov, F. N. 1951. Helminth of wolves in the Mordva National Park. *Trudi gelimint. Lab Akademii Nauk SSSR* 5:146–49.

Morton, E. S. 1975. Ecological sources of selection on avian sounds. *Am. Nat.* 109:17–34.

———. 1977. On the occurrence and significance of motivation-structural rules in some bird and mammal sounds. *Am. Nat.* 111:855–69.

Moulton, D. G. 1967. Olfaction in mammals. *Am. Zool.* 7:421–29.

Moulton, D. G., and D. A. Marshall. 1976. The performance of dogs in detecting alpha-ionone in the vapor phase. *J. Comp. Physiol., A.* 110:287–306.

Mowat, F. 1963. *Never cry wolf.* McClelland and Stewart, Toronto.

Mulders, R. 1997. Geographic variation in the cranial morphology of the wolf *(Canis lupus)* in northern Canada. Master's thesis, Laurentian University, Sudbury, Ontario. 129 pp.

Müller-Schwarze, D. 1971. Pheromones in black-tailed deer *(Odocoileus hemionus columbianus). Anim. Behav.* 19:141–52.

———. 1972. Responses of young black-tailed deer to predator odors. *J. Mammal.* 53:393–94.

Muneer, M. A., I. O. Farah, K. A. Pomeroy, S. M. Goyal, and L. D. Mech. 1988. Detection of parvoviruses in wolf feces by electron microscopy. *J. Wildl. Dis.* 24:170–72.

Munro, J. A. 1947. Observations of birds and mammals in central British Columbia. Occasional Papers of the British Columbia Provincial Museum, no. 6. 165 pp.

Munthe, K., and J. H. Hutchison. 1978. A wolf-human encounter on Ellesmere Island, Canada. *J. Mammal.* 59:876–78.

Murie, A. 1944. The wolves of Mount McKinley. U.S. National Park Service Fauna Series, no. 5. U.S. Government Printing Office, Washington, D.C. 238 pp.

———. 1963. *A naturalist in Alaska.* Doubleday, New York.

Murphy, K. M. 1998. The ecology of the cougar *(Puma concolor)* in the northern Yellowstone Ecosystem: Interaction with prey, bears, and humans. Ph.D. dissertation, University of Idaho, Moscow. 147 pp.

Murphy, K. M., P. I. Ross, and M. G. Hornocker. 1999. The ecology of anthropogenic influences on cougars. Pp. 77–101 in T. W. Clark, A. P. Curlee, S. C. Minta, and P. M. Kareiva, *Carnivores in ecosystems: The Yellowstone experience.* Yale University Press, New Haven, CT.

Murray, A. 1975. Influence of education programs on wolf conservation in Canada. Pp. 113–19 in D. H. Pimlott, ed., Wolves: Proceedings of the First Working Meeting of Wolf Specialists and of the First International Conference on the Conservation of the Wolf. Supplementary Paper no. 43. IUCN, Morges, Switzerland. 145 pp.

Musiani, M., C. Mamo, L. Boitani, C. Callaghan, C. Gates, L. Mattei, E. Visalberghi, S. Breck, and G. Volpi. In press. Human-wolf conflicts in western Canada and USA: Can fladry barriers protect livestock? *Conserv. Biol.*

Musiani, M., H. Okarma, and W. Jedrzejewski. 1998. Speed and actual distance travelled by radiocollared wolves in Bialowieza Primeval Forest (Poland). *Acta Theriol.* 43:409–16.

Musil, R. 1974. Tiergesellschaft der Kniegrotte. Pp. 30–72 in R. Feustel, ed., Die Kniegrotte, eine Magdalenien-Station in Thuringen. Veroffentlichungen des Museum fur Ur- und Fruhgeschichte Thuringens 5.

Nagy, K. A. 1994. Field bioenergetics of mammals: What determines field metabolic rates. *Aust. J. Zool.* 42:43–53.

Nagy, K. A., I. A. Girard, and T. K. Brown. 1999. Energetics of free-ranging mammals, reptiles, and birds. *Annu. Rev. Nutr.* 19:247–77.

Nash, R. 1967. *Wilderness and the American mind.* Yale University Press, New Haven, CT.

National Agricultural Statistics Board, U.S. Department of Agriculture. 1992. Cattle and calves death loss. Report, May 1, 1992, Washington, D.C.

National Geographic Explorer. 1988. *White wolf.* National Geographic Society, Washington, D.C. (Video.)

National Post. 2000. Wolf attacks camper. 4 July.

National Research Council. 1997. *Wolves, bears, and their prey in Alaska: Biological and social challenges in wildlife management.* National Academy Press, Washington, D.C.

Neff, W. D., and J. E. Hind. 1955. Auditory thresholds of the cat. *J. Acoust. Soc. Am.* 27:480–83.

Neiland, K. A. 1970. Rangiferine brucellosis in Alaskan canids. *J. Wildl. Dis.* 6:136–39.

———. 1975. Further observations on rangiferine brucellosis in Alaskan carnivores. *J. Wildl. Dis.* 11:45–53.

Neiland, K. A., and L. G. Miller. 1981. Experimental *Brucella suis* type 4 infections in domestic and wild Alaskan carnivores. *J. Wildl. Dis.* 17:183–89.

Neitz, J., T. Geist, and G. H. Jacobs. 1989. Color vision in the dog. *Vis. Neurosci.* 3:119–25.

Nelson, E., and C. Franson. 1988. Timber wolf recovery in Wisconsin: The attitudes of northern Wisconsin farmers and landowners. Research Management Findings, no. 13. Wisconsin Department of Natural Resources.

Nelson, M. E., and L. D. Mech. 1981. Deer social organization and wolf predation in northeastern Minnesota. Wildlife Monographs, no. 77. The Wildlife Society, Washington, D.C. 53 pp.

———. 1984. Observation of a swimming wolf killing a swimming deer. *J. Mammal.* 65:143–44.

———. 1985. Observation of a wolf killed by a deer. *J. Mammal.* 66:187–88.

———. 1986a. Deer population in the central Superior National Forest, 1967–1985. USDA Forest Service Research Paper NC-271. North Central Forest Experiment Station, St. Paul, MN. 8 pp.

———. 1986b. Mortality of white-tailed deer in northeastern Minnesota. *J. Wildl. Mgmt.* 50:691–98.

———. 1986c. Relationship between snow depth and gray wolf predation on white-tailed deer. *J. Wildl. Mgmt.* 50:471–74.

———. 1991. Wolf predation risk associated with white-tailed deer movements. *Can. J. Zool.* 69:2696–99.

———. 1993. Prey escaping wolves, *Canis lupus,* despite close proximity. *Can. Field Nat.* 107:245–46.

———. 1994. A single deer stands off three wolves. *Am. Midl. Nat.* 131:207–8.

———. 2000. Proximity of white-tailed deer, *Odocoileus virginianus,* home ranges to wolf, *Canis lupus,* pack homesites. *Can. Field Nat.* 114:503–4.

Nelson, R. K. 1983. *Make prayers to the raven: A Koyukon view of the Northern Forest.* University of Chicago Press, Chicago.

Neuhaus, W. 1953. Über die Riechscharfe des Hundes für Fettsauren. *Z. Vergl. Physiol.* 35:527–52.

Neuhaus, W., and E. Regenfuss. 1963. Über die Sehscharfe des Haushundes bei verschiedenen Helligkeiten. *Z. Vergl. Physiol.* 57:137–46.

Newsom, W. M. 1926. *White-tailed deer.* Scribner's, New York.

Nicolaides, N. 1974. Skin lipids: Their biochemical uniqueness. *Science* 186:19–26.

Nie, M. A. 2003. *Beyond wolves: The politics of wolf recovery and management.* University of Minnesota Press, Minneapolis. 247 pp.

Nielsen, S. W. 1953. Glands of the canine skin: Morphology and distribution. *Am. J. Vet. Res.* 14:448–54.

Niemeyer, C., E. E. Bangs, S. H. Fritts, J. A. Fontaine, M. D. Jimenez, and W. G. Brewster. 1994. Wolf depredation management in relation to wolf recovery. Pp. 57–60 in W. S. Halverson and A. C. Crabb, eds., Proceedings of the 16th Vertebrate Pest Conference (University of California, Davis).

Nikol'skii, A. A., and K. H. Frommolt. 1989. Wolf vocal activity. [In Russian, translation by T. Neklioudova.] Moscow University, Moscow. 126 pp.

Nikol'skii, A. A., K. H. Frommolt, and V. P. Bologov. 1986. Sound response of the she-wolf taking her wolf-cubs away from danger. [In Russian, translation by T. Neklioudova.] *Biulleten Moskovskogo Obshchestva Ispytatelei Prirody (Otdel Biologicheskii)* 91:53–55.

Nobis, G. 1986. Die Wildsaugetiere in der Umwelt des Menschen von Oberkassel bei Bonn und das Domestikations problem von Wolfen im Jungpalaolithifum. *Bonner Jahrb.* 1986:367–76.

Noffsinger, R. E., R. A. Laney, A. M. Nichols, D. L. Stewart, and D. W. Steffeck. 1984. Prulean Farms Wildlife Coordination Act. U.S. Fish and Wildlife Service, Raleigh, NC.

Nowak, R. M. 1972. The mysterious wolf of the south. *Nat. Hist.* 81:51–53,74–77.

———. 1978. Evolution and taxonomy of coyotes and related *Canis.* Pp. 3–16 in M. Bekoff, ed., *Coyotes: Biology, behavior, and management.* Academic Press, New York.

———. 1979. North American quaternary *Canis.* Monograph

no. 6. Museum of Natural History, University of Kansas, Lawrence. 154 pp.

———. 1983. A perspective on the taxonomy of wolves in North America. Pp. 10–19 in L. N. Carbyn, ed., *Wolves in Canada and Alaska: Their status, biology, and management.* Report Series, no. 45. Canadian Wildlife Service, Edmonton, Alberta.

———. 1991. A response to O'Brien and Mayr. *Science* 251:1187–88.

———. 1992a. The red wolf is not a hybrid. *Conserv. Biol.* 6:593–95.

———. 1992b. Wolves: The great travelers of evolution. *Int. Wolf* 2(4): 3–7.

———. 1994. Dogs and wolves—and coyotes. *Int. Wolf* 4(3): 2.

———. 1995a. Another look at wolf taxonomy. Pp. 375–97 in L. N. Carbyn, S. H. Fritts, and D. R. Seip, eds., *Ecology and conservation of wolves in a changing world.* Canadian Circumpolar Institute, Edmonton, Alberta.

———. 1995b. Hybridization: The double-edged threat. *Canid News* (Newsletter of the IUCN/SSC Canid Specialist Group) 3:2–6.

———. 1999. *Walker's mammals of the world.* Johns Hopkins University Press, Baltimore.

———. 2002. The original status of wolves in eastern North America. *Southeastern Nat.* 1:95–130.

Nowak, R. M., and N. E. Federoff. 1996. Systematics of wolves in eastern North America. Pp. 187–203 in N. Fascione and M. Cecil, eds., Proceedings of the Defenders of Wildlife's Wolves of America Conference, Albany, NY. Defenders of Wildlife, Washington, D.C.

———. 1998. Validity of the red wolf: Response to Roy et al. *Conserv. Biol.* 12:722–25.

———. 2002. The systematic status of the Italian wolf *Canis lupus. Acta Theriol.* 47:333–38.

Nowak, R. M., M. K. Phillips, V. G. Henry, W. C. Hunter, and R. Smith. 1995. The origin and fate of the red wolf. Pp. 409–15 in L. N. Carbyn, S. H. Fritts, and D. R. Seip, eds., *Ecology and conservation of wolves in a changing world.* Canadian Circumpolar Institute, Edmonton, Alberta.

Nudds, T. D. 1978. Convergence of group size strategies by mammalian social carnivores. *Am. Nat.* 112:957–60.

Obbard, M. E., J. G. Jones, R. Newman, A. Booth, J. Saterthwaite, and G. Linscombe. 1987. Furbearer harvests in North America. Pp. 1007–34 in M. Novak, J. A. Baker, M. E. Obbard, and B. Malloch, eds., *Wild furbearer management and conservation in North America.* Ontario Ministry of Natural Resources, Toronto.

O'Brien, S. J., and E. Mayr. 1991. Bureaucratic mischief: Recognizing endangered species and subspecies. *Science* 251:1187–88.

Oftedal, O. T., and J. L. Gittleman. 1989. Patterns of energy output during reproduction. Pp. 355–78 in J. L. Gittleman, ed., *Carnivore behavior, ecology and evolution.* Cornell University Press, Ithaca, NY.

Okarma, H. 1989. Distribution and numbers of wolves in Poland. *Acta Theriol.* 34:497–503.

———. 1992. The wolf—monograph of a species. [In Polish.] Bialowieza, Poland. 168 pp.

———. 1993. Status and management of the wolf in Poland. *Biol. Conserv.* 66:153–58.

———. 1995. The trophic ecology of wolves and their predatory role in ungulate communities of forest ecosystems in Europe. *Acta Theriol.* 40:335–86.

Okarma, H., B. Jedrzejewska, W. Jedrzejewski, Z. A. Krasinski, and L. Milkowski. 1995. The roles of predation, snowcover, acorn crop, and man-related factors on ungulate mortality in Bialowieza Primeval Forest, Poland. *Acta Theriol.* 40(2): 197–217.

Okarma, H., and W. Jedrzejewski. 1997. Livetrapping wolves with nets. *Wildl. Soc. Bull.* 25(1): 78–82.

Okarma, H., W. Jedrzejewski, K. Schmidt, S. Sniezko, A. N. Bunevich, and B. Jedrzejewska. 1998. Home ranges of wolves in Bialowieza Primeval Forest, Poland, compared with other Eurasian populations. *J. Mammal.* 79:842–52.

Okarma, H. K., and P. Koteja. 1987. Basal metabolic rate in the gray wolf in Poland. *J. Wildl. Mgmt.* 51:800–801.

Olivier, M., M. Breen, M. M. Binns, and G. Lust. 1999. Localization and characterization of nucleotide sequences from the canine Y chromosome. *Chromosome Res.* 7:223–33.

Olivier, M., and G. Lust. 1998. Two DNA sequences specific for the canine Y chromosome. *Anim. Genet.* 29:146–49.

Olsen, S. J. 1985. *Origins of the domestic dog: The fossil record.* University of Arizona Press, Tucson.

Olsen, S. J., and J. W. Olsen. 1977. The Chinese wolf, ancestor of New World dogs. *Science* 197:533–35.

Olson, S. F. 1938. Organization and range of the pack. *Ecology* 19:168–70.

Olsson, O., J. Wirtberg, M. Andersson, and I. Wirtberg. 1997. Wolf *Canis lupus* predation on moose *Alces alces* and roe deer in south-central Scandinavia. *Wildl. Biol.* 3(1): 13–25.

O'Neill, B. B. 1988. The law of the wolves. *Environ. Law* 18:227–40.

Oosenbrug, S. M., and L. N. Carbyn. 1982. Winter predation on bison and activity patterns of a wolf pack in Wood Buffalo National Park. Pp. 43–53 in F. H. Harrington and P. C. Paquet, eds., *Wolves of the world: Perspectives of behavior, ecology, and conservation.* Noyes Publications, Park Ridge, NJ.

Oosenbrug, S. M., L. N. Carbyn, and D. West. 1980. Wood Buffalo National Park: Wolf/Bison Studies. Progress Report no. 2. Canadian Wildlife Service.

Ortalli, G. 1973. Natura, storia e mitografia del lupo nel Medioevo. *La Cultura* 9:257–311.

Osgood, C. 1936. Contributions to the ethnography of the Kutchin. Yale University Publications in Anthropology, no. 14. 189 pp.

Ossent, P., F. Mettler, and E. Isenbugel. 1984. Retained cartilage in the ulnar metaphysis with deformation of the forelegs in two litters of captive wolves. *Zentralbl. Veterinaermed. Reihe A.* 31:241–50.

Østerberg, G. 1935. Topography of the layer of rods and cones in the human retina. *Acta Ophthalmol.* (Suppl.) 13:1–103.

Osterholm, H. 1964. The significance of distance receptors in the feeding behaviour of the fox, *Vulpes vulpes* L. *Acta Zool. Fenn.* 106:1–31.

Overall, K. L. 2001. Pharmacological treatment in behavioural medicine: The importance of neurochemistry, molecular biology and mechanistic hypotheses. *Vet. J.* 162:9–23.

Ozoga, J. J., and L. J. Verme. 1986. Relation of maternal age to fawn-rearing success in white-tailed deer. *J. Wildl. Mgmt.* 50:480–86.

Packard, J. M. 1980. Deferred reproduction in wolves *(Canis lupus).* Ph.D. dissertation, University of Minnesota, St. Paul. 347 pp.

———. 1989. Olfaction, ovulation, and sexual competition in monogamous mammals. Pp. 525–43 in J. M. Lakoski, J. R. Perez-Polo, and D. K. Rassin, eds., *Neural control of reproductive function.* Alan R. Liss, New York.

Packard, J. M., and L. D. Mech. 1980. Population regulation in wolves. Pp. 135–50 in M. N. Cohen, R. S. Malpass, and H. G. Klein, eds., *Biosocial mechanisms of population regulation.* Yale University Press, New Haven, CT.

———. 1983. Population regulation in wolves. Pp. 151–74 in F. L. Bunnell, D. S. Eastman, and J. M. Peek, eds., Symposium on Natural Regulation of Wildlife Populations. Forestry, Wildlife, and Range Experiment Station, Moscow, Idaho.

Packard, J. M., L. D. Mech, and R. R. Ream. 1992. Weaning in an arctic wolf pack: Behavioral mechanisms. *Can. J. Zool.* 70:1269–75.

Packard, J. M., L. D. Mech, and U. S. Seal. 1983. Social influences on reproduction in wolves. Pp. 78–85 in L. N. Carbyn, ed., *Wolves in Canada and Alaska: Their status, biology and management.* Report Series, no. 45. Canadian Wildlife Service, Edmonton, Alberta.

Packard, J. M., U. S. Seal, L. D. Mech, and E. D. Plotka. 1985. Causes of reproductive failure in two family groups of wolves *(Canis lupus). Z. Tierpsychol.* 68(1): 24–40.

Packer, C., L. Herbst, A. E. Pusey, D. J. Bygott, J. P. Hanby, S. J. Cairns, and M. Borgerhoff Mulder. 1988. Reproductive success of lions. Pp. 363–83 in T. Clutton-Brock, ed., *Reproductive success: Studies of individual variation in contrasting breeding systems.* University of Chicago Press, Chicago.

Packer, C., and L. Ruttan. 1988. The evolution of cooperative hunting. *Am. Nat.* 132:159–98.

Packer, C., D. Scheel, and A. E. Pusey. 1990. Why lions form groups: Food is not enough. *Am. Nat.* 136:1–19.

Palmén, J. A. 1913. Bar, Wolf and Luchs in Finland. *Zool. Beobachter* 54(3): 1–6.

Palomares, F., and T. M. Caro. 1999. Interspecific killing among mammalian carnivores. *Am. Nat.* 153:492–508.

Pamilo, P., and M. Nei. 1988. Relationships between gene trees and species trees. *Mol. Biol. Evol.* 5:568–83.

Panin, V. Ia., and L. I. Lavrov. 1962. K gel'mintofaune volkov Kazakhstana. In Parazity dikikh zhivotnykh Kasakhstana. *Trudy Inst. Zool., Akad. Nauk Kazakh. S.S.R.* 26:57–62.

Papageorgiou, N., C. Vlachos, A. Sfougaris, and S. Tsachalidis. 1994. Status and diet of wolves in Greece. *Acta Theriol.* 39:411–16.

Paquet, P. C. 1991a. Scent-marking behavior of sympatric wolves *(Canis lupus)* and coyotes *(C. latrans)* in Riding Mountain National Park. *Can. J. Zool.* 69:1721–27.

———. 1991b. Winter spatial relationship of wolves and coyotes in Riding Mountain National Park, Manitoba. *J. Mammal.* 72:397–401.

———. 1992. Prey use strategies of sympatric wolves and coyotes in Riding Mountain National Park, Manitoba. *J. Mammal.* 73:337–43.

———. 1993. Summary reference document—ecological studies of recolonizing wolves in the Central Rocky Mountains. Unpublished Report by John/Paul and Assoc. Canadian Parks Service, Banff, Alberta.

Paquet, P. C., S. Bragdon, and S. McCusker. 1982. Cooperative rearing of simultaneous litters in captive wolves. Pp. 223–37 in F. H. Harrington and P. C. Paquet, eds., *Wolves of the world: Perspectives of behavior, ecology, and conservation.* Noyes Publications, Park Ridge, NJ.

Paquet, P. C., and C. Callaghan. 1996. Effects of linear developments on winter movements of gray wolves in the Bow River Valley of Banff National Park, Alberta. Pp. 1–21 in G. L. Evink, P. Garrett, D. Zeigler and J. Berry, eds., *Trends in addressing transportation related wildlife mortality.* Proceedings of the Transportation Related Wildlife Mortality Seminar. Department of Transportation, Environmental Office, Tallahassee, FL.

Paquet, P. C., and L. N. Carbyn. 1986. Wolves, *Canis lupus,* killing denning black bears, *Ursus americanus,* in the Riding Mountain National Park area. *Can. Field Nat.* 100:371–72.

Paradiso, J. L. 1966. Recent records of coyotes, *Canis latrans,* from the southeastern United States. *Southwestern Nat.* 11:500–501.

———. 1968. Canids recently collected in east Texas, with comments on the taxonomy of the red wolf. *Am. Midl. Nat.* 80:529–34.

Paradiso, J. L., and R. M. Nowak. 1971. A report on the taxonomic status and distribution of the red wolf. Wildlife Report, no. 145. U.S. Fish and Wildlife Service, Washington, D.C.

———. 1972. *Canis rufus.* Mammalian species, 22. American Society of Mammalogists.

Parker, G. R. 1973. Distribution and densities of wolves within barren-ground caribou range in northern mainland Canada. *J. Mammal.* 54:341–48.

Parker, W. T., M. P. Jones, and P. G. Poulos. 1986. Determination of experimental population status for an introduced population of wolves in North Carolina: Final rule. *Fed. Reg.* 51: 41790–96.

Parker, W. T., and M. K. Phillips. 1991. Application of the experimental population designation to recovery of endangered red wolves. *Wildl. Soc. Bull.* 19:73–79.

Parks, H. 1950. Morphological and cytochemical observations on the circumanal glands of the dog. Ph.D. dissertation, Cornell University, Ithaca, NY.

Parmelee, D. F. 1964. Myth of the wolf. *Beaver* 295:4–9.

Parry, H. 1953. Degenerations of the dog retina. I. Structure and development of the retina of the normal dog. *Brit. J. Ophthalmol.* 37:385–404.

Parsons, D. R. 1997. Establishment of a nonessential experimental population of the Mexican gray wolf in Arizona and New Mexico: B: Final rule. *Fed. Reg.* 63:1752–72.

Parsons, D. R., and J. E. Nicholopoulos. 1995. An update of the status of the Mexican wolf recovery program in the United States. Pp. 141–46 in L. N. Carbyn, S. H. Fritts, and D. R. Seip, eds., *Ecology and conservation of wolves in a changing world.* Canadian Circumpolar Institute, Edmonton, Alberta.

Pasitchniak-Arts, M., M. E. Taylor, and L. D. Mech. 1988. Skeletal injuries in an adult arctic wolf. *Arctic Alpine Res.* 20: 360–65.

Patalano, M., and S. Lovari. 1993. Food habits and trophic niche overlap of the wolf (*Canis lupus*, L. 1758) and the red fox (*Vulpes vulpes* L. 1758) in a mediterranean mountain area. *Rev. Ecol.* (Terre Vie) 48:279–94.

Pate, J., M. J. Manfredo, A. D. Bright, and G. Tischbein. 1996. Coloradans' attitudes toward reintroduction of the gray wolf into Colorado. *Wildl. Soc. Bull.* 24:421–28.

Paul, W. J., and P. S. Gipson. 1994. Wolves. In *Prevention and control of wildlife damage.* University of Nebraska Cooperative Extension, Institute of Agriculture and Natural Resources, Lincoln.

Pavlov, I. P. 1927. *Conditioned reflexes.* [Translated by G. V. Anrep.] Oxford University Press, London.

Pavlov, M. P. 1990. Volk. Agropromizdat, Moskva, 1–350. [In Russian.]

Peace River Films, Inc. 1975. *Following the tundra wolf.* LIVE Home Video Inc., Van Nuys, CA. (Video, 45 min.)

Pearsall, M. D., and H. Verbruggen. 1982. *Scent: Training to track, search and rescue.* Alpine Publications, Inc., Loveland, CO. 225 pp.

Pearson, J. C. 1956. Studies on the life cycles and morphology of the larval stages of *Alaria arisaemoides* Augustine and Uribo, 1927, and *Alaria canis* LaRue and Fallis, 1936 (Trematoda: Diplostomidae). *Can. J. Zool.* 34:295–387.

Pedersen, S. 1982. Geographical variation in Alaskan wolves. Pp. 345–61 in F. H. Harrington and P. C. Paquet, eds., *Wolves of the world: Perspectives of behavior, ecology, and conservation.* Park Ridge, NJ: Noyes Publications, Park Ridge, NJ.

Peek, J. M., D. E. Brown, S. R. Kellert, L. D. Mech, J. D. Shaw, and V. Van Ballenberghe. 1991. Restoration of wolves in North America. Technical Review 91-1. Wildlife Society, Bethesda, MD. 21 pp.

Peek, J. M., D. L. Urich, and R. J. Mackie. 1976. Moose habitat selection and relationships to forest management in northeastern Minnesota. Wildlife Monographs, no. 48. The Wildlife Society, Washington, D.C. 65 pp.

Peichl, L. 1989. Dog retinal ganglion cells: Morphological types and breed differences. *Soc. Neurosci. Abstr.* 15:1207.

———. 1991. Catecholaminergic amacrine cells in the dog and wolf retina. *Vis. Neurosci.* 7:575–87.

———. 1992a. Morphological types of ganglion cells in the dog and wolf retina. *J. Comp. Neurol.* 324:590–602.

———. 1992b. Topography of ganglion cells in the dog and wolf retina. *J. Comp. Neurol.* 324:603–20.

Pence, D. B., and J. W. Custer. 1981. Host-parasite relationships in the wild canidae of North America. II. Pathology of infectious diseases in the genus *Canis.* Pp. 760–845 in J. A. Chapman and D. Pursley, eds., Worldwide Furbearer Conference Proceedings, vol. 2.

Pence, D. B., J. W. Custer, and C. J. Carley. 1981. Ectoparasites of wild canids from the Gulf coastal prairies of Texas and Louisiana. *J. Med. Entomol.* 18:409–12.

Person, D. K., R. T. Bowyer, and V. Van Ballenberghe. 2001. Density dependence of ungulates and functional responses of wolves: Effects on predator-prey ratios. *Alces* 37:253–73.

Persson, L. 1985. Asymmetrical competition: Are larger animals competitively superior? *Am. Nat.* 126:261–66.

Perzanowski, K. 1993. The economic aspects of wolf predation in the Bieszczady Mountains. Pp. 126–29 in C. Promberger and W. Schröder, eds., *Wolves in Europe: Status and perspectives.* Munich Wildlife Society, Ettal, Germany.

Peters, R. 1978. Communication, cognitive mapping, and strategy in wolves and hominids. Pp. 95–107 in R. L. Hall and H. S. Sharp, eds., *Wolf and man: Evolution in parallel.* Academic Press, New York.

———. 1979. Mental maps in wolf territoriality. Pp. 119–52 in E. Klinghammer, ed., *The behavior and ecology of wolves.* Garland STPM Press, New York.

Peters, R. P., and L. D. Mech. 1975a. Behavioral and intellectual adaptations of selected mammalian predators to the problem of hunting large animals. Pp. 279–300 in R. H. Tuttle, ed., Socioecology and psychology of primates. Mouton Publishers, The Hague, Paris.

———. 1975b. Scent-marking in wolves: A field study. *Am. Sci.* 63:628–37.

Peterson, E. A., W. C. Heaton, and S. D. Wruble. 1969. Levels of auditory response in fissiped carnivores. *J. Mammal.* 50: 566–79.

Peterson, E. A., W. E. Pate, and S. Wruble. 1966. Cochlear potentials in the dog: I. Differences with variations in external-ear structure. *J. Aud. Res.* 6:1–11.

Peterson, R. L. 1947. A record of a timber wolf attacking a man. *J. Mammal.* 28:294–95.

———. 1955. *North American moose.* University of Toronto Press, Toronto.

Peterson, R. O. 1974. Wolf ecology and prey relationships on Isle Royale. Ph.D. dissertation, Purdue University, Lafayette, IN.

———. 1977. Wolf ecology and prey relationships on Isle Royale. U.S. National Park Service Scientific Monograph Series, no. 11. Washington, D.C. 210 pp.

———. 1979. Social rejection following mating of a subordinate wolf. *J. Mammal.* 60:219–21.

———. 1986. Gray wolf. Pp. 951–67 in R. L. Di Silvestro, ed., *Audubon wildlife report.* National Audubon Society, New York.

———. 1995a. Wolves as interspecific competitors in canid ecology. Pp. 315–23 in L. N. Carbyn, S. Fritts, and D. R. Seip, eds., *Ecology and conservation of wolves in a changing world.* Canadian Circumpolar Institute, Edmonton, Alberta.

———. 1995b. *The wolves of Isle Royale: A broken balance.* Willow Creek Press, Minocqua, WI. 190 pp.

———. 1999. Wolf-moose interaction on Isle Royale: The end of natural regulation? *Ecol. Appl.* 9:10–16.

———. 2000. Ecological studies of wolves on Isle Royale, 1999–2000. Annual Report. Michigan Technological University, Houghton.

———. 2001. Wolves as top carnivores: New faces in new places. Pp. 151–60 in V. A. Sharpe, B. G. Norton, and S. Donnelley, eds., *Wolves and human communities.* Island Press, Washington, D.C. 321 pp.

Peterson, R. O., and D. L. Allen. 1974. Snow conditions as a parameter in moose-wolf relationships. *Nat. Can.* 101:481–92.

Peterson, R. O., A. K. Jacobs, T. D. Drummer, L. D. Mech, and D. W. Smith. 2002. Leadership behavior in relation to dominance and reproductive status in gray wolves, *Canis lupus. Can. J. Zool.* 80:1405–12.

Peterson, R. O., and R. E. Page. 1983. Wolf-moose fluctuations at Isle Royale National Park, Michigan. *Acta Zool. Fenn.* 174: 251–53.

———. 1988. The rise and fall of the Isle Royale wolves. *J. Mammal.* 69:89–99.

Peterson, R. O., R. E. Page, and K. M. Dodge. 1984. Wolves, moose, and the allometry of population cycles. *Science* 224:1350–52.

Peterson, R. O., N. J. Thomas, J. M. Thurber, J. A. Vucetich, and T. A. Waite. 1998. Population limitation and the wolves of Isle Royale. *J. Mammal.* 79:828–41.

Peterson, R. O., and J. Vucetich. 2002. Ecological studies of wolves on Isle Royale. Annual Report. Michigan Technological University, Houghton.

Peterson, R. O., J. A. Vucetich, R. E. Page, and A. Chouinard. 2003. Temporal and spatial aspects of predator-prey dynamics. *Alces.* In press.

Peterson, R. O., J. D. Woolington, and T. N. Bailey. 1984. Wolves of the Kenai Peninsula, Alaska. Wildlife Monographs, no. 88. The Wildlife Society, Bethesda, MD. 52 pp.

Pfeiffer, J. E. 1982. *The creative explosion: An inquiry into the origins of art and religion.* Harper and Row, New York.

Philips, M., and S. N. Austad. 1990. Animal communication and social evolution. Pp. 254–68 in M. Bekoff and D. Jamieson, eds., *Interpretation and explanation in the study of animal behavior,* vol. 1. Westview Press, Boulder, CO.

Phillips, D. P., W. Danilchuk, J. Ryon, and J. C. Fentress. 1990. Stereotypy of action sequence in food caching by wolves (*Canis lupus*) and coyotes (*Canis latrans*). *Am. Zool.* 30: A107.

Phillips, M. K. 1984. The cost to wolves of preying on ungulates. Proceedings of the Joint Meeting of the American Society of Mammalogists/Australian Society of Mammalogists, 9–14 July 1984. (Abstract.)

———. 1990. Measures of the value and success of a reintroduction project: Red wolf reintroduction in Alligator River National Wildlife Refuge. *Endangered Species Update* 8: 24–26.

Phillips, M. K., E. Bangs, L. D. Mech, B. Kelly, and B. Fazio. In press. Extermination and recovery of the red wolf and gray wolf in the conterminous United States. In D. Macdonald and C. Sillero, eds., *Biology and conservation of canids.* Oxford University Press, Oxford.

Phillips, M. K., and V. G. Henry. 1992. Comments on red wolf taxonomy. *Conserv. Biol.* 6:596–99.

Phillips, M. K., and J. S. Scheck. 1991. Parasitism in captive and reintroduced red wolves. *J. Wildl. Dis.* 27:498–501.

Phillips, M. K., and D. W. Smith. 1996. *The wolves of Yellowstone.* Voyageur Press, Stillwater, MN.

———. 1998. Gray wolves and private landowners in the Greater Yellowstone Area. *Trans. N. Am. Wildl. Nat. Resour. Conf.* 63:443–50.

Phillips, M. K., R. Smith, V. G. Henry, and C. Lucash. 1995. Red wolf reintroduction program. Pp. 157–68 in L. N. Carbyn, S. H. Fritts, and D. R. Seip, eds., *Ecology and conservation of wolves in a changing world.* Canadian Circumpolar Institute, Edmonton, Alberta.

Philo, L. M. 1985. Water metabolism of wolves in winter: Effect of varying food intake. *Proc. Alaska Sci. Conf.* 36:116.

———. 1987. Effects of alternate gorging-fasting. *Proc. Alaska Sci. Conf.* 38:205.

———. 1988. Water metabolism of wolves in winter: Effects of varying food intake and exercise. Ph.D. dissertation, University of Alaska, Fairbanks. 223 pp.

Pietras, R. J. 1981. Sex pheromone production by preputial gland: The regulatory role of estrogen. *Chem. Senses* 6:391–408.

Pikunov, 1981. Extent of predation of Amur tigers. Pp. 71–76 in *Rare and Endangered Land Animals of the Far East of the USSR.* DVNT's USSR, Vladivostok.

Pilgrim, K. L., D. K. Boyd, and S. H. Forbes. 1998. Testing for wolf-coyote hybridization in the Rocky Mountains using mitochondrial DNA. *J. Wildl. Mgmt.* 62:683–89.

Pimlott, D. H. 1960. The use of tape-recorded wolf howls to locate timber wolves. Presented at the 22nd Midwest Fish and Wildlife Conference, Toronto, Ontario, 5–7 December. 15 p.

———. 1961. Wolf control in Canada. *Can. Aud.* 23:145–52.

———. 1966. Review of F. Mowat's *Never Cry Wolf. J. Wildl. Mgmt.* 30:236.

———. 1967. Wolf predation and ungulate populations. *Am. Zool.* 7:267–78.

———. 1970. Predation and productivity of game populations in North America. *Trans. Int. Congr. Game Biol.* 9:63–73.

——— (ed). 1975. Wolves: Proceedings of the First Working Meeting of Wolf Specialists and of the First International Conference on the Conservation of the Wolf. Supplementary Paper no. 43. IUCN, Morges, Switzerland. 145 pp.

Pimlott, D. H., J. A. Shannon, and G. B. Kolenosky. 1969. The ecology of the timber wolf in Algonquin Provincial Park, Ontario. Research Report (Wildlife), no. 87. Ontario Department of Lands and Forests, Toronto. 92 pp.

Pinigin, A. F., and V. A. Zabrodin. 1970. On the natural nidality of brucellosis. Vest. Sel'skokhoz. Nauki (Moscow). 1970, no. 7: 96–99.

Pletscher, D. H., R. R. Ream, D. K. Boyd, D. M. Fairchild, and K. E. Kunkel. 1997. Population dynamics of a recolonizing wolf population. *J. Wildl. Mgmt.* 61:459–65.

Plimmer, H. G. 1915. Report on the deaths occurring in the Society's gardens during 1914. *Proc. Zool. Soc. Lond.* 1915:123.

———. 1916. Report on the deaths occurring in the Society's gardens during 1915. *Proc. Zool. Soc. Lond.* 1916:77.

Pliny. 1771. *Histoire Naturelle de Pline.* 12 vols.

Plotka, E. D., U. S. Seal, L. J. Verme, and J. J. Ozoga. 1983. The adrenal gland in white-tailed deer: A significant source of progesterone. *J. Wildl. Mgmt.* 47:38–44.

Ponech, C. 1997. Attitudes of area residents and various interest groups towards the Riding Mountain National Park wolf (*Canis lupus*) population, Manitoba. Master's thesis, University of Manitoba, Winnipeg. 133 pp.

Portenko, L. A. 1944. *Fauna of Anadvrskij Kray.* Chapter 3. *Mam-*

mals. Press of the Administration of the Northern Sea Route, Moscow and Leningrad, USSR.

Pospisil, J., F. Kase, and J. Vahala. 1987. Basic haematological values in carnivores. I. The canidae, the hyaenidae and the ursidae. *Comp. Biochem. Physiol. A. Comp. Physiol.* 86:649–52.

Post, E., R. O. Peterson, N. C. Stenseth, and B. E. McLaren. 1999. Ecosystem consequences of wolf behavioural responses to climate. *Nature* 401:905–7.

Post, E., N. C. Stenseth, R. O. Peterson, J. A. Vucetich, and A. M. Ellis. 2002. Phase dependence and population cycles in a large-mammal predator-prey system. *Ecology* 83(11): 2997–3002.

Potvin, F. 1988. Wolf movements and population dynamics in Papineau-Labelle reserve, Quebec. *Can. J. Zool.* 66:1266–73.

Potvin, F., L. Breton, C. Pilon, and M. Macquart. 1992. Impact of an experimental wolf reduction on beaver in Papineau-Labelle Reserve, Quebec. *Can. J. Zool.* 70:180–83.

Potvin, F., H. Jolicoeur, and J. Huot. 1988. Wolf diet and prey selectivity during two periods for deer in Quebec: Decline versus expansion. *Can. J. Zool.* 66:1274–79.

Poulle, M. L., L. Carles, and B. Lequette. 1997. Significance of ungulates in the diet of recently settled wolves in the Mercantour Mountains (Southeastern France). *Rev. Ecol.* (Terre Vie) 52:357–68.

Poulle, M. L., M. Crête, and J. Huot. 1995. Seasonal variation in body mass and composition of eastern coyotes. *Can. J. Zool.* 73:1625–33.

Poulle, M. L., T. Houard, B. Lequette, P. Havet, E. Taran, and J. C. Berthos. 1998. Wolf *(Canis lupus)* predation on moufflon *(Ovis gmelini)* and chamois *(Rupicapra rupicapra)* in the Mercantour Mountains (southeastern France) *Gibier Faune Sauvage* 15(3): 1149–59.

Poulle, M. L., B. Lequette, and T. Dahier. 1999. La recolonisation des Alpes francaises par le loup de 1992 a 1998. Office National de la Chasse, Bulletin Mensuel no. 242:4–13.

Pozio, E., A. Casulli, V. V. Bologov, G. Marucci, and G. LaRosa. 2001. Hunting practices increase the prevalence of *Trichinella* infection in wolves from European Russia. *J. Parasitol.* 87:1498–1501.

Pozio, E., G. La Rosa, P. Rossi, and R. Fico. 1989. Survival of trichinella muscle larvae in frozen wolf tissue in Italy. *J. Parasitol.* 75:472–73.

Pozio, E., G. La Rosa, F. J. Serrano, J. Barrat, and L. Rossi. 1996. Environmental and human influence on the ecology of *Trichinella spiralis* and *Trichinella britovi* in Western Eurpe. *Parasitology* 113:527–33.

Pratt, S. E., J. J. Mall, J. D. Rhoades, R. E. Hertzog, and R. M. Corwin. 1981. Dirofilariasis in a captive wolf pack. *Vet. Med. Small Anim. Clin.* 76:698–99.

Preti, G., E. L. Muetterties, J. M. Furman, J. J. Kennelly, and B. E. Johns. 1976. Volatile constituents of dog *(Canis familiaris)* and coyote *(Canis latrans)* anal sacs. *J. Chem. Ecol.* 2: 177–86.

Pritchard, J. K., M. T. Seielstad, A. Perez-Lezaun, and M. W. Feldman. 1999. Population growth of human Y chromosomes: A study of Y chromosome microsatellites. *Mol. Biol. Evol.* 16:1791–98.

Pritchard, J. K., M. Stephens, and P. Donnelly. 2000. Inference of population structure using multilocus genotype data. *Genetics* 155:945–59.

Promberger, C. 1992. Wölfe und Scavenger.—Diplomarbeit am Fachbereich Wildbiologie, Forstwissenschaftliche Fakultät der Ludwig-Maximilians-Universität München. 54 pp.

———. 1993. Research Action Plan. Pp. 120–23 in C. Promberger and W. Schröder, eds., *Wolves in Europe: Status and perspectives.* Munich Wildlife Society, Ettal, Germany.

Promberger, C., M. Dahlstrom, U. Wotschikowsky, and E. Zimen. 1993. Wolves in Sweden and Norway. Pp. 8–13 in C. Promberger and W. Schröder, eds., *Wolves in Europe: Status and perspectives.* Munich Wildlife Society, Ettal, Germany. [Based on a lecture by Petter Wabbakken.]

Promberger, C., and D. Hofer. 1994. Ein Managementplan fur Wolfe in Brandenbourg. Munich Wildlife Society, Linderhof, Germany. 99 pp.

Promberger, C., O. Ionescu, A. Mertens, M. Minca, G. Predoiu, B. Promberger-Furpab, A. Sandor, M. Scurtu, and P. Surth. 1998. Carpathian large carnivore project: Annual report 1997/1998. Munich Wildlife Society, Ettal, Germany.

Promberger, C., O. Ionescu, A. Mertens, I. Munteanu, and D. Stancu. 1995. Carpathian wolf project: Annual Report 1994/95. Munich Wildlife Society, Ettal, Germany.

Promberger, C., O. Ionescu, L. Petre, C. Roschak, P. Surth, B. Furpab, L. Todicesu, A. Sandor, M. Minca, T. Stan, H. Homm, G. Predoiu, and M. Scurtu. 1997. Carpathian large carnivore project: Annual report 1996/1997. Munich Wildlife Society, Ettal, Germany.

Promberger, C., and W. Schröder, eds., 1993. *Wolves in Europe: Status and perspectives.* Munich Wildlife Society, Ettal, Germany.

Promberger, C., C. Vogel, and M. von Loeper. 1993. Wolves in Germany. Pp. 30–34 in C. Promberger and W. Schröder, eds., *Wolves in Europe: Status and perspectives.* Munich Wildlife Society, Ettal, Germany.

Provine, R. R. 1989. Faces as releasers of contagious yawning: An approach to face detection using normal human subjects. *Bull. Psychon. Soc.* 27:211–14.

Pulliainen, E. 1965. Studies on the wolf *(Canis lupus* L.) in Finland. *Ann. Zool. Fenn.* 2:215–59.

———. 1984. Petoja ja ihmisia. 320 pp. Helsinki.

———. 1985. The expansion mechanism of the wolf *(Canis lupus)* in Northern Europe. *Rev. Ecol.* (Terre Vie) 40:157–62.

———. 1993. The wolf in Finland. Pp. 14–20 in C. Promberger and W. Schröder, eds., *Wolves in Europe: Status and perspectives.* Munich Wildlife Society, Ettal, Germany.

Pusey, A. E., and C. Packer. 1994. Non-offspring nursing in social carnivores: Minimizing the costs. *Behav Ecol.* 5:362–74.

Putman, R. J. 1984. Facts from faeces. *Mammal Rev.* 14(2): 79–97.

Queller, D. C., J. E. Strassman, and C. R. Hughes. 1993. Microsatellites and kinship. *Trends Ecol. Evol.* 8:285–88.

Quintal, P. K. M. 1995. Public attitudes and beliefs about the red wolf and its recovery in North Carolina. Master's thesis, North Carolina State University, Raleigh.

Rabb, G. B., J. H. Woolpy, and B. E. Ginsburg. 1967. Social relationships in a group of captive wolves. *Am. Zool.* 7:305–12.

Radinsky, L. B. 1981. Evolution of skull shape in carnivores, 1: Representative modern carnivores. *Biol. J. Linn. Soc.* 15:369–88.

Ragni, B., L. Mariani, A. Inverni, L. Armentano, and M. Magrini. 1985. Il lupo in Umbria. In G. Boscagli, ed., *Atti del Convegno Nazionale "Gruppo Lupo Italia,"* 1–2 May 1982, Civitella Alfedena, Italy.

Ragni, B., M. Montefameglio, and L. Ghetti. 1996. Il lupo (*Canis lupus* L.) in Umbria: Evoluzione recente della popolazione. Pp. 76–82 in F. Cecere, ed., *Atti del Convegno "Dalla parte del lupo."* Serie Atti e Studi, WWF Italia, Parma, Italy.

Ralls, K. 1971. Mammalian scent marking. *Science* 171:443–49.

Ramsay, M. A., and I. Stirling. 1984. Interactions of wolves and polar bears in northern Manitoba. *J. Mammal.* 65:693–94.

Randall, D. J., W. Burggren, and K. French. 1997. *Eckert animal physiology.* W. H. Freeman, New York.

Randi, E., F. Francisci, and V. Lucchini. 1995. Mitochondrial DNA restriction-fragment-length monomorphism in the Italian wolf (*Canis lupus*). *J. Zool. Syst. Evol. Res.* 33:97–100.

Randi, E., and V. Lucchini. 2002. Detecting rare introgression of domestic dog genes into wild wolf (*Canis lupus*) populations by Bayesian admixture analyses of microsatellite variation. *Conserv. Genet.* 3:31–45.

Randi, E., V. Lucchini, M. F. Christensen, N. Mucci, S. M. Funk, G. Dolf, and V. Loeschcke. 2000. Mitochondrial DNA variability in Italian and East European wolves: Detecting the consequences of small population size and hybridization. *Conserv. Biol.* 14:464–73.

Randi, E., V. Lucchini, and F. Francisci. 1993. Allozyme variability in the Italian wolf (*Canis lupus*) population. *Heredity* 71:516–22.

Ranson, E. W., Jr. 1981. A developmental analysis of urinary behavior in dogs. Ph.D. dissertation, University of California. Berkeley. 202 pp.

Ranson, E. W., Jr., and F. A. Beach. 1985. Effects of testosterone on ontogeny of urinary behavior in male and female dogs. *Horm. Behav.* 19:36–51.

Rausch, R. A. 1967. Some aspects of the population ecology of wolves, Alaska. *Am. Zool.* 7:253–65.

Rausch, R. A., and R. A. Hinman. 1977. Wolf management in Alaska: An exercise in futility? Pp. 147–57 in R. L. Philips and C. Jonkel, eds., Proceedings of the 1975 predator symposium. Montana Forest and Conservation Experiment Station, University of Montana, Missoula. 268 pp.

Rausch, R. L. 1953. On the status of some arctic mammals. *Arctic* 6:91–148.

Rausch, R. L. 1958. Some observations on rabies in Alaska, with special reference to wild Canidae. *J. Wildl. Mgmt.* 22:246–60.

———. 1973. Rabies in Alaska: Prevention and control. Arctic Health Research Center Report no. 111. U.S. Department of Health, Education and Welfare. 20 pp.

Rausch, R. L., B. B. Babero, R. V. Rausch, and E. L. Schiller. 1956. Studies on the helminth fauna of Alaska. XXVII. The occurrence of larvae of *Trichinella spiralis* in Alaskan mammals. *J. Parasitol.* 42:259–71.

Rausch, R. L., and F. S. L. Williamson. 1959. Studies on the helminth fauna of Alaska. XXXIV. The parasites of wolves, *Canis lupus* L. *J. Parasitol.* 45:395–403.

The Raven. 1997. The best wolf is a wild wolf (parts 1 and 2). The Raven: Visitor's Newsletter, Algonquin Provincial Park, 38(9,10), August 14, August 21.

———. 1999. A wolfian trilogy (Part 2): Thinking things through. The Raven: Visitor's Newsletter, Algonquin Provincial Park.

Raymer, J., D. Wiesler, M. Novotny, C. Asa, U. S. Seal, and L. D. Mech. 1984. Volatile constituents of wolf (*Canis lupus*) urine as related to gender and season. *Experientia* 40:707–9.

———. 1985. Chemical investigation of wolf (*Canis lupus*) anal-sac secretions in relation to breeding season. *J. Chem. Ecol.* 11:593–608.

———. 1986. Chemical scent constituents in urine of wolf (*Canis lupus*) and their dependence on reproductive hormones. *J. Chem. Ecol.* 12:297–314.

Ream, R. R., M. V. Fairchild, D. K. Boyd, and A. J. Blakesley. 1989. First wolf den in western United States in recent history. *Northwestern Nat.* 70:39–40.

Ream, R. R., M. W. Fairchild, D. K. Boyd, and D. H. Pletscher. 1991. Population dynamics and home range changes in a colonizing wolf population. Pp. 349–66 in R. K. Keiter and M. S. Boyce, eds., *The Greater Yellowstone ecosystem: Redefining America's wilderness heritage.* Yale University Press, New Haven, CT.

Ream, R. R., R. B. Harris, J. D. Smith, and D. K. Boyd. 1985. Movement patterns of a lone wolf, *Canis lupus*, in unoccupied wolf range, southeastern British Columbia. *Can. Field Nat.* 99:234–39.

Ream, R. R., and U. I. Mattson. 1982. Wolf status in the northern Rockies. Pp. 362–81 in F. H. Harrington and P. C. Paquet, eds., *Wolves of the world: Perspectives of behavior, ecology, and conservation.* Noyes Publications, Park Ridge, NJ.

Regnier, F., and M. Goodwin. 1977. On the chemical and environmental modulation of pheromone release from vertebrate scent marks. Pp. 115–33 in D. Müller-Schwarze and M. M. Mozell, eds., *Chemical signals in vertebrates.* Plenum Press, New York.

Reich, D. E., R. K. Wayne, and D. B. Goldstein. 1999. Genetic evidence for a recent origin by hybridization of red wolves. *Mol. Ecol.* 8:139–44.

Reid, R., and D. Janz. 1995. Economic evaluation of Vancouver Island wolf control. Pp. 515–21 in L. N. Carbyn, S. H. Fritts, and D. R. Seip, eds., *Ecology and conservation of wolves in a changing world.* Canadian Circumpolar Institute, Edmonton, Alberta.

Reig, S. 1993. Non aggressive encounter between a wolf pack and a wild boar. *Mammalia* 57:451–53.

Reig, S., L. de La Cuesta, and F. Palacios. 1985. The impact of human activities on the food habit of red fox and wolf in Old Castille, Spain. *Rev. Ecol.* (Terre Vie) 40:151–55.

Reig, S., and W. Jedrzejewski. 1988. Winter and early spring food of some carnivores in the Bialowieza National Park, eastern Poland. *Acta Theriol.* 33:57–65.

Reiger, I. 1979. Scent rubbing in carnivores. *Carnivore* 2:17–25.

Reilly, J. R., L. E. Hanson, and D. H. Ferris. 1970. Experimental induced predator-food chain transmission of *L. grippotyphosa* from rodents to wild Marsupialia and Carnivora. *Am. J. Vet. Res.* 31:1443–48.

Reynolds, J. C., and N. J. Aebischer. 1991. Comparison and quantification of carnivore diet by faecal analysis: A critique, with recommendations, based on a study of the fox *Vulpes vulpes. Mammal Rev.* 21(3): 97–122.

Ricker, W. E. 1954. Stock and recruitment. *J. Fish Res. Bd. Can.* 11:559–623.

Rideout, C. B. 1978. Mountain goat. Pp. 149–59 in J. L. Schmidt and D. L. Gilbert, eds., *Big game of North America: Ecology and management.* Stackpole, Harrisburg, PA.

Riley, G. A., and R. T. McBride. 1972. A survey of the red wolf (*Canis rufus*). Scientific Wildlife Report no. 162. U.S. Fish and Wildlife Service, Washington, D.C.

———. 1975. A survey of the red wolf (*Canis rufus*). Pp. 263–77 in M. W. Fox, ed., *The wild canids: Their systematics, behavioral ecology, and evolution.* Van Nostrand Reinhold, New York.

Riley, W. A. 1933. Reservoirs of *Echinococcus* in Minnesota. *Minn. Med.* 16:744–45.

Ripple, W. J., and E. J. Larsen. 2000. Historic aspen recruitment, elk, and wolves in northern Yellowstone National Park, USA. *Biol. Conserv.* 95:361–70.

Ripple, W. J., E. J. Larsen, R. A. Renkin, and D. W. Smith. 2001. Trophic cascades among wolves, elk and aspen on Yellowstone National Park's northern range. *Biol. Conserv.* 102: 227–34.

Ritter, D. G. 1981. Rabies virus assay of Alaskan canids. *Proc. Alaska Sci. Conf.* 32:22–23.

———. 1991. Rabies in Alaskan furbearers: A review. Sixth Northern Furbearer Conference, Fairbanks, Alaska.

Rivers, J. P. W., and T. L. Frankel. 1980. Fat in the diet of cats and dogs. Pp. 67–100 in R. S. Anderson, ed., *Nutrition of the dog and cat.* Pergamon Press, Oxford.

Rjabov, L. S. 1973. Wolf and dog hybrids in the Voronezh region. [In Russian.] *Bull. Mosk. obsc. ispytat. prirody* (Biol.) 78(6): 25–38.

———. 1979. New facts regarding the wolves and their hybrids with dogs in the Voronezh region. [In Russian.] *Bull. Mosk. obsc. ispytat. prirody* (Biol.) 79(4): 18–27.

Robbins, J. 1997. Return of the wolf. *Wildl. Conserv.* 100(2): 44–51, 66.

Robel, R. J., A. D. Dayton, F. R. Henderson, R. L. Meduna, and C. W. Spaeth. 1981. Relationships between husbandry methods and sheep losses to canine predators. *J. Wildl. Mgmt.* 45:894–911.

Roberts, L. H. 1975a. Evidence for the laryngeal source of ultrasonic and audible cries of rodents. *J. Zool.* 175:243–57.

———. 1975b. The rodent ultrasound production mechanism. *Ultrasonics* 13:83–88.

Robinson, D. F. 1828. *Natural history of quadrupeds.* Henry Athans, Hartford.

Robinson, J. G., and K. H. Redford. 1991. *Neotropical wildlife use and conservation.* University of Chicago Press, Chicago.

Rodman, P. S. 1981. Inclusive fitness and group size with a recon-

sideration of group size in lions and wolves. *Am. Nat.* 118(2): 275–83.

Roe, F. G. 1951. *The North American buffalo: A critical study of the species in its wild state.* University of Toronto Press, Toronto.

Rogers, L. L., and L. D. Mech. 1981. Interactions of wolves and black bears in northeastern Minnesota. *J. Mammal.* 62: 434–36.

Rogers, L. L., L. D. Mech, D. K. Dawson, J. M. Peek, and M. Korb. 1980. Deer distribution in relation to wolf pack territory edges. *J. Wildl. Mgmt.* 44:253–58.

Rook, L. 1994. The Plio-Pleistocene Old World *Canis* (*Xenocyon*) ex gr. *falconeri. Boll. Soc. Paleontol. Ital.* 33:71–82.

Root, R. B. 1967. The niche exploitation pattern of the blue-gray gnatcatcher. *Ecol. Monogr.* 37:317–50.

Rosen, W. E. 1997. Red wolf recovery in northeastern North Carolina and the Great Smoky Mountains National Park: Public attitudes and economic impacts. Technical Report. Cornell University, Ithaca, NY. [Funded by the Point Defiance Zoo and Aquarium in cooperation with the U.S. Fish and Wildlife Service.]

Rosengren, A. 1969. Experiments in color discrimination in dogs. *Acta Zool. Fenn.* 121:1–19.

Rosenzweig, M. L. 1966. Community structure in sympatric Carnivora. *J. Mammal.* 47:602–12.

———. 1968. The strategy of body size in mammalian carnivores. *Am. Midl. Nat.* 80:299–315.

Rothman, R. J., and L. D. Mech. 1979. Scent-marking in lone wolves and newly formed pairs. *Anim. Behav.* 27:750–60.

Rousset, F., and M. Raymond. 1997. Statistical analyses of population genetic data: New tools, old concepts. *Trends Ecol. Evol.* 12:313–17.

Route, W. T., and R. O. Peterson. 1991. An incident of wolf, *Canis lupus,* predation on a river otter, *Lutra canadensis,* in Minnesota. *Can. Field Nat.* 105:567–68.

Rowell, T. E. 1974. The concept of social dominance. *Behav. Biol.* 11:131–54.

Roy, L. D., and M. J. Dorrance. 1976. *Methods of investigating predators of domestic livestock.* AlbertaAgriculture, Edmonton, Alberta.

Roy, M. S., E. Geffen, D. Smith, E. A. Ostrander, and R. K. Wayne. 1994. Patterns of differentiation and hybridization in North American wolflike canids, revealed by analysis of microsatellite loci. *Mol. Biol. Evol.* 11:533–70.

Roy, M. S., E. Geffen, D. Smith, and R. K. Wayne. 1996. Molecular genetics of pre-1940 red wolves. *Conserv. Biol.* 10:1413–24.

Roy, M. S., D. J. Girman, and R. K. Wayne. 1994. The use of museum specimens to reconstruct the genetic variability and relationships of extinct populations. *Experientia* 50:551–57.

Rutberg, A. T. 1987. Adaptive hypotheses of birth synchrony in ruminants: An interspecific test. *Am. Nat.* 130:692–710.

Rutter, R. J., and D. H. Pimlott. 1968. *The world of the wolf.* J. B. Lippincott, Philadelphia.

Ryan, M. J., and S. W. Nielsen. 1979. Tonsillar carcinoma with metastases in a captive wolf. *J. Wildl. Dis.* 15:295–98.

Ryden, H. 1975. *God's dog.* Coward, McCann, Geoghegan, New York.

Ryon, C. J. 1977. Den digging and related behavior in a captive timber wolf pack. *J. Mammal.* 58:87–89.

Ryon, J., J. C. Fentress, F. H. Harrington, and S. Bragdon. 1986. Scent rubbing in wolves *(Canis lupus):* The effect of novelty. *Can. J. Zool.* 64:573–77.

Saether, B., and B. Jonsson. 1991. Conservation biology faces reality. *Trends Ecol. Evol.* 6:11–12.

Salvador, A., and P. L. Abad. 1987. Food habits of a wolf population in Leon Province, Spain. *Mammalia* 51:45–52.

Salwasser, H., and F. B. Samson. 1985. Cumulative effects analysis: An advance in wildlife planning and management. *Trans. N. Am. Wildl. Nat. Resour. Conf.* 50:313–21.

Samuel, W. M., G. A. Chalmers, and J. R. Gunson. 1978. Oral papillomatosis in coyotes *(Canis latrans)* and wolves *(Canis lupus)* of Alberta. *J. Wildl. Dis.* 14:165–69.

Samuel, W. M., S. Ramalingam, and L. N. Carbyn. 1978. Helminths in coyotes (*Canis latrans* Say), wolves (*Canis lupus* L.), and red foxes (*Vulpes vulpes* L.) of Southwestern Manitoba. *Can. J. Zool.* 56:2614–17.

Sands, M., W., R. P. Coppinger, and C. J. Phillips. 1977. Comparisons of thermal sweating and histology of sweat glands of selected canids. *J. Mammal.* 58:74–78.

Sansone-Bassano, G., and R. M. Reisner. 1974. Steroid pathways in sebaceous glands. *J. Invest. Dermatol.* 62:211–16.

Santin, L. J., S. Rubio, A. Begega, R. Miranda, and J. L. Arias. 2000. Spatial learning and the hippocampus. *Revista de Neurologia* 31:455–62.

Sapolsky, R. M. 2002. Endocrinology of the stress response. Pp. 409–50 in J. B. Becker, S. M. Breedlove, D. Crews, and M. M. McCarthy, eds., *Behavioral Endocrinology,* 2d ed. MIT Press, Cambridge, MA.

Sargeant, A. B., S. H. Allen, and J. O. Hastings. 1987. Spatial relationships between sympatric coyotes and red foxes in North Dakota. *J. Wildl. Mgmt.* 51:285–93.

Sauer, P. R. 1984. Physical characteristics. Pp. 73–90 in L. K. Halls, ed., *White-tailed deer: Ecology and management.* Stackpole, Harrisburg, PA.

Savage, C. 1988. *Wolves.* Sierra Club Books, San Francisco.

Savelli, B. G., F. Antonelli, and L. Boitani. 1998. The impact of livestock support on carnivore conservation. Large Carnivore Initiative for Europe, Institute of Applied Ecology, Rome. 170 pp.

Savolainen, P., Y. Zhang, J. Luo, J. Lundeberg, and T. Leitner. 2002. Genetic evidence for an East Asian origin of domestic dogs. *Science* 298:1610–13.

Sax, J. L. 1991. Ecosystems and property rights in Greater Yellowstone: The legal system in transition. Pp. 77–84 in R. B. Keiter and M. S. Boyce, eds., *The Greater Yellowstone ecosystem: Redefining America's wilderness heritage.* Yale University Press, New Haven, CT.

Scalia, F., and S. S. Winans. 1975. The differential projections of the olfactory bulb and accessory olfactory bulb in mammals. *J. Comp. Neurol.* 161:31–56.

Scandura, M., M. Apollonio, and L. Mattioli. 2001. Recent recovery of the Italian wolf population: A genetic investigation using microsatellites. *Mammal. Biol.* 66:321–31.

Scarce, R. 1998. What do wolves mean? Conflicting social constructions of *Canis lupus* in "Bordertown." *Human Dimensions of Wildlife* 3:26–45.

Schaefer, C. L. 2000. Spatial and temporal variation in wintering elk abundance and composition, and wolf response on Yellowstone's Northern Range. Master's thesis, Michigan Technological University, Houghton. 95 pp.

Schaeppi, U., and F. Liverani. 1979. Rod and cone components in the compound ERG of the beagle dog. *Agents and Actions* 9:294–300.

Schaller, D. T. 1996. The ecocenter as tourist attraction: Ely and the International Wolf Center. Tourism Center and Center for Urban and Regional Affairs, University of Minnesota, Minneapolis. 21 pp.

Schaller, G. B. 1967. *The deer and the tiger: A study of wildlife in India.* University of Chicago Press, Chicago.

Schaller, G. B., Li Hong, Talipu, Lu Hua, Ren Junrang, Qui Mingjiang, and Wang Haibin. 1987. Status of large mammals in the Taxkorgan Reserve, Xinjiang, China. *Biol. Conserv.* 42:53–71.

Schaller, G. B., Ren Junrang, and Qui Mingjiang. 1988. Status of the snow leopard *Panthera uncia* in Qinghai and Gannsu Provinces, China. *Biol. Conserv.* 45:179–94.

Schaller, G. B., and G. R. Lowther. 1969. The relevance of carnivore behavior to the study of early hominids. *Southwestern J. Anthropol.* 25:307–41.

Schassburger, R. M. 1978. The vocal repertoire of the wolf: Structure, function, and ontogeny. Ph.D. dissertation, Cornell University, Ithaca, NY. 343 pp.

———. 1987. Wolf vocalizations: An integrated model of structure, motivation and ontogeny. Pp. 313–47 in H. Frank, ed., *Man and wolf: Advances, issues and problems in captive wolf research.* Dr. W. Junk Publishers, Dordrecht, The Netherlands.

———. 1993. Vocal communication in the timber wolf, *Canis lupus,* Linnaeus: Structure, motivation, and ontogeny. Advances in Ethology, no. 30. Paul Parey, Berlin. 84 pp.

Schenkel, R. 1947. Ausdrucks-studien an wolfen [Expression studies of wolves]. *Behaviour* 1:81–129. [Translation from German by F. Harrington.]

———. 1967. Submission: Its features and function in the wolf and dog. *Am. Zool.* 7:319–29.

Schmidt, G. M., and D. B. Coulter. 1981. Physiology of the eye. Pp. 129–59 in K. N. Gelatt, ed., *Textbook of veterinary ophthalmology.* Lea and Febiger, Philadelphia.

Schmidt, K. P., and J. R. Gunson. 1985. Evaluation of wolf-ungulate predation near Nordegg, Alberta: Second year progress report, 1984–1985. Alberta Energy and Natural Resources Fish and Wildlife Division, Edmonton, Alberta. 53 pp.

Schmidt, P. A., and L. D. Mech. 1997. Wolf pack size and food acquisition. *Am. Nat.* 150:513–17.

Schmidt, R. H. 1986. Community-level effects of coyote population reduction. Pp. 49–65 in J. Cairns, Jr., ed., *Community toxicity testing.* Special Technical Publication no. 920. American Society for Testing and Materials, Philadelphia.

Schmitz, O. J., P. A. Hamback, and A. P. Beckerman. 2000. Trophic cascades in terrestrial systems: A review of the effects of carnivore removals on plants. *Am. Nat.* 155:141–53.

Schmitz, O. J., and G. B. Kolenosky. 1985a. Hybridization between wolf and coyote in captivity. *J. Mammal.* 66:402–5.

———. 1985b. Wolves and coyotes in Ontario: Morphological relationships and origins. *Can. J. Zool.* 63:1130–37.

Schneider, C. J., T. B. Smith, B. Larison, and C. Moritz. 1999. A test of alternative models of diversification in tropical rainforest: Ecological gradients vs. rainforest refugia. *Proc. Nat. Acad. Sci. U.S.A.* 96:13869–73.

Schoeninger, M. J., M. J. DeNiro, and H. Tauber. 1983. Stable nitrogen isotope ratios of bone collagen reflect marine and terrestrial components of prehistoric human diet. *Science* 220:1381–83.

Schotté, C. S. 1988. The development of social organization and mating patterns in a captive wolf *(Canis lupus)* pack. Ph.D. dissertation, University of Connecticut, Storrs. 434 pp.

Schotté, C. S., and B. Ginsburg. 1987. Development of social organization and mating in a captive wolf pack. Pp. 349–74 in H. Frank, ed., *Man and wolf: Advances, issues and problems in captive wolf research.* Dr. W. Junk Publishers, Dordrecht, The Netherlands.

Schröder, W., and C. Promberger. 1992. European Wolf Conservation Strategy. Pp. 2–7 in C. Promberger and W. Schröder, eds., *Wolves in Europe: Status and perspectives.* Munich Wildlife Society, Ettal, Germany.

Schultz, T. D., and J. H. Ferguson. 1974. Influence of dietary fatty acids on the composition of plasma fatty acids in the tundra wolf *(Canis lupus tundrarum). Comp. Biochem. Physiol. A. Comp. Physiol.* 49(3A): 575–81.

Schwartz, C. C. 1998. Reproduction, natality and growth. Pp. 141–71 in A. W. Franzmann and C. C. Schwartz, eds., *Ecology and management of the North American moose.* Smithsonian Institution Press, Washington, D.C.

Schwartz, C. C., and A. W. Franzmann. 1991. Interrelationship of black bears to moose and forest succession in the northern coniferous forest. Wildlife Monographs, no. 113. The Wildlife Society, Bethesda, MD. 58 pp.

Schwartz, C. C., K. J. Hundertmark, and T. H. Spraker. 1992. An evaluation of selective bull moose harvest on the Kenai Peninsula, Alaska. *Alces* 28:1–13.

Schwartz, C. C., R. Stephenson, and N. Wilson. 1983. *Trichodectes canis* on the gray wolf and coyote on Kenai Peninsula, Alaska. *J. Wildl. Dis.* 19:372–73.

Schwartz, M. 1997. *A history of dogs in the early Americas.* Yale University Press, New Haven, CT.

Scott, B. M. V., and D. M. Shackleton. 1980. Food habits of two Vancouver island wolf packs: A preliminary study. *Can. J. Zool.* 58:1203–7.

———. 1982. A preliminary study of the social organization of the Vancouver Island wolf *(Canis lupus crassodon).* Pp. 12–25 in F. H. Harrington and P. C. Paquet, eds., *Wolves of the world: Perspectives of behavior, ecology, and conservation.* Noyes Publications, Park Ridge, NJ.

Scott, H. H. 1928. Carcinoma of the tonsil in a common wolf *(Canis lupus). Proc. Zool. Soc. Lond.* 1928:43–47.

Scott, J. M., F. Davis, B. Csuti, R. Noss, B. Butterfield, C. Groves, H. Anderson, S. Caicco, F. D'Erchia, T. C. Edwards, Jr., J. Ulliman, and R. G. Wright. 1993. Gap analysis: A geographic approach to protection of biological diversity. Wildlife Monographs, no. 123. The Wildlife Society, Bethesda, MD. 41 pp.

Scott, J. P. 1961. Spectrographic analysis of dog sounds. (Abstract.) *Am. Zool.* 1:387.

———. 1967. The evolution of social behavior in dogs and wolves. *Am. Zool.* 7:373–81.

———. 1968. Evolution and domestication of the dog. *Evol. Biol.* 2:243–75.

Scott, J. P., O. S. Elliot, and B. E. Ginsburg. 1997. Man and his dog. *Science* 278:205–6.

Scott, J. P., and J. L. Fuller. 1965. *Genetics and social behavior of the dog.* University of Chicago Press, Chicago.

Scott, P. A., C. V. Bentley, and J. J. Warren. 1985. Aggressive behavior by wolves toward humans. *J. Mammal.* 66:807–9.

Seal, U. S., and T. J. Kreeger. 1987. Chemical immobilization of furbearers. Pp. 191–215 in M. Novak, J. A. Baker, M. E. Obbard, and B. Malloch, eds., *Wild furbearer management and conservation in North America.* Ontario Ministry of Natural Resources, Toronto.

Seal, U. S., and R. C. Lacy. 1989. Florida panther *(Felis concolor coryi)* population viability analysis and species survival plan. Report to the U. S. Fish and Wildlife Service. Captive Breeding Specialist Group, Species Survival Commission, IUCN, Apple Valley, MN.

Seal, U. S., and L. D. Mech. 1983. Blood indicators of seasonal metabolic patterns in captive adult gray wolves. *J. Wildl. Mgmt.* 47:704–15.

Seal, U. S., L. D. Mech, and V. Van Ballenberghe. 1975. Blood analyses of wolf pups and their ecological and metabolic interpretation. *J. Mammal.* 56:64–75.

Seal, U. S., M. E. Nelson, L. D. Mech, and R. L. Hoskinson. 1978. Metabolic indicators of habitat differences in four Minnesota deer populations. *J. Wildl. Mgmt.* 42:746–54.

Seal, U. S., E. D Plotka, L. D. Mech, and J. M. Packard. 1987. Seasonal metabolic and reproductive cycles in wolves. Pp. 109–25 in H. Frank, ed., *Man and wolf: Advances, issues and problems in captive wolf research.* Dr. W. Junk Publishers, Dordrecht, The Netherlands.

Seal, U. S., E. D. Plotka, J. M. Packard, and L. D. Mech. 1979. Endocrine correlates of reproduction in the wolf. I. Serum progesterone, estradiol and LH during the estrous cycle. *Biol. Reprod.* 21:1057–66.

Segal, M. 1994. Current American Indian attitudes towards wolves; the Kalispel and the gray wolf: A qualitative analysis. Master's thesis, University of Wisconsin, Madison.

Sears, H. J. 1999. A landscape-based assessment of *Canis* morphology, ecology and conservation in southeastern Ontario. Master's thesis, University of Waterloo, Waterloo, Ontario.

Seielstad, M., E. Bekele, M. Ibrahim, A. Touré, and M. Traoré. 1999. A view of modern human origins from Y chromosome microsatellite variation. *Genome Res.* 9:558–67.

Seielstad, M. T., E. Minch, and L. L. Cavalli-Sforza. 1998. Genetic evidence for a higher female migration rate in humans. *Nature Genet.* 1998 20:278–80.

Seip, D. R. 1991. Predation and caribou populations. *Rangifer* 11: 46–52.

———. 1995. Introduction to wolf-prey interactions. Pp. 179–86 in L. N. Carbyn, S. H. Fritts, and D. R. Seip, eds., *Ecology and conservation of wolves in a changing world.* Canadian Circumpolar Institute, Edmonton, Alberta.

Serracchiani, S. 1976. Indagine sulle opinioni nei riguardi della conservazione del lupo in Abruzzo. Laurea Thesys in Psychology. University of Rome.

Servheen, C., and R. R. Knight. 1993. Possible effects of a restored gray wolf population on grizzly bears in the Greater Yellowstone Area. Pp. 28–37 in R. S. Cook, ed., Ecological issues on reintroducing wolves into Yellowstone National Park. U.S. National Park Service Scientific Monograph Series, NPS/NRYELL/NRSM-93/22.

Servín[-Martínez], J. 2000. Duration and frequency of chorus howling of the Mexican wolf *(Canis lupus baileyi). Acta Zool. Mex.* 80:223–31.

Servín-Martínez, J. 1991. Algunos aspectos de la conducta social del lobo Mexicano *(Canis lupus baileyi)* en cautiverio. *Acta Zool. Mex.* 45:1–43.

———. 1997. El periodo de apareamiento, nacimiento y crecimiento del lobo Mexicano *(Canis lupus baileyi). Acta Zool. Mex.* 71:45–56.

Seton, E. T. 1929. *Lives of game animals.* Vol. 1: *Cats, wolves, and foxes.* Doubleday, Doran and Co., New York.

Severinghaus, C. W., and E. L. Cheatum. 1956. Life and times of the white-tailed deer. Pp. 57–186 in W. P. Taylor, ed., *The deer of North America.* Stackpole, Harrisburg, PA.

Shaffer, M. L. 1987. Minimum viable populations: Coping with uncertainty. Pp. 69–86 in M. E. Soulé, ed., *Viable populations for conservation.* Cambridge University Press, Cambridge.

Shahi, S. P. 1983. Status of the gray wolf *(Canis lupus pallipes,* Sykes) in India. *Acta Zool. Fenn.* 174:283–86.

Shaldybin, L. S. 1957. Parasitic worms of wolves in the Modrvinian ASSR. *Uch. Sap. Gor'kovsk. Gos. Ped. Inst.* 19:65–70.

Shaw, J. H. 1975. Ecology, behavior, and systematics of the red wolf *(Canis rufus).* Ph.D. dissertation, Yale University, New Haven, CT.

———. 1995. How many bison originally populated western rangelands? *Rangelands* 17:148–50.

Shelton, P. C. 1966. Ecological studies of beavers, wolves and moose in Isle Royale National Park, Michigan. Ph.D. dissertation, Purdue University, Lafayette, IN. 308 pp.

Sherman, S. M., and J. R. Wilson. 1975. Behavioral and morphological evidence for binocular competition in the postnatal development of the dog's visual system. *J. Comp. Neurol.* 161:183–96.

Shields, W. M. 1983. Genetic considerations in the management of the wolf and other large vertebrates: An alternative view. Pp. 90–92 in L. N. Carbyn, ed., *Wolves in Canada and Alaska: Their status, biology, and management.* Report Series, no. 45. Canadian Wildlife Service, Edmonton, Alberta.

Shumaker, R. W., and K. B. Swartz. 2002. When traditional methodologies fail: Cognitive studies of great apes. Pp. 335–43 in M. Bekoff, C. Allen, and G. M. Burghardt, eds., *The Cognitive Animal: Empirical and Theoretical Perspectives on Animal Cognition.* MIT Press, Cambridge, MA.

Signoret, J. P. 1970. Sexual behavior patterns in female domestic pigs *(Sus scrofa* L.) reared in isolation from males. *Anim. Behav.* 18:165–68.

Sikes, D. S. 1994. Influences of ungulate carcasses on coleopteran communities in Yellowstone National Park, USA. Master's thesis, Montana State University, Bozeman. 151 pp.

Sikes, D. S., M. A. Ivie, and L. L. Ivie. 1995. Carrion associated coleoptera on the northern range of Yellowstone National Park. P. 20 in Agenda and abstracts, Greater Yellowstone predators: Ecology and conservation in a changing landscape, September 24–27. U.S. National Park Service, Yellowstone National Park, WY.

Sikes, R. K. 1981. Rabies. Pp. 3–17 in J. W. Davis, L. H. Karstad, and D. O. Trainer, eds., *Infectious diseases of wild mammals.* Iowa State University Press, Ames.

Simberloff, D. 1998. Flagships, umbrellas, and keystones: Is single-species management passe in the landscape era? *Biol. Conserv.* 83(3): 247–57.

Simonson, E., and J. Brozek. 1952. Flicker fusion frequency: Background and applications. *Physiol. Rev.* 32:349–78.

Sinclair, A. R. E. 1989. Population regulation in animals. Pp. 197–241 in J. M. Cherrett, eds., *Ecological concepts.* Blackwell Scientific Publications, Oxford.

Singer, F. J. 1984. Some population characteristics of Dall sheep in six Alaska national parks and preserves. Proceedings of the Biennial Symposium of the Northern Wild Sheep and Goat Council 4:1–10.

Singer, F. J., and J. Dalle-Molle. 1985. The Denali ungulate-predator system. *Alces* 21:339–58.

Singer, F. J., and J. A. Mack. 1999. Predicting the effects of wildfire and carnivore predation on ungulates. Pp. 189–237 in T. W. Clark, A. P. Curlee, S. C. Minta, and P. M. Kareiva, *Carnivores in ecosystems: The Yellowstone experience.* Yale University Press, New Haven, CT.

Singleton, P. H. 1995. Winter habitat selection by wolves in the North Fork of the Flathead River Basin, Montana and British Columbia. Master's thesis, University of Montana, Missoula.

Sinnott, J. M., and C. H. Brown. 1993. Effects of varying signal duration on pure-tone frequency discrimination in humans and monkeys. *J. Acoust. Soc. Am.* 93:1541–46.

Skeel, M. A., and L. N. Carbyn. 1977. The morphological relationship of gray wolves *(Canis lupus)* in national parks of central Canada. *Can. J. Zool.* 55:737–47.

Skogland, T. 1991. What are the effects of predators on large ungulate populations? *Oikos* 61:401–11.

Skoog, R. O. 1968. Ecology of the caribou *(Rangifer tarandus)* in Alaska. Ph.D. dissertation, University of California, Berkeley. 699 pp.

Slade, V. 1983. Individual differences in social bond formation of two hand-reared prepubertal wolves. Master's thesis, University of Connecticut, Storrs.

Slatkin, M. 1987. Gene flow and the geographic structure of natural populations. *Science* 236:787–92.

Slatkin, M., and W. P. Madison. 1989. A cladistic measure of gene flow inferred from the phylogeny of alleles. *Genetics* 123: 603–13.

Slupecki, L. P. 1987. *Wilkolactwo.* Iskry, Warsaw, Poland.

Smietana, W., and A. Klimek. 1993. Diet of wolves in the Bieszczady Mountains, Poland. *Acta Theriol.* 38:245–51.

Smietana, W., and J. Wajda. 1997. Wolf number changes in Bieszczady National Park, Poland. *Acta Theriol.* 42:241–52.

Smith, C. A., R. E. Wood, L. V. Beier, and K. P. Bovee. 1987. Wolf-deer-habitat relationships in southeast Alaska. Alaska Department of Fish and Game, Division of Game, Federal Aid in Wildlife Restoration Research Final Report. Fairbanks, Alaska. 20 pp.

Smith, Debbie, T. J. Meier, E. Geffen, L. D. Mech, J. W. Burch, L. G. Adams, and R. K. Wayne. 1997. Is incest common in gray wolf packs? *Behav. Ecol.* 8:384–91.

Smith, D. W. 1997. Yellowstone Gray Wolf Restoration Project Current Wolf Population Status, August 5, 1997. N1427 (YELL). Yellowstone Center for Resources, Mammoth, WY.

———. 1998. Yellowstone Wolf Project: Annual Report, 1997. YCR-NR-98-2. U.S. National Park Service, Yellowstone Center for Resources, Yellowstone National Park, WY.

Smith, D. W., W. G. Brewster, and E. E. Bangs. 1999. Wolves in the Greater Yellowstone ecosystem: Restoration of a top carnivore in a complex management environment. Pp. 103–25 in T. W. Clark, A. P. Curlee, S. C. Minta, and P. M. Kareiva, eds., *Carnivores in ecosystems: The Yellowstone experience.* Yale University Press, New Haven, CT.

Smith, D. W., L. D. Mech, M. Meagher, W. E. Clark, R. Jaffe, M. K. Phillips, and J. A. Mack. 2000. Wolf-bison interactions in Yellowstone National Park. *J. Mammal.* 81(4): 1128–35.

Smith, D. W., K. M. Murphy, and S. Monger. 2001. Killing of a bison, *Bison bison,* calf by a wolf, *Canis lupus,* and four coyotes, *Canis latrans,* in Yellowstone National Park. *Can. Field Nat.* 115:343–45.

Smith, D. W., M. K. Phillips, and R. L. Crabtree. 1996. Interaction of wolves and other wildlife in Yellowstone National Park. Pp. 140–45 in N. Fascione and M. Cecil, eds., *Proceedings of the Defenders of Wildlife's Wolves of America Conference,* Albany, NY. Defenders of Wildlife, Washington, D.C.

Smith, D. W., D. R. Trauba, R. K. Anderson, and R. O. Peterson. 1994. Black bear predation on beavers on an island in Lake Superior. *Am. Midl. Nat.* 132:248–55.

Smith, T. B., and R. K. Wayne, eds. 1996. *Molecular genetic approaches in conservation.* Oxford University Press, New York.

Smith, T. B, R. K. Wayne, D. J. Girman, and M. W. Bruford. 1997. A role for ecotones in generating rainforest biodiversity. *Science* 276:1855–57.

Smith, W. J. 1977. *The Behavior of communicating: An ethological approach.* Harvard University Press, Cambridge, MA.

———. 1990. Communication and expectations: A social process and the cognitive operations it depends upon and influences. Pp. 234–53 in M. Bekoff and D. Jamieson, eds., *Interpretation and explanation in the study of animal behavior,* vol. 1. Westview Press, Boulder, CO.

Smythe, N. 1970. On the existence of "pursuit invitation" signals in mammals. *Am. Nat.* 104:491–94.

———. 1977. The function of mammalian alarm advertising: Social signals or pursuit invitation? *Am. Nat.* 111:191–94.

Sokolov, V. E. 1982. *Mammalian skin.* University of California Press, Berkeley.

Sokolov, V. E., and O. L. Rossolimo. 1985. Taxonomy and variability. Pp. 21–50 in D. I. Bibikov, ed., *The wolf: History, systematics, morphology, ecology.* [In Russian.] Izdatetvo Nauka, Muskva.

Sokolov, V. E., A. S. Severtsov, and A. V. Shubkina. 1990. Modelling of the selective behavior of the predator towards prey: The use of borzois for catching saigas. [Translation from Russian.] *Zool. Zh.* 69:117–25.

Solomon, N. G., and J. A. French. 1997. The study of mammalian cooperative breeding. Pp. 1–10 in N. G. Solomon and J. A. French, eds., *Cooperative breeding in mammals.* Cambridge University Press, Cambridge.

Soulé, M. E. 1980. Thresholds for survival: Maintaining fitness and evolutionary potential. Pp. 151–69 in M. E. Soulé and B. A. Wilcox, eds., *Conservation biology: An evolutionary-ecological perspective.* Sinauer Associates, Sunderland, MA.

———. 1987. *Viable populations for conservation.* Cambridge University Press, New York.

Soulsby, E. J. L. 1968. *Helminths, arthropods, and protozoa of domesticated animals.* Williams and Wilkins, Baltimore.

Southwood, T. R. E., and H. N. Comins. 1976. A synoptic population model. *J. Anim. Ecol.* 45:949–65.

Spalton, A. 2002. Canidae in the Sultanate of Oman. *Canid News* 5:1 (online). http://www.canids.org/canidnews/5/canids_in_oman.pdf.

Spaulding, R. L., P. R. Krausman, and W. B. Ballard. 1998. Summer diet of gray wolves, *Canis lupus,* in northwestern Alaska. *Can. Field Nat.* 112:262–66.

Speed, J. G. 1941. Sweat glands of the dog. *Vet. J.* 97:252–56.

Spence, C. E., J. E. Kenyon, D. R. Smith, R. D. Hayes, and A. M. Baer. 1999. Surgical sterilization of free-ranging wolves. *Can. Vet. J.* 40(2): 118–21.

Sprague, R. H., and J. J. Anisko. 1973. Elimination patterns in the laboratory beagle. *Behaviour* 47:257–67.

Stahler, D. R. 2000. Interspecific interactions between the common raven *(Corvus corax)* and the gray wolf *(Canis lupus)* in Yellowstone National Park, Wyoming: Investigations of a predator and scavenger relationship. Master's thesis, University of Vermont, Burlington. 105 pp.

Stahler, D. R., D. W. Smith, and R. Landis. 2002. The acceptance of a new breeding male into a wild wolf pack. *Can. J. Zool.* 80:360–65.

Stancampiano, L., V. Guberti, F. Francisci, M. Magi, and C. Bandi. 1993. Trichinellosis in wolves *(Canis lupus)* in Italy. 1st Naz. Fauna Selvatica Posters; no. 21. 4 pp.

Stander, P. E. 1992. Cooperative hunting in lions: The role of the individual. *Behav. Ecol. Sociobiol.* 29:445–54.

Stander, P. E., and S. D. Albon. 1993. Hunting success of lions in a semi-arid environment. *Symp. Zool. Soc. Lond.* 65:127–43.

Stanwell-Fletcher, J. F., and T. C. Stanwell-Fletcher. 1942. Three years in the wolves' wilderness. *Nat. Hist.* 49(3): 136–47.

Stardom, R. R. P. 1983. Status and management of wolves in Manitoba. Pp. 30–34 in L. N. Carbyn, ed., *Wolves in Canada and Alaska: Their status, biology, and management.* Report Series, no. 45. Canadian Wildlife Service, Edmonton, Alberta.

Steinberg, H. 1977. Behavioral responses of the white-tailed deer to olfactory stimulation using predator scents. Master's thesis, Pennsylvania State University, University Park. 49 pp.

Steinberg, R. H., M. Reid, and P. L. Lacy. 1973. The distribution of rods and cones in the retina of the cat *(Felis domesticus). J. Comp. Neurol.* 148:229–48.

Steinhart, P. 1995. *The company of wolves.* Alfred A. Knopf, Inc., New York.

Stekert, E. 1986. Folklore. Pp. 19–27 in *Wolf! A modern look.* Wolves in American Culture Committee. Northword Press, Ashland, WI.

Stenlund, M. H. 1955. A field study of the timber wolf *(Canis lupus)* on the Superior National Forest, Minnesota. Technical Bulletin no. 4. Minnesota Department of Conservation, Minneapolis. 55 pp.

Stephens, P. W., and R. O. Peterson. 1984. Wolf avoidance strategies of moose. *Holarctic Ecol.* 7:239–44.

Stephenson, R. O. 1974. Characteristics of wolf den sites. Final Report 5. Alaska Department of Fish and Game, Fairbanks.

———. 1982. Nunamiut Eskimos, wildlife biologists and wolves. Pp. 434–48 in F. H. Harrington and P. C. Paquet, eds., *Wolves of the world: Perspectives of behavior, ecology, and conservation.* Noyes Publications, Park Ridge, NJ.

Stephenson, R. O., and B. Ahgook. 1975. The Eskimo hunter's view of wolf ecology and behavior. Pp. 286–91 in M. W. Fox, ed., *The wild canids: Their systematics, behavioral ecology, and evolution.* Van Nostrand Reinhold, New York.

Stephenson, R. O., W. B. Ballard, C. A. Smith, and K. Richardson. 1995. Wolf biology and management in Alaska, 1981–92. Pp. 43–54 in L. N. Carbyn, S. H. Fritts, and D. R. Seip, eds., *Ecology and conservation of wolves in a changing world.* Canadian Circumpolar Institute, Edmonton, Alberta.

Stephenson, R. O., and D. James. 1982. Wolf movements and food habits in Northwest Alaska. Pp. 26–42 in F. H. Harrington and P. C. Paquet, eds., *Wolves of the world: Perspectives of behavior, ecology, and conservation.* Noyes Publications, Park Ridge, NJ.

Stephenson, R. O., D. G. Ritter, and C. A. Nielsen. 1982. Serologic survey for canine distemper and infectious canine hepatitis in wolves in Alaska. *J. Wildl. Dis.* 18:419–24.

Stephenson, T. R., and V. Van Ballenberghe. 1995. Defense of one twin calf against wolves, *Canis lupus,* by a female moose, *Alces alces. Can. Field Nat.* 109:251–53.

Stewart, I., and M. Golubitsky. 1992. *Fearful symmetry.* Blackwell, Cambridge, MA.

Stiles, C. W., and C. E. Baker. 1934. Key-catalogue of parasites reported for Carnivora (cats, dogs, bears, etc.). Nat. Inst. Health, U.S. Treas. Dept., Public Health Serv. Bull. 163: 913–1223.

Strain, G. M., B. L. Tedford, and R. M. Jackson. 1991. *Am. J. Vet. Res.* 52:410–15.

Strickland, D. 1983. Wolf howling in parks—the Algonquin experience in interpretation. Pp. 93–95 in L. N. Carbyn, ed.,

Wolves in Canada and Alaska: Their status, biology, and management. Report Series, no. 45. Canadian Wildlife Service, Edmonton, Alberta.

Sullivan, J. O. 1979. Individual variability in hunting behavior of wolves. Pp. 284–306 in E. Klinghammer, ed., *The behavior and ecology of wolves.* Garland STPM Press, New York.

Sumanik, R. S. 1987. Wolf ecology in the Kluane Region, Yukon Territory. Master's thesis, Michigan Technological University, Houghton. 102 pp.

Sundqvist, A. K., H. Ellegren, M. Olivier, and C. Vilà. 2001. Y chromosome haplotyping in Scandinavian wolves *(Canis lupus)* based on microsatellite markers. *Mol. Ecol.* 10:1959–66.

Sweatman, G. K. 1971. Mites and pentastomes. Pp. 3–64 in J. W. Davis and R. C. Anderson, eds., *Parasitic diseases of wild mammals.* Iowa State University Press, Ames.

Swihart, R. K., J. J. Pignatello, and M. J. Mattina. 1991. Aversive responses of white-tailed deer, *Odocoileus virginianus,* to predator urines. *J. Chem. Ecol.* 17:767–77.

Syrotuck, W. G. 1972. *Scent and the scenting dog.* Arner Publications, Rome, NY.

Szepanski, M. M., D. M. Ben, and V. Van Ballenberghe. 1999. Assessment of anadromous salmon resources in the diet of the Alexander Archipelago wolf using stable isotope analysis. *Oecologia* 120:327–35.

Tabel, H., A. B. Corner, W. A. Webster, and C. A. Casey. 1974. History and epizootiology of rabies in Canada. *Can. Vet. J.* 15:271–81.

Taberlet, P., L. Gielly, and J. Bouvet. 1996. Etude génétique sur les loups du Mercantour. Rapport pour la Direction de la Nature et des Paysages Ministère de l'Environment.

Taberlet, P., S. Griffin, B. Goossens, S. Questiau, V. Manceau, N. Escaravage, L. P. Waits, and J. Bouvet. 1996. Reliable genotyping of samples with very low DNA quantities using PCR. *Nucleic Acids Res.* 24:3189–94.

Talbot, S. L., and G. F. Shields. 1996. Phylogeography of brown bears *(Ursus arctos)* of Alaska and paraphyly within the Urisdae. *Mol. Phyl. Evol.* 5:477–94.

Taylor, R. J. 1984. *Predation.* Chapman and Hall. New York.

Taylor, R. J., and P. J. Pekins. 1991. Territory boundary avoidance as a stabilizing factor in wolf-deer interactions. *Theor. Pop. Biol.* 39:115–28.

Taylor, W. P. 1956. *The deer of North America.* Stackpole, Harrisburg, PA. 668 pp.

Taylor, W. P., and T. H. Spraker. 1983. Management of a biting louse infestation in a free-ranging wolf population. Pp. 40–41 in Proceedings of the Annual Meeting of the American Association of Zoo Veterinarians.

Tedford, R. H., B. E. Taylor, and Wang Xiaoming. 1995. Phylogeny of the Caninae (Carnivora: Canidae): The living taxa. American Museum Novitates, no. 3146. American Museum of Natural History, New York.

Tedford, R. H., and Qiu Zhanxiang. 1996. A new canid genus from the Pliocene of Yushe, Shanxi Province. *Vert. Palasiatica* 34:27–40.

Teerink, B. J. 1991. *Hair of West European mammals.* Cambridge University Press, Cambridge.

Telleria, J. L., and C. Saez-Royuela. 1989. Ecología de una poblacion iberica de lobos (*Canis lupus*). *Donana, Acta Vertebrata* 16:105–22.

Tembrock, G. 1976a. Canid vocalizations. *Behav. Processes* 1:57–76.

———. 1976b. Die Lautgebung der Caniden: Eine vergleichende Untersuchung. *MILU* 4:1–44.

Templeton, A. R., Davis, S. K., and B. Read. 1987. Genetic variability in a captive herd of Speke's Gazelle (*Gazella-Spekei*). *Zoo Biol.* 6:305–13.

Templeton, A. R., and B. Read. 1998. Elimination of inbreeding depression from a captive population of Speke's gazelle: Validity of the original statistical analysis and confirmation by permutation testing. *Zoo Biol.* 17:77–94.

Tener, J. S. 1954. A preliminary study of the musk-oxen of Fosheim Peninsula, Ellesmere Island, NWT. Wildlife Management Bulletin, Series 1, no. 9. Canadian Wildlife Service. 34 pp.

Terborgh, J., A. Estes, P. Paquet, K. Ralls, D. Boyd-Heger, B. J. Miller, and R. F. Noss. 1999. The role of top carnivores in regulating terrestrial ecosystems. Pp. 39–64 in M. E. Soulé and J. Terborgh, eds., *Continental conservation.* Island Press, Washington, D.C.

Teruelo, S., and J. A. Valverde. 1992. *Los lobos de Morla.* Circulo de Bibliofilia Venatoria, Madrid. 444 pp.

Theberge, J. B. 1973. Wolf management in Canada through a decade of change. *Nature Canada, Canadian Nature Federation* 2(1): 3–10.

———. 1975. *Wolves and wilderness.* J. M. Dent and Sons, Toronto.

———. 1983. Considerations in wolf management related to genetic variability and adaptive change. Pp. 86–89 in L. N. Carbyn, ed., *Wolves in Canada and Alaska: Their status, biology, and management.* Report Series, no. 45. Canadian Wildlife Service, Edmonton, Alberta.

———. 1990. Potentials for misinterpreting impacts of wolf predation through prey:predator ratios. *Wildl. Soc. Bull.* 18: 188–92.

———. 1991. Ecological classification, status, and management of the gray wolf, *Canis lupus*, in Canada. *Can. Field Nat.* 105:459–63.

Theberge, J. B., and T. J. Cottrell. 1977. Food habits of wolves in Kluane National Park. *Arctic* 30:189–91.

Theberge, J. B., and J. B. Falls. 1967. Howling as a means of communication in timber wolves. *Am. Zool.* 7:331–38.

Theberge, J. B., G. J. Forbes, I. K. Barker, and T. Bollinger. 1994. Rabies in wolves of the Great Lakes Region. *J. Wildl. Dis.* 30:563–66.

Theberge, J., and D. Pimlott. 1969. Observations of wolves at a rendezvous site in Algonquin Park. *Can. Field Nat.* 83: 122–28.

Theberge, J. B., and D. R. Strickland. 1978. Changes in wolf numbers, Algonquin National Park, Ontario. *Can. Field Nat.* 92:395–98.

Theberge, J. B., and M. T. Theberge. 1998. *Wolf country: Eleven Years Tracking the Algonquin Wolves.* McClelland & Stewart, Inc., Toronto. 306 pp.

Theuerkauf, J., W. Jedrzejewski, K. Schmidt, H. Okarma, I. Ruczynski, S. Sniezko, and R. Gula. 2003. Daily patterns and duration of wolf activity in the Bialowieza Forest, Poland. *J. Mammal.* 84:127–37.

Thieking, A., S. M. Goyal, R. F. Berg, K. L. Loken, L. D. Mech, R. P. Thiel, and T. P. O'Connor. 1992. Seroprevalence of Lyme disease in gray wolves from Minnesota and Wisconsin. *J. Wildl. Dis.* 28:177–82.

Thiel, R. P. 1978. The status of the timber wolf in Wisconsin 1975. *Trans. Wisconsin Acad. Sci., Arts Lett.* 66:186–94.

———. 1985. Relationship between road density and wolf habitat suitability in Wisconsin. *Am. Midl. Nat.* 113:404–7.

———. 1993. *The timber wolf in Wisconsin.* University of Wisconsin Press, Madison.

Thiel, R. P. W. H. Hall, and R. N. Schultz. 1997. Early den digging by wolves *Canis lupus* in Wisconsin. *Can. Field Nat.* 111: 481–82.

Thiel, R. P., L. D. Mech, G. R. Ruth, J. R. Archer, and L. Kaufman. 1987. Blastomycosis in wild wolves. *J. Wildl. Dis.* 23: 321–23.

Thiel, R. P., S. Merrill, and L. D. Mech. 1998. Tolerance by denning wolves, *Canis lupus*, to human disturbance. *Can. Field Nat.* 112:340–42.

Thiel, R. P., and T. P. O'Connor. 1992. Seroprevalence of Lyme disease in gray wolves from Minnesota and Wisconsin. *J. Wildl. Dis.* 28:177–82.

Thiel, R. P., and R. R. Ream. 1995. Status of the gray wolf in the lower 48 states to 1992. Pp. 59–62 in L. N. Carbyn, S. H. Fritts, and D. R. Seip, eds., *Ecology and conservation of wolves in a changing world.* Canadian Circumpolar Institute, Edmonton, Alberta.

Thiel, R. P., and T. Valen. 1995. Developing a state timber wolf recovery plan with public input: The Wisconsin experience. Pp. 169–75 in L. N. Carbyn, S. H. Fritts, and D. R. Seip, eds., *Ecology and conservation of wolves in a changing world.* Canadian Circumpolar Institute, Edmonton, Alberta.

Thompson, D. Q. 1952. Travel, range, and food habits of timber wolves in Wisconsin. *J. Mammal.* 33:420–42.

Thompson, T., and W. Gasson. 1991. Attitudes of Wyoming residents on wolf reintroduction and related issues. Wyoming Game and Fish Department, Cheyenne.

Thomson, R., J. K. Pritchard, P. Shen, P. J. Oefner, and M. W. Feldman. 2000. Recent common ancestry of human Y chromosomes: Evidence from DNA sequence data. *Proc. Nat. Acad. Sci. U.S.A.* 97:7360–65.

Thun, R., P. Watson, and G. Jackson. 1977. Induction of estrus and ovulation in the bitch, using exogenous gonadotropins. *Am. J. Vet. Res.* 38:438–86.

Thurber, J. M., and R. O. Peterson. 1991. Changes in body size associated with range expansion in the coyote (*Canis latrans*). *J. Mammal.* 72:750–55.

———. 1993. Effects of population density and pack size on the foraging ecology of gray wolves. *J. Mammal.* 74:879–89.

Thurber, J. M., R. O. Peterson, T. D. Drummer, and S. A. Thomasma. 1994. Gray wolf response to refuge boundaries and roads in Alaska. *Wildl. Soc. Bull.* 22:61–68.

Thurber, J. M., R. O. Peterson, J. D. Woolington, and J. A. Vucetich. 1992. Coyote coexistence with wolves on the Kenai Peninsula, Alaska. *Can. J. Zool.* 70:2494–98.

Thurston, L. M. 2002. Homesite attendance as a measure of alloparental and parental care by gray wolves (*Canis lupus*) in northern Yellowstone National Park. Master's thesis, Texas A&M University, College Station.

Tieszen, L. L., and T. W. Boutton. 1988. Stable carbon isotopes in terrestrial ecosystems research. Pp. 167–95 in P. W. Rundel, J. R. Ehleringer, and K. A. Nagy, eds., *Stable isotopes in ecological research.* Springer-Verlag, New York.

Tilt, W., R. Norris, and A. S. Eno. 1987. Wolf recovery in the northern Rocky Mountains. National Audubon Society and National Fish and Wildlife Foundation, Washington, D.C.

Tkachenko, A. A. 1995. Behaviour of wolves and lynxes in the Polesski Reserve. Pp. 90 in J. Gurnall, ed., 2nd European Congress of Mammalogy, 27 March–1 April 1995, Southampton, U.K.

Todd, A. W., J. R. Gunson, and W. M. Samuel. 1981. Sarcoptic mange: An important disease of coyotes and wolves of Alberta, Canada. Pp. 706–29 in J. A. Chapman and D. Pursley, eds., Worldwide Furbearer Conference Proceedings, vol. 2.

Todd, A. W., L. B. Keith, and C. A. Fischer. 1981. Population ecology of coyotes during a fluctuation of snowshoe hare. *J. Wildl. Mgmt.* 45:629–40.

Todd, S. K. 1995. Designing effective negotiating teams for environmental disputes: An analysis of three wolf management plans. Ph.D. dissertation, University of Michigan, Ann Arbor. 324 pp.

Tolman, E. C. 1955. Cognitive maps in rats and men. *Psychol. Rev.* 55:189–208.

Tolstoy, N. I., ed. 1995. Volkolak. Pp. 418–20 in *Slavic antiquities,* Vol. I. Russian Academy of Science.

Tomasello, M., B. Hare, and T. Fogleman. 2001. The ontogeny of gaze following in chimpanzees, *Pan troglodytes,* and rhesus macaques, *Macaca mulatta. Anim. Behav.* 61: 335–43.

Tompa, F. S. 1983a. Problem wolf management in British Columbia: Conflict and program evaluation. Pp. 112–19 in L. N. Carbyn, ed., *Wolves in Canada and Alaska: Their status, biology, and management.* Report Series, no. 45. Canadian Wildlife Service, Edmonton, Alberta.

———. 1983b. Status and management of wolves in British Columbia. Pp. 20–24 in L. N. Carbyn, ed., *Wolves in Canada and Alaska: Their status, biology, and management.* Report Series, no. 45. Canadian Wildlife Service, Edmonton, Alberta.

Tooze, Z. J. 1987. Some aspects of the structure and function of long-distance vocalizations of timber wolves (*Canis lupus*). Master's thesis, Dalhousie University, Halifax, NS.

Tooze, Z. J., F. H. Harrington, and J. C. Fentress. 1990. Individually distinct vocalizations in timber wolves, *Canis lupus. Anim. Behav.* 40:723–30.

Townsend, J. B. 1981. Tanaina. Pp. 623–40 in W. C. Sturtevant, ed., *Handbook of North American Indians,* Vol. 6, *Subarctic.* Smithsonian Institution, Washington, D.C.

Townsend, S. E. 1996. The role of social cognition in feeding, marking and caching in captive wolves, *Canis lupus lycaon* and *Canis lupus baileyi.* Ph.D. dissertation, University of Colorado. Boulder. 871 pp.

Trainer, D. O., F. F. Knowlton, and L. Karstad. 1968. Oral papillomatosis in the coyote. *Bull. Wildl. Dis. Assoc.* 4:52–54.

Treves, A., R. R. Jurewicz, L. Naughton-Treves, R. A. Rose, R. C. Willging, and A. P. Wydeven. 2002. Wolf depredation on domestic animals in Wisconsin, 1976–2000. *Wildl. Soc. Bull.* 30:231–41.

Trumler, E. 1990. Domesticated dogs. Pp. 79–99 in S. P. Parker, ed., *Grzimek's encyclopedia of mammals,* vol. 4. McGraw-Hill, New York.

Tsuda, K., Y. Kikkawa, H. Yonekawa, and Y. Tanabe. 1997. Extensive interbreeding occurred among multiple matriarchal ancestors during the domestication of dogs: Evidence from inter- and intraspecies polymorphisms in the D-loop region of mitochondrial DNA between dogs and wolves. *Genes Genet. Syst.* 72:229–38.

Tsutsui, T. 1975. Studies on the reproduction in the dog. VI. Ovulation rate and transuterine migration of the fertilized ova. *Jpn. J. Anim. Reprod.* 21:98–101.

Tucker, P., and D. H. Pletscher. 1989. Attitudes of hunters and residents toward wolves in northwestern Montana. *Wildl. Soc. Bull.* 17:509–14.

Turi, J. O. 1931. *Turi's book of Lapland.* London: Jonathan Cape.

Turnbull, P. F., and C. A. Reed. 1974. The fauna from the terminal Pleistocene of Palegawra Cave. *Fieldiana Anthropol.* 63:81–164.

Ulrich, R. S. 1993. Biophilia, biophobia and natural landscapes. Pp. 73–137 in S. R. Kellert and E. O. Wilson, eds., *The biophilia hypothesis.* Island Press, Washington, D.C.

U.S. Fish and Wildlife Service. 1978. Title 50. *Wildl. Fish.* 43(47): 9607–15.

———. 1982. Mexican wolf recovery plan. United States Fish and Wildlife Service, Albuquerque, NM. 103 p.

———. 1987. Northern Rocky Mountain wolf recovery plan. U.S. Fish and Wildlife Service, Denver, CO. 119 p.

———. 1989. Red Wolf recovery plan. United States Fish and Wildlife Service, Atlanta, GA.

———. 1990. Red Wolf Recovery/Species Survival Plan. U.S. Fish and Wildlife Service, Atlanta, GA.

———. 1992. Recovery plan for the eastern timber wolf. U.S. Fish and Wildlife Service, Twin Cities, MN.

———. 1994a. Establishment of a nonessential experimental population of gray wolves in Yellowstone National Park in Wyoming, Idaho, and Montana and central Idaho and southwestern Idaho B: Final rule. *Fed. Reg.* 59:60252–81.

———. 1994b. The reintroduction of gray wolves to Yellowstone National Park and central Idaho: Final environmental impact statement. U.S. Fish and Wildlife Service, Helena, MT.

———. 1998. Availability of draft reassessment of the interim wolf control plan for the northern Rocky Mountains for review and comment. *Fed. Reg.* 63(78): 20212.

———. 2000. Endangered and threatened wildlife and plants; proposal to reclassify and remove the gray wolf from the list of endangered and threatened wildlife in portions of the conterminous United States; proposal to establish three special regulations for threatened gray wolves; proposed rule. *Fed. Reg.* 65(135): 43450–96.

————. 2002. Mexican wolf reintroduction update, December 1–15. U.S. Fish and Wildlife Service, Albuquerque, NM.

U.S. Fish and Wildlife Service, Nez Percé Tribe, U.S. National Park Service, and USDA Wildlife Services. 2002. Rocky Mountain Wolf Recovery 2001 Annual Report. T. J. Meier, ed. USFWS, Ecological Services, Helena, MT. 43 pp.

Vainisi, S. J., H. F. Edelhauser, E. D. Wolf, E. Cottlier, and F. Reeser. 1981. Nutritional cataracts in timber wolves. *J. Am. Vet. Med. Assoc.* 179:1175–80.

Van Ballenberghe, V. 1974. Wolf management in Minnesota: An endangered species case history. *Trans. N. Am. Wildl. Nat. Resour. Conf.* 39:313–22.

————. 1977. Physical characteristics of timber wolves in Minnesota. Pp. 213–19 in R. L. Phillips and C. Jonkel, eds., Proceedings of the 1975 predator symposium. Montana Forest and Conservation Experiment Station, University of Montana, Missoula. 268 pp.

————. 1983a. Extraterritorial movements and dispersal of wolves in south-central Alaska. *J. Mammal.* 64:168–71.

————. 1983b. Two litters raised in one year by a wolf pack. *J. Mammal.* 64:171–72.

————. 1987. Effects of predation on moose numbers: A review of recent North American studies. *Swed. Wildl. Res.* 1 (Suppl.): 431–60.

————. 1991. Forty years of wolf management in the Nelchina basin, south-central Alaska: A critical review. *Trans. N. Am. Wildl. Nat. Resour. Conf.* 56:561–66.

Van Ballenberghe, V., and W. B. Ballard. 1994. Limitation and regulation of moose populations: The role of predation. *Can. J. Zool.* 72:2071–77.

Van Ballenberghe, V., A. W. Erickson, and D. Byman. 1975. Ecology of the timber wolf in northeastern Minnesota. Wildlife Monographs, no. 43. The Wildlife Society, Washington, D.C. 44 pp.

Van Ballenberghe, V., and L. D. Mech. 1975. Weights, growth, and survival of timber wolf pups in Minnesota. *J. Mammal.* 56:44–63.

Van Camp, J., and R. Gluckie. 1979. A record long-distance move by a wolf *(Canis lupus). J. Mammal.* 60:236.

Vandenbergh, J. G. 1988. Pheromones in mammalian reproduction. Pp. 1679–98 in E. Knobil and J. D. Neill, eds., *The physiology of reproduction,* vol. 2. Raven Press, New York.

Van Heerden, J. 1981. The role of integumental glands in the social and mating behaviour of the hunting dog *Lycaon pictus* (Temminck, 1820). *Onderstepoort J. Vet. Res.* 48:19–21.

Van Hooff, J. A. R. A. M., and J. A. B. Wensing. 1987. Dominance and its behavioral measures in a captive wolf pack. Pp. 219–52 in H. Frank, ed., *Man and wolf: Advances, issues and problems in captive wolf research.* Dr. W. Junk Publishers, Dordrecht, The Netherlands.

Vannelli, G. B., and G. C. Balboni. 1982. On the presence of estrogen receptors in the olfactory epithelium of the rat. Pp. 279–83 in W. Breipohl, ed., *Olfaction and endocrine regulation.* IRL Press, London.

Van Valkenburgh, B. 1988. Trophic diversity in past and present guilds of large predatory mammals. *Paleobiology* 14:155–73.

Van Valkenburgh, B., and K.-P. Koepfli. 1993. Cranial and dental adaptations to predation in canids. *Symp. Zool. Soc. Lond.* 65:15–37.

Van Valkenburgh, B., and C. B. Ruff. 1987. Canine tooth strength and killing behaviour in large carnivores. *J. Zool.* 212:379–98.

Veitch, A., W. Clar, and F. Harrington. 1993. Observations of an interaction between a barren-ground black bear, *Ursus americanus,* and a wolf, *Canis lupus,* at a wolf den in northern Laborador. *Can. Field Nat.* 107:95–97.

Verme, L. J. 1962. Mortality of white-tailed deer fawns in relation to nutrition. National White-Tailed Deer Disease Symposium: Proceedings 1:15–38.

Verme, L. J. 1963. Effect of nutrition on growth of white-tailed deer fawns. *Trans. N. Am. Wildl. Conf.* 28:431–43.

Vest, J. H. C. 1988. The medicine wolf returns: Traditional Blackfeet concepts of *Canis lupus. Western Wildlands* 14:28–33.

Victor, P. E., and J. Lariviere. 1980. *Les loups.* F. Nathan, Paris. 191 pp.

Vignon, V. 1995. Analyse de la predation des ongules par les loups *(Canis lupus)* dans un massif des monts Cantabriques (Asturies, Espagne). [In French with English summary.] *Cahiers d'Etholgie* 15(1):81–92.

Vilà, C., I. R. Amorim, J. A. Leonard, D. Posada, J. Castroviejo, F. Petrucci-Fonseca, K. A. Crandall, H. Ellegren, and R. K. Wayne. 1999. Mitochondrial DNA phylogeography and population history of the grey wolf *Canis lupus. Mol. Ecol.* 8:2089–2103.

Vilà, C., J. Castroviejo, and V. Urios. 1993. The Iberian wolf in Spain. Pp. 104–9 in C. Promberger and W. Schröder, eds., *Wolves in Europe: Status and perspectives.* Munich Wildlife Society, Ettal, Germany.

Vilà, C., J. E. Maldonado, and R. K. Wayne. 1999. Phylogenetic relationships, evolution, and genetic diversity of the domestic dog. *J. Hered.* 90:71–77.

Vilà, C., P. Savolainen, J. E. Maldonado, I. R. Amorim, J. E. Rice, R. L. Honeycutt, K. A. Crandall, J. Lundeberg, and R. K. Wayne. 1997. Multiple and ancient origins of the domestic dog. *Science* 276:1687–89.

Vilà, C., A. K. Sundqvist, O. Flagstad, J. Seddon, S. Bjornerfeldt, I. Kojola, A. Casulli, H. Sand, P. Wabakken, and H. Ellegren. 2002. Rescue of a severely bottlenecked wolf *(Canis lupus)* population by a single immigrant. *Proc. R. Soc. Lond. B* (online).

Vilà, C., V. Urios, and J. Castroviejo. 1990. Ecología del lobo en la Cabrera (León) y la Carbaleda (Zamora). Pp. 95–108 in J. C. Blanco, L. Cuesta, and S. Reig, eds., *El lobo* (Canis lupus) *en España: Situacion, problematica y apuntes sobre su ecología.* Istituto Nacional para la Conservacion de la Naturaleza, Madrid.

————. 1993. Tooth losses and anomalies in the wolf *(Canis lupus). Can. J. Zool.* 71:968–71.

————. 1994. Use of faeces for scent marking in Iberian wolves *(Canis lupus). Can. J. Zool.* 72:374–77.

Vilà, C., U. Vicente, and J. Castroviejo. 1995. Observations on the daily activity patterns in the Iberian wolf. Pp. 335–40 in

L. N. Carbyn, S. H. Fritts, and D. R. Seip, eds., *Ecology and conservation of wolves in a changing world.* Canadian Circumpolar Institute, Edmonton, Alberta.

Vilà, C., C. Walker, A.-K. Sundqvist, O. Flagstad, Z. Andersone, A. Casulli, I. Kojola, H. Valdmann, J. Halverson, and H. Ellegren. 2003. Combined use of maternal, paternal and biparental genetic markers for the identification of wolf-dog hybrids. *Heredity* 90:17–24.

Vilà, C., and R. K. Wayne. 1999. Hybridization between wolves and dogs. *Conserv. Biol.* 13:195–98.

Vogler, A. P., C. B. Knisley, S. B. Glueck, J. M. Hill, and R. DeSalle. 1993. Using molecular and ecological data to diagnose endangered populations of the puritan tiger beetle *Cicindela puritana. Mol. Ecol.* 2:375–83.

Voigt, D. R. 1973. Summer food habits and movements of wolves (*Canis lupus*) in central Ontario. Master's thesis, University of Guelph, Ontario.

Voigt, D. R., and W. E. Berg. 1987. Coyote. Pp. 344–57 in M. Novak, J. A. Baker, M. E. Obbard and B. Malloch, eds., *Wild furbearer management and conservation in North America.* Ontario Ministry of Natural Resources, Toronto.

Voigt, D. R., G. B. Kolenosky, and D. H. Pimlott. 1976. Changes in summer foods of wolves in central Ontario. *J. Wildl. Mgmt.* 40:663–68.

Volterra, V. 1928. Variations and fluctuations of the number of individuals in animal species living together. *Journal du Conseil International pour l'Exploration de la Mer* 3:3–51.

Von Schantz, T. 1984. Spacing strategies, kin selection, and population regulation in altricial vertebrates. *Oikos* 42:48–58.

Vormbrock, J. K., and J. M. Grossberg. 1988. Cardiovascular effects of human-pet dog interactions. *J. Behav. Med.* 11:509–17.

Vos, J. 2000. Food habits and livestock depredation of two Iberian wolf packs (*Canis lupus signatus*) in the north of Portugal. *J. Zool.* 251:457–62.

Vucetich, J. A., R. O. Peterson, and C. L. Schaefer. 2002. The effect of prey and predator densities on wolf predation. *Ecology* 83(11): 3003–3013.

Vucetich, J. A., R. O. Peterson, and T. A. Waite. 1997. Effects of social structure and prey dynamics on extinction risk in gray wolves. *Conserv. Biol.* 11:957–65.

Wabakken, P., H. Sand, O. Liberg, and A. Bjärvall. 2001. The recovery, distribution, and population dynamics of wolves on the Scandinavian peninsula, 1978–1998. *Can. J. Zool.* 79:710–25.

Wabakken, P., O. J. Sorensen, and T. Kvam. 1983. Wolves (*Canis lupus*) in southeastern Norway. *Acta Zool. Fenn.* 174:277.

Wade, D. A., and J. E. Brown. 1982. *Procedures for evaluating predation on livestock and wildlife.* Texas Agricultural Extension Service, San Angelo 42 pp.

Wagner, F. H. 1994. Scientist says Yellowstone Park is being destroyed. *High Country News,* 30 May, 14–15.

Wagner, K. K., R. H. Schmidt, and M. R. Conover. 1997. Compensation programs for wildlife damage in North America. *Wildl. Soc. Bull.* 25:312–19.

Waite, C. 1994. Age determination of gray wolves from Isle Royale National Park. Master's thesis, Michigan Technological University, Houghton. 49 pp.

Waits, L. P., S. L. Talbot, R. H. Ward, and G. F. Shields. 1998. Mitochondrial DNA phylogeography of the North American brown bears and implications for conservation. *Conserv. Biol.* 12:408–17.

Walker, D. N., and G. C. Frison. 1982. Studies on Amerindian dogs, 3: Prehistoric wolf/dog hybrids from the northwestern plains. *J. Archaeol. Sci.* 9:125–72.

Walker, R. F., D. Riad-Fahmy, and G. F. Read. 1978. Adrenal status assessed by direct radioimmunoassay of cortisol in whole saliva or parotid saliva. *Clin. Chem.* 24:1460–63.

Walls, G. L. 1942. *The vertebrate eye and its adaptive radiation.* (Reprint 1963.) Hafner Publishing, New York.

Walters, C. 1986. *Adaptive Management of Renewable Resources.* Macmillan, New York. 384 pp.

Walters, C. J., M. Stocker, and G. C. Haber. 1981. Simulation and optimization models for a wolf-ungulate system. Pp. 317–37 in C. W. Fowler and T. D. Smith, eds., *Dynamics of large mammal populations.* John Wiley and Sons, New York.

Walther, F. 1961. Einige Verhaltensbeobachtungen am Bergwild des Georg von Opel Freigeheges. *Jahrb. G. v. Opel Freigehege. Tierforsch.* 1960–1961:53–89.

Walton, L. R., H. D. Cluff, P. C. Paquet, and M. A. Ramsay. 2001. Movement patterns of barren-ground wolves in the central Canadian Arctic. *J. Mammal.* 82:867–76.

Waser, P. M., and M. S. Waser. 1977. Experimental studies of primate vocalization: Specializations for long-distance propagation. *Z. Tierpsychol.* 42:239–63.

Wassle, H., and B. B. Boycott. 1991. Functional architecture of the mammalian retina. *Physiol. Rev.* 71:447–80.

Watkins, A. 1997. Mind-body pathways. Pp. 1–25 in A. Watkins, ed. *Mind-Body Medicine: A Clinician's Guide to Psychoneuroimmunology.* Churchill Livingstone, New York.

Wayne, R. K. 1986. Cranial morphology of domestic and wild canids: The influence of development on morphological change. *Evolution* 40:243–61.

———. 1992. On the use of morphologic and molecular genetic characters to investigate species status. *Conserv. Biol.* 6:590–92.

———. 1993. Molecular evolution of the dog family. *Trends Genet.* 9:218–24.

———. 1995. Red wolves: To conserve or not to conserve. *Canid News* (Newsletter of the IUCN/SSC Canid Specialist Group) 3:7–12.

———. 1996. Conservation genetics in the Canidae. Pp. 75–118 in J. C. Avise and J. L. Hamrick, eds., *Conservation genetics: Case histories from nature.* Chapman and Hall, New York, London.

Wayne, R. K., and D. M. Brown. 2001. Hybridization and conservation of carnivores. Pp. 145–62 in J. L. Gittleman, S. Funk, D. W. Macdonald, and R. K. Wayne, eds., *Carnivore conservation.* Cambridge University Press, Cambridge.

Wayne, R. K., E. Geffen, D. J. Girman, K. P. Koepfli, L. M. Lau, and C. R. Marshal. 1997. Molecular systematics of the Canidae. *Syst. Biol.* 46:622–53.

Wayne, R. K., D. A. Gilbert, A. Eisenhawer, N. Lehman, K. Hansen, D. Girman, R. O. Peterson, L. D. Mech, P. J. P. Gogan, U. S.

Seal, and R. J. Krumenaker. 1991. Conservation genetics of the endangered Isle Royale gray wolf. *Conserv. Biol.* 5: 41–51.

Wayne, R. K., and J. L. Gittleman. 1995. The problematic red wolf. *Sci. Am.* 273:36–39.

Wayne, R. K., and S. M. Jenks. 1991. Mitochondrial DNA analysis supports extensive hybridization of the endangered red wolf (*Canis rufus*). *Nature* 351:565–68.

Wayne, R. K., N. Lehman, M. W. Allard, and R. L. Honeycutt. 1992. Mitochondrial DNA variability of the gray wolf: Genetic consequences of population decline and habitat fragmentation. *Conserv. Biol.* 6:559–69.

Wayne, R. K., N. Lehman, and T. K. Fuller. 1995. Conservation genetics of the gray wolf. Pp. 399–407 in L. N. Carbyn, S. H. Fritts, and D. R. Seip, eds., *Ecology and conservation of wolves in a changing world.* Canadian Circumpolar Institute, Edmonton, Alberta.

Wayne, R. K., J. A. Leonard, and A. Cooper. 1999. Full of sound and fury: The recent history of ancient DNA. *Annu. Rev. Ecol. Syst.* 30:457–77.

Wayne, R, K., W. G. Nash, and S. J. O'Brien. 1987a. Chromosomal evolution of the Canidae. I. Species with high diploid numbers. *Cytogenet. Cell Genet.* 44:123–33.

———. 1987b. Chromosomal evolution of the Canidae. II. Divergence from the primitive carnivore karyotype. *Cytogenet. Cell Genet.* 44:134–41.

Wayne, R. K., and S. J. O'Brien. 1987. Allozyme divergence within the Canidae. *Syst. Zool.* 36:339–55.

Wayne, R. K., M. S. Roy, and J. L. Gittleman. 1998. Origin of the red wolf: Response to and Federoff and Gardner. *Conserv. Biol.* 12:726–29.

Weaver, J. 1978. The wolves of Yellowstone. Natural Resources Report no. 14. U.S. National Park Service, Washington, D.C. 38 pp.

Weaver, J. L. 1993. Refining the equation for interpreting prey occurrence in gray wolf scats. *J. Wildl. Mgmt.* 57:534–38.

———. 1994. Ecology of wolf predation amidst high ungulate diversity in Jasper National Park, Alberta. Ph.D. dissertation, University of Montana, Missoula. 183 pp.

Weaver, J. L., C. Arvidson, and P. Wood. 1992. Two wolves, *Canis lupus,* killed by a moose in Jasper National Park, Alberta. *Can. Field Nat.* 106:126–27.

Weaver, J. L., R. E. F. Escano, and D. S. Winn. 1987. A framework for assessing cumulative effects on grizzly bears. *Trans. N. Am. Wildl. Nat. Resour. Conf.* 52:364–76.

Weaver, J. L., and S. H. Fritts. 1979. Comparison of coyote and wolf scat diameters. *J. Wildl. Mgmt.* 43:786–88.

Weaver, J. L., and S. W. Hoffmann. 1979. Differential detectability of rodents in coyote scats. *J. Wildl. Mgmt.* 43:783–86.

Webb, S. D. 1974. Chronology of Florida Pleistocene mammals. Pp. 5–31 in S. D. Webb, ed., *Pleistocene mammals of Florida.* University Presses of Florida, Gainesville.

———. 1984. Ten million years of mammal extinctions in North America. Pp. 189–210 in P. S. Martin and R. G. Klein, eds., *Quaternary extinctions: A prehistoric revolution.* University of Arizona Press, Tucson.

Weeden, R. B. 1985. Northern people, northern resources and the dynamics of carrying capacity. *Arctic* 38:116–20.

Weibel, E. R., C. R. Taylor, J. J. O'Neil, D. E. Leith, P. Gehr, H. Hoppeler, V. Langman, and R V. Baudinette. 1983. Maximal oxygen consumption and pulmonary diffusing capacity: A direct comparison of physiologic and morphometric measurements in canids. *Respir. Physiol.* 54:173–88.

Weiler, G. J., G. W. Garner, and D. G. Ritter. 1995. Occurrence of rabies in a wolf population in northeastern Alaska. *J. Wildl. Dis.* 31:79–82.

Weiner, J. 1989. Metabolic constraints to mammalian energy budgets. *Acta Theriol.* 34:3–35.

Weise, T. F., W. L. Robinson, R. A. Hook, and L. D. Mech. 1975. An experimental translocation of the eastern timber wolf. Audubon Conservation Report 5. National Audubon Society, New York. 28 pp.

———. 1979. An experimental translocation of the eastern timber wolf. Pp. 346–419 in E. Klinghammer, ed., *The behavior and ecology of wolves.* Garland STPM Press, New York.

Wells, M. C. 1978. Coyote senses in predation: Environmental influences on their relative use. *Behav. Processes* 3:149–58.

Wells, M. C., and M. Bekoff. 1981. An observational study of scent-marking in coyotes, *Canis latrans. Anim. Behav.* 29:332–50.

Wells, M. C., and P. M. Lehner. 1978. The relative importance of the distance senses in coyote predatory behaviour. *Anim. Behav.* 26:251–58.

Wen, G., J. Sturman, and J. Shek. 1985. A comparative study of the tapetum, retina and skull of the ferret, dog and cat. *Lab. Anim. Sci.* 35:200–210.

Wheler, C. L., and A. M. Pocknell. 1996. Ectopic perianal gland tumor in a timber wolf. *Can. Vet. J.* 37:41–42.

Whitacre, F. E., and B. Barrera. 1944. War amenorrhea. *J. Am. Med. Assoc.* 124:399–403.

White, G. C. 2000. Population viability analyses: Data requirements and essential analyses. Pp. 288–331 in L. Boitani and T. K. Fuller, eds., *Research techniques in animal ecology: Controversies and consequences.* Columbia University Press, New York.

White, K. S., H. N. Golden, K. J. Hundertmark, and G. R. Lee. In press. Predation by wolves, *Canis lupus,* on wolverines, *Gulo gulo,* and an American marten, *Martes americana,* in Alaska. *Can. Field Nat.*

White, P. A, and D. K. Boyd. 1989. A cougar, *Felis concolor,* kitten killed and eaten by gray wolves, *Canis lupus,* in Glacier National Park, Montana. *Can. Field Nat.* 103:408–9.

White, P. S., O. L. Tatum, L. L. Deaven, and J. L. Longmire. 1999. New, male-specific microsatellite markers from the human Y chromosome. *Genomics* 57:433–37.

Whitten, W. K., and A. K. Champlin. 1973. The role of olfaction in mammalian reproduction. Pp. 109–23 in R. O. Greep and E. B. Astwood, eds., *Handbook of physiology,* section 7: *Endocrinology,* vol. 2: *Female reproductive system,* part 1. American Physiological Society, Washington, D.C.

Wicker, K. J. 1996. An analysis of public testimonies on the reintroduction of wolves to the Greater Yellowstone ecosystem. Master's thesis, Texas A&M University, College Station.

Willems, R. A. 1995. The wolf-dog hybrid: An overview of a controversial animal. *Anim. Welfare Inform. Newsl.* 5(4): 3–8.

Williams, C. K., G. Ericsson, and T. A. Heberlein. 2002. A quantitative summary of attitudes toward wolves and their reintroduction (1972–2000). *Wildl. Soc. Bull.* 30:575–84.

Williams, G. C. 1966. *Adaptation and natural selection.* Princeton University Press, Princeton, NJ.

Williams, L. W., R. L. Peiffer, and K. N. Gelatt. 1977. Cataract resorption in a young wolf. *Vet. Med.* 72:419–21.

Williams, T. M. 1990. Summer diet and behaviour of wolves denning on barren-ground caribou range in the Northwest Territories, Canada. Master's thesis, University of Alberta, Edmonton. 86 pp.

Williamson, L. L., and R. D. Teague. 1971. The role of social sciences in wildlife management. Pp. 34–37 in *A manual of wildlife management.* The Wildlife Institute, Washington D.C.

Wilson, E. O. 1975. *Sociobiology.* Harvard University Press, Cambridge, MA.

———. 1984. *Biophilia: The human bond with other species.* Harvard University Press, Cambridge, MA.

———. 1992. *The diversity of life.* Harvard University Press, Cambridge, MA.

———. 1993. Biophilia and the conservation ethic. Pp. 31–41 in S. R. Kellert and E. O. Wilson, eds., *The biophilia hypothesis.* Island Press, Washington, D.C.

Wilson, E. O., and W. H. Bossert. 1963. Chemical communication among animals. *Recent Prog. Horm. Res.* 19:673–716.

Wilson, M. A. 1997. The wolf in Yellowstone: Science, symbol, or politics? Deconstructing the conflict between environmentalism and wise use. *Soc. Nat. Resour.* 10:453–68.

———. 1999. Public attitudes towards wolves in Wisconsin. In Wisconsin Wolf Management Plan (draft). Wisconsin Department of Natural Resources.

Wilson, M. A., and T. A. Heberlein. 1996. The wolf, the tourist, and the recreational context: New opportunity or uncommon circumstance. *Human Dimensions of Wildlife* 1(4): 38–53.

Wilson, P. J., S. Grewal, I. D. Lawford, J. N. M. Heal, A. G. Granacki, D. Pennock, J. B. Theberge, M. T. Theberge, D. R. Voigt, W. Waddell, R. E. Chambers, P. C. Paquet, G. Goulet, D. Cluff, and B. N. White. 2000. DNA profiles of the eastern Canadian wolf and the red wolf provide evidence for a common evolutionary history independent of the gray wolf. *Can. J. Zool.* 78:2156–66.

Wisconsin Department of Natural Resources. 1999. Wisconsin wolf management plan. Wisconsin Department of Natural Resources, Madison.

Witter, J. F. 1981. Brucellosis. Pp. 280–87 in J. W. Davis, L. H. Karstad, and D. O. Trainer, eds., *Infectious diseases of wild mammals.* Iowa State University Press, Ames.

Wobeser, G. 1992. Traumatic, degenerative, and developmental lesions in wolves and coyotes from Saskatchewan. *J. Wildl. Dis.* 28:268–75.

Wobeser, G., W. Runge, and R. R. Stewart. 1983. *Metorchis conjunctus* (Cobbold, 1860) infection in wolves *(Canis lupus)* with pancreatic involvement in two animals. *J. Wildl. Dis.* 19: 353–65.

Wolfe, M. L., and D. L. Allen. 1973. Continued studies of the status, socialization and relationships of Isle Royale wolves, 1967–70. *J. Mammal.* 54:611–33.

Wolstenholme, R. 1996. Attitudes of residents toward wolves in a rural community in Northwestern Montana: Summary and results. Master's thesis, University of Montana, Missoula.

Woollard, T., and S. Harris. 1990. A behavioural comparison of dispersing and non-dispersing foxes *(Vulpes vulpes)* and an evaluation of some dispersal hypotheses. *J. Anim. Ecol.* 59:709–22.

Woolpy, J. H. 1967. Socially controlled systems of mating among wolves and other gregarious mammals and their implications for the genetics of natural populations. Ph.D. dissertation, University of Chicago, Chicago, IL.

———. 1968. The social organization of wolves. *Nat. Hist.* 77: 46–55.

Woolpy, J. H., and I. Eckstrand. 1979. Wolf pack genetics, a computer-simulation with theory. Pp. 206–24 in E. Klinghammer, ed., *The behavior and ecology of wolves.* Garland STPM Press, New York.

Worley, D. E., D. S. Zarlenga, and F. M. Seesee. 1990. Freezing resistance of a *Trichinella spiralis nativa* isolate from a gray wolf, *Canis lupus,* in Montana, with observations on genetic and biological charcteristics of the biotype. *J. Helminth. Soc.* 57:57–60.

Wozencraft, W. C. 1993. Order Carnivora. Pp. 279–348 in D. E. Wilson and D. E. Reeder, eds., *Mammal species of the world: A taxonomic and geographic reference.* Smithsonian Institution Press, Washington, D.C.

Wurster-Hill, D. H., and W. R. Centerwall. 1982. The interrelationships of chromosome banding patterns in canids, mustelids, hyena, and felids. *Cytogenet. Cell Genet.* 34:178–92.

Wydeven, A. P., T. K. Fuller, W. Weber, and K. MacDonald. 1998. The potential for wolf recovery in the northeastern United States via dispersal from southeastern Canada. *Wildl. Soc. Bull.* 26:776–84.

Wydeven, A. P., and R. N. Schultz. 1993. Management policy for wolf den and rendezvous sites. Wisconsin Department of Natural Resources, Madison. 12 pp.

Wydeven, A. P., R. N. Schultz, and R. P. Thiel. 1995. Monitoring of a gray wolf *(Canis lupus)* population in Wisconsin, 1979–1991. Pp. 147–56 in L. N. Carbyn, S. H. Fritts, and D. R. Seip, eds., *Ecology and conservation of wolves in a changing world.* Canadian Circumpolar Institute, Edmonton, Alberta.

Wysocki, C. J. 1979. Neurobehavioral evidence for the involvement of the vomeronasal system in mammalian reproduction. *Neurosci. Biobehav. Rev.* 3:301–41.

Wywialowski, A. P. 1994. Agricultural producers' perceptions of wildlife-caused losses. *Wildl. Soc. Bull.* 22:370–82.

Young, S. P. 1944. History, life habits, economic status, and control. Pp. 1–385 in S. P. Young and E. A. Goldman, eds., *The wolves of North America.* American Wildlife Institute, Washington, D.C. 636 pp.

———. 1946. *The wolf in North American history.* Caxton Printers, Ltd., Caldwell, OH. 149 pp.

———. 1970. *The last of the loners.* Macmillan, New York.

Young, S. P., and E. A. Goldman. 1944. *The wolves of North America.* American Wildlife Institute, Washington, D.C.

Young, S. P., and H. H. T. Jackson. 1951. *The clever coyote.* Stackpole Press, Harrisburg, PA.

Yudakov, A. G., and I. G. Nikolaev. 1987. *Ecology of the Amur tiger.* Nauka, Moscow.

Yudin, V. G. 1992. The wolf of the far east of Russia. Russian Academy of Sciences, Far Eastern Branch Biology and Soil Institute, Blagoveshchensk, USSR.

Yukon Wolf Management Planning Team. 1992. The Yukon wolf conservation and management plan. Yukon Department of Renewable Resources, Whitehorse. 17 pp.

Zamenhof, S., E. Van Marthens, and L. Grauel. 1971. DNA (cell number) in neonatal brain: Second generation (F_2) alteration by maternal (F_0) dietary protein restriction. *Science* 172:850–51.

Zarnke, R. L., and W. B. Ballard. 1987. Serological survey for selected microbial pathogens of wolves in Alaska, 1975–1982. *J. Wildl. Dis.* 23:77–85.

Zarnke, R. L., J. Evermann, J. M. V. Hoef, M. E. McNay, R. D. Boertje, C. L. Gardner, L. G. Adams, B. W. Dale, and J. Burch. 2001. Serologic survey for canine coronavirus in wolves from Alaska. *J. Wildl. Dis.* 37:740–45.

Zarnke, R. L., and T. H. Spraker. 1985. *Trichodectes canis* louse infestation of wolves in Alaska. *Proc. Alaska Sci. Conf.* 36:112.

Zarnke, R. L., D. E. Worley, J. M. Ver Hoef, and M. E. McNay. 1999. *Trichinella* sp. in wolves from interior Alaska. *J. Wildl. Dis.* 35:94–97.

Zasukhin, D. N., K. S. Vel'iaminov, and L. P. D'Iakonov. 1979. Sarcocystosis of animals. [In Russian.] *Veterinariia* 1979(1): 49–55.

Zheleznov, N. 1991. Helminthofauna of wolf (*Canis lupus* L.) in northern Asia. *Trans. Int. Congr. Game Biol.* 20(2): 784–91.

———. 1992. Ecology of the grey wolf *Canis lupus* L. on Chukotka. *Trans. IUBG Cong.* 18:381–84.

Zimen, E. 1971. *Wolfe und Koenigspudel: Vergleichende Verhaltungsbeobachtungen.* R. Piper & Co. Verlag, Munich. 257 pp.

———. 1975. Social dynamics of the wolf pack. Pp. 336–62 in M. W. Fox, ed., *The wild canids: Their systematics, behavioral ecology, and evolution.* Van Nostrand Reinhold, New York.

———. 1976. On the regulation of pack size in wolves. *Z. Tierpsychol.* 40:300–341.

———. 1978. *Der Wolf: Mythos und Verhalten.* Meyster Verlag, Muenchen. 373 pp.

———. 1981. *The wolf: A species in danger.* Delacorte Press, New York.

———. 1982. A wolf pack sociogram. Pp. 282–322 in F. H. Harrington and P. C. Paquet, eds., *Wolves of the world: Perspectives of behavior, ecology, and conservation.* Noyes Publications, Park Ridge, NJ.

Zimen, E., and L. Boitani. 1975. Number and distribution of wolves in Italy. *Z. Saugetierk.* 40:102–12.

———. 1979. Status of the wolf in Europe and the possibilities of conservation and reintroduction. Pp. 43–83 in E. Klinghammer, ed., *The behavior and ecology of wolves.* Garland STPM Press, New York.

Zimmerman, W. J. 1971. Trichinosis. Pp. 127–39 in J. W. Davis and R. C. Anderson, eds., *Parasitic diseases of wild mammals.* Iowa State University Press, Ames.

Author Index